International Conference on Boilers and Heat Exchangers (ICBHE2019)

IOP Conference Series: Materials Science and Engineering
Volume 721

Hangzhou, China
1 - 4 November 2019

ISBN: 978-1-7138-2580-7
ISSN: 1757-8981

Printed from e-media with permission by:

Curran Associates, Inc.
57 Morehouse Lane
Red Hook, NY 12571

CURRAN ASSOCIATES INC. proceedings.com

Some format issues inherent in the e-media version may also appear in this print version.

Institute of Physics
Dirac House, Temple Back
Bristol BS1 6BE UK

Phone: 44 1 17 929 7481
Fax: 44 1 17 920 0979

techtracking@iop.org

Additional copies of this publication are available from:

Curran Associates, Inc.
57 Morehouse Lane
Red Hook, NY 12571 USA
Phone: 845-758-0400
Fax: 845-758-2633
Email: curran@proceedings.com
Web: www.proceedings.com

TABLE OF CONTENTS

IOP Conf. Series: Materials Science and Engineering **721** (2020) 011001 doi:10.1088/1757-899X/721/1/011001

Preface

2019 International Conference on Boiler and Heat Exchangers (ICBHE2019), co-organized by China Boiler and Boiler Water Treatment Association and China Jiliang University, was successfully held on November 1-3, 2019 in Hangzhou, China.

ICBHE2019 was to bring together innovative academics and industrial experts in the field of boilers, heat exchangers, energy and environment, measurement, certification and evaluation to a common forum. The conference aims to promote and contribute research and developmental advancement in the industries of boilers, heat exchangers and heat transfer technology. And another goal is to promote scientific information interchange between researchers, developers, engineers, students, and practitioners working all around the world.

It was wonderful to have around one hundred participants representing various universities at home and abroad attended this conference. The scientific program of the conference mainly consists of themed forums, special seminars and academic inspection activities. And we are honored to have Dean Guangxi Yue representing Tsinghua University, China, Prof. Yidong Chen from University of Nevada Las Vegas, USA, Prof. Xianming Dai from The University of Texas at Dallas, USA, and Prof. Wuqiang Yang from University of Manchester, UK as our distinguished keynote speakers. During the conference, these professors have shared their valuable and remarkable speeches on the following topics: 1. Research progress of industrial circulating fluidized bed boilers in China; 2. Numerical analysis of air condenser; 3. Application of New Bionic Nanomaterials in Thermal Energy Transmission; 4. Electrical capacitance tomography and industrial applications.

We are pleased to introduce the Proceedings of ICBHE2019 which contains a selection of papers from contributed presentation in the conference. The Proceedings features recent result and contemporary research in boilers, heat exchangers, energy and environment and measuring technique. All the papers have been through rigorous review and process to meet the requirements of International publication standard.

On behalf of the committee, we would like to express our gratitude to the distinguished speakers who have come to share their valuable findings and knowledge, to the chairman and all participants, who contributed a lot and made this conference a great success. We would also like to thank the contributors for submitting their papers and patiently revised them according to the suggestions of the designated reviewers.

Finally, we hope that all the articles in this conference can provide valuable summary and reference for researchers in this field.

The Organizing Committee of ICBHE2019

IOP Conf. Series: Materials Science and Engineering **721** (2020) 011001 doi:10.1088/1757-899X/721/1/011001

Committee member

Academic Committee

Chairman

Prof. Guangxi Yue, Energy and Power Engineering Department Tsinghua University, China
Co-Chairman:
Prof. Yitung Chen, Department of Mechanical Engineering, University of Nevada Las Vegas, USA
Prof. Guoli Qi, China Special Equipment Inspection and Research Institute, China
Prof. Zuohe Chi, The Institute for Energy Engineering, China Jiliang University, China
Committee:
Prof.Shaohua Wu, The Combustion Engineering Research Institute (CERI) of Harbin Institute of Technology (HIT) ,China
Prof. Xianming Dai, Mechanical Engineering,The University of Texas at Dallas,USA
Prof.Wuqiang Yang, School of Electrical and Electronic Engineering, University of Manchester, UK
Prof. Zhongxiao Zhang, School of Mechanical Engineering, Shanghai Jiao Tong University, China
Prof. Linhua Liu, School of Energy and Power Engineering, Shandong University, China
Prof. Bo Yu, Department of oil & gas storage and transportation engineering, Beijing Institute of Petrochemical Technology, China
Prof. Lyu Junfu, Department of Energy and Power Engineering, Tsinghua University, China
Prof. Jian Guan, China Special Equipment Inspection and Research Institute, China
Prof. Bin Chen, School of Environment,Beijing Normal University,China
Prof. Zhiguo Qu, School of Energy & Power Engineering, Xi'an Jiaotong University, China
Prof.Tong Zhu, School of Mechnical Engineering,Tongji University, China
Prof.Yongdong Chen, Hefei General Machinery Research Institute Corporation Ltd,China
Prof.Yong Dong, School of energy and power engineering, Shandong university, China
Prof.Jianmin Gao, School of Energy Science and Engineering , Harbin Institute of Technology, China
A.Prof.Jiaoling Wang, China Boiler and Boiler Water Treatment Association, China
A.Prof.Hua Guo, China Boiler and Boiler Water Treatment Association, China
Prof.Defang Cheng, Zhejiang Provincial Special Equipment Inspection And Research Institute, China

Organizing Committee

Chairman:

A.Prof.Jiaoling Wang, China Boiler and Boiler Water Treatment Association, China
Prof.Jinhui Cai, School of Measurement and Testing Enginering, China Jiliang University,China

Co-Chairman:

Baijun Xie, Zhejiang new zhonggang thermoelectricity co.,ltd

Youth Academic Committee

Chairman:

Prof. Zhiguo Qu, School of Energy & Power Engineering, Xi'an Jiaotong University, China

Co-Chairman:

Prof.Jianmin Gao, School of Energy Science and Engineering , Harbin Institute of Technology, China
Dr. Yan Yao, The Institute for thermal detection and automatic control, China Jiliang University,China
Dr.Yang Zhang, Department of Energy and Power Engineering, Tsinghua University, China

Secretary-general:

Dr.Xuemin Liu, Energy Saving Technology Center, China Special Equipment Inspection and Research Institute, China

IOP Conf. Series: Materials Science and Engineering **721** (2020) 011001 doi:10.1088/1757-899X/721/1/011001

2019 International Conference on Boilers and Heat Exchangers（ICBHE2019）
2019年锅炉与换热器国际学术会议
中国・杭州 2019.11.02

IOP Conf. Series: Materials Science and Engineering **721** (2020) 011002 doi:10.1088/1757-899X/721/1/011002

Peer review statement

All papers published in this volume of *IOP Conference Series: Materials Science and Engineering* have been peer reviewed through processes administered by the proceedings Editors. Reviews were conducted by expert referees to the professional and scientific standards expected of a proceedings journal published by IOP Publishing.

Effect of technology application of NO_x removal by O_3 on fly ash concentration

Wang Jianfeng[1], Gong Lei[2], Liu Peiqi[1], Hu Ximing[1], Zhu Yue[1]

[1]HuaDian Electric Reaesrch Institute Co. LTD, Hangzhou, Zhejiang, 310030, China

[2]Demister (shanghai) environmental technology Co. LTD，shanghai, 201620, China

*jianfeng-wang@chder.com, 545893518@qq.com, peiqi-liu@chder.com, ximing-hu@chder.ocm@chder.com, yue-zhu@chder.com

Abstract. An experimental study was carried out on the concentration of dust in flue gas from the outlet of desulphurization system under the conditions of different ammonia and ozone recharge, When the dust samples were dried at different temperatures, the mass concentration of dust was quite different. Analysis of samples by TG-DSC and XPS, It was concluded that the great change in the mass concentration of soot is mainly due to the influence of the combined denitrification process of SNCR and O_3 on the composition of the original flue gas. The formation of Soluble salt was carried by droplets in the flue gas and retained on the filter membrane. It led to a great change in the concentration of dust. The main substances of the soluble salt were carbonic acid, ammonium nitrate and calcium nitrate.

1. Introduction

At present, in order to implement the Circular of the General Office of the State Council on the issuance of the Strategic Action Plan for Energy Development (2014-2020) (issued by the State Office (2014) No. 31). more than 90% of the thermal power units in China had completed the transformation of ultra-low emissions. The emission standards could meet the emission limits of NO_x, SO_2 and dust emission concentrations 50, 35 and 10mg/m^3 (standard, dry base, 6%O_2) specified in the notice[1]. In order to meet the NO_x emission target value, Low nitrogen Combustion Technology (LNB) and selective Catalytic reduction Technology (SCR) were generally used in Coal-fired generating sets of large Power Plant boilers[1,2], Selective non-catalytic reduction technology (SNCR) was generally used in small pulverized coal units or circulating fluidized bed units. More than 90% of NOx was NO in flue gas of coal-fired units[3-5], Because of the very low water solubility of NO, it could not be effectively absorbed by desulphurization absorption liquid. According to the very low water solubility of NO, Some small coal-fired units adopt a new denitrification process to meet the emission requirements of NO_x after using SNCR. Such as non-thermal plasma oxidation[6], electron beam flue gas treatment[7], direct oxidation of O_3 and $NaClO_2$ oxidants[8-12] to promote the absorption of NO in flue gas. Among them, O^3 oxidants had the advantages of high oxidation efficiency, strong oxidation selectivity and no secondary pollution of oxidation products at low temperature[13]. However, the application of these new denitrification technologies only took into account the oxidative absorption of NO or the synergistic removal of NO_x, SO_2 and mercury, and did not study and discuss the effect of the application of this technology on the emission of fly ash concentration.

This paper took a small boiler as the research object, using ammonia as reducing agent, using SNCR denitrification process and desulphurization inlet using O_3 oxidation process. the concentration

IOP Conf. Series: Materials Science and Engineering **721** (2020) 012001 doi:10.1088/1757-899X/721/1/012001

of dust in flue gas was measured, and the filter membrane was analyzed by comprehensive thermal analysis (TG-DSC) and X-ray photoelectron spectroscopy (XPS), The composition of dust was qualitatively analyzed after SNCR and O_3 denitrification, and the effect of ozone denitrification on soot emission was discussed.

2. Test object and method

2.1Brief introduction of Test Unit

This paper mainly aimed at the 210t/h circulating fluidized bed of a chemical enterprise in Zhejiang Province as the test object, The unit used environmental protection facilities as shown in the following figure.

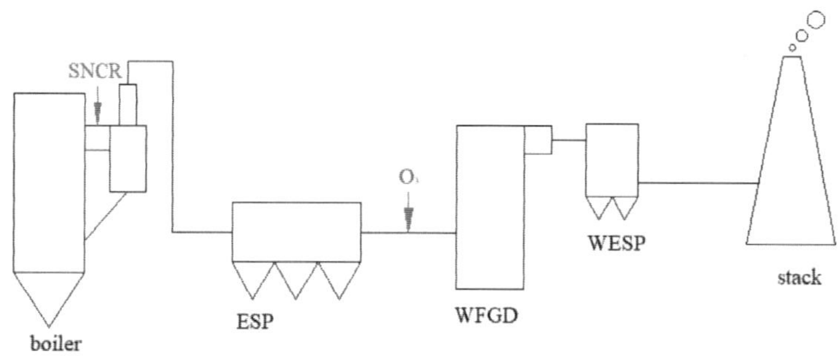

Figure 1. Process flow chart of test unit.

The unit adopt SNCR and ozonation process, high efficiency limestone-gypsum wet desulfurization process, high efficiency dedusting device and wet electric dedusting process. In the denitrification process, SNCR was arranged at the inlet of the cyclone separator, and ozone was injected into the inlet of the desulfurization absorption tower for denitrification.

2.2Analytical instruments and methods

Under different ozone consumption conditions, samples of dust were done at the outlet of desulfurization absorption Tower by using LaoYing 3012H Dust Sampler. The filter membrane was dried at 105℃, 175℃ and 250℃. The dried samples were weighed after constant weight, And analyzed by TG-DSC and XPS.

3. Results and discussion

3.1Filter membrane weight at different drying temperatures

Table 1. Statistics of filter membrane weight under different conditions.

number	Inlet flue gas temperature	Outlet flue gas temperature	flue gas volume	Ammonia consumption per hour	Ozone consumption per hour	Weight after drying at 105℃-- mgm^{-3}			Weight after drying at 175℃-- mgm^{-3}			Weight after drying at 250℃- mgm^{-3}		
	℃	℃	m3/h	kg/h	m3/h	①	②	③	①	②	③	①	②	③
1st dsy	100	54	200000	108	90-120	5.2	8.7	5.9	0.8	2.1	0.9	/	/	/
2nd dsy	125	49	200000	262	150-170	35.2	30.1	25.0	25.5	21.4	12.6	2.06	/	2.7
3rd day	145	49	442996	488	281	9.7	8.3	/	/	/	/	4.5	3.7	/

Remarks: Numbers①、②、③ were three samples, the quality of the same sample at different drying temperatures.

According to *Method of performance tests for electrostatic precipitator* (GB 13931-2017) and *Stationary source emission-Determination of mass concentration of particulate matter at low concentration-Manual gravimetric method*(HJ 836-2017), the sampling heads should be placed at a constant temperature of 105°C and 110°C, dried for 2h and 1h, and then cooled in constant temperature and humidity equipment, and weighed after meeting the corresponding prescribed conditions.

From the comparison of the data on the 1st and 2nd days in Tab.1, Under the condition that the load of the unit remained unchanged and the dedusting system was running normally, When the inlet flue gas temperature increased, the amount of ammonia and ozone recharge in the SNCR region increased, and the smoke concentration at the desulphurization outlet increased from 5.2mg/m^3, 8.7mg/m^3, 5.9mg/m^3 to 35.2mg/m^3, 30.1mg/m^3, 25.0mg/m^3. The average value increased from 6.6mg/m^3 to 30.1mg/m^3, and the absolute value of weight increased by about 78%. The increase of the soot concentration was mainly due to the chemical reaction between the escape ammonia and the unreacted NO in the SNCR region and the ozone supplied at the inlet of the desulfurization system, resulting in the formation of substances that were easily soluble in water. After desulphurization and washing, the soot sampler captures the droplets in the flue gas and leaves the soluble salt and insoluble or slightly soluble substances on the filter membrane, which lead to a significant increase in the concentration of soot.

From the comparison of the data on the 1st and 3rd days in Tab.1, Under the condition that the load of the unit remained unchanged and the dedusting system was running normally, With the increase of load, the amount of ammonia and ozone recharge in SNCR region increased, and the smoke concentration at the outlet of desulphurization increased from 5.2mg/m^3, 8.7mg/m^3 and 5.9mg/m^3 to 9.7mg/m^3 and 8.3mg/m^3. The average value rose from 6.6mg/m^3 to 9.0mg/m^3, and the absolute value of weight increased by about 7%. The smaller increase in fly ash concentration was mainly due to the increase in the load of the unit, although the amount of ammonia and ozone recharge increases. However, with the increase of unit load, the initial NO_x concentration increased greatly, and the ammonia water consumption and ozone consumption increased. The reaction degree of unreacted NO and escape ammonia equal to desulphurization inlet ozone was lower than that of excess recharge reaction. The formation of soluble matter decreased, so the concentration of smoke and dust increased slightly.

3.2Filter membrane composition analysis

Under the condition of different load, different consumption of ammonia and ozone. There was a great difference in the concentration of dust at the outlet of the desulfurization system. It was mainly due to the existence of a large number of decomposing substances in the residual substances of the filter membrane. In view of this phenomenon, the qualitative analysis of filter membrane was carried out by using TG-DSC and XPS.

3.2.1Analysis of existing substances in filter membrane. In the denitrification, desulphurization and dedusting process adopted by the unit, the participation of ozone led to the change of flue gas composition in the desulfurization system and the change of ions in the slurry of the desulfurization system.

Table 2 lists several major elementary reactions in the process of O_3 oxidation of industrial flue gas. The kinetic parameters of the reaction were taken from the NIST (National Institute of Standards and Technology) database of the American Institute of Standards. The reaction rate constant was:

$$k=AT^b\exp(-E_a/RT)$$

Table 2. Reaction kinetic parameter [14-16].

Reaction	A	b	Ea/(cal/mol)
$O_3+NO==NO_2+O_2$	8.43×10^{11}	0	2600

Reaction	A	b	Ea/(cal/mol)
$O_3+NO_2==NO_3+O_2$	8.40×10^{10}	0	4910
$NO_2+NO_3==N_2O_5$	7.98×10^{17}	-3.9	0
$O_3+SO_2==O_2+SO_3$	1.81×10^{12}	0	13910
$O_3+CO==O_2+CO_2$	6.02×10^{2}	0	0
$O_3+H_2O==O_2+H_2O_2$	6.62×10	0	0
$O_3+HCl==HOCl+O_2$	2.83	0	0

By the reaction parameters, In the process of O_3 oxidation of NO_x, the formation rate of NO_2 and N_2O_5 was higher than that of NO_3, The generated NO_3 was easy for NO_2 to react to generate N_2O_5, the final products of O_3 oxidation of NO were NO_2 and N_2O_5. The resulting N_2O_5 was easily soluble in water to form HNO_3. The reaction rate of O_3 oxidation of NO_x was higher than that of O_3 oxidation of SO_2, So O_3 had a certain selectivity [17].

According to the above reaction, the following main ions might exist in the slurry of desulphurization system due to denitrification escape ammonia and other reasons.

Cation: Ca^{2+}, NH_4^+, H^+

Anion: CO_3^{2-}, NO_3^-, SO_4^{2-}, OH^-, Cl^-

3.2.2TG-DCS result analysis.

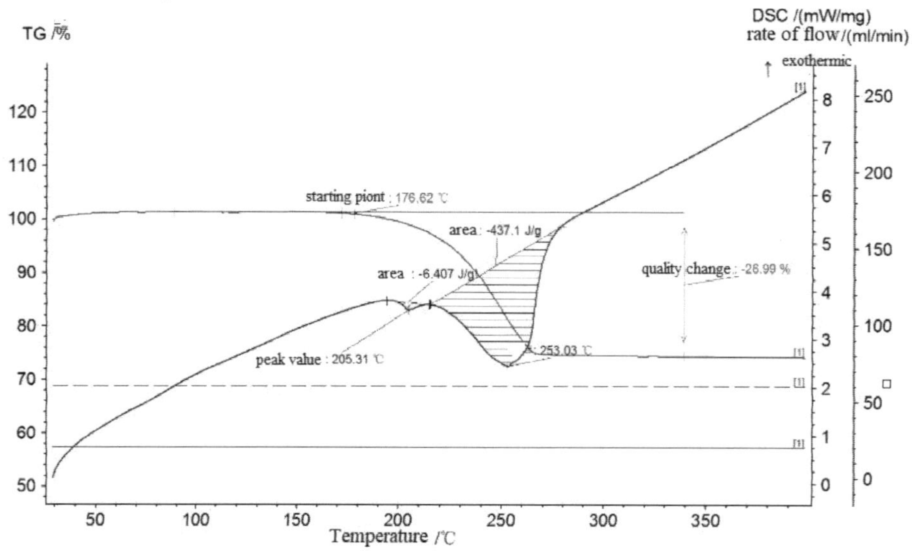

Figure 2. Analytical atlas and test results.

According to Fig.2, The sample began to decompose at 176.62℃ and stopped to decompose at 260.01℃, The weight loss rate was 26.99%.

Table 3. Decomposition temperature and Solubility of substances in slurry.

Number	substances	Decomposition temperature -℃	Solubility
1	$CaCO_3$	800	Slight dissolution

Number	substances	Decomposition temperature -℃	Solubility
2	$Ca(NO_3)_2$	132	Soluble
3	NH_4HCO_3	30	Soluble
4	NH_4NO_3	110	Soluble
5	$(NH_4)_2SO_4$	280	Soluble
6	NH_4Cl	100	Soluble
7	dust	/	Difficult to tolerate

According to Fig.2 and Tab.3, The starting decomposition substances should be mainly NH_4HCO_3, NH_4NO_3, NH_4Cl, $Ca(NO_3)_2$ and $(NH_4)_2SO_4$ at 176.62℃, After treatment at 260.01℃, The remaining substances should be $CaSO_4$, $CaCO_3$ and fly ash.

3.2.3XPS result analysis

Table 4. X-ray photoelectron spectroscopy analysis and test results.

	1st specimen		**2nd specimen**	
	After treatment at 105℃	**After treatment at 250℃**	**After treatment at 105℃**	**After treatment at 250℃**
Name	**Atomic %**	**Atomic %**	**Atomic %**	**Atomic %**
Al2p	5.81	6.21	5.5	5.33
C1s	26.31	19.14	29.61	23.74
Ca2p	1.02	0.79	0.74	0.67
Cl2p	3.14	0.34	1.18	0.4
Cr2p	0.6	0.57	0.58	0.55
F1s	0.89	0.85	0.88	0.52
Fe2p	1.39	1.258	1.23	0.77
I3d	0.08	0.09	0.1	0.1
K2p	0.14	0.19	0.28	0.13
Mg1s	0.67	0.3	0.5	0.45
N1s	5.12	3.92	3.39	2.23
Na1s	1.43	0.95	1.37	1.28
O1s	39.23	47.61	40.07	45.75
P2p	0.73	0.77	0.74	0.75
S2p	1.69	3.63	1.43	2.74
Si2p	11.07	12.79	11.8	14.07

	1st specimen			2nd specimen
	After treatment at 105℃	After treatment at 250℃	After treatment at 105℃	After treatment at 250℃
Name	Atomic %	Atomic %	Atomic %	Atomic %
Ti2p	0.69	0.59	0.6	0.54

According to Tab.4, After calcination at different temperatures, the content of oxygen was maintained at 40 to 50%, Main reason: The matrix itself contains oxygen, In the process of sample preparation, oxygen existed in the microscopic gap in the process of sample preparation, mainly by physical adsorption. Oxygen adsorption occurred in the process of entering XPS[18].

According to Tab.4, The sample decomposed after calcination at 250℃. The main substances were nitrate and carbonate, and a small amount of hypochlorite.

4. Conclusion

Through the analysis of dust concentration at the outlet of wet desulphurization after using SNCR and ozone combined with denitrification process, Denitrification combined with ozone denitrification process directly affected the concentration of smoke and dust at the outlet of desulfurization system after traditional denitrification process. According to the requirements of the national and environmental standards, the soot concentration contains the soluble salts carried in the droplets, and the main substances were carbonic acid and nitrate ammonium salt and nitrate calcium salt.

Based on the results of the above analysis, In view of the combination of the traditional denitrification process and the ozone denitrification process unit, the ozone replenishment should be strictly controlled, Prevent excessive ozone emission while preventing excessive ozone from reacting with the original components in flue gas. After the reactant was dissolved in water, it reacted with escape ammonia and other substances to form a large number of soluble salt, resulting in difficulty in dehydration of gypsum. At the same time, under the condition of high droplet control index, the droplets carried soluble salt and caused the desulphurization outlet smoke to exceed the standard.

Acknowledgements

We appreciate the financial support from the National Key Research and Development Program (NO. 2017YFB0603201).

References

[1] Ministry of Environmental Protection, National Development and Reform Commission, National Energy Administration. Fully implement the work plan of ultra-low emission and energy-saving transformation at coal burning point [Z]. 2015.

[2] BRUGGEMANN T C, KEIL F J. Theoretical investigation of the mechanism of the selective catalytic reduction of nitric oxide with ammonia on H-form zeolites[J]. Journal of Physical Chemistry C, 2008, 115(112): 17378-17387.

[3] LIU Q Y, LIU Z Y, LI C Y. Adsorption and activation of NH_3 during selective catalytic reduction of NO by NH_3[J]. Chinese Journal of Catalysis, 2006, 27(7): 636-646.

[4] ZHU T Y, LIU Q, LI Y R, *et al.* Emission characteristics of multiple pollutants from iron-steel sintering flue gas and review of control technologies[J]. Science & Technology Review, 2014, 32(33): 51-56.

[5] GóMEZ-GARCIA M A, PITCHON V, KIENNEMANN A. Pollution by nitrogen oxides: an approach to NOx abatement by using sorbing catalytic materials[J]. Environment International, 2005, 31(3): 445-67.

[6] DURME J V, DEWULF J, LEYS C, *et al.* Combining non-thermal plasma with heterogeneous catalysis in waste gas treatment: a review[J]. Applied Catalysis B Environmental, 2008, 78(3/4): 324- 333.
[7] RAJANIKANTH B S, ROUT S. Studies on nitric oxide removal in simulated gas compositions under plasma-dielectric/catalytic discharges[J]. Fuel Processing Technology, 2001, 74(3): 177-195.
[8] BASFAR A A, FAGEEHA O I, KUNNUMMAL N, *et al.* Electron beam flue gas treatment (EBFGT) technology for simultaneous removal of SO_2 and NO_x from combustion of liquid fuels[J]. Fuel, 2008, 87(8): 1446-1452.
[9] DESHWAL B R, LEE S H, JUNG J H, *et al.* Study on the removal of NO_x from simulated flue gas using acidic $NaClO_2$ solution[J]. Journal of Environmental Sciences, 2008, 20(1): 33-38.
[10] BYUN Y C, HAMILTON I P, XU T, *et al.* Formation of chlorinated species through reaction of SO_2 with $NaClO_2$ powder and their role in the oxidation of NO and Hg0[J]. Environmental Science & Pollution Research, 2014, 21(13): 8052-8058.
[11] ZHAO Y, HAO R L, QI M. Integrative process of preoxidation and absorption for simultaneous removal of SO_2, NO and Hg0[J]. Chemical Engineering Journal, 2015, 269: 159-167.
[12] BYUN Y C, KO K B, CHO M, *et al.* Reaction pathways of NO oxidation by sodium chlorite powder[J]. Environmental Science & Technology, 2009, 43(13): 5054-5059.
[13] DAI S K, XU W Q, TAO W L, *et al.* Experimental research for the application of O_3 oxidation in simultaneous removal of SO_2, NO and Hg in coal-fired flue gas[J]. Environmental Engineering, 2014, 32(10): 85-89.
[14] WEI L S, ZHOU J H, WANG Z H, *et al.* Kinetic modeling of homogeneous low-temperature multi-pollutant oxidation by ozone[J].Ozone Science & Engineering, 2007, 29(3): 207-214.
[15] WANG H Q, ZHOU K, SUN C L, *et al.* Numerical evaluation of the effectiveness of NO_2 and N_2O_5 generation during the NO ozonation process[J]. Journal of Environmental Sciences, 2016, 41(3): 51-58.
[16] WANG Z H. Mechanism study on multi-pollution control simultaneously during coal combustion and direct numerical simulation of reaction jets flow[D]. Hangzhou: Zhejiang University, 2005.
[17] Tong Ming. Simultaneous desulfurization and denitration of flue gas by ozone oxidation-calcium absorption process [J]. ENVIRONMENTAL PROTECTION OF CHEMICAL INDUSTRY. 2017, 32(2):223-226.
[18] Wang S X, He W J, Huang J J. *et al.* X-ray Photoelectron Spectroscopy of Hard Alloy with Multi-ion Implantation[J]. Acta Scientiarum Naturalium Universitatis Nankaiensis. 2019, 52(1):10-14.

Experimental Study on High Temperature Pyrolysis Characteristics of Biomass Tar in Biomass Gasification-reburning Technology

Peng He[1], Qian Zhang[1], Bing Zhang[2*], Jinhua Liu[1], Guohong Gong[1] and Dong Li[1]

[1]Shanxi Boiler and Pressure Vessel Supervision Inspection Institute, Taiyuan, Shanxi Province, 030012, China

[2]Wenzhou Special Equipment Inspection and Research Institute, Wenzhou, Zhejiang Province, 325007, China

[*]Corresponding author's e-mail: 326109580@qq.com

Abstract. In order to solve the problem of high-tar content of gasification products in biomass gasification-reburning technology, this thesis proposes an experimental scheme for high temperature thermal cracking of biomass tar, using representative rice hull tar as a sample, and self-designed pyrolysis experiment. On the stage, the reaction temperature of the gasification chamber was measured by a high-temperature thermocouple. At the same time, the biomass tar cracking rate was measured at different temperatures with the thermobalance by appropriately changing pyrolysis temperature, and the Agilent HP5973 GC/MS was employed to analyze the components of pyrolysis products of biomass tar under different temperature conditions. The experimental results show that the total thermal cracking rate of tar increased with the increasing of experimental temperature. In the temperature range of 550°C~655°C, the increasing rate of thermal cracking becomed slower; when it reached between 655°C and 790°C, the thermal-cracking rate showed a significant growth trend; while entering 790°C~853°C, the growth rate slowed down. The experimental and analytical results reveal the basic laws of high-temperature pyrolysis of biomass tar, and provide a certain technical basis for the industrial production of low-tar and high-quality biomass gas.

1. Introduction

While the global fossil fuel reserves are gradually declining, and the problems of environmental changes or pollution brought by fossil fuels combustion are becoming increasingly prominent, and biomass reburning technology has gradually attracted attention. Biomass mainly refers to agricultural waste such as straw, plant and forestry processing waste, etc. It is a kind of renewable and clean energy. Biomass resource in China is abundant. According to statistics, the annual output of crop straw alone is about 700 million tons, and the annual output of forestry and wood processing waste is about 900 million tons, equivalent to approximately 800 million tons of standard coal [1]. Due to its seasonality, scattered distribution, and low energy density, and high transportation and storage costs, the large-scale application of biomass energy is greatly limited. However, considering the maximization of economic and environmental benefits, the effective use of biomass can alleviate energy tension to a certain extent, and it is still the most popular technology.

Generally, there are two ways on biomass reburning technology, including direct reburning and reburning with gasification [2]. Direct reburning is realized by the drying and pulverization of biomass as a variety of secondary fuel; reburning with gasification is a way to secondary fuel produced by biomass pyrolysis gasification [3]. The main components of biomass fuel are C, O, low content of N, S, Na, K, etc. It is easy to catch fire because of its high volatile content and low calorific value; its combustion mainly produces CO_2, Sox and a little of NO_x. While CO_2 can be absorbed during plant photosynthesis, biomass burning can achieve zero CO_2 emissions from a large cycle. Meanwhile, Na and K can be utilized properly as catalysts to promote the forward reaction progress and promote the efficiency of de-NO; but it can reduce the ash melting point and cause ash to be fouled in heated areas [4].

At present, the high tar content of biomass gasification products is one of the problems that restrict the development of biomass gasification technology. The tar removal of biomass gasification gas is divided into physical method and chemical method. Although tar can be removed effectively with the physical method, it only undergoes a whole of phase change and is not really removed. The tar-containing wastewater must be treated to prevent environmental pollution; and energy contained in tar can not be utilized, which reduces the gasification efficiency [5]. Chemical methods [6] can be divided into catalytic cracking and pyrolysis. Cartin Myren and other foreign scholars have done a lot of work around tar catalytic cracking. The catalysts are used commonly such as dolomite, silica, and nickel base [7]. In China, Hou Bin [8] et al. in Tsinghua University, Zhou Jinsong [9] et al in Zhejiang University, Lv Pengmei [10] et al. in Guangzhou Energy Institute have done a lot of research in biomass tar catalytic cracking. In terms of high-temperature pyrolysis, YG Pan [11] et al. believe that a certain amount of secondary air is injected under the operating conditions of 830~850°C in the fluidized bed biomass gasification experiment, which is beneficial to eliminating tar in the product gas. For the problem of high biomass tar content, most scholars at home and abroad are devoted to the research of tar catalytic cracking, and there are relatively fewer studies on high temperature pyrolysis of tar. In this thesis, a high temperature secondary pyrolysis experiment is carried out on rice husk tar, and the effects of pyrolysis temperature variation on the pyrolysis ratio of rice hull tar and the composition and content of pyrolysis products are studied. The fixed bed reactor used in this experiment is small in volume, and the gas residence time can be strictly controlled, and the temperature distribution can be adjusted, which is beneficial to increasing the conversion rate and selectivity of chemical reactions, and is beneficial to the high-temperature secondary cracking of tar.

2. Experimental Devices and Methods

2.1Experimental Materials

Sample: The experimental sample selected from rice husk tar was taken from a farm in Heilongjiang Province. The sample was ventilated and dried at 105°C for 4 hours in a dry box to remove moisture, and sealed and stored in the dark. Its elemental analysis was tested by the commissioned National Coal and Coal Chemical Product Quality Supervision and Inspection Center. The composition analysis results are shown in Table 1.

Table 1. Elemental analysis of biomass tar.

Sample	Water content (%)	Ash content (%)	C_{ar} (%)	H_{ar} (%)	O_{ar} (%)
Rice husk tar	16.50	0.06	39.8	7.3	52.9

2.2Experimental bench design

As shown in Figure 1, the self-designed biomass tar high-temperature pyrolysis test-bed is adopted in this experiment, which is mainly composed of gas supply system, combustion system, thermal cracking system, temperature measurement and temperature-control system and gas-cooling system. The main reactor of the thermal cracking system has a total length of 700 mm, and the reaction section of the bed is a stainless steel tube with an inner diameter of 48 mm, and the length of the reaction section is 600 mm. In order to minimize the heat loss of reactions, the gap between the periphery of the reactor and the protective iron sheet is filled with insulating quartz sand. The manual quantitative one-time feeding is used in the main reactor, and carbon-steel material is used in the upper exhaust pipe, and the other components are made of stainless steel. In order to achieve the purpose on high-temperature secondary pyrolysis of biomass tar, the test a two-stage burner design is adopted, as well as a high-temperature-resistant concrete to insulate the combustion system. The high-temperature flue gas produced by the propane burner provides the heat required for the biomass tar pyrolysis. The uppermost part of the reactor is equipped with a quick-response armored thermocouple to measure the inlet gas temperature, and the reaction temperature is controlled in the range from 400°C to 1300°C. In order to ensure the accuracy of the thermocouple temperature measurement, the working end of the thermocouple should be kept at the center of the reactor, so as to achieve real-time monitoring of the cracking and gasifi-cation temperature in the reactor. The gas generated by biomass-tar pyrolysis is condensed by a condenser, and the tap water is used as a condensing medium, and is recycled by a water pump. Non-condensable gas is emptied or sampled for analysis.

2.3Experimental process

The biomass tar is dosed to the inlet of the pyrolysis reactor and falls to the bottom stainless steel mesh tray. On the one hand the tray acts as a filter during the thermal cracking process, and on the other hand extends the reaction residence time of the tar. During the experiment firstly the combustible gas supplied by the propane cylinder and the air provided by the air compressor is fully burned in the secondary burner, and Laoying3026 Infrared flue gas analyzer is used to monitor the oxygen content in the high temperature flue gas at the tail to ensure that the tar-amount reduction is due to high-temperature pyrolysis instead of combustion; the high-temperature flue gas discharged from the secondary burner enters the thermal-cracking reactor; after the temperature is stabilized, the valve of the bypass line will be opened to make the high-temperature flue gas by the first-stage burner (combustion of propane and air) enter the biomass-tar pyrolysis main-bed (high-temperature flue gas requires a slight excess of fuel to ensure that the biomass tar is pyrolyzed under anoxic conditions); the tar undergoes thermal cracking, and the solid product after thermal cracking remains on the mesh disk, and the gaseous product along with other residues, flows out through the pores of the disk. At the same time, the gaseous products and residues produced by once-cracking are subjected to secondary-cracking by the higher temperature flue gas generated by the secondary burner; and the cracked gas is condensed into liquid through the condenser and collected in the vessel; and the cooled permanent gas flows through the detention barrel and then is vented or collected for sampling analysis. Experiments have confirmed that the tar content in gas after cooling is extremely low and can be ignored. Within the test the system maintained a micro-positive pressure condition to prevent air leakage and ensure the pyrolysis conditions inside the reactor.

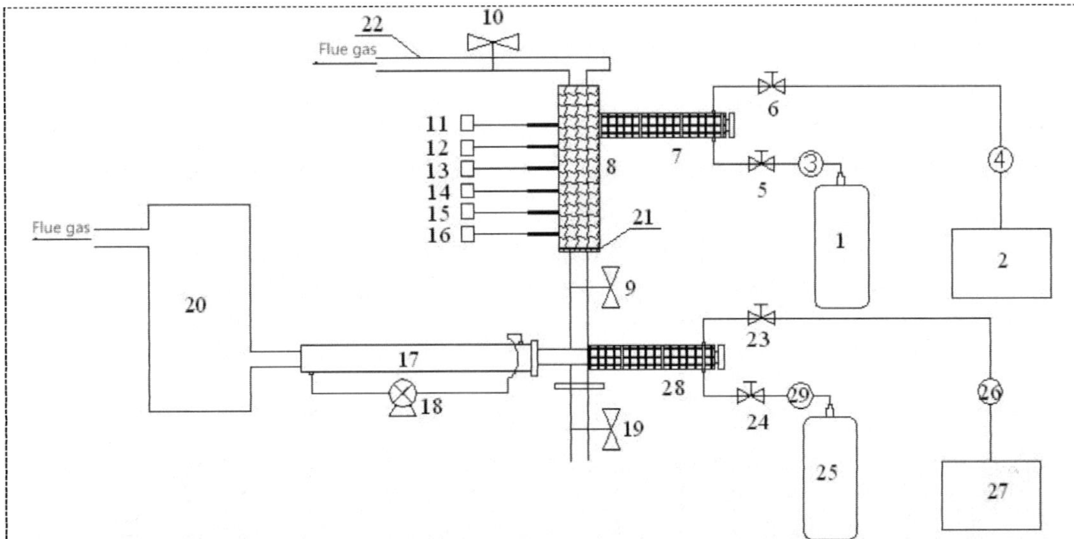

Figure 1. Schematic diagram of biomass-tar high-temperature pyrolysis test bench with components including: propane cylinders of 1 and 25; air compressors of 2 and 27; barometer of 3, 4, 26 and 29; regulating valve of 5, 6, 23 and 24; burners of 7 and 28; fixed bed reactor of 8; ball valves of 9, 10 and 19; thermocouple and digital display tables of 11, 12, 13, 14, 15, 16, 30, 31 and 32; condensation tube of 17; water pump of 18; retention barrel of 20; air distribution plate of 21; reactor upper exhaust tube of 22.

3. Experimental contents and results analysis of biomass-tar high-temperature pyrolysis

3.1Effect of pyrolysis temperature on biomass tar pyrolysis

Biomass-tar pyrolysis is the secondary cracking of biomass tar at higher temperature levels, which is converted into smaller molecular weight gaseous compounds and other products. The temperature required for this method is relatively high, generally reaching 1000~1200°C. The tar is sufficiently decomposed twice at high temperature to make the content extremely low. During the biomass tar pyrolysis, it is heated under an inert atmosphere to cause decomposition of the polymer to produce biomass charcoal, condensable liquid and gaseous products. The pyrolysis conversion reaction of biomass tar is an endothermic reaction. The pyrolysis temperature plays a decisive role in the pyrolysis effect of biomass tar, which not only has a great influence on the pyrolysis ratio of biomass tar, but also directly affects the relevant components of pyrolysis products.

Researchers have added water vapor to reduce tar content and promote tar conversion. The principle is that water vapor reacts with certain components in tar to produce CO, H_2 and CH_4, and reduces the formation of carbon black. The addition of some oxide can be used because the oxide react with some components of the tar, resulting in a tar conversion [12]. In this experiment, the high-temperature flue gas used in the pyrolysis biomass taris provided by propane combustion, and there is no shortage of water vapor, which is also beneficial to the tar content reduction in the tar cracking gas to some extent.

Biomass tar pyrolysis is usually a process of thermal decomposition under an inert atmosphere. It is affected by many aspects, but the reaction temperature plays a decisive role, which directly determines the pyrolysis ratio of biomass tar, that is to say, the degree of biomass tar pyrolysis.

In order to characterize the cleavage effect of biomass tar, this experiment defines the biomass tar pyrolysis ratio as follows:

Pyrolysis ratio of biomass tar represents the percentage of the total mass for the mass of biomass tar pyrolysis during a certain temperature rise, that is:

$$X_{tar} = \frac{M_{tar1} - M_{tar2}}{M_{tar1}} \times 100\% = \frac{\Delta m_{tar}}{M_{tar1}} \times 100\% \tag{1}$$

where X_{tar} denotes the pyrolysis ratio of the biomass tar sample (%); M_{tar1} and M_{tar2} respectively mean the mass of biomass tar before and after cracking (g); Δm_{tar} is the mass reduction of the biomass tar sample after undergoing a certain temperature-rise process with $\Delta m_{tar} = M_{tar1} - M_{tar2}$ (g).

Experiments have proved that acetone has good solubility and high solubility for tar. The condenser tube can be washed with acetone. The mass difference before and after the acetone rinsing is the mass of the biomass tar pyrolysis product. And make sure to recover all solvents as much as possible to reduce systematic errors. Thus, the pyrolysis ratio is calculated as shown in Table 2.

Table 2. Effect of pyrolysis temperature on pyrolysis ratio.

primary/secondary pyrolysis temperature (°C)	mass of tar sample before pyrolysis (g)	mass of tar sample after pyrolysis (g)	Pyrolysis mass of tar samples (g)	pyrolysis ratio of tar samples (%)	Condenser inlet/outlet flue gas temperature (°C)
694/708	31.746	7.395	24.351	76.71	280/31
500/790	24.397	4.138	20.259	83.04	470/41
660/924	23.985	1.970	22.015	91.79	483/43
715/1009	24.841	1.726	23.115	93.05	541/46
715/1100	27.536	1.537	25.999	94.42	644/50
678/1203	27.272	0.836	26.436	96.93	800/71

Experimental results are analyzed as follows:

1) Figure 2 shows the variation of rice husk tar with pyrolysis temperature under an inert atmosphere. It can be seen that under the pyrolysis condition, X_{tar} grows faster than the pyrolysis temperature increasing, within the primary pyrolysis temperature range of 500~715°C and the secondary pyrolysis temperature range of 708~1203°C, and reaches up to over 96% at the primary/secondary pyrolysis temperature of 678°C/1203°C. The analysis of reasons is that the secondary pyrolysis reaction of the tar in the experimental system is aggravated by the increase of the pyrolysis temperature, and some tar components will be further converted into gaseous light hydrocarbons or liquid products or other small molecules or charcoal etc by demethylation, intramolecular bridges and hydrogen bond cleavage and so on, causing an overall reduction of tar content in the pyrolysis gas. By contrast, it is explained that the proper increase of the temperature level during the tar pyrolysis contributes to the deepening of the pyrolysis reaction of biomass tar, as well as the decrease of the tar content in pyrolysis gas products. Pyrolysis at temperatures above 1200°C has not been studied due to experimental conditions, but it is foreseeable that the trend of further increase with the increasing of pyrolysis temperature will gradually become flat.

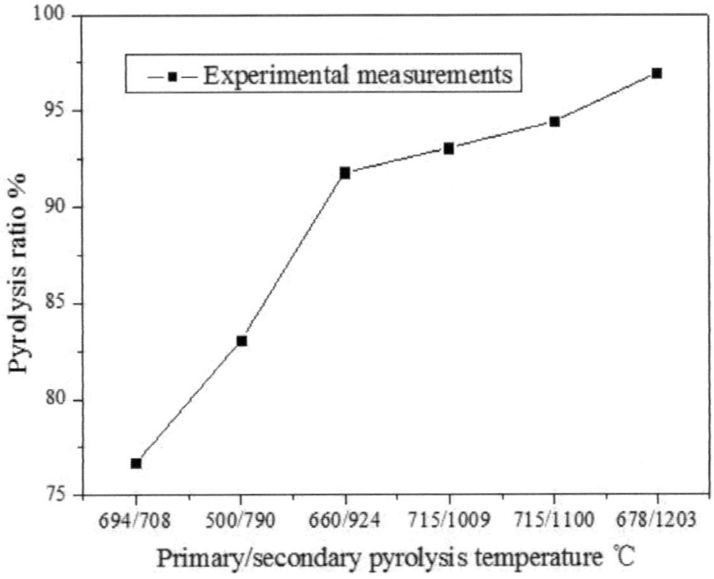

Figure 2. Relationship between primary/secondary pyrolysis temperature and pyrolysis ratio.

2) In the experimental conditions, X_{tar} can reach up to 96.93% and increase with the increasing of the pyrolysis temperature. As can be seen from table 2, the pyrolysis ratio of the tar sample increased from 76.71% to 91.79% with the primary pyrolysis temperature approaching 700°C and the secondary pyrolysis temperature increasing from 708°C to 924°C. The pyrolysis ratio of the tar sample after 924°C grows with temperature rise and slows down. This shows that high-temperature secondary pyrolysis has a significant effect on biomass tar pyrolysis, and it is very difficult to attempt to achieve the same pyrolysis ratio variation under high temperature compared with lower temperatures.

3) There must be components which are quite difficult to pyrolyze in the remaining pyrolysis products, so the substances which are continuously increasing s as the pyrolysis temperature rise are components which are more difficult to pyrolyze in the GC-MS analysis.

4) When the pyrolysis temperature is in the range of 600~700°C, and the secondary pyrolysis temperature increases, the tail flue discharges rapidly a large amount of non-condensable gas, which reflects the pyrolysis rate is relatively high under the temperature conditions.

5) Considering that some of biomass tar pyrolysis products may be carried by the flue gas into the detention tank to be decelerated and not collected by the condenser, the residue in the detention tank needs to be collected after each experiment. There is mainly water, in addition to a small amount of impurities; the mass of impurities obtained with filter paper, corresponding to the secondary pyrolysis temperatures from high to low, is 0.203g, 0.239g, 0.313g, 0.475g, 0.513g and 0.561g. Even if the impurities are all pyrolysis products, their mass is much smaller than the total amount, and so it can be ignored.

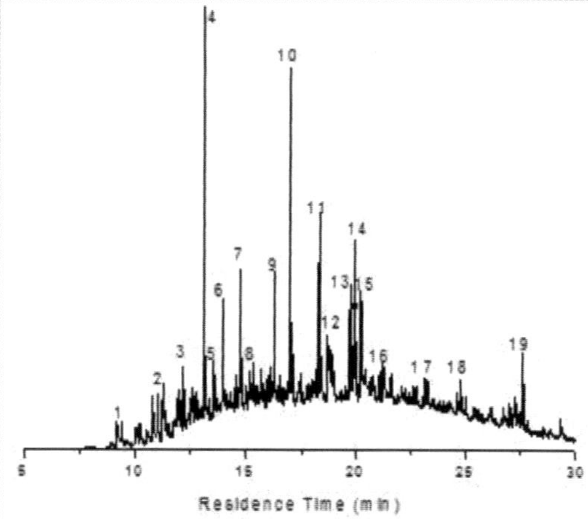

Figure 3. Total ion chromatogram of tar pyrolysis at the pyrolysis temperature of 694/708°C.

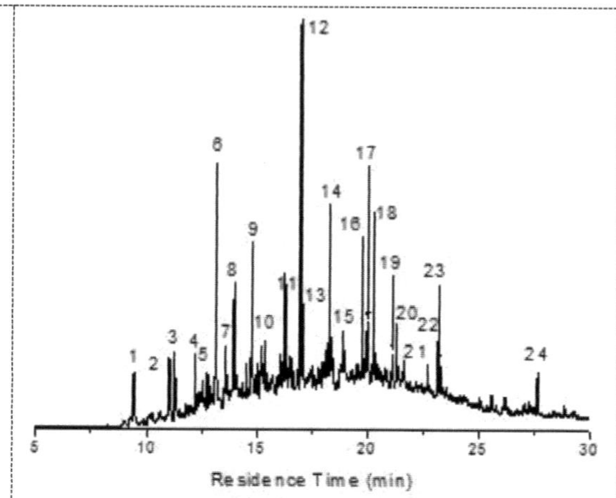

Figure 4. Total ion chromatogram of tar pyrolysis at the pyrolysis temperature of 500/790°C.

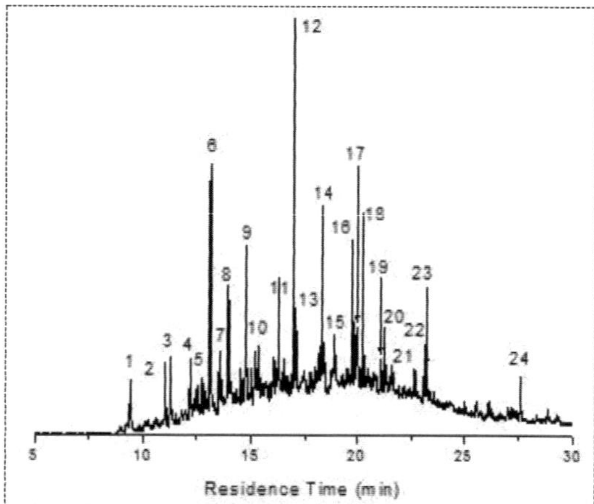

Figure 5. Total ion chromatogram of tar pyrolysis at the pyrolysis temperature of 660/924°C.

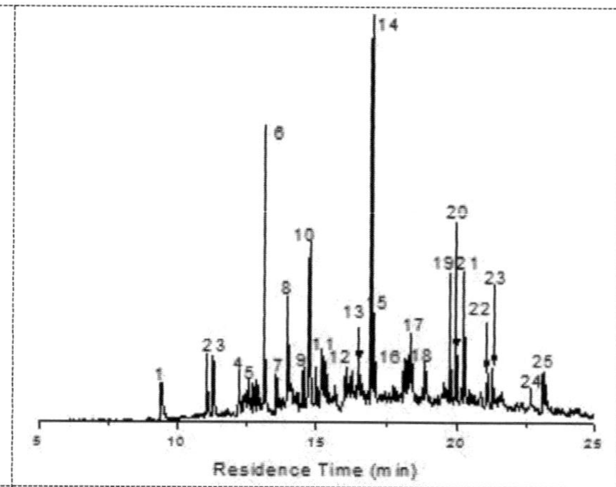

Figure 6. Total ion chromatogram of tar pyrolysis at the pyrolysis temperature of 715/1009°C.

3.2GC-MS Analysis of High Temperature Pyrolysis of Biomass Tar

Instrument: American Agilent HP5973GC/MS GC/MS.

Chromatographic conditions: HP-5 trace analysis column is used with the capillary column of 30m ×0.25mm×0.25μm; carrier gas is helium of high purity, and its gas flow is 1.0mL/min,and its nominal initial pressure is 7.89psi, and its average flow rate is 36cm/s, and the split ratio is 5:1; the column temperature (initial temperature) is 55°C, and the equilibrium time is 0.5min and it rises to 300°C at the rate of 10°C/min and stay 22 minutes; the inlet temperature is 320°C.

Mass spectrometry conditions: ionization source is EI, and electron bombardment energy is 70eV, and mass range m/z is 15-650amu, and scanning time is 3min, and mass scanning mode is SCAN.

These small molecular products are carried into the column by carrier gas (usually ruthenium) for separation, and then enter the mass spectrometry system; the structure identification of the cleavage products is carried out, and finally the pyrolysis components of the biomass tar are analyzed. Analysis of the cleavage products can characterize the pyrolysis of one kind of material.

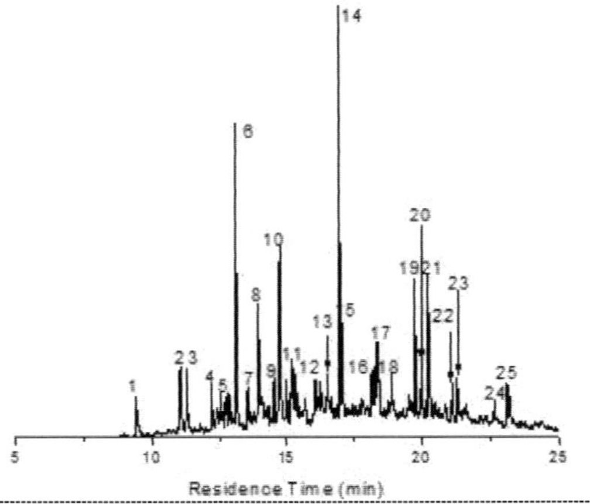

Figure 7. Total ion chromatogram of tar pyrolysis at the pyrolysis temperature of 715/1100°C.

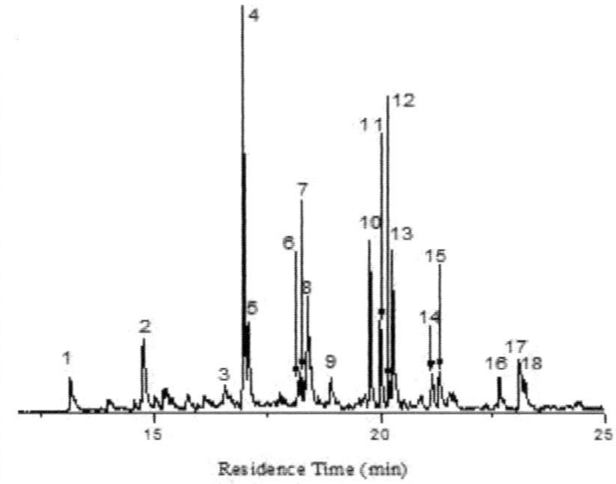

Figure 8. Total ion chromatogram of tar pyrolysis at the pyrolysis temperature of 678/1203°C.

Figure 3 shows the total ion chromatogram of the pyrolysis of rice husk tar at the pyrolysis temperature of 694°C/708°C. Figures 4, 5, 6, 7 and 8 are respectively the total ion chromatograms of tar pyrolysis at 500°C/790°C, 660°C/924°C, 715°C/1009°C, 715°C/1100°C and 678°C/1203°C. The main components and relative contents of biomass tar pyrolysis products under different primary/secondary pyrolysis temperature combinations are shown in Tables 3 to 7. Summarily the biomass tar is gradually cracked as the reaction temperature rises, and the types of pyrolysis product gradually increase, and the types of pyrolysis products decrease when it rises to 1300°C. The main products of high-temperature secondary pyrolysis of biomass tar are naphthalene, phenanthrene, anthracene, anthracene, anthracene, dibenzofuran, fluoranthene, biphenyl, biphenylene and its derivatives. And with the increasing of pyrolysis temperature the total contents of main cleavage products of tar showed an increasing trend; the contents of several main components such as lanthanum, phenanthrene, anthracene, fluoranthene and their derivatives increased significantly with the increasing of pyrolysis temperature; the increase amplitudes were 19.540%, 17.792%, 20.389% and 13.518% respectively. The product components whose contents are increasing are mainly aromatic organic compounds and polycyclic aromatic hydrocarbons.The analysis of reasons is that: on the one hand, some components in the tar are converted into aromatic organic compounds or polycyclic aromatic hydrocarbons under high temperature conditions; on the other hand, because the aromatic group has a lower reaction rate than other groups, it is expressed as the content of the aromatic has shown an increasing trend compared with other ethnic groups and its final increase amplitude was higher than the others. This further demonstrates that as the pyrolysis proceeds, the treatment of the tar pyrolysis products will become more and more difficult.

Sample analysis results: After instrumental analysis the total ion chromatograms can be obtained. According to them, the relative contents of main compounds that can be identified are listed below:

Table 3. The relative contents of the distinguishable major compounds of tar pyrolysis products at the primary/secondary pyrolysis temperature of 694°C/708°C.

Serial Number	Residence Time (min)	Name	Molecular Formula	Molecular Weight	Relative Content (%)
1	9.173	Phenol, 4-ethyl-	$C_8H_{10}O$	122	1.033
2	10.811	4-Hydroxy-3-methylacetophenone	$C_9H_{10}O_2$	150	1.182
3	11.320	Phenol, 2-methoxy-4-propyl-	$C_{10}H_{12}O_2$	164	1.213
4	11.920	Naphthalene, 2, 3-dimethyl-	$C_{12}H_{12}$	156	1.374

5	12.184	Biphenyl	$C_{12}H_{10}$	154	1.015
6	12.570	Dibenzofuran	$C_{12}H_8O$	168	1.290
7	13.100	Fluorene	$C_{13}H_{10}$	166	7.221
8	13.540	4-Methyl-5-penta-1, 3-dienyltetrahy drofuran-2-one	$C_{10}H_{14}O_2$	166	1.378
9	13.608	Dibenzofuran, 4-methyl-	$C_{13}H_{10}O$	182	1.412
10	13.939	Pyridine, 4-(2-(pyridyl-4) ethenyl)-, trans-	$C_{12}H_{10}N_2$	182	2.281
11	14.520	Cyclopropane, 1methyl-2-pentyl-	C_9H_{18}	126	1.085
12	14.736	Anthracene	$C_{14}H_{10}$	178	2.433
13	15.337	Phenanthrene, 2-methyl-	$C_{15}H_{12}$	192	1.139
14	15.683	n-Hexadecanoic acid	$C_{16}H_{32}O_2$	256	1.141
15	16.057	8, 11-Octadecadienoic acid, methylester	$C_{19}H_{34}O_2$	294	1.152
16	16.257	Fluoranthene	$C_{16}H_{10}$	202	2.500
17	16.947	11H-Benzo[b]fluorene	$C_{17}H_{12}$	216	4.806
18	17.055	Fluoranthene, 2-methyl-	$C_{17}H_{12}$	216	1.573
19	18.279	1-Propene, 3-(2-cyclopentenyl)-2-methyl-1, 1-diphenyl-	$C_{21}H_{22}$	274	2.307

The phenolic materials present at the pyrolysis temperature combination of 694°C/708°C is absent at higher temperatures, indicating that the reaction temperature is increased, the secondary pyrolysis of the tar is enhanced, and the intramolecular bridges of tar components are broken. The polycyclic substance is subjected to a ring-opening reaction by heat and is converted into monocyclic substances.

Table 4. The relative contents of the distinguishable major compounds of tar pyrolysis products at the primary/secondary pyrolysis temperature of 500°C/790°C.

Serial Number	Residence Time (min)	Name	Molecular Formula	Molecula r Weight	Relative Content (%)
1	9.391	Azulene	$C_{10}H_8$	128	7.612
2	11.022	Naphthalene, 2-methyl-	$C_{11}H_{10}$	142	5.915
3	11.259	Naphthalene, 1-methyl-	$C_{11}H_{10}$	142	6.547
4	12.189	Biphenyl	$C_{12}H_{10}$	154	3.051
5	12.525	Naphthalene, 1, 5- dimethyl-	$C_{12}H_{12}$	156	1.533
6	12.711	Naphthalene, 1, 6-dimethyl-	$C_{12}H_{12}$	156	1.338
7	12.754	Naphthalene, 2, 7-dimethyl-	$C_{12}H_{12}$	156	1.415
8	13.105	Biphenylene	$C_{12}H_8$	152	9.716
9	13.541	1-Isopropenylnaphthalene	$C_{13}H_{12}$	168	2.009
10	13.942	Dibenzofuran	$C_{12}H_8O$	168	4.036
11	14.736	Fluorene	$C_{13}H_{10}$	166	5.974

IOP Conf. Series: Materials Science and Engineering **721** (2020) 012002 doi:10.1088/1757-899X/721/1/012002

12	15.18	Dibenzofuran, 4-methyl-	$C_{13}H_{10}O$	182	2.304
13	15.337	[1, 1'-Biphenyl]-4-carboxaldehyde	$C_{13}H_{10}O$	182	2.009
14	16.060	9H-Fluorene, 1-methyl-	$C_{14}H_{12}$	180	1.599
15	16.253	3H-Benz[e]indene, 2-methyl-	$C_{14}H_{12}$	180	3.648
16	16.947	Phenanthrene	$C_{14}H_{10}$	178	10.958
17	17.055	Anthracene	$C_{14}H_{10}$	178	3.191
18	18.278	Pentadecanoic acid, 14-methyl-, methylester	$C_{17}H_{34}O_{12}$	270	2.033
19	18.357	4H-Cyclopenta [def] phenanthrene	$C_{15}H_{10}$	190	2.181
20	18.872	1, 2, 4, 8-Tetramethy lbicyclo [6.3.0] undeca-2, 4-diene	$C_{15}H_{24}$	204	1.211
21	19.981	Fluoranthene	$C_{16}H_{10}$	202	8.342
22	20.239	Pyrene	$C_{16}H_{10}$	202	7.616
23	21.269	1-Naphthalenamine, N-phenyl-	$C_{16}H_{13}N$	219	1.351
24	23.080	Cyclopenta[cd]pyrene	$C_{18}H_{10}$	226	3.448

Table 5. The relative contents of the distinguishable major compounds of tar pyrolysis products at the primary/secondary pyrolysis temperature of 660°C/924°C.

Serial Number	Residence Time (min)	Name	Molecular Formula	Molecular Weight	Relative Content (%)
1	9.400	Naphthalene	$C_{10}H_8$	128	2.332
2	11.028	Naphthalene, 2-methyl-	$C_{11}H_{10}$	142	3.412
3	11.268	Naphthalene, 1-methyl-	$C_{11}H_{10}$	142	3.645
4	12.189	Biphenyl	$C_{12}H_{10}$	154	1.915
5	12.714	Naphthalene, 1, 6-dimethyl-	$C_{12}H_{12}$	156	1.212
6	12.755	Naphthalene, 2, 6-dimethyl-	$C_{12}H_{12}$	156	1.365
7	13.107	Biphenylene 1-	$C_{12}H_8$	152	8.688
8	13.544	Isopropenylnaphthalene	$C_{13}H_{12}$	168	2.009
9	13.942	Dibenzofuran	$C_{12}H_8O$	168	4.209
10	14.522	L-Valine, N-[(9H-fluoren-9-ylmethoxy) carbonyl]-	$C_{20}H_{21}NO_4$	339	1.277
11	14.737	Fluorene	$C_{13}H_{10}$	166	6.872
12	14.985	1H-Phenalene	$C_{13}H_{10}$	166	1.735
13	15.178	Dibenzofuran, 4-methyl-	$C_{13}H_{10}O$	182	2.181
14	15.338	3-(2-Naphthyl) acrylaldehyde	$C_{13}H_{10}O$	182	1.568
15	16.058	9H-Fluorene, 1-methyl-	$C_{14}H_{12}$	180	1.849
16	16.256	1-Heptene, 2-isohexyl-6-methyl-	$C_{14}H_{28}$	196	4.716
17	16.508	1, 1'-Biphenyl, 4-ethenyl-	$C_{14}H_{12}$	180	1.802

18	16.948	Anthracene	$C_{14}H_{10}$	178	10.193
19	17.055	1, 10b(2H)-Dihydropyrano[3, 4, 5-jk] fluorene	$C_{15}H_{12}O$	208	2.721
20	18.125	Anthracene, 9-methyl-	$C_{15}H_{12}$	192	1.045
21	18.279	Pentadecanoic acid, 14-methyl, methyl ester	$C_{17}H_{34}O_2$	270	4.714
22	18.361	4H-Cyclopenta[def]phenanthrene	$C_{15}H_{10}$	190	1.691
23	18.872	5, 16[1', 2']: 8, 13[1", 2"]-Dibenzen odibenzo[a, g] cyclododecene, 6, 7, 14, 15-tetrahydro-	$C_{32}H_{24}$	408	1.526
24	18.915	Benzenamine, 4-methoxy-N-(3-pyridinylmethylene)-	$C_{13}H_{12}N_2O$	212	1.095
25	19.747	Fluoranthene	$C_{16}H_{10}$	202	4.774
26	19.941	9-Octadecenoic acid (Z)-, methylester	$C_{19}H_{36}O_2$	296	1.311
27	19.981	Benzene, 1, 1'-(1, 3-butadiyne-1, 4-diyl) bis-	$C_{16}H_{10}$	202	2.938
28	20.240	Pyrene	$C_{16}H_{10}$	202	4.131
29	21.111	Fluoranthene, 2-methyl-	$C_{17}H_{12}$	216	1.872

Table 6. The relative contents of the distinguishable major compounds of tar pyrolysis products at the primary/secondary pyrolysis temperature of 715°C/1009°C.

Serial Number	Residence Time (min)	Name	Molecular Formula	Molecula r Weight	Relative Content (%)
1	11.038	Naphthalene, 1-methyl-	$C_{11}H_{10}$	142	5.989
2	11.274	Naphthalene, 2-methyl-	$C_{11}H_{10}$	142	6.996
3	12.198	Biphenyl	$C_{12}H_{10}$	154	4.696
4	13.112	Biphenylene 1-	$C_{12}H_8$	152	13.042
5	13.551	Isopropenylnaphthalene	$C_{13}H_{12}$	168	3.751
6	13.949	Dibenzofuran	$C_{12}H_8O$	168	8.744
7	14.742	Fluorene	$C_{13}H_{10}$	166	12.135
8	15.186	Dibenzofuran, 4-methyl-	$C_{13}H_{10}O$	182	2.476
9	16.953	Phenanthrene	$C_{14}H_{10}$	178	18.783
10	17.065	Anthracene	$C_{14}H_{10}$	178	6.340
11	18.284	Pentadecanoic acid, 14-methyl-, methylester	$C_{17}H_{34}O_2$	339	3.908
12	19.755	Fluoranthene	$C_{16}H_{10}$	202	6.041
13	14.985	Pyrene	$C_{16}H_{10}$	202	7.098

The contents of naphthalene, biphenyl, biphenylene, dibenzofuran and their derivatives have all risen first and fallen later; and the contents of naphthalene and its derivatives reached maximum at the pyrolysis temperature of 500°C/790°C; and the contents of biphenyl, biphenylene and dibenzofuran reached maximum at the combination of pyrolysis temperature of 715°C/1009°C and then decreased rapidly. This is indicated that the high temperature above 1000°C is not conducive to the presence of these compounds. The reason for this may be that the substances containing -OH, C-O and C=O are cracked due to their thermal instability under high temperature conditions; meanwhile, the side rings of these materials are easy to break under high temperature conditions to form other substances, resulting in a sudden content increase of naphthalene, biphenyl, biphenylene, dibenzofuran and its derivatives. Substances existing at lower temperatures such as Pyridine, 4-(2-(pyridyl-4) ethenyl)-, trans-, have N element and heterocyclic compounds containing an element such as N or S are also thermally unstable, so they are liable to be broken when the temperature is further increased.

Table 7. The relative contents of the distinguishable major compounds of tar pyrolysis products at the primary/secondary pyrolysis temperature of 715℃/1100℃.

Serial Number	Residence Time (min)	Name	Molecular Formula	Molecular Weight	Relative Content (%)
1	9.409	Naphthalene	$C_{10}H_8$	128	2.465
2	11.032	Naphthalene, 1-methyl-	$C_{11}H_{10}$	142	3.276
3	11.268	Naphthalene, 2-methyl-	$C_{11}H_{10}$	142	2.933
4	12.196	Naphthalene, 2-ethenyl-	$C_{12}H_{10}$	154	1.915
5	12.718	Naphthalene, 2, 6-dimethyl-	$C_{12}H_{12}$	156	2.404
6	12.853	Biphenyl	$C_{12}H_{10}$	154	1.161
7	13.108	Biphenylene	$C_{12}H_8$	152	9.391
8	13.552	1-Isopropenylnaphthalene	$C_{13}H_{12}$	168	1.696
9	13.947	Dibenzofuran	$C_{12}H_8O$	168	5.750
10	14.527	1H-Phenalene	$C_{13}H_{10}$	166	1.702
11	14.686	Fluorene-9-methanol	$C_{14}H_{12}O$	196	1.178
12	14.740	Fluorene	$C_{13}H_{10}$	166	15.117
13	15.345	Dibenzofuran, 4-methyl-	$C_{13}H_{10}O$	182	2.502
14	16.261	9H-Fluorene, 1-methyl-	$C_{14}H_{12}$	180	5.020
15	16.518	9H-Fluoren-9-one	$C_{13}H_8O$	180	2.749
16	16.949	Phenanthrene	$C_{14}H_{10}$	178	11.348
17	17.059	Anthracene	$C_{14}H_{10}$	178	4.058
18	18.287	Phenanthrene, 1-methyl-	$C_{15}H_{12}$	192	4.944
19	18.364	4H-Cyclopenta[def]phenanthrene	$C_{15}H_{10}$	190	3.393
20	18.877	1, 2, 4, 8-Tetramethylbicyclo[6.3.0] undeca-2, 4-diene	$C_{15}H_{24}$	204	2.025
21	19.984	Fluoranthene	$C_{16}H_{10}$	202	6.501
22	20.243	Pyrene	$C_{16}H_{10}$	202	2.968
23	21.114	Pyrene, 1-methyl-	$C_{17}H_{12}$	216	1.891
24	21.275	11H-Benzo[b]fluorene	$C_{17}H_{12}$	216	1.788
25	23.113	Triphenylene	$C_{18}H_{12}$	228	1.042

Table 8. The relative contents of the distinguishable major compounds of tar pyrolysis products at the primary/secondary pyrolysis temperature of 678°C/1203°C.

Serial Number	Residence Time (min)	Name	Molecular Formula	Molecular Weight	Relative Content (%)
1	13.137	Acenaphthylene	$C_{12}H_8$	152	3.724
2	14.758	Fluorene	$C_{13}H_{10}$	166	6.357
3	16.544	9H-Fluoren-9-one	$C_{13}H_8O$	180	2.478
4	16.954	Anthracene	$C_{14}H_{10}$	178	20.881
5	17.067	Phenanthrene	$C_{14}H_{12}$	178	7.095
6	18.142	Anthracene, 1-methyl-	$C_{15}H_{12}$	192	1.092
7	18.204	Phenanthrene, 2-methyl-4H-	$C_{15}H_{12}$	192	2.035
8	18.368	Cyclopenta[def]phenanthrene	$C_{15}H_{10}$	190	9.801
9	18.890	1, 2, 4, 8-Tetramethylbicyclo [6.3.0]undeca-2, 4-diene	$C_{15}H_{24}$	204	2.371
10	20.164	Fluoranthene	$C_{16}H_{10}$	202	17.519
11	20.247	Pyrene	$C_{13}H_{10}$	202	11.108
12	21.129	Pyrene, 2-methyl-	$C_{17}H_{12}$	216	3.024
13	21.295	11H-Benzo[a]fluorene	$C_{17}H_{12}$	216	3.132
14	23.100	Cyclopenta[cd]pyrene	$C_{18}H_{10}$	226	6.257
15	23.224	Triphenylene	$C_{18}H_{12}$	228	3.124

4. Conclusions

Pyrolysis products were obtained by high-temperature secondary pyrolysis of biomass tar (represented by rice hull tar) through a self-designed experimental bench, and the pyrolysis ratios of tar at different temperatures were recorded and compared, and simultaneously the relevant indicators are analyzed and the conclusions are as follows:

- The pyrolysis ratio of tar increases with the increase of experimental temperature, and the pyrolysis ratio increases slowly between 550°C and 655°C; the pyrolysis ratio exhibites a large upward trend between 655°C and 790°C and the rising trend slowes down between 790°C and 853°C. The above phenomenon indicates that the pyrolysis ratio will increase with the increase of the reaction temperature, but the trend will gradually slow down combined with the measurement results of the thermal balance. Compared with the experimental data of the thermal balance, the tar pyrolysis ratio from the experimental platform has been lower than that of the thermobalance because the amount of tar pyrolysis from the thermobalance is smaller than the bench, and the heat supplied at a certain temperature can be more fully contacted with tar, resulting in a higher pyrolysis ratio.
- The changes of the main components of pyrolysis products represented by naphthalene, anthracene, phenanthrene, 4-ethyl-phenol and n-hexadecanoic acid show that while the reaction temperature is increased, the secondary pyrolysis of tar is enhanced and the breakage of intramolecular bridge bonds are enhanced; the polycyclic substances are converted into monocyclic ones by ring-opening reactions and the increase of the chain-like substances is caused by the ring-opening reactions of a part of the monocyclic substances by heat, and the chain-like radical is converted. In combination with other simple free radicals, such as carboxyl and hydroxyl groups, a chain-like substance is

formed. At the same time, it can be concluded that the phenol derivative and the materials of chain structure have better thermal stability.

- Combined with the GC-MS analysis results of pyrolysis products at reaction temperatures of 423°C, 550°C, 655°C, 790°C and 853°C and the pyrolysis ratio at reaction temperatures, it can be concluded that the pyrolysis ratio varies with the reaction temperature rising and gradually increases due to the reduction of polycyclic aromatic hydrocarbons such as naphthalene, anthracene and phenanthrene.
- Using the American Agilent HP5973GC/MS to analyze pyrolysis products of biomass tar under different pyrolysis temperature combinations, the main components of of biomass tar pyrolysis at different temperatures are naphthalene, phenanthrene, anthracene, pyrene, pyrene, and benzofuran, fluoranthene, biphenyl, biphenylene and their corresponding derivatives. The pyrolysis temperature has a great influence on the pyrolysis products of biomass tar. The yield of solid products gradually decreases with the increase of temperature, and gradually becomes a fixed value; the yield of the gas products decreases with the increase of temperature. The yield of non-condensable pyrolysis gas increases with the increase of reaction temperature.
- At the pyrolysis temperatures of 550°C, 1050°C and 1300°C, the types of pyrolysis components undergo a process of increasing first and then decreasing while the total mass fraction of pyrolysis components shows a process of decreasing from high to low. That is from 85.397% to 83.198%, and finally to 75.483%, a drop of 10%.
- The main pyrolysis products of tar show different changes: the mass fraction of pyrolysis components containing side chains, heterocycles and long chains decreases with temperature rise, while the mass fraction of aromatic organic compounds and polycyclic aromatic hydrocarbons increases with temperature rise; and the total increase is as high as 15.535%.

Acknowledgments
This research is vigorously supported by the project team of Application Research on Boiler Energy Efficiency and Environmental Protection Testing Analysis System in our unit. We are thankful to Dr. Guoli Qi from China Special Equipment Inspection And Research Institute and Dr. Ming Zhai from Harbin Institute of Technology for useful suggestions.

References

[1] He J L. (2011) Study on characteristics of biomass reduction of nitrogen oxides under reburning conditions. Master's Thesis of Huazhong University of Science and Technology, Wuhan [BJ1].
[2] Li X L. (2008) Numerical simulation of direct reburning of biomass. Master's Thesis of Harbin Institute of Technology, Harbin.
[3] Gong Z Q. (2008) Experimental study on direct reburning and denitrification of biomass. Master's Thesis of Shandong University, Jinan.
[4] Ye Y J. (2007) Study on the characteristics of biomass ash and its slagging mechanism. Master's Thesis of Huazhong University of Science and Technology, Wuhan.
[5] Tao J. (2016) Research progress on removal method of biomass gasification tar. Biochemical Engineering, 2 (6): 79-80.
[6] Liu X W. (2008) Removal method and optimization of biomass gasification tar. Master's thesis of North China Electric Power University, Baoding.
[7] Myren C, Hornell C. (2002) Catalytic tar decomposition of biomass pyrolysis gas with a combination of dolomite and silica. Biomass and Bioenergy, 23(3): 217-227.
[8] Hou B, Lü Z, Li X H, et al. (2001) Catalytic cracking of tar derived from biomass pyrolysis. Journal of Fuel Chemistry and Technology, 29(1): 70-75.
[9] Lü P M, Chang J, Wu C Z. (2005) An investigation on the kinetics of biomass fast catalytic pyrolysis. Acta Energiae Solaris Sinica, 26(5): 647-653.
[10] Zhou J S, Liu Y J, Luo Z Y, et al. (2005) Effects of solid acid and alkali catalysts on catalytic

cracking of biomass tar. Journal of Zhejiang University, 39 (7): 1047-1051.
[11] Pan Y G, Roca X. (1999) Removal of tar by secondary air in fluidised bed gasification of residual biomass and coal. Fuel, 78(14): 1702-1710.
[12] Hepola J, Mccarty J, Krishnan G. (1999) Elucidation of behavior of sulfur on Nickel-based hot gas cleaning catalysts. Appl Catal B: Environ, 20: 191-203.

IOP Conf. Series: Materials Science and Engineering **721** (2020) 012003 doi:10.1088/1757-899X/721/1/012003

Prediction method of coal slagging property in layer-fired boiler

Wang Zhao[1], Zhang Song-song[1], Guo Li[2] and Qi Guo-li[1,*]

[1]China Special Equipment Inspection and Research Institute, 100029 Beijing, China

[2]School of Energy Science and Engineering, Harbin Institute of Technology, Harbin, 150001, China

[*]Corresponding author's e-mail: wenyuhe0616@126.com

Abstract. In this article, different coal samples are selected and their physicochemical properties are analyzed. Study and comprehensively analyze various factors affecting slagging. Prediction of coal slagging properties using 12 influencing factors, include: softening temperature, the difference between deformation temperature and flow temperature, characteristic of char residue, sodium oxide content, silicon ratio, silica-alumina ratio, iron-calcium ratio, acid-base oxide ratio, sulfur slagging index, contamination index, chlorine content, inorganic mineral content. The fuzzy clustering analysis method was used to classify and quantify all the influencing factors, and the coal slagging prediction method of layer fired boiler was proposed and verified by application.

1. Introduction

The slagging, ashing and ash accumulation of the layer fired boiler is a comprehensive change process of coal combustion and pyrolysis and melting of minerals in coal. The type of grate, the atmosphere in the boiler, the physical and chemical properties of the coal will cause mutual coupling. It is also directly related to the design and operating conditions of the overall boiler. Therefore, the problem of slagging in a layer-fired boiler is a practical problem of cross-disciplinary and infiltration. These factors have seriously affected the safety and economic and environmental protection of the boiler. The discriminative indicators for predicting slagging, such as slag characteristics, ash fusibility and other physical and chemical indicators, failed to comprehensively summarize all the factors, leading to slagging in the actual operation process. At present, the fuel of boiler equipment is usually not a single type of coal, which deepen the difficulty in judging the performance of slagging. Therefore, combined with internal factors and external factors can make comprehensive prediction of coal slag combustion slagging more reasonable. The multi-factor weights are used and eventually classified as a single indicator to reflect the degree of coal slagging. In order to reduce the deviation caused by the change of threshold value, the fuzzy judgment method is introduced into the discriminating range of the influencing factors, so that the final discriminant index is more realistic and more convincing, making the prediction result more objective.

2. Study on the physicochemical properties of coal samples

The 13 kinds of coal samples in the operation of the layer-fired boiler were selected, the industrial analysis of the samples was analyzed, the elemental analysis was carried out, and the thermogravimetric analyzer was used to study the weight loss process of the coal pyrolysis, the heat absorption and desorption process, and the weight loss with temperature. The composition and content of minerals in coal were analyzed by X-ray fluorescence spectrometer and X-ray diffractometer.

The weight loss of the coal sample is analyzed by thermogravimetric analysis, and the coal does not burn in an inert atmosphere. The main weight loss is derived from the precipitation, pyrolysis and phase transformation of the volatile matter. In the interval of 300°C to 600°C, the weight loss is obvious, mainly the release of volatiles in coal. At 500°C, the chloride begins to volatilize, accompanied by pyrolysis and phase transformation. Carbonate pyrolysis at 600-1000°C, forming free alkali metal compound, while releasing CO_2. In the range of 900°C to 1200°C, alkali metal compounds volatilized, there is about 2% weight loss, through the curve can be found. At 1000°C, the residual part of the basic and coal quality analysis of the volatilization the content of the fraction is close.

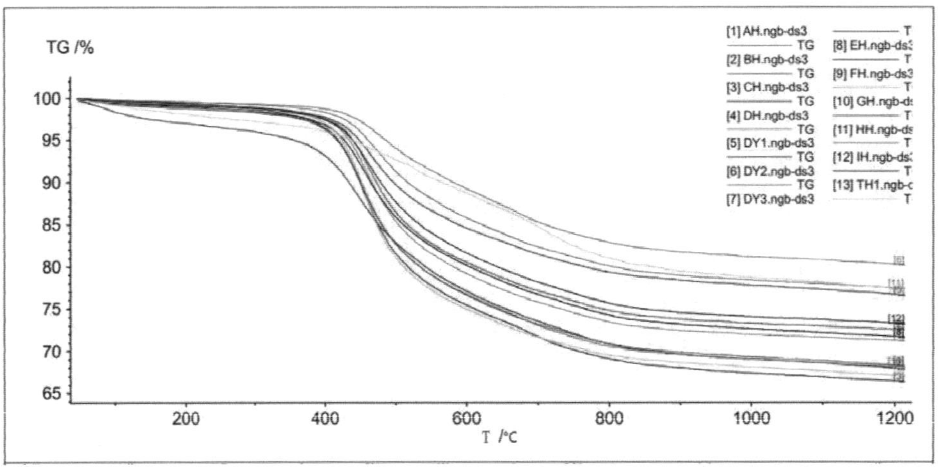

Figure 1. Weight loss in an inert atmosphere.

Coal contains clay minerals, carbide minerals, carbonate minerals, sulfate minerals, chloride minerals and silicate minerals. These minerals will undergo pyrolysis and eutectic combustion in the combustion process of a layer-fired boiler. Quantitative analysis of minerals determines the mineral composition of the coal sample. Through data and theory, the eutectic of acidic oxides and basic oxides, SiO_2 and Al_2O_3, the melting point of Fe oxides in different atmospheres, and the low-melting alkali metal compounds such as Na_2O and K_2O are the effects of minerals on the ash deposition of coal ash. The key to the factors.

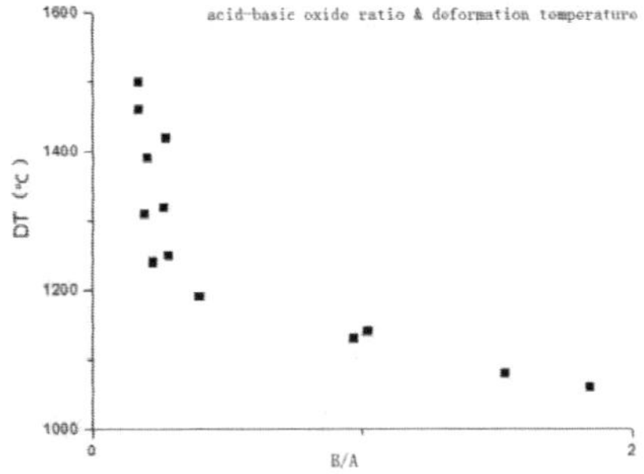

Figure 2. Relationship between silicon ratio and deformation temperature

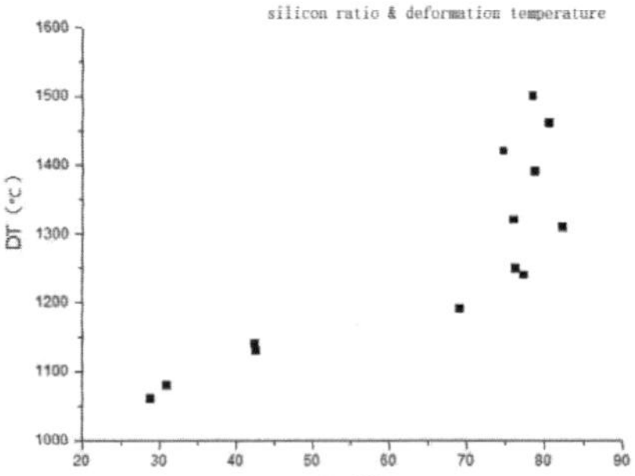

Figure 3. Relationship between acid-base ratio and deformation temperature

Through different indicators to explore the impact on the slagging between them, it is found that the index prediction of each coal sample indicates that the coal sample is severely slagging, and the

IOP Conf. Series: Materials Science and Engineering **721** (2020) 012003 doi:10.1088/1757-899X/721/1/012003

other indicators predict that the coal sample is slightly slagging or not slag, they are contradictory.

Table 1. Prediction results of various influencing factors

Sample	ST	ΔT	CB	Na	G	Si/Al	Fe/Ca	B/A	S	Rf	Cl	MM
AH	M	M	L	N	M	S	L	M	L	S	M	L
BH	M	L	L	N	M	M	M	M	S	S	S	L
CH	S	S	L	N	S	M	M	S	N	L	M	L
DH	M	S	L	N	L	S	M	L	S	S	S	L
EH	L	N	L	N	M	S	L	M	M	S	M	L
FH	N	N	M	N	M	L	M	L	S	S	M	M
GH	S	M	L	N	S	S	M	S	N	S	S	L
HH	L	S	L	N	M	M	L	L	S	S	L	S
IH	M	S	L	N	M	S	L	M	S	S	S	L
DY1	S	M	L	M	M	S	M	M	M	S	L	S
DY2	L	N	L	N	L	M	M	L	S	S	L	S
DYH	S	S	L	N	S	S	M	S	M	S	L	M
TH1	S	S	L	N	S	S	M	S	L	L	L	M

(S= severe slagging, M= medium slagging, L= light slagging, N= non-slagging)

3. Establish fuzzy clustering analysis

The limit value of the traditional discriminant index is a step function. When the calculated value of the index is near the critical value, a small amount of change will affect the level of the slag level. In order to reduce the deviation caused by the change of threshold value, the fuzzy judgment method is introduced into the discriminating range of the influencing factors, so that the final discriminant index is more realistic and more convincing, making the prediction result more objective. Using fuzzy clustering analysis method, the slagging factor becomes a mathematical quantitative form, and then the relationship between these factors is judged. Through the membership function of each factor, the evaluation limit is judged, and finally an overall evaluation value is obtained.

3.1 Establish fuzzy factor sets and evaluation levels

Combining the influencing factors of the slagging evaluation object $x=\{x_1,x_2,\cdots x_j\}, j=1,2,\cdots,p$

Using the physical and chemical experimental data, 12 discriminating factors were selected as follows:

1) Softening temperature of ash meltability (ST): a comprehensive indicator for measuring grate and heating surface slagging;

2) The difference between deformation temperature and flow temperature (ΔT): measuring the ability of ash from melting to cooling;

3) Characteristic of char residue (CB): measure the shape of coal char after staying at 900°C;

4) Sodium oxide content (Na): measure the Na_2O content that is most prone to ash formation; equivalent $Na_2O=(Na_2O+0.659K_2O)\ A/100$

5) Silicon ratio (G): the components in the denominator are mostly fluxes, and the higher the silicon ratio means less adhesion.; G=SiO_2/ (equivalent Fe_2O_3+SiO_2+MgO+CaO) equivalent $Fe_2O_3=Fe_2O_3+1.11$

6) Silica-alumina ratio (Si/Al): SiO_2 and Al_2O_3 are the main acidic substances in coal ash, and this ratio can judge the melting point of the acidic oxide eutectic;

7) Iron-calcium ratio (Fe/Ca): judging the slagging factors of bituminous coal ash;

8) Acid-basic oxide ratio (B/A): judging according to the tendency of the melting point of the acidic oxide and the basic oxide; $B/A=(Fe_2O_3+CaO+MgO+K_2O+Na_2O_2)$

9) Sulfur slagging index (S): considering that pyrite in coal ash is a flux, it is easy to reduce the ash melting point; $Rs = A/B \times S_d$

10) Contamination index (R$_f$): considering the degree of coal bonding; $R_f = A/B \times (Na_2O)$ $Na_2O = (Na_2O + K_2O)$

11) Chlorine content (Cl): factors in determining fusible alkali metal oxides;

12) Inorganic mineral content (MM): as a factor in the quality of slag in coal; $MM_d = 1.129A_d + 0.35S_d - 0.2$

The set V of the various evaluation levels divided by the judged object according to the discriminant limit; the predictive index is divided into non-slagging, light slagging, medium slagging and severe slagging according to the foregoing. When calculating the dispersion standard value (the center value of the indicator divided by the maximum value of each indicator), the center value of the range is required and determined. The discriminant limits and center values are summarized, as shown in Table 2:

Table 2. Judging level

	Index	non-slagging	light slagging	medium slagging	severe slagging
ST	central value	1500.00	1445.00	1325.00	1260.00
	Discriminant limit	> 1500	1390-1500	1260-1390	< 1260℃
ΔT	central value	200	175	125	75
	Discriminant limit	> 200	150-200	100-150	< 100
CB	central value	1.00	2.00	4.00	7.00
	Discriminant limit	1、2	2、3	4、5	6、7
Na	central value	0.30	0.38	0.53	0.67
	Discriminant limit	< 0.3	0.3-0.45	0.45-0.6	> 0.6
G	central value	78.80	75.40	69.05	63.15
	Discriminant limit	> 78.8	78.8-72	72-66.1	< 66.1
Si/Al	central value	2.00	1.94	2.26	3.04
	Discriminant limit	< 1.87	2-1.87	2.65-1.87	> 2.65
Fe/Ca	central value	0.30	3.00	1.35	1.35
	Discriminant limit	< 0.3	> 3	0.3-3	
B/A	central value	0.15	0.15	0.31	0.49
	Discriminant limit	< 0.206	0.1-0.206	0.206-0.4	> 0.4
S	central value	0.50	1.30	2.30	2.90
	Discriminant limit	< 0.6	0.6-2.0	2.0-2.6	2.60
Rf	central value	0.13	0.35	0.75	1.25
	Discriminant limit	< 0.2	0.2-0.5	0.5-1	> 1
Cl	central value	0.10	0.25	0.40	0.60
	Discriminant limit	< 0.2	0.2-0.3	0.3-0.5	> 0.5
MM	central value	2.5	7.25	15	25
	Discriminant limit	0-5	5-10	10.01-20	> 20

3.2 Determine membership functions and specify clustering rights

Determine the membership function: list the triangle membership function for the factors with the evaluation limit, and divide the slagging degree into four grades. Take the softening temperature as an example. The expression is as follows:

$$\text{Non-slagging: } f_{j11}(t) = \begin{cases} 1, \ t > 1500 \\ \frac{t-1445}{55}, \ 1000 < t \leq 1500 \\ 0, \ t < 1445 \end{cases} \tag{3.1}$$

$$\text{Light slagging: } f_{j12}(t) = \begin{cases} 0, & 1385 < t, \; t > 1500 \\ \frac{t-1445}{60}, & 1500 < t < 1445 \\ \frac{t-1385}{60}, & 1385 < t < 1445 \\ 1, & t = 1445 \end{cases} \quad (3.2)$$

$$\text{Medium slagging: } f_{j13}(t) = \begin{cases} 0, & 1445 < t, \; t > 1195 \\ \frac{t-1325}{130}, & 1325 < t < 1445 \\ \frac{t-1195}{130}, & 1195 < t < 1325 \\ 1, & t = 1325 \end{cases} \quad (3.3)$$

$$\text{Severe slagging: } f_{j13}(t) = \begin{cases} 0, & 1325 < t \\ \frac{t-1325}{130}, & 1195 < t < 1325 \\ 1, & t < 1195 \end{cases} \quad (3.4)$$

Calculating clustering coefficients δ_{jk}:

$$\delta_{jk} = \sum_{j=1}^{P} f_{jk}(X_{ij})\eta_{jk} \quad \text{In formula, } i = 1,2\cdots n;\; k = 1,2\cdots m$$

(n: number of samples, m: number of grades);

(3.5)

3.3 Determine the dispersion standard value and cluster weight value

According to the discrimination of classification boundaries and the judgment of single results, in order to eliminate the difference in the magnitude of different indicators, the data of the grading standards are processed, and the central value of each indicator is divided by the maximum value of each indicator.

$$M_j = i^{max}(x_{ij}) \cdots\cdots i=1,2\cdots 13 \quad (3.6)$$

$$\lambda_{ij} = \frac{x_{ij}}{M_j} \cdots\cdots j = 1,2,3,4 \quad (3.7)$$

According to different evaluation levels, determine the relative weight of each level of the corresponding level

$\eta_{jk} = \lambda_{jk} / \sum_{j=1}^{P} \lambda_{jk}$ ......In formula, λ_{jk} Standard value, $j = 1,2\cdots p$ Number of factors, $k = 1,2\cdots m$ Number of grades.

(3.8)

4. Result analysis

Normalize the four evaluation index values of each clustering coefficient (the sum of the clustering coefficients of the four evaluation indicators is 1), and determine the classification according to the maximum principle (if there are two numerical differences less than 0.15, take the more serious prediction index), finally get Table 3

Table 3. Prediction result

sample	non-slagging	light slagging	medium slagging	severe slagging	forecast result
AH	0.12	0.38	0.35	0.14	medium slagging
BH	0.20	0.24	0.34	0.20	medium slagging
CH	0.21	0.05	0.21	0.53	severe slagging
DH	0.43	0.04	0.19	0.35	severe slagging
EH	0.43	0.24	0.17	0.16	non-slagging
FH	0.43	0.22	0.28	0.07	non-slagging

GH	0.46	0.07	0.17	0.31	severe slagging
HH	0.44	0.19	0.12	0.25	non-slagging
IH	0.14	0.49	0.14	0.23	light slagging
DY1	0.00	0.10	0.49	0.41	medium slagging
DY2	0.47	0.09	0.18	0.26	non-slagging
DYH	0.16	0.08	0.11	0.65	severe slagging
TH1	0.18	0.13	0.00	0.69	severe slagging

5. Conclusion

In this paper, fuzzy clustering analysis method is established to predict the slagging property of coal-fired boilers. The prediction index is divided into non-slagging, slight slagging, medium slagging and severe slagging. The determination of the boundary and the determination of the membership function relationship are finally predicted. After this method, an overall evaluation value is obtained, which makes the prediction result more accurate and objective. The discriminating factors of coal-fired slagging of layered boilers are difficult to classify and difficult to quantify into clear and simple mathematical quantitative indicators.

In practical engineering applications, using the single-index experimental data available at the site, this fuzzy clustering analysis method is used to provide a new way to predict coal slagging performance of layer-burning boilers. There are still many shortcomings in the research of coal slag after combustion. If combined with the analysis of coal slag after combustion at different temperatures, the choice of predictive factors will be more objective and accurate.

Acknowledgements

This study was Supported by the Science and Technology Planning Project of AQSIQ (No. 2017QK178).

References

[1] Wang Cai-hong, Lin Xiong-chao, Yang Sa-sha. (2016) Evaluation of the thermal and rheological characteristics of minerals in coal using SiO_2-Al_2O_3-CaO-FeO_x quaternary system. Journal of Fuel Chemistry and Technology, 44(9), 1025-1033.

[2] Ni J, Zhou Z, Yu G, et al. (2010) Molten slag flow and phase transformation behaviors in a slagging entrained-flow coal gasifier. Industrial & Engineering Chemistry Research, 49(23): 12302-12310.

[3] Cham, ST, Sahajwalla, V, Sakurovs, R. (2004) Factors influencing carbon dissolution from cokes into liquid iron. ISIJ International, NO.11, 1835-1841.

[4] Vargas S, Frandesen F J, Dam-Johansen K. (2001) Properties of high-temperature melts of coal ashes and other silicates. K Progress in Energy and Combustion Science, 27(3):237-429.

[5] Chen Fan-min, Wang Jia-rui, Zhao Bing. (2015) Effect of mineral form in coal on combustion characteristics and fusibility of coal ash. Journal of Fuel Chemistry and Technology, 1th, 27-33.

[6] Wu Xiao jiang, Zhang Xiang, Chen Nan. (2017) Effect of Adding Clay on Ash Fouling/ Slagging Characteristic of Xinjiang High Alkali Coal. Boiler Technology, 48(2): 5-9.

[7] Seggiani M, Pannocchia G. (2003) Prediction of coal ash thermal properties using partial least-squares regression[J]. Industrial & engineering chemistry research, 42(20): 4919-4926.

[8] Su S, Pohl J H, Holcombe D, et al. (2001) Slagging propensities of blended coals. Fuel, 80(9): 1351-1360.

[9] Xie Liang-cai, Li Feng-hai, Xue Zhao-min. (2016) Influence of coal blending on ash fusion characteristics for coal with high ash fusion temperature. Journal of Fuel Chemistry and Technology, 12:1430-1439.

[10] Zhu Hong-qing, Luo Ming-gang, Xiang Ming-rui. [2016] Coal combustion characteristics research and dynamic analysis based on thermogravimetric experiments. China Safety Science Journal, 4: P40-45.

IOP Conf. Series: Materials Science and Engineering **721** (2020) 012004 doi:10.1088/1757-899X/721/1/012004

The NOx of Grate-Fired Furnaces Reduction strategy

Baibing Yang[1], Songsong zhang[1*]

[1]China Special Equipment Inspection And Research Institute, Beijing, 100029, P.R.China

[*]zhangsong0536@163.com

Abstract. According to the operating characteristics of Coal Grate-Fired Furnaces, the causes of nitrogen oxides formation in Coal Grate -Fired Furnaces were analyzed, the measures to reduce nitrogen oxides emissions were introduced in principle, and according to the actual situation, the engineering measures for reducing nitrogen oxides in Coal Grate-Fired Furnaces were proposed.

1. Introduction

According to the "Statistical Bulletin on National Economic and Social Development", in 2017, China's crude coal output was 3.52 billion tons, it is 3.3% points higher than 2016, the number of coal exports decreased by 7.0 percentage points, and the imports increased by 6.1 percentage points, and raw coal consumption increased significantly compared with 2016. However, more than 80% of coal resources are used for direct combustion of power plant boilers and industrial boilers. The chemical elements such as nitrogen, sulfur and mercury in coal will form pollutants such as nitrogen oxides, sulfur dioxide and mercury compounds after direct combustion. Nitrogen oxides, as one of the main pollutants, have a great impact on human health and environment[1-2].

By the end of 2016, there are about 412,000 coal-fired industrial boilers in use in China, the concentration of NOx emitted by industrial boilers is 400-800 mg/m^3, and millions of tons of NOx are emitted from industrial boilers every year, which causes serious environmental pollution. The number of layer-fired chain grate boilers occupies an important proportion in coal-fired industrial boilers, GB13271-2014"Discharge Standard of Boiler Air Pollutant"clearly puts forward the limit value of nitrogen oxide emission concentration of layer-fired boilers at 300 mg/m3, the emission concentration limit of nitrogen oxides is 200 mg/m^3 in key areas. There are great differences in furnace structure, combustion mode and combustion temperature between chain boilers and utility boilers, the flue gas denitrification technology of utility boilers can not be directly applied to chain boilers, so it is of great significance to develop low nitrogen combustion technology for chain boilers.

2. The Introduction of Chain Boiler

Chain boilers are generally divided into two types: double-drum horizontal type and single-drum vertical type. They have the advantages of high mechanization, stable combustion, easy operation, wide adaptability of coal, low concentration of boiler smoke and dust. There is no relative movement between fuel and grate, and the coal seam is ignited from top to bottom in time gradient. On the grate length gradient, the combustion, coke combustion and burnout of fuel are analyzed by drying and volatilization. The combustion process of chain boilers has no periodicity, but it has section character.

3. Main Formation Mechanism of NOx in Chain Boiler

Nitrogen oxides produced during coal combustion mainly exist in the form of NO and NO_2, and their proportion is related to combustion temperature. In the normal operation of chain boiler, the ignition of coal is homogeneous, and the combustion temperature is generally about 1350 C. The proportion of NO in the nitrogen oxides produced is larger.

Nitrogen oxides can be generated in three ways: Thermal NOx, Prompt NOx and Fuel NOx. Thermal NOx: Nitrogen oxides formed by oxidation of N_2 in the air during combustion at high temperature. Because the decomposition of nitrogen molecule requires large activation energy, the most influential factor on the formation of Thermal NOx is temperature, followed by the concentration of oxygen and reaction time (residence time). Prompt NOx: hydrocarbon fuels are formed under the condition of $\alpha < 1$, which is mainly affected by fuel characteristics, temperature and other factors. Fuel NOx: oxidized products of nitrogen-containing functional groups in fuels after thermal decomposition, mainly affected by temperature, oxygen concentration, fuel properties and other factors.

Because the combustion temperature of chain boilers is relatively low, fuel-type nitrogen oxides are the main source of nitrogen oxides. Fuel-type nitrogen oxides are the products of the interaction between nitrogen elements in coal and oxygen elements in air during coal combustion, the formation process of fuel-type nitrogen oxides is shown in Fig.1.

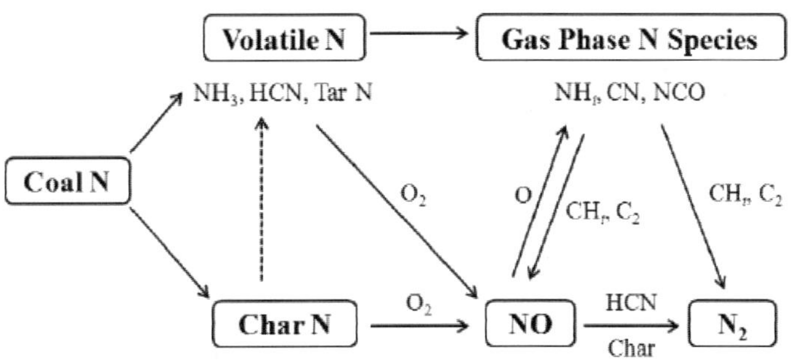

Fig.1 Schematic diagram of transformation path of nitrogen in coal[3]

4. The status of Laminar-Fired Boilerse

Coal bed combustion has a long history, and experimental research was the main method in the early stage. Rosin et al. studied the ignition of chain boiler through experimental system. The factors affecting the ignition were divided into upper radiation and lower airflow. The upper radiation was mainly affected by radiation temperature, air velocity and particle size. The lower airflow was mainly affected by air temperature, flow velocity, particle size and moisture[4]. On the basis of Rosin, Dunningham et al. increased the influence of ignition by fire source. It was found that under the condition of open fire ignition, the ignition events were greatly shortened[5]. Grumell et al. divided the bed into ignition zone, combustion zone and burnout zone by measuring the parameters of bed atmosphere and temperature. As shown in Figure 2, the main factors affecting the bed combustion were analyzed, including fuel type, ash content, moisture content, primary air speed, coal expansion, bed resistance and grate temperature[6].

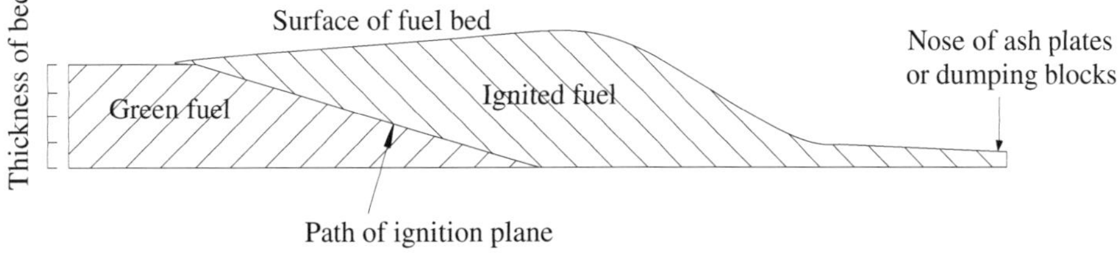

Fig. 2 Schematic diagram of combustion zoning in bed [6]

By combining the analysis of gases in the fuel layer, the measurement of air velocity under the fuel layer and the measurement of the temperature field of the fuel itself, Kenori divided the combustion process of the fuel layer on the grate into three areas: volatilization analysis, coke gasification and coke combustion. The bed combustion scheme is shown in Fig. 3. O_1N is the isotherm of volatilization analysis, O_2K is the isotherm of pure coke precipitation, $O_2O_3O_4O_2$ is the bed gasification zone, $O_3O_4O_6O_5WKO_3$ is the bed oxidation zone, $WO_5O_6O_4DCW$ is the ash burnout zone[7].

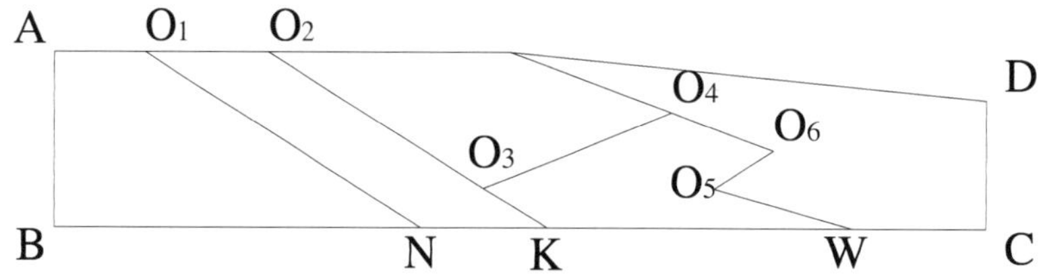

Fig.3 Schematic diagram of bed combustion scheme[7]

Luo Yonghao's team of Shanghai Jiaotong University proposed a double arch structure and studied the effects of combustion characteristics of different coal types, air staged combustion and fuel staged combustion on the emission of nitrogen oxides. It was found that different coal types had different exhaustion time and burning intensive area in the furnace, and both air staged combustion and fuel staged combustion could be significantly reduced. The effect of low NOx emission is that the air staged combustion is affected by the residence time in the reduction zone and the secondary air volume, and the fuel staged combustion is affected by the amount of reburning fuel, the residence time in the reburning zone and the excess air coefficient in the main combustion zone [8-10].

Gao Jianmin's team of Harbin University of Technology found that the unit test-bed can simulate the combustion process and NOx generation process of chain grate boiler, and the combustion and NOx generation processes are basically the same[11]. The team analyzed the effects of coking residence time, coking temperature and catalyst loading on NO reduction from 1.7-2.8mm coal coke. The combustion process and NO release of coke layer at different coke/O2 ratios and temperatures were studied. It was found that the reduction layer of coke could better inhibit NO emission, but was affected by heat transfer and temperature [12-13].

5. Measures to Reduce NOx Emission

5.1Measures to Reduce NOx in Principle

The formation of fuel-type NOx is affected by temperature, oxygen concentration and fuel properties. Therefore, the main methods to reduce the formation and emission of NOx are as follows:

(1) Reducing oxygen concentration: reducing excess air coefficient in furnace, reducing oxygen concentration in ignition zone.

(2) Sufficient residence time: Nitrogen in fuel is not easy to produce NOx at low oxygen concentration, and the NOx generated is reduced by homogeneous or multi-phase reaction.

(3)Reduce the peak temperature: When oxygen is sufficient, the hot air temperature can be lowered or the NOx generation can be reduced by flue gas recirculation.

(4) Adding reductant: generating CO, NH_3 and other reducing gases, reducing and decomposing NOx.

5.2Engineering Practice of Chain Boiler

Large-scale boilers usually adopt the method of injecting reducing agent to the furnace or tail heating surface to reduce NO X. There are mainly two kinds of reduction methods: selective catalytic reduction of ammonia (SCR) and selective non-catalytic reduction (SNCR). However, the two methods require higher flue gas temperature conditions. Because of the large load variation, the temperature window is unstable and other factors, result in the reduction denitrification method is not suitable for industrial chain boiler. There are several ways of engineering renovation application:

(1) Air Classified Combustion: Air Classified Combustion is a technology used in almost all combustion modes to reduce NOx. The aim is to avoid reducing excess air coefficient and NOx generation in areas where the temperature is too high and the formation of NOx is easy. The principle is shown in Figure 4.

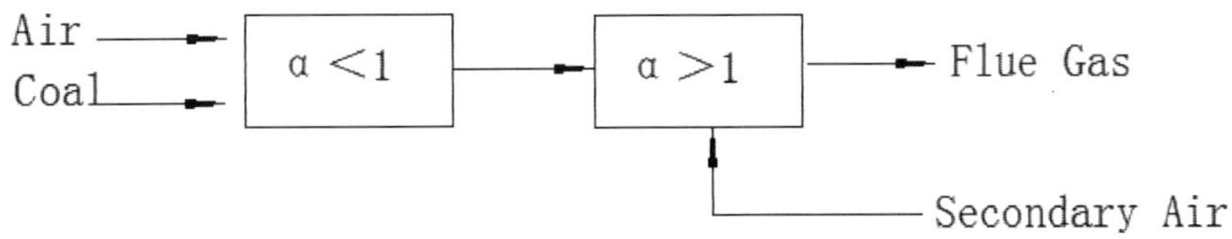

Fig. 4 Air classification schematic diagram

(2) Fuel-staged combustion: also known as "reburning method", which uses three-stage combustion, the furnace is divided into main combustion zone, reburning zone (or reduction zone) and burnout zone. The primary fuel which accounts for about 80%-85% of the total fuel is fed into the main combustion zone, and NOx is generated under the condition of excess air ratio> 1. The remained secondary fuel of 15%-20% is fed into the reburning zone, which ensures excess air ratio < 1. Under the experimental conditions, the emission reduction of 50%-60% NOx can be achieved[3-4]. The principle is shown in Fig. 5.

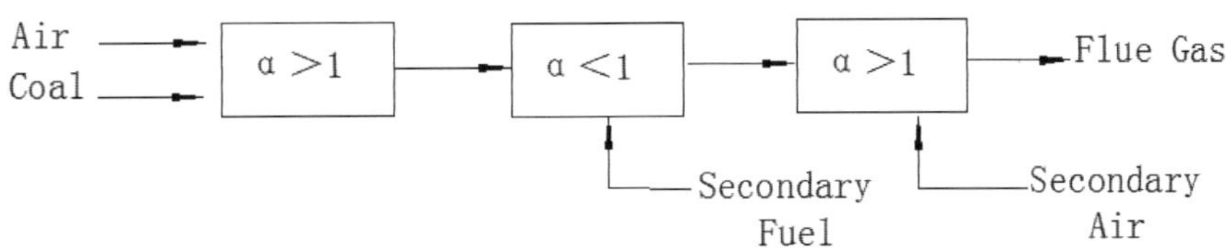

Fig.5 The Fuel-staged combustion schematic diagram

(3) Flue gas recirculation: Part of low-temperature flue gas is directly fed into the furnace, or mixed with air (primary or secondary air) and fed into the furnace. Because the flue gas absorbs heat and dilutes the concentration of oxygen, the combustion speed and furnace temperature are reduced, and the thermal NOx is reduced. If fuel classification is used, flue gas recirculation is generally used to transport secondary fuel. The reduction degree of NOx by flue gas recycling is mainly affected by the ratio of flue gas recycling and excess air coefficient. In operation, the circulation ratio should not be too high or too low, and the effect of too low is not obvious. Excessive high NO recycling will increase the amount of NO and lead to the enrichment of NOx.

(4) Coke reduction: The coke particles in the flue gas are sent back to the combustion chamber to reduce NOx and incomplete combustion loss. By collecting coke particles in tail flue and nitriding with urea, the reduction capacity of coke is strengthened and the emission of NOx is reduced.

6. Conclusions

(1) The causes of different types of NOx formation are analyzed. Because of the low temperature in the furnace, the fuel type is the main type of NOx formation in the laminated-fired boiler.

(2) In principle, the method of reducing NOx is analyzed, and the renovation measures for reducing NOx emission of stratified-fired boilers are put forward, such as Air Classified Combustion, Fuel-staged combustion, Flue gas recirculation and Coke reduction.

Acknowledgement
This work was financially supported by the project supported by National Key R&D Program of China" (2017YFF0209806)

References

[1] M Kampa,E.Castanas. Human health effects of air pollution.Environmental Pollution[J],2008, 151(2): 362-367..

[2] Wang, X.R. Nitrogen Oxide Hazards and Treatment Methods[J].Technology Application and Research， 2018,10:116-117.

[3] Tomas K M.The release of nitrogen oxides during char combustion[J].Fuel,1997,76(6):457-473.

[4] P.O.Rosin, H.R.Fehling, The Ignition of Coal on a Grate[J].Journal of the Institute of Fuel, 1937, 11:102-117.

[5] A.C.Dunningham, E.S.Grumell, Addendum to Contribution to the Study of the Ignition of Fuel Beds[J]. Journal of the Institute of Fuel, 1937, 11:129-133.

[6] E.S.Grumell, A.C.Dunningham, Combustion of Fuel on a Travelling Grate[J]. Journal of the Institute of Fuel 1938 12 87-95.

[7] Knorieh write,Yuyi Ma,Weizhen Hu translate.Boiler combustion process[D],Electric power industry press,1956first edition:144.

[8] Xu H.D.Experimental Study on Combustion Characteristics and Low NOx Combustion Mechanism of Laminated Furnace[D]. Shanghai Jiaotong University.2001.

[9] Ji,J.J., Luo, Y.H., Lu, F., Xu, H.D. Study on the Effect of Staged Combustion on NOx Emission Reduction in Layer Combustion Furnace[J].Boiler Technology,2006,37: 50-53.

[10] Feng,Y.L., Luo, Y.H., Lu, F., Ji, J.J. Experimental study on low NOx reburning technology for stoker furnace[J]. Power Engineering. 2004,24(1):29-32.

[11] Gao,J.M., Zhao L.F., Xu, L., Du, Q., Gao, J.H., Luo, J.Y., Wu, S.H., Zhao, G.B. Characteristics of NOx generation in typical bituminous coal seam combustion process [J]. Industrial Boiler.2012, (02):1-5.

[12] Xu,L.,Wei,Z.Z.,Gao,J.M,etc.Effect of Coking Conditions and Catalysts on NO Reduction by Large Particle Coal Char[J].Journal of Harbin Institute of Technology.2016,48(7): 52-57.

[13] Xu L,Gao J,Zhao G，et al,Influenc of Oxidation-Reduction Layering on Fuel Nitrogen Oxide Emissions during a Char Grate-fired Process[J].Energy＆Fuel,2017,31(9):9736-9744.

IOP Conf. Series: Materials Science and Engineering **721** (2020) 012005 doi:10.1088/1757-899X/721/1/012005

Effects of Friction Pressure on Microstructures and Mechanical Properties of Friction Welded T92/Super304H Dissimilar Steel Weld Joints

Debiao Li, Yuan Lu*, Mo Chen, Yaqin Sun, Lining Wang, Cong Ding

Xi'an Special Equipment Inspection Institute, Xi'an, 710065, China

*Corresponding author's e-mail: luyuan19801104@163.com

Abstract. In this paper, the T92 martensitic steel and the Super304H austenitic steel were welded by continuous drive friction welding technique. The friction welding joints exhibited good mechanical and metallurgical properties. The influence of friction pressure on the microstructures and mechanical properties of T92/Super304H dissimilar steel joints was investigated. With an increase in the friction pressure, the grains in the welding zone and the heat affected zone didn't grow up and the number of second-phase particles precipitates increased. The Vickers hardness increased and the impact toughness decreased, the tensile fracture and fracture location didn't change.

1. Introduction

Due to the problems of the earth environment such as the increase of carbonic acid gas and the fuel consumption, improvement of thermal efficiency is required in fossil fired power plants. Ultra super critical boilers that are operated under the condition of a temperature over 873K and steam pressure over 24MPa are expected as one of the higher heat efficiency plants. However, with the increase of steam parameters, requirements for the materials applied in the Ultra-super critical boilers components are becoming higher [1, 2]. In order to fulfill the tough parameters of the ultra-super critical power generation, traditional heat-resistant steels used in the subcritical and supercritical units have been replaced by those new heat-resistant steels. The recently developed Super304H (0.1C-18Cr-9Ni-3Cu-Nb-N) austenitic steel has shown considerable promise due to its high oxidation resistance, corrosion resistance and creep resistance. The Super304H has higher strength at elevated temperatures are required for super heater tubes in the boilers. This excellent creep rupture strength is based on finely precipitated particles such as Nb(C, N) and NbC. Thus, it is widely used for super-heaters and re-heaters, which have the abominable service environment in Ultra-super critical boilers. The T92 (9Cr-0.5Mo-2-W-V-Nb) martensitic steel approximately similar to T91 but with a little modification in chemical compositions for preferable high temperature properties. In comparison with T91 steel, the resistance to creep of T92 is about 30% higher. The T92 martensitic steel will certainly have a broad application prospect in forth coming ultra-super boilers. In this case, the welding between T92 and Super304H steels will be necessary.

At present, the T92 martensitic steel and Super304H austenitic steel are welded through the gas tungsten arc welding (GTAW) technique usually in engineering applications. The GTAW technique is mature and suitable for field welding. Nevertheless, the T92/Super304H joints welding through the conventional fusion welding still exhibit inferior mechanical properties. The reasons are: (i) The base metal near the fusion line during the fusion welding process is prone to form the wide heat affected zone with coarse grains; (ii) A transition layer (fusion zone) will appear in the welding zone near the weld line; (iii) The migration of carbon atoms occured though the fusion line, and carbonization layers and decarburization layers are formed at the welding zone. The friction welding is a solid state welding process, and the welding temperature was under the melting temperature. This welding process nullifies adverse effects of the uncertainties in the filler metal selection by GTAW technique. The friction welding has a number of advantages over the conventional fusion welding process. The major advantage relies on the direct conversion of mechanical energy into thermal energy at the joint interface in contrast to fusion welding. The very high temperature gradient available in the joint area accounts for a very small heat affected zone. Owing to the narrow the heat affected zone, the welding distortion is kept minimum. Additionally, because it is a solid state process, the defects associated with melting solidification phenomenon such as porosity and slag are not present [3-5]. The aim of this paper is to fabricate a combination with higher better mechanical properties between the Super304H austenitic steel and the T92 martensitic steel by using continuous drive friction welding. The effect of friction pressure on the microstructures and mechanical properties of the welded joints were thus investigated.

2. Experimental procedures

The pipes of Super304H austenitic steel and T92 martensitic steel that in the sizes of Φ44.5mm x 9mm were used as base metals. The chemical compositions and room mechanical properties of the base metals were listed in Tables 1 and 2. The pipes of Super304H and T92 were friction welded by a continuous drive friction welding machine. In the process of friction welding, the friction speed was 1500 rpm. The friction pressure was 100 MPa of Process 1, 150 MPa of Process 2 and 200 MPa of Process 3. The upset pressure was 200 MPa and the friction time was 5s respectively. The metallographs and fractographs of the welding joints were performed by an optical microscope and a scanning electron microscope respectively. The X-ray XRD patterns were examined by X-ray Diffraction. The Vickers hardness values of the joints were tested using a Vickers hardness testing instrument. The tensile samples were tested using a testing instrument. The impact samples were tested using an impacting instrument.

Table 1 Chemical composition of Super304H and T92

	Chemical composition										
	C	Si	Mn	Cr	Ni	Mo	Cu	Nb	N	Al	B
T92	0.13	0.28	0.43	8.474	0.239	0.985	0.043	0.062	-	0.0057	0.002
Super304H	0.07	0.21	0.81	18.37	9.06	0.42	2.93	0.52	0.10	0.01	0.002

Table 2 Mechanical properties of Super304H and T92

	$\sigma_{0.2}$/MPa	σ_b/MPa	δ/%	A_{KV}/J	Hardness/ HB
T92	≥440	≥620	≥20	≥40	≤250
Super304H	≥235	≥590	≥35	–	–

3. Results and Discussion

Figure 1 showed the X-ray diffraction profiles for the Super304H side and the T92 side of the welding joints welded by Process 2 respectively. It could be sure that the peaks with the highest intensity correspond to the γ-Fe phase and the precipitated phases $Cr_{23}C_6$ and NbC for the Super304H side. Therefore, the main phase of the Super304H side were the γ-Fe phase and the precipitated phases of $Cr_{23}C_6$ and NbC. The X-ray result indicated that the precipitated phases included the Cr and Fe elements [6]. The Cr element had a tendency to form carbides. So some C element and Cr element were precipitated from matrix and formed the precipitated phases $Cr_{23}C_6$ at the grain boundary preferentially. Some N element replaces the element C, and a part of the NbC is converted into Nb(C, N). It was difficult to index the Nb(C, N) due to lower intensities peaks. Therefore, $Cr_{23}C_6$ and NbC were denoted precipitated phases in the X-ray diffraction profiles. It can be sure that the peaks with the highest intensity correspond to the α-Fe phase and the precipitated phases $Cr_{23}C_6$ for the T92 side of the welding joints. Therefore, the main phase of the T92 side are the α-Fe phase and the precipitated phases of $Cr_{23}C_6$ which were precipitated from supersaturated martensite grains. Because there were some trace elements of Nb and V in the T92 side. So the MX carbonitrides containing Nb and V should be precipitated theoretically. But it was difficult to index the MX carbonitrides due to lower intensities peaks.

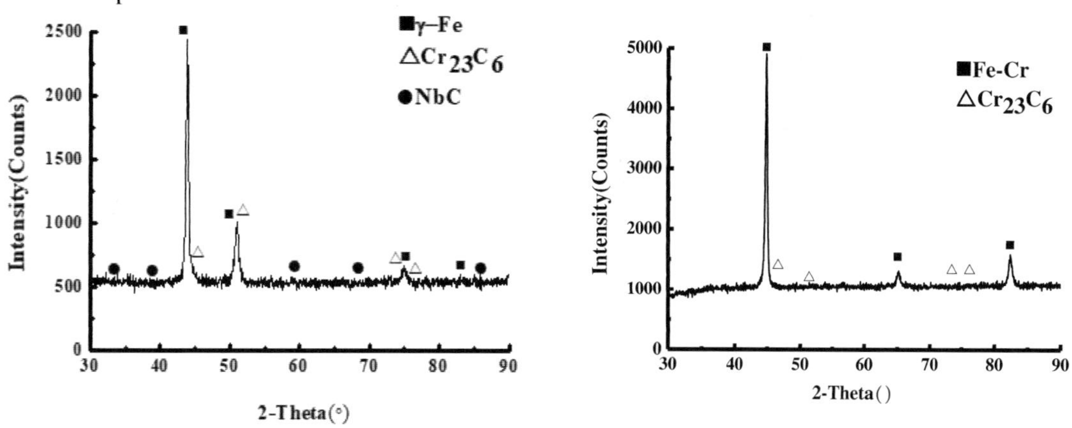

(a) the Super304H side (b) the T92 side

Figure 1 the XRD patterns of for the Super304H side and the T92 side

Figure 2, 3 and 4 shows the metallography of the welding joints with different friction pressure respectively. During the friction welding process, the welded zone produced agglutinate and shear tear behaviour. It led to the deformation of the grains in the welding zone, and the dynamic recrystallization driving force and lattice distortion energy increased. The thermoplastic deformation temperature and unit volume free energy of recrystallized grains decreased, which produced a large number of recrystallized nucleation [7, 8]. The welding zone included a welding interface and the areas (about 50μm from the welding interface) on both sides of the welding interface. The centre of the welding zone was the welding interface that was straight and clear. Due to agglutinate and shear tear behaviour, there was the T92 lays appear in the Super304H side of the welding zone occasionally and the T92 lays were parallel to the welding interface. The Super304H side of the welding zone was mainly composed of fine equiaxed austenite grains because of grain refinement. The martensite grains which in the T92 side of the welding zone was austenitized due to the higher temperature generated by friction process, than it produced a large number of fine equiaxed austenitic grains by dynamic recrystallization. In the cooling process after welding, the recrystallization of the equiaxed austenitic grains takes place, resulting in the formation of the fine plate martensite grains. The welding zone of T92/Super304H dissimilar steel welded by fusion welding process had the typical coarse δ-phase grains. The coarse δ-phase grains were not formed in the welding zone by the friction welding. Because that the austenite grains and plate martensite grains on both sides of the welding interface was fine, so the welding zone of the Super304H/T92 welding joints exhibit excellent microstructures and mechanical properties. The friction welding was solid state welding techniques, the melting process did not occur in the welding zone and the heat affected zone. So the heat input during the friction welding process was much less than the heat input during the fusion welding process. The width of the heat affected zone caused by friction welding process were much smaller than the width of the heat affected zone caused by the fusion welding process. The tendency of grains growth during the friction welding process was less obvious than the tendency of grains growth during the fusion welding process. So the heat affected zone of the welding joints with the fine grains showed excellent microstructure and mechanical properties. With the distance from the welding zone increased, the size of the austenitic grains and the martensite grains decreased in the both heat affected zone, until the size of grains was equal to the size of grains in the base metal. The overheated zone with coarse δ-phase grains were not founded in the T92 heat affected zone too [9, 10]. So the friction welded Super304H/T92 joints with the fine grains in the welding zone and heat affected zone showed excellent microstructure and mechanical properties.

(a) (b) (c)

Figure 2 The metallography of welded joints for Process 1 (a) the T92 heat affected zone (b) the welding zone, (c) the Super304H heat affected zone

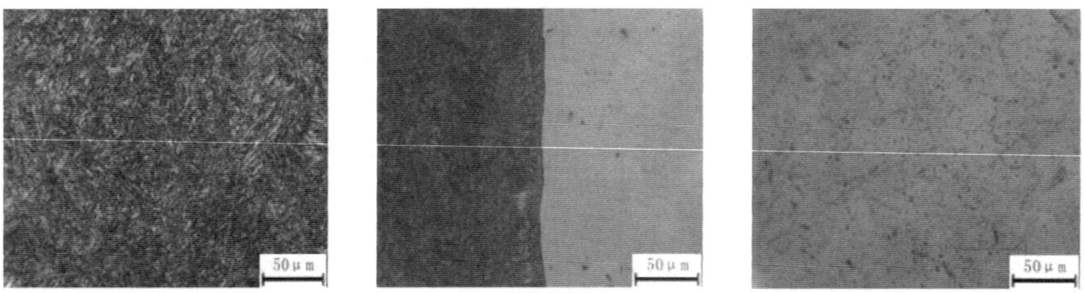

(a) (b) (c)

Figure 3 The metallography of welded joints for Process 2 (a) the T92 heat affected zone (b) the welding zone, (c) the Super304H heat affected zone

Figure 4 The metallography of welded joints for Process 3 (a) the T92 heat affected zone (b) the welding zone, (c) the Super304H heat affected zone

The surface temperature of the welding joints was equal to the temperature of the welding heat source in friction welding process. It affected the heating temperature, temperature distribution, deformation and diffusion of the metal of the welding joints directly. So the surface temperature of the welding joints had a great influence on the microstructure and mechanical properties of the welding joints. The welding heat source was considered to be a linearly propagated planar heat source in friction welding process. If the heat dissipation to the surrounding space was not considered, the surface temperature of the welding joints was calculated according to the Eq (1)

$$T(t) = qt/(\pi\lambda C) \quad (1)$$

T was referred to the surface temperature of the welding joints， *t* was referred to friction time and *q* was referred to friction heating power. It was calculated according to the Eq (2)

$$q = (\pi n T)/30 \quad (2)$$

T was referred to the friction torque， *n* was referred to the friction speed, and *q* was referred to the friction heating power. The friction heating power was mainly depended on the friction torque and the friction speed. When the friction speed was constant during the friction welding process, as the friction torque increased, the friction heating power and the surface temperature of the welding joints increased correspondingly. The friction torque was mainly depended on the friction pressure and the coefficient of friction, so the friction torque was proportional to the friction pressure. Therefore, as the friction pressure increased, the friction heating power and the surface temperature of the welding joints increased. Because that melting process did not occur in the welding zone, although the surface temperature of the welding joints increased, the size change of the grains in the welding zone was not detectable. As the frictional pressure increased, the heat input during the friction welding process increased in the heat affected zone, and the temperature of the heat affected zone increased accordingly. The size change of the austenite grains in the Super304H heat affected zone and the martensite grains in the T92 heat affected zone was not detectable, but the amount of the precipitated phases in the both heat affected zone increased obviously [11, 12].

Table 3 depicted the variation of the yield strength and tensile strength as a function of the friction pressure. It was detected that the fracture position of the welding joints was in the Super304H base metal for the different friction pressure. The fractographs of the tensile strength samples were shown in figure 4. As the friction pressure increased, the tensile strength change of the welding joints was not obviously and regularly. The fracture morphology was characterized by typical ductile fracture, and a lot of fine and shallow dimples distributed on the facture surface. Table 3 depicted the variation of the impact toughness as a function of the friction pressure. The grains size had a great influence on the impact toughness of each zone of the welded joints. When the grains size was smaller, the impact toughness was higher. The grain size of the heat affected zone with coarse grains was the larger, so the impact toughness of the heat affected zone was the lower. As the friction pressure increased, the dislocation density and lattice distortion energy in the welding zone increased, and the impact toughness of the welding zone decreased. The precipitation of the $Cr_{23}C_6$ carbides were main reason that the impact toughness of the heat affected zone decreased with the friction pressure increased. The fractographs of the impact toughness samples were shown in figure 6, 7 and 8. The fracture mode of the impact toughness samples was ductile fracture with the aggregation dimples on the facture surface. In the welding zone, the dimples were distributed densely and deeply, and the shapes of dimples were round. In the heat affected zone, the dimples were distributed sparsely and shallow, and the shapes of dimples were flat. Table 3 depicted the variation of the Vickers hardness as a function of the friction pressure. When the grains size was smaller, the Vickers hardness was higher. The grains size of the welding zone with fine grains was the smaller, so the Vickers hardness of the welding zone was the higher. The dislocation density and lattice distortion energy in the welding zone increased as the friction pressure increased, and the Vickers hardness of the welding zone increased. The precipitation of the $Cr_{23}C_6$ carbides was the reason that the Vickers hardness of the heat affected zone increased with the friction pressure increased.

Table 3 The mechanical properties of Super304H/T92 welded joints

Tensile strength /MPa			Impact toughness /J		Vickers hardness / Hv	
$R_{p0.2}$	R_m	Fracture position	Welding zone	Heat affected zone (Super304H/T92)	Welding zone /Hv	Heat affected zone (Super304H/ T92)

IOP Conf. Series: Materials Science and Engineering **721** (2020) 012005 doi:10.1088/1757-899X/721/1/012005

Process 1	422.38	653.15	Super304H base metal	78.5	69.0/59.1	312.5	224.1/321.5
Process 2	413.11	644.41	Super304H base metal	76.9	64.2/56.2	349.6	248.9/358.9
Process 3	435.71	644.43	Super304H base metal	70.6	58.7/49.6	379.9	282.9/381.4

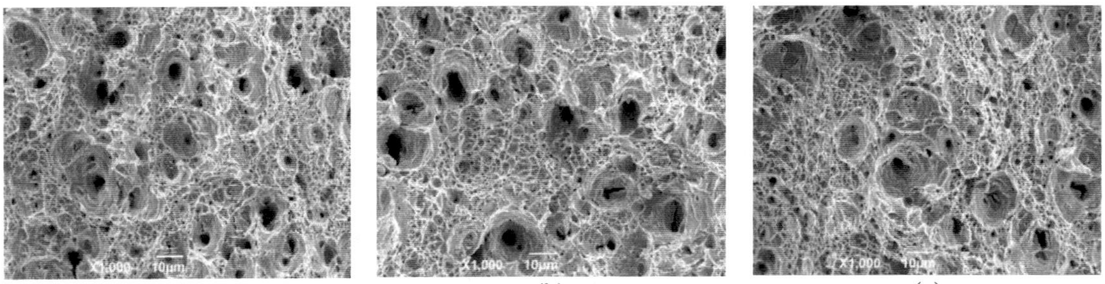

(a) (b) (c)

Figure.5 The fracture appearance of the tensile strength samples (a) Process 1, (b) Process 2, (c) Process 3

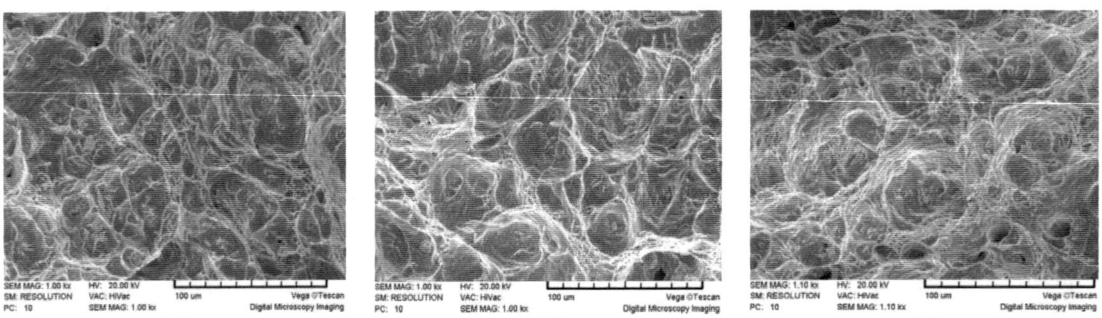

(a) (b) (c)

Figure.6 The fracture appearance of the impact toughness samples for Process 1 (a) the T92 heat affected zone, (b) the welding zone, (c) the Super304H heat affected zone

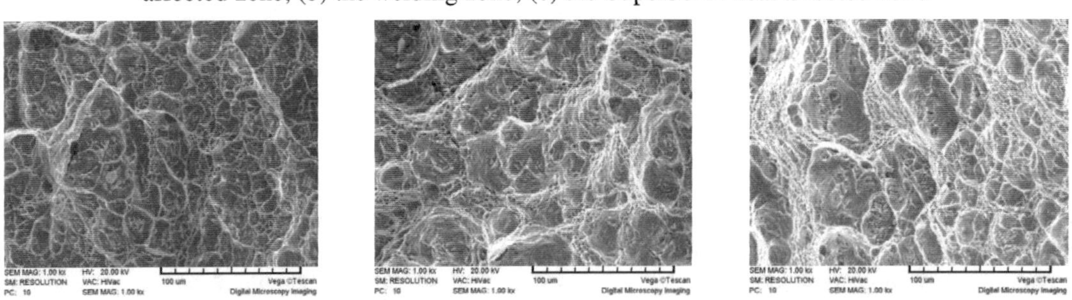

(a) (b) (c)

Figure.7 The fracture appearance of the impact toughness samples for Process 2 (a) the T92 heat affected zone, (b)the welding zone, (c) the Super304H heat affected zone

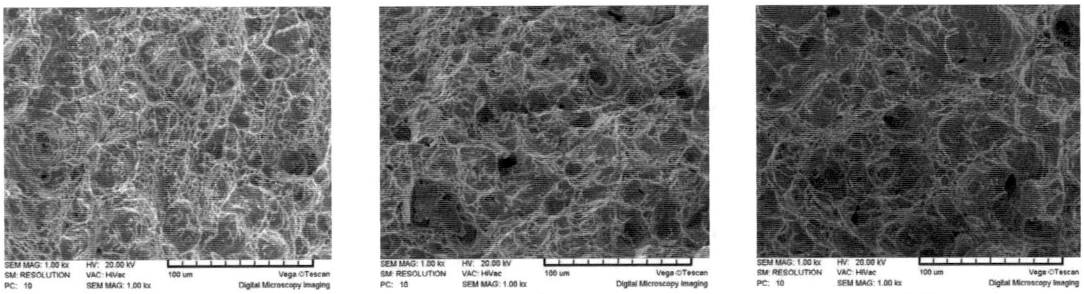

(a) (b) (c)

Figure.8 The fracture appearance of the impact toughness samples for Process 3 (a) the T92 heat affected zone, (b)the welding zone, (c) the Super304H heat affected zone

4. Conclusions

The T92/Super304H dissimilar steel welded joints with excellent properties were welded by continuous drive friction welding technique. The welding interface was well combined, and the austenite grains and plate martensite grains in the welding zone was fine. The fracture position of the welding joints was in the Super304H base metal. The influence of friction pressure on the structure and properties of T92/Super304H dissimilar steel joints was investigated. With an increase in the friction pressure, the grains of weld zone didn't grow up and the number of second-phase particles precipitates increased. The Vickers hardness increased and the impact toughness decreased, the tensile fracture didn't change.

References

[1] X M Li, Y Zou and Z W Zhang. (2010) Microstructure Evolution of a Novel Super304H Steel Aged at High Temperatures. Mater. Trans, 608: 164-173

[2] J Cao, Y Gong and Z G Yang. (2011) Creep fracture behavior of dissimilar weld joints between T92 martensitic and HR3C austenitic steels. Int. J. Press. Ves. Pip, 88: 94–98.

[3] P Li, J L Li and M Salman. (2014) Effect of friction time on mechanical and metallurgical properties of continuous drive friction welded Ti6Al4V/SUS321 joints. Materials. &. Design, 56: 649-656.

[4] P Li, J Li, X Li. (2015) A study of the mechanisms involved in initial friction process of continuous drive friction welding. J. Adhes. Sci. Technol, 29: 1246-1257.

[5] X Li, J Li, Z Liao. (2018) Effect of rotation speed on friction behavior and radially non-uniform local mechanical properties of AA6061-T6 rotary friction welded joint. J. Adhes. Sci. Technol, 32: 1987-2006.

[6] G H Chen, Y M Song, J Q Wang. (2012) High-temperature short-term tensile test and creep rupture strength prediction of the T92/TP347H dissimilar steel weld joints. Eng. Fail. Anal, 26:220-229.

[7] K MyungYeon, K Sukchul and C InSuk. (2014) High-temperature tensile and creep deformation of cross-weld specimens of weld joint between T92 martensitic and Super304H austenitic steels. Mater. Charact. 97: 161-168.

[8] J Cao, Yi Gong and Zhenguo Yang. (2011) Microstructural analysis on creep properties of dissimilar materials joints between T92 martensitic and HR3C austenitic steels. Mater. Sci. Eng. A, 528:6103-6111.

[9] J Cao, Y Gongan and K Zhu. (2011) Microstructure and mechanical properties of dissimilar materials joints between T92 martensitic and S304H austenitic steels. Materials. &. Design, 32:2763-2770

[10] G H Chen, Q Zhang and J J Liu. (2013) Microstructures and mechanical properties of T92/Super304H dissimilar steel weld joints after high-temperature ageing. Materials. &. Design, 44:469-475

[11] A Zielinski. (2012) Structure and properties of Super 304H steel for pressure elements of boilers with ultra-supercritical parameters. Journal of Achievements in Materials and Manufacturing Engineering, 55:403-409

[12] S Indrani, E Amankwaha, S Kumara. (2011) Microstructure and mechanical properties of annealed SUS 304H austenitic stainless steel with copper. Mater. Sci. Eng. A, 528: 4491-4499.

IOP Conf. Series: Materials Science and Engineering **721** (2020) 012006 doi:10.1088/1757-899X/721/1/012006

Assessment of Creep Rupture Properties of Friction Welded Super304H/T92 Steel Weld Joints

Yuan Lu, Jianlong Zhang, Debiao Li, Cheng Bi, Fuyang Sun, Yang Yang

Xi'an Special Equipment Inspection Institute, Xi'an, 710065, China

Abstract. The T92 martensitic and Super304H austenitic dissimilar steels welded joints were always encountered in the super critical boilers. The isothermal method was used to assess the creep rupture strength of weld joints. With the higher loading stress, the transcrystalline fracture was the rupture mode of creep specimens, and results in forming dimples. With the higher loading stress, the transcrystalline fracture was the rupture mode of creep specimens and results in forming dimples. With the lower loading stress, creep cavities joined each other along the grain boundary, the creep rupture was the rupture mode of creep specimens and results in cleavage rupture.

1. Introduction

Because that troubles of the environment issue such as the fuel consumption, the power plants required of improvement of thermal efficient. Ultra super critical boiler was one of the higher heat efficient plants. However, the requirements of the steels applied in the Ultra-super critical boilers are becoming higher [1]. The new developed Super304H austenitic steel showed a good future for not only excellent resistance to high-temperature corrosion and steam oxidation, but also superior creep strength than martensitic steels. The T92 martensitic steel approximately similar to T91 but with a little modification in chemical compositions showed preferable higher temperature properties. There are T92 martensitic and Super304H austenitic dissimilar welded joints inevitably, and the creep performance of the joints plays a significant role for Ultra-super critical boilers in service [2]. The Super304H/T92 dissimilar steels welded joints though friction welding performed excellent performance. The friction welding can be widely used for welding of martensitic steel and austenitic steel for higher temperature application [3,4].

2. Experimental procedures

In the process of friction welding, the friction speed was 2000 rpm and the friction pressure was 200 MPa. The upset pressure was 250 MPa and the friction time was 10s respectively. The creep experimental temperature was 625°C and the loading stress were 100 MPa, 120 MPa, 140 MPa, 160 MPa and 180 MPa respectly. The rupture appearance and the metallographs of the creep specimens were observed and analyzed by the scanning electron microscopy with an energy dispersive spectroscopy (EDS) system.

3. Results and Discussion

The creep rupture locations of the weld joints are the T92 heat affected zone for the different loading stress. Table 1 shows the creep strength experiment results of Super304H/T92 weld joints. With the loading stress increased, the rupture time reduced. The fracture surface of the creep specimen for loading stress of 180 MPa was figure 1(a) (d). There is a significant plastic deformation occurred in the creep specimen. Adjacent dimples in some areas are connected to grow up, and form small cracks. So the creep specimen under the higher stress is the plastic fracture caused by the connection of the dimples. There were some wedge cracks appeared, and the wedge cracks often appeared at the intersection of grains boundaries. The fracture surface of the creep specimen for loading stress of 140 MPa was figure 1 (b) (e). The fracture surface change with the loading stress decreased, the number of the dimples decreased and the shapes of the dimples were different. Under the higher loading stress, the dimples were distributed densely and deeply, and the shapes of dimples were round. Under the lower loading stress, the dimples were distributed sparsely and shallow. There was the larger spacing between the adjacent dimples, and the shapes of dimples were flat. The fracture surface of the creep specimen under the loading stress of 180 MPa is observed in figure 1 (c) (f). With the loading stress decreased, the fracture surface of the creep specimen was smoother and the rupture mode was intergranular brittle fracture. There was no obvious plastic deformation for the creep specimen. The fracture surface of the specimen consists of many small, flat and shining cleavage surfaces, and indicating that the fracture was along the grains boundaries.

Table 1 The creep rupture strength of 304H/T92weld joints

Sample No.	Test temperature/°C	Loading stress /MPa	Rupture time/h	Rupture location
1		100	5182.3	
2		120	2883.2	
3	625	140	1726.9	T92 heat affected zone
4		160	912.8	
5		180	537.1	

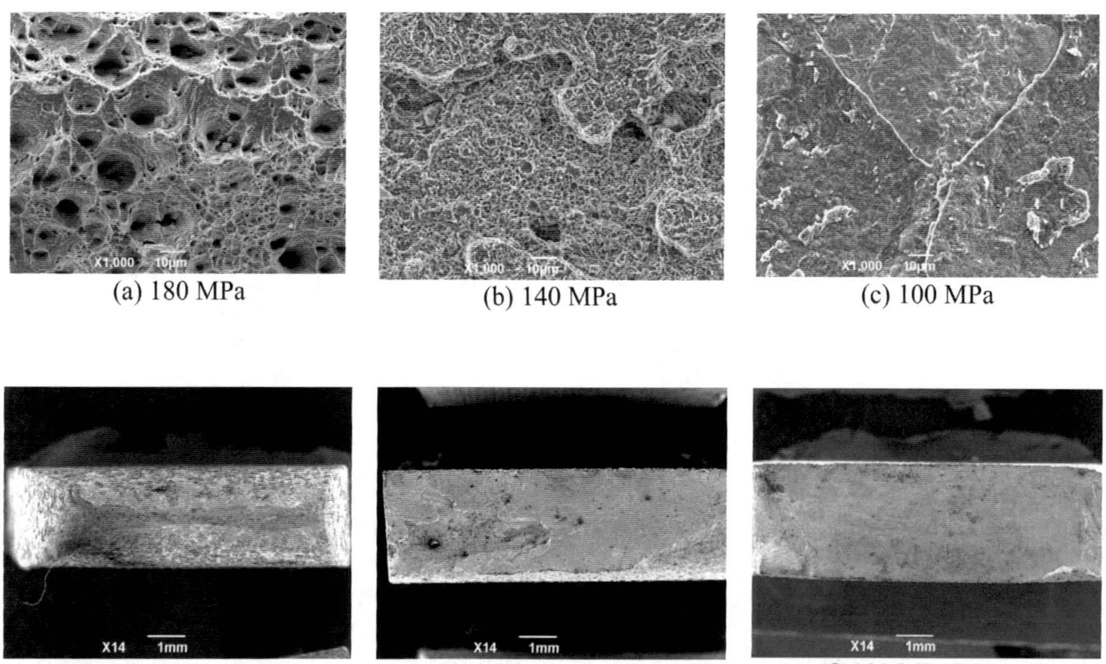

(a) 180 MPa (b) 140 MPa (c) 100 MPa

(d) 180 MPa (Macro) (e) 140 MPa (Macro) (f) 100 MPa(Macro)

Figure 1 The creep rupture appearance of welded joints

With the loading stress decreased and the creep rupture time increased, the number of dimples was reduced significantly and the number of the cleavage surfaces was increased gradually. The fracture surface of the creep specimen was more flat. Therefore, under the higher loading stress, the transcrystalline plastic fracture was the rupture mode of creep specimens and results in forming dimples. Under the lower loading stress, the intergranular brittle fracture was the rupture mode of creep specimens and results in cleavage rupture [5].

Figure 2 (a) shows metallograph of creep specimen for loading stress of 180 MPa. The matrix feature of the T92 heat-affected zone was the plate martensite. After several hundred hours under the higher loading stress, the shapes of the martensite grains were greatly deformed and some martensites grains were change into elliptical. It indicates that the grains showed a large slip along the grains boundary before the fracture of the creep specimen, and the transcrystalline plastic fracture was the rupture mode of creep specimens. Figure2 (b) shows metallograph of creep specimen for loading stress of 100 MPa. There were many creep cavities in the T92 heat-affected zone, and the creep cavities were distributed along the grains boundaries with little spacing between each other [6].

Table 2 exhibited the EDS analysis results of the precipitated phases. Figure 3 (a) shows the shape of precipitated phases of the creep specimen for loading stress of 180 MPa, the precipitated phase A is $Cr_{23}C_6$ that distributed along the martensite grains boundaries. The fine precipitate phase B is MX carbonitride that distributed in the interior of martensite grains. There was no significant coarsening of precipitate phase, and the creep rupture of the creep specimens was not caused by coarsening of the precipitated phase. Figure 3 (b) shows the shape of precipitated phases of the creep specimen under the loading stress of 180 MPa. The precipitated phase A is $Cr_{23}C_6$ and the precipitated phase B is the Laves phases. The $Cr_{23}C_6$ and the Laves phases were distributed along the grains boundaries. The aggregation and growth of the precipitated phase were obviously, which leads to creep rupture of the creep specimens.

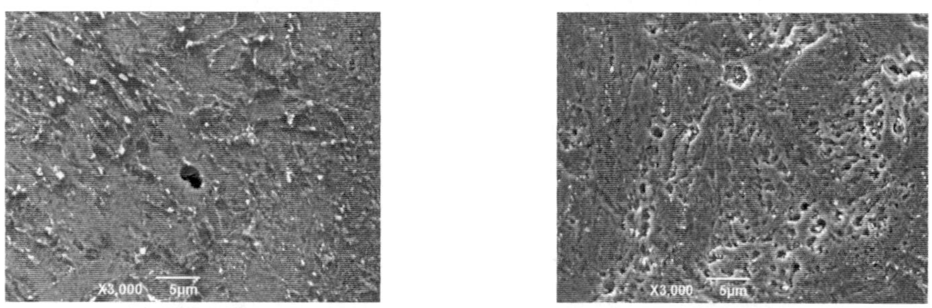

(a) 180 MPa (b) 100 MPa

Figure 2 The metallograph of the T92 heat-affected zone

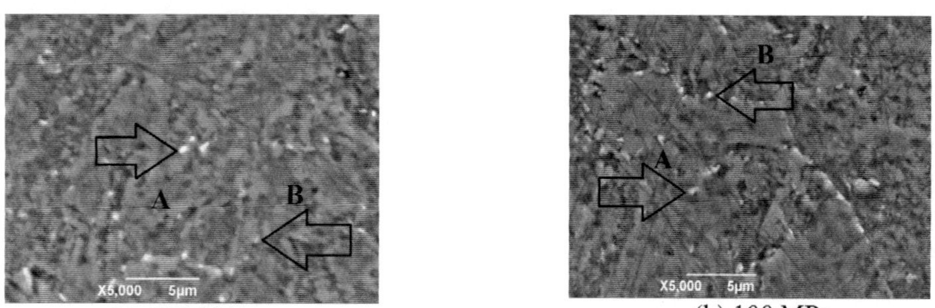

(a) 180 MPa (b) 100 MPa

Figure 3 The shape of precipitated phases

Table 2 The energy spectrum analysis of the precipitated phase (at%)

	C	N	Si	V	Cr	Fe	Nb	Mo	W
The A (180 MPa)	8.34/28.90	1.5/4.46	-	-	11.86/9.55	75.47/56.26	-	0.9/0.39	1.93/0.44
The B (180 MPa)	8.6/28.95	5.9/17.04	-	4.2/2.87	14.32/11.13	37.59/27.22	29.39/12.79	-	-
The A (100 MPa)	23.63/59.19	-	-	-	9.71/6.02	63.27/34.08	-	1.02/0.32	2.37/0.39
The B (100 MPa)	7.78/31.21	-	5.83/10.0	0.13/0.12	6.64/6.15	48.95/42.22	0.68/0.35	6.06/3.04	22.65/5.93

Figure 4(a) shows the cracks of creep specimens for loading stress 180 MPa. The martensite strips in the T92 heat-affected zone were subjected to compression deformation. In some areas the fragmentation of the martensite strips generated a subgrains boundaries, and the size and shape of the precipitates were not changed significantly. In the creeping process of the T92 heat-affected zone, the dislocations of the plate martensite grains occurred slipping and climbing. When the dislocations have encountered the MX carbonitrides that precipitated in the matrix dispersedly, and the MX carbonitrides prevent the movement of dislocations, which led to the stress concentrations. As the creeping process continued, these areas with stress concentrations will develop into cracks sources gradually that leading to the formation of creep rupture eventually. Figure 4(b) shows the cracks of creep specimens for loading stress of 100 MPa. The martensite strips which through a long creep process in the T92 heat-affected zone were restored, and the sub-grains boundaries between the martensite strips became blurred. Because of the long-term higher temperature, the C, Cr, Fe, W and other elements of the martensite grains moved to the grains boundaries with higher grain boundary energy. The $M_{23}C_6$ that at the original austenite grains boundaries and sub-grains boundaries are aggregated and growed up. The Laves phases were precipitated at the grains boundaries intersection. The nucleation and growth of creep cavities was caused by the coarse $Cr_{23}C_6$ carbide and the Laves phases. The creep crack was formed by the unioned creep cavities, which were characterized by intergranular brittle fracture. Therefore, the creep rupture of the T92 heat-affected zone under the higher loading stress was caused by the interaction between the dislocations and the precipitation phases in the interior of grains. The creep rupture of the T92 heat-affected zone under the lower loading stress was caused by the coarsening process of $M_{23}C_6$ and the Laves phase that distributed along the grains boundaries. With the loading stress reduced continuously, the precipitated phase are aggregated and growed up further along the grains boundaries, and the tendency of the intergranular brittle fracture for the creep specimens is more obvious.

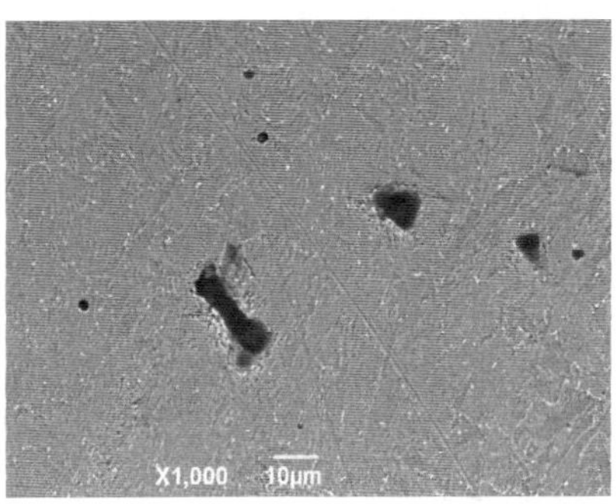

(a) 180MPa

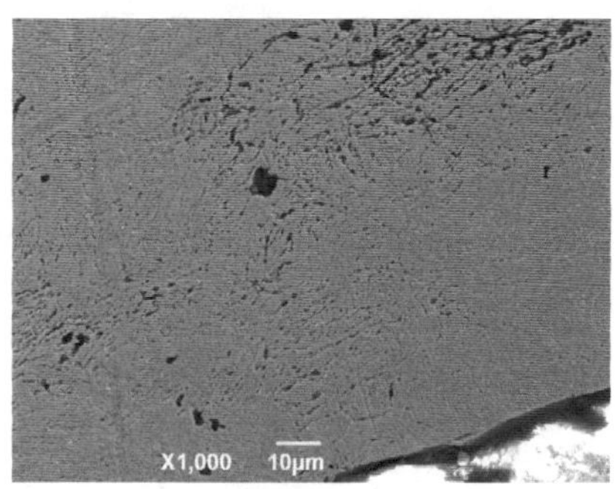

(b) 100MPa

Figure 4 The cracks of the T92 heat-affected zone

4. Conclusions

The creep rupture strength is evaluated using isothermal method. With the higher loading stress, the transcrystalline fracture was the rupture mode of creep specimens, and results in forming dimples. With the lower loading stress, the creep rupture was the rupture mode of creep specimens, and results in cleavage rupture.

References

[1] X M Li, Y Zou and Z W Zhang. (2010) Microstructure Evolution of a Novel Super304H Steel Aged at High Temperatures. Mater. Trans, 608: 164-173

[2] J Cao, Y Gong and Z G Yang. (2011) Creep fracture behavior of dissimilar weld joints between T92 martensitic and HR3C austenitic steels. Int. J. Press. Ves. Pip, 88: 94–98.

[3] G H Chen, Y M Song, J Q Wang. (2012) High-temperature short-term tensile test and creep rupture strength prediction of the T92/TP347H dissimilar steel weld joints. Eng. Fail. Anal, 26:220-229.

[4] J Cao, Yi Gong and Zhenguo Yang. (2011) Microstructural analysis on creep properties of dissimilar materials joints between T92 martensitic and HR3C austenitic steels. Mater. Sci. Eng. A, 528:6103-6111.

[5] J Cao, Y Gongan and K Zhu. (2011) Microstructure and mechanical properties of dissimilar materials joints between T92 martensitic and S304H austenitic steels. Materials. &. Design, 32:2763-2770

[6] G H Chen, Q Zhang and J J Liu. (2013) Microstructures and mechanical properties of T92/Super304H dissimilar steel weld joints after high-temperature ageing. Materials. &. Design, 44:469-475

IOP Conf. Series: Materials Science and Engineering **721** (2020) 012007 doi:10.1088/1757-899X/721/1/012007

Effects of Friction Speed on Microstructures and Mechanical Properties of Friction Welded Super304H Welding Joints

Xu Yang, Yuan Lu, Debiao Li, Cheng Bi, Jianlong Zhang, Fuyang Sun, Bo Cao

Xi'an Special Equipment Inspection Institute, Xi'an, 710065, China

Abstract. The Super304H austenitic steel were welding though friction welding technique. The Super304H joints showed excellent microstructure and mechanical properties. The effects for friction speed on the microstructure and mechanical properties of Super304H welding joints was investigated. With an increase in the friction speed, the grains didn't grow up and second-phase particles precipitates increased in welding joints. The tensile fracture location was in the Super304H heat affected zone. When the friction speed was increased, the tensile strength increased and impact toughness was decreased.

1. Introduction

Under the ultra-supercritical condition, the heat efficiency of ultra-supercritical boilers was about 45%. Many ultra-supercritical (USC) boilers have been built and successfully operated all over the world. At present, the worldwide aspirations for ensuring energy safety with simultaneous tightening of the requirements of the act on permissible emission of pollutants into the atmosphere have contributed to the development of materials that allow the improvement of steam working parameters resulting in units designed for ultra-supercritical steam parameters. The boilers between 550 and 760°C requires the use of new materials with higher mechanical parameters and more complex manufacturing process for making the components of boiler pressure section out of them [1]. Among the critical boiler pressure section components, there should be distinguished, among other things, the thin-wall pipes of steam superheaters whose materials can be operated at up to 600°C and 25-30 MPa in systems with supercritical parameters. The materials more often used for components of steam super-heater with so high parameters are mainly based on high-temperature corrosion resistant steels of austenitic structure, which included the Super304H austenitic steel. The Super304H austenitic exhibited both excellent microstructures and creep rupture strength. The Super304H employed precipitation strengthening through finely dispersed NbCrN and are designed to prevent degradation of properties finely though dispersed Cu rich phase [2]. At present, the Super304H austenitic steel are welded through the GTAW technique usually for engineering applications. The GTAW technique was mature and suitable for field welding. Nevertheless, the Super304H joints welding through the conventional fusion welding still showd inferior mechanical properties, because that the heat affected zone of the fusion welding is prone to form coarse grains.[3, 4]. This paper was to fabricate a combination the Super304H austenitic steel by using friction welding with better mechanical properties. The effect of friction speed on the microstructure and mechanical properties of the welded joint was discussed.

2. Experimental procedures

The chemical compositions and mechanical properties of the Super304H austenitic steel were listed in Tables 1 and 2. Figure 1 shows the outward appearance and the metallography. The metallography of Super304H showed fine austenite grains with a small amount of carbide precipitation phase. The friction speed was 1200 rpm of Process 1, 1500 rpm of Process 2 and 1800 rpm of Process 3 respectively for the process of friction welding. The metallography and fractographs of the welding joints were performed by the optical microscope and the scanning electron microscope respectively. The X-ray XRD patterns were examined by X-ray Diffraction. The tensile samples were tested using a testing instrument, and the impact samples were tested using an impacting instrument.

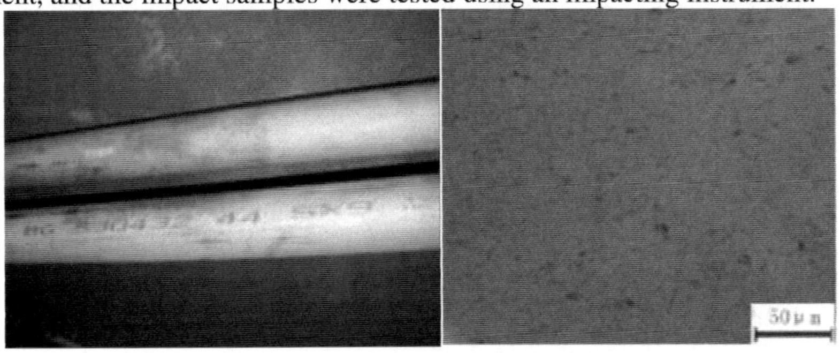

(a) (b)

Figure 1 The appearance and metallography of Super304H (a) the appearance, (b) the metallography

Table 1 Chemical composition of Super304H

	Chemical composition										
	C	Si	Mn	Cr	Ni	Mo	Cu	Nb	N	Al	B
Super304H	0.07	0.21	0.81	18.37	9.06	0.42	2.93	0.52	0.10	0.01	0.002

Table 2 Mechanical properties of Super304H

	$\sigma_{0.2}$/MPa	σ_b/MPa	δ/%	A_{KV}/J	Hardness/ HB
Super304H	≥235	≥590	≥35	–	–

3. Results and Discussion

Figure 2 showed the X-ray diffraction profiles for the Super304H welded by Process 2. The main phases were the γ-Fe phase and the precipitated phases of $Cr_{23}C_6$ and NbC. The Cr element had a tendency to form carbides, and some C element and Cr element were precipitated from matrix and formed the precipitated phases $Cr_{23}C_6$ at the grain boundary preferentially. Therefore, $Cr_{23}C_6$ and NbC were denoted precipitated phases in the the Super304H welding joints [5]. Figure 3, 4 and 5 exhibited the metallography for different friction speed. The dynamic recrystallization driving force and lattice distortion energy increased in the friction welding process. The welding zone included a welding interface, the areas (about 50μm from the welding interface) on both sides of the welding interface. The welding interface that was straight and clear, so the welding interface is well combined. The welding zone was mainly composed of fine equiaxed austenite grains because of grain refinement. The welding zone for the fusion welding was typical dendritic morphology with large grains. So the welding zone exhibit excellent microstructure and mechanical properties [6]. The width of the heat affected zone caused by friction welding process were much smaller than the width of the heat affected zone caused by the fusion welding process. The tendency of grains growth was less obvious than the tendency of grains growth. So the welding joints with the fine grains showed excellent microstructure and mechanical properties. With the distance from the welding zone increased, the size of the austenitic grains decreased, until the size of grains was equal to the size of grains in the base metal. So the friction welded Super304H joints with the fine grains showed excellent microstructure and mechanical properties.

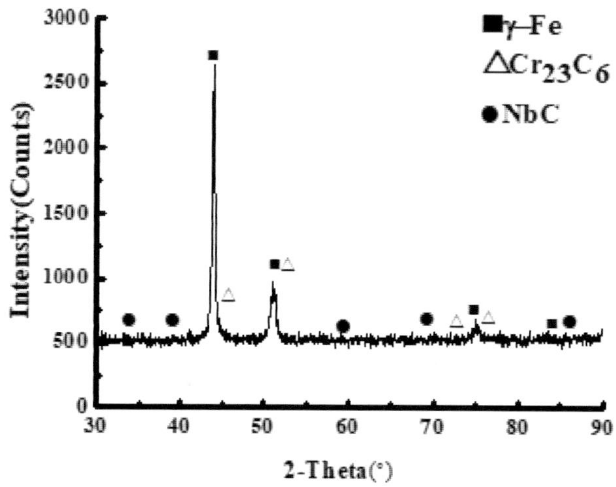

Figure 2 XRD patterns of for the Super304H

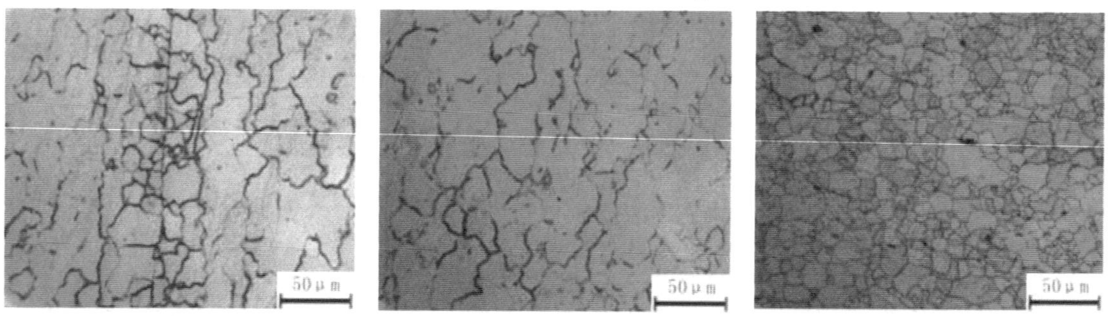

(a) (b) (c)

Figure 3 The metallography for Process 1 (a) the welding zone,
(b) the heat affected zone (c) the base metal

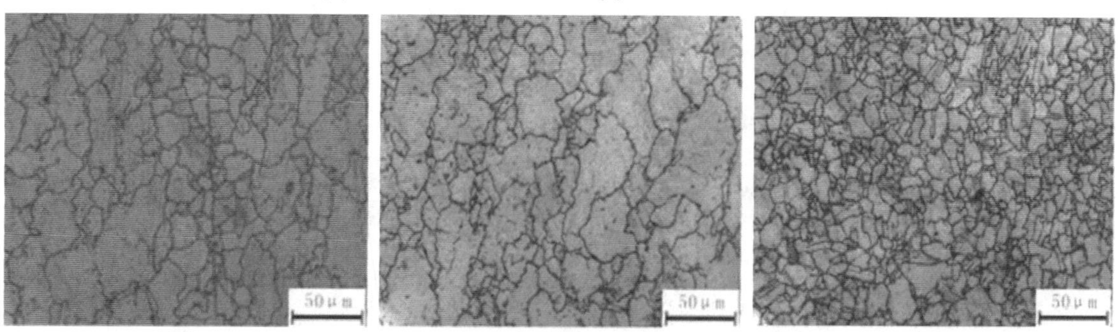

(a) (b) (c)

Figure 4 The metallography for Process 2 (a) the welding zone,
(b) the heat affected zone (c) the base metal

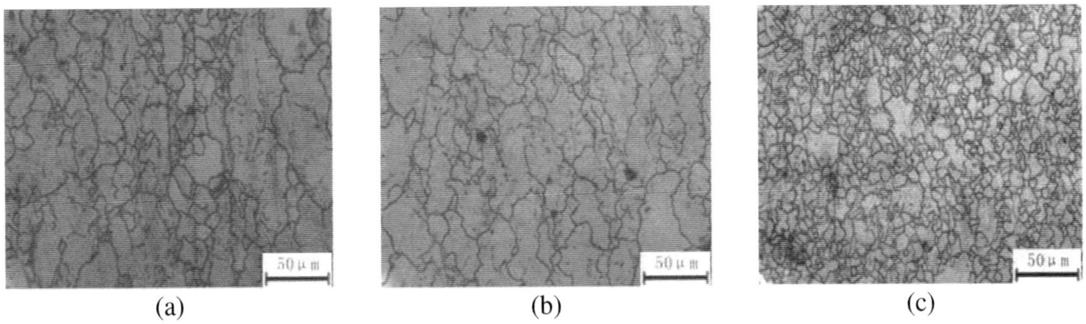

Figure 5 The metallography for Process 3 (a) the welding zone,
(b) the heat affected zone (c) the base metal

The surface temperature of the welding joints was equal to the temperature of the welding heat source, The welding heat source was considered to be a linearly propagated planar heat source in friction welding process. If the heat dissipation to the surrounding space was not considered, the surface temperature of the welding joints was calculated according to the Eq (1)

$$T(t) = qt/(\pi\lambda C) \tag{1}$$

T was referred to the surface temperature of the welding joints, *t* was referred to friction time and *q* was referred to friction heating power. It was calculated according to the Eq (2)

$$q = (\pi n T)/30 \tag{2}$$

T was referred to the friction torque, and *n* was referred to the friction speed. The friction heating power was mainly depended on the friction torque and the friction speed. When the welding torque was constant , as the friction speed increased, the friction heating power and the surface temperature of the welding joints increased correspondingly. Although the surface temperature of the welding joints increased, the change about the grains in the welding zone was ignored. As the friction speed increased, the heat input and the temperature of the heat affected zone increased accordingly. The size change of the austenite grains was not detectable, but the amount of the precipitated phases $Cr_{23}C_6$ and NbC in the both heat affected zone increased obviously [7,8].

Table 3 depicted the change of the tensile strength as for the different friction speed. The fracture position of the welding joints was in the heat affected zone for the different friction speed. The fractographs of the samples were shown in figure 6. As the friction speed increased, the tensile strength was increased gradually. When the friction speed was increased, the friction heating power and the temperature was increased during the friction welding process. The amount of the precipitated phases $Cr_{23}C_6$ and NbC increased obviously, the dispersion strengthening effect of the $Cr_{23}C_6$ and NbC carbides was main reason of the tensile strength increased. The fracture morphology was characterized by typical ductile fracture, and a lot of dimples distributed on the facture appearance. Table 2 depicted the variation of the impact toughness as a function of the friction speed. The heat affected zone with coarse grains was the larger, so the impact toughness of the heat affected zone was the lower. As the friction speed increased, the dislocation density and lattice distortion energy in the welding zone increased, and the impact toughness of the welding zone decreased. Because of the precipitation of the $Cr_{23}C_6$ and NbC carbides, the impact toughness of the heat affected zone decreased with the friction speed increased. The fractographs of the impact toughness samples were shown in figure 7, 8 and 9. The fracture mode of the impact toughness samples was ductile fracture with the aggregation dimples. In the welding zone, the dimples were distributed densely and deeply, and the shapes of dimples were round. In the heat affected zone, the dimples were distributed sparsely and shallow, and the shapes of dimples were flat .

Table 3 The mechanical properties of Super304H welded joints

	Tensile strength /MPa		Impact toughness /J	
	$R_{p0.2}$	R_m	Welding zone	Heat affected zone
Process 1	412.54	715.13	96.77	81.56
Process 2	439.71	744.98	85.42	74.95
Process 3	470.91	771.11	72.88	64.44

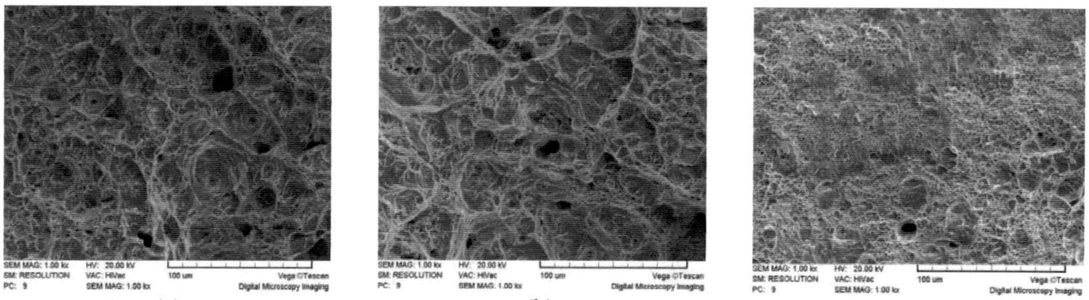

(a) (b) (c)

Figure 6 The fracture appearance of the tensile strength samples (a) Process 1, (b) Process 2, (c) Process 3

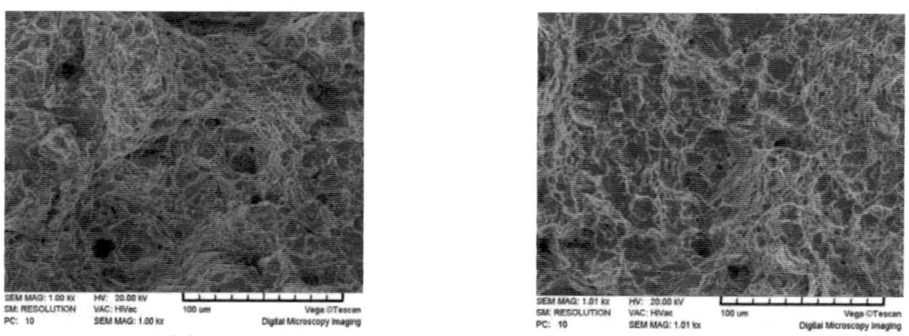

(a) (b)

Figure 7 The fracture appearance of the impact toughness samples for Process 1 (a) the welding zone (b) the heat affected zone

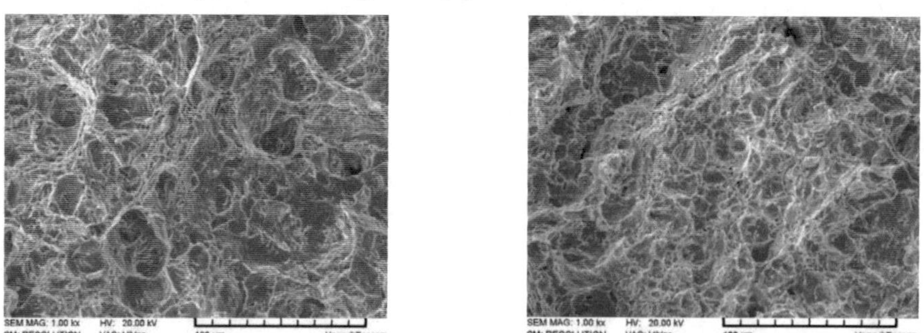

(a) (b)

Figure 8 The fracture appearance of the impact toughness samples for Process 2 (a) the welding zone (b) the heat affected zone

IOP Conf. Series: Materials Science and Engineering **721** (2020) 012007 doi:10.1088/1757-899X/721/1/012007

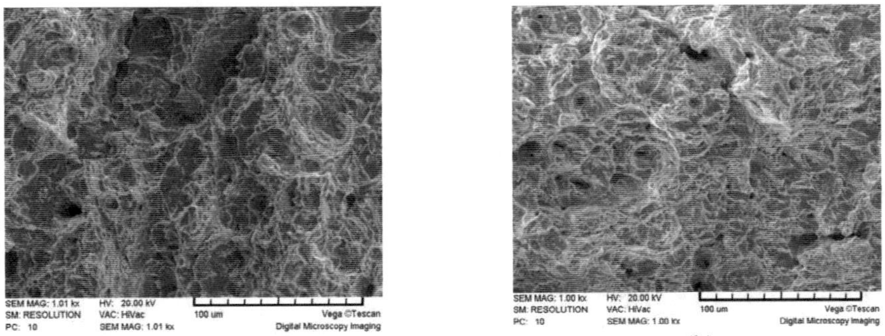

(a) (b)

Figure 9 The fracture appearance of the impact toughness samples for Process 3
(a) the welding zone (b) the heat affected zone

4. Conclusions

Super304H austenitic steel welded joints with excellent properties by friction welding. The welding interface was well combined and the the welding zone grains was fine. The effects of friction speed for the structure and mechanical properties of Super304H steel joints was investigated. The friction speed was increased, the welding zone grains didn't grow up and the second-phase particles precipitates was increased. The friction speed was increased, the tensile strength was increased and the impact toughness was decreased.

References

[1] X M Li, Y Zou and Z W Zhang. (2010) Microstructure Evolution of a Novel Super304H Steel Aged at High Temperatures. Mater. Trans, 608: 164-173

[2] J Cao, Y Gong and Z G Yang. (2011) Creep fracture behavior of dissimilar weld joints between T92 martensitic and HR3C austenitic steels. Int. J. Press. Ves. Pip, 88: 94–98.

[3] P Li, J Li, X Li. (2015) A study of the mechanisms involved in initial friction process of continuous drive friction welding. J. Adhes. Sci. Technol, 29: 1246-1257.

[4] X Li, J Li, Z Liao. (2018) Effect of rotation speed on friction behavior and radially non-uniform local mechanical properties of AA6061-T6 rotary friction welded joint. J. Adhes. Sci. Technol, 32: 1987-2006.

[5] J Cao, Y Gongan and K Zhu. (2011) Microstructure and mechanical properties of dissimilar materials joints between T92 martensitic and S304H austenitic steels. Materials. &. Design, 32:2763-2770

[6] G H Chen, Q Zhang and J J Liu. (2013) Microstructures and mechanical properties of T92/Super304H dissimilar steel weld joints after high-temperature ageing. Materials. &. Design, 44:469-475

IOP Conf. Series: Materials Science and Engineering **721** (2020) 012008 doi:10.1088/1757-899X/721/1/012008

Uncertainty Research of Particles Concentration Measurement in Continuous Emission Monitoring System of Thermal Power Plants

Tianqi Liang[1], Tianlin Zheng[2], Guangxue Zhang[1], Luyao Ma[1] and Hong Xu[1*]

[1] College of Metrology and Measurement Engineering, China Jiliang University, Hangzhou, Zhejiang, 310018, China

[2] Hangzhou Zunbang Technology Co., Ltd, Hangzhou, Zhejiang, 310018, China

[*] Corresponding author's e-mail: xuhonghappy@cjlu.edu.cn

Abstract. In order to monitor the reliability of the CEMS operation data of a thermal power plant in Hangzhou, the data of its operation process were collected. The components of uncertainty influencing on the particles concentration measurement, including zero drift, range drift, relative accuracy and measuring repeatability, were analyzed in the system. The confidence interval (2.80%) and the allowable interval (8.94%) were introduced to ensure the reliability of the system data. The result of evalution shows that the related expanded uncertainty of particles concentration measurement is 9.26% when coverage factor is k=2 and level of confidence is 95%. Measuring repeatability is the main factor affecting the uncertainty.

1. Introduction

With the continuous development of China's economy, the national demand for energy is increasing. Energy consumption of coal-based thermal power plants is one of main causes of air pollution. Curbing pollution emissions from the source has become the focus of the environmental industry. At present, particles are one of the major pollutant emissions from thermal power plants. It has become an important task for enterprises to control the concentration of pollutants to meet environmental emission standard. Flue gas Continuous Emission Monitoring System (CEMS) is a device for continuous monitoring of stationary pollution sources. In recent years, CEMS has been widely used in thermal power plants for continuous monitoring of pollution sources. In order to ensure the implementation of 'Hangzhou Emission Standard of Air Pollutants for Boilers', that is, the particles concentration do not exceed 5mg/Nm^3, regular calibration and testing of the power plant's CEMS is essential. Meanwhile, it is necessary to analyze and evaluate the uncertainty to ensure the accuracy and reliability of the measuring results [1-2].

Uncertainty in measurement analyzes the dispersion of the measured values based on the information provided [3]. Each factor in the system may introduce corresponding uncertainty, which can evaluate the credibility and acceptability of results of measurement [4]. In the paper, the operation data are monitored after ultra-low emission renovation project is completed at a thermal power plant in Hangzhou, and uncertainty of the data is analyzed and evaluated according to 'Guide to the Evaluation and Expression of Uncertainty in Measurement' (GB/T 27418-2017).

2. Ultra-low emission renovation project in thermal power plant

There are three 130t/h high temperature and high pressure circulating fluidized bed boilers at a thermal power plant in Hangzhou. To meet the requirements of ultra-low emission, low-nitrogen combustion and wet electro-static precipitator (WESP) have been added in the plant, on the basis of retaining semi-dry absorption tower, electrostatic precipitator and fabric filter. The flue gas of the boiler is desulfurized by desulfurization tower, then is purifed by WESP at the top of the tower, finally is discharge from the chimney [5]. Figure 1 shows the technological process of flue gas treatment after ultra-low emission renovation project in the power plant.

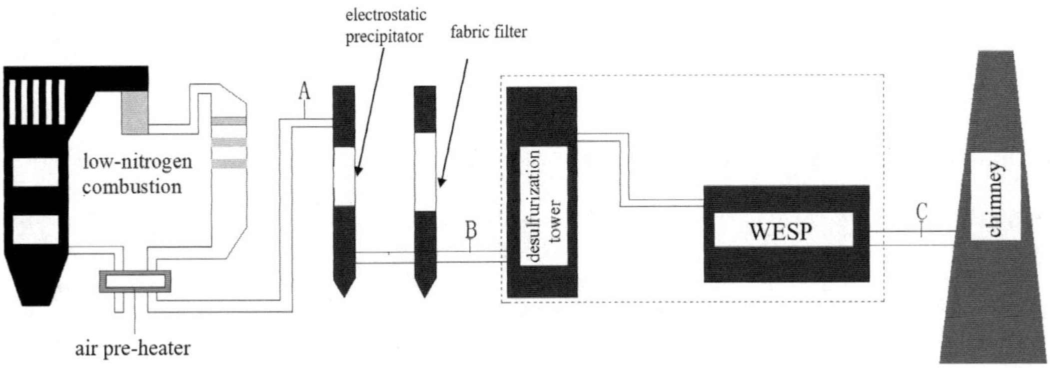

Figure 1. Flue gas purification system process and sampling points

The flue gas sampling points were located at A and B. CEMS were installed at points C as shown in Figure 1, which is in the monitoring room of the chimney wall of the power plant. Flue gas is monitored online by CEMS to ensure that emissions meet national environmental standards.

2.1Wet electro-static precipitator

The main working principles of WESP are that the water mist is sprayed to the discharge electrode and corona area. Under the action of the barbed electrode, the water mist forms a strong phenomenon of collision, interception and adsorption in the corona area. Driven by electric field force, dust particles reach the dust collecting pole and are thus captured by the water mist. The water mist forms a continuous water film on the dust collector to flush the captured dust into the ash hopper and discharge it.

The WESP not only has a good removal effect on PM2.5, but also on acid fog (SO_3), toxic heavy metals such as mercury and other pollutants removal according to domestic and foreign studies [6].

2.2Test results and evaluation of dust removal facilities

2#boiler was selected for test to evaluate dust removal efficiency of the renovation. The inlet and outlet particles concentration of electrostatic precipitator, fabric filter and WESP were monitored separately. The concentration at the sampling point during the test is summarized in the table below. Test data came from two working conditions, high load (90%) and low load (55-65%).

As shown in Table 1, under high boiler load >90%, the dust removal efficiency of electrostatic precipitator and fabric filter is about 99.99%, while the combine removal efficiency of desulfurization tower and WESP 63.24%. Under low boiler load 55-65%, the dust removal efficiency of two combined devices mentioned above is about 99.95% and 89.57% respectively. The integration of desulfurization and WESP can control the particles emission concentration at a lower level under both boiler loads, which fully meets national environmental emission standards of thermal power plants.

Table 1. Test results of dust removal efficiency

Boiler load	Dust removal facilities	Inlet particles concentration (mg/Nm^3)	Outlet particles concentration (mg/Nm^3)	Efficiency (%)
>90%	Electrostatic precipitator and fabric filter	1.22×10^4	1.36	99.99
	Desulfurization tower and WESP	1.36	0.50	63.24
55-65%	Electrostatic precipitator and fabric filter	1.01×10^4	4.70	99.95
	Desulfurization tower and WESP	4.70	0.49	89.57

3. Analysis of factors affecting uncertainty of particles concentration measurement

3.1Sources of uncertainty of particles concentration measurement

Figure 2 shows the sources of uncertainty of particles concentration measurement. The uncertainty introduced by zero drift, range drift and relative accuracy is mainly determined by CEMS itself; the uncertainty introduced by measuring repeatability is affected by multiple measurements.

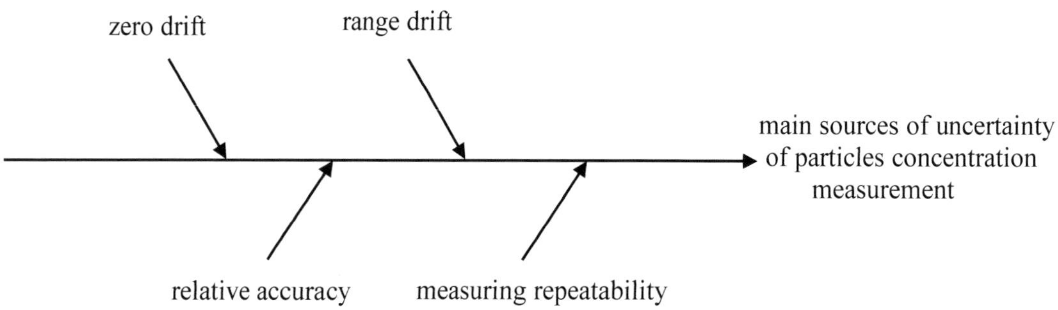

Figure 2. Sources of uncertainty of particles concentration measurement

The uncertainty of particulates concentration measurement includes two types, A and B. Uncertainty of type A is introduced by measurement repeatability, which can be calculated by Bessel formula [2]. Uncertainty of Type B is mainly obtained according to the instrument parameters and calibration nameplate. However, since the CEMS monitoring of flue gas is a dynamic process, different influencing factors will also change with the continuous changes in the working conditions of the measurement process. Therefore, it is necessary to collect and analyze the experimental data during operation of the system, so as to have more accurate feedback on the influencing factors of measurement uncertainty.

The technical indexes of CEMS include the requirements of the half width of the confidence interval, the half width of the allowable interval and correlation coefficient, which also reflects the uncertainty of particles concentration measurement of CEMS and ensures the data quality and credibility of flue gas monitoring process [7].

3.2 Mathematical model of calculating standard uncertainty

Particles concentration can be directly obtained by CEMS. The zero drift, range drift, relative accuracy and measuring repeatability of the instruments and equipment are independent of each other and have no correlation [8], so the correlation coefficient is 0 and the sensitivity coefficient is 1 for each component. According to the calculation method of uncertainty, the combined standard uncertainty can be obtained as follows:

$$u_c = \sqrt{\sum_{i=1}^{n}\left(\left(\frac{\partial f}{\partial x_i}\right)^2 (u_{xi})^2\right)} = \sqrt{\sum_{i=1}^{n}\left((u_{xi})^2\right)} \quad (1)$$

Where u_c is combined standard uncertainty; $\partial f / \partial x_i$ is sensitivity coefficient; u_{xi} is standard uncertainty of each component.

3.3 Analysis of factors affecting uncertainty of particles concentration measurement

3.3.1 Data analysis of zero drift and range drift in CEMS. After the stable Operation of the analytical instrument to be tested, the zero calibration department will adjust to zero and record the stable zero reading of the instrument as Z_0; Then switch to the range calibration unit and record the reading as L_0. Repeat the above operation after 24 hours of equipment operation, record the zero reading and range reading Z_n and L_n, and continue the above operation for 3 days. According to equations (2) and (3), the values of zero drift and range drift are calculated respectively.

$$Z_d = \frac{|\Delta Z|}{R} \times 100\% \quad (2)$$

$$L_d = \frac{|\Delta L|}{R} \times 100\% \quad (3)$$

Where Z_d is zero drift; ΔZ is absolute error of zero drift; R is full range of instrument; L_d is range drift; ΔL is absolute error of range drift. Table 2 shows the test data of zero drift and range drift.

Table 2. Test data of zero and range drift in CEMS

Test location		Monitoring room		Principle of measurement		High temperature and dilution lazer	
Response value of calibration		49.3mg/Nm^3		Range		50mg/Nm^3	
Unit of measurement (mg/Nm^3)							
Z_0	Z_n	$\Delta Z = Z_n - Z_0$	Z_d	L_0	L_n	$\Delta L = L_n - L_0$	L_d
0.02	0.04	0.02	0.04	49.7	49.3	-0.40	0.80
0.13	0.16	0.03	0.06	49.0	49.7	0.70	1.40
1.10	0.98	-0.12	0.24	48.8	48.5	-0.30	0.60
Maximum absolute error of zero drift			-0.12	Maximum absolute error of range drift			0.70
Zero drift			0.24	Range drift			1.40

From the data of table 2, the zero drift and range drift in CEMS is 0.24% and 1.40%, better than 2%F.S. (full span) and 5%F.S. required by equipment parameters.

3.3.2 Data analysis of relative accuracy of particles concentration measurement in CEMS. The evaluation index of particles concentration measurement in CEMS introduces the half width of the confidence interval and the half width of the allowable interval, which does not directly affect the uncertainty from the data level, but also reflects the quality of the measured data of particles concentration. The CEMS data of particles were taken and compared with the concentration values measured by reference method [9] at the same time period. Manual measurement data have converted the cross-section concentration of particles under standard conditions into the cross-section concentration value of particles under actual flue gas.

Table 3. Comparison of particles concentration measured by reference method and CEMS

Serial number	Indicating value of CEMS (X_i)	Indicating value of reference method (Y_i)	Serial number	Indicating value of CEMS (X_i)	Indicating value of reference method (Y_i)
1	2.86	2.85	9	0.96	0.86
2	3.39	3.19	10	0.13	0.21
3	3.33	3.22	11	1.14	1.01
4	3.63	3.53	12	2.03	2.05
5	3.19	3.23	13	1.96	2.01
6	2.98	2.88	14	2.1	1.96
7	1.66	1.56	15	2.48	2.22
8	0.48	0.28		$\overline{X}$:2.15	$\overline{Y}$:2.07

The linear equation is:

$$Y = aX + b \tag{4}$$

Where Y is the particles concentration predicted by reference method; X is measuring concentration of particles of CEMS; a is slope of curve; b is intercept of curve.

According to relevant calculation standards [10], the slope is 0.989, the intercept is -0.060, the correlation coefficient is 0.996. The linear equation is $Y = 0.989X - 0.060$.

The equation for calculating the half width of the confidence interval as follows:

$$CI{=}t_{df,1-a/2}S_E\sqrt{\frac{1}{n}} \tag{5}$$

$$S_E = \sqrt{\frac{1}{n-2}\sum_{i=1}^{n}\left(Y_i - Y_i\right)^2} \tag{6}$$

$$CI\%{=}\frac{CI}{EL}\times 100\% \tag{7}$$

Where CI is the half width of the confidence interval at the average value; S_E is dispersion or deviation of the calibration curve. Refer to the relevant table, $t_{df,1-a/2}$ is 2.160, so S_E is 0.104, CI is 0.058. EL is the average value measured by reference method during the detection period. Therefore, $CI\%$ is 2.80%, which is better than 10%, the half width of the confidence interval required by the indicator.

The equation for calculating the half width of the allowable interval as follows:

$$TI = u_{n'}V_{df}S_E \tag{8}$$

$$TI\% = \frac{TI}{EL}\times 100\% \tag{9}$$

Where TI is the half width of the allowable interval at the average value; refer to the relevant paper, $u_{n'}$ is 1.198, V_{df} is 1.485, TI is 0.185. Then $TI\%$ is 8.94%, which is better than 25%, the half width of the allowable interval required by the indicator.

The confidence interval is: $\hat{Y}{=}0.989X - 0.060 \pm 0.058$

The allowable interval is: $\hat{Y}{=}0.989X - 0.060 \pm 0.185$

Figure 3 shows the relationship between correlation calibration curve, the confidence interval and the allowable interval.

IOP Conf. Series: Materials Science and Engineering **721** (2020) 012008 doi:10.1088/1757-899X/721/1/012008

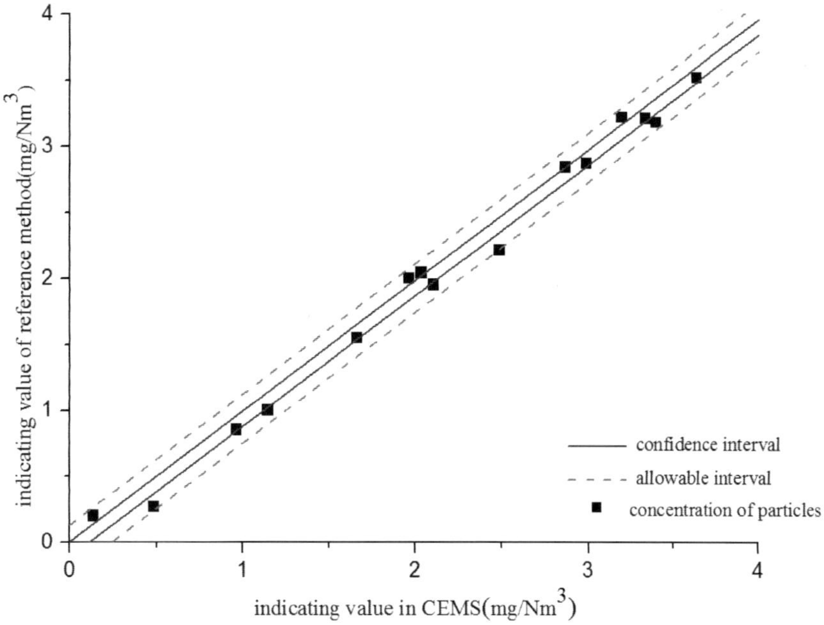

Figure 3. Sketch map of correlation calibration curve, confidence interval and allowable interval

Combined with the above calculations and figures, it can be explained that the calibration curve of CEMS measured values detected by reference method is within the confidence interval and the allowable interval, and the calibration of the two indexes is far lower than the requirements of instrument performance indexes, further demonstrating that the system has higher reliability in monitoring patricles concentration.

4. Evaluation results of standard uncertainty of particles concentration measurement

4.1 Uncertainty introduced by zero drift and range drift

According to the results, the zero drift and range drift are respectively 0.24% and 1.40%. The evalution equation of uncertainty of type B as follows:

$$u_B = \frac{\alpha}{k} \tag{10}$$

Where α is relative error of influencing factor, k is coverage factor. On the basis of uniform distribution, $k=\sqrt{3}$, $u_{Bzero} = \alpha / k = 0.24\% / \sqrt{3} = 0.14\%$. $u_{Brange} = \alpha / k = 1.40\% / \sqrt{3} = 0.81\%$.

4.2 Uncertainty introduced by measuring repeatability

According to the real monitoring data, the hourly data of the day are summarized in the following table.

Table 4. Monitoring data of particles concentration one day

Test location: monitoring room			Unit: mg/Nm^3		
Time	Concentration	Time	Concentration	Time	Concentration
0~1	2.02	8~9	1.58	16~17	1.73
1~2	1.85	9~10	3.14	17~18	1.94
2~3	2.01	10~11	1.94	18~19	1.99
3~4	2.18	11~12	1.77	19~20	2.04

4~5	2.21	12~13	2.11	20~21	1.83
5~6	3.01	13~14	3.37	21~22	1.85
6~7	2.02	14~15	2.14	22~23	3.02
7~8	1.68	15~16	2.12	23~24	2.03
Average value:2.15					

Uncertainty introduced by measuring repeatability taking evalution method of type A, on the basis of Bessel formula, the related standard uncertainty is u_A=4.56%.

4.3 Related combined standard uncertainty of particles concentration measurement
According to the measurement and caculation above, the evaluation results of related standard uncertainty of each component are summarized in the following table.

Table 5. Summary of uncertainty of each component

Serial number	Source of uncertainty	Component of uncertainty	Related standard uncertainty
1	Zero drift	u_{Bzero}	0.14%
2	Range drift	u_{Brange}	0.81%
3	Measuring repeatability	u_A	4.56%

In the operation process of the coal-fired power plant, in addition to the above factors affecting the uncertainty, there are also voltage fluctuations, ambient temperature and other factors [11-12]. Since the uncertainty introduced by such factors is relatively small, the calculation can be omitted in this paper. According to the above component combination formula (1), the related combined standard uncertainty of particles concentration measurement is $u_c = 4.63\%$.

The measuring error of particles concentration in CEMS follows the normal distribution. When coverage factor is k=2 and level of confidence is 95%, the related expanded uncertainty is $U = k \times u_c = 9.26\%$.

4.4 Summary
Among the factors affecting the uncertainty of particles concentration measurement in CEMS, the measuring repeatability has the greatest influence. First of all, because the paticles concentration monitoring is dynamic, the monitoring value fluctuates up and down with time, and its standard deviation will be relatively high. Secondly, since the monitoring point is located behind the WESP, the velocity and moisture content of flue gas will affect the monitoring of pollutants by the probe, resulting in obvious drift of repeated measurement values of flue gas by the CEMS, and the introduced uncertainty component is large.

The measurement principle of portable flue gas analyzer used by reference method is fixed potential electrolysis method, which is different from the Fourier infrared absorption method of CEMS, resulting in deviation of monitored data. Therefore, it is necessary to introduce the confidence interval and allowable interval to evaluate the measurement results.

5.Conclusion

(1) In this paper, the uncertainty of particles concentration measurement of CEMS in a thermal power plant in Hangzhou is researched and analyzed. The results show that the related expanded uncertainty of particulates concentration measurement is 9.26%. Measurement repeatability is the main influencing factor.

(2) The uncertainty evalution of particles concentration measurement introduces two indexes. The confidence interval half width is 0.48% and the allowable interval half width is 1.54%, which are far

less than the required value of device, further demonstrating that CEMS has higher reliability in monitoring particles concentration.

Acknowledgements
This paper was supported by National Key Research and Development Program of China (2018YFC1802100).

References

[1] Zhang, X.H., Schreifels, J. (2011) Continuous emission monitoring systems at power plants in China: improving SO_2 emission measurement. Energy Policy, 39: 7432-7438.

[2] Wang, Z.Y., Liu, Z.M., Xia, X.T. (2008) Measuring Error and Uncertainty Evalution. Science Press, Beijing.

[3] Wang, X. (2014) Interpretation of the current definition of uncertainty measurement. China Metrology, 12: 67-69.

[4] Ma, K., Li, X.J., Cui, M.M. (2015) Uncertainty evaluation for the determination of amaranthus red in wine samples by high performance liquid chromatography. Acta Metrologica Sinica, 35: 102-106.

[5] Zhang, X.F., Ruan, X. (2010) Key parameters influencing the efficiency of desulfurization inwet flue gas desulfurization systems. Energy Environmental Protection, 24: 41-44.

[6] Yang, L., Zhang, B., Wang, K.H. (2019) Conversion characteristics of combustible particles from coal-fired flue gas in WFGD and WESP. Environmental Science, 40: 121-125.

[7] Ni, Y.C. (2009) Practical Uncertainty of Measurement Evalution. China Metrology Press, Beijing.

[8] Song, M.S., Chen, Y.H., Tao, J.S. (2005) Risk of the estimating measurement uncertainty in artificial ignored correlative case. Acta Metrologica Sinica, 26: 90-92.

[9] Ministry of Environmental Protection of thePeople's Republic of China. (2017) HJ 836 Stationary source emission—Determination of mass concentration of particulate matter at low concentration—Manual gravimetric method.

[10] https://wenku.baidu.com/view/793ae37a66ec102de2bd960590c69ec3d5bbdbe7.html.Ministry of Environmental Protection of thePeople's Republic of China. (2017) HJ 76 Specifications and test procedures for continuous emission monitoring systems for SO_2, NO_X and particulate matter in flue gas emitted from stationary sources. https://www.docin.com/p-2204910993.html.

[11] Wang, H.G., Yang, J., Wang, G.P. (2007) Evalution uncertainty of measurement for sculpture dioxide in flue gas. Instrumentation and Analysis monitoring, 4: 42-44.

[12] Dong, F., Li, X.T., Li, J.H. (2008) Evaluation and analysis of uncertainty in measurement of SO_2 automatic monitor. In: 2008 International Conference on Machine Learning and ybernetics. Kunming. pp. 1958-1961.

Design and Uniformity Analysis of Fully Premixed Natural Gas Burner Venturi Mixer

Sun Zhenglu[1] Wang Yongtang[1*] Liu Hong[1] Dong Aihua[1] Shi Yunxin[1] Yu Bin[1]

[1] Harbin electric corporation central research institute, Harbin, Heilongjiang province, 150028, China

[*]Corresponding author's e-mail:wyongt@harbin-electric.com

Abstract. The venturi mixer is one of the key structures of the fully premixed surface natural gas burner. It has the function of uniformly mixing gas and air in a specific fluid space. In order to improve the burn-up rate of the combustion, and make the combustion temperature lowered to reduce the emission of NOx pollutants, a two-channel venturi mixer was designed, according to the venturi mixer performance of pressure loss and uniform. The mixer structure was analyzed by numerical simulation with the evaluation of the uniform performance parameters from the downstream following isometric characteristic cross-sectional. The results show that the venturi shrinkage angle and diffusion angle have little effect on the uniformity of mixer. The number or area of the gas injection holes is related to the uniform performance of the mixer and the pressure loss of the gas. The gas injection holes on the venturi mixer can be adjusted to balance two key parameters and it is an important factor in the design of venturi mixer structure of the fully premixed surface natural gas burner.

1. Introduction

As the country's requirements for energy conservation and emission reduction increase, and the urgent need for energy structure reform. It can be foreseen that the domestic demand for natural gas will continue to grow in the next ten years[1]. The fully premixed surface natural gas burner is a kind of high-efficiency combustion and ultra-low emission natural gas burner, which has gradually become the focus of enterprises, research institutions and developers. Premixing degree and pressure loss are key design parameters for the development and design of fully premixed surface natural gas burners. Therefore, designing a gas/air mixer with high efficiency, compactness, good uniform performance and stability determines premixed surface burner development level.

2. Theory of premixed mixers

2.1Low emission theory of fully premixed surface natural gas burner

As a sulfur-free fuel, methane (CH_4) is widely considered to be a highly efficient clean fuel[2]. Non-premixed atmospheric burners usually have concentrated concentration areas during combustion. Not only will there be insufficient combustion of the area, so carbon monoxide will be producted, but the inner flame will have less contact with oxygen, which will form a rapid NO_x[3], simultaneously, the combustion temperature is too high, the outer flame is in sufficient contact with oxygen. According to the Zeldovich mechanism, it also causes the oxidation of N_2 in the air to form thermal NO_x[4]. In general, the thermal NO_x dominates pollutant emissions of natural gas burner[5].

$$\frac{dc_{NO}}{dt} = 3\times10^{14}\times c_{N_2} c_{O_2}^{0.5} e^{(-542000/RT)} \tag{1}$$

$$N + OH \Leftrightarrow NO{+}H \tag{2}$$

Where c_{N_2}, c_{O_2}, c_{NO} is the concentration of N_2, O_2, NO. R is the gas constant (8.314 J/mol·K) and T is the temperature. It can be seen from the equal mechanism that the thermal NO_x will be produced in a large amount under high temperature conditions, and the upward trend has grown substantially. Another factor for the formation of the thermal NO_x is the oxygen contact concentration during combustion, and the amount of thermal NOx produced is proportional to the square root of the oxygen contact concentration. As a ultra-low NO_x emission burner, the fully premixed surface natural gas burner is from the above principles to reduce pollutants. There are two means, one is the surface combustion method. Surface burning uses mesh metal fiber as a porous medium and is laid on the surface of the burner's head. The burning on the surface of the metal fiber mesh belongs to the micro-flame combustion, the overall heat release is uniform, which leads to the reduction of the overall combustion temperature. The other is full premixing. The premixing method belongs to lean fuel premixing, and the excess air coefficient is >1. Although the excess air increases the oxygen contact concentration during combustion, the low temperature air will also carry a large amount of heat during combustion. The reduction of thermal NO_x caused by flame cooling is far greater than the increase of thermal NO_x caused by the increase of oxygen concentration.

2.2Mixers

Normally, when the gas is freely diffused in the air, it takes a long time to mix evenly. Therefore, the gas and air need a sufficiently long mixing distance in the state of absolute flow. But the long mixing distance cannot exist in a burner. For cost reduction and efficiency, a compact mixer is required to design for deep mixing of gas and air. At present, the mixing technology mainly has the structure of dispersing tube type, venturi type, swirling type, and pore network type. The structure of venturi type has the characteristics of high efficiency, compactness and uniform performance.

3. Design of gas/air mixing structure

3.1Premix chamber and venturi mixer

The mixer is a mechanical device for enhancing lean fuel and rich air mixing, while the premix chamber provides gas and air closed compact mixing space and achieves gas and air a relatively uniform. The premixing chamber and the venturi mixer installed in the premixing chamber are shown in Fig.1. The gas enters from the gas inlet pipe and the air enters from the upstream of premixing chamber. After passing through the venturi structure, the gas and air are evenly mixed in a 365mm×250mm mixing section. The narrowest air flow cross-section area of venturi structure is 61575.8mm^2.

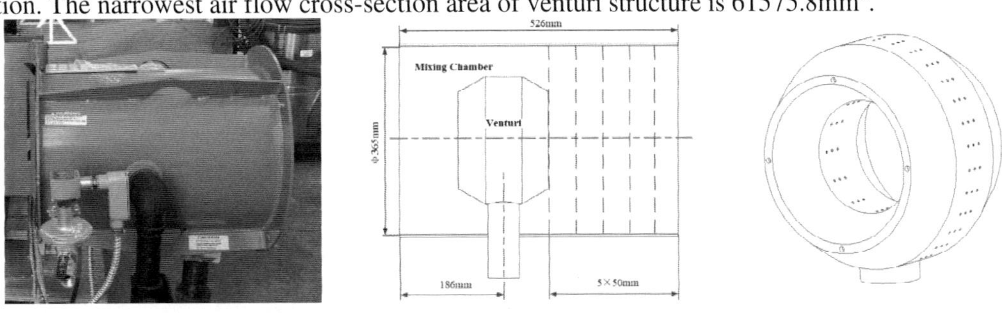

(a) (b) (c)

Fig.1 Premix chamber and venturi mixer

The gas injection hole is a key factor in the design of venturi mixer. According to Kitti Wichettapong[6], the gas diffusion velocity in the air through the venturi structure is in relation to gas pressure. Based on this point, six kinds of open-hole venturi structures are designed, which are 40, 56, 72, 88, 104, and 120, φ3mm gas injection hole models.

In order to compare the effect of verification standard and non-standard shrinkage and diffusion angle on the uniform performance of venturi structure, three shrinkage and diffusion angle models are built: standard venturi shrinkage and diffusion angle (21.5°), and non-standard shrinkage and diffusion angle (38.0°, 50.0°). According to the research of Readerharris M.J.[7] and the design requirements of venturi, the non-standard shrinkage and diffusion angle will lead to the decrease of the flow coefficient and the increase of the flow resistance. The structure requires less tolerance to air flow resistance loss and higher requirements for uniform performance under actual conditions.

3.2Entrance condition

The simulated burner uses 1.4MW as the rated load power of the burner. The volume ratio of fuel and air was 1:14.15, that is, the excess air ratio is about 1.4. The entry condition parameters are shown in Tab.1.

Table 1. Entry condition parameters

	Gas	Air
Volume flow rate（m3/s）	0.0459375	0.65
Temperature/K	298	298
Density（kg/m3）	0.7831	1.1850
Mass flow rate（kg/s）	0.03597	0.77025
Pressure（Pa）	4000	4000
Diameter of Inlet（mm）	60	365

4. Numerical simulation analysis of mixer models

4.1Mixer models

To verify the effect of the designed venturi mixer, the following models are established, a total of 8 mixer models, as shown in Tab.2.

Table 2. Mixer models for CFD simulation

Name	Angle	Number of Holes
Mixer I	21.5°	120
Mixer II	38°	120
MixerIII	50°	120
MixerIV	50°	104
Mixer V	50°	88
MixerVI	50°	72
MixerVII	50°	56
MixerVIII	50°	40

4.2Numerical simulation analysis method

In the performance verification of the mixer, the pressure loss and non-uniformity analysis method are used. The undisturbed smooth mixing section at the end of the venturi structure and the downstream

250mm space, and 6 characteristic cross-sectionals are equally divided. The characteristic cross-sectional is used as a mixer performance evaluation section, each section being 50 mm apart.

Numerical evaluation of the uniformity of each characteristic cross-sectional as an evaluation index for the uniform performance of the mixer.

$$SMD = \frac{\iint \left| f - \bar{f} \right| dA}{A\bar{f}} \tag{3}$$

Where SMD is the Spatial Mixing Deficiency, f is the equivalent concentration of fuel, $\bar{f}$ is the average concentration of fuel on the characteristic cross-sectional, and A is the area of the characteristic cross-sectional.

4.3Calculation results

Fig.2 shows the comparison of the standard venturi shrinkage and diffusion angle (MixerI), and the non-standard shrinkage and diffusion angle (MixerII, MixerIII) on the SMD non-uniformity of the mixer. It can be seen from the figure that the standard shrinkage and diffusion angle is the contribution of venturi structure mixing uniformity is slightly higher than the non-standard shrinkage and diffusion angle. Observing the data trend, from the first characteristic cross-sectional to the exit characteristic cross-sectional, the downward trend of each structure is basically the same as the amplitude. Explain that the mixing stroke has a good contribution to gas/air mixing uniformity (SMD decreases by about 0.3 in a 250mm mixing stroke, and still has a downward trend). The shrinkage and diffusion angle do not fundamentally change the uniformity of the mixer within a certain range(21.5°~50°), and the non-uniformity is slightly increased only when the shrinkage and diffusion angle is increased. Considering the axial structure size of the mixer, the smaller the shrinkage and diffusion angle and the larger the axial structure size, the axial structure size of MixerI, MixerII, MixerIII are 304mm, 150mm, 100mm. It can be seen that the increase of the axial size and the increase of the mixing distance will reduce the mixing non-uniformity of the mixer.

The change of the shrinkage and diffusion angle leads to the change of the axial size, and the mixing stroke determines the non-uniformity of the mixer to a certain extent. Both weaken the non-uniformity due to the decrease of the shrinkage and diffusion angle.

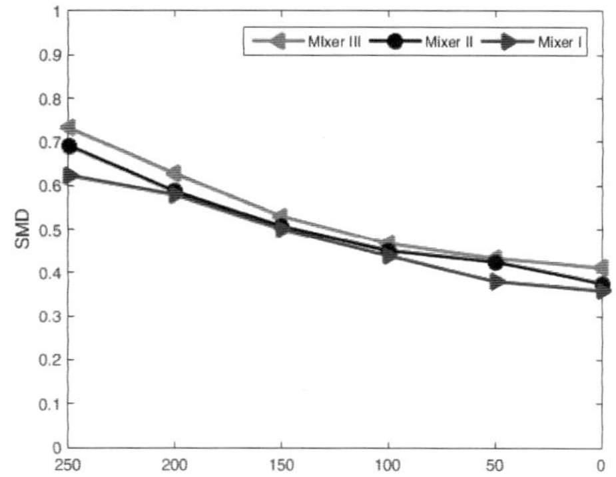

Fig.2 SMD of non-standard angle

Fig.3 shows the CH_4 concentration profile over the 250mm equidistant 6 characteristic cross-sectionals downstream of the mixer. From MixerVIII to MixerIII, the CH_4 concentration distribution on the characteristic cross-sectionals show that the center and the edge have a concentration concentration zone to a ring concentration concentration zone, and gradually the edge concentration dead zone. It is because the reduction of the opening area leads to a decrease in the gas flow velocity, which causes the gas to be injected at the small hole without sufficient initial kinetic energy, and during the injecting process, the air at a certain velocity is sucked. It causes the radial flow velocity too low, and the concentration dead zone formed, before the edge of the mixing fluid domain has not been touched by gasflow.

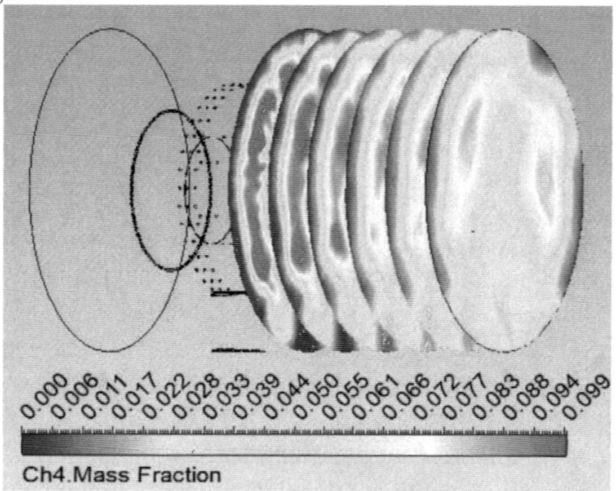

(a) MixerIII

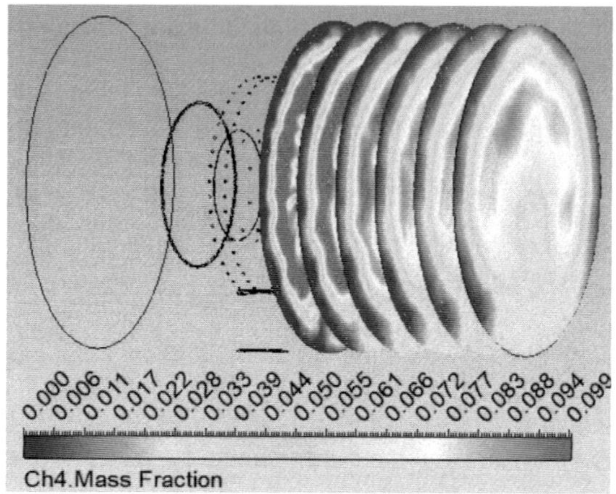

(b) MixerIV

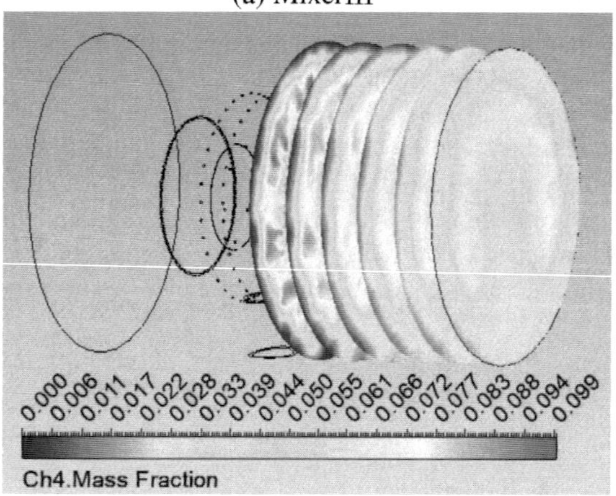

(c) Mixer V

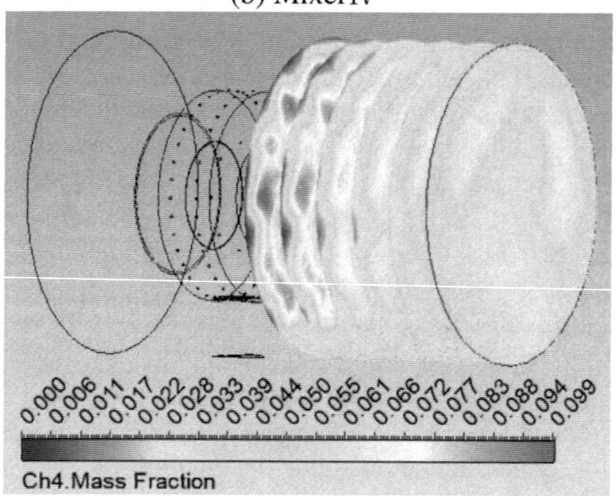

(d) MixerVI

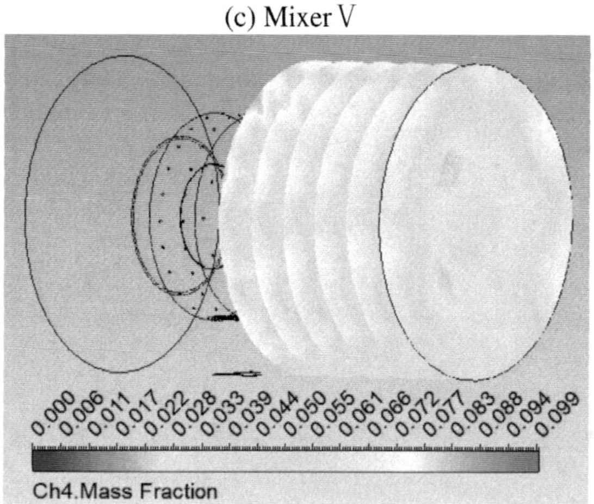

(e) MixerVII

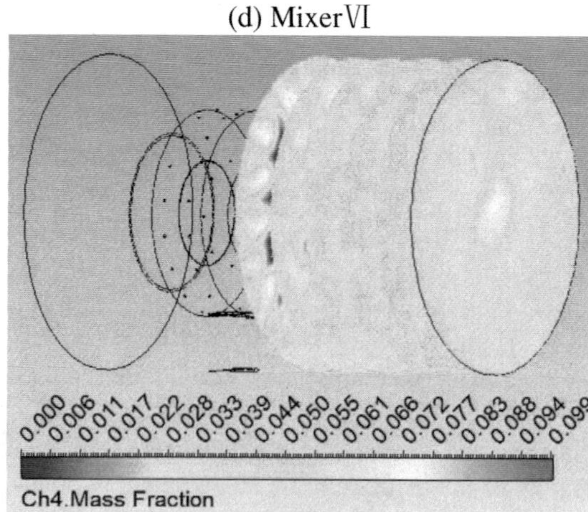

(f) MixerVIII

Fig.3 Mass fraction of CH_4 on the characteristic cross-sectionals

According to the calculation, the SMD evaluation values of six characteristic cross-sectionals downstream of the six venturi mixers are obtained, as shown in Fig.4. The number of openings has a great influence on the uniform performance of the venturi mixer. Within a certain range (40~120 holes), the smaller the number of openings (or the smaller the area), the smaller the non-uniformity of the cross-sectional of the mixer outlet. The number of openings has little effect on the air pressure characteristics, but has an exponential effect on the gas pressure characteristics. The smaller the number of openings (or the smaller the area), the greater the pressure loss on the gas side of the overall structure, and the greater the gas pressure required to stably maintain the gas inlet. According to the continuity theorem and the local resistance formula (e.q.4), small opening number (or area) has high injection velocity and high local resistance loss.

$$\Delta P_f = \zeta \frac{\rho u^2}{2} \tag{4}$$

Uniformity improvement and high gas pressure loss are caused by the velocity of the injection. In the modelⅧ, most of the gas injected through the outer holes hits the boundary of the fluid domain, that is, the wall of the mixing space, and the gas injected through the inner holes is collected at the center of the inner fluid channel. The flows form a collision, and the collision leads to uniform diffusion. Simultaneously the larger the average injection velocity at the gas injection holes, the greater the partial pressure loss (e.q.4). According to the calculation results, the static pressure loss of gas is 21524.773Pa, and the dynamic pressure loss is 236Pa. At full load, the inlet pressure of the main gas pipe must be maintained at 23044.773Pa.

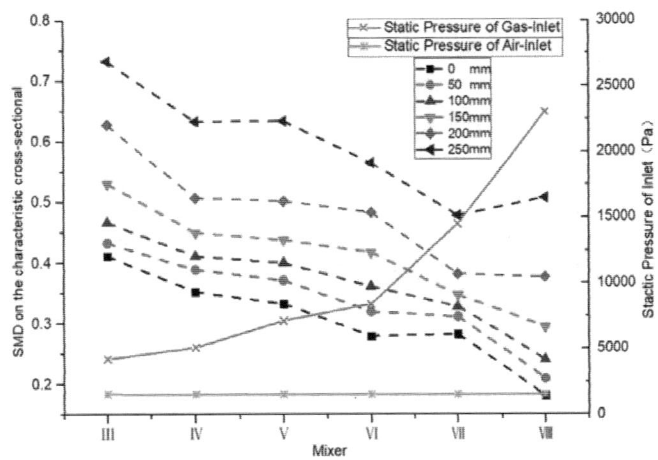

Fig.4 The uniformity and pressure characteristics of the six kinds of gas injection hole structure

According to the actual operation of the fully premixed surface natural gas burner in the experiment and investigated, the gas inlet pressure before the gas valve group is generally less than 20000Pa, and the boiler back pressure is between 500Pa~2000Pa. The gas inlet pressure is generally guaranteed to be between 3000Pa and 5000Pa under full load state. In this pressure range, it can not only ensure the continuity of the gas input, but also filter the back-propagation fluctuation of the gas flow and combustion. Therefore, if the pressure loss is too large, the target effect of the actual operation of the mixer cannot be satisfied.

5. Conclusion

According to the theoretical analysis and numerical simulation results, the guiding conclusions of the design of the full premixed surface burner mixer are obtained:

a. The opening number (or area) of the gas injection holes of the mixer is the key factor for the uniform mixing design of the venturi mixer. The small opening number or area can increase the injection velocity, and increases the gasflow collision dispersion, which is favorable for the gas/air mixing uniformity.

b. The pressure loss of the mixer is a joint factor in the design of the characteristic parameters of the venturi mixer, which is mainly determined by the opening number (or area) of the gas injection holes of the mixer, the smaller the opening area, the greater the local pressure loss, and the greater the overall gas pressure loss of the mixer. vice versa. Mixer gas pressure loss is also a key design factor that determines burner reliability and versatility. High gas pressure loss will require a stable input at the burner gas inlet, reducing combustor versatility and increasing unstable combustion risk.

c. The mixer uniformity design and pressure characteristic design need to be balanced with each other, usually with the stability target as the precondition. Therefore, the gas pressure loss should be taken as the first design target. The uniformity of sacrifice can be compensated for by a structure with enhanced mixing that increases the small pressure loss, such as a large pressure angle swirling vane, a spoiler bluff body etc.

d. The shrinkage and diffusion angle of the venturi mixer has little effect on the uniformity of the mixer in the range of 21.5° to 50°, and the angle can be appropriately increased to reduce the axial dimension of the mixed structure, thereby improving the compactness of the overall structure.

References

[1] Natural gas industry. (2018) China's natural gas demand will reach 600 billion cubic meters in 2030. J. Natural Gas Industry., 38(09): 87.

[2] Yan, J., Sun, F., Chou, S.K., Desideri, U., Li, H., Campana, P., Xiong, R. (2017) Impacts of emission reduction and external cost on natural gas distribution. J. Applied Energy. 207.

[3] Ryoji, H., Koji, H. (1995) Study on NOx formation mechanism in a Bunsen flame. J. Nippon Kikai Gakkai Ronbunshu. Japan Society of Mechanical Engineering. 61(583): 1177-1182.

[4] Wu, X.S., Cheng, L.M., Yan, K., Zhang, W.G. (2018) Experimental study of industrial low nitrogen porous media burner. J. Journal of Zhejiang University (Engineering Science). 52(11): 2136-2141+2170.

[5] Wu, P., Cao, T.Z., Zhang, C.X., Li, M.J. (2015) CFD study on combustion properties and NOx emission of reboiling furnace for xylene tower. J. Journal of Engineering for Thermal Energy and Power, 30(02): 180-186+314-315.

[6] Readerharris, M.J., Brunton, W.C., Hodges, D., et al. (2002) Discharge coefficients of Venturi tubes with standard and non-standard convergent angles. J. Flow Measurement & Instrumentation. 12(2): 135-145.

[7] Minghou, X., Caiyuan H. (1995) Numerical Simulation on a Gas Particulate Flow of the Trail in an Orthogonal Jet. J. Journal of Wuhan University of Hydraulic & Electric Engineering.

IOP Conf. Series: Materials Science and Engineering **721** (2020) 012010 doi:10.1088/1757-899X/721/1/012010

Design and optimization of distribution structure of fully premixed surface burner

Dong Aihua[1] Qiu Penghua[2] Liu Hong[1] Liu Li[2] Sun Zhenglu[1] Wang Hui[2] Wang Yongtang[1*] Liu Guichuang[2] Shi Yunxin[1]

[1]Harbin electric corporation central research institute, Harbin, Heilongjiang province, 150028, China

[2]Combustion engineering institute, School of energy science and engineering, Harbin institute of technology, Harbin, Heilongjiang province, 150001, China

[*]Corresponding author's e-mail:wyongt@harbin-electric.com

Abstract. In response to China's energy policy and the increasingly stringent NOx emission standards, premixed surface burners have gradually entered the field of vision as an ultra-low NOx emission gas burner. In this paper, the numerical simulation software ANSYS FLUENT was used to analyze the airflow distribution structure of a certain type of premixed surface burner, and the distribution characteristics of the premixed gas flow in the distribution structure during actual operation were also obtained. At the same time, the parameters of the distribution structure were optimized according to the problems appearing in the distribution of airflow by numerical simulation. The results showed that the design and optimization of the distribution structure of the premixed surface burner were reasonable and effective.

1. Introduction

As a high-quality, high-efficiency, clean and low-carbon energy source, natural gas can be controlled in a very low range compared with conventional coal powder if it is equipped with a suitable combustion method. According to the requirements of the 13th Five-Year Plan of the National Development and Reform Commission[1], by 2020, natural gas will account for 10% of primary energy consumption, and the natural gasification rate of urban population will reach 57%. In the Beijing-Tianjin-Hebei, Yangtze River Delta, Pearl River Delta, and other regions, the "coal to gas" project is being implemented on a large scale[2], and has achieved certain results, which will certainly promote the rapid growth of natural gas demand.

Beijing Municipality's "Boiler Air Pollutant Emission Standard" DB11/139-2015[3], which was revised in 2015, requires a new boiler NOx emission limit of 80 mg/m3 before March 31, 2017. The new boiler NOx emission limit from April 1, 2017 is 30mg/m3.

Low-NOx gas burners currently being developed by European and American companies primarily use semi-premixed combustion technology to achieve low NOx emissions through fuel/air grading and flue gas internal circulation[4]. When implementing more stringent emission standards, such as emission targets of 30 mg/m3, European brands mostly use the technical route of staged combustion plus external flue gas recirculation. This is also the mainstream technology for ultra-low emission burners on the market in China. However, such technologies cannot achieve emission targets below 30mg/m3 for boilers with smaller boiler sizes, while surface fully premixed burners can Achieve the goal of low NOx combustion in small and medium boilers.

The surface combustion technology is characterized by fully premixed combustion, which achieves complete uniform premixing of gas and air, while the excess air coefficient in the combustion process is small, and the heat released by combustion can be quickly transferred through radiation and convection. The flame temperature is lower, which can achieve lower NOx levels than other burners[5].

2. Numerical Simulation Study on a Mixture Structure of a Burner

The gas/air mixing technology and the gas distribution technology are two key technical points of the surface fully premixed burner. Bao Tong et al[6] numerically simulated the gas/air mixing structure of a brand surface burner. The results show that the uniformity of gas and air mixing will directly affect the burner combustion requirements and boiler temperature and emissions of nitrogen oxides. The distribution technique is designed to make the mixed gas evenly distributed on the surface of the combustion head, and to avoid uneven distribution of local flame due to uneven flow. Several common distribution structures are shown in Fig.1-Fig.4[7-10]. It can be seen that the characteristics of the distribution structure are mainly concentrated in the following aspects: the number of layers of the distribution structure, the holes distribution and the hole form of the distribution structure, The shape of the burner head.

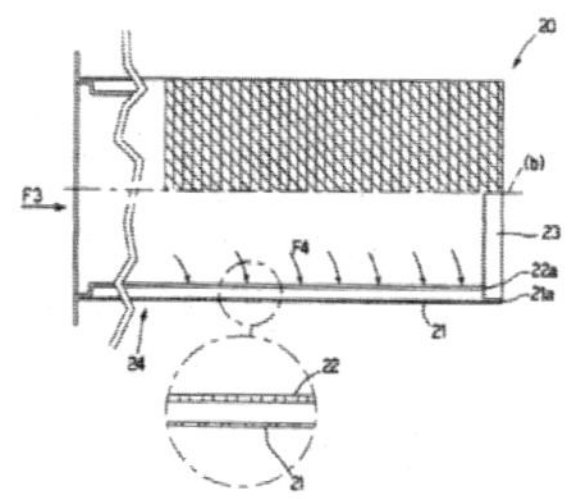

Fig.1 Double orifice plate distribution structure

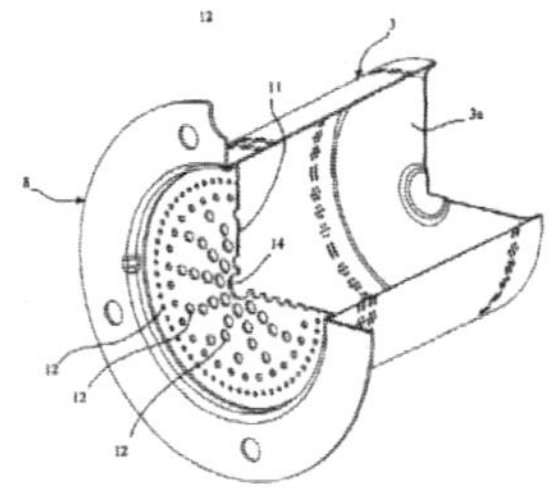

Fig.2 Distribution structure with uniform circular holes at the entrance of the combustion head

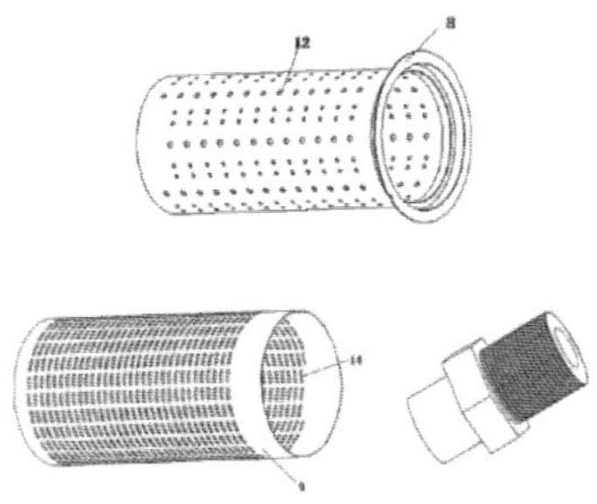

Fig.3 Multi-layer mixing distribution structure

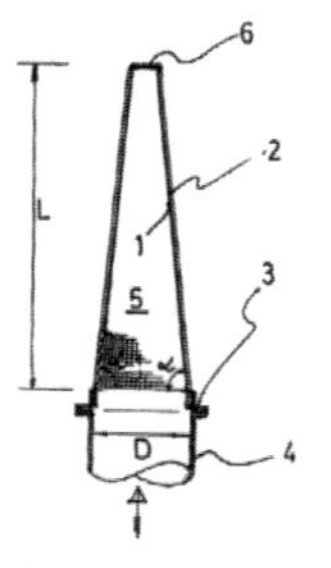

Fig.4 Conical burner head

At present, there are few related studies on the distribution structure of surface burners.

3. New distribution structure

A type of burner is used as a simulation object. The burner distribution structure is located inside the combustion head. It is mainly divided into: the inner layer distribution cylinder with scaly holes, as shown in Fig.5; The outer layer distribution cylinder, and the sintered metal fiber[11] wrapped around the strip hole cylinder, as shown in Fig.6.

Fig.5 Inner layer cylinder in-kind

Fig.6 Metal fiber and outer layer cylinder in-kind

In order to accurately simulate the flow of the mixed gas flow inside the burner, the simulation of this paper starts from the inlet of the combustion head, that is, the starting end of the distribution structure, and simulates the flow state of the airflow when both the fish scale hole and the strip hole exist.

The fish scale hole cylinder of one of the distribution structures is an approximate positive 45-sided shape, and about 100 fish scale holes are opened in the longitudinal direction of each side, and about 4,500 fish scale holes are opened in the entire fish scale hole cylinder.

Due to the large number of holes in the distribution structure, in order to reduce the number of meshes and ensure the grid quality to save calculation time, the model adopts a radial thin plate-like unitization model at the center of the fluid domain. Considering the symmetry of the fluid in the entire circumferential direction, only the fish scale holes which are opened in the circumferential direction on one side of the horizontal cross section 45 and the strip holes corresponding to the upper portion thereof are drawn. As shown in Fig.7, the outer cylindrical area is the fluid domain structure inside the actual combustion head, and the red area is based on The fluid domain region as used herein above.

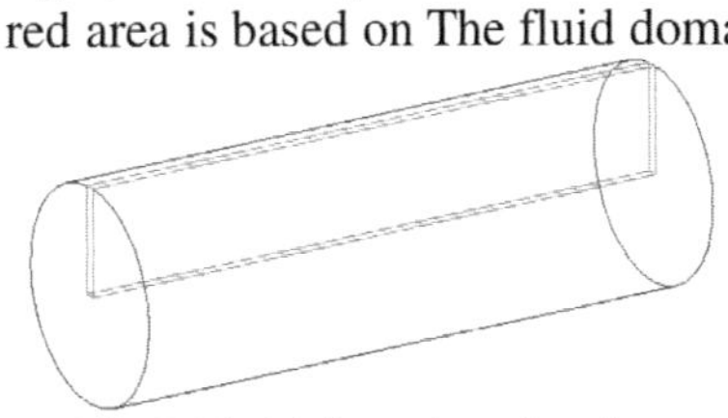

Fig.7 Fluid domain selection

The fish scale holes are opened in the inner cylinder wall, and the number of fish scale holes is 100, and the fluid domain model of the inner cylinder and the outer cylinder fluid domain is built, and The bottom end of the inner cylinder is at a distance of 37 mm from the bottom end of the outer cylinder. The structure is as shown in Fig.8 Shown.

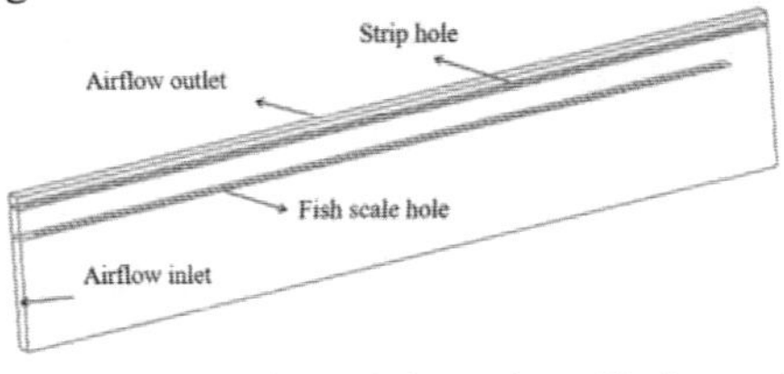

Fig.8 Fluid domain model of the fish scale cylinder and strip cylinder

The gas actually flowing in the burner distribution structure should be a mixed gas of natural gas and air. Table.1 shows the relevant physical parameters of natural gas and air under operating conditions.

Tab.1 Gas and air physical parameters

Fluid	Temperature(K)	Density(kg/m^3)
Air	300	1.185
Gas	300	0.7174

Since the air volume flow rate during the mixing process of this type of burner is about 14 times that of natural gas volume flow, it can be seen from Table.1 that under the same environmental conditions, the air density is about twice the density of natural gas. Therefore, during the simulation, we set all the gases involved in the distribution to be air. The import and export models used in the simulation process are shown in Table.2.

Tab.2 Boundary conditions

Name	Type	Velocity(m/s)	Pressure(Pa)
Mix Inlet	Velocity-Inlet	25	--
Outlet	Pressure-Outlet	--	0

In order to facilitate the analysis of the obtained simulation results, as shown in Fig. 9, a plane passing through the center of the fluid domain as indicated by the arrow is established, and the relevant results analyzed are derived from the sections obtained by this position.

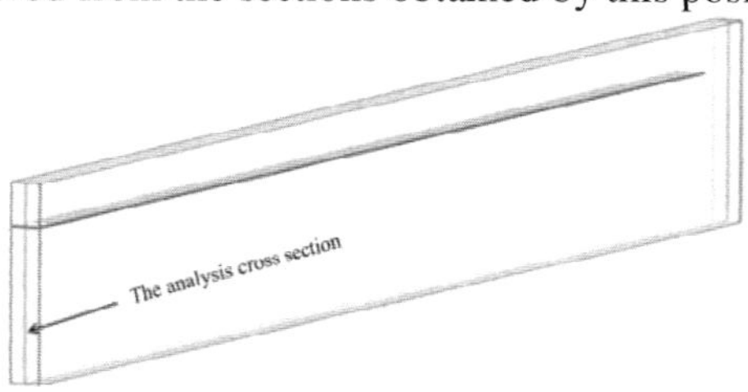

Fig.9 The position of the analysis cross section

The ANSYS FLUENT software was used for the simulation calculation. The five points in the flow field were randomly monitored during the calculation. After the simulation is completed, the velocity nephogram and the local velocity vector diagram on the analysis plane are obtained. As shown in Fig.10 and Fig.11, respectively.

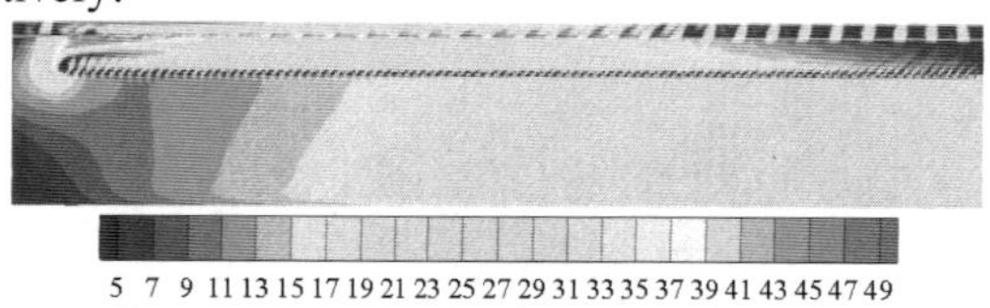

Fig.10 The velocity nephogram of the distribution structure analysis cross section

Fig.11 Local velocity vector of the distribution structure analysis cross section

It is easy to see from the velocity nephogram that at the end of the combustion head, that is, at the end of the flow of the airflow, there is a region where the airflow velocity is sharply increased due to the sudden decrease in the resistance. However, at the airflow inlet position, the air flow between the inner and outer cylinder is small. The area where the airflow is concentrated in the middle and rear parts of the overall distribution structure. The occurrence of the airflow concentration zone may be caused by the sudden decrease of the resistance of the tail of the combustion head, resulting in a large amount of airflow not flowing out of the fish scale hole and the strip hole, and the reason for the area of small airflow may be due to the special structural form of the fish scale hole. The tendency of the airflow to flow internally causes less flow out of the fish scale hole near the inlet.

4. Numerical Simulation and Optimization of New Distribution Structure

According to the simulation results of a certain type of burner distribution structure, in actual operation, there may be a case where the flow velocity at the inlet end is small and the flow velocity at the tail end is large, which may be a special opening type and scale of the fish scale hole of a certain type of burner. The hole is related to the arrangement of the strip hole. Because the structure of the strip hole is difficult to change under the current conditions, this paper mainly considers the optimization of the air inlet mode and the relevant parameters of the fish scale hole, in order to obtain a more reasonable one.

4.1Analysis of fluid domain model for increasing airflow inlet

Since a type of burner has a gas flow cell between the distribution interlayers at the inlet, in order to optimize this area, first attempts to improve the low flow area between the fish scale cylinder and the strip cylinder sandwich at the inlet in a manner that increases the inlet area of the gas stream. Fig.12 shows an optimized structure with several circular holes at the inlet of the fish scale to increase air flow at the entrance of the interlayer.

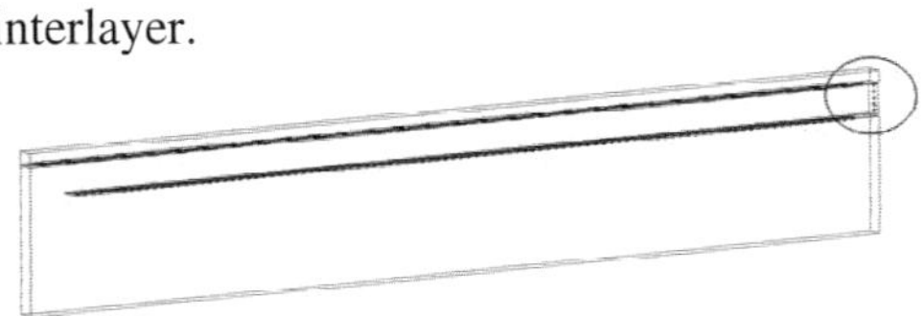

Fig.12 A fluid domain structure model with airflow flowing into the orifice at the inlet mezzanine

The fluid domain meshing and boundary conditions are the same as above. After the simulation is completed, the velocity nephogram distribution on the new reference plane and the local magnification speed nephogram at the entrance position are obtained, as shown in Fig.13 and 14.

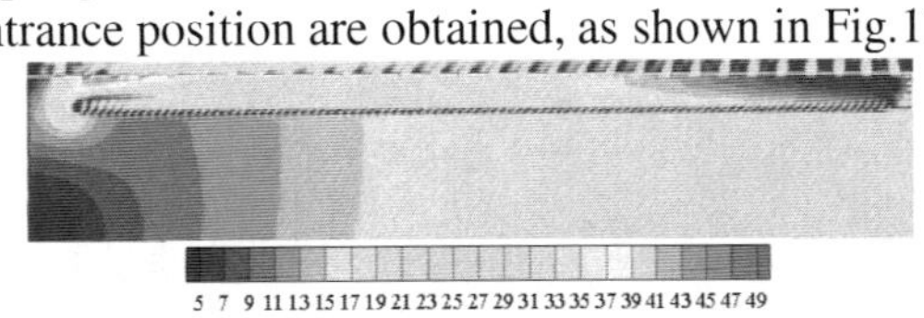

Fig.13 Distribution structure velocity map of the mezzanine with airflow flowing into the small hole

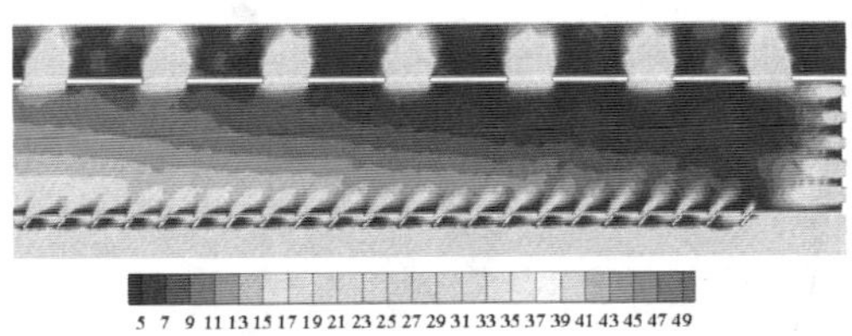

Fig.14 Enlarged cloud at the entrance of the distribution structure

It can be seen from Fig. 13 and Fig. 14 that the flow of air into the small hole at the entrance of the two-layer distribution structure is more effective when the airflow just flows in, but as the flow deepens, it improves the overall low flow area of the front section of the interlayer. The effect is not very significant, which may be related to the size of the opening, the arrangement of the holes and other factors.

In order to further verify the effect of increasing the inflow area of the airflow to improve the low flow area, an extreme model is set up, that is, the entrance end, that is, the initial end of the interlayer is also set as the airflow inlet, as shown in Fig. 15, after the simulation is completed, a new one is obtained. The velocity nephogram on the reference plane is shown in Figure 16.

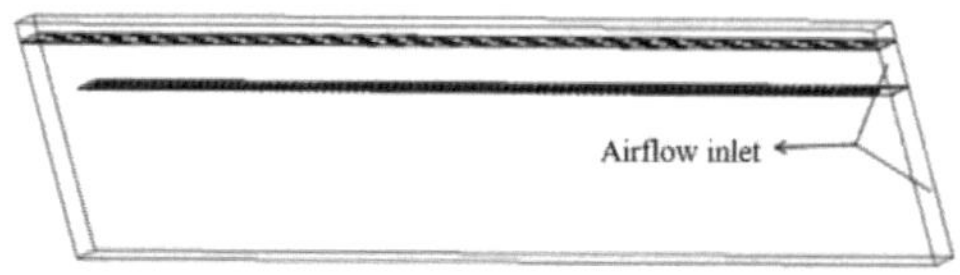

Fig.15 Fluid domain structure model with inlet section added

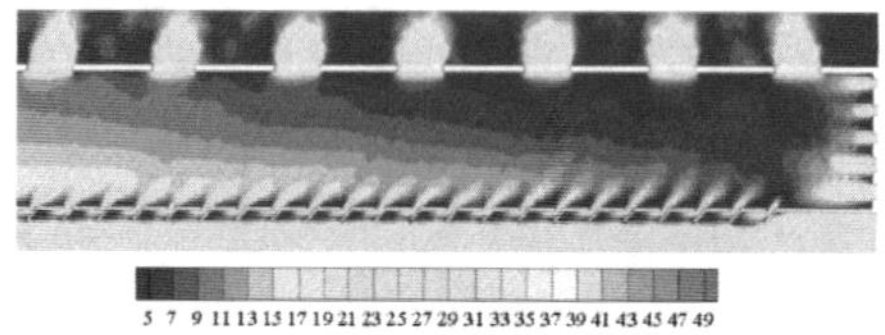

Fig.16 Increased distribution profile of the inlet section

It can be seen from Fig. 16 that the front section of the interlayer is set as the airflow inlet, which also plays a partial improvement role. Due to the effect of the interlayer opening, the low flow area originally located in the front section moves backwards, but the overall has not been reduced. The trend, thus showing that increasing the airflow inlet area is not a better way to improve the low velocity flow zone.

4.2Fluid Domain Model Analysis of Changing the Opening Area at the Entrance of Fish Scale Holes

One of the reasons for the low flow area between the fish scale cylinder plate and the strip cylinder plate in the inlet section is that there is less airflow from the fish scale hole at the inlet. For this purpose, the structure is increased the opening area of the fish scale hole at a certain length of the entrance of the fish scale hole plate by about twice the original, and aims to increase the outflow of the fish scale hole at the inlet of the orifice plate further serves to improve the flow rate of the inlet. The velocity nephogram distribution and the local velocity of the entrance location are obtained on the reference plane, as shown in Fig.17 and Fig.18.

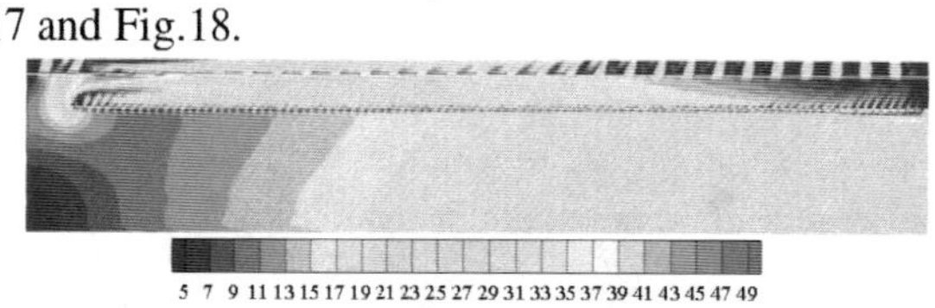

Fig.17 Overall distribution structure velocity map with increased opening area at the entrance

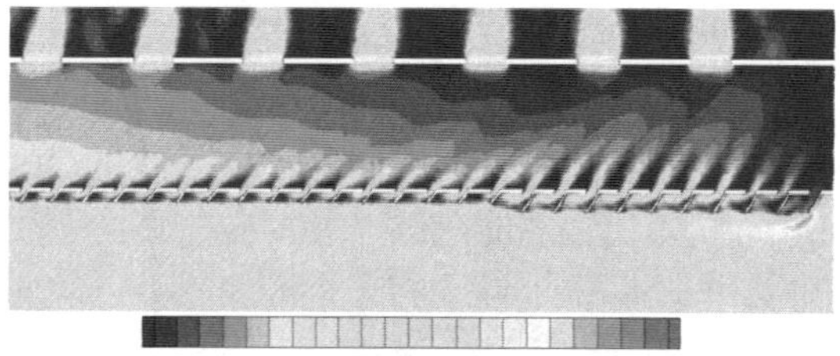

Fig.18 Entrance position partial magnification speed nephogram

It can be seen from Fig.17 and Fig.18 that after the opening area of the front fish scale hole is increased, the air flow rate in the low flow area is increased to some extent, so the structure has a certain improvement effect on the expected airflow distribution, but it is still necessary to carry out comparative analysis by numerical simulation to increase the opening area and increase the length of the opening.

4.3Fluid Domain Model Analysis of Reversed Fish Scale Holes

The direction of the opening of the fish scale hole of a certain type of burner will form a flow aid the flow of air flowing into the inside of the combustion head, and the flow direction of the fish scale hole is set to the opposite direction. The influence of the structure on the flow distribution of the gas flow between the interlayers, and then the relevant conclusions regarding the "the effect of the flow-through effect of the fish scale hole opening form". Fig.19 shows the fluid domain model under this structural condition. After the simulation is completed, the velocity nephogram distribution and local velocity vector on the new reference plane are obtained. As Fig.20 and Fig.21 show.

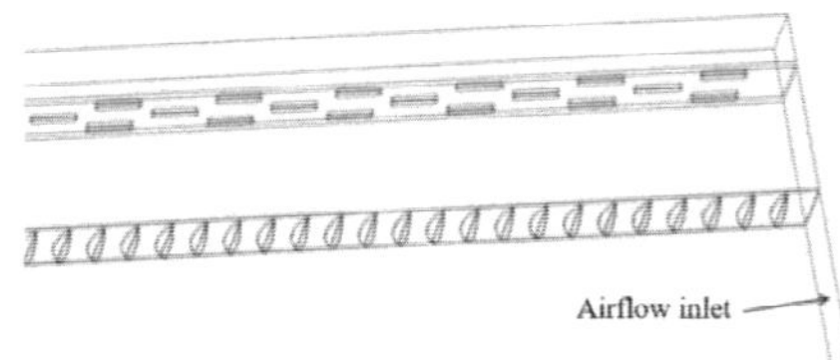

Fig.19 Fluid domain structure model of fish scale hole opening reverse

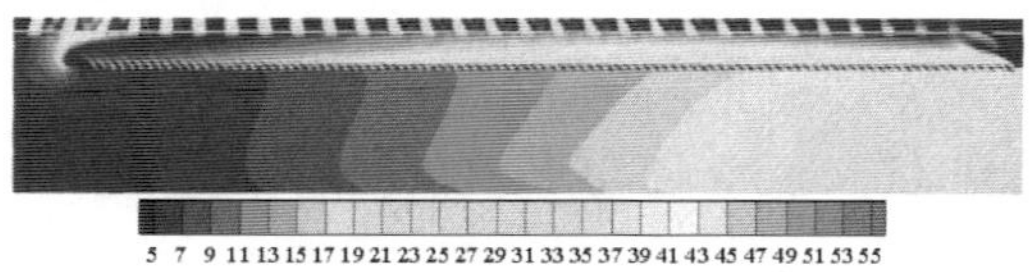

Fig.20 Fish scale hole opening reversed distribution structure analysis section velocity nephogram

Fig.21 Fish scale hole opening reversed distribution structure analysis section local velocity vector

As can be seen from Fig. 20, when the fish scale hole openings are all reversed, the front end portion of the combustion head no longer has a very small velocity region, but since most of the gas flows out from the fish scale hole at the front end, and the characteristics of the opening of the fish scale hole can be seen from the velocity vector of Fig.21. The airflow flowing from the fish scale hole at the front end has a tendency to converge toward the middle of the combustion head, so that the flow of air between the distribution interlayers, that is, the middle of the combustion head is the largest. And there is still a certain speed surge phenomenon due to the decrease of the tail resistance.

This kind of structure indicates that the relevant conclusions about the influence of the opening direction of the fish scale hole of a certain type burner on the flow of the inflow airflow are correct. This result also provides a certain idea and direction for the optimization design of the fish scale hole structure.

4.4Fluid domain model analysis of new fish scale hole opening structure

For the analysis results in Section 4.3, a new type of fish scale hole opening method is proposed, as shown in Fig.22. The structural change of the scale hole is structurally changed. The punched portion is completely perpendicular to the plate surface, and the opening is set as a rectangular opening. The single opening area is $12mm^2$, which is consistent with the opening area of the fish scale hole of a certain type of burner, and the rectangular hole size is 6mm × 2mm. The velocity nephogram of the simulation results is shown in Fig.23, and the local velocity vector diagram is shown in Fig.24.

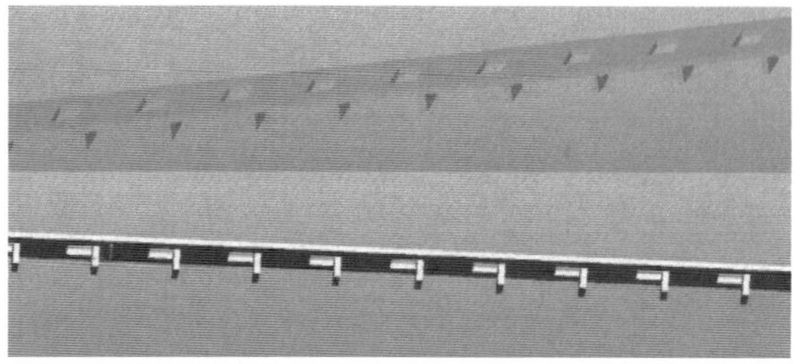

Fig.22 New fish scale hole opening structure view

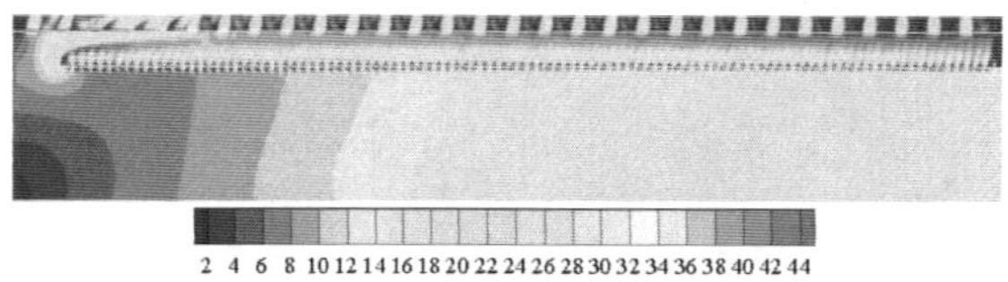

Fig.23 New open hole distribution structure analysis section velocity nephogram

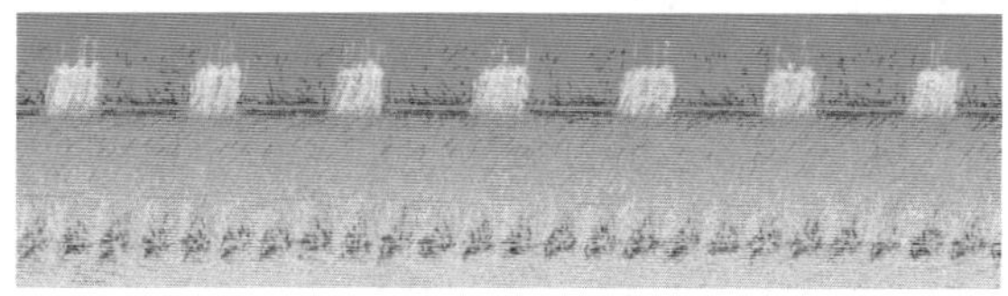

Fig.24 New opening distribution structure analysis section local velocity vector

It can be seen from Figures 22 and 23 that under such opening conditions, the velocity distribution between the distribution interlayers is relatively uniform, the flow rate of the inlet location disappears, and although the velocity surge region still exists in the tail, the airflow velocity can be seen from the nephogram. It is not as obvious as a certain type of burner. This is due to the fact that under the opening structure, the flow of gas flows out more in the length of the whole scale of the fish scale, and the flow of the air flowing out of the tail is less. When the resistance suddenly decreases, although the speed increases rapidly, the whole trend is not big.

This structure can not only prove the principle of the influence of the opening form of the fish scale hole of a certain type of burner on the flow of the airflow, but also provide a reasonable airflow distribution structure.

5. Conclusion

Based on the simulation results of a certain type of burner, this paper conducts an in-depth simulation of the overall distribution structure under the same boundary conditions for the actual situation, and carries out related improvement experiments and analysis on the distribution structure according to the successive optimization experience.

Through the above four sets of simulation results, it can be seen that, without changing the structure of the fish scale hole and the strip hole, attempting to open the holes between the fish scale cylinder and the strip cylinder entrance can improve the distribution interlayer to a certain extent. But the improvement effect is not obvious and may cause tempering under operating conditions. Without changing the strip hole structure, only the structure of the fish scale hole is optimized, and the opening area of the local fish scale hole is appropriately increased, can achieve a certain airflow flow optimization effect but the overall optimization effect is not obvious.

The open hole form of the fish scale hole plays an important role in the flow. According to this conclusion, a new type of fish scale hole opening structure is proposed, and the numerical simulation results are also reasonable. It provides a reasonable idea for the design study of the distribution structure in the future. However, it should be noted that the specific optimization parameters of the new structure still need to carry out a large number of numerical simulations and compare the results to obtain a more realistic value.

References

[1] Gas & Heat. (2017) Notice of the National Development and Reform Commission on Printing and Distributing the National Energy 13th Five-Year Plan. J. Gas & Heat. 37(02):1-16.
[2] Fan, H. (2017) Discussion on Some Important Issues of "Coal to Gas" in China. N. 002.
[3] Beijing Municipal Ecological Environment Bureau. (2015) The standard of boiler air pollutant discharge (DB11/139-2015). Z.
[4] Wang, Z.X., Liao, X. Yang, J.J. (2011) Research on fully premixed metal fiber burner. J. Gas & Heat. (06):27-31.
[5] Hu, B., Lin, L.C., Deng, Y.C., Feng, L., Qi, Y.T. (2015) Study on NOx Emission Characteristics of Metal Fiber Premixed Combustion. J. Shanghai Gas. (06):21-23+26.
[6] Bao, T., Zhang, H.L., Miao, Y.W. (2017) Numerical Simulation of Cold Flow Field of Metal Fiber Gas Burner. J. Shanghai Gas. (06):27-31.
[7] Lonardi, A., Lovato, A. (2005) Premix burner combustion head: EP, EP 1538395 A9. P.
[8] Puzzato, D., Scribano, G., Badoka, L., et al. (2008) Burner, especially premixed burner: CN 101861496 A. P.
[9] Liu, F.G., Zhang, R., Zhang, B.L., et al. (2014) Fully premixed gas burner: CN 203687038 U. P.
[10] Dewaegheneire, G., Marrecau, W., Vansteenkiste, P. (2000) Conical surface burner: US, US 6065963 A. P.
[11] Li, L.J., Tao, Z., Pang, J.L., Long, W.J. (2014) Progress in the application of metal fiber burners in the utilization of low concentration coalbed methane. J. 28(09):35-40

IOP Conf. Series: Materials Science and Engineering **721** (2020) 012011 doi:10.1088/1757-899X/721/1/012011

Design and optimization of mixing structure of fully premixed surface burner

Wang Yongtang[1*] Wang Hui[2] Sun Zhenglu[1] Liu Li[2] Shi Yunxin[1] Qiu Penghua[2] Dong Aihua[1] Zhao Kaixiang[2] Liu Hong[1]

[1]Harbin electric corporation central research institute, Harbin, Heilongjiang province, 150028, China

[2]Combustion engineering institute, School of energy science and engineering, Harbin institute of technology, Harbin, Heilongjiang province, 150001, China

[*]Corresponding author's e-mail:wyongt@harbin-electric.com

Abstract. This paper disassembles a brand of fully premixed surface burners, designs a new type of air-fuel mixing structure through literature survey, and uses Fluent software to simulate the prototype and new type of mixing structure through numerical simulation. The simulation of the mixing was carried out. Through comparative analysis, the relevant structural parameters were changed and the new mixing structure was optimized and verified by numerical simulation. The results show that the new mixing structure proposed in this paper has a better effect on mixing, and some rules affecting the mixing are obtained.

1. Introduction

Surface combustion technology is a new type of combustion technology that has been developed in recent years with high efficiency and low pollution[1]. At present, the environmental protection emission standards of developed provinces (municipalities or districts) are becoming stricter. The fully premixed surface burner can achieve the goal of low NOx combustion, and will be the mainstream of ultra low emission combustion technology for small and medium-sized combustion equipment.

Fuel and air mixing technology is the core technology of a fully premixed surface burner. NOx emissions are directly related to mixing unevenness. As the mixing unevenness increases, the NOx emission increases sharply[2], especially when the mixing unevenness is >5% and the adiabatic flame temperature is higher than 1800K. Only the uniform mixing of fuel and air can achieve the combustion economy, low emission and tempering safety of the burner. Therefore, the mixing technology of fully premixed surface burners is crucial. Fric[3] combined theoretical analysis and experimental research on the fuel/air mixing process in premixed nozzles, and proposed the concept of premixing inhomogeneity, and gave the calculation method. Leong[4] et al. achieve uniform fuel/air mixing by varying the fuel jet angle, position, and fuel/air momentum ratio. Sun Baocheng[5] used numerical simulation to study the gas/air mixing law, and summarized the factors including the turbulence, the fuel/air momentum ratio and other factors used to predict the mixing unevenness of the cylindrical premixing section. Xie Gang [6-7] studied the fuel/air mixing process and pollutant emission performance in the R0110 gas turbine combustor through simulation and experimental methods.

IOP Conf. Series: Materials Science and Engineering **721** (2020) 012011 doi:10.1088/1757-899X/721/1/012011

2. Numerical Simulation Study on a Mixture Structure of a Burner

2.1Meshing and irrelevance verification

According to the survey results, the mixing structure was drawn by ProE software. The model was slightly simplified, the igniter and the fire detection fixture were removed. The fluid domain obtained after the mixed structure model is shown in Fig.1.

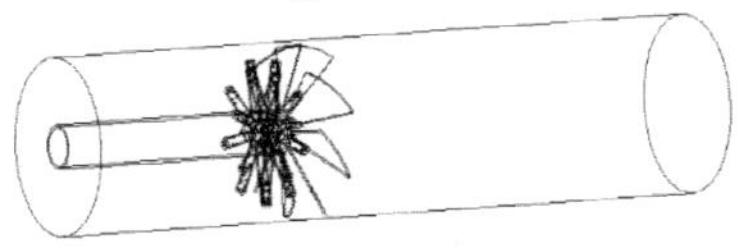

Fig.1 Prototype prototype mixing model

Use ICEM to mesh, in order to ensure the quality of the grid, while reducing the total number of grids, define the global size parameter mapping grid in two sections, and connect the two parts of the grid with interface. Using a tetrahedral unstructured grid, the mesh quality is above 0.3, which can meet the accuracy requirements.

Three grids of 2.50 million, 3.99 million and 4.86 million are taken for calculation. The first section is taken in the flow direction of the blade trailing edge of 100 mm, and another section is placed every 50 mm, and a total of six sections are taken. Spatial Mixing Velocity-unevenness (SMV) of 6 sections is taken out by Fluent calculation result. The curve is shown in Fig. 2. It can be seen from the graph that the results of 2.50 million, 3.99 million and 4.86 million are basically coincident. It can be considered that the grid increase has little effect on the result. In order to save the calculation cost, 2.50 million grids were selected to calculate the prototype model.

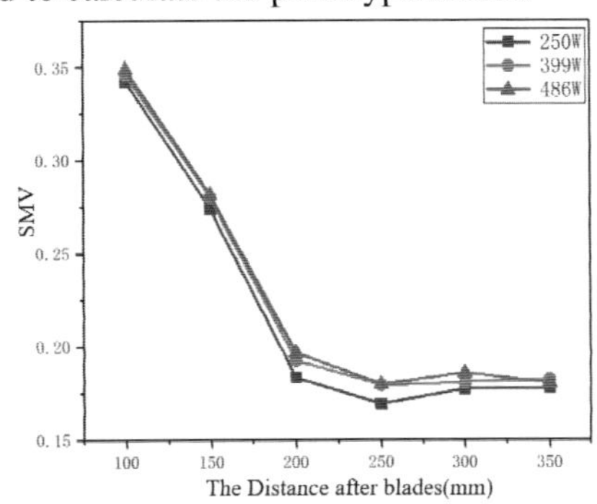

Fig.2 SMV distribution with three grid numbers

2.2Numerical simulation and result analysis

The numerical calculation uses the standard realizable turbulence model and the methane-air component model. The air and gas inlets are both set as mass flow inlets and the outlet is pressure outlet, as shown in Table 1.

Table 1. Boundary conditions

	Fuel	Air	Outlet
Volume flow rate（m3/s）	0.0459375	0.65	-
Temperature/K	298	298	298
Density（kg/m3）	0.7831	1.1850	-
Mass flow rate（kg/s）	0.03597	0.77025	-
Pressure（Pa）	13789	2074	1600

IOP Conf. Series: Materials Science and Engineering **721** (2020) 012011 doi:10.1088/1757-899X/721/1/012011

The formula for calculating the concentration unevenness [8] is as follows:

$$SMD = \frac{\iint \left| f - \bar{f} \right| dA}{A\bar{f}} \tag{1}$$

Where A is the cross-sectional area of the selected section.
Approximately, the speed unevenness is calculated as follows:

$$SMV = \frac{\iint \left| v - \bar{v} \right| dA}{A\bar{v}} \tag{2}$$

The distribution of the six sections CH4 is shown in Figure 3.

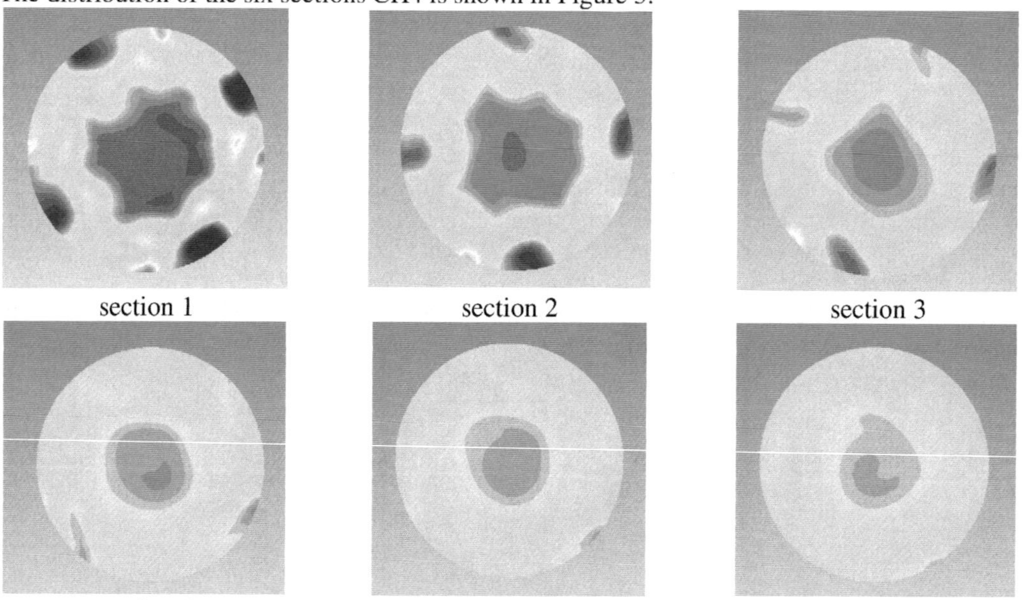

section 1 section 2 section 3
section 4 Section 5 section 6

Fig.3 Cloud map of each section CH4 distribution

The uniformity of the distribution of air and CH4 gas increases, indicating that the mixing section promotes the uniform mixing of the gas. However, the gas was not completely mixed, and the CH4 concentration was extremely low near the center. It shows that the prototype has certain defects, there is no gas flowing behind the central gas pipe, resulting in a dead zone with extremely low concentration, which is difficult to mix evenly. The structure needs to be improved.

3. New mixing structure

After analyzing the patent research mixing technology [9-11], the following mixing scheme is proposed and have applied for a patent. As shown in Figure 4, the mixing structure mainly includes a premixing chamber, an internal gas pipe, a ring gas chamber and swirl blades. The inner horizontal tube is longer than the outer horizontal tube, and the inner and outer two sections are connected to form a truncated cone shape. The annular gas chamber divides the air passage into an inner and outer air passage, and the outer air passage is in a venturi structure. Gas injection holes are formed in the inner and outer sides of the annular cavity, and gas is respectively injected into the inner and outer air passages, and the swirling blades are bolted to the end of the annular gas chamber.

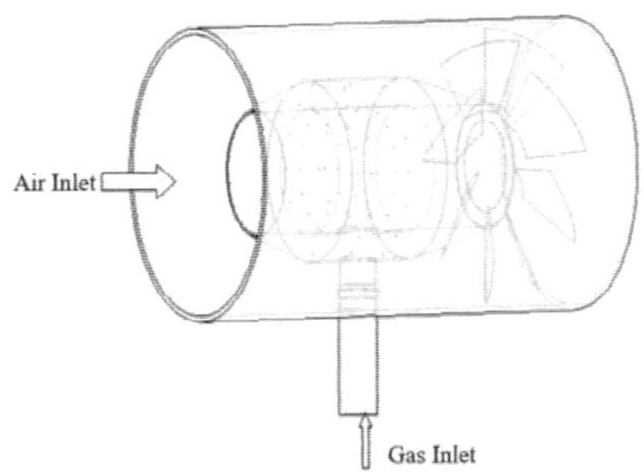

Fig.4 Schematic diagram of new mixing structure

4. Numerical Simulation and Optimization of New Mixing Structure

Similar to section 2.1, the model and fluid domain were established. Simultaneously, the model was meshed, and the independence was verified. 3.89 million grids were selected for calculation.

4.1Effect of the ratio of the number of holes in the gas chamber to the mixing

The gas hole diameter is 3 mm, and the ratio of the inner and outer holes is set to (1:1), (2:3), (1:2), (1:4). Numerical simulation is performed to study the effect of the number of holes in the cavity on the mixing.

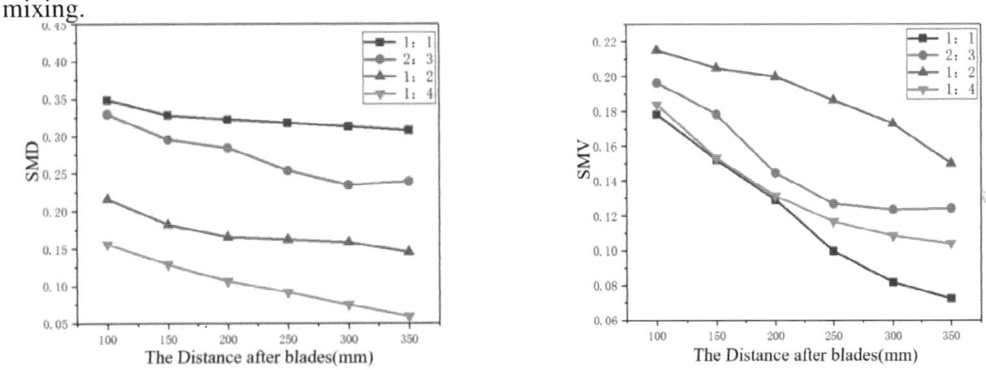

Fig.5 SMD and SMV distribution different internal and external aperture ratios

It can be seen from Fig.5 that the mixing result has the lowest concentration unevenness under the ratio of the inner and outer holes is 1:4. However, from Figure 7, the speed unevenness does not match, and there is no obvious law.

The ratio of the cross-sectional area of the inner and outer air ducts is close to 1:4. Therefore, it can be concluded that after the gas enters the chamber, the flow rate to the gas holes of each inner and outer air passages is approximately the same, so the ratio of the inner and outer gas holes determines the flow ratio. When the cross-sectional area of the air duct is very different, the difference between the two gas concentrations will be large, and there will be a large concentration unevenness. Therefore, the ratio of the inner and outer holes should be consistent with the cross-sectional area of the inner and outer air ducts. The gas flow ratio of the duct is consistent with the cross-sectional area ratio, which is conducive to uniform concentration. The distribution has no obvious relationship with the speed, so the speed uniformity does not conform to this law.

4.2Effect of gas pore size on mixing

The ratio of the number of holes in the chamber is 1:4 (8:32), and the diameter of gas holes is set to 3mm, 4mm, and 5mm respectively. Numerical simulation is carried out to study the effect of gas pore size on the mixing.

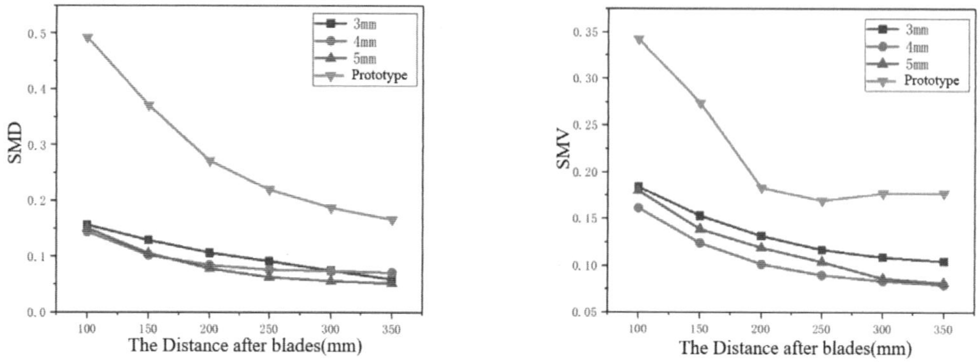

Fig.6 SMD and SMV distribution under different gas pore sizes

It can be seen from Fig.6 that in the three cases, the mixing result is the best at 5mm aperture and the SMD is the smallest, but it can be seen that the three results are relatively close, and the difference is not very large. Compared with the prototype burner, the three results are much better. The analysis shows that the size of the gas pore size affects the gas exit velocity, which in turn affects the jet depth, thus affecting the mixing. It can be seen from Fig.6 that the SMV and SMD distributions have similar consistency, and the three cases are smaller than the results of the prototype.

4.3Effect of the number of gas holes on the mixing

The ratio of the inner and outer holes of the gas chamber is 1:4, the gas hole diameter is 3mm, and the number of gas holes is 8+32, 12+48, 16+64, respectively. Numerical simulation is carried out to study the effect of the number of holes on the mixing.

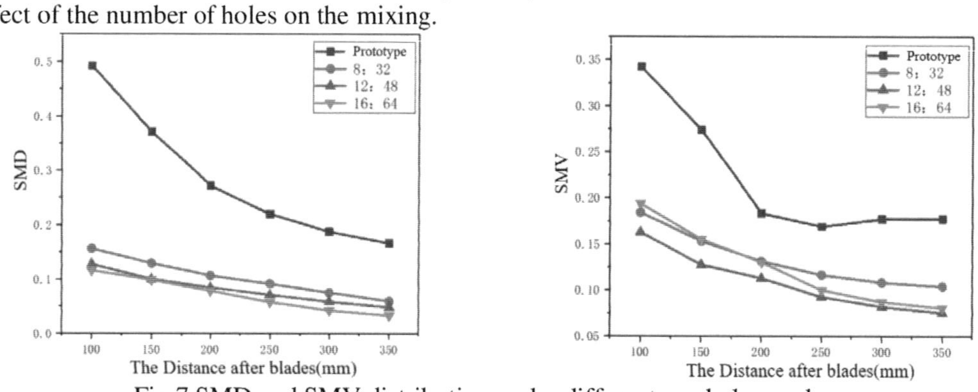

Fig.7 SMD and SMV distribution under different gas hole numbers

It can be seen from Fig.7 that in the three cases, the mixing result is the best at 16:64, and the SMD is the smallest, but it can be seen that the three results are relatively close, and the difference is not very large. Compared with the prototype burner, the three results are better. The number of gas holes also affects the gas outlet speed, which in turn affects the jet depth, thus affecting the mixing. It can be seen from Fig.7 that the SMV and SMD distributions have similar consistency.

4.4Effect of the number of gas holes on the mixing

The ratio of the inner and outer holes of the gas chamber is 1:4 (8:32), the gas hole diameter is 3 mm, and the length of blades is taken as 98 mm, 70 mm, 45 mm and 0 mm (no blade) respectively.

IOP Conf. Series: Materials Science and Engineering **721** (2020) 012011 doi:10.1088/1757-899X/721/1/012011

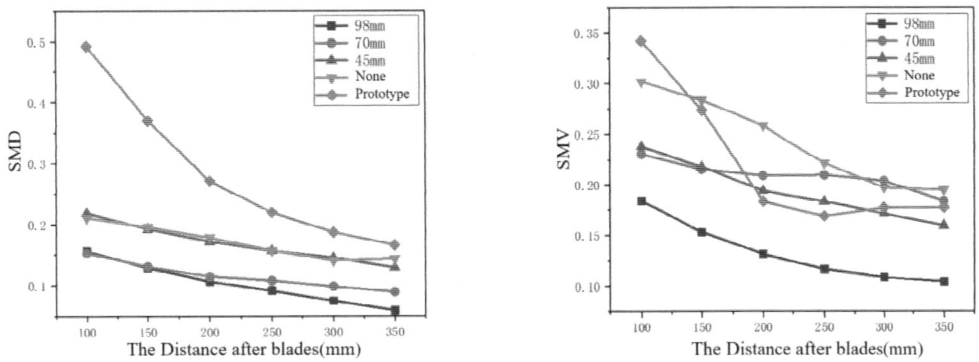

Fig.8 SMD and SMV distribution with different blade lengths

It can be seen from Fig.8 that the swirling blade has a certain promoting effect on the gas mixing. The SMD becomes larger as the length of the blade decreases. The SMV has the same tendency as SMD. From this, it can be concluded that the blade has a significant effect on promoting the mixing, especially for the velocity uniformity, and the gas will generate a swirling flow through the swirling blade, thereby improving the uniformity of the gas concentration and the uniformity of the velocity.

5. Conclusion

In this paper, the numerical simulation method is used to simulate the mixing structure and the new mixing structure of a certain type of burner, and the number of inner and outer holes, the size of the gas hole, the number of gas holes and the swirling flow of the new mixing structure are changed. The effect of the blade length on the mixing is compared with a mixing structure of a burner. Some conclusions are as follows:

a. There is a certain defect in the mixing structure of a certain type of burner. There is no gas flowing through the central duct and the blade, and there is a dead zone of the airflow, resulting in a dead zone with extremely low concentration in the vicinity of the central axis of the blade, even a long distance from the center. The CH4 concentration is very low and it is difficult to mix evenly.

b. The ratio of the inner and outer gas holes determines the flow ratio, so the ratio of the inner and outer holes should be cut with the inner and outer air passages. The gas flow ratio of the inner and outer air ducts is consistent with the cross-sectional area ratio, which is conducive to uniform concentration.

c. The size of the gas pore size affects the gas exit velocity, which in turn affects the jet depth, thereby affecting the mixing.

d. The number of gas holes also affects the gas exit velocity, which in turn affects the jet depth. However, when the jet depth can be reached, the number of gas holes is large, and the holes are denser, which is more favorable for mixing.

e. The blade has a significant effect on promoting the mixing, especially for the speed uniformity. The gas will generate a swirling flow through the swirling blade, which improves the gas concentration uniformity and speed uniformity.

References

[1] Tang, H.P., Xi, Z.P. (2006) Development Status of Surface Burners for Metal Porous Materials. J. Rare metal materials and engineering. S2: 423-427.
[2] Shao, W.W., Zhao, Y., Liu, Y. (2015) Study on Gas Fuel/Air Blending Uniformity of Gas Turbine Combustion Chamber Premixed Burner. J. Proceedings of The Chinese Society for Electrical Engineering. 37.
[3] Fric, T.F. (1993) Effects of fuel-air unmixedness on NOxemissions. J. Journal of Propulsion and Power., 9(5): 708-713

[4] Leong, M.Y., Samuelsen, G.S., Holdeman, J.D. (1999) Mixing of jetair with a fuel-rich, reacting crossflow. J. Journal of Propulsion and Power, 15(5): 617-622.
[5] Sun, B.C. (2007) Numerical study on gas turbine fuel/air mixing. D. Beijing: Tsinghua University.
[6] Feng, C., Kuang, H.Y., Xie, G., etc. (2011) Analysis of fuel/air premixing uniformity of dry low NOx gas turbine combustor. J. Proceedings of The Chinese Society for Electrical Engineering. 31(17): 9-19.
[7] Xie, G., Kuang, H.Y., Li, Y.H., etc. (2010) Study on Pollution Emission Performance of R0110 Heavy Duty Gas Turbine Combustor. J. Proceedings of The Chinese Society for Electrical Engineering. 30(20): 51-57.
[8] Zhao, Y. (2016) Study on Gas-filling and Combustion Characteristics of Natural Gas Lean Premixed Burners. D. Institute of Engineering Thermophysics, Chinese Academy of Sciences.
[9] Zhao, R.X., Wang, Y.Q. (2015) Gas-grading combustion ultra-low NOx burner: CN202483751U. P.
[10] Jiang, Y.K., Yang, J., Guo, Y. (2012) Venturi tubular gas mixer: CN202483751U. P.
[11] Qiu, P.H., Gu. X.L., Liu, L. High-speed injection combustion device adapted to multiple fuels: CN105509049A. P.

IOP Conf. Series: Materials Science and Engineering **721** (2020) 012012 doi:10.1088/1757-899X/721/1/012012

Numerical investigations of the flow distributions with a transient three-dimensional multi-component ejector model

Zhanrui Liu, Kui Jiao, Zirong Yang and Qing Du*.

State Key Laboratory of Engines, Tianjin University, Tianjin, 300350, China

*Corresponding author's e-mail: duqing@tju.edu.cn

Abstract. Ejector is a promising hydrogen recirculation for proton exchange membrane fuel cell (PEMFC) systems since it does not require parasitic power consumption. However, the complicated transport mechanisms inside the ejector remain unclear. In this study, the heat and mass transport processes are investigated by a transient three-dimensional multi-component model. The distribution of velocity, temperature, and pressure are presented. To avoid water vapor condensation, the maximum primary flow pressure should be limited to prevent the formation of a low temperature region at the nozzle throat. Compared with the converging-diverging nozzle, the convergent nozzle is recommended since it increases the condensation of water vapor. To predict the lowest temperature, the relationship between the relative pressure and the temperature is further obtained by curve fitting methods.

1. Introduction

Ejector is a promising device for the hydrogen recirculation in proton exchange membrane fuel cell (PEMFC) systems. Meanwhile, it also recycles the heat and moisture, enhancing energy utilization and anode humidification. However, several technical challenges remain to be solved, such as narrow working ranges, inaccurate flow control, and complicated transport mechanisms.

In the past decades, both numerical and experimental researches have been conducted. Experiments were mainly focused on the effects of fluid component, operating conditions, and structural designs. Li et al.[1] established an ejector test facility to study the effects of water vapor on the hydrogen recycle entrainment ratio. Arun et al.[2] used the schlieren flow visualization method to investigate the velocity distributions. It was found that shock waves were generated and influenced by velocity differences between the primary flow and the secondary flow. As regards modeling studies, Mazzelli et al.[3] comprehensively compared the two-dimensional (2D) model with a three-dimensional (3D) model. It was concluded that the 3D model generated more accurate results such as mass flow rate and entrainment ratio. Yin et al.[4] optimized the ejector geometric parameters to improve the entrainment ratio with a 3D steady-state model. To investigate the transient processes, Ankit et al.[5] proposed the inertial effect, which was found to be negatively correlated with the primary flow pressure. Lijo et al.[6] investigated the transient phenomena within a vacuum ejector. Ferrari et al.[7] developed a 3D ejector model for the SOFC recirculation. The model was validated against their experimental data. Despite aforementioned researches, the distribution of parameters (velocity, temperature, and pressure) inside the ejector are still not fully revealed. Water vapor condensation inside ejectors can cause severe damages to the performance and reliability. However, the studies to prevent vapor condensation are hard to find in literature. Investigating the transport processes inside ejectors not only helps improving performances but also gives advice on the prevention of water vapor condensation.

In the study, a transient 3D multi-component ejector model is developed since there are multiple components in the fuel cell anode exhaust. The distribution of velocity, temperature, and pressure along the primary flow direction are presented. The differences between two turbulence models are studied. For guiding the choice of the working conditions, the relationship between primary flow pressure and the lowest temperature in ejector is investigated.

2. Mathematical method

2.1 Geometric parameters

As shown in Figure 1, the high-pressure hydrogen enters the ejector through the primary flow tube. The velocity of the primary flow increases dramatically. Meanwhile, the pressure declines rapidly due to the convergent nozzle. The vacuum region is formed at the nozzle throat. Therefore, the secondary flow can be drawn into the ejector by the pressure difference between secondary flow and the vacuum region. Subsequently, the primary flow and the secondary flow are fully mixed in the mixing tube. At the diffuser section, mixed gases restore the pressure and temperature to meet the fuel cell anode inlet requirements. In this study, geometric parameters are set the same as Dadvar et al. [8] and the operation conditions are adopted from reference [4].

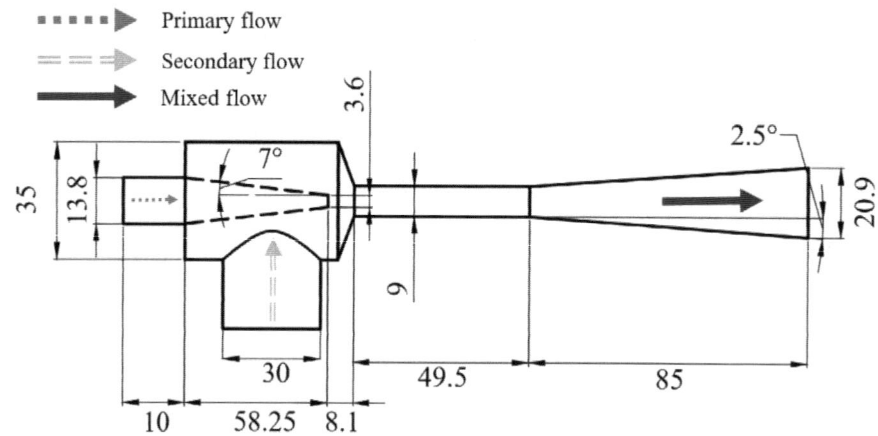

Figure 1. Structural parameters of the ejector.

2.2 Governing equations

Continuity equation:

$$\frac{d\rho}{dt}+\nabla\cdot\left(\rho\vec{U}\right)=0 \tag{1}$$

Where ρ is density (kg m^{-3}), t is time (s) and $\vec{U}$ is body velocity (m s^{-1}).

Momentum equation:

$$\frac{d\rho\vec{U}}{dt}+\nabla\cdot\left(\rho\vec{U}\vec{U}\right)+\nabla p=0 \tag{2}$$

Where p is pressure (Pa).

Energy equation:

$$\frac{d\rho E}{dt}+\nabla\cdot\left(\vec{U}\rho E\right)+\nabla\cdot\left(\vec{U}p\right)=0 \tag{3}$$

Where E is total energy (J kg^{-1}).

Species transport equation:

$$\frac{d}{dt}\left(\rho Y_i\right)+\nabla\cdot\left(\rho\vec{U}Y_i\right)=-\nabla\cdot\vec{J}_i \tag{4}$$

$$\vec{J}_i=-\rho D_{diff,i}\nabla Y_i \tag{5}$$

Where Y_i is mass fraction of i species (hydrogen, water vapor, nitrogen), $\vec{J}_i$ is the diffusion flux (kg $m^{-2}s^{-1}$) of i species, $D_{diff,i}$ is the diffusion coefficient of i species.

In this study, the ejector is mainly designed for fuel cell systems. The operating conditions are listed in Table 1. The ejector is initially filled with pure hydrogen. The initial temperature is the same as ambient temperature. The RNG k-ε turbulence model is adopted for all cases.

Table 1. Ejector operating conditions.

Parameters	Values
Primary flow temperature T_p (k)	293
Secondary flow temperature T_s (k)	353
Primary flow pressure P_p (kPa)	450/325
Secondary flow pressure P_s (kPa)	300
Secondary flow water vapor mass fraction	0.55
Secondary flow nitrogen mass fraction	0.05
Secondary flow hydrogen mass fraction	0.4

2.3 Numerical procedures

The simulations are conducted based on the open-source software OpenFOAM, and the *rhoCentralFoam* solver is adopted for calculation. However, the *rhoCentralFoam* solver in OpenFOAM cannot solve the flow problems with multiple components [9]. Therefore, the species transport equations are added to the solver owing to the existence of various gas species (including hydrogen, nitrogen, and water vapor) in the fuel cell system. The variable time step is used to accelerate the convergence speed and improve the calculation stability. Since the solver is a transient one, the steady-state solution does not exist. Therefore, when the relative error of mass rate between time points is less than 0.03, a relatively stable state is considered to be reached. All simulations are conducted based on the Tianhe-HPC1 system at the National Supercomputer Center in China.

3. Results and discussions

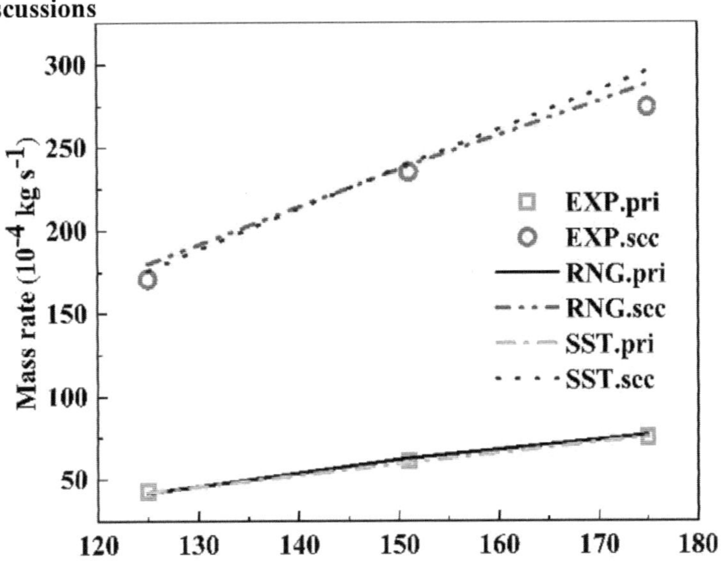

Figure 2. Comparison between the experimental data [10] and two different turbulence models.

The numerical results are validated with experimental data [10] in two different turbulent models, which are *k-ε* RNG and *k-ω* SST, respectively. Figure 2 illustrates the comparison of mass flow rates with different turbulence models. It is seen that the two different turbulence models can predict the flow rate of the primary flow well. When it comes to the secondary flow, both models overestimate the mass flow rate. The k-ω SST model predicts more accurately when the pressure is lower (125kPa), and the error increases significantly when the pressure is higher (175kPa). In general, the fluctuation of k-ε RNG is smaller than k-ω SST. Therefore, the k-ε RNG turbulence model is finally selected.

Figure 3 shows the distribution of velocity, pressure, and temperature along the primary flow direction. The differences between two different primary flow pressures (325kPa, 450kPa) are presented. The abscissa represents the relative position in the z-direction, 0 represents the inlet of primary flow, and 1 represents the outlet. The result shows that the velocity gradually increases in the nozzle (0.05-0.3) and the pressure gradually decreases in the nozzle. In addition, the maximum value of velocity and the minimum value of pressure occur at 0.3, where is the nozzle throat position. Note that the maximum velocity of 450kPa is much larger than 325kPa (Figure 3 (a) shown) and the vacuum region pressure of 450kPa is lower than 325kPa (Figure 3 (b) shown).

The Figure 3(c) shows the temperature variation inside the ejector. It is found that the minimum of temperature occurs at the nozzle throat. Higher primary flow pressure results in lower temperature at the nozzle throat. The lowest temperature is about 268K with the primary flow pressure being 450kPa. Compared to the inlet region of primary flow, the temperature drops by nearly 25K. However, the secondary flow from fuel cell anode exhaust contains water vapor, which will condense due to the low temperature at the nozzle throat position. If the primary flow pressure is 325kPa, the lowest temperature drops a little (only 3-4K). Therefore, the inlet primary flow pressure of ejectors for fuel cell systems must be limited to prevent water vapor condensation. Since the converging-diverging nozzle results in lower temperature than the converging nozzle [11], the converging-diverging nozzle is not suitable for PEMFC systems.

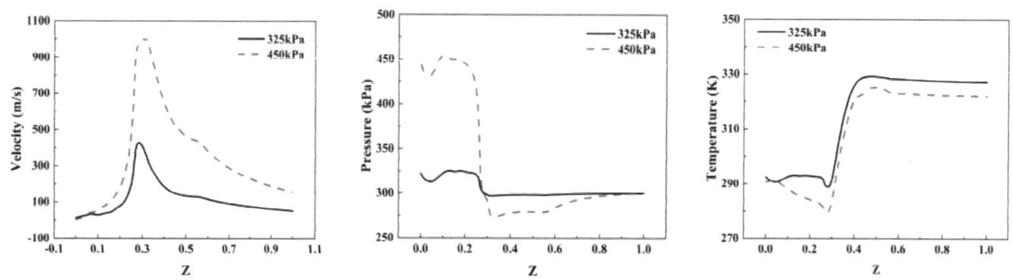

(a) (b) (c)

Figure 3. the trends along the primary flow direction: (a) velocity (b) pressure (c) temperature.

For working fluids including water vapor, predicting the lowest temperature helps prevent water vapor condensation, which improves the ejector stability. As shown in Figure 4, the relationship between the relative pressure of the primary flow and the lowest temperature inside the ejector is investigated. A linear fitting equation is proposed for predicting the lowest temperature inside the ejector. The details of linear fit equation are shown in Table 2. It can be seen that there is a linear negative correlation between the relative pressure of the primary flow and the lowest temperature. The lowest temperature of the ejector inside can be predicted by this equation which can guide the choice of the primary inlet pressure.

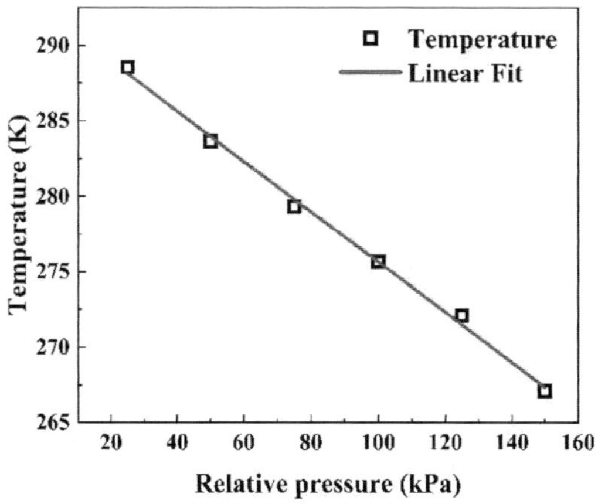

Figure 4. Linear fit of relative pressure to temperature.

Table 2. The parameters of the linear fit equation.

Description	Types and value
Equation	y = a + b*x
Plot	Temperature
Weight	No Weighting
Intercept	342.17277 ± 1.80675
Slope	-0.16631 ± 0.00463
Residual Sum of Squares	0.9397
Pearson's r	-0.99845
R-Square(COD)	0.9969
Adj.R-Square	0.99613

4.Conclusions

In this study, a transient three-dimensional multi-component ejector model is established for investigating the heat and mass transport processes inside the ejector. The *k-ω* SST model better predicts the ejector performance than *k-ε* RNG if the primary flow pressure is lower than 125kPa. Conversely, the predictive ability of *k-ω* SST is better than *k-ε* RNG at high pressure. The distribution of velocity, pressure, and temperature along the primary flow direction are presented. The results indicate that the most dramatic change occurs in the nozzle throat. The velocity rises significantly while the pressure and temperature drops. The maximum value of the primary flow pressure of ejectors for fuel cell systems should be limited to prevent water vapor condensation caused by low temperature region. The diverging nozzle is more suitable than the converging-diverging nozzle. A linear fitting equation of primary flow relative pressure to lowest temperature is proposed to prevent condensation inside the ejector.

Acknowledgments

This work was supported by the National Key Research and Development Program of China (2018YFB0105500).

References

[1] F. Li, J. Du, L. Zhang, J. Li, G. Li, G. Zhu, M. Ouyang, J. Chai, H. Li, (2017), Experimental determination of the water vapor effect on subsonic ejector for proton exchange membrane fuel cell (PEMFC), International Journal of Hydrogen Energy 42(50) 29966-29970.

[2] K.M. Arun, S. Tiwari, A. Mani, (2019), Experimental studies on a rectangular ejector with air, International Journal of Thermal Sciences 140 43-49.

[3] F. Mazzelli, A.B. Little, S. Garimella, Y. Bartosiewicz, (2015), Computational and experimental analysis of supersonic air ejector: Turbulence modeling and assessment of 3D effects, International Journal of Heat and Fluid Flow 56 305-316.

[4] Y. Yin, M. Fan, K. Jiao, Q. Du, Y. Qin, (2016), Numerical investigation of an ejector for anode recirculation in proton exchange membrane fuel cell system, Energy conversion and management 126 1106-1117.

[5] A. Mittal, G. Rajesh, V. Lijo, H.D. Kim, (2014), Starting Transients in Vacuum Ejector-Diffuser System, Journal of Propulsion and Power 30(5) 1213-1223.

[6] V. Lijo, H.D. Kim, G. Rajesh, T. Setoguchi, (2010), Numerical simulation of transient flows in a vacuum ejector-diffuser system, Proceedings of the Institution of Mechanical Engineers, Part G: Journal of Aerospace Engineering 224(7) 777-786.

[7] M. Ferrari, M. Pascenti, A. Massardo, (2018), Validated ejector model for hybrid system applications, Energy 162 1106-1114.

[8] M. Dadvar, E. Afshari, (2014), Analysis of design parameters in anodic recirculation system based on ejector technology for PEM fuel cells: A new approach in designing, international journal of hydrogen energy 39(23) 12061-12073.

[9] C.J. Greenshields, H.G. Weller, L. Gasparini, J.M. Reese, (2010), Implementation of semi - discrete, non - staggered central schemes in a colocated, polyhedral, finite volume framework, for high - speed viscous flows, International journal for numerical methods in fluids 63(1) 1-21.

[10] K. Nikiforow, P. Koski, H. Karimäki, J. Ihonen, V. Alopaeus, (2016), Designing a hydrogen gas ejector for 5 kW stationary PEMFC system–CFD-modeling and experimental validation, International Journal of Hydrogen Energy 41(33) 14952-14970.

[11] NAKAGAWA Masafumi, M.S. BERANA, KISHINE Akinori, (2009), Supersonic two-phase flow of CO2 through converging–diverging nozzles for the ejector refrigeration cycle, International Journal of Refrigeration 32(6) 1195-1202.

Study on dynamic measurement of fouling thermal resistance in a boiler

Kairan Zou[1], Ning Qiao[1*], jiaoling Wang[2], Rong Si[2]

[1]Laboratory of Electrochemical Process and Technology for Materials, Beijing, 100029, China

[2]China Boiler and Boiler Water Treatment Association, Beijing, 100029, China

[*] Corresponding author's e-mail: qn_buct@163.com

Abstract. A fouling thermal resistance method is introduced, for evaluating water quality in real time through measuring fouling thermal resistance on heat transfer surface. A simple device as a boiler heat transfer model was designed for measuring the deposition of dirt by the fouling thermal resistance method. The change in the thermal resistance of the fouling was measured with regard to samples with different water quality. The temperature-time curves were plotted with respect to the measured data of various temperatures over the whole process of temperature rise and fall to examine the feasibility of the method in evaluating the thermal resistance of the fouling contributed by the hardness of the water, finding that the fitted curve can characterize the fouling deposition of high hardness water. It has been shown that this experimental device can be improved further in the hope of being applied in practical use.

1. Introduction

Fouling is an undesirable substance (sediment) accumulated on the surface of heat exchanger. From the basic point of view, fouling can be described as a comprehensive transfer problem of unsteady heat, mass and momentum of heat exchanger fluid and surface, which is closely related to the operating conditions of heat exchanger. Undesirable substances may be crystals, sediments, inorganic salts, biological growth and corrosion products [1, 2]. Fouling is usually invisible from the outside of the processing equipment and can only be reflected by its significant influence on the equipment[3], such as, an increases in the pressure drop, a reduction the heat transfer efficiency. Investigation of fouling dynamics is of great importance since the deposit layer grows with time and may not develop a steady layer having a certain thickness. It is fairly difficult to measure the rate of growth of fouling layers directly due to their soft and readily deformable nature [4] and the challenge in immobilizing a measurement tool in the flow.

The gravimetric method and the direct measurement of thickness most commonly used to measure the degree of fouling are discontinuous or invasive in nature [5]. The gravimetric method evaluates the degree of fouling by weighing the amount of fouling deposited on the heat transfer surface. Without corrosion on the heat transfer surface, preferably the surface state is consistent [6], otherwise there is no point in evaluating in view of nonrepeatability of the test data. The thickness measurement method uses a high-frequency sound wave to pass through the material [7] to measure the wall thickness, with measuring the time when the echo returns to the probe, or recording the amplitude generated when the resonance is produced as a signal source. This method is often adopted to detect the internal defect, the corrosion damage, and the wall thickness of the equipment and piping, however, the disadvantage of which is difficulty tracking changes in the recorded wall thickness for lack of sufficient sensitivity.

In the study of fouling thermal resistance, a commonly adopted approach is to measure the heat transfer performance of the system and then to relate changes in the thermal resistance to the amount of fouling or properties of the deposit such as its thickness and thermal conductivity. An analytical model and monitoring device is put forward in this paper, which is used to simulate the actual operation process of boiler and monitor the change of temperature caused by fouling thermal resistance, thereby determining the degree of scaling. This analytical model based on a simple concept of heat transfer resistance can accurately simulate the deposition process without the assistance of highly sensitive and expensive measurement sensors. Therefore, this device is highly suitable for boiler industry, to monitor the degree of scale formation degree of in a boiler.

2. Experiments

2.1 Configuration of device

Dynamic monitoring device for fouling thermal resistance mainly includes a heating pipe, a temperature controller and a thermometer. The heating pipe is made of stainless steel to eliminate the corrosion effects caused by the open system during the temperature control process, simulating the fouling process in the boiler. As shown in figure 1, the temperature controller is connected with the u-shaped heating pipe and Pt1000 temperature probe, which are all placed into a beaker with 2L water to be tested. The heating pipe is vertical relative to the liquid level and positioned at a fixed depth. Another thermometer is fixed on the wall of the u-shaped heating pipe to measure the wall temperature.

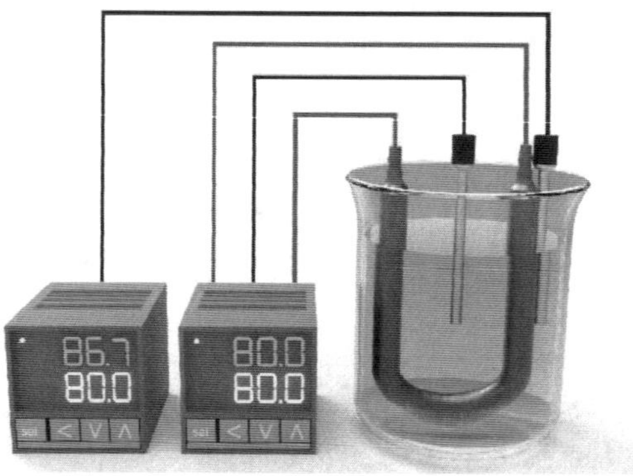

Figure 1 Dynamic monitoring device for fouling thermal resistance

2.2 Process of measurement

The test water sample was prepared with a standard concentration of calcium chloride solution, adjusting the hardness. Two liters of experimental water was added into the beaker, marking the liquid level. The experiment was performed with the designed fouling resistance measuring device (Figure 1), and the experimental temperature was set to 80 °C. The heating pipe and the temperature probe for the water sample were fixed in position, and the other temperature probe was fixed in the pipe wall groove to measure the wall temperature. An experimental water sample was added every 1 hour to supplement the loss of water by evaporation. The change of temperature was recorded every 2 hours. As the device is an open system, the temperature controller has a limited sensing range of temperature with sensitivity of the sensor. The temperature controller starts up to heat from its lowest temperature. The heating does not stop until the water temperature reaches its highest temperature. Thus, the whole experiment is a process of temperature change via a cycle of from heating up to cooling down. Therefore, the actual temperature control range of the temperature controller is 80 ± 1 °C. The minimum wall temperature and the maximum wall temperature were recorded at 2-hour intervals and the period of time for heating and cooling (the liquid level was kept consistent whenever the temperature was recorded). According to the heat transfer model described in 3.1, the change of the thermal resistance of the fouling was calculated, and then the degree of fouling was evaluated.

3. Analysis and discussion

3.1 Heat transfer model

According to Chao Shen's uncertainty analysis on the design of the heat exchanger fouling experimental device, higher surface heat transfer performance and low water flow rate are beneficial to reduction in the uncertainty of the device. The fouling deposition has little effect on the heat flux density of the water [8]. The design is simplified to static scaling, minimizing the effect of heat transfer at the flow rate.

For the sake of calculation of the effect of fouling on heat transfer efficiency, the following assumptions are made:

(1) The presence of fouling on the heat transfer surface has negligible effect on the convective heat transfer coefficient of the wall and fluid;

(2) The heating pipe power is constant and the heating is a continuous process;

(3) The effect of hardness on specific heat capacity of the water to be measured can be ignored. According to the heat transfer theory, the total fouling thermal resistance of the surface of the heating pipe during the whole experiment is [9]:

$$R_f = \frac{1}{U_f} - \frac{1}{U_c} \tag{1}$$

Where U_f and U_c are the total heat transfer coefficient in the presence of fouling on the heat transfer surface and in the clean state, respectively. According to Fourier's law, the thermal conductivity k can be expressed as [10, 11]:

$$k = |q| / \frac{\partial T}{\partial x} \tag{2}$$

Where q is the heat flux on the heat transfer surface, $\partial T/\partial x$ is the temperature gradient in the continuous temperature field (projection in the x direction). When the position of the heating pipe is fixed, the temperature field is fixed and, the heat flow field is also fixed during the whole heating-cooling process, and then the temperature becomes a function of space. Then the heat Q flowing over the area A through a certain point at t time can be written as:

$$Q = \int_0^t dt \int_A k|gradT|dA \tag{3}$$

During the experiment, the temperature of the heating pipe is measured by a thermometer. It is assumed that only one side of the heat transfer surface is fouled. The amount of water is constant every time the wall temperature is recorded. The amount of water required to rise from the lowest temperature to the highest temperature is constant, that is, the total heat flux during the heating time is constant, which can be regarded as a constant heat flux model [12, 13]. As shown in Figure 2, T_l is the temperature of the experimental water sample in the beaker, and T_w is the wall temperature of the heating pipe. In the distance between $x=0$ and $x=a$, there is a temperature gradient grad T between T_w and T_l, and the heat transfers from the heating pipe to the water sample. As the degree of fouling continues to increase and the deposit of fouling on the wall of the heating pipe causes the temperature of the pipe wall to rise, the thermal resistance of the dirt at any one time is:

$$R_f(t) = \frac{1}{U_f(t)} - \frac{1}{U_c} = \left(\frac{T_w(t) - T}{q_c} - \frac{T_w(t=0) - T}{q_c}\right) = \frac{T_w(t) - T_w(t=0)}{q_c} \tag{4}$$

Where $T_w\ (t)$ is the average wall temperature during the temperature rise and fall for any measurement time, $T_w(t=0)$ is the initial wall temperature, and q_c is the heat flux which is a constant for the experimental device. According to equation (4), the thermal resistance of the fouling at each measurement point is calculated directly from the measured wall temperature.

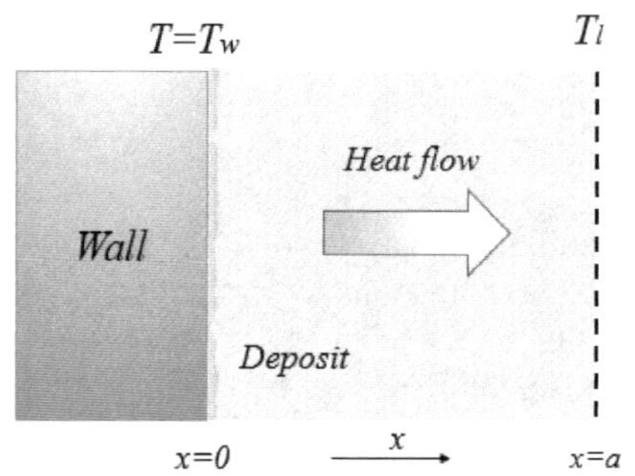

Figure 2 Constant heat flux heat transfer model

During the experiment, by comparing the thermal resistance of the different experimental water samples with regard to the same fouling time, the degree of fouling can be determined, as shown in equation (5):

$$\eta = \frac{R_{fa}(t)}{R_{fb}(t)} = \frac{T_{wa}(t)-T_{wa}(t=0)}{T_{wb}(t)-T_{wb}(t=0)} \tag{5}$$

Where η is the ratio of the degree of fouling of the a experimental water sample to the b experimental water sample at the fouling time t, and $R_{fa}(t)$ and $R_{fb}(t)$ are the thermal resistance of the fouling during the scaling time. The remaining terms are the same as equation (4).

3.2 Temperature measurement with three-point method

The whole experiment is a process of warming and cooling. With a random number of points at the same interval, it is hard to accurately describe the wall temperature change caused by fouling of the heating pipe, but it is easy to record the temperature value on the temperature detector by a simple device. In the course of the experiment, every two hours, the extreme point of the temperature rising and cooling process, that is, the initial temperature T_0 at which the temperature rise starts, the highest temperature T_1 for temperature rise, and the lowest temperature T_2 for temperature drop are recorded. At the same time, the time t_1, t_2 taken by the temperature rise and temperature drop is recorded by a stopwatch. Due to the assumption in (2) of 3.1, the measured data are fitted into the time-temperature curve $T(t)$ by Newton interpolation, and then the function curve is time-integrated. The integrated area is proportional to the heat flux, and then the temperature integral is averaged. The number represents the temperature value during this time period. This method measures three temperature values, referred to as the three-point method, as shown in equation (6) [14]:

$$\bar{T} = \frac{\int_0^{t_2} T(t)\,dt}{t_2} \tag{6}$$

Deepening of the degree of fouling, will affect the heat transfer efficiency of the heat transfer surface. In the heating and cooling processes, the wall temperature of the heating pipe will change due to the change of the thermal resistance of the fouling. Therefore, the average temperature value obtained by this calculation method is related to the heat flux during the whole time period, for which the thermal resistance of the fouling is responsible. The average temperature matches up with the temperature that causes a change in the temperature rise and temperature fall process.

3.3 Evaluation of device stability

The deionized water was subjected to a 24-hour temperature test on the designed device. The stability of the device was verified in accordance with to the wall temperature curve (Figure 3) in the clean state of the heating pipe.

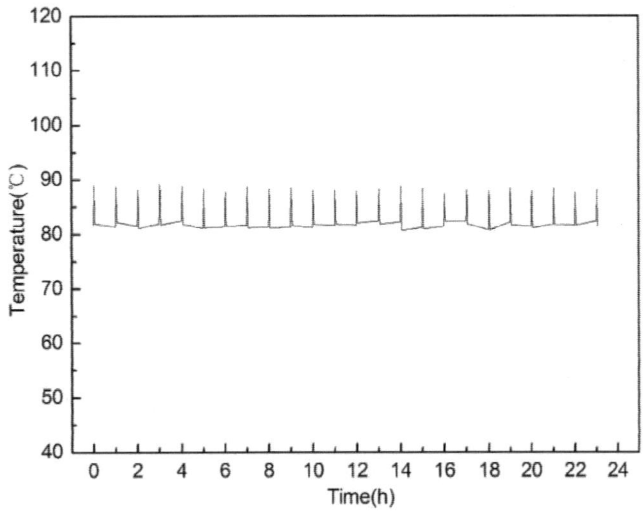

Figure 3 temperature - time curve of deionized water

In terms of the data of the measured deionized water, the temperature change in the temperature-time curve is smooth. The temperature peak fluctuates around 1°C, and the temperature fluctuates between 6 and 7°C during the whole experiment. It has shown that the device has good stability for the impurity-free deionized water test. The temperature data of the long-term scaling experiment were integrated and calculated to obtain the following table 1.

Table 1 Data analysis of different water sources

Water sample	Water hardness (mmol/L)	Mean heating time (s)/ Standard deviation	mean highest tem perature (℃) / Standard deviation	Fitted slope of the line(*10^{-2})
Deionized water	0	94/13.19	88.28/0.41	0.07
Municipal water supply	1.43	119/26.54	84.91/1.06	0.58
Formulation water sample	3.00	107/29.75	85.13/1.12	1.39

As can be seen from the above table 1, the average heating time obtained for deionized water as the experimental water sample is 94 seconds, which is less than the heating time for scaling precipitation. The reasons are that scaling leads to the reduction in heat transfer coefficient, and in turn the heat transfer efficiency decreases, with taking more of time. The comparing between the standard deviation of the highest temperature and the slope of the fitted curve, has displayed that the temperature rise-fall process of the device in the clean state is basically the same, with good stability.

3.4 Measurement and feasibility analysis of fouling thermal resistance

A scale thermal resistance test was performed using a municipal water with a hardness of 1.43mmol/L and a water with a hardness of 3.00mmol/L on a fouling thermal resistance measuring device. Assuming that the wall temperature of the heating pipe rises evenly during the heating period, and using the average value of the integral of each set of temperature data over time as the average temperature within the two hours, all the temperatures of the entire test are plotted and then curve-fitted. Figure 4 shows the variation of the wall temperature of the heating pipe caused by the thermal resistance of the fouling:

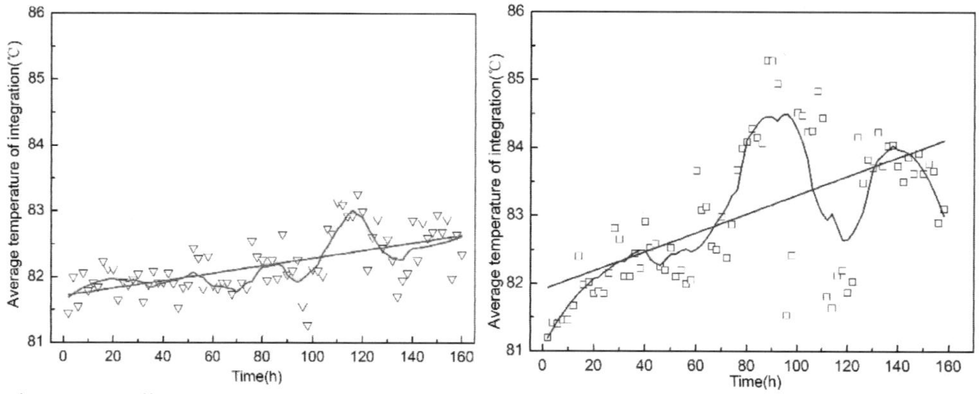

Figure 4 Fouling temperature curve (left: municipal water; right: a water with a hardness of 3.00mmol/L)

Seen from the left figure, the fouling time of municipal water is within 100 hours. The figures show that the temperature fluctuates little, and thus temperature rises slowly. It is learned from this group of temperature data that in a relatively short period of time (within 100 hours), the degree of fouling does not exert a significant effect on temperature, It is noted that the degree of scaling becomes greater as the concentration of calcium ions in the water sample increases, and the temperature changes significantly. The temperature rises and falls sharply at a higher temperature in the later stage. After 200hours of temperature control, the municipal water used in the experiment is gradually concentrated with evaporation, and the hardness ions gradually increase. The concentration of chloride ion in the final concentrated water sample is 0.0141mol/L, and the chloride ion concentration of Municipal water is $6.31*10^{-4}$mol/L. This indicates that the water sample was concentrated by a factor of 22 during the experiment. The high concentration of hardness ions contributes to the scale driving force to a certain extent, facilitating the process of fouling. This also explains that the temperature changes obviously after 100hours.

In this test, the hardness of the experimental water (3.00mmol/L) is more than two times that of the municipal water (1.43mmol/L). From the figure on the right, it can be observed that after 60hours temperature control, the temperature curve shows a steep climbing pattern. The temperature data of the whole temperature control process were fitted into a straight line with a slope of 0.0139, which was remarkably higher than the fitted slope of 0.0058 for the municipal water as experimental water. In the time of 160 hours of fouling, the ratio of the two fitted slopes in the two experiments was 2.398 adequately reflected the change in the degree of fouling based on formula (5) during in the same period of time with different water samples. According to the theoretical model of fouling thermal resistance, this is the direct evidence of the increase in fouling thermal resistance. This test has demonstrated that this designed method can be used to predict the degree of fouling by dynamically monitoring the wall temperature change caused by the thermal resistance of the boiler.

According to the above study of the thermal resistance of the boiler by the wall temperature method, the degree of fouling in the boiler can be measurably detected by the change in the wall temperature of the boiler. However, for the temperature-time curve of the experimental water sample with a hardness of 3.00mmol/L, the temperature changes over the later period greatly. As the amount of water in the temperature control system is constant, the heat flux of the heating pipe has a certain range for maintaining the temperature of system in the temperature range of from 79 °C to 81 °C. As the thermal resistance increases, the heat transfer efficiency decreases, resulting in a longer heat transfer time. The system heat transfer is slow, causing the wall temperature of the heating pipe to rise with the increase in the degree of fouling, characterized by the temperature rise and fall process overtime. However, such phenomenon of temperature rise and fall is strictly subject to the nature of the test system. According to equation (3) in 3.1, the heat required to maintain 2L water at 80 °C is constant. When the temperature continues to increase due to the increase in the thermal resistance of the fouling, the temperature difference between the heating pipe at a higher temperature and the experimental water sample at a temperature of 80 °C will keep heat transferring on. Thus, the period of time for temperature rise becomes shorter over the whole heating-cooling process, while the period of time for the temperature drop remains constant, or becomes longer. In short, the change of the temperature curve at the latter stage is not as obvious as the earlier and middle stages. Moreover, the fouling thermal resistance device was designed as a static scaling mode for simplicity, and the temperature was controlled at 80 °C. In this circumstance, due to the influence of various factors, including, among other things, bubbling, heat flow, will cause the entire experimental system to reach a certain degree of fouling, finally striking a sediment-desorption balance of the scale [15, 16], and making it difficult to continue to increase the scale.

Taking into account the amount of water and the degree of fouling, the device, designed to determine the fouling thermal resistance of a boiler by simulating the fouling of a boiler has its limitations. When the temperature reaches a certain limit, it is difficult to further increase due to the deposition of the scale. However, for an industrial boiler, a higher temperature limit is allowable depending on the heat amount of the boiler and its mode of operation. Compared to the simulated device designed in this experiment, the boiler wall temperature in practice is relatively easy to measure. Thus, in actual application, it is only necessary to determine the slope of the temperature-time curve or the wall temperature change arising out of a certain degree of fouling to watch over scale formation in the boiler for the purpose of avoiding any safety accident by taking timely actions of cleaning.

4. Conclusion

The device designed for fouling thermal resistance in this experiment has good stability and can simulate the operation of a boiler to some extent. The proposed three-point wall temperature method has succeeded in measuring the thermal resistance of the fouling and the fitted slope by data analysis can roughly reflect the temperature change in a scaling process. The measurement of fouling thermal resistance for the designed device has its own restrictions, determined by the total heat required to maintain a certain temperature of the system as the entire experimental system to reach a sediment-desorption balance of the scale. Compared with the simulation equipment used in the experiment, it is of great significance in industry that the degree of scale formation in a boiler can be simply monitored by recording different temperatures of the system as a larger measurable temperature rise can be expected during the actual operation.

Acknowledgement

This work was supported by China Boiler and Boiler Water Treatment Association.

References

[1] T.R. Bott, Fouling Notebook. The Institution of Chemical Engineers, Rugby, England, 1990.

[2] Lister V Y, Davidson J F, Wilson D I. Calculating thermal fouling resistances from dynamic heat transfer measurements[J]. Chemical Engineering Science, 2012, 84(none):772---780.

[3] Markowski M, Trafczynski M, Urbaniec K. Validation of the method for determination of the thermal resistance of fouling in shell and tube heat exchangers[J]. Energy Conversion and Management, 2013, 76:307-313.

[4] Chew, Y.M.J., Paterson, W.R., Wilson, D.I., 2004. Fluid dynamic gauging for measuring the strength of soft deposits. J. Food Eng. 65, 175–187.

[5] Tuladhar, T. R., Paterson, W. R., Macleod, N., & Wilson, D. I. (2000). Development of a novel noncontact proximity gauge for thickness measurement of soft deposits and its application in fouling studies. The Canadian Journal of Chemical Engineering, 78, 935–947.

[6] Li G, Gao J, Liu Y. Research on effects of residual polymer on the corrosion and scaling of the oilfield produced water [J]. Chemical Engineering of Oil & Gas, 2012, 41(6):612-615.

[7] Cao J H, Yan G B, Han X F. (2002) Principle of thickness measurement of electromagnetic

ultrasound and its application —— A new ultrasound thickness measurement [J]. Journal of Zhejiang University(Engineering Science),36(1):89-91

[8] Shen C, Cirone C, Wang X. Uncertainty analysis: Design of a fouling test device for the liquid-to-refrigerant heat exchangers [J]. Applied Thermal Engineering, 2015, 85:148-159.

[9] FENG Dianyi, Fan Jia, Sun Shouren (2009) Research on the Real time Prediction of the Fouling Resistance of a Waste Heat Boiler/ Journal of Engineering for Thermal Energy & Power. - 2009, 24(2). – 196-199

[10] Albert F, Augustin W, Scholl S. Roughness and constriction effects on heat transfer in crystallization fouling [J]. Chemical Engineering Science, 2011, 66(3):499-509.

[11] Incropera, Dewitt F P /, Bergman D P /, et al. Foundations of Heat Transfer [J]. John Wiley & Sons, 2012.

[12] Lister V Y, Davidson J F, Wilson D I. Calculating thermal fouling resistances from dynamic heat transfer measurements [J]. Chemical Engineering Science, 2012, 84(none):772---780.

[13] Sarafraz M M, Nikkhah V, Madani S A , et al. Low-frequency vibration for fouling mitigation and intensification of thermal performance of a plate heat exchanger working with CuO/water nanofluid[J]. Applied Thermal Engineering, 2017, 121:388-399.

[14] Johansen T A, Foss B A. Empirical modeling of a heat transfer process using local models and interpolation[C]// American Control Conference. 1995.

[15] Mwaba M G, Golriz M R, Gu J. A semi-empirical correlation for crystallization fouling on heat exchange surfaces [J]. Applied Thermal Engineering, 2006, 26(4):440-447.

[16] Lacmann R. Crystallization, Third Edition. J. W. MULLIN, Butterworth-Heinemann, Oxford 1997, 527 Seiten, zahlr. Abb. und ISBN 0-7506-3759-5[J]. Chemie Ingenieur Technik - CIT, 2010, 70(11):1468-1468.

IOP Conf. Series: Materials Science and Engineering **721** (2020) 012014 doi:10.1088/1757-899X/721/1/012014

Selection of typical wet desulfurization system and economic analysis of variable load operation for industrial coal-fired boilers

Ximei Li[1], Jianmin Gao[1*], Yinuo Liu[1], Qian Du[1], Xuemin Liu[2], Guoli Qi[2]

[1]School of Energy Science and Engineering, Harbin Institute of Technology, Harbin, Heilongjiang Province, 150001, China

[2]Institute of Energy Conservation Technology, China Special Equipment Inspection and Research Institute, Beijing, 100029, China

[*]Corresponding author's e-mail: yagjm@hit.edu.cn

Abstract. In this paper, the operation costs of wet flue gas desulfurization based on MgO, CaO and $CaCO_3$ under basic, special and ultra-low emission limits are analyzed from the unit price of desulfurizer and electricity price. Guidance about desulfurizer selection in Liaoning Province under different electricity prices and different loads is given. The results show that the desulfurization cost of MgO is greatly affected by the desulfurizer price, while the $CaCO_3$ is more sensitive to the fluctuation of electricity price, and the CaO is between the two. In Liaoning Province, the overall economy of $CaCO_3$ is relatively poor in the whole range of electricity price, while CaO and MgO perform better. The former has obvious advantages at low electricity price, while the latter is preferred at high. With the emission standard becoming stricter, the dominant range of CaO is expanded. When the boiler load changes, the power consumption cost and desulfurizer cost also change accordingly under different emission requirements. When the electricity price of Liaoning Province is 0.8 RMB/kWh, no matter under which emission limit, the operation cost of MgO is always the lowest under the considered loads, followed by CaO, and the economy of $CaCO_3$ is the worst.

1. Introduction

As energy conversion equipment, industrial boilers have been playing an important role in industrial manufacturing and society development. Different from power plant boilers, industrial boilers are used in a large number of areas and widely distributed throughout the country. At present, about 600,000 industrial boilers have been put into operation in China, which can be divided into coal-fired,oil-gas-fired, electric heating and biomass industrial boilers according to the different fuels. Among them, the number of coal-fired boilers is the largest, about 500,000, accounting for about 80% of the total [1].

The widespread use of coal-fired industrial boilers and the consequential SO_2 emission from coal combustion have contributed to the serious environmental contamination [2, 3], which are also some of the main reasons for the serious haze that has continued to occur in recent years in China [4]. Moreover, industrial boilers are mostly located in the city or suburb with usually low chimneys, so the impact on the environment and human health should not be overlooked [5]. Faced with this situation, China has become more stringent in the formulation of pollutant emission standards for coal-fired boilers, and the ultra-low emission has been proposed as the strictest emission standard currently. It is required by the ultra-low emission standard that after processing, the concentration of SO_2 produced from coal-fired boilers should be less than 35 mg/m^3, for NOx, it's 50 mg/m^3, and PM is 5 mg/m^3 [6].

For its low operation cost, high desulfurization efficiency and almost no water consumption, wet flue gas desulfurization is the most universal and mature desulfurization technology [7], which accounts for about 85% of the desulfurization units built in the world [8]. However, as a complex system bringing huge environmental benefits, the wet flue gas desulfurization unit is also a high energy consumption system. Therefore, for the existing wet flue gas desulfurization system, the energy-saving operation is of great significance to improve its economic performance. At present, there are many studies on energy consumption and economic analysis of wet flue gas desulfurization, and the influence of desulfurization materials, operation of equipment, coal quality and etc. on energy consumption and cost are discussed in depth. Zhou et al. [9] carried out the on-the-spot investigation on more than ten power plants, summarized the common problems of energy consumption and operation of desulfurization system, and gave the optimum feasible scheme according to the actual operation of the power plant. Xue et al. [10] analyzed the energy consumption of desulfurization system from three aspects: energy consumption of desulfurization equipment, selection of coal types and chemical technology, and gave suggestions on optimization of energy consumption. Xu et al. [11] investigated the operation characteristics of various equipment in desulfurization system, and established mathematical models such as fitting curve to study their characteristics, and provided guidance for reducing the energy consumption of desulfurization system from quantitative analysis. Lu [12] established the energy consumption model of the main equipment used in the wet flue gas desulfurization system and optimized the operation of the main energy-consuming equipment such as oxidation fan and circulating slurry pump. Zhang [13] focused on the cost and benefit of the wet flue gas desulfurization system and discussed the influence of generation hours, electricity price and raw material cost on the economy. Wang [14] put forward a weighted model of desulfurization cost by analyzing the factors affecting the cost of desulfurization. Shi [15] took unit capacity, sulfur content, desulfurization efficiency, annual utilization time, furnace type, coal quality, and regional difference into consideration, and established a cost-benefit calculation model of the limestone-gypsum method. Wang [16] of Shandong University proposed a calculation method of energy consumption based on power consumption and studied the influence of operation parameters, flue gas parameters and technical parameters on the energy efficiency of desulfurization tower under ultra-low emission condition.

Although there have been many studies on the energy consumption and economy of wet flue gas desulfurization, greater challenges for its economic operation have been brought by ultra-low emission standard[17]. The discharge concentration of pollutants are greatly reduced by ultra-low emission, however, the energy consumption of the unit is also increased in the process of transformation and emission reduction. Nowadays, the implementation of the ultra-low emission standard is mainly realized by upgrading the existing technology and equipment or optimizing the system, and the specific implementation process is also related to the coal quality, boiler units and the region where it's located [18]. In this situation, there are few energy efficiency studies and economic optimization for wet flue gas desulfurization of coal-fired industrial boilers. At the same time, due to the difference of electricity cost, original pollutant emission concentration, and treatment methods of products between industrial boilers and utility boilers, the research results of coal-fired boilers in thermal power plants can't be directly applied to industrial ones. Thus, it is necessary to establish energy consumption and economic evaluation of wet flue gas desulfurization system for coal-fired industrial boilers.

In this paper, the economic analysis of three typical desulfurizers for wet flue gas desulfurization of coal-fired industrial boilers will be carried out, and the effects of desulfurizer price, electricity price and load variation on operation cost under different emission standards will be discussed, so as to provide guidance for energy saving and emission reduction of industrial coal-fired boiler enterprises.

2. Typical wet flue gas desulfurization system for industrial boilers

Typical wet flue gas desulfurization system for industrial boilers usually consists of flue gas subsystem, SO_2 absorption subsystem, absorbent reserve and supply subsystem, product treatment subsystem, feed water subsystem, and wastewater treatment subsystem.

Flue gas is introduced into the wet desulfurization system by the flue gas subsystem so that SO_2 can react with desulfurizer in the system. A booster fan or induced draft fan is usually used to boost pressure of the flue gas before it enters absorber. Then the liquid drops in the flue gas after SO_2 removal are eliminated by demister and discharged through the chimney.

SO_2 absorption subsystem including absorber, demister, circulating slurry pump and oxidation fan is the main place of SO_2 reacting with desulfurizer. Different liquid-gas ratio (L/G) are adopted according to the different desulfurization efficiencies, and 2~4 spray layers are usually used for a certain circulating slurry. Because the flow rate through the nozzle and atomization pressure are fixed values, one pump is usually used to correspond to one spray layer, or one pump is used to control two layers by combining frequency conversion and throttling adjustment. In this paper, the first method is adopted.

Typical wet flue gas desulfurization processes used in industrial boilers generally do not have a pulverizing system, and desulfurizers are purchased in powder. Desulfurizers are usually alkaline compounds used in SO_2 and SO_3 removal, including lime (the main component is CaO), slaked lime (the basis is $Ca(OH)_2$), limestone (the base is $CaCO_3$), magnesium oxide(MgO), magnesium hydroxide($Mg(OH)_2$), soda ash(Na_2CO_3), carbide slag(the basis is $Ca(OH)_2$) and so on. In this paper, three different desulfurizers of CaO, $CaCO_3$ and MgO will be studied.

The solid content of slurry discharged from the outlet of desulfurization circulating pump is about 10%~15%. The slurry flows through the slurry pipe to the sedimentation tank, where slurry clarified liquid after sedimentation is pumped back to the circulating pool, and the precipitated product is discharged. If the desulfurization products are to be used, further dehydration can be carried out. The product dehydration subsystem generally includes first and second vacuum dehydration. After treatment, the desulfurization products with water content less than 10% are generally stored in the storage room, which is convenient for transportation. But not all industrial boilers dehydrate and reuse the desulfurization products, and some by-products which are difficult to be used may be discarded.

The feed water subsystem is mainly responsible for the water supply during the operation of the wet flue gas desulfurization system. Generally, the water is introduced from the original water supply system of the boiler and then sent to users through pumps, such as the flushing water of the demister during operation, and start-up and replenishment water of the slurry pond. The flushing water for all slurry conveying and storage equipment is provided by the feed water subsystem.

Normally, the wastewater produced in the operation of the system is lower in pH value and contains many heavy metals, which don't meet the environmental protection requirements of discharge. It must be treated to reduce environmental hazards. Wastewater can be mixed with alkaline ash slurry in the ash removal system, or directly introduced into professional wastewater treatment equipment. After professional physical sedimentation, chemical neutralization and other processes, the wastewater whose pH value and element content reach certain standard can be reused in the process water subsystem.

3. Test conditions

In this paper, the basic, special and ultra-low emission standards are considered, and the limits for SO_2 under the three standards are 400 mg/m^3, 200 mg/m^3, and 35 mg/m^3, respectively.

The following information got from a heating industrial coal-fired boiler located in Shenyang City are used in the next investigation. The boiler capacity is 70 MW with 0.29% sulfur content of the designed coal. And the flue gas volume is 63.89 m^3/s and the original SO_2 emission concentration is 644.96 mg/m^3 as the exhaust temperature is 150 ℃ at the 100% boiler load. When the basic, special and ultra-low emission standards are considered respectively, the SO_2 removal amounts are correspondingly 56.34 kg/h, 102.34 kg/h, and 140.29 kg/h. Due to the different consumption and unit price of desulfurizers and the electricity cost fluctuation caused by the change of flow rate and circulating pump's power, the total cost of desulfurization will change subsequently. Therefore, the economy of the system under different desulfurizer price, electricity price and changing load is analyzed.

4. Result and discussion

4.1 The effect of desulfurizer price on the cost

The influence of different desulfurizers on the technical parameters of desulfurization is shown in table1. It's can be seen from table1 that when the desulfurization efficiency is assumed as 90%, the L/G of different desulfurization methods are quite different. Because the dissolving alkalinity of magnesium ion is stronger than that of calcium ion, the reaction ability of which can be more than ten times of that of calcium ion, the reaction intensity of MgO with SO_2 is higher. The molecular mass of MgO is 40 g/mol, CaO and $CaCO_3$ are 56 g/mol and 100 g/mol, respectively. Therefore, the amount of MgO is 73% of CaO and 40% of $CaCO_3$ when reacting with the same amount of SO_2.

Table 1 Technical indicators of desulfurization methods under different emission standards

Emission standard	Desulfurization efficiency	Desulfurization method	L/G (L/m^3)	Ca(Mg)/S	The pH of circulating slurry
		Lime	>5	<1.10	5.0~7.0
Basic	>90%	Magnesium	>2	<1.05	5.0~7.0
		Limestone	>10	<1.05	5.0~6.0
		Lime	>6	<1.10	5.0~7.0
Special	>90%	Magnesium	>2.5	<1.05	5.0~7.0
		Limestone	>12	<1.05	5.0~6.0
		Lime	>8	<1.10	5.0~7.0
Ultra-low	>90%	Magnesium	>4	<1.05	5.0~7.0
		Limestone	>16	<1.05	5.0~6.0

The desulfurizer price is highly related to the desulfurizer storage and transportation cost caused by geographical location. The prices of desulfurizers in different regions are shown in table 2. And according to table2, the price range of MgO, CaO and $CaCO_3$ is 650~1350 RMB/t, 500~600 RMB/t and 200~300 RMB/t, respectively.

Table 2 Unit price of desulfurizers in different provinces

Desulfurizer	Heilongjiang (RMB/t)	Jilin (RMB/t)	Liaoning (RMB/t)	Shandong (RMB/t)
MgO	850	780	1350	650
CaO	600	580	500	550
$CaCO_3$	300	300	250	200

The desulfurizer costs are calculated under the basic, special and ultra-low emission limits and the results are shown in figure 1, figure2 and figure3.

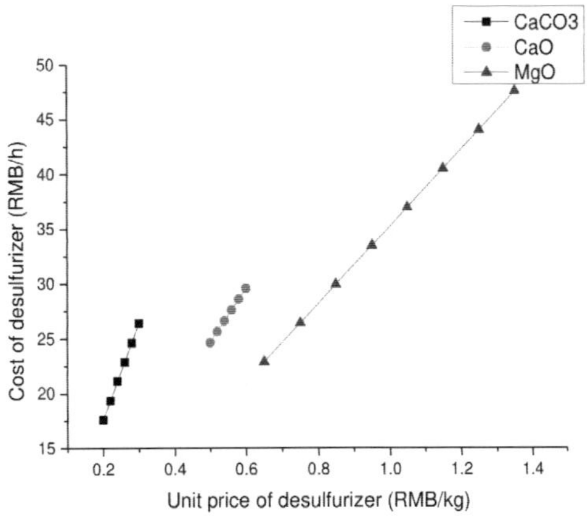

Figure 1 Desulfurizer cost under basic emission standard

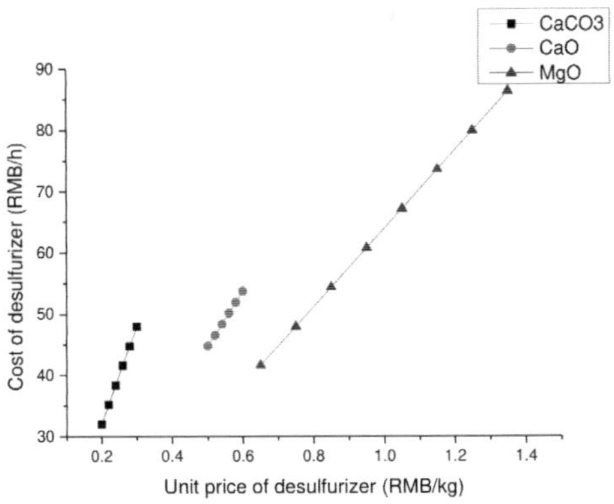

Figure 2 Desulfurizer cost under special emission standard

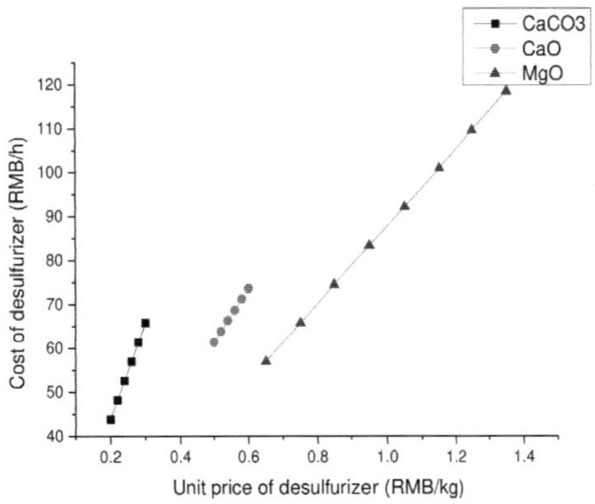

Figure 3 Desulfurizer cost under ultra-low emission standard

As can be seen from the figures, the desulfurizer costs of different desulfurizers increase with the stringent emission standards, but the slope of the same desulfurizer is the same under different emission standards. Because the quality of desulfurizer reacting with the same quality of SO_2 is $CaCO_3$ > CaO > MgO, the slope of $CaCO_3$ is the largest, followed by CaO, and MgO is the smallest. Under the three emission standards, the desulfurizer costs of the three desulfurizers all increase. But because of the highest unit price of MgO, its desulfurizer cost is relatively high, ranging from 22.89~47.54 RMB/h under the basic emission standard to 41.58~86.35 RMB/h under the special, until 56.99~118.37 RMB/h under the ultra-low. The unit price of $CaCO_3$ is the cheapest, so its desulfurizer cost of 17.61~65.76 RMB/h is relatively low and that of calcium oxide of 24.65~73.65 RMB/h is medium.

4.2The effect of electricity price on the cost

For improving the utilization rate of equipment and energy efficiency and reducing electricity consumption, many factories and enterprises choose to make use of the peak-valley tariff. When different SO_2 emission standards are taken into consideration, the value of L/G will change and the power consumption of the desulfurization system may be variable. The power consumptions of different desulfurizers under the three emission limits are shown in table3, table4 and table5.

Table3 Electricity consumption under basic emission limits

Desulfurizer	Circulating slurry volume (m^3/s)	Pump power (kW)	The resistance of flue gas side (Pa)	Fan power (kW)	Total power (kW)

MgO	127.78	36.03			112.70
CaO	319.45	90.08	1200	76.67	166.75
$CaCO_3$	638.9	180.17			256.84

Table4 Electricity consumption under special emission limits

Desulfurizer	Circulating slurry volume (m^3/s)	Pump power (kW)	The resistance of flue gas side (Pa)	Fan power (kW)	Total power (kW)
MgO	159.72	45.04			140.88
CaO	383.34	108.10	1500	95.84	203.94
$CaCO_3$	766.68	216.20			312.04

Table5 Electricity consumption under ultra-low emission limits

Desulfurizer	Circulating slurry volume (m^3/s)	Pump power (kW)	The resistance of flue gas side (Pa)	Fan power (kW)	Total power (kW)
MgO	255.56	72.07			187.09
CaO	511.12	144.13	1800	115.02	259.15
$CaCO_3$	1022.24	288.27			403.29

Considering the large range of electricity price in factories and enterprises, the electricity costs of different desulfurizers are calculated in the scope of 0.2~1 RMB/kWh, and the results are shown in figure4, figure5 and figure6.

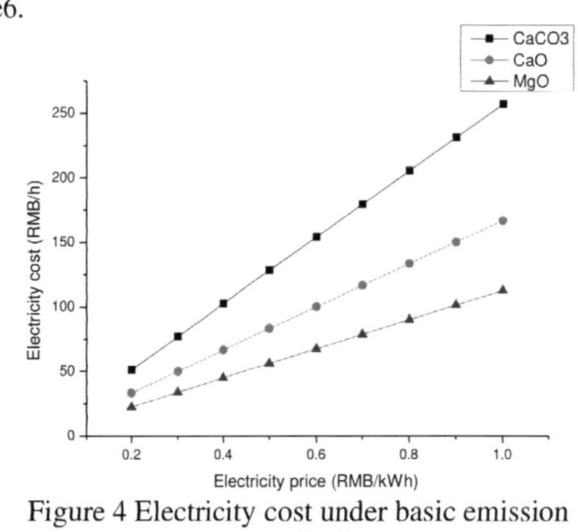

Figure 4 Electricity cost under basic emission standard

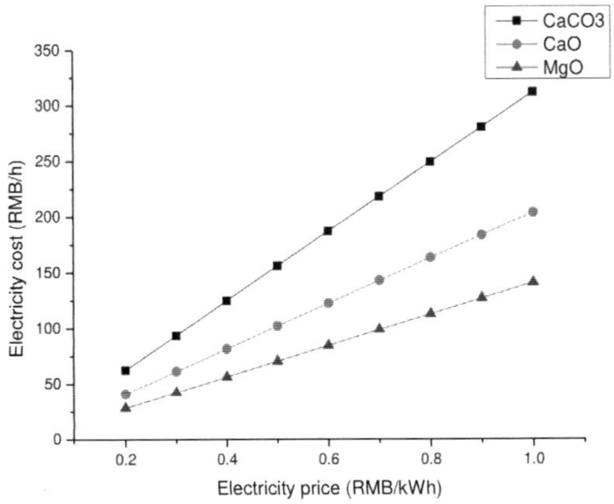

Figure 5 Electricity cost under special emission standard

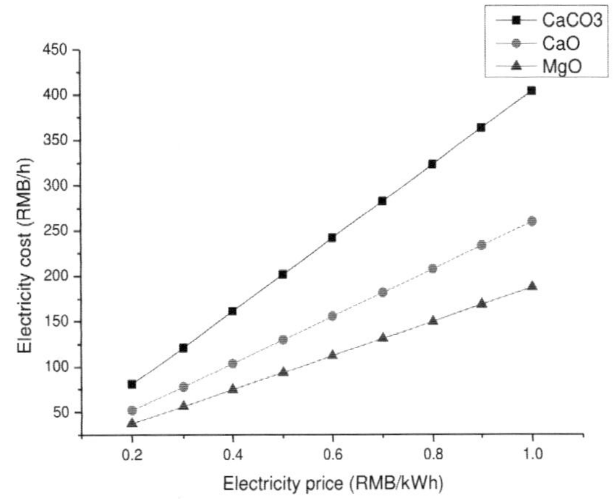

Figure 6 Electricity cost under ultra-low emission standard

Figure4, figure5 and figure6 demonstrate that with the strictness of emission standards, the electricity costs of MgO, CaO and $CaCO_3$ all increase substantially. Among them, the electricity cost of MgO increases the most with the strictness of emission standards, the increasing rate of which from the basic to the special is 25.01%, and 32.80%from the special to the ultra-low, and the total growth rate was 66.01%. The electricity cost for CaO increases by 55.41% from the basic to the ultra-low, and for $CaCO_3$, it's 57.02%. Under the basic, special and ultra-low emission limits of SO_2, the electricity costs of the three desulfurization methods grow with the electricity price climbing, and the range of the $CaCO_3$ is the largest, which indicates that the electricity cost of the limestone method is the most sensitive to the change of electricity price. And the electricity cost of MgO has a minimum increase with the rise in electricity price, showing that the magnesium method is the least sensitive to electricity price and has better adaptability. With the fluctuation of electricity price, the electricity cost of MgO is always the lowest among the three methods no matter which standard is carried out for its smallest circulating slurry volume and the least power consumption.

4.3The operation cost of the desulfurization system

The desulfurizer cost and the electricity cost constitute the total operation cost of the desulfurization system. For the ultra-low emission standard, the total cost of the desulfurization system is given according to the change of desulfurizer cost and electricity cost, and the total costs of the three desulfurizers are shown in Figure 7.

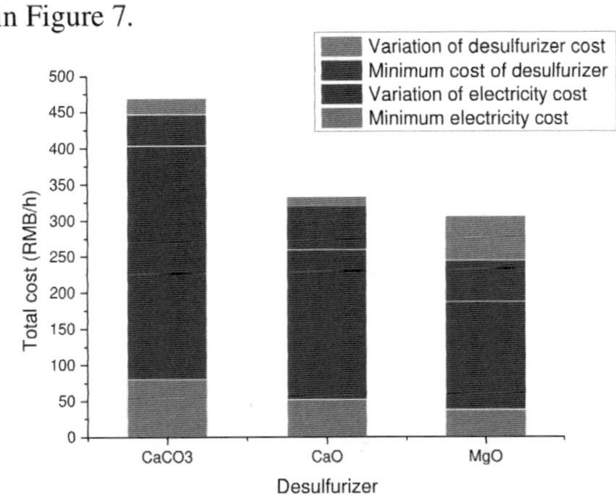

Figure 7 Total cost under ultra-low emission standard

Figure 7 illustrates that it's $CaCO_3$ that has the highest minimum electricity cost and electricity cost variation, which is related to the maximum pump power of the limestone method. Because of the low power consumption of MgO, the minimum electricity cost and the electricity cost variation of MgO are the lowest, and CaO is between the two. However, the minimum desulfurizer cost is not completely arranged according to the unit price of desulfurizer. Because the desulfurizer cost equals the unit price of desulfurizer times the consumption of desulfurizer, it is obvious that the inhibition of the unit price of $CaCO_3$ on the desulfurizer cost exceeds the influence of the desulfurizer consumption, thus, the minimum of desulfurizer cost of $CaCO_3$ is the lowest. While the consumption of desulfurizer of CaO and the unit price of CaO are both relatively higher, which results in its highest minimum value of desulfurizer cost. The variation of desulfurizer cost is equal to the unit price change of desulfurizer times the consumption change of desulfurizer. Therefore, the variation of desulfurizer cost in MgO is maximum should be contributed to its highest unit price change. And it should be the most relevant to the change of consumption that the desulfurizer cost variation in $CaCO_3$ ranks the next. Because the unit price change and the consumption change of desulfurizer in CaO are both smaller, its desulfurizer cost variation is the lowest.

4.4The operation cost of desulfurization under different electricity prices

Based on the provinces referred in section 4.1, when the electricity price is 0.2 RMB/kWh, the unit prices of MgO, CaO and $CaCO_3$ are 650 RMB/t, 500 RMB/t and 200 RMB/t, the total cost of desulfurization system is the smallest. The desulfurizer cost and electricity cost are shown in table 6.

When the electricity price is 0.8 RMB/kWh, the unit prices of MgO, CaO and $CaCO_3$ are 650 RMB/t, 500 RMB/t and 200 RMB/t, the desulfurizer cost and electricity cost are shown in table 7.

When the electricity price is 0.8 RMB/kWh, the unit prices of MgO, CaO and $CaCO_3$ are 1350 RMB/t, 600 RMB/t and 300 RMB/t, the desulfurizer cost and electricity cost are shown in table 8.

Table 6 Proportion of desulfurizer cost and power consumption cost

Desulfurizer	Desulfurizer cost (RMB/h)	Electricity cost (RMB/h)	The proportion of desulfurizer cost (%)	The proportion of Electricity cost (%)
MgO	56.99	37.42	60.37	39.63
CaO	61.37	51.83	54.22	45.78
$CaCO_3$	43.84	80.66	35.21	64.79

Table 7 Proportion of desulfurizer cost and power consumption cost

Desulfurizer	Desulfurizer cost (RMB/h)	Electricity cost (RMB/h)	The proportion of desulfurizer cost (%)	The proportion of Electricity cost (%)
MgO	56.99	149.67	27.58	72.42
CaO	61.37	207.32	22.84	77.16
$CaCO_3$	43.84	322.63	11.96	88.04

Table 8 Proportion of desulfurizer cost and power consumption cost

Desulfurizer	Desulfurizer cost (RMB/h)	Electricity cost (RMB/h)	The proportion of desulfurizer cost (%)	The proportion of Electricity cost (%)
MgO	118.37	149.67	44.16	55.84
CaO	73.65	207.32	26.21	73.79
$CaCO_3$	65.76	322.63	16.93	83.07

According to the results, when the unit prices of desulfurizers are the same, if the electricity cost is 0.2RMB/kWh, the desulfurizer cost is more than the electricity cost, and if the electricity price is 0.8 RMB/kWh, the opposite has occurred. When the electricity price is fixed, the proportions of desulfurizer costs are also changed when the unit prices of desulfurizers are the most expensive and the cheapest. Therefore, when the electricity price and desulfurizer price fluctuate, there exists a balanced electricity price which makes the total costs of different desulfurizers equal.

For Liaoning Province, the unit price of MgO is 1350 RMB/t, CaO is 500 RMB/t, and $CaCO_3$ is 250 RMB/t. The balanced point of the electricity price affecting the total cost is found and shown in table 9.

Table 9 Total operation cost of different electricity prices under different emission standards

Electricity price RMB/h	Basic			Special			Ultra-low		
	MgO	CaO	$CaCO_3$	MgO	CaO	$CaCO_3$	MgO	CaO	$CaCO_3$
0.2	70.08	58.00	73.38	114.53	85.56	102.39	155.79	113.21	135.46
0.3	81.35	74.64	99.06	128.62	105.96	133.59	174.50	139.12	175.79
0.4	92.62	91.35	124.74	142.70	126.35	164.79	193.21	165.04	216.12
0.5	103.89	108.02	150.43	156.79	146.74	196.00	211.92	190.95	256.45
0.6	115.16	124.70	176.11	170.88	167.14	227.20	230.63	216.87	296.78
0.7	126.43	141.37	201.80	184.97	187.53	258.41	249.34	242.78	337.11
0.8	137.70	158.05	227.48	199.06	207.93	289.61	268.04	268.70	377.43
0.9	148.97	174.72	253.16	213.14	228.32	320.81	286.75	294.61	417.76
1.0	160.24	191.40	278.85	227.23	248.71	352.02	305.46	320.53	458.09

Under the basic, special and ultra-low emission limits, the total costs of different desulfurizers are shown in figure 8, figure 9 and figure 10.

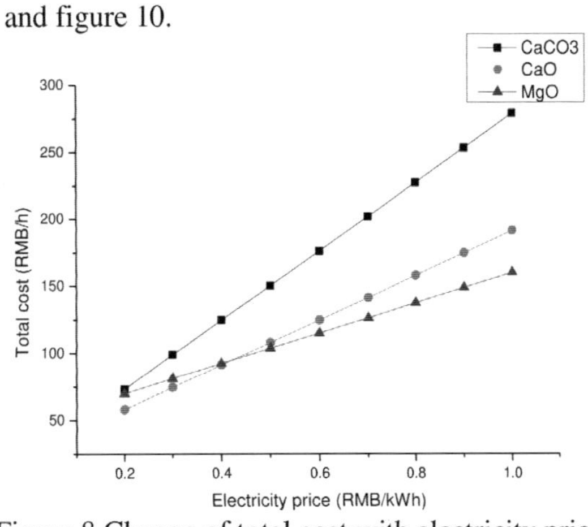

Figure 8 Change of total cost with electricity price under basic emission

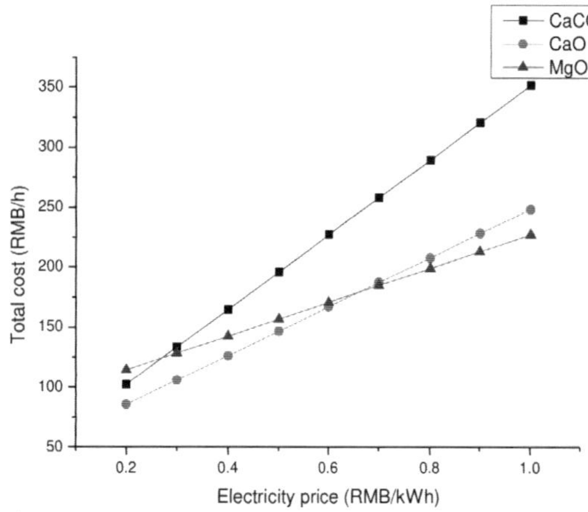

Figure 9 Change of total cost with electricity price under special emission

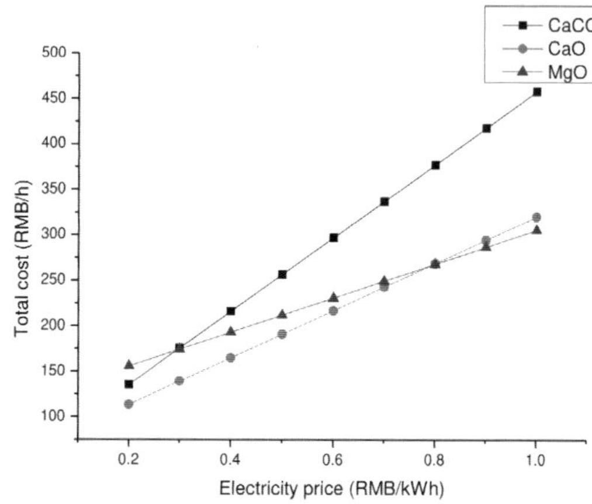

Figure 10 Change of total cost with electricity price under ultra-low emission

It can be seen from figure 8 that when the basic emission limit is implemented in Liaoning Province, the total cost of CaO is the lowest at the scope of lower than 0.42 RMB/kWh, and beyond this value, using MgO is more economical. Within the range of 0.2 RMB/kWh to 1.0 RMB/kWh, the total cost of $CaCO_3$ is always the highest. As figure 9 shows, under the special emission limit, CaO is optimal when the electricity price is lower than 0.66 RMB/kWh, and MgO is more favorable as it exceeds 0.66 RMB/kWh. When the electricity price is beyond 0.27 RMB/kWh, $CaCO_3$ performs worst economically. Figure 10 depicts the total costs of ultra-low emission and it illustrates that 0.79 RMB/kWh is the demarcation point between CaO and MgO and the less costly region lower than 0.79 RMB/kWh is dominated by CaO and otherwise is MgO. The $CaCO_3$ shows obvious disadvantage compared with the other two desulfurizers at the range of 0.29 RMB/kWh to 1 RMB/kWh.

Generally speaking, although the cost of $CaCO_3$ is lowest, its electricity cost is relatively high, especially when the electricity price rises, as it needs a larger circulating slurry volume and a higher pump power. Compared with $CaCO_3$, both CaO and MgO perform better economically. The former has an obvious advantage at low electricity price, while the latter is preferred at high. As the emission limit gets lower, the electricity price level with the same total cost of both increases gradually, from 0.42 RMB/kWh at basic emission to 0.66 RMB/kWh at special, and until 0.79 RMB/kWh of ultra-low, and the range where CaO shows better economy is becoming wider.

4.5 The operation cost of desulfurization under variable load conditions

Under basic, special and ultra-low emission limits, there will be two, three and four spray layers respectively, so the spray flow rate of circulating slurry volume under basic emission limit can be assigned as 50% or 100%, while under the special it can be set as 33.3%, 66.7% or 100%, and under the ultra-low it can be 25%, 50%, 75% or 100%. Taking Liaoning Province into consideration, the unit price of MgO is 1350 RMB/t, CaO is 500 RMB/t and $CaCO_3$ is 250 RMB/t. The total costs under the three emission limits are discussed at four changing loads (100%, 70%, 50% and 25%) when the electricity price is 0.8 RMB/kWh.

When the load is 70%, three layers are opened under the ultra-low emission standard, and all of the layers should be fully opened under both basic and special emission limits. Assuming the load to be 50%, two spray layers should be opened under the special and ultra-low emission standards and one under the basic. Then it comes to 25% load, there should be only one layer to be opened no matter under which limits. The calculation results of electricity cost and desulfurizer cost under different loads are shown in table10, table11, table12 and table13.

Table 10 Desulfurizer cost and electricity cost at 100% load

Emission standard	Desulfurizer	Pump power (kW)	Fan power (kW)	Total power (kW)	Electricity cost (RMB/h)	Desulfurizer cost (RMB/h)	Total cost (RMB/h)
	MgO	36.03	76.67	112.70	90.16	47.54	137.70
Basic	CaO	90.08	76.67	166.75	133.40	29.58	162.98
	$CaCO_3$	180.17	76.67	256.84	205.47	22.01	227.48
	MgO	45.04	95.84	140.88	112.70	86.35	199.06
Special	CaO	108.10	95.84	203.94	163.15	53.73	216.88
	$CaCO_3$	216.20	95.84	312.04	249.63	39.98	289.61
	MgO	72.07	115.02	187.09	149.67	118.37	268.04
Ultra-low	CaO	144.13	115.02	259.15	207.32	73.65	280.97
	$CaCO_3$	288.27	115.02	403.29	322.63	54.80	377.43

Table 11 Desulfurizer cost and electricity cost at 70% load

Emission standard	Desulfurizer	Pump power (kW)	Fan power (kW)	Total power (kW)	Electricity cost (RMB/h)	Desulfurizer cost (RMB/h)	Total cost (RMB/h)
	MgO	36.03	53.67	89.70	71.76	33.28	105.04
Basic	CaO	90.08	53.67	143.75	115.00	20.71	135.71
	$CaCO_3$	180.17	53.67	233.84	187.07	15.41	202.48
	MgO	45.04	67.09	112.13	89.70	60.45	150.15
Special	CaO	108.10	67.09	175.19	140.15	37.61	177.76
	$CaCO_3$	216.20	67.09	283.29	226.63	27.98	254.62
	MgO	54.05	80.51	134.56	107.65	82.86	190.51
Ultra-low	CaO	108.10	80.51	188.61	150.89	51.56	202.45
	$CaCO_3$	216.20	80.51	296.71	237.37	38.36	275.73

Table 12 Desulfurizer cost and electricity cost at 50% load

Emission standard	Desulfurizer	Pump power (kW)	Fan power (kW)	Total power (kW)	Electricity cost (RMB/h)	Desulfurizer cost (RMB/h)	Total cost (RMB/h)
	MgO	18.02	38.34	56.36	45.09	23.77	68.86
Basic	CaO	45.04	38.34	83.38	66.70	14.79	81.49
	$CaCO_3$	90.08	38.34	128.42	102.74	11.00	113.74
	MgO	30.18	47.92	78.10	62.48	43.18	105.66
Special	CaO	72.43	47.92	120.35	96.28	26.86	123.14
	$CaCO_3$	144.85	47.92	192.77	154.22	19.99	174.20
	MgO	36.04	57.51	93.55	74.84	59.19	134.03
Ultra-low	CaO	72.07	57.51	129.58	103.66	36.83	140.49
	$CaCO_3$	144.14	57.51	201.65	161.32	27.40	188.72

Table 13 Desulfurizer cost and electricity cost at 25% load

Emission standard	Desulfurizer	Pump power (kW)	Fan power (kW)	Total power (kW)	Electricity cost (RMB/h)	Desulfurizer cost (RMB/h)	Total cost (RMB/h)
	MgO	18.02	19.17	37.19	29.75	11.88	41.64
Basic	CaO	45.04	19.17	64.21	51.37	7.39	58.76
	$CaCO_3$	90.08	19.17	109.25	87.40	5.50	92.90
	MgO	14.86	23.96	38.82	31.06	21.59	52.64
Special	CaO	35.67	23.96	59.63	47.70	13.43	61.14
	$CaCO_3$	71.35	23.96	95.31	76.25	9.99	86.24

	MgO	18.02	28.76	46.77	37.42	29.59	67.01
Ultra-low	CaO	36.03	28.76	64.79	51.83	18.41	70.25
	$CaCO_3$	72.01	28.76	100.82	80.66	13.70	94.36

As shown in figure 11, the total costs under 100%, 70%, 50% and 25% of loads are compared.

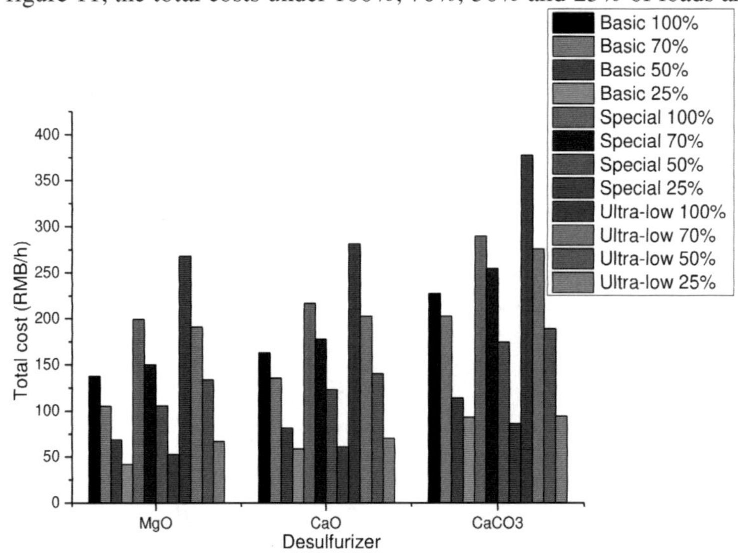

Figure 11 Comparison of total cost under four load conditions and different emission standards

Under the three standards, with the decrease of load, the consumption of desulfurizer and power both decrease, and the total cost decreases gradually. For MgO and CaO, the total cost increases with the strictness of emission standards under any load, but for $CaCO_3$, the total cost under special emission standards is lower than that under the basic at 25% load. When the load is 25%, the flow rate of the spray layer can only be assigned as 50% under the basic limit, and 33.3% under the special, so the power consumption of the pump decreases by a much greater extent than that of the fan, resulting in the reduction of the total power consumption and a significant decrease of the power consumption cost, offsetting the increase in the desulfurizer cost, thus reducing the total cost. For the three desulfurizers, no matter under which discharge standard and which load, the operation cost of MgO is always the lowest, followed by CaO, and the economy of $CaCO_3$ is the worst.

When the boiler load is 70%, the spray layer of the absorber can only be set to 100% for meeting the basic and special emission limits, while the spray layer can be set to 75% for ultra-low emission, which plays an important role in energy conservation. The situation is similar when the boiler load is 25%. Therefore, when the boiler load changes, there will be some dynamic changes in the cost of electricity and desulfurizer under different emission requirements. For industrial enterprises, desulfurizer should be selected reasonably according to desulfurizer price, electricity price, emission standard and load range.

5. Result and discussion

According to the actual situation of the Northeast Three Provinces and Shandong Province, this paper analyses the influence of the unit price of desulfurizer and electricity price on the operation cost of three typical wet desulfurization systems under different emission limits. Taking Liaoning Province as an example, guidance is given for the selection of desulfurizer under different electricity prices and boiler loads. Because of the high unit price of MgO, the operation cost of magnesium method is greatly affected by unit price compared with electricity price. The higher circulating slurry volume of limestone method determines that it is more sensitive to the fluctuation of electricity price, and the lime method is between the two. When electricity price fluctuates, the overall economy of $CaCO_3$ in Liaoning Province is relatively poor, while the CaO and MgO are more economical. The former has obvious advantages in lower electricity price, while the latter is preferred at higher electricity price. With the strictness of emission standards, the electricity price level with the same total cost of the two methods gradually increases, from 0.42 RMB/kWh at the basic emission to 0.66 RMB/kWh at the special and until 0.79 RMB/kWh at the ultra-low, the range where the operation cost of CaO is lowest becomes much wider. As the boiler load changes, there will be some dynamic changes in the electricity cost and desulfurizer cost under different emission requirements. When the electricity price of Liaoning Province is 0.8 RMB/kWh, under the four load conditions discussed, the operation cost of MgO is always the lowest, followed by CaO, and $CaCO_3$ is the most costly. When the boiler load is 70%, the spray amount can only be set to 100% for basic and special emission limits, while it can be set to 75% for ultra-low emission, thus realizing energy conservation. The situation is similar when the boiler load is 25%. Therefore, under ultra-low emission, the electricity cost can be reduced by flexibly adjusting the spray amount to meet the desulfurization requirements under different loads.

Acknowledgments
This work was financially supported by the National Key Research and Development Project (grand number: 2017YFF0209805).

References

[1] Wu Z B, Weng X L, Wang Y J. (2017). Ultra-low Emission Control Technology for Coal-fired Industrial Boilers [J]. China Science and Technology Achievements,18(1):24-24.
[2] Cui SH, Hao RL, Fu D. (2019). An integrated method of non-thermal plasma combined with catalytical oxidation for simultaneous removal of SO2 and NO [J]. Fuel, 246:365-374.
[3] Wang M, Du Q, Li YP, Xu JJ, Gao JM, Wang H. (2019). Effect of steam on the transformation of sulfur during demineralized coal pyrolysis[J]. Journal of Analytical and Applied Pyrolysis, 140:161-169.
[4] Caiazzo G, Nardo A D, Langella G, et al. (2012). Seawater scrubbing desulfurization: A model for SO2 absorption in fall-down droplets[J]. Environmental Progress & Sustainable Energy. 31(2):277-287.
[5] Dilinuer A. (2016). Causes of haze pollution in cities and its control measures [J].Business, 3:57.
[6] National Development and Reform Commission, Ministry of Environmental Protection of the People's Republic of China, National Energy Administration. (2015). Circular on the Issuance of the Action Plan for the Upgrading and Renovation of Coal and Electricity Energy Conservation and Emission Reduction (2014-2020). http://bgt.ndrc.gov.cn/zcfb/201409/t20140919_626242.html
[7] Cheng G, Zhang C X, Zhang X M, Zhang Z L.(2018). Preparation of lignite-based activated carbon and its flue gas desulfurization performance. https://iopscience.iop.org/article/10.1088/2053-1591/aaebeb
[8] Desulfurization and Denitrification Committee of China Environmental Protection Industry Association. (2013). Summary of the development of desulfurization and denitrification industry in China in 2012[J]. Environmental Protection Industry in China, 7: 8-20.
[9] Zhou X H, Luo W B, Liao Y J. (2009). Operation Analysis and Optimization Suggestions for Flue Gas Desulfurization Devices of Thermal Generating Units [J]. Guangdong Electric Power, 2(8):52-56.
[10] Xue J M, Liu T, Xu Y Y, et al. (2009). Key factors affecting energy saving in wet flue gas desulfurization system[J]. China Electric Power, 42(8):50-52.
[11] Xu G, Yang Y, Wang N, et al. (2010). Analysis on Energy Consumption and Optimal Operation of FGD System in Power Plant[C]. In: Power&Energy Engineering Conference. Chengdu. pp. 185-193.
[12] Lu S Y. (2011). Energy-saving Analysis of the flue gas desulfurization system and research on

Comprehensive Performance Evaluation Method in Large Coal-fired Units [D]. North China Electric Power University (Beijing).
[13] Zhang C Q. (2012). Analyzing on environmental quality improving and economical efficiency by wet flue gas desulfurization system in power plant [D]. North China Electric Power University.
[14] Wang X. (2013). The efficiency and cost of desulfurization multi-objective optimization method and application research [D]. Central South University.
[15] Shi J Y. (2015). Cost-benefit analysis of flue gas desulfurization and denitrification technology in coal-fired power plants [D]. Zhejiang University.
[16] Wang Y C. (2018). Study on energy efficiency of wet flue gas desulphurization tower under ultra-low emission conditions [D]. Shandong University.
[17] Stanislaus A, Marafi A, Rana M S. (2010). Recent advances in the science and technology of ultra-low sulfur diesel (ULSD) production[J]. Catalysis Today, 153:1-68.
[18] Xin S W. (2017).Study on modification of SO2 ultra-low emission in CFB boiler[J]. Electric Power Technology & Environmental Protection, 33(4):10-13.

Discussion on Relevant Issues of Thermal performance test code for industrial boilers

Chang Yongqiang[1*], Li Tiankui[2] and Li Peng[1]

[1]China Special Equipment Inspection and Research Institute, Beijing 100029

[2]Qingdao Special Equipment Inspection and Testing Institute, Qingdao 266071

[*]Corresponding author's e-mail: chayoqi@163.com

Abstract: In the study and application of GB/T 10180-2017 "Thermal performance test code for industrial boilers", the author found some unreasonable or error-prone areas in the rules. In this paper, the value of boiler thermal efficiency, the test method of condensing boiler, the correction method of boiler thermal efficiency and the calculation of desulfurization thermal efficiency are discussed, and personal suggestions are put forward.

1. Preface

GB/T 10180-2017 "Thermal performance test code for industrial boilers" (here in after referred to as GB/T 10180-2017) was issued in 2017 and implemented on February 1, 2018. The standard specifies the terms and definitions, symbols and units, general principles, test preparation, test requirements, measurement items and test instruments, test methods, calculation of boiler thermal efficiency and test reports in the thermal performance test of industrial boilers [1]. In the process of learning and using the regulations, some unreasonable or error-prone places in the regulations have been found. This paper discusses these problems.

2. Discussion on Relevant Issues

2.1Boiler thermal efficiency adopts the average value of direct and indirect balance

Direct balance measurement is a method of directly measuring input heat and output heat to determine boiler thermal efficiency, also known as direct measurement or input-output method. According to the definition of direct balance measurement method, we can conclude that the advantages of direct balance measurement method are: direct measurement to determine the main parameters of efficiency (input and output); Less measurement is required; There is no need to estimate the loss that cannot be measured. The disadvantages of direct balance measurement method are: accurate measurement of fuel quantity, fuel calorific value, steam flow rate and properties is needed to minimize uncertainty and cannot help analyze the reasons for low efficiency; It is required to revise the experimental results to the standard or guarantee conditions, which can only be done by using the energy balance method.

Indirect balance measurement method is a method to determine the boiler thermal efficiency by measuring the heat loss of various combustion products and the heat dissipation loss of the boiler. It is also called indirect measurement method, heat loss method or energy balance method. According to the definition of the indirect balance measurement method, we can conclude that the advantages of the indirect balance measurement method are: it can measure the main quantities (flue gas analysis and flue gas temperature) very accurately; The experimental results can be corrected to standard or guarantee conditions. Since the measured (various losses) only account for a small share of the total energy, the uncertainty of the efficiency of the experiment is small. The error between the secondary measured value and the estimated value has little influence. The source of the larger loss can be confirmed. The disadvantages of the indirect balance measurement method are: more measurement is needed; Evaporation and output data cannot be obtained automatically; Some losses are actually unmeasurable and their values must be estimated.

Section 5.4 of GB/T 10180-2017 stipulates that the boiler thermal efficiency shall be measured by both the direct balance measurement method and the indirect balance measurement method. The boiler thermal efficiency value shall take the average value of the results measured by the direct balance measurement method and the indirect balance measurement method. However, from the above analysis of the direct balance measurement method and the indirect balance measurement method, it can be seen that the direct balance measurement method and the indirect balance measurement method are two different test methods, and the test contents and test principles are different, so the boiler thermal efficiency tested by the two different test methods cannot be compared. In addition, according to the uncertainty analysis of boiler thermal efficiency made by Zhu Chunmei [2], it can be seen that the uncertainty result of boiler thermal efficiency calculated by the direct balance method is 4.9783, the uncertainty result of boiler thermal efficiency calculated by the indirect balance method is 0.4785, and the uncertainty result of boiler thermal efficiency calculated by the direct balance method and the uncertainty result of boiler thermal efficiency calculated by the indirect balance method are obviously not at the same level. Therefore, no matter from the perspective of testing principle or uncertainty analysis, the author believes that the boiler thermal efficiency should not be the average value of the results measured by the direct balance method and the indirect balance measurement method.

2.2Test Methods for Condensing Boilers

Section A.2 of Appendix A to GB/T 10180-2017 stipulates that the boiler thermal efficiency calculation for flue gas condensation and heat recovery within the boundary of the boiler thermal balance system is obtained by adding the thermal efficiency $\eta 1$ of the boiler body and the thermal efficiency $\eta 2$ of the condenser, wherein the thermal efficiency $\eta 1$ of the boiler body is calculated in accordance with Chapter 10 of GB/T 10180-2017, and the thermal efficiency $\eta 2$ of the condenser is calculated by the direct balance method. The method solves the problem that the indirect balance test method cannot be used to test the thermal efficiency of the system where flue gas condensation occurs within the boundary of the boiler thermal balance system in GB/T 10180-2003 "Industrial Boiler Thermal Performance Test Regulations", but there are still some problems in actual application.

This section analyzes the problems of this method from the following aspects:

*2.2.1*The thermal efficiency $\eta 1$ of the boiler body is calculated according to Chapter 10 of GB/T 10180-2017, which requires the flue gas temperature and flue gas composition to be measured at the outlet flue of the boiler body. However, the excess air coefficient at the exhaust flue required by TSG G0002-2010 "Supervision Administration Regulation on Energy Conservation Technology for Boiler" shall be the test data at the boiler outlet flue, and the test position shall be arranged behind the last heating surface of the boiler. Therefore, in order to ensure that the boiler test data meet the requirements of TSG G0002-2010 "Supervision Administration Regulation on Energy Conservation Technology for Boiler" during the test of boiler type products, we need to test the composition of boiler flue gas again in the flue behind the last heating surface of the boiler. This content is not explained in detail in GB/T 10180-2017, and many testers did not notice this problem during the test.

*2.2.2*According to the actual situation encountered in the field test, many gas-fired boilers (especially cast aluminum boilers which are currently used more frequently) have condensation phenomenon when the flue gas comes out of the boiler body without installing condensers. In this case, the boiler thermal efficiency test still cannot be carried out by the indirect balance test method, and cannot meet the relevant provisions of subsection (5.4 c) of GB/T 10180-2017;

*2.2.3*In this calculation method, the thermal efficiency η1 of the boiler body is the average value of the thermal efficiency of the boiler measured by the direct balance measurement method and the indirect balance measurement method, and the thermal efficiency η2 of the condenser is measured by the direct balance measurement method. The calculation method has the problem of chaotic use of two test methods with different levels of uncertainty.

2.3boiler thermal efficiency correction method

2.3.1Correction of inlet air temperature deviation from design value
GB/T 10184-2015 "Performance test code for utility boiler" derives the correction formula of exhaust gas temperature when the inlet air temperature deviates from the design value based on the heat balance of the air heater (assuming that the flue gas side efficiency of the air heater does not change when the inlet air temperature of the air heater changes), and then corrects the boiler thermal efficiency according to the corrected exhaust gas temperature[3].However, GB/T 10180-2017 does not consider the correction of inlet air temperature deviation from the design value.

2.3.2Correction of inlet flue gas temperature deviation from design value
At present, GB/T 10184-2015 "Performance test code for utility boiler" aims at the correction of the deviation of inlet flue gas temperature from the design value [4]. When the inlet feed water temperature deviates from the design value, the exhaust gas temperature is corrected, and then the boiler thermal efficiency is corrected according to the corrected exhaust gas temperature.

However, GB/T 10180-2017 gives an empirical correction method. If the deviation between the feed water temperature of the steam boiler and the design temperature exceeds -20 K, the thermal efficiency of the boiler with a difference of -60 K between the two will be converted by one percentage point. For coal-fired hot water boilers without air heater, the difference between the outlet water temperature and the rated temperature -15 K boiler thermal efficiency decreases by 1 percentage point; For oil-fired and gas-fired hot water boilers without air heater, the difference between the outlet water temperature and the rated temperature is -25 K, and the thermal efficiency of the boiler decreases by one percentage point. At present, the author has not found the experimental basis and theoretical basis of the empirical correction method.

2.3.3Fuel correction
In GB/T 10184-2015 "Performance test code for utility boiler",the corrected losses are calculated by substituting the ultimate analysis and lower heating value of design fuel for corresponding values in all computations. Corrections for fuel are not mentioned in GB/T 10180-2017.

2.4 q3 Calculation in Simple Test of Thermal Efficiency under Operating Conditions
Appendix B.1 of GB/T 10180-2017 refers to table 1 for the values of incomplete combustion loss of gas in the simple test of thermal efficiency of boiler operating conditions.

Table 1. Heat Loss of Incomplete Combustion of Gas (q3)

Project	Unit	Value		
CO	%	CO≤0.05	0.05<CO≤0.1	0.1<CO
q_3	%	0.2	0.5	1

In the actual measurement process, many boilers, especially gas-fired boilers, due to poor air distribution adjustment, result in small air distribution volume and large amount of CO production, many reaching more than 1%, some even reaching 10%, and the actual incomplete combustion loss of gas reaches more than 1%, some even more than 10%. However, according to the values in the table, the incomplete combustion loss of gas can only be taken as 1%, which is quite different from the actual value. In addition, when CO is 0, the incomplete combustion loss of gas in the table is 0.2%, but it is actually 0.There is a big error in calculating the incomplete combustion loss of gas by using the value method, which cannot accurately reflect the actual combustion condition of the boiler[5]. It is suggested to use empirical formula to calculate:

$$q_3 = 0.11(\alpha_{py} - 0.6) \times 30.2CO \tag{1}$$

The formula is adopted for calculation, which not only does not increase the workload, but also can more accurately reflect the actual operation of the boiler.

2.5Calculation of Desulfurization Thermal Efficiency

During the calculation of circulating fluidized bed with desulfurizer added according to appendix I of GB/T 10180-2017, the author found that the excess air coefficient calculated by excess air coefficient after adding desulfurizer in appendix I.3 of GB/T 10180-2017 was wrong. After consulting relevant documents, it is found that the calculation method of excess air coefficient after adding desulfurizer in GB/T 10180-2017 comes from ASME PTC4-2013 Fired Steam Generators Performance Test Code[6].After comparing and analyzing the relevant calculation formulas in the two standards, it is found that the formulas (I.18), (I.20) and (I.21) in GB/T 10180-2017 are incorrect.

Among them, formula (I.18) is a citation error. ASME PTC4 standard uses excess air rate to represent excess air. the actual delivered air quantity minus the theoretical required air quantity divided by the theoretical air quantity and multiplied by 100 is used as excess air rate. GB/T 10180-2017 uses excess air coefficient to represent excess air, and the ratio of actual used air quantity to theoretical air quantity is used as excess air coefficient. The calculation of excess air coefficient in appendix I of GB/T 10180-2017 confuses the excess air rate with the excess air coefficient, and no relevant conversion has been carried out in the process of quoting the ASME PTC4 formula. Therefore, the correct expression of formula (I.18) should be (2). If there are publishing errors in formula (I.20) and formula (I.21), the correct expressions should be (3) and (4).

$$\alpha_{ds} = \frac{O_2\left(mo_{gycw} + 0.795mo_{lk}\right)}{mo_{lk}\left(20.95 - O_2\right)} + 1 \tag{2}$$

$$mo_{tlgy} = \frac{CaCO_{3,ar}}{10008.6}\frac{B_{shs}}{B} \tag{3}$$

$$mo_{lk} = \frac{0.1151C_{ar}^{r} + 0.3430H_{ar} + 0.0431S_{ar}\left(1 + 0.5\eta_{SO2}\right) - 0.0432O_{ar}}{28.963} \tag{4}$$

3. Conclusion

According to the actual working conditions, this paper discusses the problems of boiler thermal efficiency value, condensing boiler test method, boiler thermal efficiency correction method, q3 calculation in simple test of thermal efficiency under operating conditions and desulfurization thermal efficiency calculation in GB/T 10180-2017 "Thermal performance test code for industrial boilers", and puts forward personal suggestions. The author thinks that the research on the testing and calculation methods of boiler energy efficiency should be strengthened so that the testing and calculation results can more truly reflect the actual situation of the boiler and better guide the energy saving work of the boiler.

Acknowledgments
This project was supported by CSEI Research Program.

References

[1] GB/T 10180-2017.Thermal performance test code for industrial boilers[S]. General Administration of Quality Supervision, Inspection and Quarantine of the People's Republic of China.
[2] Zhu C.M. The Economic Performance Index Calculation of Thermal Power Based On Uncertainty Analysis [D]. North China Electric Power University(HeBei),2010.
[3] Zhu C.W, Yan Z.S. (2011) Discussion on Improvement of Correction Method of Outgoing Gas Temperature in GB10184-88 Codes[J]. Boiler Technology.42(05):56-59.
[4] Chang Y.Q,Liu X.M,Qi G.L, et al. (2018)Comparison Research of Chinese Boiler Performance Test Codes and ASME PTC 4-2013[J].Journal Of Chinese Society Of Power Engineering,2018,38(08):610-616+632.
[5] Cheng J, Xie C.H, Wu J.Q, Chen H.(2013)Relevant Issues of "energy efficiency test and evaluation regulation for industrial boiler"[J].China Special Equipment Safety,29(06):50-51.
[6] The American Society of Mechanical Engineers. ASME PTC 4-2013.Fired Steam Generators Performance Test Codes[S]. New York:The American Society of Mechanical Engineers,2014.

Prediction of flue gas acid dew point temperature distribution range in boilers

Cheng Bi[1]*, Debiao Li[1], Xu Yang[1], Yuan Lu[1], Jianlong Zhang[1, 2]

[1]Xi'an Special Equipment Inspection Institute, Xi'an, Shaanxi, 710065, P.R. China

[2]College of Mechanical Engineering, Xi'an University of Science and Technology, Xi'an, Shaanxi, 710054, P.R. China

*Corresponding author's e-mail: chengbi_doc@163.com

Abstract. Prediction of the acid dew point temperature of flue gas has a great effect on the heat exchangers design or control as recover energy from the boiler exhaust gas. The present study focuses on the prediction of the flue gas acid dew point temperature as well as its distribution interval, which can further help to prevent and resist the low temperature corrosion from the flue gas. Popular prediction models of acid dew point are reviewed. The 1973 version of Soviet Union calculation standard model has a priority to calculate the mean acid dew point prediction in coal fired boilers, and the Okkes model is also recommended for different types of boilers. Several groups of experimental and published data are substituted to different prediction models for analysis, and the results show that the Verhoff & Banchero model and Haase & Borgmann model can be explored to calculate the upper and lower limits of the distribution interval of the acid dew point.

1. Introduction

The direct feature of boiler convective exchangers, such as economizer and air pre-heater, is to reduce the temperature of the exhaust gas for energy recovery [1-3]. Generally, a 10-15 ℃ temperature drop of boiler exhaust gas can lead to an increase of boiler thermal efficiency by 1% [4]. However, it is impossible to install too many exchangers for a perfect energy saving, because that the acid gas in the flue gas together with water vapor will generate liquid acid when the flue gas temperature is cooled by the exchanger below the acid dew point temperature (t_{ADP}), and then the liquid acid will further result in a corrosion on the surface of the heat exchanger, which is called Low Temperature Corrosion (LTC) and directly leads to serious decrease in the heat exchangers safety and life [5-7]. Therefore, the surface temperature of exchangers should be controlled higher than the t_{ADP} to avoid LTC [8].

The t_{ADP} of the boiler flue gas cannot be obtained easily since the gas components are too much complex and changeable, and thus the t_{ADP} often depends on the experimental data as well as theirs fitting curve formula and empirical prediction models [9-16]. The experimental test in the present study is called direct method, and the t_{ADP} form prediction model is called indirect method. For direct method, its test system is often complicated because of the changeable gas components, fly ash, water vapor condensation, sensor precision, etc., which often make the experimental system overpriced. Instead, developing indirect method will need less consumption due to the existing thermodynamics theory and experimental data. For decades of studies, there are dozens of empirical or semi-empirical formulas have been proposed to predict the t_{ADP}, and these methods can be divided into two categories: (1) the t_{ADP} is derived from the fuel composition [8, 9, 17-18], and (2) the t_{ADP} is predicted from the gas contents of SO_2/SO_3 and H_2O in the combustion products [10-16]. In recent years, the development of numerical simulation also creates numerical methods to predict the T_{ADP} with high prediction accuracy, such as CFD simulations [2, 19-21] and numerical iterations [1, 3, 7, 22]. However, these studies almost focus on the mean t_{ADP}, while few have paid attention to the t_{ADP} distribution range. As described in ref. [23], the t_{ADP} always varies in an interval with about 30~60 °C width at a certain SO_3 content. For this phenomenon, we can conclude that good temperature control of exchangers against LTC not only needs an accuracy prediction of mean t_{ADP} but also its distribution range.

In the present study, we focus on both the t_{ADP} and its distribution range predictions on the basis of the indirect methods. We choose three groups of tested data from operating boilers to predict t_{ADP} and its distribution range for analysis, and some published data are also employed for the comparison. From the analysis, we shall recommend effective t_{ADP} prediction methods for a further control of the boiler exchanger temperature.

2. Review on acid dew point prediction models

2.1Prediction models derived from fuel composition

The Soviet Thermodynamic Calculation Standard (1973 edition) formula (here we called it as Soviet Standard Model 1973) [9] is one of the most classic models to calculated t_{ADP}, expressed by,

$$t_{ADP} = \frac{\beta \cdot \sqrt[3]{S^{\Pi}}}{1.05^{\alpha_{ash} \cdot A^{\Pi}}} + t_{DP} \tag{1}$$

where t_{DP} (°C) is the dew point temperature of the water vapor in the boiler exhaust gas, S^{Π} and A^{Π} the equivalent sulfur content and ash content (%, mass percentage), respectively, α_{ash} the fly ash coefficient, and β the empirical coefficient which often equals to 125. The t_{DP} can be obtain from the saturated water vapor thermal properties with its partial pressure (p_{H_2O}, Pa), or predicted by fitting formula, here we shall provide a simple equation cited from ref. [24], given by,

$$t_{DP} = 42.4332(p_{H_2O})^{0.13434} - 100.35 \tag{2}$$

The parameters of S^{Π} and A^{Π} are converted according to the sulfur content S_{ar} and ash content A_{ar} as received basis (%, mass percentage), expressed by,

$$S^{\Pi} = 4187 S_{ar} / Q_{ar,net}, \quad A^{\Pi} = 4187 A_{ar} / Q_{ar,net} \tag{3}$$

where $Q_{ar,net}$ is the fuel net calorific value as received basis, kJ/kg. The last parameter α_{ash} denotes the ratio of the fly ash mass to the total ash mass in the combustion products, and one can obtain α_{ash} by experimental test. In addition, we shall introduce some empirical data of coal fired boilers for reference, which are recommended from Ref. [4], as shown in Table 1.

Table 1. Empirical value of α_{ash} in coal fired boilers

Combustion mode	α_{ash}
Inclined reciprocating grate	0.1 ~ 0.2
Travelling grate stoker	0.1 ~ 0.2
Spreader stoker	0.2 ~ 0.3
Circulating fluidized bed	0.4 ~ 0.5
Pulverized coal fired boiler	0.8 ~ 0.9
Coal-water slurry	0.7 ~ 0.8

Some notes of the Soviet Standard Model 1973 are as follows: (1) it is mostly used to predict the coal fired boilers as well as solid fuel fired boilers, but for oil-fired boiler it sometimes presents underestimated predictions [23], (2) For high percentage of alkaline oxide in the ash and A_{ar} >35%, Eq.

(1) may overestimate the t_{ADP} [9], (3) it is not recommended when $S^{\Pi}<0.2$ [25], (4) once the desulfurization measures are carried out during the boiler operation, Eq. (1) should be modified by the SO_3 desulfurization rate (η_{SO_3}), i.e. replace S^{Π} in Eq. (1) with $S^{\Pi}(1-\eta_{SO_3})$ [26], and (5) it can be used without boiler running, and thus is mostly selected for the design of boiler as well as its exchangers.

There are also some modified models based on Eq. (1) for special cases, in which good predictions cannot be obtained from Eq. (1). Li *et al.* replace the following formula expressed by [18]

$$t_{\mathrm{ADP}}=\frac{1.1\beta\cdot\sqrt[10]{S^{\Pi}}}{1.05^{\alpha_{\mathrm{ash}}\cdot A^{\Pi}}}+t_{\mathrm{DP}} \tag{4}$$

to deal with the underestimated prediction from Eq. (1). To consider the effect of alkaline oxide in the ash, А.Ф Гаврцлов also proposed a modified model, given by [23, 25],

$$t_{\mathrm{ADP}}=\frac{\beta\cdot\sqrt[3]{S^{\Pi}}}{K}+t_{\mathrm{DP}} \tag{5}$$

where K is ash characteristic coefficient as the functions of CaO & MgO mass contents compared to the total ash mass. In addition, a simple empirical formula, expressed by,

$$t_{\mathrm{ADP}}=120+17(S^{\Pi}-0.25),\qquad S^{\Pi}>0.25 \tag{6}$$

is describe in Ref. [8], which is a rough prediction model with low accuracy. This model is often used to quickly estimate the magnitude of the t_{ADP} under the condition that the ash content A_{ar} is unknown.

2.2Prediction models derived from exhaust gas contents

The prediction model derived from exhaust gas contents is simpler than that from fuel composition. Once the components of gaseous SO_2/SO_3 and H_2O are measured, the results can be fast obtained. Since the predictions need the components of the combustion products, the experimental test must combine with the boiler running, therefore, the second kind model is mostly selected for the temperature control of the boiler exchangers. The widely used models are listed in Table 2.

Table 2. Prediction models of t_{ADP} (°C) derived from exhaust gas contents

Model	Expression	Dimension	Index
Muller Model [10]	$t_{\mathrm{ADP}}=116.55+16.06\log V_{\mathrm{SO_3}}+1.05(\log V_{\mathrm{SO_3}})^2$	V, ppm	Eq. (7)
Halstead Model [11]	$t_{\mathrm{ADP}}=113.02+15.08\lg V_{\mathrm{H_2SO_4}}+2.10(\lg V_{\mathrm{H_2SO_4}})^2$	V, ppm	Eq. (8)
Okkes Model [12] (Version A)	$t_{\mathrm{ADP}}=10.88+27.6\lg p_{\mathrm{H_2O}}+10.83\lg p_{\mathrm{SO_3}}+1.06(\lg p_{\mathrm{SO_3}}+2.9943)^{2.19}$	p, Pa	Eq.(9a)
Okkes Model [12] (Version B)	$t_{\mathrm{ADP}}=203.25+27.6\lg p_{\mathrm{H_2O}}+10.83\lg p_{\mathrm{SO_3}}+1.06(\lg p_{\mathrm{SO_3}}+8)^{2.19}$	p, atm	Eq.(9b)
Verhoff & Bancher [13] (Version A)	$\frac{1000}{t_{\mathrm{ADP}}+273.15}=1.784+0.027\lg p_{\mathrm{H_2O}}-0.103\lg p_{\mathrm{SO_3}}+0.034\lg p_{\mathrm{H_2O}}\lg p_{\mathrm{SO_3}}$	p, atm	Eq.(10a)
Verhoff & Bancher [13] (Version B)	$\frac{1000}{t_{\mathrm{ADP}}+273.15}=2.988-0.1376\lg p_{\mathrm{H_2O}}-0.267\lg p_{\mathrm{SO_3}}+0.03287\lg p_{\mathrm{H_2O}}\lg p_{\mathrm{SO_3}}$	p, Pa	Eq.(10b)
Japan Power Model [14]	$t_{\mathrm{ADP}}=20\lg V_{\mathrm{SO_3}}+a-80$	V, ppm	Eq. (11)

Table 2 (cont.). Prediction models of t_{ADP} (°C) derived from exhaust gas contents

Model	Expression	dimension	Index

Exponential Model [15]	$t_{ADP} = t_{DP} + B \cdot (p_{H_2SO_4})^n$	p, atm	Eq.(12)
Варанова Model [8, 15]	$t_{ADP} = 186 + 20\lg V_{H_2O} + 26\lg V_{SO_3}$	V, %	Eq.(13)
Haase & Borgmann [16]	$t_{ADP} = 255 + 18.7\lg p_{H_2O} + 27.6\lg p_{SO_3}$	p, atm	Eq.(14)

Remark:

(1) Eq. (8) is based on V_{H_2O}= 11%, for V_{H_2O} < 9% or > 13%, $t_{ADP} = t_{ADP}$ -3 or +3 ℃, respectively.

(2) Calculations by Eqs. (9a) and (9b) are equal, and Eqs. (10a) and (10b) also have identical predictions.

(3) In Eq. (11), a = 184, 194, 201 for water vapor contents = 5%, 10%, 15%, respectively

(4) Eq. (12) includes two versions, named Version A and Version B for different parameters of B and n, and details of the B and n are listed in Refs. [7, 23].

2.3Calculation of SO₃ content

The volume content of SO_3 is too low (~1.0 ppm or ~10 ppm) to directly detect by the equipment of smoke analyzer. Therefore, the SO_3 content is often predicted from the SO_2 content which can be easily tested by apparatus. Zhang [23] proposed a fitting formula out of some published data to predict the conversion ratio of the SO_2 to the SO_3, expressed by,

$$\begin{cases} V_{SO_3} = K_{SO_3} V_{SO_2} \\ K_{SO_3} = m(\alpha - 1) v_{exhaust} S_{ar}^{-0.1} \end{cases} \tag{15}$$

where K_{SO_3} is the conversion ratio of (SO_2 -> SO_3), α the excess air coefficient, $v_{exhaust}$ the volume of 1.0 kg exhaust gas (m^3/kg), and m the coefficient dependent with α. For pulverized coal combustion, m = 1.0 with α < 0.135, and for $\alpha \geq 0.135$, $m = 0.143/(\alpha\text{-}1)$. For oil combustion, m = 3.94 with α < 0.135, and for $\alpha \geq 0.135$, $m = 0.53/(\alpha\text{-}1)$.

Equation (15) includes $v_{exhaust}$ term which is complicated to calculate, and we prefer to substitute the element balance of v_{SO_2}=0.007S_{cal}, where S_{cal} denotes the calculating sulfur mass content and usually can be replaced by S_{ar}, into Eq. (15) for convenience, and then we have,

$$K_{SO_3} = 0.007m(\alpha - 1)S_{ar}^{0.9} / V_{SO_2} \tag{16}$$

In Eq. (16), all the input parameters can be directly obtained, and thus significantly reduce the difficult of the calculations.

Although Eq. (15) is an explicit formula, the m parameter is still insufficient to predict the K_{SO_3} of other types of boilers, and thus we usually prefer the empirical value for the calculations. To simplify the predictions, we shall conclude a fast prediction method as follows: (1) the volume content of water vapor in the flue gas often distributes during 10~14%, and we can choose 12% for an average estimation [8, 15], (2) 500 ppm of SO_2 in the flue gas could be corresponding to 1% of S_{ar}, and this relation can be assumed as a linear relationship [8], i.e. $V_{SO_2} \approx 500S_{ar}$, and (3) the value of K_{SO_3} mostly distributes in the range of 0.5-3% (the Maximum is about 5%), and Blanco and Pena [27] used K_{SO_3} = 1% for calculation and obtained a good prediction curve as the V_{SO_3} < 5 ppm). Generally, K_{SO_3} = 3% would be enough to get safe predictions against the low temperature corrosion [8].

3. Result and discussion

3.1Importance of acid dew point distribution range

During the boiler operation, once the exhaust gas flows across the tube banks of the exchangers, complicated flow field and temperature field may occur, such as boundary effect, backflow, vortex, fly ash effect, and these factors inevitably have significant effect on the SO_2/SO_3 and water vapor content, finally making the t_{ADP} always change. Figure 1 shows the statistics of t_{ADP} distribution for the coal fired boiler. From Fig.1, we can find that the t_{ADP} generally varies in a special interval when the V_{SO_3} is fixed, in which the upper limit is about 30-60 °C higher than the lower limit as the V_{SO_3} > 10 ppm. For instance, when V_{SO_3} = 20 ppm, t_{ADP} distributes between 70 and 110 °C, and for V_{SO_3} = 40 ppm, t_{ADP} varies during 100 and 160 °C. In this study, we shall assume 50 °C as the average interval width.

Once the t_{ADP} distribution range is obtained, the interval upper and lower limits will be two critical temperature points for the exchanger design and control. If the Minimum surface temperature of the exchanger is higher than the upper limit temperature, i.e. right zone of the shadow in Fig. 1, low temperature corrosion will not occur. Conversely, once the Maximum surface temperature of the exchanger is in the left zone of the shadow, the low temperature corrosion must emerge. If the surface temperature of the exchanger distributes in the shadow, we should take necessary protective measures against the low temperature corrosion.

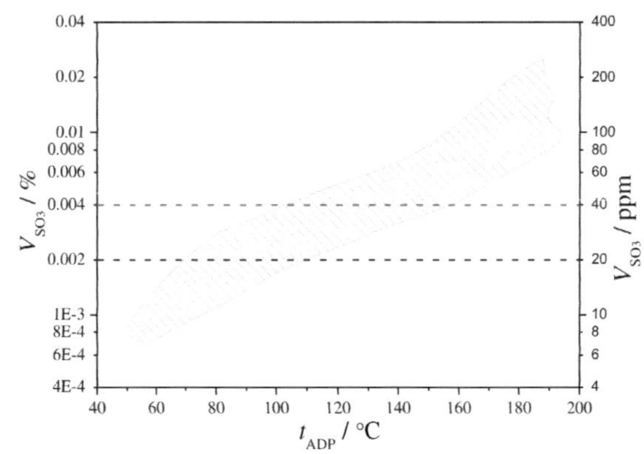

Fig. 1 Acid dew point distribution of the coal fired boilers. (The data are digitized from Ref. [23])

3.2Prediction of acid dew point and its distribution range

Figure 2 shows the results predicted by the present experimental data which is tested from three operating boilers in the city of Xi'an. The types of the three boilers are DHL140-1.6/110/70-AII, QXL70-1.6/13/70-AII, and SHL20-13-A, of which the boiler capacity is 140MW (~200 t/h), 70MW (~100t/h), and 20 t/h (~14 MW). The tested boiler capacities cover two magnitudes, so that the test data would be more representative for the comparisons. The index number means: 1- Soviet Standard Model 1973, 2- Haase and Borgmann Model, 3- Баранова Model, 4- Okkes Model, 5- Verhoff & Bancher Model, 6- Japan Power Model, 7- Exponential Model A, 8- Exponential Model B, 9- Muller Model, and 10- Halstead Model.

From Fig. 2, we can find most predictions distribute above the blue dash line (lower limit of the interval, corresponding to Haase & Borgmann Model) and below the wed dash line (upper limit of the interval, corresponding to Verhoff & Banchero Model), and the interval range width is not great than 50 °C, which denotes that the assumption in Sec. 3.1 is reasonable and can be effectively to predict the acid dew point and its distribution range. The predictions by Exponential Model A (index number = 7) are not in the t_{ADP} distribution interval, while the calculation by Exponential Model A can meet the requirement.

For the comparison of the mean t_{ADP} predictions (Note that, we choose K_{SO_3} = 1.5% for the prediction of SO_3 content. The exhaust gas pressure is fitted as 1.0 atm, and the water vapour content is assumed 10% if it is not available by the equipment test), the Soviet Standard Model 1973 is firstly to be recommended for the designing calculation of the coal fired boiler. For controlling the temperature of the exchangers, most of the models listed in Table 2 (except Exponential Model A) can be used, and in the present study we shall recommend Okkes Model for the prioritized method since this formula is a modified model based on Muller Model, of which includes the water vapour effect and can be use for different types of boilers [18, 24-26].

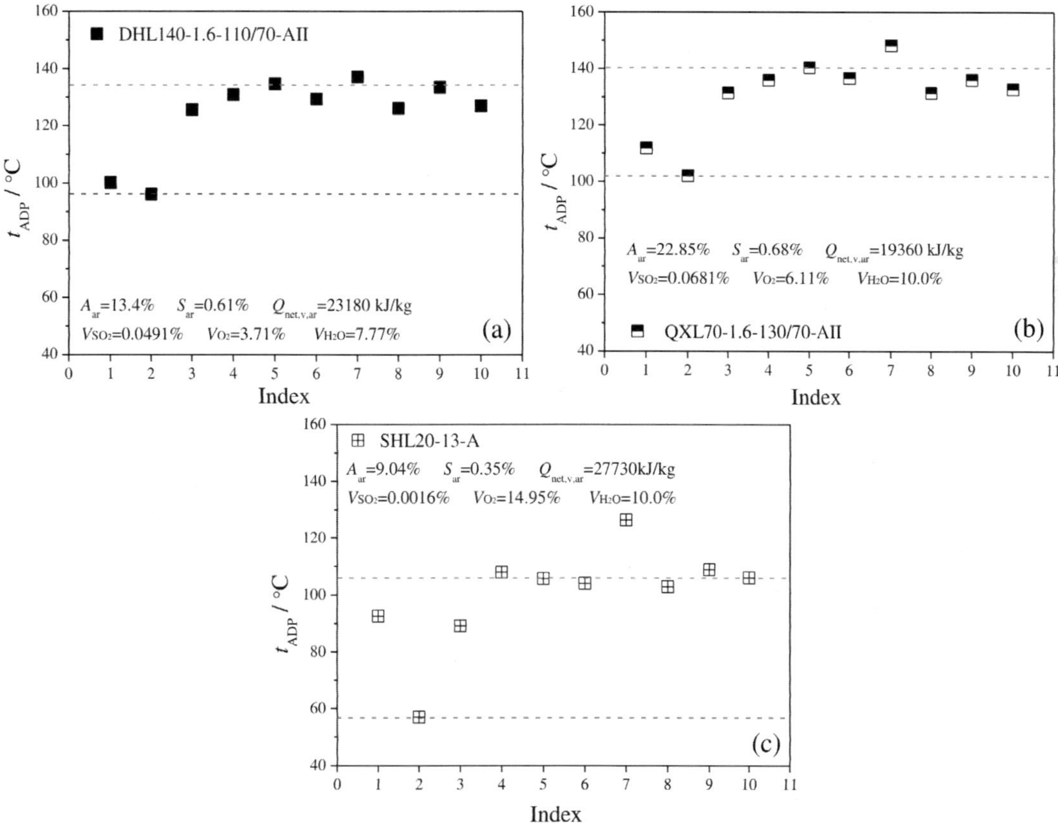

Fig.2 Comparison of t_{ADP} predictions among different models with the present test data. (a) 140MW (~200t/h). (b) 70MW (~100t/h). (c) 20t/h (~14MW).

Figure 3 depicts the predictions of the t_{ADP} corresponding to the published data of a coal fired boiler [23], an oil fired boiler [23] and a circulating fluidized bed boiler [28], and the flue gas content of the fluidized bed boiler is artificially controlled with a fixed O_2 content, which is called artificial controlled flue gas in this study. The meanings of the number index are consistent with that in Fig. 2.

In Fig. 3, Haase & Borgmann Model and Verhoff & Banchero Model are used to evaluate the lower and upper limits of the t_{ADP} distribution interval. We can find that most of the predictions are in the distribution interval except two points, i.e. Point 7 in Fig. 3(b) and Point 1 in Fig.3(c), and the two exceptions can be explained as follows, (1) the former is calculated with Exponential Model A which is not suitable for the prediction as proved in the Ref. [23, 26] and by the above examples, and (2) the latter is predicted by the Soviet Standard Model 1973 which is not recommended in the case of S^{Π}<0.2 as mentioned in Section 2.1 (here, the value of S^{Π} is about equal to 0.164, with an assumption of $Q_{ar,net}$ = 20000 kJ/kg). In addition, the t_{ADP} of point 1 in Fig. 3(b) is slightly lower than the lower limit, and this exception can be attributed to the liquid fuel, i.e. oil, of which the ash content is too low to use the Soviet Standard Model 1973 for calculations.

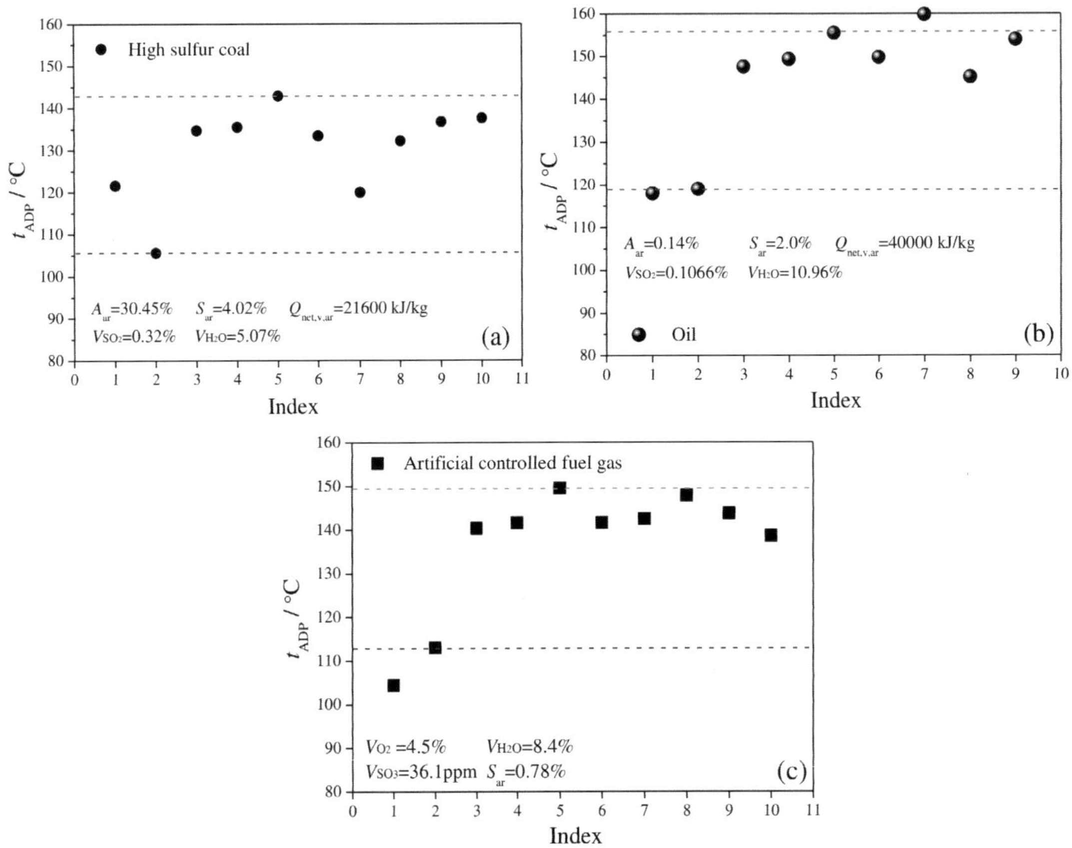

Fig. 3 Comparison of t_{ADP} predictions among different models with published data.

4. Conclusions

The prediction methods of the acid dew point temperature of the flue gas in boilers are studied, and the widely used prediction models are also reviewed. The existing models are mostly used to calculate the mean acid dew point instead of its distribution range. The present study focus on the prediction of the upper and lower limits of the changeable acid dew point, and the purpose is to further prevent and resist the low temperature corrosion of the exchangers. Six group data are introduced to predict the acid dew point for comparisons. The Soviet Standard Model 1973 is more suitable for the coal fired boilers on the condition of the equivalent sulfur content higher than 0.2%, and the Okkes model can be used for many types of boilers. The Soviet Standard Model 1973 is preferred for the designing calculation of the exchanger temperature, while the Okkes Model is recommended for the control of the exchanger temperature. The Verhoff & Banchero model and Haase & Borgmann model is recommended to calculate the upper and lower limits of the acid dew point distribution range, respectively.

Acknowledgments

This work was supported by Shaanxi Province Quality and Technology Supervision Bureau Research Project under Grant No. 2016KY02.

References

[1] Bahadori A. (2011) Estimation of combustion flue gas acid dew point during heat recovery and efficiency gain. Appl. Therm. Eng., 31: 1457-1462.

[2] Han H., He Y.L. Tao W.Q. (2013) A numerical study of the deposition characteristics of sulfuric acid vapor on heat exchanger surfaces. Chem. Eng. Sci., 101: 620-630.
[3] Wei W., Sun F.Z., Shi Y.T., Ma L. (2017) Theoretical prediction of acid dew point and safe operating temperature of heat exchangers for coal-fired power plants. Appl. Therm. Eng., 123: 782-790.
[4] Song J.H., Wang X.L., Leng H., et al. (2016) Technologies of energy efficiency test for boiler. Chemical Industry Press, Beijing, pp. 104-124. (in Chinese)
[5] Shi Y.T., Zhang X.Y., Li F., Ma L. (2014) Engineering acid dew temperature: the limitation for flue gas heat recovery. Chin. Sci. Bull., 59 (33): 4418-4425.
[6] Song, J., Yuan, H., Deng, J. (2018) Experiment comparison between engineering acid dew point and thermodynamic acid dew point. In E3S Web of Conferences, EDP Sciences, 38, pp: 04008.
[7] Xiang, B., Tang, B., Wu, Y., Yang, H., Zhang, M., and Lu, J. (2016) Predicting acid dew point with a semi-empirical model. Appl. Therm. Eng., 106: 992-1001.
[8] Qin K.F., Fan J.R., Chi Z.H., et al. (1994) Boiler and heat exchanger fouling, slagging, abrasion and corrosion prevention principles and calculation. Science Press, Beijing, pp: 377-401 (in Chinese).
[9] Boiler unit thermal calculation standard method//Translation of Soviet thermodynamic calculation standard (1973 edition), Mechanical Industry Press, Beijing, 1976. (in Chinese)
[10] P. Müller. (1959) Contribution to the problem of the action of sulfuric acid on the dew point temperature of flue gases. Chem. Eng. Technol., 31: 345-350.
[11] Halstead W.D., Talbot J. (1980) Sulphuric acid dew point in power station flue gases. J. Energy Inst. 53: 142-145.
[12] Okkes A.G., Badger B.V. (1987) Get acid dew point of flue gas. Hydrocarbon Process, 66 (7): 53-55.
[13] Verhoff F.H., Banchero J.T. (1974) Predicting dew points of flue gases. Chem. Eng. Prog., 70 (8): 171-172.
[14] Xu J. G. (1990) Determination of flue gas dew point temperature. Energy Saving, 6: 28-30. (in Chinese)
[15] Rong L.E., Yuan T.F., Liu Z.M., Tian Z.P. (1997) Power plant boiler principle. China Electric Power Press, Beijing, pp: 206-207. (in Chinese)
[16] Haase R., Borgmann H.W. (1963) Präzisionsmessungen zur ermittlung vonsauertaupunkten. Korrosion, 15: 47-49. (in German)
[17] Boiler unit thermal calculation standard method//Translation of Soviet thermodynamic calculation standard (1998 edition), Mechanical Industry Press, Beijing, 1976. // Sa Y.L. (1994) Boiler heating surface external process. Water Power Press, Beijing, pp: 152-154. (in Chinese)
[18] Li P.F., Tong H.L. (2009) Comparison and analysis on the calculation methods of acid dew point of flue gas. Boiler Tech., 40 (6): 5-8. (in Chinese)
[19] Wang Y.C., Tang G.H. (2014) Acid condensation and heat transfer characteristics on H-type fin surface with bleeding dimples and longitudinal vortex generators. Chin. Sci. Bull., 59(33): 4405-17. http://dx.doi.org/10.1007/s11434-014-0564-3.
[20] Wang Y.C., Tang G.H. (2016) Prediction of sulfuric acid dew point temperature on heat transfer fin surface. Appl. Therm. Eng., 98: 492- 501.
[21] Karvounis N., Pang K.M., Mayer S., Walther J. H. (2018) Numerical simulation of condensation of sulfuric acid and water in a large two-stroke marine diesel engine. Appl. Ene., 211: 1009-1020.
[22] ZareNezhad B., Aminian A. (2010) A multi-layer feed forward neural network model for accurate prediction of flue gas sulfuric acid dew points in process industries. Appl. Therm. Eng., 30: 692-696.

[23] Zhang J.Z. (2003) Discussion on calculation of SO_x displacement of flue gas, prediction of flue gas dew point temperature before and after desulfurization and evaluation of flue gas corrosion. Heat Eng. Tech., B11: 84-99. (in Chinese)
[24] Jia M.S., Ling C.M. (2003) Factors of affecting the flue gas acid dew point temperature and its way of calculation. Industrial Boiler, 6: 31-35. (in Chinese)
[25] Zhang J.Z., Wang J.W. (2012) Review on current calculating method of gas acid dew point temperature for coal fired boilers and inquire into its optimization and improvement. Total Corrosion Control, 26 (10): 1-6. (in Chinese)
[26] Tang Z.Y. Investigation on corrosion behaviours of downstream components of power plant following the WFGD. Dissertation of Doctoral Degree of Southeast University, Nanjing, pp: 26-36. (in Chinese)
[27] Blanco J.M., Pena F. (2008) Increase in the boiler's performance in terms of the acid dew point temperature: Environmental advantages of replacing fuels. Appl. Therm. Eng., 28: 777-784.
[28] Xiang B.X., Xing W.C., Li J.F., Li Y.Z., Yang H.R., Lu J.F. (2014) The measurement and correction of correlation formulas of flue gas acid dew point. Boiler Technol., 45 (1): 1-4 (in Chinese).

IOP Conf. Series: Materials Science and Engineering **721** (2020) 012017 doi:10.1088/1757-899X/721/1/012017

Validation of Unipolar Diffusion Charging Models for Spherical and Agglomerated Nanoparticles

Lipeng Su[1,2], Qian Du[1,*], Jiwang Wen[1], Qisheng Ou[2,*], David Y.H. Pui[2], Jianmin Gao[1], Peng Dong[1]

[1]School of Energy Science and Engineering, Harbin Institute of Technology, 92 Xidazhi Street, Harbin, 150001, China.

[2]Particle technology Laboratory, University of Minnesota, Minnesota 55455, USA

[*]Duqian@hit.edu.cn (Qian Du)

Abstract. An ion-jet diffusion charger was used in this study to investigate the diffusion charging of spherical and agglomerated nanoparticles, and validate corresponding theoretical models. Results show that the limiting-sphere model and White's theory could preferably predict the diffusion charging of spherical particles, compared with other models. As to agglomerated particles, it is affected by primary particle size and sintered morphology to some extent. Especially for loose agglomerate which is point-to-point contact, obviously more charges could be acquired at same mobility diameter in unipolar diffusion charger. In addition, it was validated that all of Chang's theory, normalized capacitance model and collision kernel model are in good agreement with experiments results. Chang' theory and normalized capacitance model, however, are more concise and could be used to evaluate diffusion charging of non-spherical particles in real occasions.

1. Introduction

Charging properties of nanoparticles play important roles in particle collection, aerosol purification and measurement [1-3]. In the field of coal-fired and industrial plant, the electrostatic precipitator (ESP) is one of the most popular techniques to remove particles in exhausted gas. For this method, the collection efficiency of nanoparticle largely depends on its diffusion charging in corona discharge, so investigating the charging mechanism is of great significance to the optimization and prediction of collection efficiency [4]. In another field, aerosol measurement, particle charging is used in many instruments, such as electrical low pressure impactor, to classify and characterize particles [5]. Therefore, an appropriate model to evaluate particle charging, including spherical and agglomerated particles, is important for the further development of such instruments.

Given that the complexity of charging process, it is difficult to describe diffusion charging in the whole size range by just utilizing one theoretical model. Diffusion-mobility model, which was first applied in diffusion charging of spherical particles, was reported to have a good accuracy in the continuous regime [6]. Subsequently, Fuchs limiting-sphere theory and BKG model were developed to predict diffusion charging in transition regime, while White's theory was applied in free-molecule regime [6]. With regard to non-spherical particles, Laframboise [7] first proposed a theoretical model which is based on diffusion mobility theory, analytical solution of electric field near the charged particle and ion flux equation. Afterwards, Chang [8] established a model to be applied in continuous and free-molecule regimes. On this basis, many scholars [9,10] acquired the charge distribution of primary particles in an agglomerate, which further enriched charging models of non-spherical particles. Gopalakrishnan [11] proposed an ion-particle collision kernel model in the whole range of Knudsen number to solve particle charge distribution, which utilized dimensional analysis, Brownian diffusion dynamics and molecular dynamic methods. However, this model has not been directly verified by experiments. Additionally, Biskos [6] simulated particle diffusion charging by Monte Carlo method and compared with other charging models. Even though popular charging models used in early stage were discussed and analyzed in Biskos'study [6], effective experiments are still needed to verify current theoretical models.

In this study, an ion-jet diffusion charger was used to evaluate and verify the most of charging models for both spherical and agglomerated particles. The charger adopts an ion-particle impinge mode, which minimizes the effect of external electric field on particle charging and loss. By means of this setup, theoretical models for spherical particles were first validated. Afterwards, agglomerated particles were produced in tandem tube-furnace and the influence of furnace temperature on particle charging was expatiated. Finally, multiple non-spherical charging models were validated by experimental results.

2. Theoretical models

2.1. *Spherical particle*

Diffusion charging models for spherical particle mainly include White'theory, Fuchs limiting-sphere theory, BKG model and diffusion-mobility theory, which are shown in equation (1) ~ (4), respectively.

$$\frac{dq_p}{dt} = \frac{\pi}{4} d_p^2 v_r \rho_i \exp\left(\frac{-q_p e}{2\pi\varepsilon_0 d_p kT}\right), \tag{1}$$

where q_p is charges carried by single particle, t is charging time, d_p is particle diameter, v_r is the thermal velocity of ions, ρ_i is ion density, ε_0 is vacuum dielectric constant, k is Boltzmann constant and T is the absolute temperature.

Fuchs limiting-sphere theory assumes that the outer space of a particle is divided into two parts by a concentric sphere. In the gap of hypothetical sphere and particle surface, the movement of gas ions is dominated by thermal diffusion and ion-particle interaction potential, while it is controlled by macroscopic diffusion-mobility theory in the outer space of hypothetical sphere. Based on similar flux theory, BKG model and diffusion-mobility theory are formulated as equation (3) and (4), respectively.

$$J = \frac{\pi\gamma v_r \delta^2 \rho_i \exp(-\phi(\delta)kT)}{1+\exp(-\frac{\phi(\delta)}{kT})\gamma v_r \delta^2 / 4D_i \int_{\infty}^{r} 1/r^2 \exp(-\phi(r)kT)dr}, \tag{2}$$

$$J = \frac{\pi d_p^2 v_r \rho_i E_0}{4(1+\lambda E_1 / 2\sqrt{\pi} E_0)}, \tag{3}$$

$$J(r) = -4\pi r^2 (D_i \frac{d\rho_i}{dr} - Z_i \rho_i E(r)), \tag{4}$$

where γ is the probability of ions that enter into the hypothetical sphere and transfer charges to the particle, δ is the radius of hypothetical sphere, $\phi(\delta)$ is the potential at the location, δ . D_i is ion diffusion coefficient, E_0 is the zeroth-order correction of molecule-flux flux, E_1 is first-order correction, and λ is the self-defined parameter. In equation (4), Z_i is ion mobility, $E(r)$ is the field strength at the distance r from particle.

2.2. *Agglomerated particle*

Diffusion charging models for agglomerated particle primarily involve Laframboise's theory [7], Chang's theory [8], normalized capacitance model [9] and ion-particle collision model [11], which are formulated in equation (5), (6) ~ (8), (9) and (10), respectively. Laframboise's theory was also established based on flux theory, while the normalized capacitance model was acquired by dimensionless surface potential and self-defined normalized capacitance. For Chang' theory, equation (6) and (7) are applicable to the continuous regime, whereas the equation (8) corresponds to free-molecule regime. Additionally, Monte Carlo method could be also applied to the calculation of particle charging in diffusion charger, and its effectiveness is also validated in section 4.2.3.

$$J = 4\pi\rho_i q_p \frac{L}{In(2L)} \frac{\varphi_p \exp(-\varphi_p)}{1-\exp(-\varphi_p)+Kn_e\varphi_p}, \tag{5}$$

$$q_p \cong (C_p\rho_i D_i t)/\varepsilon_0, (\phi_p < 0.1), \tag{6}$$

$$q_p \cong \frac{2C_p kT}{e^2}[(\frac{e^2}{kT}\frac{\rho_i D_i t}{\varepsilon_0}+1)^{0.5}-1], (\phi_p < 1), \tag{7}$$

$$q_p \cong \frac{2C_p kT}{e^2} In(1+\frac{e^2\rho_i v_r S_p t}{4kTC_p}), \tag{8}$$

$$q_p = \frac{\varphi_p kT}{e^2}C_p = \frac{\varphi_p kT}{e^2}2\pi\varepsilon_0 d_p C_p^*, \tag{9}$$

$$\frac{dq_p}{dt} = R_{p-1,i} - R_{p,i} = \rho_i(\beta_{p-1,i}n_{p-1} - \beta_{p,i}n_p) \tag{10}$$

where L is the ratio of polar radius to midline radius, φ_p is dimensionless surface potential, Kn_e is the effective Knudsen number, C_p is particle capacitance, C_p^* is normalized capacitance, S_p is particle geometric surface area, $R_{p,i}$ is the charging rate of a particle carried with p elementary charges, $\beta_{p,i}$ and n_p are corresponding collision kernel and particle number concentration, respectively.

3. Experiments

Figure 1 shows the setup diagram for unipolar diffusion charging, which involves aerosol generation for both spherical and agglomerated particles, particle classification and measurement, unipolar charging and detection. Spherical particles are first produced in a collision-type atomizer and subsequently flow into a diffusion dryer and a neutralizer. 0.01% K_2SO_4 solution and Di-Ethyl-Hexyl-Sebacate (DEHS) liquid are individually filled in the atomizer to produce polydisperse salt-particles and oil-particles, respectively. After moisture absorption in dryer and charge neutralization in neutralizer, polydisperse particles are classified into monodisperse particles by a DMA (TSI, 3081) followed by a neutralizer and a CPC (TSI, 3776). For non-spherical particles, tandem furnaces and an agglomeration chamber were used to produce agglomerate and aggregate in different fractal dimensions. Firstly, solid silver is placed in the centre of first tube-furnace with 1150°C, then gaseous silver is produced continuously at this temperature. At the exit of first tube-furnace, gaseous silver gradually condenses into a larger number of nanoparticles due to the decrease of gas temperature. Secondly, these nanoparticles collide into loose agglomerates in agglomeration chamber by random Brownian diffusion. After that, loose agglomerates are sintered in the second tube-furnace (25°C~600°C) into aggregates in different fractal dimensions and neutralized to reach Boltzmann equilibrium distribution in a neutralizer.

Aerosol flow produced by atomizer or tandem furnaces is then divide into two branches after classification by a DMA. The one flows into a CPC to detect particle concentration while the other enters into a unipolar charging system. At the inlet of this system, the aerosol is split into two portion again. One portion of 1.5L/min directly flows into a mixing chamber while the other portion of 1L/min is filtered and introduced into a diffusion charger. In this charger, needle electrode is connected to the positive potential (+2.5kV), then a large number of positive ions are produced and carried into the mixing chamber. Injected ions within this mixing chamber impinge with particles, so that particles could acquire charges. After that, charged particles exit from mixing chamber and pass through a needle-tube ESP (20V) followed by an electrometer to detect induced electric current. Charges carried by single particle could be calculated by equation (11):

$$q_p = I / efc, \tag{11}$$

where I is electric current detected by the electrometer, f is the aerosol flowrate and c is particle number concentration。

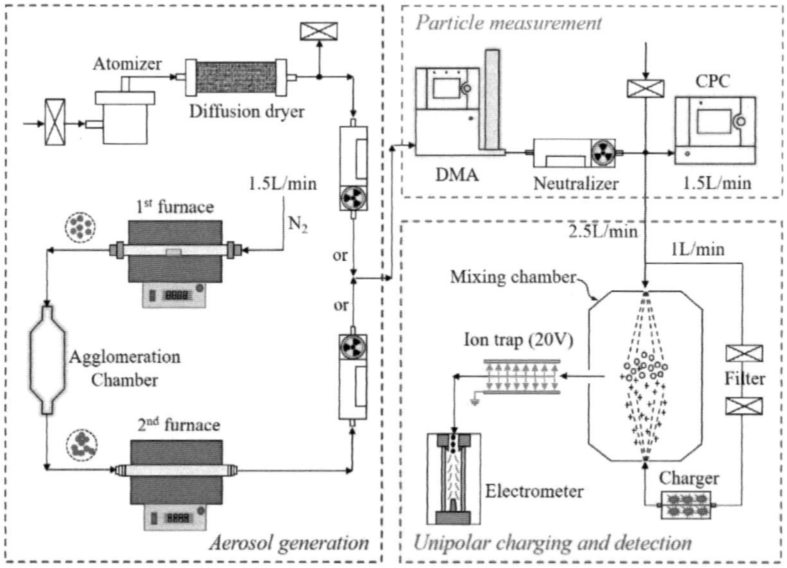

Figure 1. Setup diagram used in this study

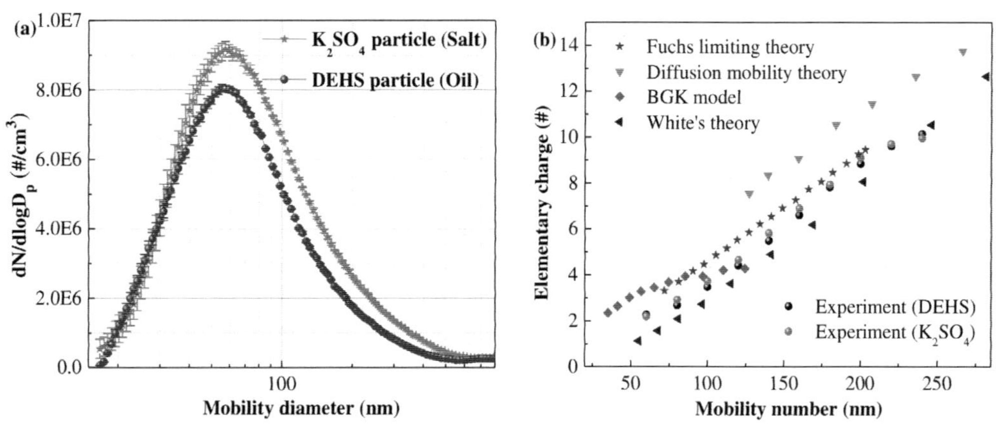

Figure 2. Validation results for spherical particles (a) particle size distribution, (b) comparison with multiple theoretical models

4. Results and discussion

4.1. Spherical particle

Particle size distribution and charges carried by single particle in diffusion charging are shown in figure 2(a) and 2(b), respectively. Similar distribution, peak size and absolute concentration for K_2SO_4 and KCl aerosols could be seen in figure 2(a), just expect for material composition. In figure 2(b), diffusion charges carried by single particle increase with mobility diameter, which is because particle surface area is an important factor in diffusion charging at same charging time and ion concentration. As a contrast, material composition has a marginal effect on diffusion charging, which agrees well with limiting-sphere theory and shin's study [12]. In order to validate current charging models, Figure 2(b) also shows the comparison of theoretical models with experimental results. Major parameters used in models include the product of ion concentration and charging time ($\rho_i t$ =2.5×10^{13}s/cm^3) and ion mobility (Z_i=1.15cm^2/V×s). For all theoretical models, both Coulomb force and image force are considered in the process of ion-particle collision. Compared with experimental results, all models predict a reasonable charging trend with mobility diameter. However, gaps for absolute level also occur in figure 2(b). BKG model and diffusion-mobility theory slightly deviate from testing results while limiting-sphere theory and White's theory are in accordance with experimental results.

4.2. Agglomerated particle

4.2.1. Effect of evaporation temperature on agglomerate diffusion-charging. Effects of evaporation temperature on agglomerate size distribution and diffusion charging are shown in figure 3(a) and 3(b), respectively. Unimodal distribution trend could be seen in figure 3(a) and the peak size gradually increases with temperature, stating that average diameter also increases with temperature of the first furnace. A reasonable explanation for this phenomenon is that the amount of gaseous silver would quickly increase once evaporation temperature is improved, so that both size and number of primary particle produced by condensation increase [13]. As a result, these primary particles coalesce into the larger agglomerate with point-to-point structure. Figure 3(b) shows that charges acquired by agglomerates just slightly increase with evaporation temperature at same mobility diameter, suggesting that the primary particle in this study has a marginal effect on diffusion charging of agglomerate. According to Chang' theory [8], particle surface area and capacitance are crucial factors in diffusion charging. Whereas increasing evaporation temperature could not change agglomerate structure in a large extent, so just minor increase occurs in figure 3(b).

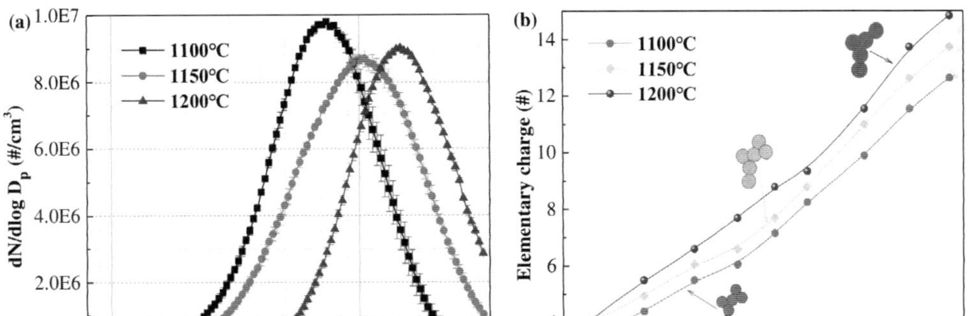

Figure 3. Effect of evaporation temperature on size distribution and diffusion charging (a) particle size distribution, (b) charges acquired in diffusion charger.

4.2.2 Effect of sintering temperature on agglomerate diffusion-charging. Effects of sintering temperature on particle size distribution and diffusion charging are shown in figure 4(a) and 4(b), respectively. With an increase of sintering temperature, peak size in figure 4(a) shifts from 80nm to 55nm and the size range becomes narrow. In room temperature, particles produced by tandem furnaces are regarded as loose agglomerates with fractal dimension of 1.78 [14]. When sintered in high temperature, agglomerates gradually become dense ellipsoid or nearly spherical particles, meanwhile the fractal dimension increases from 1.78 to 3 approximately. As particle morphology varies, the drag force applied to the particle and mobility diameter selected by DMA also change, which agrees well with experiment in this study. In figure 4(b), sintering temperature has a significant influence on particle charges acquired in diffusion charger, with more charges in the higher temperature. In a specific charger, particle charges in steady stage depend on collision frequency of ion-particle and corresponding adsorption efficiency. It is generally assumed that charges transfer to conductive particle immediately once the event of ion-particle collision occurs in a diffusion charger [11]. Consequently, only ion-particle collision probability is the main parameter to determine unipolar diffusion charging. At same mobility diameter, loose structure of agglomerate could reduce the distance of ion-particle collision to some extent, meanwhile the large surface area could increase the collision probability and storage more charges. Therefore, the geometric structure of particle itself is the one of important factors to affect diffusion charging, and large surface area could significantly promote ion adsorption on particle surface.

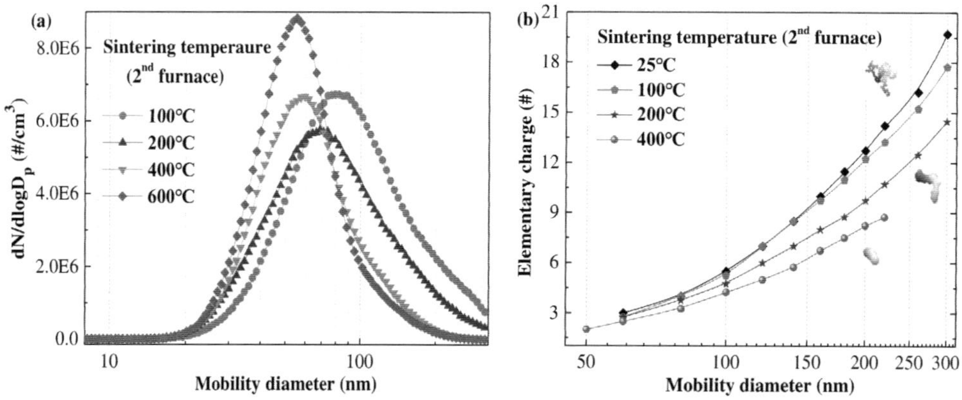

Figure 4. Effect of sintering temperature on size distribution and diffusion charging (a) particle size distribution, (b) charges acquired in diffusion charger.

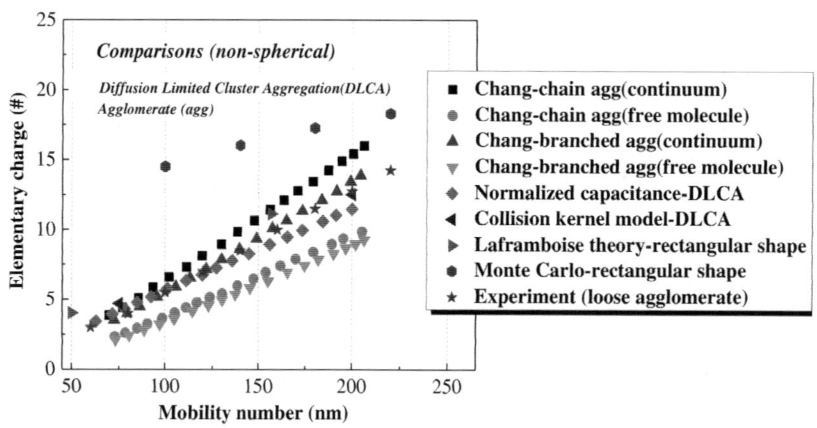

Figure 5. Comparison with multiple theoretical models.

4.2.3 Comparison with theoretical models. Figure 5 shows the comparison of experimental data with modeling results predicted by multiple theoretical models. It should be noted that experimental data in figure 5 correspond to charges carried by loose agglomerates which are assumed in most of models. In a contrast, results provided by Monte Carlo simulation are quite different from that of other models as well as experiment. However, it should be attributed to the modeling rectangular particle in Monte Carlo simulation to some extent. For this shape, the surface area just varies a little at different mobility diameters but the aspect ratio changes considerably. Compared with other models, it could be found that the aspect ratio has a significant effect on diffusion charging, which is mainly because aspect ratio affects the potential energy distribution on particle surface and probability of ion-particle collision. As an overall comparison in figure 5, Chang's theory, normalized capacitance method and collision kernel model could provide a good prediction on diffusion charging. Ulteriorly, collision kernel model can describe the charging process in detail, while Chang's theory and normalized capacitance model are more concise to be applied in real conditions.

5. Conclusion

An ion-jet diffusion charger was adopted in this study to investigate the diffusion charging of both spherical and agglomerated particles, and validate multiple theoretical models. Results show that charges acquired by spherical particles in diffusion charger increase with an increase of mobility diameter. This experiment also verifies that Fuchs limiting-sphere theory and White's theory could provide a good prediction for spherical particles. For agglomerated particles, diffusion charging is mainly affected by particle morphology and primary particle number while primary particle size is the secondary factor. With increasing sintering temperature, particle fractal dimension also increases, whereas charges acquired by particles decrease. Similarly, multiple theoretical models were validated by experiments in this study, indicating all of Chang's theory, normalized capacitance model and collision kernel model agree well with testing data. Further, Chang's theory and normalized capacitance model are more concise and applicable to real occasions.

References

[1] Arif, S., Branken, D.J., Everson, R.C., Neomagus, H.W.J.P., Le Grange, L.A., Arif, A. (2016) CFD modeling of particle charging and collection in electrostatic precipitators. Journal of Electrostatics, 84:10-22.

[2] Zheng, C., Duan, D., Chang, Q., Liu, S., Yang, Z., Liu, X., Weng, W., Gao, X. (2019) Experiments on Enhancing the Particle Charging Performance of an Electrostatic Precipitator. Aerosol and Air Quality Research, 19(6):1411-1420.

[3] Schriefl, M.A., Bergmann, A., Fierz, M. (2018) Design principles for sensing particle number concentration and mean particle size with unipolar diffusion charging. IEEE sensors journal, 19(4):1392-1399.

[4] Du, Q., Su, L., Dong, H., Gao, J., Zhao, Z., Lv, D., Wu, S. (2016) The experimental study of a water-saving wet electrostatic precipitator for removing fine particles. Journal of Electrostatics, 81:42-47.

[5] Su, L., Ou, Q., Cao, L.N., Du, Q., Pui, D.Y.H. (2019). A new instrument prototype to measure the geometric surface area of nanoparticles with a time resolution of 1s. Journal of Aerosol Science, 132:32-43.

[6] Biskos, G., Mastorakos, E., Collings, N. (2004) Monte-Carlo simulation of unipolar diffusion charging for spherical and non-spherical particles. Journal of aerosol science, 35(6): 707-730.

[7] Laframboise, J.G., Chang, J.S. (1977) Theory of charge deposition on charged aerosol particles of arbitrary shape. Journal of Aerosol Science, 8(5): 331-338.

[8] Chang, J.S. (1981) Theory of diffusion charging of arbitrarily shaped conductive aerosol particles by unipolar ions. Journal of Aerosol Science, 12(1): 19-26.

[9] Shin, W.G., Wang, J., Mertler, M., Sachweh, B., Fissan, H., Pui, D.Y.H. (2010). The effect of particle morphology on unipolar diffusion charging of nanoparticle agglomerates in the transition regime. Journal of Aerosol Science, 41(11): 975-986.

[10] Cao, L.N., Wang, J., Fissan, H., Pratsinis, S.E., Eggersdorfer, M.L., Pui, D.Y.H. (2015) The capacitance and charge of agglomerated nanoparticles during sintering. Journal of Aerosol Science, 83: 1-11.

[11] Gopalakrishnan, R., Thajudeen, T., Ouyang, H., Hogan Jr, C.J. (2013) The unipolar diffusion charging of arbitrary shaped aerosol particles. Journal of Aerosol Science, 64: 60-80.

[12] Shin, W.G., Qi, C., Wang, J., Fissan, H., Pui, D.Y.H. (2009) The effect of dielectric constant of materials on unipolar diffusion charging of nanoparticles. Journal of Aerosol Science, 40(5): 463-468.

[13] Kim, S.C., Wang, J., Emery, M.S., Shin, W.G., Mulholland, G.W., Pui, D.Y.H. (2009) Structural property effect of nanoparticle agglomerates on particle penetration through fibrous filter. Aerosol Science and Technology, 43(4): 344-355.

[14] Sorensen, C.M. (2011) The mobility of fractal aggregates: a review. Aerosol Science and Technology, 45(7): 765-779.

Research on the detection of fouling and slagging in cold state based on capacitance principle

Jingqi Peng[1], Jiajie Zhang[1*], Shaojun Meng[1] and Jiansheng Zhang[2]

[1]College of Electrical and Power Engineering, Taiyuan University of Technology, Taiyuan, Shanxi, 030024, China

[2]Department of Thermal Engineering, Tsinghua University, Beijing, 10084, China

[*]Corresponding author's e-mail: zhangjiajie@tyut.edu.cn

Abstract. Aiming at the requirement of on-line monitoring for the fouling and slagging in the syngas cooler of coal gasifier, this paper proposes a detection method of deposition thickness based on the capacitance principle. A capacitance measurement rig for deposition detection is built to conduct a cold state simulation experiment. The coal ash and syrup are selected as medium to simulate the fouling and slagging in cold state. Two kinds of electrodes including single-electrode and double-electrode probes are used to measure the capacitance of deposition. The results show that the measured capacitance increases linearly with the increase of the deposition thickness for all the cases. The linear correlations between the capacitance value and the deposition thickness are obtained for ash and syrup under two kinds of electrodes, respectively. Verification tests show that the predicted deposition thicknesses by employing the linear correlations are basically consistent with the real thickness, and the maximum error is 5.15%. The detection error with single-electrode is less than that with double-electrode due to more deposition amount between two electrodes.

1. Introduction

Coal gasification is one of the main methods on high efficient and clean utilization of coal resources. The syngas from gasification chamber has a high temperature about 1200 ~ 1600 ℃. The sensible heat of high temperature syngas generally accounts for more than 15% of the total heat in coal [1]. Therefore, the waste heat recovery for the high temperature syngas can effectively improve the overall efficiency of coal gasification. In current gasification systems, the syngas cooler is the most commonly used equipment to recovery the syngas heat, including radiant syngas cooler (RSC) and convection syngas cooler (CSC) [2, 3]. RSC and CSC are used to recovery the waste heat of high temperature syngas (>800℃) and low temperature syngas (<800℃), respectively. They have the similar functions as radiant heat exchanger and convection heat exchanger in utility boilers.

However, compared with high temperature gas in utility boilers, the syngas of coal gasification contains a large number of slag and ash particles, and there is no internal heat source in syngas cooler. It results in that the fouling and slagging on the heating surface of syngas cooler are more prominent. Generally, the slagging occurs in RSC when the molten slags in the high temperature syngas contact with the heating surface and then stick on it. The fouling happens much in CSC where the fly ash particles in the low temperature syngas deposit on the heating surface. The fouling and slagging weaken the heat transfer of syngas cooler and decrease the unit efficiency. Excessive fouling and slagging can lead to the ash or slag blocking the syngas passage, and even unit shutdown accident [4, 5]. Therefore, on-line monitoring about ash and slag deposition on the heating surface of syngas cooler is very necessary to ensure the safe operation of gasification unit. In most present gasifiers, the fouling and slagging in syngas cooler is indirectly detected based on the heat transfer parameters such as heat flux, furnace temperature and syngas outlet temperature. However, due to the complex environment with high temperature, high pressure, and gas/liquid/solid multiphase flow in the syngas cooler, these indirect monitoring methods are unreliable to detect the fouling and slagging in practical process.

In recent years, some novel approaches are developed to detect the pollutant deposition in heat exchangers, such as the sonic wave, infrared visualization and electrical principle [6]. These new detection methods have been the research focuses, especially in food industry [7, 8]. Among, the electrical detection principle is that, one or more electrodes is installed on the surface of heat exchanger. Thus, the fouling condition can be detected by monitoring the electrical parameters (e.g. resistance, capacitance), which will be changed when the fouling occurring on the heat transfer surface. Because the measuring electrode directly contacts with the pollutant (namely, an in-situ monitoring), this method has the advantages of high accuracy, sensitivity, and little influence from environment.

Considering the complex and changeable internal environment in syngas cooler of gasifier, the electrical monitoring for the fouling and slagging in syngas cooler has a potential research value. Based on this, the present study builds a deposition detection rig with capacitance principle. A cold state simulation experiment is carried out to study the relation of measured capacitance with the deposition thickness.

2. Detection principle of capacitive

The detection of deposition thickness in current study is based on the coaxial capacitor principle [9], as shown in Figure 1. A metal probe is placed in the center of container, where there are some measured mediums being deposited. Thus, the metal probe, the medium, and the container wall constitute a capacitor. The metal probe and the container wall are as the two electrode of the capacitor. The capacitance between the two electrodes can be written as:

$$C_t = C_1 + C_2 + C_3 \tag{1}$$

Where, C_t is the total capacitance, C_1 is the equivalent stray capacitance, C_2 is the capacitance caused by air between two electrodes, and C_3 is the capacitance caused by the measured medium.

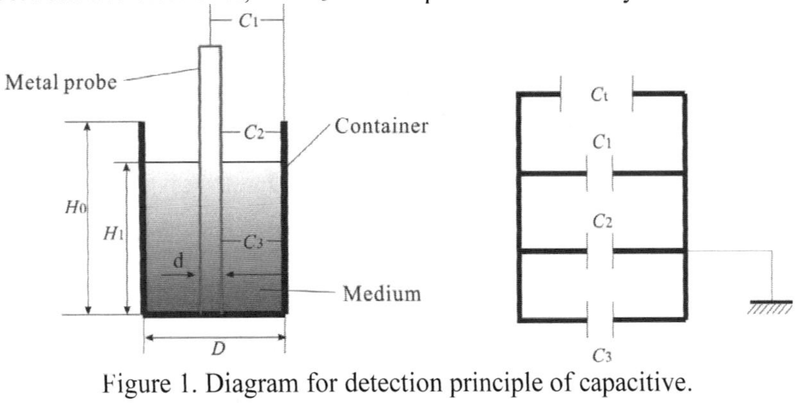

Figure 1. Diagram for detection principle of capacitive.

According to the calculation formula of coaxial capacitance [10], the capacitances of air and medium between two electrodes are obtained as follows:

$$C_2 = \frac{2\pi\varepsilon_0 (H_0 - H_1)}{\ln(D/d)} \tag{2}$$

$$C_3 = \frac{2\pi\varepsilon_1 H_1}{\ln(D/d)} \tag{3}$$

Where, ε_0 and ε_1 are the dielectric constants of air and measured medium, respectively. H_0 is the height of container and H_1 is the thickness of measured medium. D and d are the diameters of container and metal probe, respectively.

Substituting Eqs. (2) and (3) into Eq. (1), the total capacitance can be written as:

$$C_t = C_1 + \frac{2\pi\varepsilon_1 H_0}{\ln(D/d)} + \frac{2\pi(\varepsilon_1 - \varepsilon_0)H_1}{\ln(D/d)} \quad C_3 = \frac{2\pi\varepsilon_1 H_1}{\ln(D/d)} \tag{4}$$

Therefore, Eq. (4) can be abbreviated as:

$$C_t = C_0 + KH_1 \tag{5}$$

Where, C_0 and K are two constants, which are related to the dielectric constant of the medium and the structure of the container. It can be seen that the measured total capacitance C_t is correlated linearly with the deposition thickness H_1 of the medium. Therefore, if the constants of C_0 and K can be determined, the thickness of the measured medium can be obtained according to the measured capacitance based on the Eq. (5).

3. Experimental apparatus

3.1 Test section

Figure 2 shows the diagram of the test section in the experimental apparatus. The test section consists of the electrode probe and the deposition chamber. The deposition chamber adopts a separated structure, which contains a stainless open cavity, a stainless board as the bottom, and an acrylic board as the top. The center of bottom board is drilled with a threaded holes of 20mm to install the electrode probe. The top board is cut with a square hole to add the medium into chamber. The bottom and top boards are clamped with the screws to avoid the measured medium to leak.

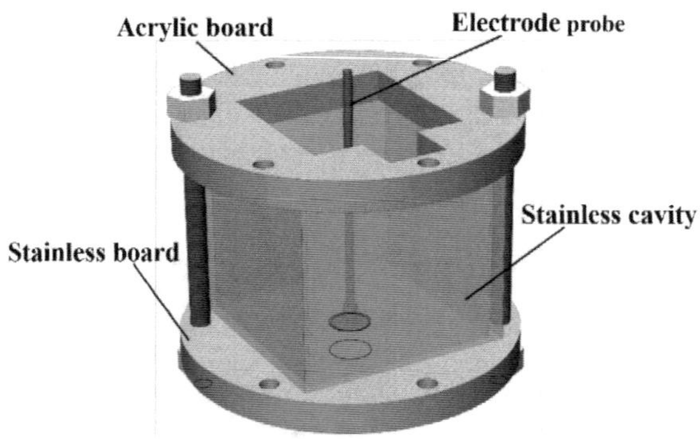

Figure 2. Diagram of test section.

3.2 Measured mediums

The fouling occurring in the CSC is mainly caused by the fly ash particles entrained in syngas, therefore, the current experiment selects the coal ash [Figure 3 (a)] as medium to simulate the fouling in cold state.

(a) Coal ash (b) Potential substitutes for slag

Figure 3. Mediums.

The slagging in the RSC is happened when the molten slags contact the water wall, solidify and stick to the wall. Since it is hard to realize the high temperature environment (>1000℃) of RSC under laboratory conditions, some substitutes are selected to simulate the slagging process in the open literatures, such as syrup and paraffin [11,12]. Based on these, this study chooses the three kinds mediums of syrup, paraffin, rosin as the potential substitutes of slag, as shown in Figure 3 (b). Among, the slag is obtained from the gasifier in Yangmei Fengxi Fertilizer Industry.

Due to the dielectric constant being a key parameter to determinate the capacitance of measured medium, the dielectric constants of three substitutes and the real slag are measured by employing a broadband dielectric spectrometers (Model: Novocontrol GmbH, Germany). The test data are shown in Figure 4. It is shown that the dielectric constant of slag decreases with the increase of test frequency. The dielectric constant of slag is obviously higher than that of other mediums in the low testing frequency range (< 100 kHz). However in the high frequency range (> 100 kHz), the dielectric constant of syrup is basically same to that of slag, while the dielectric constants of paraffin and rosin are obviously smaller. Therefore, this study selects the syrup as an alternative medium to simulate the slag in cold state.

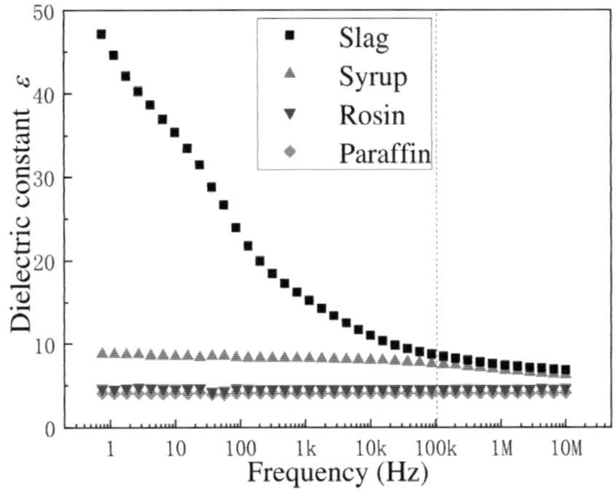

Figure 4. Dielectric constants of mediums.

3.3 Test electrode

Two kinds of electrode probes are used in this experiment, including the single-electrode probe and the double-electrode probe, as shown in Figure 5. When using the single-electrode probe for testing, the probe and cavity wall are used as two electrodes of the capacitor (as shown in Figure 2). While using the double-electrode probe, two probes are used as two electrodes of capacitor. The capacitance value for the capacitor is recorded by a high precision digital electric bridge (Model: Victor-4080C, China). Figure 6 shows the photograph of the experimental apparatus.

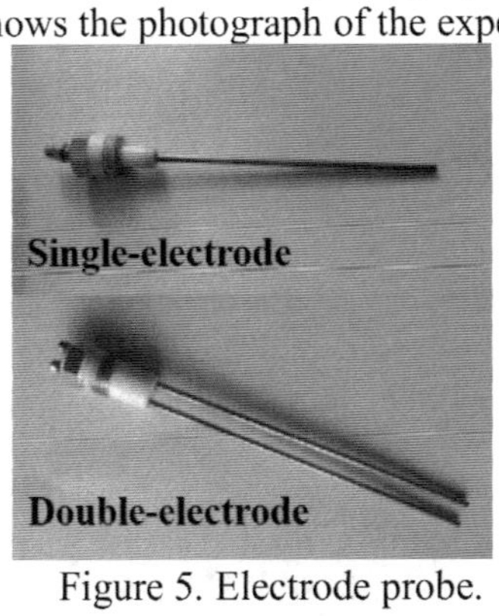

Figure 5. Electrode probe.

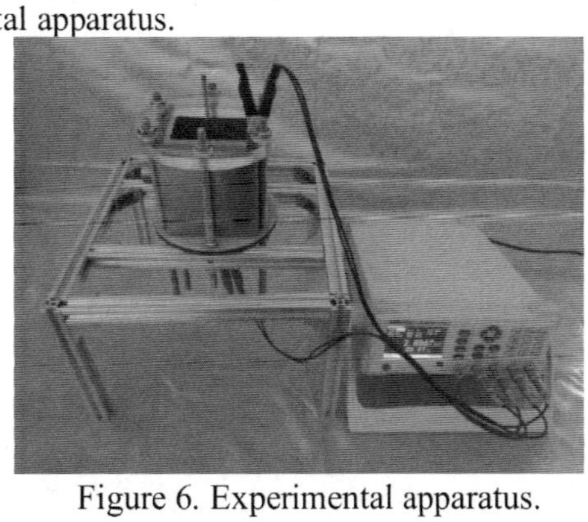
Figure 6. Experimental apparatus.

During the test process for ash deposition, the coal ash is firstly naturally deposited in the test chamber, then a spatula is used to shave the ash surface to be flat, and the capacitance of ash deposition is recorded. For the test of syrup deposition, the syrup is firstly heated to 180 ℃ for melting, and the molten syrup is deposited in the test chamber. When the molten syrup solidifies and its temperature decreases to the ambient temperature, the capacitance is measured. The ambient temperature is maintained at 25 ~ 26 ℃ by air conditioning during all test processes.

4. Results and discussion

4.1 Effect of test frequency on capacitance value

Figures 7 (a) and (b) show the effects of test frequency on capacitance value for the single-electrode probe, with ash deposition and syrup deposition, respectively. It can be seen that, under a fixed frequency, the test capacitance increases with the deposition thickness of ash/syrup increasing. For a given deposition thickness, the capacitance value has a fluctuant variation with the test frequency changing, and the variation trends are basically consistent for different deposition thicknesses. Figures 8 (a) and (b) show the effects of test frequency for the double-electrode probe. The phenomena is consistent with that of single-electrode test.

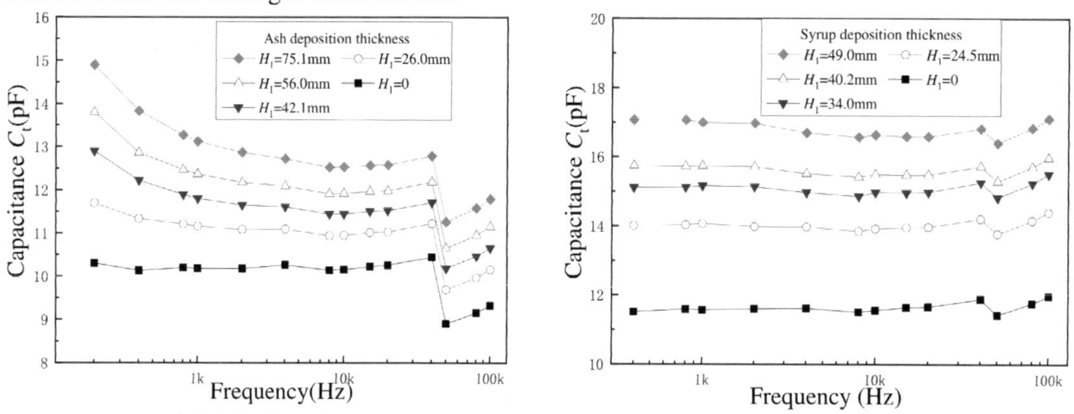

(a) Ash deposition (b) Syrup deposition

Figure 7. Effect of test frequency on capacitance value for the single-electrode probe.

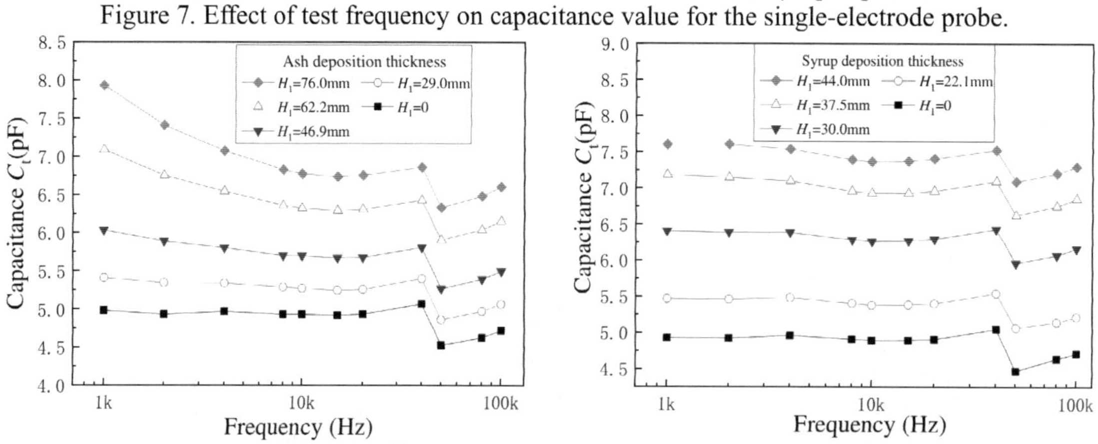

(a) Ash deposition (b) Syrup deposition

Figure 8. Effect of test frequency on capacitance value for the double-electrode probe.

4.2 Effect of deposition thickness on capacitance value

According to the results in Figure 4, the dielectric constant of syrup is basically consistent to that of slag at high frequencies. Meanwhile, considering that the frequency bands of capacitive sensors in industrial applications are mostly near 100kHz [13,14], this paper chooses 100 kHz as the test frequency to study the relationship between the deposition thickness of medium and the measured capacitance value.

4.2.1 Ash deposition

Figure 9 displays the variations of test capacitance with ash deposition thickness varying, under the single-electrode probe and double-electrode probe, respectively. It shows that with the increase of the ash deposition thickness, the measured capacitance value increases gradually. Especially, the capacitance value has a significant linear relation with the deposition thickness. This is consistent with the linear law presented in Eq. (5). By fitting the test data, the linear correlations between the capacitance and the ash deposition thickness are obtained as follows: C_t=0.033H_1+9.309 for the case of single-electrode probe, C_t=0.033H_1+4.069 for the case of double-electrode probe. As can be noticed, the slopes of the two correlations are same, but the intercept for the single-electrode is about twice than that for double-electrode. It can be attributed to the different ash content between the two electrodes for two kinds of test electrodes. Under the same ash deposition thickness, the amount of ash between single-electrode probe and the cavity wall (as the other electrode) is much larger than that between the two probes with double-electrode probe.

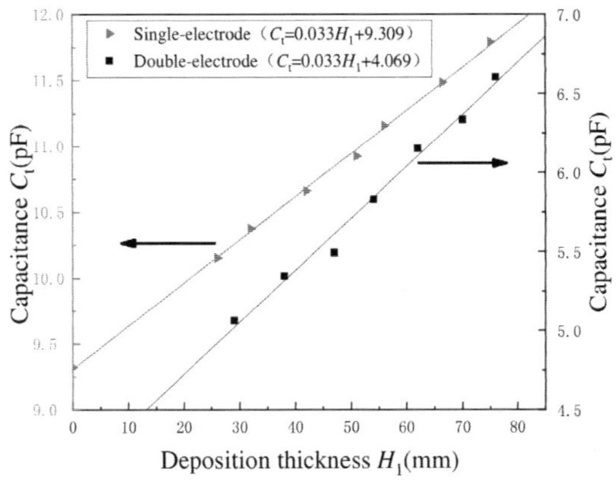

Figure 9. Variations of test capacitance with ash deposition thickness varying.

4.2.2 Syrup deposition

Figure 10 displays the variations of test capacitance with syrup deposition thickness varying. There is also a linear relationship between the test capacitance and the syrup deposition thickness. The linear correlations for single-electrode and double-electrode are C_t=0.103H_1+11.943 and C_t=0.093H_1+3.272, respectively. The slopes of two correlations are similar, but the intercept of former is obviously larger than that of latter.

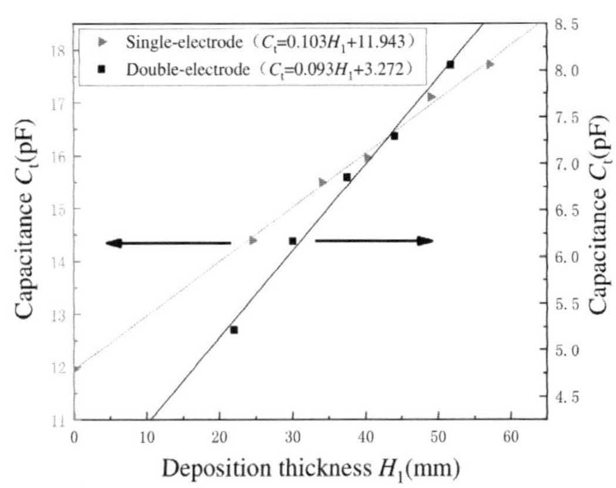

Figure 10. Variations of test capacitance with syrup deposition thickness varying.

4.3 Verification tests

Based on the linear correlations above, the verification tests are conducted as follows. By setting random deposition thickness of ash/syrup and reading the measured capacitance value, thus, the predicted deposition thickness may be obtained by employing the above linear correlations. Tables 1-2 display the comparisons of the predicted deposition thickness and the real thickness. The real thickness is obtained by using a vernier caliper.

Table 1. Comparisons for ash deposition with single-electrode and double-electrode.

Case	Single-electrode					Double-electrode				
	1	2	3	4	5	1	2	3	4	5
Capacitance value /pF	10.205	10.517	10.758	10.959	11.209	5.203	5.412	5.512	5.992	6.319
Predicted thickness /mm	27.30	36.70	43.97	50.03	57.61	34.27	40.64	43.67	58.21	68.21
Real thickness /mm	27.9	36.5	43.5	49.0	58.0	33.9	42.1	44.9	57.0	69.3
Relative error	2.14%	0.54%	1.08%	2.10%	0.68%	1.10%	3.48%	2.75%	2.13%	1.57%

Table 2. Comparisons for syrup deposition with single-electrode and double-electrode.

Case	Single-electrode					Double-electrode				
	1	2	3	4	5	1	2	3	4	5
Capacitance value /pF	13.963	15.463	16.548	17.265	17.812	5.574	6.223	7.023	7.731	8.496
Predicted thickness /mm	19.58	34.15	44.73	51.72	56.96	24.71	31.70	40.30	47.94	56.22
Real thickness /mm	19.9	34.5	43.2	51.0	58.5	23.5	33.1	41.5	49.6	57.5
Relative error	1.60%	1.03%	3.54%	1.41%	2.63%	5.15%	4.23%	2.89%	3.36%	2.23%

As seen from the data in the tables, the predicted deposition thicknesses by applying the linear correlations agree well with the real thicknesses. The maximum errors (absolute value) for ash and syrup deposition with the single-electrode are 2.14% and 3.54%, respectively. These are lower than those with the double-electrode, which are 3.48% and 5.15%, respectively. It can be explained that the capacitance measured by the single-electrode is decided by the deposition between the probe and the cavity wall, while the capacitance for double-electrode is decided by the deposition between two probes. The deposition amount of former is much larger than that of latter. As a result, the capacitance for the double-electrode is lower and the maximum error is larger, compared to the single-electrode. However, the double-electrode has an advantage that it does not need to use the cavity wall as an electrode of capacitor. It means that the double-electrode is more flexible to install in practical application and is not limited by the shape and material of the cavity.

5. Conclusion

This paper conducts an experiment study on the detection of deposition in cold state based on capacitance principle. The coal ash and syrup are selected as mediums to simulate the fly ash and slag. Two kinds of electrode probes are used to measure the capacitance value. The results show that the syrup has a similar dielectric constant with the real slag in high frequency range (> 100 kHz). The capacitance values fluctuate with the test frequency changing, under a given deposition thickness. The capacitance increases linearly with the deposition thickness increasing. The linear fitting correlations of ash deposition by single/double-electrode are C_t=0.033H_1+9.309 and C_t=0.033H_1+4.069, respectively; the correlations of syrup are C_t=0.103H_1+11.943 and C_t=0.093H_1+3.272, respectively.

Verification tests show that the predicted deposition thickness by applying the above correlations is basically consistent with the real thickness, and the maximum error is 5.15% for the current cases. The prediction error of the deposition thickness by single-electrode is less than that by double-electrode. These findings can provide beneficial guidance for the capacitance detection to be applied for the on-line monitoring of the fouling and slagging in the syngas cooler.

Acknowledgments
This work was supported by the National Key R&D Program of China (2017YFB0602703) and Shanxi Province Postgraduate Innovation Project (No. 110).

References

[1] Ni, J.J. (2011) Multiphase flow in entrained-flow gasifier and its radiation waste boiler, study on heat transfer and slag behavior. East China University of Science and Technology, Shanghai.
[2] Zhang, L., Wang, G.B., Xie, D.S., et al. (2011) Simulation of whole GE water-coal slurry gasification process. J. Chemical Engineering, 39(7):78-82.
[3] Deng, G., Han, L., Fan, Y., et al. (2014) Influence of syngas sensible heat recovery on performances of IGCC power plant based on coal water slurry gasification. J. Journal of Chinese Society of Power Engineering, 34(12):985-989 and 996.
[4] Wang, C., Liu, X.R. (2016) Cause analysis of slagging in waste boiler of Texaco gasifier. J. Shandong Chemical Industry, 45 (20): 121-122.
[5] Jing. Y.H., Yang, L., Liu, H.G., et al. (2013) Causes and improvement of slagging in radiation waste boiler of Texaco whole waste boiler process. J. Clean Coal Technology, 19 (2): 106-109.
[6] Wallhäußer, E., Hussein, M., Becker, T. (2012) Detection methods of fouling in heat exchangers in the food industry. J. Food Control, 27 (1): 1-10.
[7] Guérin, R., Ronse, G., Bouvier, L., et al. (2007) Structure and rate of growth of whey protein deposit from in situ electrical conductivity during fouling in a plate heat exchanger. J. Chemical Engineering Science, 62 (7): 1948-1957.
[8] Chen, X.D., Li, D.X.Y., Lin, S.X.Q., et al. (2004) On-line fouling/cleaning detection by measuring electric resistance—equipment development and application to milk fouling detection and chemical cleaning monitoring. J. Journal of Food Engineering, 61 (2): 181-189.
[9] Fan, X.C. (2005) Detection technology and system. Beijing University of Aeronautics and Astronautics Press, Beijing. pp. 215~227
[10] Sha, X. (2006) Research on radio frequency admittance capacitive level measuring instrument. D. Harbin Institute of Technology, Harbin.
[11] Zhang, J.F. (2011) Study on slag flow characteristics at slag mouth of gasifier. D. East China University of Science and Technology, Shanghai.
[12] Yuan, H.Y., Zhai, H.G., Ren, H.P., et al. (2005) Experimental study on slag deposition in entrained-flow gasifier. J. Journal of East China University of Science and Technology, 31(3):393-397.
[13] Xue, Y., Dou, Y.K., Chang, X.M., et al. (2015) Research on capacitance induction ice coating thickness sensor for overhead transmission lines. J. Practice and Cognition of Mathematics, 45(1):151-157.
[14] Dou, Y.K. (2010) Research on mechanism and application of fixed-point ice thickness detection based on capacitance induction technology. Taiyuan University of Technology, Taiyuan.

IOP Conf. Series: Materials Science and Engineering **721** (2020) 012019 doi:10.1088/1757-899X/721/1/012019

Efficiency Calculation and Economic Analysis of Straight Fins with Various Shapes Based on Steady-state Heat Conduction

Jiying Chen[1], Shenghao Jiang[2], Jiayi Hu[3] and Jiaxin Ji[1,*]

[1]College of New Energy, China University of Petroleum, Qingdao, Shandong, 266580, China

[2]School of Engineering and Applied Sciences, Harvard University, Cambridge, Massachusetts, 02138, America

[3]International School, Jiangxi University of Finance and Economics, Nanchang, Jiangxi, 330000, China

*Corresponding author's e-mail: Ying1606050115@163.com

Abstract. Fins are an important means of heat transfer enhancement, which can increase the heat transfer area, reduce the heat resistance of convection heat transfer and enhance the heat transfer capacity of equipment. The analytical solutions of rectangular fins, trapezoidal fins and triangular fins in one dimension are given by using the direct solution method. Then the heat conduction of fins is discretized by the finite volume method (FVM). By solving the discrete linear equations, the numerical solutions of temperature distribution on the surface of fins with constant property and variable property are obtained. Based on the law of temperature distribution, the relationship between the length, the height and the width of fins and fin efficiency is explored, and its economic efficiency is discussed. The results show that the temperature gradient of rectangular fins is the smallest, and that of triangular fins is the most significant. For various shapes of fins, the most effective way to increase heat dissipation is to increase the width of the fin, and for the fin whose machining length is 1 unit, the efficiency of the first 20%~30% part of the maximum length changes significantly, which can be used as the focus of cost optimization.

1. Introduction

With the continuous progress of productivity, the energy problem in social development is becoming more and more serious. Energy conservation and emission reduction has become an important goal in industrial production. Although China is a country with large reserves of resources, there are still many problems to be solved urgently in energy utilization. Improving energy utilization rate is the key to production. Heat exchanger is the key equipment in the thermal system. Depending on its structural characteristics, heat transfer in different fluid media can be realized. Fins are widely used in heat exchangers, such as condensers, radiators and air heaters of refrigerating plant [1-4], which is a measure to enhance heat transfer by increasing the heat transfer area. Optimum design of fins with different shapes to improve the heat transfer efficiency of products is an effective measure to promote energy saving. The fins have various shapes. At present, the shapes of fins are mainly rectangular fins, trapezoidal fins, triangular fins and so on (Fig. 1). In special occasions, there will be circular finned tube and elliptical finned tube [5-6]. This study mainly focuses on the efficiency of rectangular fins, trapezoidal fins and triangular fins under reasonable thermodynamic assumptions in engineering.

IOP Conf. Series: Materials Science and Engineering **721** (2020) 012019 doi:10.1088/1757-899X/721/1/012019

2. Overview of physical model

Assuming that the fin is placed in the external environment with temperature T, the temperature at the end of the fin is constant T_W, as the first boundary condition; (2) the end of the fin is adiabatic (heat flux $q = 0$), as the second boundary condition. Because most of the fins are made of metal with good heat conduction, it is assumed that the temperature of the fins does not change in the direction of width and height, that is, the heat transfer resistance 1/h on the surface of the fins is much greater than the thermal-conduction resistance δ / λ in the fins. Therefore, the temperature of the fins can be considered uniform on any cross section. The total heat dissipation of fins can be calculated by accumulating the total heat dissipation ϕ of each infinitesimal element, and the heat dissipation of each infinitesimal element at the base-temperature of fins is ϕ_0. Based on these assumptions, the research problem is simplified to one-dimensional steady-state heat conduction without internal heat source.

The formula for calculating fin efficiency η is:

$$\eta = \frac{\phi}{\phi_0} \tag{2-1}$$

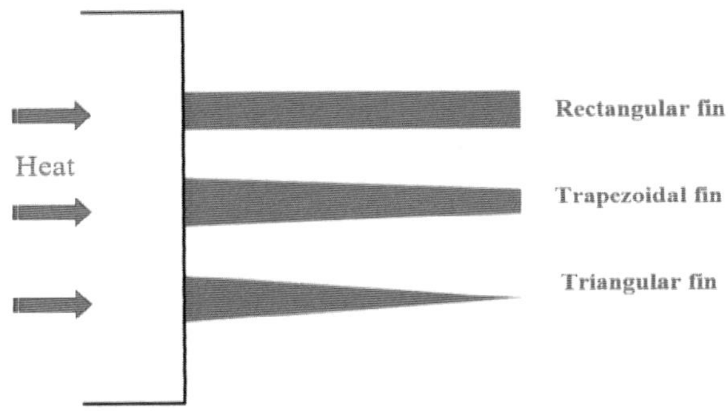

Figure 1. Side view of fins with various shapes

3. Establishment of mathematical model

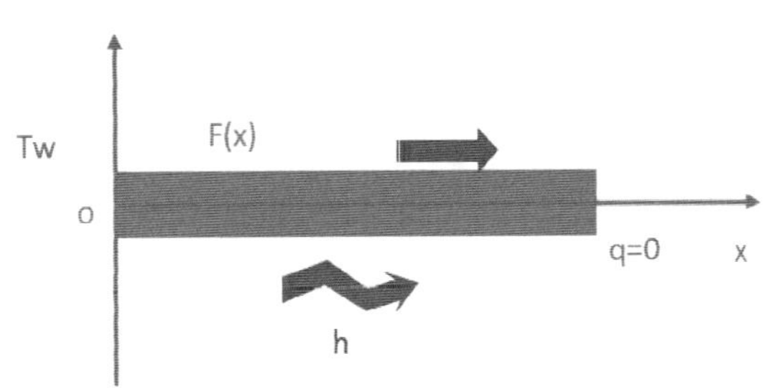

Figure 2. Schematic diagram of thermodynamic conditions for one-dimensional steady-state heat conduction

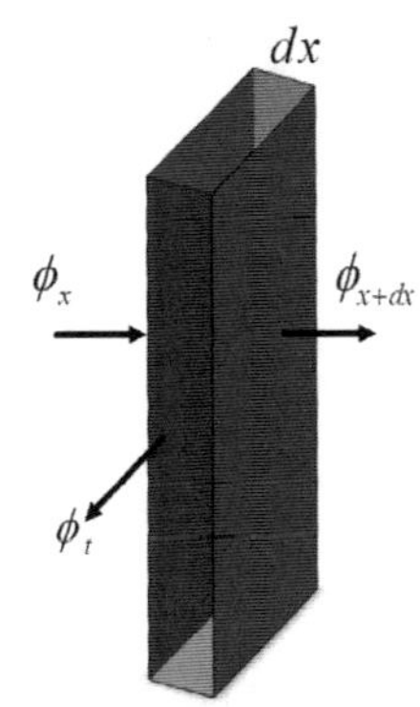

Figure 3. Analysis of infinitesimal element energy (rectangular, trapezoidal and triangular are all applicable. Here is the trapezoidal fin infinitesimal element.)

In order to establish a mathematical description of the temperature distribution of fins, the coordinates shown in Fig. 2 are established.

As shown in Fig. 3, under steady-state conditions, the infinitesimal element satisfies the following energy conservation:

$$\phi_x = \phi_t + \phi_{x+dx} \tag{3-1}$$

By Fourier's law:

$$\phi_x = -\lambda A(x)\frac{dT}{dx} \tag{3-2}$$

$$\phi_{x+dx} = \phi_x + \frac{d\phi_x}{dx} \tag{3-3}$$

Where,λ: The thermal conductivity of fin materials;A(x): The cross-sectional area of the fin at x.

According to the formula of Newton's law of cooling, the heat dissipation through the side of the fin is as follows:

$$\phi_t = hF(T - T_f) \tag{3-4}$$

Where, h: Local surface heat transfer coefficient of fluid with temperature T_f and fin surface at location x; F: Side superficial area of infinitesimal element.

And because:

$$F = Uds = U(x)\sqrt{1+f'^2}dx \tag{3-5}$$

Where, U(x): The perimeter of infinitesimal element at x; ds: The molded line of infinitesimal element fins; f: The molded line equation of fins.

Substituting (3-5) into (3-4), the following formula is obtained:

$$\phi_t = hU(x)\sqrt{1+f'^2}(T - T_f)dx \tag{3-6}$$

By substituting (3-2), (3-3) and (3-6) into the energy balance equation (3-1), it is concluded that:

$$\frac{d}{dx}\left(\lambda A(x)\frac{dT}{dx}\right) - hU(x)\sqrt{1+f'^2}(T - T_f) = 0 \tag{3-7}$$

By introducing excess temperature $\theta = T - T_f$, the complete mathematical model is as follows:

$$\begin{cases} \frac{d}{dx}\left(\lambda A(x)\frac{d\theta}{dx}\right) - hU(x)\sqrt{1+f'^2}\theta = 0 \\ \theta|_{x=0} = T_w - T_f \\ -\lambda\frac{d\theta}{dx}|_{x=L} = 0 \end{cases} \tag{3-8}$$

For rectangular fins, the reference analysis solution is [7]:

$$\theta = \theta_0 \frac{ch[m(x-H)]}{ch(mH)} \tag{3-9}$$

Where:

The cosine function $ch(x) = \frac{e^x + e^{-x}}{2}$.

To calculate trapezoidal and triangular straight fins, simply replace all the molded line expressions:

$$U(x) = 2b + 4f(x) \tag{3-10}$$

Substituting (3-10) into (3-8), the following formula is obtained:

$$2\lambda b\frac{d}{dx}\left[f(x)\frac{d\theta}{dx}\right]=h(2b+4f(x))\sqrt{1+f'^2}\theta \tag{3-11}$$

Concluded as:

$$f(x)\frac{d^2\theta}{dx^2}+f'(x)\frac{d\theta}{dx}=\frac{h(b+2f(x))}{\lambda b}\sqrt{1+f'^2}\theta \tag{3-12}$$

Assuming $\theta=\theta_1=e^{\int g(x)dx}$

Making

$$\varphi(x)=\frac{h(b+2f(x))}{\lambda b}\sqrt{1+f'^2} \tag{3-13}$$

Concluded as:

$$g'(x)+\frac{f'(x)}{f(x)}g(x)+[g(x)]^2=\frac{\varphi(x)}{f(x)} \tag{3-14}$$

The solution of equation (2-14) must contain two linear independent special solutions. By using undetermined coefficient method and constant variation method, the general solution of (2-14) is obtained as follows:

$$\theta(x)=(1+C_1\int\frac{e^{-\int 2\frac{1}{f(x)}\left[(\int\varphi(x)dx+C_2)+\frac{1}{\left(\int\frac{dx}{f(x)}+C_3\right)}\right]dx}}{f(x)}dx)e^{\int\frac{1}{f(x)}\left[(\int\varphi(x)dx+C_2)+\frac{1}{\left(\int\frac{dx}{f(x)}+C_3\right)}\right]dx} \tag{3-15}$$

4. The numerical solution

4.1Computational domain and equation discretization

The inner node method is used to discretize the computational domain, and the discretized object of the equation is the energy conservation equation of the infinitesimal element (3-7).Take the control volume as shown in Figure 4. Integrate the control volume P:

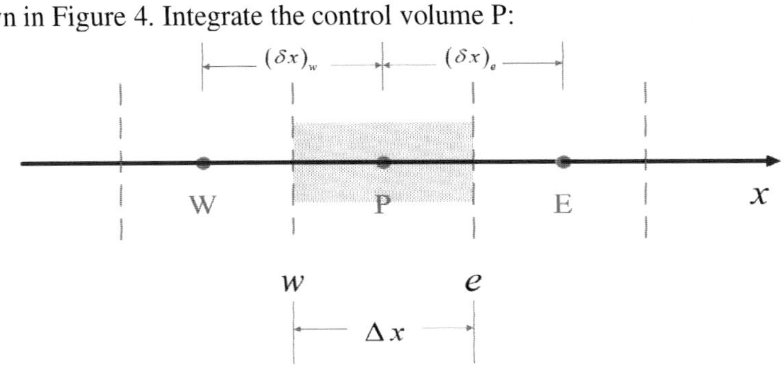

Figure 4. The selection of control volume

$$\int_w^e\frac{d}{dx}\left(\lambda A(x)\frac{dT}{dx}\right)dx=\int_w^e hU(x)\sqrt{1+f'^2}(T-T_f)dx \tag{4-1}$$

Suppose that the T molded line is linear [8], then (4-1) on the left side is:

$$\int_w^e \frac{d}{dx}\left(\lambda A(x)\frac{dT}{dx}\right)dx = \left(\lambda A(x)\frac{dT}{dx}\right)_e - \left(\lambda A(x)\frac{dT}{dx}\right)_w = \lambda_e A_e(x)\frac{T_E - T_P}{(\delta x)_e} - \lambda_e A_e(x)\frac{T_E - T_P}{(\delta x)_e} \quad (4\text{-}2)$$

Assuming that T is a stepped change in space

$$\int_w^e hU(x)\sqrt{1+f'^2}(T-T_f)dx = h\int_w^e U(x)\sqrt{1+f'^2}(T-T_f)dx \quad (4\text{-}3)$$

By substituting (4-2) into (4-1), we can get:

$$\left(\frac{\lambda_e A_e(x)}{(\delta x)_e} + \frac{\lambda_w A_w(x)}{(\delta x)_w} + h\int_w^e U(x)\sqrt{1+f'^2}dx\right)T_P = \frac{\lambda_w A_w(x)}{(\delta x)_w}T_w + \frac{\lambda_e A_e(x)}{(\delta x)_e}T_E + h\int_w^e U(x)\sqrt{1+f'^2}dxT_f \quad (4\text{-}4)$$

Formula (4-4) is rearranged into

$$a_P T_P = a_W T_W + a_E T_E + h\int_w^e U(x)\sqrt{1+f'^2}dx \quad (4\text{-}5)$$

The temperature distribution can be obtained by solving the linear equations (4-5) obtained by the discrete method. Actual heat dissipation:

$$\phi = \sum_{i=2}^{N-1} hF(x_i)(T_i - T_f) \quad (4\text{-}6)$$

The heat dissipation of the whole fin at the base-temperature of fins:

$$\phi_0 = \phi = \sum_{i=2}^{N-1} hF(x_i)(T_i - T_w) \quad (4\text{-}7)$$

5. Analysis of solution results

Assuming that the fins are adiabatic at the end are placed in the air, the convection heat transfer coefficient on the surface of the fins is $10.55W/(m^2 \cdot K)$, the ambient air temperature is 62℃, the length of the fins perpendicular to the vertical wall is L=0.10m, the width of the joints between the fins and the vertical wall is b=0.01m, and the constant temperature at the joints is Tw=80℃. Fig.5 shows the temperature of rectangular, trapezoidal and triangular fins varying with position.

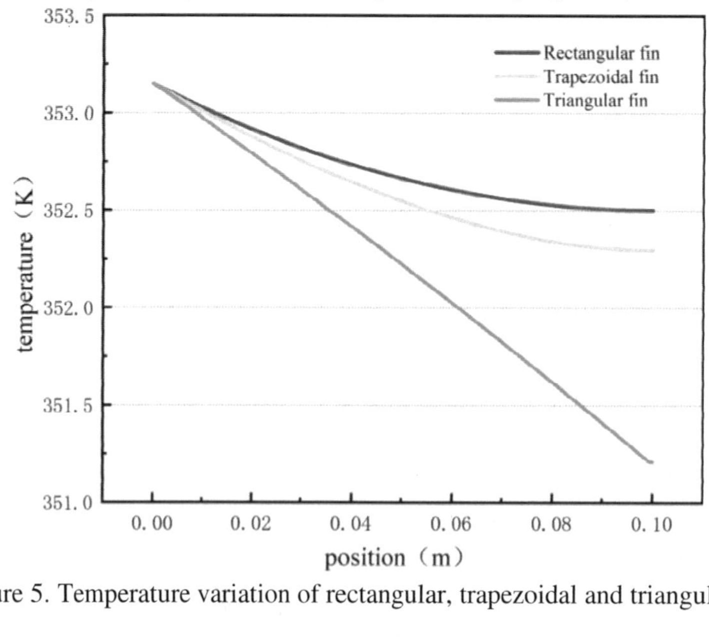

Figure 5. Temperature variation of rectangular, trapezoidal and triangular fins

As can be seen from Fig. 5, under the same thermodynamic conditions, the temperature of rectangular fins decreases most slowly along the length of fins, while that of triangular fins decreases most quickly. And the temperature drop rate of triangular fins tends to increase. For components requiring rapid heat transfer, triangular fin is a more preferable choice.

Take the length of fins from zero to one meter as research object. Define the heat dissipation efficiency coefficient e_i corresponding to node i in numerical calculation:

$$e_i = \frac{\varphi_i}{\varphi_{\max[1,2,3...n]}} \tag{5-1}$$

e_i denotes the ratio of the heat dissipation of corresponding control volume of each node to the maximum heat dissipation of all nodes. This method of comparing each value with regional maximum to normalize can be used to denote the heat conduction of the corresponding infinitesimal element relative to the whole fin after each segment of fin length is added. The data is plotted in Fig. 6.

From Fig. 6, it can be seen that the fin efficiency increases with the increase of fin length. To further illustrate this trend, the marginal cost theory in economics is more appropriate. In economics and finance, marginal cost refers to the cost that a manufacturer increases per unit of output. This concept indicates that the cost per unit of product is related to the total volume of product. Mathematically, marginal cost (MC) is expressed as partial derivative ($\frac{\partial Tc}{\partial Q}$) of total cost (TC) and quantity (Q) [9-10]. For the research of fins, heat dissipation and fin efficiency are our concerned research objectives, and fin length is the "product" that we can increase or decrease freely. Therefore, the coefficient ζ can be defined:

$$\zeta = \frac{\Delta\varphi}{\Delta L} \tag{5-2}$$

Where: φ is the heat dissipation of infinitesimal element, and L is the length of fins. ζ essentially represents the slope of the curve in Fig. 6.

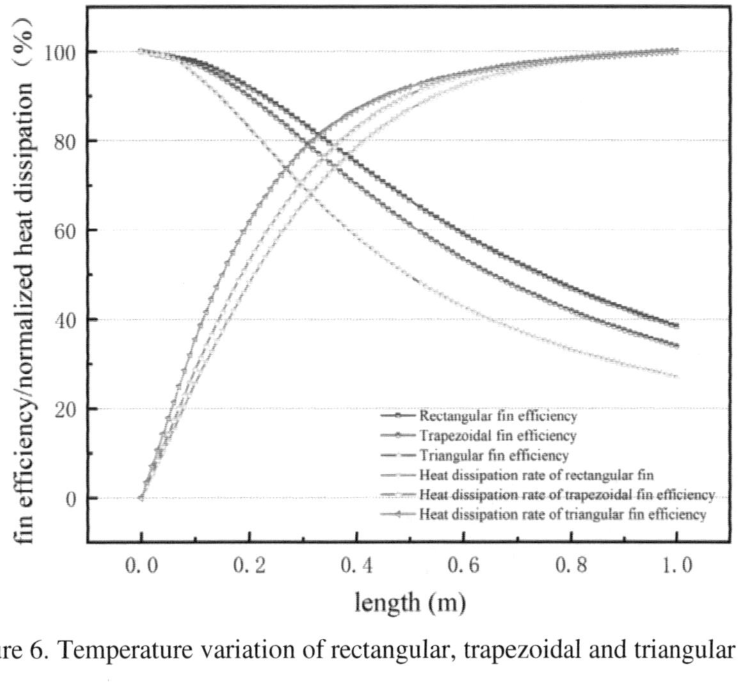

Figure 6. Temperature variation of rectangular, trapezoidal and triangular fins

According to the change trend of image slope: increasing the length of fins at the initial stage can significantly increase heat dissipation and maintain high material efficiency. When the length is increased to about 0.3 m, the change of heat dissipation will slow down when the length is increased. In order to find the appropriate balance point between fin efficiency and heat dissipation, the length of fins should be selected between 0.2 and 0.4m in the image. The intersection point of heat dissipation and fin efficiency curve is also a reasonable choice in this case. The corresponding normalized heat dissipation and fin efficiency at the intersection point are both more than 70%.
The relations between the fin efficiency of rectangular, trapezoidal and triangular fins and the width and height of the fins are shown in Fig. 7 and Fig. 8.

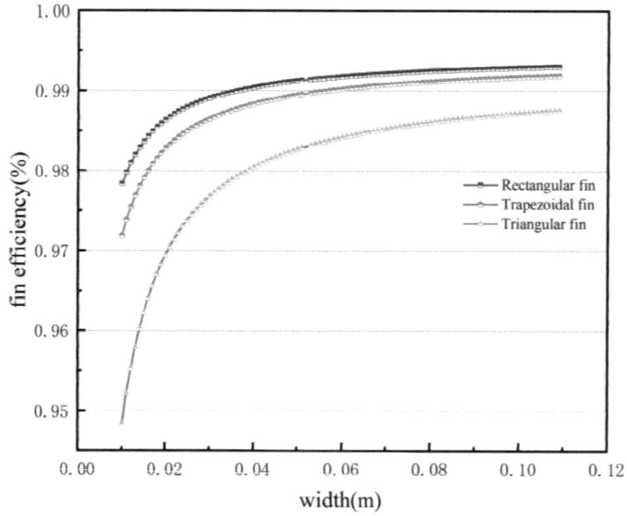

Figure 7. The relationship between fin efficiency and the width of fins

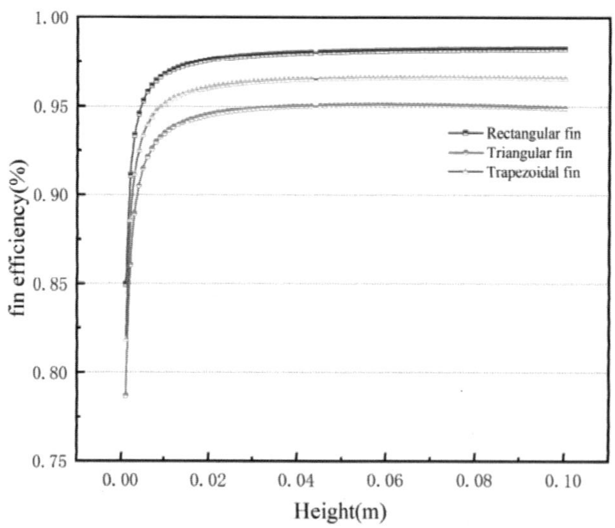

Figure 8. The relationship between fin efficiency and the height of fins

It can be seen from Fig. 7 and Fig. 8 that the increase of the height and width of fins is accompanied by the increase of thermal efficiency. Compared Fig. 6, Fig. 7 and Fig. 8, the effect of increasing the height of fins on efficiency is the most significant. Similar to the variation law of the length of fins, increasing the height and width of fins also has the phenomenon of marginal cost decreasing. Especially in the study of fin width, after experiencing short and steep efficiency fluctuation, increasing fin width has little effect on the fin efficiency of all kinds of fins. The efficiency of rectangular fins is the highest under the same fin height compared with the other two.

6. Conclusion

1) The temperature of the fins decreases with the length of the fins, and the speed of the decrease varies with the shape of fins. Compared with rectangular and trapezoidal fins, triangular fins have the fastest temperature drop rate. When the fin is long enough, the temperature of the fin will drop to the fluid temperature. By analyzing the molded line equation, it can be further summarized: with the increase of the absolute value of the slope of fin molded line, the temperature of fins decreases faster.

2) With the increase of the length of fins, the fin efficiency decreases, and the decreasing speed is getting slower and slower. For different shapes of fins, changing the length of fins, the decreasing speed of the fin efficiency increases with the increase of absolute value of fin molded line. From Fig. 7, it can be seen that with the increase of the width of fins, the fin efficiency increases rapidly and finally approaches one. For the same width of fins, the efficiency of rectangular fin is the highest, trapezoidal fin is the second, and triangular fin is the smallest. Fig. 8 shows that with the increase of the height of fins, the fin efficiency will increase rapidly, and finally approaches one. For the same height of fins, the efficiency of triangular fin is less than that of rectangular fin.

3) From Figs. 6, 7 and 8, it can be seen that the heat dissipation of fins increases with the increase of the length, height and width of fins, but the fin efficiency decreases with the increase of the length of fins. Compared Figs. 7 and Fig. 8, the most effective way to increase the heat dissipation of both rectangular fins and triangular fins is to increase the width of fins.

References

[1] W.T. Ji., C.Y. Zhao, J. Lofton., Z.Y. Li, D.C., Zhang, Y.L. He, W.Q. Tao (2018) Condensation of r134a and r22 in shell and tube condensers mounted with high-density low-fin tubes. Journal of Heat Transfer, 140:091503

[2] N. Lewpiriyawong, K. L. Khoo, C. Sun, P. S. Lee. (2019) Thermal and hydraulic analysis of aluminium oblique-tube condenser coils with plain fins manufactured by controlled atmosphere brazing. International Journal of Refrigeration,101:81-89

[3] Priyam A, Chand P (2019) Experimental investigations on thermal performance of solar air heater with wavy fin absorbers. Heat and Mass Transfer,55:2651–2666

[4] Sathish Kumar, T. R., Jegadheeswaran, S., Chandramohan, P. (2018). Performance investigation on fin type solar still with paraffin wax as energy storage media. Journal of Thermal Analysis and Calorimetry,136:101-112

[5] Abdulkerim O., Ali P., Ali B. Olcay, M., Hilmi A (2018) An experimental, computational and flow visualization study on the air-side thermal and hydraulic performance of louvered fin and round tube heat exchangers. International Journal of Heat and Mass Transfer,121:153-169

[6] S. S., Yogesh, A.S. Selvaraj., D.K. Ravi., T.K.R. Rajagopal (2018) Heat transfer and pressure drop characteristics of inclined elliptical fin tube heat exchanger of varying ellipticity ratio using CFD code. International Journal of Heat and Mass Transfer,2018,119:26-39

[7] Shiming Yang， Wenquan Tao(2006)Heat Transfer. Higher Education Press, Beijing

[8] Payam H., Hassan K. B., Mehdi M., Navid B., Emad H. M.(2019) Numerical modeling of nanofluid flow and heat transfer in a quartered gearwheel-shaped heat exchanger using FVM. Chinese Journal of Physics,59:591-605

[9] Siverskog, J., Henriksson, M. (2019). Estimating the marginal cost of a life year in sweden's public healthcare sector. The European Journal of Health Economics.20: 1-12.

[10] Bryon J.P, Allen M.F. (2019) A comparison of parametric and nonparametric estimation methods for cost frontiers and economic measures. Journal of Applied Economics 22:59-84

IOP Conf. Series: Materials Science and Engineering **721** (2020) 012020 doi:10.1088/1757-899X/721/1/012020

Study on the effect of molten ash on the combustion process for biomass

Yao Xu, Xun Zou, Shuai Jin, Ming Zhai*, Peng Dong

School of Energy Science and Engineering, Harbin Institute of Technology, Harbin, Heilongjiang, 150001, China

*Corresponding author's e-mail: zhaiming@hit.edu.cn

Abstract. In this study, the high temperature visual test system (HTVTS) and high temperature thermogravimetric test system (HTTTS) were set up to observe the formation process of ash, thereby analyzing the mechanism of biomass melt combustion. Then, the effect of the molten ash on the combustion process was investigated. Results show that during the combustion, once the local temperature of the biomass particle is higher than the ash melting temperature, the generated ash could be melted and contracted into a spherical shape by the surface tension. The molten ash located in the interior of the biomass particle adhere to the char surface, thereby preventing the contact between char and oxygen. The molten ash located on the surface of biomass particle could drop down from the surface of biomass particle. When the conversion rate is higher than 45%, the effect of molten ash on the reaction rate becomes more obvious.

1.Introduction

One of the biggest challenges facing in 21st century is overcoming the shortage of fossil fuels and controlling greenhouse gas emissions [1].This challenge has prompted industry and academia to study the feasibility of using renewable fuels[2-3]. In general alternative energy sources, biomass has the advantages of renewability, cleanliness and abundant sources, and it has become a potential energy source to replace fossil fuels [4-5,6].

Combustion is the most mature and widely used technology in biomass utilization. Many scholars have studied the combustion characteristics of biomass particles from different perspectives [7-9]. However, these studies are mainly focus on the temperature range below 1000 ° C, corresponding to the operating temperature range of fluidized bed and fixed bed. In the recent retrofit of a suspended-fired boiler, biomass was first granulated and then pulverized into millimeter-sized particles [10]. Abbas et al. [11] found that when milled biomass is less than 1 mm, energy demand is significantly increased. This indicates that it is important to study the combustion characteristics of biomass particles larger than 1 mm in the higher temperature range.

Biomass combustion can be divided into a pyrolysis stage and a char conversion stage, in which the char conversion stage is considered as a rate limiting process [13]. During the char conversion stage, ash is formed on the surface of the char. Due to the large amount of alkali metals and alkaline earth metals (AAEMs), these ash can be melted and fused, thereby changing the particle structure[14-16].It is not difficult to imagine that as the particle size of biomass particles increases, the char conversion stage is mainly controlled by the gas diffusion. Therefore, the effect of ash fusion on char conversion stage is more pronounced. However, similar studies are rarely mentioned in the combustion of biomass pellets.

In this study, the high temperature visual test system (HTVTS) and high temperature thermogravimetric test system (HTTTS) were set up to observe the formation process of ash, thereby analyzing the mechanism of biomass melt combustion. Then, the effect of the molten ash on the combustion process was investigated, which could provide the experimental references for the numerical modeling of biomass particles.

2.Experimental part

The biomass (corn stalk) used in this work comes from Heilongjiang province, China. Its fuel characteristics are shown in table 1. Before the experiment, the biomass particles are polished into the regular samples. High-temperature combustion experiments are carried out on the self-designed high temperature visual test system (HTVTS) (figure 1(a)) and high temperature thermogravimetric test system (HTTTS) (figure 1(b)). The furnace temperature is increased to the setting temperature at the rate of 20°C/min firstly. Then starts the air intake device. The flow rate is controlled to 5 L/min. For HTVTS, when the morphology of the sample is no longer changed, the experiment is stopped. After that, the thermocouple is buried in the particle center and quickly placed into the furnace of HTTTS. When the temperature and mass are no longer changed, the experiment is finished.

Table 1. Proximate analysis and AFT test

Sample	Proximate analysis					AFT test			
	M_{ad}/%	V_{ad}/%	A_{ad}/%	FC_{ad}/%	$Q_{net,ar}$/kJ/kg	IDT/℃	ST/℃	HT/℃	FT/℃
Corn stalk	7.24	71.69	12.2	14.5	17.5	1045	1101	1152	1185

M: Moisture. V: Volatile. A: Ash. FC: Fixed carbon. a ad: Air drying base. b daf: Dry ash free base

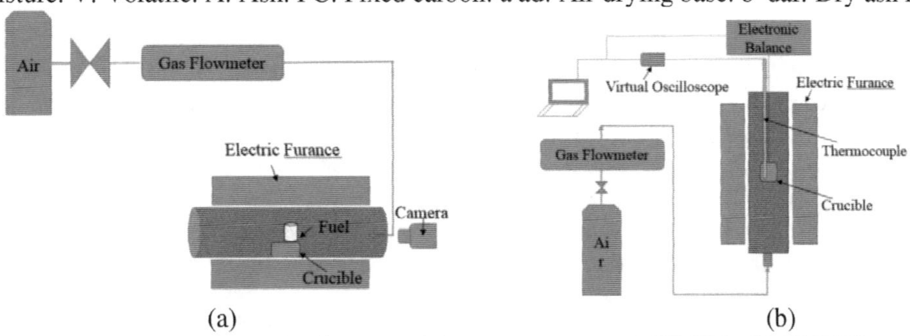

Figure 1. Schematic diagram of the test system: (a) HTVTS; (b) HTTTS.

3. Results and discussion

Figure 2(a) shows the morphology changes of biomass particle during combustion at different ambient temperatures. It can be found that the biomass combustion process could be divided into the three stages: the combustion stage of volatiles, the co-combustion stage of volatiles and char, as well as the combustion stage of char. When the ambient temperature are 920°C and 1053°C, biomass particle could maintain the initial cylindrical structure during the combustion. Once the temperature increases to 1192°C and 1345°C, the particle shape of biomass changes significantly. Firstly, the smaller spherical particles are formed on the surface of biomass particle. Then, the neighboring spherical particles are coalesced into the large spherical particle. Under the action of gravity, the large spherical particles drop down from the surface of biomass particle. Figure 2(b) shows the changes of the section area during the combustion. It can be seen that the particle size of biomass decreases with the progress of the reaction. Moreover, the decreasing tendency increases with the ambient temperature. It is important to note that there is a critical point at 1192°C and 1345°C. After the critical point, the particle size decreases rapidly. Combined with figure 2(a), the critical point corresponds to the time

IOP Conf. Series: Materials Science and Engineering **721** (2020) 012020 doi:10.1088/1757-899X/721/1/012020

range of large spherical particles appearing on the particle surface. This means that the obvious change of particle size at 1192°C and 1345°C could be attributed to the merging and peeling of adjacent spherical particles.

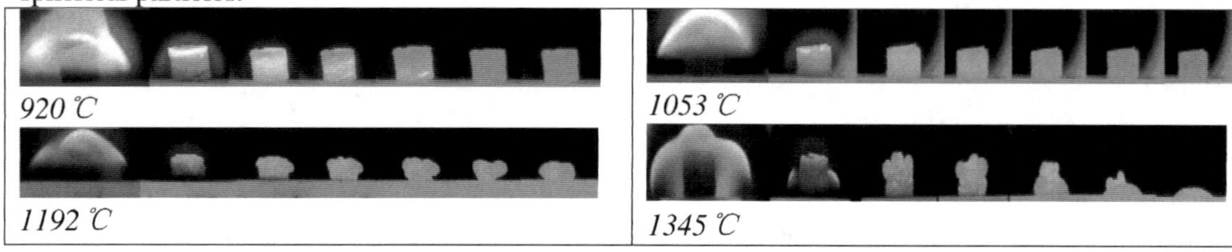

(a)

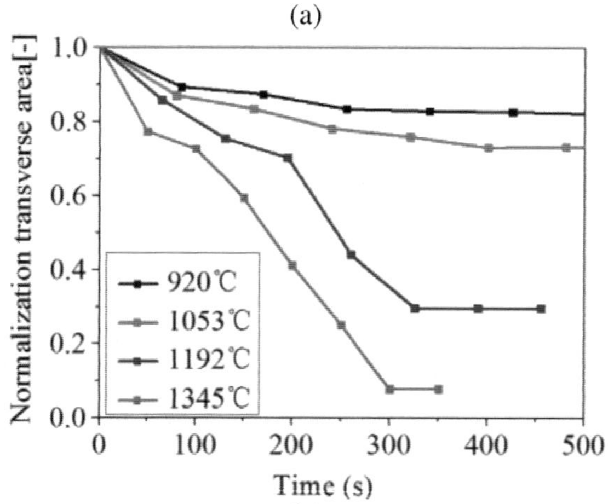

(b)

Figure 2. Combustion process: (a) direct observation; (b) evolution of particle size.

Figure 3 shows the results of SEM-DES analysis. When the ambient temperature is 920°C, the biomass particle exhibits an irregular loose structure. Meanwhile, the clusters of spherical particles could be observed. The loose structure of the particle contains a large amount of Si, C and O elements, and a small amount of Al, Mg, Fe, K, Ca. While, the spherical particle cluster contains Si, O, Al, Mg, Fe, K and a small amount of C elements. Studies have shown that Si, Al and Fe elements are stable and retained elements in the molten ash. During the combustion, due to the exothermic heat of combustion, the local temperature of the biomass particle is higher than the ash melting temperature. At this time, the generated ash is melted and further contracted into a spherical shape by the surface tension. When the carbon is present, the spherical particles exists in the form of clusters of particles. As the decrease of the carbon, the spherical particles exist in form of the individuals. When the temperature rises to 1053 ° C, the structure of the biomass particle becomes smoother. The reason may be caused by the melting and plastic transformation of the cell structure. When the ambient temperature is increased to 1192°C and 1345°C, the biomass particle is almost completely covered by the independent spherical particles. Meanwhile, the C element is not detected in the spherical particles, which means that the ash melting and peeling could not cause char loss.

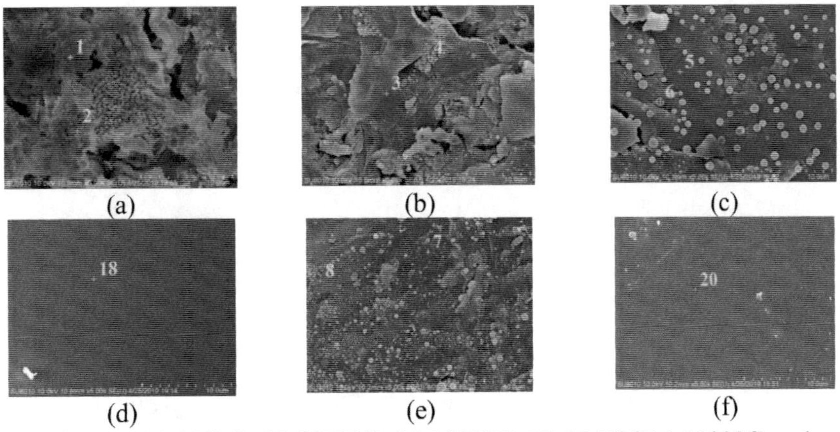

Figure 3. SEM image: (a) 920°C; (b) 1053°C; (c) 1192°C; (d) 1345°C(e) 1192°C molten sphere; (f) 1345°C molten sphere.

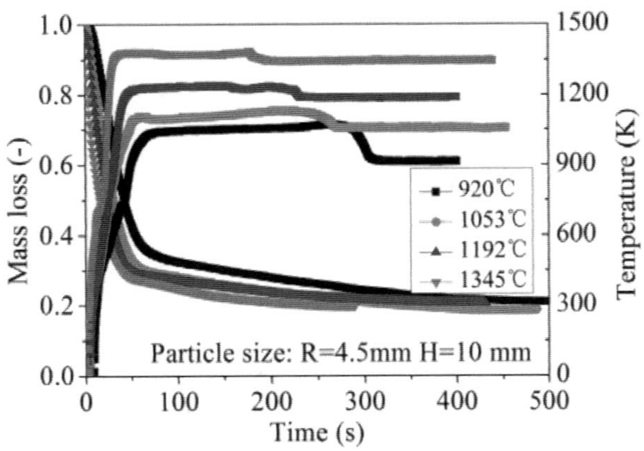

Figure 4. Center temperature and mass of biomass particle as a function of time.

Fig.4 shows the change of the center temperature and mass for biomass particle during the combustion. When the biomass particle is exposed to a high-temperature environment, the external heat transfers to the biomass particle by the radiation and convection. Therefore, the temperature of the particle rises rapidly because of the large initial temperature difference between the particle and the environment. In addition, the volatiles in the particle sharply release and react, which leads to the particle temperature is higher than the ambient temperature. As the reaction proceeds, the biomass reaction gradually changes from the volatiles combustion to char combustion, thereby reducing the rate of heat release. At the same time, the heat dissipation of the particle and the ash fusion also suppress the increase of the particle temperature. When the heat released by the combustion is less than the heat dissipation of the particles and the heat absorption of ash melting, the particle temperature decreases and finally is consistent with the ambient temperature. The initial temperature difference between the environment and the biomass particle increases with the ambient temperature, so that the weight loss curve and the temperature curve move to the upper left. However, the increase of the particle temperature results in the increase of the heat transfer between the particle and the environment. Therefore, the heating rate of the second stage decreases with the increase of the ambient temperatures.

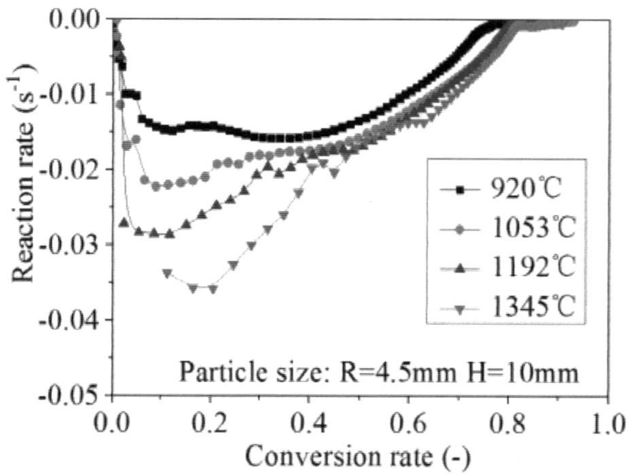

Figure 5. Reaction rate as a function of conversion

Fig.5 shows the reaction rate of biomass particle as a function of conversion. It can be found that the reaction rate of the biomass increases firstly and then decreases with the conversion rate. When the conversion rate is lower than 45%, rising ambient temperature could increase the reaction rate, which is because that the release of volatiles is mainly controlled by the heat transfer. When the conversion rate is higher than 45%, with the increase of the ambient temperature from 920°C to 1053°C, the reaction rate increases. However, the reaction rate at 1192°C is nearly same as that at 1053°C, which means that the reaction stage is in the diffusion control area. Compared to the reaction rate at 1192°C, when the temperature rises to 1345°C, the reaction rate of the particle is further raised. The reason can be explained by that rising temperature speeds up the ash melting. The molten ash falls off the particle surface of the biomass, which decreases the diffusion resistance of the oxygen.

4.Conclusions

- During the combustion, once the local temperature of the biomass particle is higher than the ash melting temperature. The generated ash could be melted and contracted into a spherical shape by the surface tension. When the carbon is present, the spherical particles exists in the form of clusters of particles. As the decrease of the carbon, the spherical particles exist in form of the individuals.
- The molten ash located in the interior of the biomass particle adhere to the char surface, thereby preventing the contact between char and oxygen. The molten ash located on the surface of biomass particle could drop down from the surface of biomass particle.
- When the conversion rate is higher than 45%, the effect of molten ash on the reaction rate becomes more obvious.

Acknowledgment
This work was supported by National Natural Science Foundation of China (Grant No.: 51976049).

References

[1] Dhamodaran G, Esakkimuthu GS, Pochareddy YK, Sivasubramanian H. Investigation of n-butanol as fuel in a four-cylinder MPFI SI engine. Energy, 2017, 125: 726-735.
[2] Haseli Y, Oijen JAV, Goey LPHD. A detailed one-dimensional model of combustion of a woody biomass particle. Bioresource Technology, 2011, 102(20): 9772-9782.
[3] Singh A, Pant D, Korres NE, Nizami AS, Prasad S, Murphy JD. Key issues in life cycle assessment of ethanol production from lignocellulosic biomass: challenges and perspectives. Bioresource Technology, 2010, 101(13): 5003-5012.
[4] Shan F, Lin Q, Zhou K, Wu Y, Fu W, Zhang P, Song L, Shao C, Yi B. An experimental study of ignition and combustion of single biomass pellets in air and oxy-fuel. Fuel, 2017, 188: 277-284.

[5] Liang F, Wang R, Jiang C, Yang X, Zhang T, Hu W, Mi B, Liu Z. Investigating co-combustion characteristics of bamboo and wood. Bioresource Technology, 2017, 243: 556-565.

[6] Werther J, Saenger M, Hartge E U, et al. Combustion of agricultural residues. Progress in Energy Combustion Science, 2000, 26(1):1-27.

[7] Shan F, Lin Q, Zhou K, et al., 2017. An experimental study of ignition and combustion of single biomass pellets in air and oxy-fuel[J]. Fuel, 2017, 188:277-284.

[8] Kuo, J. T., Hsi, C. L., 2005. Pyrolysis and ignition of single wooden spheres heated in high-temperature streams of air. Combustion and Flame, 142 (4), 401-412.

[9] Roy, M. M., Dutta, A., Corscadden, K., 2013. An experimental study of combustion and emissions of biomass pellets in a prototype pellet furnace. Applied Energy, 108, 298-307.

[10] Momenikouchaksaraei M. Fundamental Study of Single Biomass Particle Combustion. 2013.

[11] Abbas T, Costen P G, Lockwood F C. Solid fuel utilization: From coal to biomass. Symposium on Combustion, 1996, 26(2):3041-3058.

[12] Momeni M, Yin C, K? R S K , et al. Experimental Study on Effects of Particle Shape and Operating Conditions on Combustion Characteristics of Single Biomass Particles[J]. Energy & Fuels, 2013, 27(1):507-514.

[13] Kim YT, Seo DK, Hwang J. Study of the Effect of Coal Type and Particle Size on Char-CO2 Gasification via Gas Analysis. Energy & Fuels, 2011, 25(11):5044-5054.

[14] Damoe A J, Jensen P A, Frandsen F J, et al. Fly Ash Formation during Suspension-Firing of Biomass. Effects of Residence Time and Fuel-Type. Energy Fuels, 2017, 31(1): 555-570.

[15] Liu M, Shen Z, Liang Q, et al. Morphological evolution of single char particle with low ash fusion temperature during the whole gasification process. Energy & Fuels, 2018, 32, 1550-1557.

[16] Lin S, Hirato M, Horio M. The characteristics of coal char gasification at around ash melting temperature Energy & Fuels, 1994, 8, 598-606.

IOP Conf. Series: Materials Science and Engineering **721** (2020) 012021 doi:10.1088/1757-899X/721/1/012021

Review of flow regime in CFB standpipe and circulation rate measurement

Zhao Yang [1,2], Xuemin Liu [2*], Guoli Qi [2], Yong Dong [1*]

[1]Shandong University, Jinan, Shandong Province, 250000, China

[2]China Special Equipment Inspection and Research Institute, Beijing, 100029, China

*liuxuemin@csei.org.cn

*dongy@sdu.edu.cn

Abstract. The classification and characteristics of flow regimes in a circulating fluidized bed standpipe were briefly described. Based on the flow characteristics, several measuring methods of circulating flow rate in standpipe were introduced, including accumulation method, time-of-descent measurements, particle motion detection method, flowmeter method, and correlation method. Advantages and disadvantages of these methods were compared. It can be seen that at present, all measurement methods have some defects, which need to be further studied and improved.

1. Introduction

Circulating flow rate is one of the most important design parameters of circulating fluidized bed (CFB) boiler, which can reflect the gas-solid flow state in the furnace, and has an important influence on the mass and heat transfer characteristics, so it is necessary to measure the circulating flow rate. The measurement of circulating flow rate is mostly carried out in the standpipe. It is the basis of realizing the measurement of circulating flow rate to clarify the flow regime and flow characteristics in the standpipe. As an important part of CFB, the flow characteristics and stability of gas-solid two-phase flow in the standpipe are very important to the operation of CFB. The reverse pressure gradient flow of granular material in the CFB standpipe leads to the complexity of gas-solid two-phase flow. There are a large number of possible flow states in the standpipe, and each of them has different characteristics. This paper briefly describes the classification and characteristics of flow regimes in the CFB standpipe, and compares and analyzes the commonly used measurement methods of circulating flow rate.

2. Flow regime classification

In order to study the flow characteristics in CFB standpipe, the researchers classified the flow regimes in standpipe according to different criteria.

The first qualitatively phase regime diagram was proposed by Zenz [1]. Later, Kojabashian [2] made a quantitative analysis of the flow regime in the standpipe and proposed the flow pattern assuming a linear relationship between void fraction and gas-solid slip velocity, and he used the gas-solid slip velocity and u_{mf}/ε_{mf} to divide the gas-solid flow regime into fluidized solid flow and non-fluidized solid flow.

- Fluidized solid flow, $u_{sl}>u_{mf}/\varepsilon_{mf}$, $\varepsilon>\varepsilon_{mf}$, and the particle is suspended;

- Non-fluidized solid flow, $u_{sl} \leqq u_{mf}/\varepsilon_{mf}$, $\varepsilon \leqq \varepsilon_{mf}$, and the particles move as a whole, with little relative motion. Non-fluidized solid flow is also called packed bed flow, moving bed flow or slip-stick flow.

Kojabashian divided the fluidized flow into three regions according to the direction of gas flow in the standpipe and the positive and negative of $(\partial\varepsilon/\partial P)_{Ws,Wg}$, and further divided each region into two subgroups according to the existence of bubbles; non-fluidized flow is subdivided into the transition packed bed flow where voidage increases with slip velocity and the packed bed flow where voidage is not affected by slip velocity.

Table 1. Flow regime classification by Kojabashian.

Flow regimes		Characterization formula	
Non-fluidized solid flow	Transition packed bed flow	$u_{sl}<u_{mf}/$ $_{mf}$, $=f(u_{sl})$	
	Packed bed flow	$u_{sl}<0$, is constant	
Fluidized solid flow	Flow area I The gas is flowing upwards	$(\partial\varepsilon/\partial P)_{Ws,Wg}<0$	No bubble
			Bubbles
	Flow area II The gas is flowing downwards	$(\partial\varepsilon/\partial P)_{Ws,Wg}>0$	No bubble
			Bubbles
	Flow area III The gas is flowing downwards	$(\partial\varepsilon/\partial P)_{Ws,Wg}<0$	No bubble
			Bubbles

In the table, u_{sl} is slip velocity, u_{mf} is minimum fluidization velocity, is voidage, $_{mf}$ is minimum fluidization void fraction, P is pressure, W_s is solid mass velocity, and W_g is gas mass velocity.

Leung and Jones [3] proposed the flow pattern based on a more realistic relationship between slip velocity and voidage, and the flow pattern is similar to that proposed by Kojabashian. As for Kojabashian's classification of fluidized flow in the standpipe, Leung et al. [4] thought that area I was very rare in reality; the voidage of area I and III decreased along the standpipe, while the voidage of area II increased along the standpipe; therefore, area I and III were uniformly classified as Dense Phase Fluidized Solid Flow (DENFLO), and area II was classified as Lean Phase Fluidized Solid Flow (LEANFLO).

Leung et al. [5] modified and simplified the flow pattern classification of Kojabashian and proposed their own classification method.

Table 2. Flow regime classification by Leung et al.

Flow regimes		Characterization formula
Non-Fluidized Regime	Transition Packed Bed Flow (TRANPACFLO)	$0<u_{sl}<u_{mf}/\varepsilon_{mf}$ $\varepsilon<\varepsilon_{mf}$ $\varepsilon=f(u_{sl})$
	Packed Bed Flow (PACFLO)	$u_{mf} \leqq 0$ $\varepsilon=\varepsilon_{p}$, ε is constant
Fluidized Regime	Dense Phase Fluidized Solid Flow (DENFLO)	$u_{sl} \geqq u_{mf}/\varepsilon_{mf}$ $\varepsilon \geqq \varepsilon_{mf}$ $\left(\frac{\partial u_g}{\partial \varepsilon}\right)_{W_s}>0$

IOP Conf. Series: Materials Science and Engineering **721** (2020) 012021 doi:10.1088/1757-899X/721/1/012021

Lean Phase Fluidized Solid Flow (LEANFLO)	$u_{sl} \geqq u_{mf}/\varepsilon_{mf}$ $\varepsilon \geqq \varepsilon_{mf}$ $\left(\frac{\partial u_g}{\partial \varepsilon}\right)_{W_S} < 0$

In the table, ε_p is packed bed voidage, and W_S is solids phase velocity, u_g is superficial gas velocity in standpipe.

Knowlton [6] named a dilute phase flow with a high voidage as streaming flow, which was classified into the third flow regime besides fluidized flow and non-fluidized flow. But this regime was very rare in reality[7]. Fluidized flow was classified into bubbling fluidized flow and non-bubbling fluidized flow. Bubbling fluidization was further divided into four types.

- Emulsion gas flow up, bubble flow up, net gas flow up;
- Emulsion gas flow down, bubble flow up, net gas flow up;
- Emulsion gas flow down, bubble flow up, net gas flow down;
- Emulsion gas flow down, bubble flow down, net gas flow down.

Geldart et al. [8] continued Knowlton's classification method by dividing the fluidized regime into bubbling fluidized flow and non-bubbling fluidized flow.

3. Flow regime in CFB standpipe

Depending on the amount and direction of air intake at the bottom of the standpipe, the material in the standpipe may be in a moving bed or a fluidized bed[9-10].

At present, most models in literature assume that the flow state in standpipe is moving bed [11, moving packed bed [19], minimum fluidization [20-21] or minimum fluidization/moving bed [22-23].

In fact, there may be many different flow systems in the standpipe, and it is even possible to coexist two different flow systems in the same standpipe. Therefore, the judgment of the flow state in the standpipe has always been controversial in the academic circle.

Kunii and Levenspiel [24] described a coexistence of LEANFLO in the top and PACFLO (or TRANPACFLO) in the bottom of standpipe. Leung and Wilson [25] described a coexistence of DENFLO in the top and PACFLO (or TRANPACFLO) in the bottom of standpipe. Judd and Rowe [26] described a coexistence of LEANFLO in the top and DENFLO in the bottom of standpipe.

Geldart et al. [27] observed the pressure profile of the standpipe and found that the apparent density increased from top to bottom. Wang et al. [28] found that flow in the standpipe can be divided into three regions, namely inlet, dilute and dense flow, through observation of the axial pressure profile. The inlet region is characterized by that pressure gradient is nearly zero, and the dense phase flow can be divided into fluidized (preferred) or packed.

Ji et al. [29] carried out experiments on the gas-solid two-phase flow rule in the standpipe under different working conditions in a large-scale cold mold. It was found that the flow regime is a coexistence of the upper dilute phase fluidization and the lower dense phase fluidization. When the gas volume continued to rise, the gas node would appear in the standpipe. When the gas volume decreased, the flow pattern in dense phase tended to be non-fluidized.

Li et al. [30] thought that the direction of net gas flow in the standpipe varies with the solid circulation rate according to his experimental results, and there is a critical solid circulation rate. Wei et al. [9] confirmed that the flow state in the standpipe is related to the particle mass flow rate G_s through establishment of the calculation model and experimental research. There is a critical particle mass flow rate (G_{sc}). When the particle mass flow rate G_s is lower than G_{sc}, the flow pattern in the standpipe is composed of dilute phase flow in the upper zone and dense phase flow in the lower zone. With the increase of G_s, the dilute-dense phase interface moves down; unitl $G_s > G_{sc}$, the dilute-dense phase interface disappears, and the flow pattern changes into a single concentrated phase transport regime, namely DENFLO.

4. Circulation rate measurement method

Circulating flow rate is one of the most important design parameters of CFB boiler. By associating gas velocity u_f with circulation rate G_s, Yue et al. [31] proposed Fluidization State Specification (FSS) block diagram for the design of CFB boiler. It is necessary to measure the circulating flow rate, as it can reflect the gas-solid flow state in the furnace. At present, the main methods to measure the circulation rate include accumulation method, time-of-descent method, particle motion detection method, flowmeter method and correlation method, etc.

4.1. Accumulation method

The accumulation method is a method of measuring the accumulation rate of solid particles by stopping the flow. The principle is to close the permeation valve and form a fixed bed with a continuous accumulation of height. Solid particles accumulate in a short enough time (assuming that accumulation and consumption of solids in different parts of the system do not affect the overall behavior). Then the circulation rate can be obtained by measuring the solid accumulation rate at the butterfly valve and linking it with the pressure change.

Previous researchers have used porous butterfly valves [32-33], a purged slide valve [34] or diversion into a collecting vessel [35] to accumulate solids over times[36].Harris et al. [37] collected solid particles by using a slot flow meter (SFM), which was calibrated by an in situ technique, and this technique was based on the dynamic response to change of input flow. Bodelin et al. [38] installed a weighing funnel on the standpipe, allowing solids to accumulate in the funnel by shutting down mechanical valves. Kreuzeder et al. [39] measured the circulation rate by tracking the rate of solid accumulation height change per unit time in the L-shaped section of the transmission pipe under closed fluidized air condition.

However, the technique is highly intrusive and can cause variations in standpipe operation. The system pressure and fluid flow will change as the solid particles no longer return to the standpipe, which will result in system errors. It was found that the pressure drop of the riser changed during the experiment process when using this method [40]. In addition, the accumulation method can only be used under a stable operating condition, and is not suitable for high temperature environment.

4.2. Time-of-descent measurement

The principle of time-of-descent measurement is to determine the circulation rate by measuring the falling time of an identifiable particle in the standpipe.

Patience et al. [41] measured the circulating flow rate by time-of-descent measurement, and studied the relationship between the riser pressure drop and circulating flow rate. Muir et al. [42] installed a settler in the standpipe, which was dragged down by the solid particles. The circulating flow rate was calculated by measuring the time that the settler took to move a specified distance.

Although time-of-descent measurement has no interference to the circulation system, it is greatly affected by the accuracy of the timing device and the material of the measuring section.

4.3. Particle motion detection method

In addition to the time-of-descent measurements,there are some other particle motion detection methods, such as optics, tracers, acoustics, electricity and so on.

4.3.1. Optical method. Oki et al. [43] developed a method for measuring the velocity of solid particles by using a fiber optic probe. Dong et al. [44] used an optical fiber probe to measure the radial distribution of particle velocity and the circulation rate of particles. Song et al. [45] measured the circulating flow rate at room temperature with a special high-temperature fiber optic probe. Using optical techniques, differential pressure techniques and particle extraction techniques, Medrano et al. [46] measured the circulating flow rate between the air reactor and the membrane assisted fuel reactor in a two-dimensional interconnected reactor system.

4.3.2. Tracer method. Kuramoto et al. [47] added fluorescent dye-coated tracers into the circulating system and detected their movement with two fiber-optic probes at different heights in the standpipe. Both Wei et al. [48] and Li et al. [49] used gas tracers, the difference is that Wei connected the bottom of the standpipe to the dense phase zone of the riser, while Li artificially isolated the standpipe from the overall system[50]. Bhusarapu et al. [51] tracked the movement of one radioactive tracer particle in the CFB standpipe operated at ambient pressure and temperature. Guio-Perez et al. [52] used ferromagnetic particles as tracers and used inductive coils to track them.

4.3.3. Acoustic method. The dynamic system of particles produces acoustic interference due to its own nature, which can be used to provide information of process monitoring and control. The particles also respond positively to acoustic signals applied to them.

Davies et al. [53] tried to use sound pressure wave to obtain the solid flow rate, and the results showed that this method was relatively accurate for the prediction of large pariticles, while the prediction of small particles was quite different from the reality. Ellis et al. [54] used acoustic emission sensors to measure the solid flow rate at room temperature. Chorpening et al. [55] used a microwave Doppler system to detect sliding or intermittent flows of particles.

4.3.4. Electrical method. Spenik et al. [56] used a piezoelectric pressure sensor to measure the size and density of particles to obtain the mass flow rate in the riser, which can be referred to in the measurement of the standpipe.

4.4. Flowmeter method

4.4.1. Impacting flowmeter. In many cases, flow in standpipe is non-fluidized bed flow, also known as plug flow[57]. The solid-phase velocity of the plug flow is considered uniform throughout the standpipe. The bulk density of solids changes little with solid phase velocity, and the average velocity of solids in the standpipe is approximately equal to the local velocity[58]. The gas-solid mixture in the standpipe can be regarded as a quasi-fluid[59]. The impacting flowmeter is more effective for particles in the state of free fall or non- pneumatic transport. The principle of impacting flowmeter method is to calculate the circulating flow rate by measuring the force of falling particles on the target.

Judd and Bernhardt [60] placed a cylinder perpendicular to the flow direction in the dense phase of the standpipe and obtained the particle flow rate by measuring the drag force on the cylinder. Wu et al. [61] installed a double-ended wire mesh baffle with a rotating shaft on the top of the standpipe.

Wu et al. [62] developed a kind of flowmeter which worked by measuring the torque of a hinged plate when falling solids impacted the plate. Similarly, Hu et al. [63] designed an impacting flowmeter to measure the impact force of solid particles by means of bending moment and strain, and realized online measurement.

4.4.2. Turbine flowmeter. Compared with impacting flowmeter, turbine flowmeter is favored by many researchers because of its high sensitivity and less interference to particle flow.

Liu et al. [64] placed an impeller flowmeter in the dense phase section of the standpipe. The special impeller would rotate under the push of materials, and the volume flow rate of particles through the standpipe could be obtained by measuring the rotating speed of the impeller. Ludlow et al. [65] installed a rotating spiral blade made of glass fiber in the standpipe, determined the average velocity of solid flow by recording the solid accumulation rate under the condition of stopping the solid circulation, and measured the rate of pressure drop change to determine the solid cycle rate.

4.5. Correlation method

It is difficult to measure the flow state of the standpipe directly, so some researchers have shifted the direction to such easily measurable parameters as pressure, and obtained the circulating flow rate through correlation method.

Based on the related literature and experimental data of a cold dual fluidized bed, Lim et al. [66] established an correlation relationship by using four dimensionless numbers Fr, Re_p, u_0/u_t, ρ_s/ρ_g as the key parameters to estimate the solid circulatin rate under high solid-gas density ratio. The correlation was verified by the literature data of a dual fluidized bed steam gasifier at high temperature, and the results show that the maximum deviation of the estimate is 25%.

In a circulating fluidized bed, control of the solid circulation rate can be achieved by changing the opening of the mechanical valve at the bottom of the standpipe. Grieco and Marmo [67] modified the predictive equation proposed by Jones and Davidson [68] and Cheng [69] and established the relationship between the pressure drop of the mechanical valve and the flow of solids through the control valve at room temperature.

Monazam and Shadle [70] adopted a transient method: stopping the solid phase flow into the standpipe when the standpipe was in the state of full dense transportation, then analyzing the derivative of the transient pressure drop of the standpipe with respect to time, and correlating the solid circulation rate with the axial pressure distribution and the solid holdup of the riser.

4.6. *Others*

In addition, Burkell et al. 366] also studied the modified orifice meter and calorimetric method, and conducted comparative analysis with the permeable valve accumulation method, time-of-descent measurement and impacting meter mentioned above. The results showed that the pressure drop obtained by the modified orifice meter was too small to be recorded under low circulation rate, and it was also difficult to measure under high circulation rate because of the great fluctuation of pressure drop. The problem of calorimetric method was that the pipe had radial temperature gradient and the heat loss in the test section could not be accurately determined, so the accuracy was low, and the measurement would cause additional heat loss. These two methods have many defects and are difficult to be widely used.

4.7. *Summary*

In general, the accumulation method is simple to operate, but it has a great disturbance to the system, and the changing conditions cannot be measured. Time-of-descent measurement has no interference to the flow in the tube, but it requires more experimental equipment and measuring equipment, and cannot be measured online. Optics, tracers, acoustics, electricity and other methods for particle motion detection use relatively advanced technologies, the specific accuracy of which remains to be studied. The flowmeter method has the characteristics of online measurement and low interference to flow. However, the friction resistance brought by the rotating elements in the turbine flowmeter may influence the experiment and the impacting flowmeter has a slightly insufficient measuring capacity under high circulation rate and high temperature. And it is inevitable that the correlation method has errors.

Therefore, the measurement method and technology of circulating flow rate still need to be further studied and improved.

5. Conclusion

The flow pattern in CFB standpipe is generally divided into fluidized flow and non-fluidized flow. Fluidized flow can be divided into dense phase fluidized solids flow and lean phase fluidized solids flow, or bubbling fluidized flow and non-bubbling fluidized flow. Non-fluidized flow can be divided into transition packed bed flow and packed bed flow.

The gas-solid two-phase flow in the standpipe presents a variety of flow state superposition, and it is generally assumed to be moving bed flow in the simulation.

An ideal measurement technology of circulating flow rate in standpipe should have the characteristics of online measurement, no interference to the steady flow in standpipe, large range and applicable to high temperature operation. The main methods to measure the circulation rate include accumulation method, time-of-descent measurement, particle motion detection method, flowmeter method and correlation method, etc. However, all of these measurement methods have some defects at present. Therefore, the measurement methods and techniques need to be further studied and improved.

Acknowledgments
This study was supported by the National Key R&D Program of China (No. 2018YFF0216000).

References

[1] Zenz, F. A. (1953) Pet. Refiner, 32 123.
[2] Kojabashian, C. (1958). Properties of dense-phase fluidized solids in vertical downflow (Doctoral dissertation, Massachusetts Institute of Technology, Department of Mechanical Engineering).
[3] Leung, L. S., & Jones, P. J. (1978). Coexistence of fluidized solids flow and packed bed flow in standpipes. In Fluidization: Proceedings of the Second Engineering Foundation Conference, Trinity College, Cambridge, England 2-6 April 1978 (Vol. 2, p. 115). CUP Archive.
[4] Leung, L. S., & Jones, P. J. (1978). Flow of gas—Solid mixtures in standpipes. A review. Powder Technology, 20(2), 145-160.
[5] Jones, P.J., Leung, L.S. (1985) Down flow of Solids through Pipes and Valves [M]. London: Academic Press p. 293-329
[6] Knowlton, (1986) T.M. "Standpipes," in D. Geldart. Gas Fluidization Technology.Chichester, New York: John Wiley and Sons Ltd.
[7] Sarra, A. M. (2001). Particle-wall shear stress measurements within the standpipe of a circulating fluidized bed.
[8] Geldart, D., & Radtke, A. L. (1986). The effect of particle properties on the behaviour of equilibrium cracking catalysts in standpipe flow. Powder technology, 47(2), 157-165.
[9] Wei, Y.D., Liu, R.H., Shi, M.X. (2006) A model of the Axial Voidage Distribution of Gas-solid Two-phase Flow in the Standpipe under Negative Pressure Gradient[C]. In: Dept.Chem.Eng.University of Petroleum,Beijing 102200,China.
[10] Li, H., Kwauk, M. (1989) Vertical Pneumatic Moving-bed Transport: II.Experimental Findings [J]. Chem. Eng. Sci., 44(2): 261-271.
[11] Rhodes, M. J., & Geldart, D. (1987). A model for the circulating fluidized bed. Powder Technology, 53(3), 155-162.
[12] Yang, W. C. (1988). A model for the dynamics of a circulating fluidized bed loop. In Circulating Fluidized Bed Technology (pp. 181-191). Pergamon.
[13] Breault, R. W., & Mathur, V. K. (1989). High-velocity fluidized-bed hydrodynamic modeling. 1. Fundamental studies of pressure drop. Industrial & engineering chemistry research,28(6), 684-688.
[14] Hannes, Svoboda, J., Renz, K. , Bleek, U. , & D, C. M. V. . (1993). Mathematical modelling of CFBC - an overall modular programming frame using a 1. International Conference on Fluidized Bed Combustion. ASME.
[15] Basu, P., & Cheng, L. (2000). An analysis of loop seal operations in a circulating fluidized bed.Chemical Engineering Research and Design,78(7), 991-998.
[16] Dutta, A., & Basu, P. (2003). An investigation on heat transfer to the standpipe of a circulating fluidized bed boiler.Chemical Engineering Research and Design,81(8), 1003-1014.
[17] Ludlow, J. C., Panday, R., & Shadle, L. J. (2013). Standpipe models for diagnostics and control of a circulating fluidized bed.Powder technology,242, 51-64.
[18] Heng, F. E. N. G., Qinghai, L. I., Chao, G. A. N., & Aihong, M. E. N. G. (2015). One-dimensional hydrodynamic model of the recycling valve in a circulating fluidized bed.Journal of Tsinghua University (Science and Technology),54(2), 229-234.
[19] Yang, X.P. (2009) Research on characteristics of gas-solid flow in air nozzle area in the furnace and U-valve in CFB boiler[D]. Shanghai： Shanghai Jiao Tong University .
[20] Mori, S., Yan, Y., Kato, K., Matubara, K. and Liu, D. (1991). Hydrodynamics of Circulating

Fluidized Bed. In Circulating Fluidized Bed Technology III; Basu, P., Horio, M., Hasatani, M., Eds.; Pergamon Press: Oxford, England; p 113.

[21] Kim, S. W., Kim, S. D., & Lee, D. H. (2002). Pressure balance model for circulating fluidized beds with a loop-seal. Industrial & engineering chemistry research,41(20), 4949-4956.

[22] Lei, H., & Horio, M. (1998). A comprehensive pressure balance model of circulating fluidized beds.Journal of chemical engineering of Japan,31(1), 83-94.

[23] Lim, K. S., & PeelerP, C. R. (1999). EstimationofSolids CirculationRateinCFBfrom PressureLoopProfiles.

[24] Kunii, D., & Levenspiel, O. (1969). Fluidization Engineering, Wiley, New York, p.372.

[25] Leung, L.S. and Wilson, L.A. (1973). Powder Technol.vol. 7 343 - 349.

[26] Judd, M.R. and Rowe, D.N. (1978). Dense phase flow of a powder down a standpipe.inJ.F.Davidson(Ed.),Proc. International Fluidization Conference, Cambridge Univ. Press, Cambridge.

[27] Geldart, D., Broodryk, N., & Kerdoncuff, A. (1993). Studies on the flow of solids down cyclone diplegs. Powder technology,76(2), 175-183.

[28] Wang, J., Bouma, J. H., & Dries, H. (2000). An experimental study of cyclone dipleg flow in fluidized catalytic cracking.Powder technology,112(3), 221-228.

[29] Ji, H., & Shi, M. (1997). Study on Gas-Solid Flow in Constrained Dipleg of Cyclone Separator in Fluidized Beds.PETROLEUM REFINERY ENGINEERING,27, 35-38.

[30] Li, Y., Lu, Y., Wang, F., Han, K., Mi, W., Chen, X., & Wang, P. (1997). Behavior of gas-solid flow in the downcomer of a circulating fluidized bed reactor with a V-valve.Powder technology,91(1), 11-16.

[31] Yue, G., Lu, J., Zhang, H., Yang, H., Zhang, J., Liu, Q., ... & Jaud, P. (2005). Design theory of circulating fluidized bed boilers. In 18th International Fluidized Bed Combustion Conference (pp. 18-21).

[32] Yerushalmi, J., D.H. Turner and A.M. Squires (1976). Ind. Eng. Chem. Proc. Des. Dev., 15, 47-53.

[33] Hartge, E.U., Y. Li and J. Werther (1986). In Circulating Fluidized Bed Technology, P. Basu (Ed.), Pergamon, Toronto, pp. 153-160.

[34] Fusey, I., C.J. Lim and J.R. Grace (1986). In Circulating Fluidized Bed Technology, P. Basu (Ed.), Pergamon Press, Toronto, pp. 409-416.

[35] Arena, U., A. Cammarota and L. Pistone (1986). In Circulating Fluidized Bed Technology, P. Basu (Ed.), Pergamon, Toronto, pp. 119-125.

[36] Burkell, J.J., Grace, J.R., Zhao, J., et al. (1988). Circulating fluidized bed technology :vol .II[M] .Oxford:Pergamon Press, 501-509 .

[37] Harris, B. J. , Davies, C. E. , & Davidson, J. F. . (1997). The slot flow meter: a new device for continuous solids flow measurement. Chemical Engineering Science, 52(24), 4637-4648.

[38] Bodelin, P., Malodtsof, F., & Delebarre, A. (1994). Flow structure investigations in a CFB In circulating fluidized bed technology IV A. Aviden (Eds), AIChE NY, 151.

[39] Kreuzeder, A., Pfeifer, C., & Hofbauer, H. (2007). Fluid-dynamic investigations in a scaled cold model for a dual fluidized bed biomass steam gasification process: Solid flux measurements and optimization of the cyclone. International Journal of Chemical Reactor Engineering, 5(1).

[40] Lim, K.S. Peeler, P., Joyce, T., Zakhari, A., Close, R. (2002) 7th Int'l CFB Conference, Niagara Falls, New York, p. 169

[41] Pateence, G. S., Chaouki, J., & Grandjean, B. P. A. (1990). Solids flow metering from pressure drop measurement in circulating fluidized beds. Powder Technology, 61(1), 95-99.

[42] Muir, J. R., Brereton, C. M. H., Grace, J. R., & Lim, C. J. (2004). Line-and sinker measurement of solids circulation rate in a CFB combustor. In Fluid. XI, Engineering Conferences International, New York (pp. 315-322).

[43] Oki, K. T., Akehata, T., & Shirai, T. (1973). A method for measuring the velocity of solid particles with fiber optic probe. Kagaku Kogaku, 37, 965.

[44] Dong, Z.B. (1992).Experimental Study on Gas-Solid Downflow in a CFB: [Master Degree Thesis]. ICCCAS.
[45] Song, X., Bi, H., Lim, C. J., Grace, J. R., Chan, E., Knapper, B., & McKnight, C. (2004). Hydrodynamics of the reactor section in fluid cokers. Powder technology, 147(1-3), 126-136.
[46] Medrano, J. A., Nordio, M., Manzolini, G., van Sint Annaland, M., & Gallucci, F. (2016). On the measurement of solids circulation rates in interconnected fluidized beds: Comparison of different experimental techniques. Powder Technology, 302, 81-89.
[47] Kuramoto, M., Kunii, D., & Furusawa, T.(1986). Flow of dense fluidized particles through an opening in a circulation system, Powder Technol. 47 (2) 141–149.
[48] Wei, Y.D., Liu, R.H., Sun, G. G., et al.(2003).Fluidization Behaviors of Gas-Solid Two-phase Flow in the Standpipe under Negative Pressure Gradient[J]. Chin J Proc Eng, 3(5):385-389.
[49] Li, Y., Lu, Y., Wang, F., Han, K., Mi, W., Chen, X., & Wang, P. (1997). Behavior of gas-solid flow in the downcomer of a circulating fluidized bed reactor with a V-valve. Powder technology, 91(1), 11-16.
[50] Yao, X. (2010). Study on Hydrodynamic and Particles Segregation in the Loop of Circulating Fluidized Bed[D].Tsinghua University.
[51] Bhusarapu, S., Fongarland, P., Al-Dahhan, M. H., & Duduković, M. P. (2004). Measurement of overall solids mass flux in a gas–solid circulating fluidized bed. Powder technology, 148(2-3), 158-171.
[52] Guío-Pérez, D. C., Dietrich, F., Cala, J. N. F., Pröll, T., & Hofbauer, H. (2017). Estimation of solids circulation rate through magnetic tracer tests. Powder technology, 316, 650-657.
[53] Davies, C. E., Tallon, S. J., & Webster, E. S. (2010). Applications of active acoustics in particle technology. Particuology, 8(6), 568-571.
[54] N. Ellis, C.J. Lim, P.A. Reyes, J.I. Soletti, J.R. Grace.(2012). Acoustic emissions method for solids mass flux measurements, in: 21st Int. Conf. Fluid. Bed Combust., Naples, 2012: pp. 681–688.
[55] B.T. Chorpening, M. Spencer, J. Charley, R.C. Stehle, D.W. Greve.(2016). Microwave Doppler Sensing of Sliding or Intermittent Particle Flows, 2016 Multiph. Flow Sci. Work.
[56] Spenik, J. L. , & Ludlow, J. C. . (2010). Use of piezoelectric pressure transducers to determine local solids mass flux in the riser of a cold flow circulating fluidized bed. Powder Technology, 203(1), 86-90.
[57] Jones, P.J., Leung, L.S. ,Davidson, J.F. Clift, R. and Harrison, D. (1985) Fluidization,Academic Press, London, 2nd edn., p. 294.
[58] Liu, J., & Huan, B. (1995). Turbine meter for the measurement of bulk solids flowrate. Powder technology, 82(2), 145-151.
[59] Wei, W., & Youchu, L. (2000). Progress of the simulation of particle-fluid two-phase flow. PROGRESS IN CHEMISTRY-BEIJING-, 12(2), 208-217.
[60] Judd, M.R. and Bernhardt, H.W.(1990).Circulating fluidized bed technology:vol.III[M].Oxford:Pergamon Press,621.
[61] Wu, C.Z., Zhu, J.X., Yin, X.L. (2009) A measuring method applied to a circulating fluidized bed[J]. Guangzhou Institute of Energy Conversion, Chinese Academy of Sciences.
[62] Wu, W., Gerhart, A. L., Chen, Z., Dellenback, P. A., & Agarwal, P. K. (2001). A device for measuring solids flowrate in a circulating fluidized bed, Powder Technol. 120(3), 151-158.
[63] Hu N,Yao X,Chao J N,Yang H R,LU J F. (2011).Impacting Method for On-line Measurement of Solid Circulating Rate in CFB.Journal of Chinese Society of Power Engineering,Vol.31 No .7.
[64] Liu, J., & Huan, B. (1995). Turbine meter for the measurement of bulk solids flowrate. Powder technology, 82(2), 145-151.
[65] Ludlow, J. C., Monazam, E. R., & Shadle, L. J. (2008). Improvement of continuous solid circulation rate measurement in a cold flow circulating fluidized bed. Powder Technology, 182(3), 379-387.

[66] Lim, M. T., Saw, W. L., & Pang, S. (2015). Estimation of solids circulation rate from risers at high solids-to-gas density ratios. Chemical Engineering Communications, 202(3), 356-360.
[67] Grieco, E., & Marmo, L. (2006). Predicting the pressure drop across the solids flow rate control device of a circulating fluidized bed. Powder technology, 161(2), 89-97.
[68] Jones, D. R. M. , & Davidson, J. F. . (1965). The flow of particles from a fluidised bed through an orifice. Rheologica Acta, 4(3), 180-192.
[69] Cheng, L. , Basu, P. , & Cen, K. . (1998). Solids circulation rate prediction in a pressurized loop-seal. Chemical Engineering Research & Design, 76(6), 761-763.Monazam, E. R., & Shadle, L. J. (2004). A transient method for characterizing flow regimes in a circulating fluid bed. Powder technology, 139(1), 89-97.
[70] Monazam, E. R., & Shadle, L. J. (2004). A transient method for characterizing flow regimes in a circulating fluid bed. Powder technology, 139(1), 89-97.

IOP Conf. Series: Materials Science and Engineering **721** (2020) 012022 doi:10.1088/1757-899X/721/1/012022

Structure optimization and thermal field analysis of biogas derived methane fueled Solid Oxide Fuel Cell

Ruobing Zhan, Yang Wang, Qing Du* and Kui Jiao*

State Key Laboratory of Engines, Tianjin University, Tianjin, 300350, China

*Corresponding author's e-mail: kjiao@tju.edu.cn (K. Jiao); duqing@tju.edu.cn (Q. Du)

Abstract. The suitability for using various gaseous fuels makes the Solid Oxide Fuel Cell (SOFC) the most promising heat-electricity cogeneration device. Thermal field distribution of SOFC directly determines whether the cell is operating in sufficient catalytic activity and weakened by thermal stress. For biogas derived methane fueled SOFC, the internal reforming reactions can cause the inhomogeneous temperature distribution. So it is crucial to optimize the thermal field to achieve optimal cell operation state. In this paper, a 3D muti-physics model is developed to study the improvements of a new cathode flow field structure. When adopting new cathode structure, the power density increases by over 11%, and the maximum temperature gradient can be evidently reduced.

1. Introduction

Among many energy conversion devices, Solid Oxide Fuel Cell (SOFC) possesses an excellent fuel flexibility, which can be fueled by H_2, coal syngas and various alcohol fuels [1]. Besides,after purification and upgrading, the biogas can be refined as high purity methane which is a non-polluting high heating value fuel. For methane fueled SOFC, the cell thermal management is mainly carried out by adjusting the cathode air flow rates. The temperature distribution and the maximum temperature gradient must be controlled within a reasonable range which insures a high catalyst activity without excessive thermal stress [2].

There are many studies on the flow field optimization [3-5]. Among them, Wei et al. [6] replaced the cathode flow channel as a porous cathode current collector, and the relevant simulations are performed. Zielke et al. [7] employed a Cu-Mn alloy made metal foam as the cathode contact material to enhance the conductivity of cathode electrode. The experimental results show that this new type of cathode structure can significantly improve the cell output performance, and the endurance of this new structure against long-time oxidation is verified.

In this study, inspired by Zielke's experiment, an optimized cathode structure of SOFC is presented. And the cell performance and thermal field analysis are conducted by developing a 3D multi-physics SOFC model. Compared to traditional straight channel flow field, the improvements of optimized structure are discussed in detail.

2. Modeling development

2.1 Cell geometry

The anode-supported planar SOFC are adopted in this study. For typical anode-supported SOFC, the cell is consist of interconnects, anode/cathode channels and the positive electrode-electrolyte-negative electrode (PEN). The partly pre-reformed methane is supplied as the fuel to the anode inlet. And the chemical and electrochemical reactions proceeds inside the porous electrode inside PEN. The cell geometries of two kinds of flow fields are shown in Figure 1. For the new cathode structure, different from the straight channel flow field, the porous media (normally metal foam) is placed in the position between cathode interconnect and PEN as the air channel.

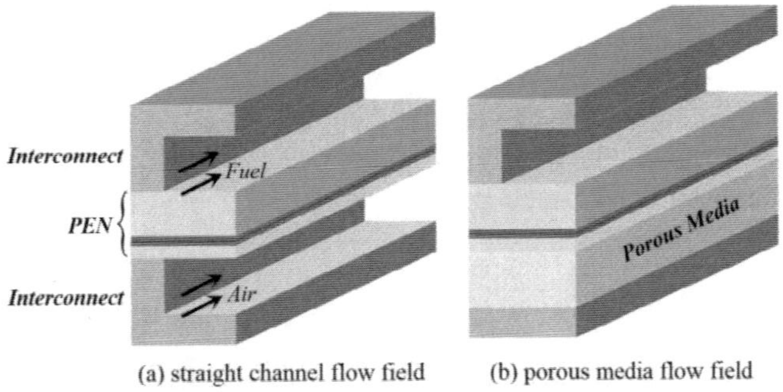

Figure 1. Cell geometry (a) Straight channel flow field (b) Porous media flow field.

2.2 Governing equation

The heat and mass transfer, fluid flow and the chemical/electrochemical reactions are highly coupled in present model. And the conservation equations are listed in Table 1. The multi-component diffusion process (molecular diffusion and Knudsen diffusion) are considered in this model. The effective multi-component diffusion coefficients D_{eff} and the related source terms (S) can be found in Ref [8].

Table 1. Conservation equations.

Description	Conservation Equations
Mass	$\frac{\partial(\varepsilon\rho)}{\partial t}+\nabla\cdot(\varepsilon\rho v)=S_{\mathrm{m}}$
Momentum	$\frac{\partial}{\partial t}(\varepsilon\rho v)+\nabla\cdot(\varepsilon\rho vv)=-\varepsilon\nabla p+\nabla\cdot\left[\varepsilon\mu\left(\nabla v+(\nabla v)^{\mathrm{T}}\right)\right]+S_{\mathrm{u}}$
Species	$\frac{\partial}{\partial t}(\varepsilon\rho y_{\mathrm{i}})+\nabla\cdot\left(-\rho y_{\mathrm{i}}\sum_{\mathrm{j}\neq\mathrm{i}}^{\mathrm{n}}D_{\mathrm{eff,ij}}\nabla x_{\mathrm{j}}+\rho v y_{\mathrm{i}}\right)=S_{\mathrm{i}}$
Energy	$\frac{\partial}{\partial t}(\varepsilon\rho c_{\mathrm{p}}T)+\nabla\cdot(\varepsilon\rho c_{\mathrm{p}}vT)=\nabla\cdot(\kappa_{\mathrm{eff}}\nabla T)+S_{\mathrm{T}}$
Electronic charge	$0=\nabla\cdot\left(\sigma_{\mathrm{ele}}^{\mathrm{eff}}\nabla\varphi_{\mathrm{ele}}\right)+S_{\mathrm{ele}}$
Ionic charge	$0=\nabla\cdot\left(\sigma_{\mathrm{ele}}^{\mathrm{eff}}\nabla\varphi_{\mathrm{ele}}\right)+S_{\mathrm{ele}}$

2.3 Chemical and electrochemical model

When the methane contained gas is supplied to SOFC, several internal reforming reactions occurs simultaneously. Thereinto the methane steam reforming reaction (MSR) and the water gas shift reaction (WGSR) play a leading role, and can be described by following reactions:

MSR: $$CH_4+H_2O\leftrightarrow CO+3H_2 \tag{1}$$

WGSR: $$CO+H_2O\leftrightarrow CO_2+H_2 \tag{2}$$

The strongly endothermic reaction, MSR, will have an impact on the cell temperature gradient distribution. Thus, in this study, these two reactions are both considered with the same reaction rate expressions as Haberman's model [9]. The related equations (reaction rates and reaction heats) can be found in our previous study [8].

The cell output voltage is calculate as the following expression:

$$V_{\text{output}} = E_{\text{rev}} - \eta_{\text{act}} - \eta_{\text{ohm}} \tag{3}$$

where the reversible voltage (E_{rev}) can be obtained as:

$$E_{\text{rev}} = -\frac{\Delta G^0}{2F} + \frac{RT}{2F}\ln\frac{p_{\text{H}_2/\text{CO}}\, p_{\text{O}_2}^{0.5}\, p_0^{-0.5}}{p_{\text{H}_2\text{O}/\text{CO}_2}} \tag{4}$$

We take account of the coupled H_2/CO electrochemical oxidation which is assumed to occur in anode catalyst layer. The expressions of anode and cathode electrochemical reaction rates (J_a and J_c) are determined by the Butler-Volmer equations:

$$J_{\text{a}} = j_{0,\text{a}} A_{\text{v}}\left(\exp\left(\alpha\frac{2F\eta_{\text{act,a}}}{RT}\right) - \exp\left(-(1-\alpha)\frac{2F\eta_{\text{act,a}}}{RT}\right)\right) \tag{5}$$

$$J_{\text{c}} = j_{0,\text{c}} A_{\text{v}}\left(\exp\left(\beta\frac{4F\eta_{\text{act,c}}}{RT}\right) - \exp\left(-(1-\beta)\frac{4F\eta_{\text{act,c}}}{RT}\right)\right) \tag{6}$$

where A_v is the electrochemical active area (m^2 m^{-3}), j_0 the exchange current density (A m^{-2}), α and β the transfer coefficient， η_{act} the activation overpotential (V). It should be mentioned that the exchange current density of CO is supposed to be 1/3 of that of H_2 due to the CO's low catalyst activation.

2.4 Numerical procedures

Th commercial software Fluent solver is employed to solve the conservation equations. And the in-house C code are compiled as the user defined functions (UDF) to simulate the multi-component diffusion process and the chemical/electrochemical reactions. Besides, the mesh-independence tests are carefully conducted to guarantee the error caused by mesh is in allowable range.

3. Results and discussions

3.1Experimental comparison

The experiment validation is conducted in different operating temperature. The operating conditions and cell parameters are consistent with the experiment [10]. In order to validate the coupled electrochemical oxidation effects of H_2 and CO, two kinds of fuels (pure H_2 and coal derived syngas) are selected, and the C-V curves are shown in Figure 2, in which the simulation results fit the experiment data well.

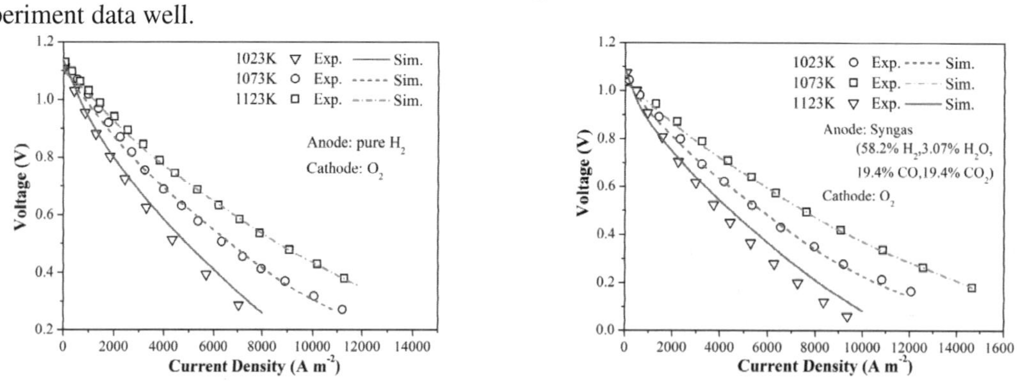

(a) (b)

Figure 2. Experimental comparison: (a) pure H_2 (b) Syngas.

3.2Structure optimization

In order to investigate the effects of new cathode structure, a six-channel single cell and the corresponding single cell with the porous material flow field are simulated. The Current-Voltage (C-V) curves and Power-Voltage (P-V) cures of two kinds of cells are shown in Figure 3. As the results shown, the new cathode structure can significantly improve the cell output performance, and the increment can reach to over 11%. Figure 4 shows the electrochemical reaction rates at the interface of cathode catalyst layer (CL) and electrolyte (ELE). For straight channel flow field, there are some oxygen-deficient areas which will result in zero reaction rates. In contrast, the new cathode structure can make the reaction rates concentrate on a small range. The same tendency can be found in the species distribution field. Besides, the porous material flow field completely contacts with the porous electrode, which markedly decreases the contact resistance. All these factors contribute to improve the cell output performance.

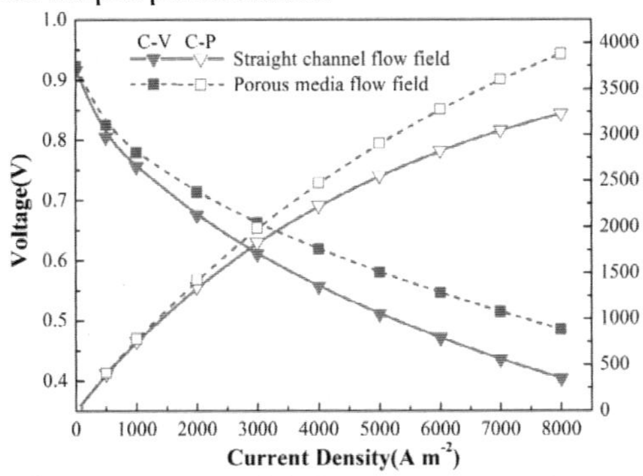

Figure 3. C-V and C-P curves comparisons.

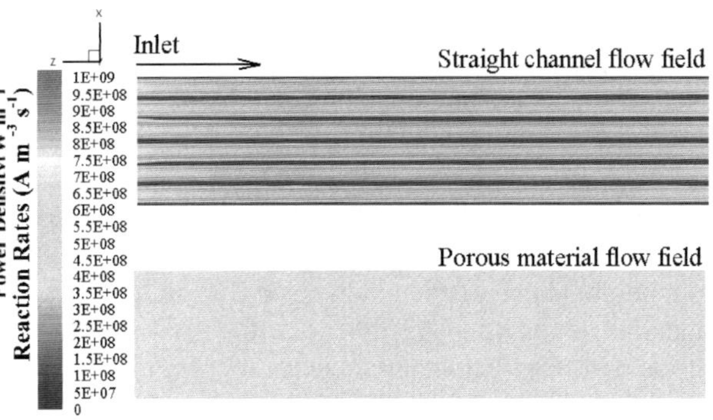

Figure 4. Electrochemical reaction rates at the interface of CL/ELE.

3.3Thermal analysis

For SOFC, the extra heat is mainly taken away by the air. It should be mentioned that the air amount supplied to cathode inlet remains unchanged for all cases. When the fuel concluding methane is fed in the cell anode, the MSR mainly proceeds in front of the cell. In some cases, this endothermic reaction may make the cell inlet temperature less than the inlet gas temperature. But as the heat accumulates along the flow direction, the temperature of the end of the cell is raised greatly. For these reasons, the temperature gradient is normally large.

Figure 5 shows the temperature distribution of two structures. The maximum temperature gradient of the optimized flow field is 105K which is much less than the traditional one. Table 2 lists each heat source terms of two structures. It is observed that the main difference lies in the ohmic heat. As mentioned above, the porous material flow field can sufficiently contact with the cathode porous electrode, which means that this structure can break through the limitation of the channel-rib structure of straight channel. The electrons are directly transfer to the interconnect through the porous material without the horizontal transfer process inside porous electrode. This process can decrease the electron transfer distance, and reduces the ohmic loss. Then, the ohmic heat is reduced.

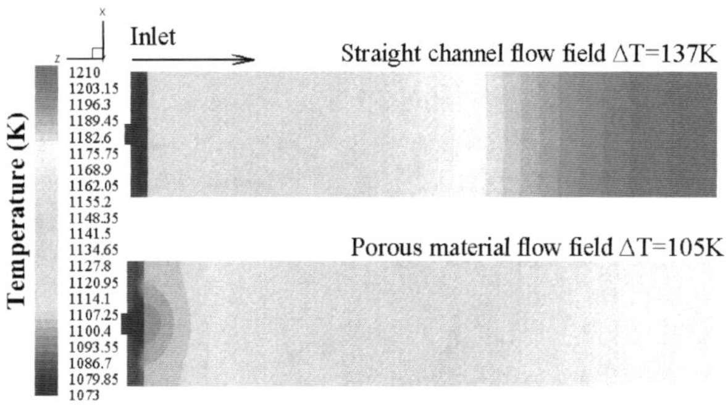

Figure 5. Temperature distribution.

Table 2. Heat source terms (Unit W).

	Straight Channel	Porous Material
Ohmic	0.628	0.250
Reversible (anode)	3.860	3.773
Reversible (cathode)	-1.967	-1.919
Irreversible	1.624	1.487
Chemistry	-1.432	-1.384
Total	2.713	2.207

4. Conclusions

We developed a multi-physics biogas derived methane fueled SOFC model to investigate the effects of the optimized cathode structure on the cell performance. Compared with the straight-channel cell, the cell adopting the porous material flow field shows a larger output voltage. Meanwhile, the distribution of the electrochemical reactions gets more uniform and concentrates within a small range. Besides, through the contrastive study on the heat source terms, we found the main influence factor on the maximum temperature gradient is the ohmic loss. And the sufficient connection between porous material flow field and porous electrode reduces the contact resistance. The total extra heat production of the optimized structure decreases of almost 20%. The maximum temperature gradient drops by 32K.

Acknowledgments

This work is supported by the National Key Research and Development Program of China (2017YFB0601904).

References

[1] Bao, C., Wang, Y., Feng, D., et al. (2018) Macroscopic modeling of solid oxide fuel cell (SOFC) and model-based control of SOFC and gas turbine hybrid system. Progress in Energy and Combustion Science, 66: 83-140.

[2] Grew, K.N., Chiu, W.K.S., (2012) A review of modeling and simulation techniques across the length scales for the solid oxide fuel cell. Journal of Power Sources, 199: 1-13.

[3] Bhattacharya, D., Mukhopadhyay, J., Biswas, N., et al. (2018) Performance evaluation of different bipolar plate designs of 3D planar anode-supported SOFCs. International Journal of Heat and Mass Transfer, 123:382-396.

[4] Saied, M., Ahmed, K., Nemat-Alla, M., Ahmed, M., El-Sebaie, M. (2018) Performance study of solid oxide fuel cell with various flow field designs: numerical study. International Journal of Hydrogen Energy, 43(46):20931-20946.

[5] Duhn, J.D., Jensen, A.D., Wedel, S., et al. (2016) Optimization of a new flow design for solid oxide cells using computational fluid dynamics modelling. Journal of Power Sources, 336:261-271.

[6] Wei, S.S., Wang, T.H., Wu, J.S. (2014) Numerical modeling of interconnect flow channel design and thermal stress analysis of a planar anode-supported solid oxide fuel cell stack. Energy, 69: 553-561.

[7] Zielke, P., Wulff, A.C., Sun, X., et al. (2017) Investigation of a Spinel-forming Cu-Mn Foam as an Oxygen Electrode Contact Material in a Solid Oxide Cell Single Repeating Unit. Fuel Cells, 17(5).

[8] Wang, Y., Zhan, R., et al. (2018) Three-dimensional modeling of pressure effect on operating characteristics and performance of solid oxide fuel cell. International Journal of Hydrogen Energy, 43(43): 20059-20076.

[9] Haberman, B.A., Young, J.B. (2004) Three-dimensional simulation of chemically reacting gas flows in the porous support structure of an integrated-planar solid oxide fuel cell. International Journal of Heat and Mass Transfer, 47(17-18): 3617-3629.

[10] Li, W., Shi, Y., Luo, Y., et al. (2015) Elementary reaction modeling of solid oxide electrolysis cells: Main zones for heterogeneous chemical/electrochemical reactions. Journal of Power Sources, 273:1-13.

Slagging resistance mechanism of precursor ceramic coating for solid fuel-fired boilers

Guangxin Wang[1], Bing Zhang[2], Jinqing Wang[1*], Zuohe Chi[1], Guangxue Zhang[1], Jin li[1], Jie Wang[1], Peihao Qian[1] and Songjie Yang[1]

[1]Institute of Energy Engineering, China Jiliang University, Hangzhou, Zhejiang, 310018, China

[2]Wenzhou Special Equipment Inspection and Research Institute, Wenzhou, Zhejiang, 325007, China

*Corresponding author's e-mail: jqwang@cjlu.edu.cn

Abstract. Slag formation on the heating surface in solid fuel-fired boilers is a common problem. In this paper, with polysilazane used as the precursor, three different ceramic coatings were prepared by adding hexagonal boron nitride (h-BN), ZrO_2, and Al_2O_3 inert fillers. The coatings were applied to a TP347 substrate by the Czochralski method. The spreading and solidification experiments of the molten ash were performed and matched with numerical simulation. The results show that compared with the steel sheet, the molten ash has a large contact angle on the coating, a small spreading coefficient and the same spreading time, indicating that the coating has the slagging resistance property. The coating with the addition of h-BN has better slagging resistance properties than the other two coatings. The numerical simulation results are consistent with the experimental results.

1. Introduction

Boiler slagging is common in solid fuel-fired boilers such as coal-fired boilers and waste incinerators. The slagging on the heating surface of the boiler will reduce the thermal efficiency of the boiler and increase the emission of nitrogen oxides. More seriously, some accidents such as superheater bursting, furnace stalling, and water wall damage may occur, which will threaten the safe and economic operation of the boiler[1],[2]. To resist the slagging problem, many measures have been proposed, including a reasonable design of the combustion system[3], installing soot blower[4], blending weakly slag coal[5], adding additives[6], spraying protective coating and so on. While, there are few studies on protective coatings compared to general measures.

Naganuma et al.[7],[8]prepared different ceramic materials on the heating surfaces by thermal spraying technology, and found that the bonding strengths of the molten ash to different materials were different. Typically, the nickel alloy coating could effectively slow the slagging. Chen et al.[9] also found that it can reduce the adhesion strength of molten ash by thermal spraying a nickel-containing coating. Formanek et al.[10] sprayed a ceramic coating containing aluminum oxide and chromium oxide on the water wall to reduce the contamination and corrosion of the heated surface. However, thermal spraying technology has the disadvantages of high cost, low utilization rate of spraying materials and it is difficult to operate. These limit its promotion and development[10]. Shan et al.[11] synthesized a ceramic coating with good impact resistance and oxidation resistance on the chrome-manganese alloy substrate by slurry method. Wang et al.[12],[13],[14] prepared a composite ceramic coating for the high-sodium coal, high-sulfur coal, high-alkali coal and other easily-slagging coal by slurry method. In the laboratory, it was found that the coating with h-BN added had good adsorption resistance, shock resistance and stain resistance. They have been applied to a Zhundong coal boiler and achieved good results. However, this composite coating produces some cracks that shorten the life.

Precursor ceramic coating technology can alleviate the above problems. The technology has advantages of low energy consumption and simple spraying. With the features of organic polymer and ceramic materials, the coating is not easy to crack, owning high hardness, good wear resistance and hydrophobicity. However, during the pyrolysis of the precursor, the volume shrinkage due to gas escaping, which leads to defects, cracks and even delamination of the coating. For single precursor coating, the critical thickness is too small and not conducive to the use of the coating in harsh environments. Greil et al.[15] first proposed the addition of active fillers in the precursors to alleviate the problems caused by the volume shrinkage of the precursors, and established a physical model for the volumetric shrinkage during the cracking process. Schütz et al.[16]used glass frit as a filler to densify the coating for the increase of coating thickness and adhesion. The addition of a low surface energy material to the precursor can improve the slag resistance of the coating. Different from other coating measures, the precursor ceramic coating technology first uses a chemical method to synthesize a precursor cross-linked polymer, which is then pyrolyzed to form a ceramic coating. The coating can be sprayed on the heated surface by a general spraying method.

In this paper, three kinds of ceramic coatings were prepared by respectively adding the inert fillers hexagonal boron nitride (h-BN), ZrO_2 and Al_2O_3 with the precursor method. The three types of coatings were subjected to a melt diffusion and solidification test using a slagging test apparatus. The slag resistance of the coatings was evaluated and compared by studying the contact angle, diffusion coefficient and temperature of the droplet and the coating, combined with numerical simulation.2. Experiment and simulation

1.1 Preparation of coating

1.1.1 Preparation of slurry. First, the slurry was prepared with the precursor method. The organopolysilazane was used as the precursor, butyl acetate was used as the solvent, dicumyl peroxide was used as the free radical initiator. Three inert fillers h-BN, ZrO_2 and Al_2O_3 were added separately. The components were uniformly mixed with a stirrer for 30 minutes to obtain three composite ceramic slurries. The proportions of the three slurries components are shown in Table 1.

Table 1. Proportion of slurry ingredients.

Polysilazane (Vol.%)	h-BN/ ZrO_2/ Al_2O_3 (Vol.%)	Butyl Acetate (Vol.%)	Dicumyl peroxide (Vol.%)
40	30	27	3

1.1.2 Preparation of coating samples. The coated steel piece was made of TP347 steel and cut into squares of 20 mm × 20 mm × 1 mm. In order to facilitate the embedding of the thermocouple, a 1 mm hole is made in the center, and a groove of 0.5 mm in depth and 1 mm in diameter is dug at the right side of the back, as shown in Figure 1.

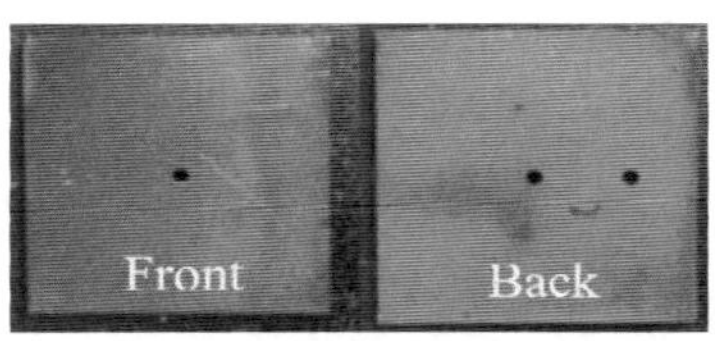

Figure 1. Schematic diagram of the opening of the sample piece.

To begin with, it was sandblasted with quartz sand having an average particle diameter of 40 μm for coating adhesion and sintering. After that, the surface oil was cleaned with ultrasonic acetone. Finally, it was washed with alcohol and dried.

Thereafter, the slurry was applied to the steel sheet by the Czochralski method, and the critical thickness of the coating was about 20 μm. In order to prevent bubbles or cracks in the coating caused by direct heating, the steel sheet covered with the coating was first placed in a dry oven and heated from room temperature to 85°C at a heating rate of 3°C/min, dried for 1 h. After drying, it was placed in a muffle furnace and heated to 600°C at a heating rate of 3°C/min, kept for 2 h. Finally, it is taken out after cooling along with the furnace.

1.1.3 Comparative steel sheet specimen. During the actual slagging process, the the object of adhesion of the molten ash is mainly derived from the oxide film layer of the steel. Therefore, for comparison, the tp347 steel sheet was subjected to the same washing and drying step, placed in a muffle furnace for high-temperature oxidation, with an oxidation temperature of 500 ° C and an oxidation time of 24 hours, to form a dense oxide film on the surface.

1.2 Molten ash spreading and solidification experiment

In the slagging testing device (Figure 2), the simulated ash was poured into a corundum tube. The simulated ash composition was as shown in Table 2. First, the simulated ash was melted into liquid in a high-temperature frit furnace (1300°C), and the corundum tube was quickly lowered to 3 cm from the sample piece. In a high temperature electric furnace (700 °C), the droplets were dropped into the center of the sample piece on the cooling and temperature measuring device (Figure 3) by means of a syringe at the top of the pressurized corundum tube. In order to avoid too high content of the oxygen, N_2 is introduced into the furnace to provide a reducing atmosphere. The high-speed camera shoots through the high-temperature glass observation hole provided in the lower part of the high-temperature electric furnace. The wettability of the coating was analyzed by measuring the contact angle. The pressure of the compressed air (0.2 MPa) introduced into the cooling device was controlled by a pressure regulating valve to maintain the upper sample piece of the cooling device at 500°C. When the droplet drops into the center of the sample, the thermocouple in the temperature measurement hole can measure the temperature of the droplet, and the temperature of the droplet can be transmitted by the data acquisition card (DAQ) to the computer for the temperature change recording.

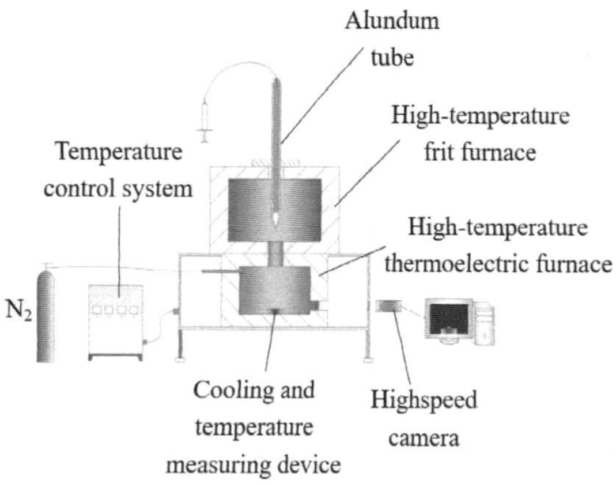

Figure 2. Diagram of slagging testing device.

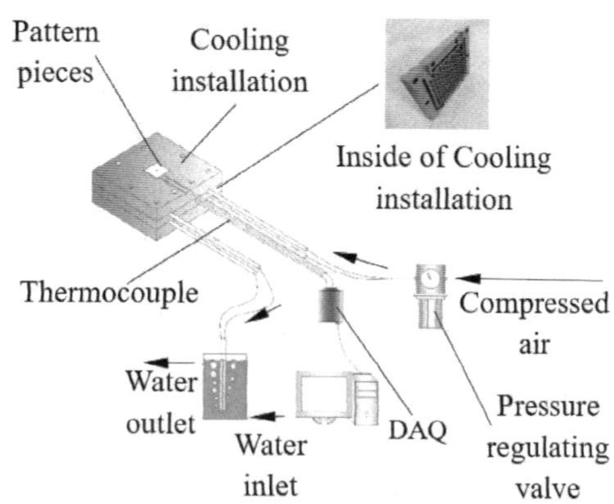

Figure 3. Diagram of cooling and temperature measuring device.

Table 2. Simulated ash composition.

Na_2O/%	BaO/%	SiO_2/%	CrO_3/%	Al_2O_3/%	Softening Temperature /°C
5	40	30	15	10	680

In order to avoid the influence of the placement of the thermocouple on the measuring point and the temperature in the vicinity thereof, the groove and the temperature measuring hole are buried. In order to measure the temperature of the droplets passed through the through holes in the center of the sample piece, two thermocouples with ceramic sleeves were placed in the grooves of the cooling device, The thermocouple probe that measures the temperature of the sample piece is placed in the groove on the right side of the sample piece to measure the temperature of the sample piece.

The contact angle is the angle between the tangent of the gas-liquid interface and the boundary between the liquid and the solid-liquid, expressed by θ. The JC2000 was used to measure the contact angle of the spreading coagulation process recorded by the high-speed camera. During the measurement, multiple steel plate samples and three coating samples were taken and averaged to obtain the final contact angle data. Since the solid surface has a roughness r, the measurement results need to be corrected according to the Wenzel equation[17]. θ_w in the formula is the true contact angle.

$$\cos \theta_w = r \cos\theta \tag{1}$$

The dynamic diameter of the contact surface of the droplet to the sample piece is d. The ratio of the dynamic diameter d and the equivalent diameter D of the droplet is the spreading coefficient β, which can be used to characterize the wettability of the liquid to the solid.

$$\beta = d/D \tag{2}$$

During the measurement, the spreading coefficients of the steel sheet and the three coated samples were each taken from a plurality of data and averaged to obtain the final spreading coefficient.

1.3 Numerical simulation of molten ash

From the above experiment, it was found that the h-BN coating has the best resistance to the slagging effect. However, the information such as the temperature distribution inside the droplet, the spreading and solidification process of the droplet, could not be obtained. Thus, the numerical simulation is further carried out. By analyzing the shape and temperature field changes of the droplets, and measuring the parameters such as contact angle and spreading coefficient, the numerical simulation is well instructive compared with the experimental results.

The multi-phase flow VOF method is used to track the liquid-gas interface, and the DO model is used to solve the radiation heat transfer problem during the droplet movement. The solidification/melting model was combined with the VOF model to solve the solidification problem after the droplets contacted the substrate. The physical property parameters are shown in Table 3.

Table 3. Physical parameters.

Materials	Simulation Ash	Air	Sample Piece
Density(kg/m^3)	2500	0.362	7850
Specific Heat (J/kg·K)	700	1135	460
Thermal Conductivity (W/m·K)	1.1	0.1671	61(steel sheet) 59(h-BN)
Viscosity (Pa·s)	0.25	4.15e-5	/
Solidus Temperature (K)	953.15	0	/
Liquidus Temperature (K)	953.15	0	/
Surface Tension (N/m)	0.56	/	/
Initial Contact Angle (°)	/	/	91.574(sheet steel) 126.142(h-BN)

The initial contact angle of the two sample pieces are the final contact angles which was measured by the melt ash spreading and solidification experiment.

2. Results and discussion

2.1 Morphology analysis
The simulated droplet diameter was 7 mm, the collision speed was 0.75 m/s, and the initial position was the point contacted with the sample piece. The simulated picture of the morphological change of the molten ash on the steel sheet and the h-BN coating were spliced with the experimental photograph, as shown in Figure 4. It can be seen that the simulation results are almost identical to the experimental results. After the droplets on the steel sheet begin to deposit, they gradually spread on the substrate, and the radius increases continuously. However, due to the high temperature of the droplets and the low temperature of the ambient air and the substrate, the heat exchange between the three is very intense, resulting in short spreading process time. The leading edge of the intermediate droplet solidifies in a short time, and the droplet spread is insufficient. When the droplets are deposited on the substrate, the air just below the droplets is drawn into the droplets because it is too late to be discharged. The deposition process of the droplet on the h-BN coating is short, the contact area of the droplet with the substrate is small, and after the speed is reduced, it is almost no longer spread.

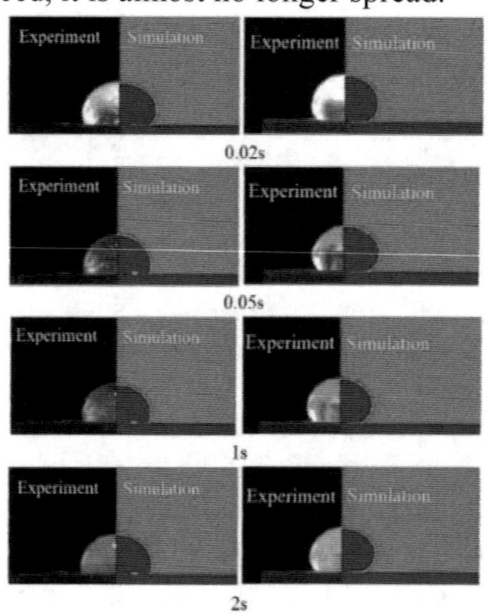

Figure 4. Diagram of morphological comparison. In the experiment and simulation, the morphological changes of the droplets on the steel sheet and the h-BN coating were compared.

2.2 Contact angle
The change in the contact angle of the molten ash on each sample piece with time in the experiment and simulation is shown in Figure 5.

According to the change of the contact angle of the three coatings and the steel sheet, the contact angle of the molten ash on the steel sheet gradually decreases with the increase of time. On the coating, the contact angle first decreases and then increases with time. Mainly because the molten ash spreads slowly under the action of the coating, and the contact angle gradually decreases. When the contact line stops spreading, the upper part of the molten ash continues to deposit downward under the action of inertial force and gravity, so the contact angle gradually increases. The contact angle of the final molten ash on the three coatings is larger than that of the steel sheet. Among the three coatings, the contact angle of the molten ash in the h-BN coating is the largest, which is 34.7° larger than that of the steel sheet. It indicates that the coating weakens the wettability of the molten ash, and the addition of the h-BN filler to the precursor is more effective.

In the numerical simulation, the contact angle of the droplet on the steel sheet gradually decreases with time, and the simulated value of the contact angle is larger than the experimental value. After the droplet solidification stops spreading, the numerically calculated contact angle is 93°, and the final angle is 91.3° through experimental measurement. The simulated value of the contact angle of the droplet on the h-BN coating is generally smaller than the experimental value. After the droplet deposition is completed, the experimental value is 1° smaller than the simulated value. The simulation results agree well with the experimental results.

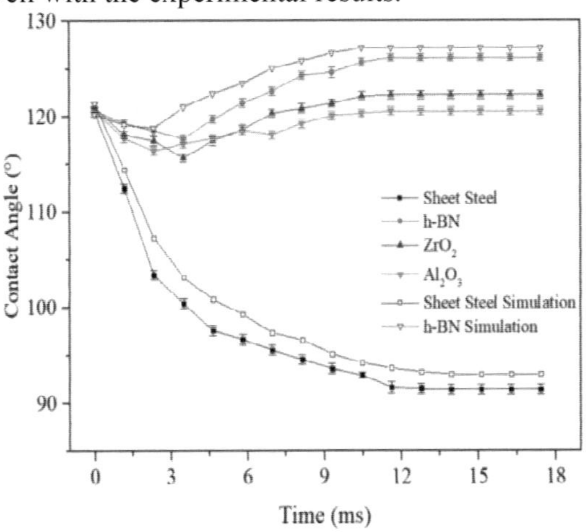

Figure 5. Curves of contact angle versus time.

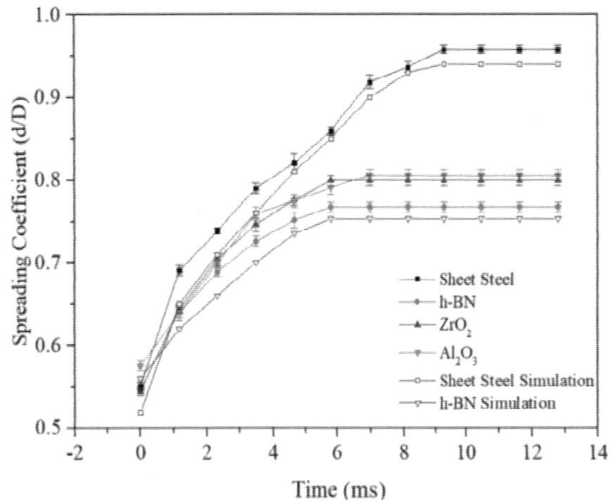

Figure 6. Curves of spreading coefficients versus time.

2.3 Spreading coefficient

Figure 6 shows the curves of the spreading coefficients of the molten ash on steel sheets and coatings with time in experiments and simulations. It can be seen that the spreading coefficient of the molten ash increases with time. Due to the higher initial molten ash temperature, the viscosity is smaller and the spreading is faster. As time increases, the temperature at the bottom of the molten ash and the spreading speed decrease gradually. The molten ash reaches a spread balance at 9 ms and no longer spreads with a spreading coefficient of 0.96.

The spreading coefficient of the molten ash on the h-BN coating reaches a maximum value of 0.77 at 6 ms. The molten ash on the ZrO_2 coating no longer spreads at 6 ms, and the spreading coefficient is 0.8. On the Al_2O_3 coating, the spreading coefficient reaches a maximum value of 0.81 at 7 ms.

As can be seen from the figure, when the molten ash no longer spreads, all the spreading coefficients on the coating are smaller than those on the steel sheet. The molten ash needs a shorter time to reach the spread balance on the coating. The results indicate that the coating weakens the wettability of the molten ash. After the spreading stops, the spreading coefficient of the molten ash on the h-BN coating is the smallest, which is 0.19 smaller than that of the steel sheet. It shows that the wettability of molten ash on h-BN coating is the worst. This results in a smaller contact area of the sample piece, which makes the adhesion of the molten ash difficult and finally achieves the effect of slagging resistance.

The simulated value of the spreading coefficient of the droplet on the steel sheet is slightly smaller than the experimental value. After spreading, the simulated value is 1.84% smaller than the experimental value. The experimental and simulated spreading coefficients of the droplet on the h-BN coating differ by 1.82%, which is in good agreement.

2.4 Temperature change

During the simulation, the temperature monitoring point is set at the same position as the experiment, in the center of the droplet. This shows the temperature change of the droplet on the steel sheet and the h-BN coating. The comparison between the experimental and simulation results is shown in Figure 7. The temperature changes are similar. When the molten ash is dropped onto the steel sheet and the coating,

IOP Conf. Series: Materials Science and Engineering **721** (2020) 012023 doi:10.1088/1757-899X/721/1/012023

the temperature of the sample piece is about 860°C. From the analysis of the spreading coefficient, it can be seen that the spreading of molten ash stops at 7 ms, the temperature of the molten ash measured at 840°C is higher than the solidification temperature, indicating that the front edge of the molten ash has been solidified. With the passage of time, the molten ash gradually cools from the bottom, the molten ash temperature reaches 680°C at 1 s, which is the solidification temperature, and the molten ash is gradually solidified. Therefore, most of the solidification during the spreading process occurs after the solid-liquid-vapor contact line stops, and the local solidification at the molten ash contact line controls the stop point of the contact line, which plays a central role in determining the final shape of the deposited droplet.

During the simulation, the temperature of the droplets rapidly decreases when spreading on the substrate. Compared with the experimental data, the temperature decreases rapidly. It is stable at 500°C after 8 s. This is because thermocouple temperature measurement requires a certain response time during the experiment.

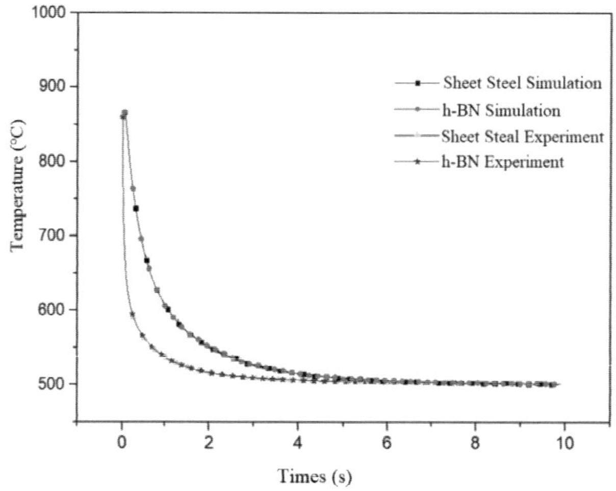

Figure 7. Curves of temperature versus time of steel sheet and h-BN coating in the experiments and simulations.

2.5 Slagging resistance mechanism of precursor coating

The slagging resistance mechanism is shown in Figure 8. The melted or partially melted particles carried by the flue gas are cooled and solidified on the heating surface, resulting in slagging of the boiler. When the molten ash diffuses and impacts on the precursor ceramic coating, the contact angle of the droplets is greater than 90° due to the wettability of the coating, resulting in residual air between the droplets dispersed on the coating. The area between the molten ash and the heating surface becomes smaller. When the molten ash accumulates into a slag, the adhesion between the molten ash and the heating surface is insufficient, causing it to fall off. Then the slag prevention effect is achieved.

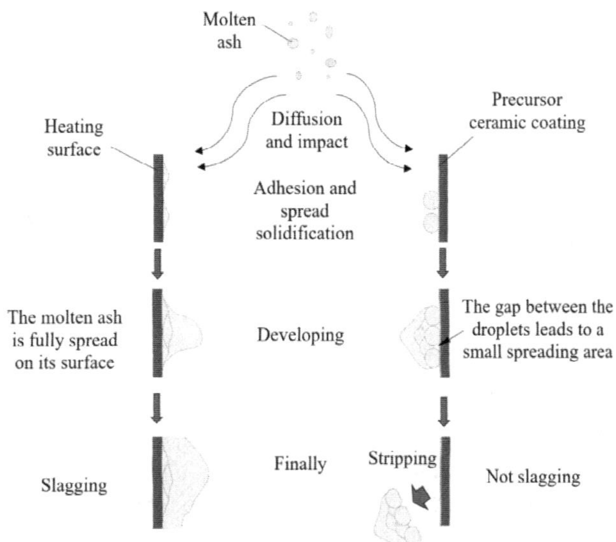

Figure 8. Diagram of the slagging resistance mechanism.

3. Conclusions

(1) The molten ash has poor wettability on the surface of the coating, and the h-BN coating is the worst. The contact angle of molten ash on the coating is more than 90°, and the spreading coefficient is much smaller than 0.96 of the steel sheet. The contact angle of the h-BN coating surface is 34.7°. The spreading coefficient is reduced by 0.19 compared with the surface of the steel sheet. The spreading coefficients of the ZrO_2 coating and the Al_2O_3 coating are also reduced by 0.16 and 0.15, respectively. The results show that adding h-BN to the coating has better slag resistance characteristics than the other two fillers.

(2) The temperature changes of the droplets between the steel sheet and the coating are very similar. The solidification process of the droplets is independent of the sample. The difference between the simulated and experimental values of the contact angle is not more than 2°, and the difference between the simulated and experimental values of the diffusion coefficient is not more than 1.84%, which is in good agreement. The simulation results show that the wettability of the droplets on the coating is poor, which is consistent with the experimental results.

(3) Slagging resistance mechanism of the precursor coating is to reduce the wettability of the heating surface, so that it cannot be completely spread and reduce the contact surface. Eventually it will fall off due to the insufficient adhesion.

Acknowledgments

This work was supported by National Key Research and Development Project of China (grant number 2017YFF0205000), the National Natural Science Foundation of China (grant number 51876197), the Zhejiang Natural Science Foundation (grant number LQ20E060017).

References

[1] Martinson, C.A., van Schoor, G., Uren, K.R. (2017) Energy and exergy analyses of a subcritical pulverised coal-fired boiler based on the effects of slagging and fouling. IFAC Papers Online, 50(1): 2017-2022

[2] Wang, J. (2009) Study on the fuzzy mathematics model for the comprehensive judgment of slagging characteristics of utility boilers. 2009 International Conference on Machine Learning and Cybernetics. Hebei China.

[3] Ma, L., Fang, Q., Yin, C., Wang, H., Zhang, C., Chen, G. (2019) A novel corner-fired boiler system of improved efficiency and coal flexibility and reduced NO_x emissions. Applied Energy, 238:

453-465
[4] Andrei,V., Aleksandr, S., Zavorin, Aleksandr, V.,Starchenko. (2019) Numerical investigation of the combustion process for design and non-design coal in T-shaped boilers with swirl burners. Energy,186: 115844
[5] Li, H., Qiu, X., Tang, Y. (2008) Ash melting behavior by fourier transform infrared spectroscopy. Journal of China University of Mining and Technology, 18(2):245-249.
[6] Stam, A. F., Brem, G. (2019) Fouling in coal-fired boilers: biomass co-firing, full conversion and use of additives – a thermodynamic approach. Fuel, 239: 1274-1283.
[7] Naganuma, H., Ikeda, N., Ito, T., Matsuura, M., Nunome, Y., Ueki, Y., Yoshiie, R., Naruse, I. (2013) Reduction mechanisms of ash deposition in coal and/or biomass combustion boilers, Fuel, 106: 303–309
[8] Naganuma, H., Ikeda, N., Ito, T., Satake, H., Matsuura, M., Ueki, Y., Yoshiie, R., Naruse, I. (2013) Control of ash deposition in solid fuel fired boiler, Fuel Process. Technol. 105: 77–81
[9] Chen, J., Ninomiya, Y., Naganuma, H., Sasaki, Y., Noguchi, M., Cho, H., Ueki, Y., Yoshiie, R., Naruse, I. (2016) Development of thermal spraying materials through several corrosion tests for heat exchanger tube of incinerators, Fuel Processing Technology, 141: 216–224
[10] Formanek, B., Szymański, K., Szczucka-Lasota, B., Włodarczyk, A. (2005) New generation of protective coatings intended for the power industry, J. Mater. Process. Technol. 164–165:850–855
[11] Shan, X., Wei, L.Q., Zhang, X.M., Li, W. H., Tang, W. X., Liu, Y., Tong, J., Ye, S. F., Chen, Y. F. (2015) A protective ceramic coating to improve oxidation and thermal shock resistance on CrMn alloy at elevated temperatures, Ceramics International, 41: 4706–4713
[12] Wang, J., Yuan, Y., Chi, Z., Zhang, G. (2017) Development and application of anti-fouling ceramic coating for high-sodium coal-fired boilers. Journal of the Energy Institute.91(6):962-969.
[13] Wang, J., Yuan, Y., Chi, Z., Zhang, G. (2018) High-temperature sulfur corrosion behavior of h-BN-based ceramic coating prepared by slurry method. Materials Chemistry and Physics, 206: 186-192.
[14] Wang, J., Zhang, G., Li, F., Chi, Z. (2019) Lyophobicity and slagging resistance mechanism of h-BN based coating for coal-fired boilers. Fuel Processing Technology, 188: 43-50.
[15] Greil, P. (1995) Active-filler-controlled pyrolysis of preceramic polymers. Journal of the America Ceramic Society, 78(4): 835-848.
[16] Schütz, A., Günthner, M., Motz, G., et al. (2012) Characterisation of novel precursor-derived ceramic coatings with glass filler particles on steel substrates. Surface & Coatings Technology, 207(20): 319-327.
[17] Wenzel, R.N. (1936) Resistance of solid surfaces to wetting by water. Industrial & Engineering Chemistry Research, 28(8): 988—994.

IOP Conf. Series: Materials Science and Engineering **721** (2020) 012024 doi:10.1088/1757-899X/721/1/012024

Fracture Analysis on Turbine Intermediate Regulating Valve of an Electric Generator Unit

Tianpeng Wang[1*], Yuqiu Sui[2], Yunan Wang[1,3], Hongqiang Li[1], Mingcheng Sun[1], Li Dai[1] and Nan Wu[1]

[1]Electric Power Research Institute of State Grid Liaoning Electric Power Co., Ltd, Shenyang, Liaoning province, 110006, PR China

[2]State Grid Liaoning Electric Power Supply Co. Ltd., Shenyang, Liaoning province, 110006, PR China

[3]Liaoning Dongke Electric Power Co., Ltd, Shenyang, Liaoning province, 110006, PR China

[*]Corresponding author's e-mail: wtp0619004@gmail.com

Abstract. A turbine intermediate regulating valve fractured in a power plant, and the reasons for the fracture were analysed by using macro morphology observation, mechanical properties test, chemical composition analysis and microstructure observation, such as scanning electron microscopy (SEM) analysis or metallographic analysis. The results indicated that the chemical composition and strength properties were in accordance with the standard requirements, but the elongation after fracture and the impact toughness value were all lower than the required values. Precipitation of σ phase further weakened the matrix and reduced the strength of the structure. Fretting friction and collision appeared between the valve stem and the anti-rotating pin, and additional stress was generated by the reduction of radial clearance between the pin and valve. Then cracks and brittle expansion occurred in the stress concentration areas close to the anti-rotating pin holes and finally fractured.

1. Introduction

A staff member of a power plant carried out the partial stroke activity test of the No. 2 combined steam valve as schedule on the unit 4 in March 2016, (part of the journey activity test once a day, and once a week for the whole trip activity test). He found that jamming occurred in the intermediate pressure regulating valve (8% valve position jamming) and estimated that the jamming location was in the regulating valve (the total journey of the valve is 257.9±1.5mm and the regulating valve's journey accounted for about 16% of the total trip). Slight oxide skin was discovered on the valve rod after disassembling, whereas the valve bushing remained intact without fracture. After removing the valve cover and lifting out the regulating valve, fracture was found in the stress concentration areas close to the anti-rotating pin holes.

The No. 4 unit in this power plant was put into operation on May 18th, 2015. The steam was reportedly held at a temperature of 600 ℃ under a pressure of less than 8 MPa. The valve's fracture site was not affected by external stress during the process of operation. A pin, whose material is 2Cr12NiMo1W1V, was mainly applied to prevent the rotation of the valve disc, and a certain amount of clearance exists between the valve rod and pin. The material of the regulating valve was designed as a nickel-based high temperature deformation alloy (GH901) [1,2], which is based on Fe-43Ni-12Cr with reinforced element such as titanium, aluminum, molybdenum and a small amount of boron and carbon. The reasons for the fracture were analyzed in this paper through the investigation of the fracture morphology, mechanical properties, chemical composition and microstructure of the material, and some suggestions on the maintenance and management of valves were put forward.

2. Experiment analysis

2.1Macro morphology analysis

The macro morphology of the regulating valve sent for inspection is shown in figure 1. It can be observed that the fracture position is located in the middle of the stem with anti-rotating pin installed. Further analysis of the macro morphology of the fracture surface is shown in figure 2. Two crack sections can be observed on the fracture surface (named section 1 and section 2, respectively), which have been severely oxidized, suggesting that the fracture surface had been exposed to the environment for a significant period of time. Both of the two crack initiations are located in the center of the contacting area between the pin hole and pin. Figure 2b) indicates the propagation directions of the crack towards the edge and the final rupture regions after the formation of two crack initiations during the operation of the valve stem. A fiber region, a radiation region, and a final fracture region, which were formed centrally outwardly across the fracture, are characteristic of a typical brittle fatigue fracture. The crack initiations of the section 1 and section 2 are not in the same horizontal plane, which has about 20 mm difference in height. The fracture section of the section 1 is generally flat, whereas the cross section of the section 2 has an obvious convex platform. The source areas of section 1 and section 2 have a large number of oxide skins in connection with the pin holes, while no oxide skin is observed at the bottom of the pin holes.

Figure 1. Macro morphology of the regulating valve

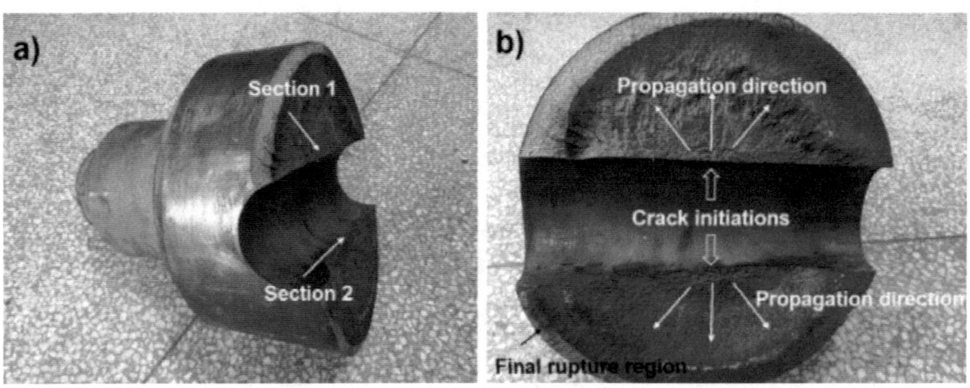

Figure 2. Macro morphology of the fracture surface
a) Crack sections picture; b) Crack initiations, propagation directions and final rupture regions

2.2Chemical composition analysis

A specimen was excised from the valve stem for spectral inspection to determine the chemical composition of the part, and the test execution standard was GB/T 4336-2016. The instrument model was DV-6 quantitative direct-reading spectrometer.

The results of the chemical composition analysis are shown in table 1. The required values of the China Aeronautical Materials Handbook [3] (hereinafter referred as handbook) are also listed in table 1. It can be seen that all chemical components in the fracture valve meet the requirements of GH901. There are no anomalies present in the composition of the valve stem that will be considered contributory to the failure of the part.

Table 1. Chemical analysis results (weight percent)

element	Ni	Cr	Mo	Ti	C	Si	Mn	P	S	B	Al	Cu
Fracture sample	44.17	12.33	6.49	3.01	0.033	0.01	0.024	0.002	0.004	0.01	0.191	0.036
Handbook requirement	40.0 ~45.0	11.0 ~14.0	5.00 ~6.50	2.80 ~3.10	0.02 ~0.06	≤0.40	≤0.50	≤ 0.020	≤ 0.008	0.010~ 0.020	≤0.30	≤0.20

2.3Mechanical property analysis

2.3.1Hardness. The hardness test of the valve's cross section was carried out by 320HBS-3000 brinell hardness tester. The measured hardness values were 368HB, 374HB and 370HB, respectively. The hardness of the GH901 nickel based high temperature deformed alloy introduced in the handbook is 302-388HB. So it can be concluded that the material's hardness meets the material requirements, but is close to the upper limit.

2.3.2Tensile properties at room temperature. Two Φ10 mm rod tensile samples along the longitudinal and transverse directions were sampled respectively from the fracture valve stem. CSS-1120 electronic universal testing machine was applied to measure the room temperature tensile properties, and the results are listed in table 2, along with the standard values of the tensile properties for the GH901 alloy written in the handbook. It can be seen that all the tensile strengths and yield strengths along the valve's two directions meet the standard requirement. Brittle intergranular fracture can be observed along the fracture section, and almost no necking appears. The elongation and contraction of the cross sectional area are far lower than the required value.

Table 2. Tensile testing results at room temperature

No.	Tensile strength (N/mm^2)	Yield strength (N/mm^2)	Elongation (%)	Reduction in area (%)	Remark
1	1194	961	7.0	7.84	longitudinal
2	1204	949	8.0	9.75	longitudinal
3	1153	946	2.0	3.96	transverse
4	1156	964	3.6	3.96	transverse
Standard value	≥1130	≥810	≥9%	≥12%	—

2.3.3Impact performance. Three U-notched impact specimens were sampled from each of the longitudinal and transverse directions. NI300C pendulum instrumented impact tester was used and the results were listed in table 3. It is required in the handbook that the impact energy of the GH901 alloy is not less than 56J, but the actual impact energy of the 6 u-notched samples is far below the requirement.

Table 3. Impact performance results

Sampling direction	Impact energy/J
Longitudinal direction	12.0; 11.1; 11.0
Transverse direction	6.9; 6.5; 7.7

2.4Mechanical property analysis

A sample was taken at the source of the crack for microscopic examination, as shown in figure 3-5. It can be seen that both the crack initiation and the matrix are twin austenite, but the grain size is large, which can be classified as grade 3 in accordance with GB/T 6394-2002. It can be seen from figure 3 that there is a serious oxidation phenomenon at the location of the crack source. The thickness of the oxide layer is about 100μm, and there are microcracks extending along the grain boundaries.

To further investigate the microstructure distribution of the corrosion zone, a SEM observation of the area around the corrosion layer is shown in figure 4. It can be seen that the corrosion zone has a multi-layered structure: the oxide skin is divided into two layers (layer 4 and layer 5), and the total thickness is about 0.2mm; an oxidation film (layer 3) with no more than 50μm can be seen close to the layer 4; a transition layer (layer 2) with a thickness of 30-50μm exists between the oxidation film (layer 3) and the valve substrate. Energy disperse spectrum (EDS) analysis was carried out at different regions marked in the figure 4, and the result is shown in table 4. The EDS results shows that O, Si, V, Cr, Mn, Fe, Ni, Mo, W and other elements are detected in the oxide skin close to the pin side (layer 5), but only three elements such as O, Cr, Fe can be detected in the oxide layer (layer 4). Considering that the pin material is 2Cr12NiMo1W1V, it is inferred that the oxide skin near the pin side is constituted by the oxidation of the pin material remaining and sticking to the inner surface of the pin hole during operation. More O element can be detected at layer 3, which is further confirmed as the oxide film. The EDS analysis of transition layer and substrate (positions 2 and 1) shows that the elements of the transition layer and substrate are the same.

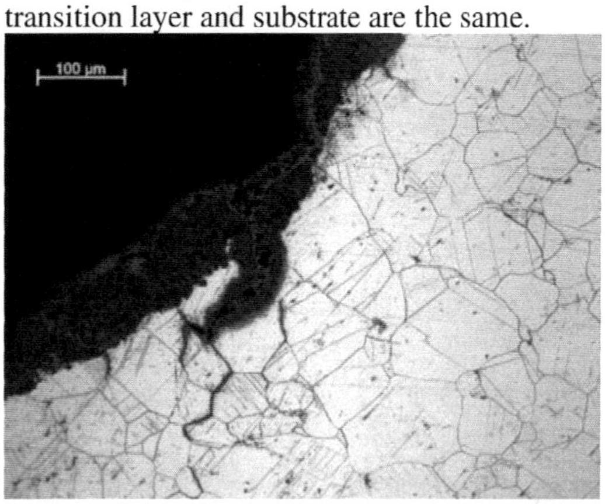

Figure 3. Metallographic structure around the crack initiation

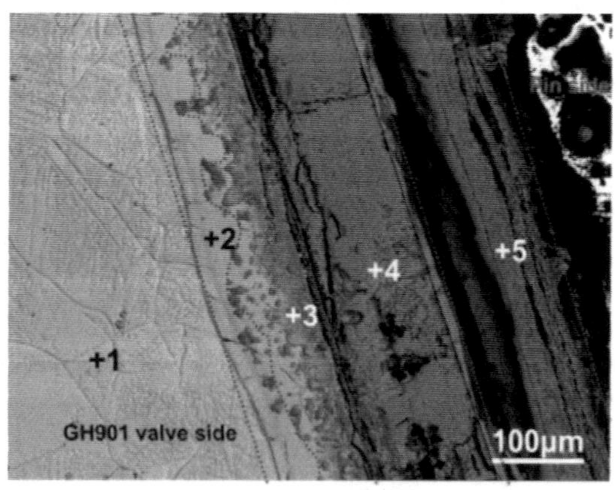

Figure 4. SEM picture of the area around the corrosion layer

Table 4. EDS analysis results of different regions in figure 4

Position	Element content (weight %)									
	O	Si	V	Cr	Mn	Fe	Ni	Mo	W	Ti
1	—	—	—	13.51	—	31.56	35.30	5.20	—	2.41
2	—	—	—	12.77	—	32.97	36.29	5.80	—	2.32
3	15.80	—	—	18.65	1.02	20.55	27.21	8.72	—	4.14
4	13.31	—	—	0.65	—	89.04	—	—	—	—

5	22.13	0.39	0.49	16.09	1.18	44.05	1.64	3.02	1.58	—

3. Discussion

The mechanical performance test result shows that the chemical component and tensile strength of the failure valve meet the standard requirements. However, the elongation and impact toughness values are far below the required value. Brittle intergranular fracture can be observed along the tensile and impact fracture section, and the material's toughness is very poor. Macro morphology observation indicates that the crack originated from the position of both sides of the valve's center close to the anti-rotating pin and expanded outward. The section is rough and there is no obvious plastic deformation, indicating a typical brittle fracture[4]. Microstructure investigation shows that the microstructure of the valve is large twin austenitic, and precipitations appear, further weakening the substrate and reducing the strength of the component[5,6].

The component was run under the condition of 600 ℃ for a long time, the valve stem material elongation decreased significantly because of the existence of the GH901 material's aging embrittlement[7]. There are a lot of microcracks on the inner surface of pin hole close to the crack source, and the cracks have extended along the grain boundary to the inside of the valve and have penetrated the entire diffusion layer. There are a large number of oxide skins accumulated in connection with the pin holes, while no accumulated or peeling oxide skin is observed at the bottom of pin holes. So we can know that there was a strong interaction force between the pin and the valve. The stress mainly comes from two aspects[8]: firstly, the pin is mainly used to prevent the rotation of the valve, therefore, micro friction is generated inevitably between the pin and the valve rod in the running process. The material strength decreases under high temperature, especially for the 2Cr12NiMo1W1V[9]. Then the two contacting parts produced caking phenomenon because of the micro friction. The caking part produces adhesive particles, which are more easily oxidized under high temperature condition. With continuous operation of the unit, more and more adhesive oxidations accumulate. When the adhesive oxidations reach a certain thickness, they no longer produce a bond phenomenon due to the poor binding force. Secondly, the designed maximum assembly clearance between the pin and valve is 0.05mm, which will become smaller at high temperature due to the expansion of the pin and valve[10]. As the thickness of the adhesive oxidation increases, the clearance may be completely eliminated, causing the pin to bite to death in the pin hole. Thus a considerable expansion force is generated between the pin hole and valve stem. As that the material's toughness is very poor, and the constraint force on the edge is less than that in the core, micro cracks generate in the center of pin hole when the expansion stress reaches a certain extent. Then the crack sources initiate on the both sides. With the continuous oxidation along the grain boundary and the collision between the valve stem and the anti-rotating pin, the crack can be expanded and even cracked.

To sum up, supervision job should be done strictly by the power plant during the process of the valve's producing, assembly, inspection or test. The quality of the assembly should be specifically considered. The problem should be eliminated in the germ state, thus a good quality can be ensured. In addition, the inspection and maintenance of the intermediate regulating valve should follow the correct and reasonable maintenance process during the process of maintenance.

4. Conclusion

1) The fracture mechanism of the intermediate regulating valve belongs to brittle fracture;

2) The chemical composition and tensile strength of the valve meet the standard requirements, whereas the elongation and impact toughness are all lower than the required values;

3) Fretting friction and collision between the valve stem and the anti-rotating pin, additional stress generated by the reduction of radial clearance between the pin and valve, and lower plastic toughness account for the crack formation, extension and final fracture of the valve.

Acknowledgments
This research was financially supported by the National Key R&D Program of China NO. 2017YFB0902100.

References

[1] Duan S., Liu J. (2017) Effect of forging process on wear resistance of GH901 high temperature alloy turbine shaft forgings. Hot Working Technolo., 46: 51-55.
[2] Zhang H., Li Q., Guo Z.J., He J.G., Wen G.P. Li W.J. (2015) Effects of heat treatment on microstructure and hardness of GH901 alloys. Foundry Technolo., 18: 25-31.
[3] Yan M.G. (2001) China Aeronautical Materials Handbook, volume 2, Wrought Superalloy and Cast Superalloy. China Standards Press, Beijing.
[4] Schroeder C.J., (2015) Metallurgical failure analysis of a fractured steam control valve stem. J. Fail. Anal. Prevention, 15: 1-9.
[5] Zhou J., (2015) Analysis of surface cracking on stepped shaft of GH901 alloy and it's process optimization. Special Steel Technolo., 21: 18-20.
[6] Jerenz M., Martin U., Schurack F., Schmidt U., Vöhringer O., Oettel H. (1999) The microstructure and precipitation behaviour of superalloys with thermally insulating coating systems. Praktische Metallographie, 36: 13-21.
[7] Park K.H., LaSalle J.C., Schwartz L.H., Kato M. (1986) Mechanical properties of spinodally decomposed Fe-30 wt% Cr alloys: Yield strength and aging embrittlement. Acta Metall., 34: 1853-1865.
[8] Hopf R., Sündermann S.H., Born S., Ruiz C.E., Van Mieghem N.M., De Jaegere P.P., Maisano F., Falk V., Mazza E. (2017) Postoperative analysis of the mechanical interaction between stent and host tissue in patients after transcatheter aortic valve implantation. J. Biomech., 53: 15-21.
[9] Wang M., Chen M., Rui W., Wang G. (2016) High temperature flow stress model and hot processing map for 2Cr12NiMo1W1V supercritical steel. J. Cent. S. Univ., 47: 25-30.
[10] Reiter P., Kaschnitz E. (2002) Measurement of thermal expansion at high temperature by a transient interferometric technique. Int. J. Thermophys., 23:1327-1338.

Numerical investigations on entropy generation in cross-wavy channels

Xusheng Shi[1, 2], Yongwei Wang[1, 2*] and Xiulan Huai[1, 2*]

[1]Institute of Engineering Thermophysics, Chinese Academy of Sciences, Beijing, 100190, China

[2]University of the Chinese Academy of Sciences, Beijing, 100049, China

[*]Corresponding author: wangyongwei@iet.cn (Y. Wang), hxl@iet.cn (X. Huai)

Abstract. The numerical study to research the influence of ratio of channel amplitude to pitch of cross-wavy on entropy generation is conducted in this paper. In Reynolds number range of 200-1200, five types of cross-wavy channel are analyzed. The simulation results show the thermal entropy generation rising rate is fast when Reynolds number less than 700 and then flat. In addition, it is smaller with increasing channel amplitude to pitch value. The frictional entropy generation increases almost linearly with Reynolds number. However, it is much less than thermal entropy generation under the same conditions. In brief, the channels which have larger ratio of amplitude to pitch can reduce the total irreversibility to the least under the same flow conditions.

1. Introduction

In recent years, distributed power generation system develops rapidly and micro gas turbine is one of the core power equipment. For micro gas turbine, the thermal efficiency is only about 20% with simple cycle, while over 30% with recuperated cycle [1]. Therefore, recuperator which is the key part of gas turbine is often used in recuperated cycle to recover waste heat of exhaust gas and increase the inlet temperature of combustion. Cross-wavy (CW) primary surface heat exchangers have the advantages of high heat exchange efficiency, low pressure loss, compacted structure and light weight, and so on [2]. Therefore, it is often used as recuperator in gas turbine. Due to its compacted structure and small hydraulic diameter, the investigation of thermal-hydraulic performance in CW primary surface is useful.

Recently, many scholars have done lots of interesting study on CW primary surface by the method of experimental study and numerical simulation. Xiao et al. [3] reviewed recuperators used in micro gas turbines. The research provides a comprehensive understanding about optimization methods of recuperators. Stasiek and Ciofalo et al. [4, 5] researched thermal-hydraulic performance of corrugated passages. They obtained the local and average Nusselt number distribution data over channel surface. Utriainen and Sunden [6] studied thermal-hydraulic performance of cross-wavy duct by utilizing numerical simulation method. The results found amplitude of the duct have the great influence on heat transfer enhancement. Du et al. [7] implemented a three-dimensional numerical model of CW channel and investigated the influence of the value of the ratio L/A (wave length L, unit cell amplitude A) on comprehensive performance of channel.

Heat exchanger as an exchanging heat experiment is widely used to high energy-consuming sectors. The entropy generation analysis provides an insight that cannot be achieved by the energy analysis.

IOP Conf. Series: Materials Science and Engineering **721** (2020) 012025 doi:10.1088/1757-899X/721/1/012025

There is little research about the entropy generation in CW channels. In this paper, the influences of ratio of amplitude to pitch (*A/P*) on entropy generation are studied. It is meaningful to reduce energy consumption and improve the performance of recuperators.

2. Mathematical model

CW primary surface recuperator consists of many CW plates which are stacked and welded, as shown in figure 1. In the flow direction, the channels are corrugated. The hot channel and cold channel are alternating and this makes the structure more compact. The geometry parameters of the hot and cold channel, such as amplitude *A*, unit length *L*, pitch *P* and channel inner height *H*, are shown in figure 1.

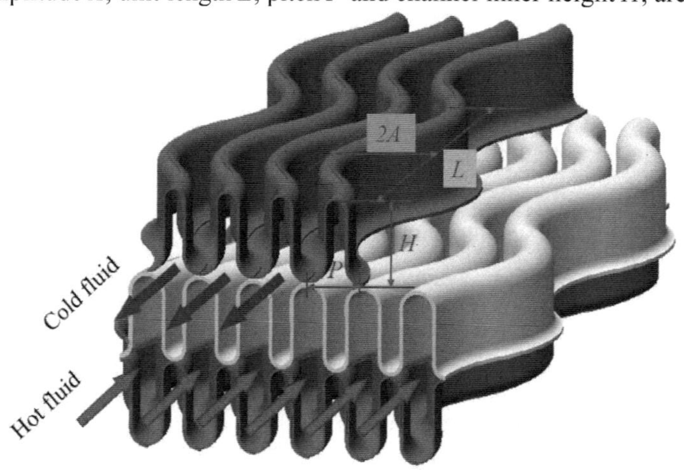

Figure 1. CW primary surface heat exchanger.

To reduce the computational complexity, three flow units of air channel are created as the numerical model, as shown in figure 2. Due to the flow and structure are periodic, the periodic boundary conditions can be used. It can be applied in the S1-1 and S1-2, S2-1 and S2-2, S3-1 and S3-2, S4-1 and S4-2, S5-1 and S5-2, S6-1 and S6-2. More details about multi-periodic boundary conditions can be found in literatures[8].

Given the influence of geometry parameters on distribution of the flow field of the channel, the dimensionless parameters *A/P* is used as variable to research entropy generation performance of CW channel. The values of *A/P* for different CW channel are 0.2333, 0.3333, 0.5000, 0.6667 and 0.8333, respectively.

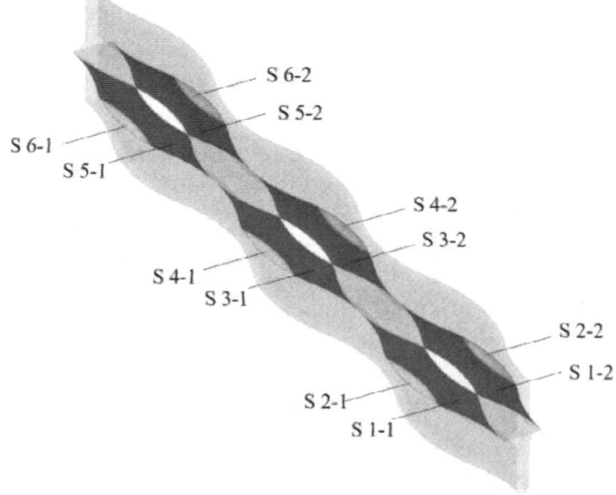

Figure 2. Numerical model of CW channel.

2.1 Governing equations

Zhang and Che [9] researched the performance of cross-corrugated plates using various turbulence models, and the results of low Reynolds number turbulence (LBKE) model agree with experiments. Therefore, the LBKE turbulent model is used.

In this model, the continuum equation, momentum equation and energy equation are respectively expressed as [10]:

$$\frac{\partial(\rho u_i)}{\partial x_i}=0 \tag{1}$$

$$\frac{\partial(\rho u_i u_j)}{\partial x_j}=\frac{\partial}{\partial x_j}\left(\mu_{eff}\left(\frac{\partial u_j}{\partial x_i}+\frac{\partial u_i}{\partial x_j}\right)-\frac{2}{3}\mu_{eff}\frac{\partial u_k}{\partial x_k}\right)-\frac{\partial p_i}{\partial x_i}+\rho g_i \tag{2}$$

$$\frac{\partial(\rho u_i c_p T)}{\partial x_i}=\frac{\partial}{\partial x_i}\left(\lambda\frac{\partial T}{\partial x_i}\right)+\Phi \tag{3}$$

where Φ is the energy dissipation due to viscosity, μ_{eff} is effective viscosity, which is equal to the sum of molecular viscosity μ and turbulent viscosity μ_t.

2.2 Calculation of entropy generation rate

The thermal and frictional volumetric entropy generation rate are respectively defined as[11]:

$$S'''_{g,\Delta T}=\frac{\lambda_{eff}}{T^2}\left[\left(\frac{\partial T}{\partial x}\right)^2+\left(\frac{\partial T}{\partial y}\right)^2+\left(\frac{\partial T}{\partial z}\right)^2\right] \tag{4}$$

$$S'''_{g,\Delta P}=\frac{\mu_{eff}}{T}\left(2\left[(\frac{\partial u}{\partial x})^2+(\frac{\partial v}{\partial y})^2+(\frac{\partial w}{\partial z})^2\right]+\left[\frac{\partial u}{\partial y}+\frac{\partial v}{\partial x}\right]^2+\left[\frac{\partial w}{\partial y}+\frac{\partial v}{\partial z}\right]^2+\left[\frac{\partial u}{\partial z}+\frac{\partial w}{\partial x}\right]^2\right) \tag{5}$$

Where μ_{eff} is effective viscosity, λ_{eff} is effective thermal conductivity which is equal to the sum of thermal conductivity λ and turbulent conductivity $C_p\mu_t/\sigma_t$.

The total volumetric entropy generation:

$$S'''_g=S'''_{g,\Delta T}+S'''_{g,\Delta P} \tag{6}$$

The corresponding dimensionless entropy generation rate [12]:

$$N_{g,\Delta T}=\iiint_V\frac{S'''_{g,\Delta T}}{\lambda/D_h^2}dV \tag{7}$$

$$N_{g,\Delta P}=\iiint_V\frac{S'''_{g,\Delta P}}{\lambda/D_h^2}dV \tag{8}$$

$$N_g=\iiint_V\frac{S'''_g}{\lambda/D_h^2}dV \tag{9}$$

where V is fluid region.

3. Grid independence and validation

Different grid systems are used to test the grid independence for model of type 2 with $Re = 451$. The results shown average Nusselt number changes only 1.38% when grid number ranged from 7,262,830 to 9,230,730. Therefore, the grid number of 7,262,830 is adopted.

To demonstrate the accuracy of present numerical method, numerical results and experimental data studied by Ma et al. [13] are shown in Figure 3. From the figure, the maximum error of *Nu* and *f* are about 13.5% and 10.6%, respectively. The results agree well with the experimental data.

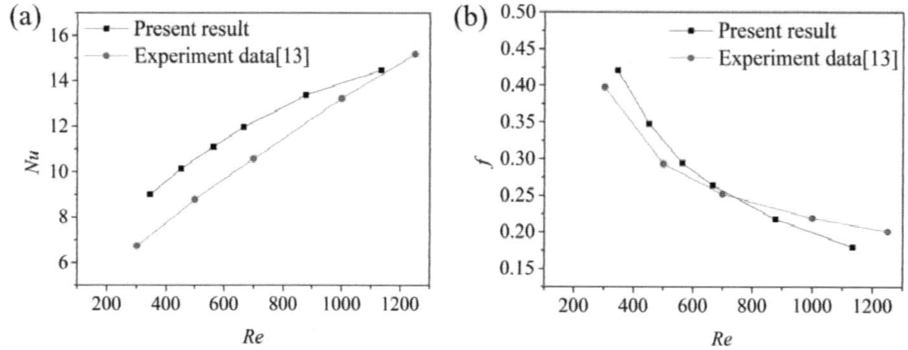

Figure 3. Numerical method validation.

4. Results and discussion

4.1 Volumetric entropy generation distribution

Figure 4 shows $S_{g,\Delta T}$ distributions of type 2. At $z = 0.5L$, the maximum of $S_{g,\Delta T}$ appeared at the channel center. However, at $z = L$, the value of $S_{g,\Delta T}$ near the channel wall is higher than channel center. This is because the fluid mixing of adjacent passages is intense and the temperature gradient increases.

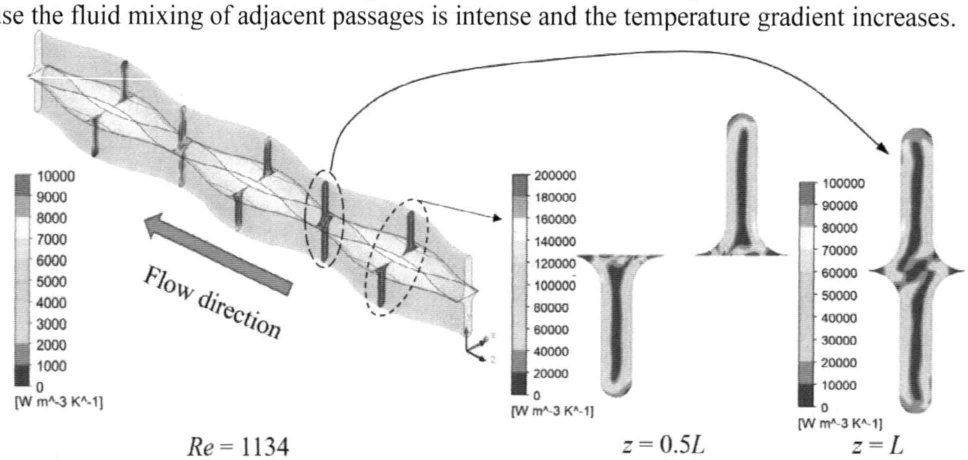

Figure 4. Thermal entropy generation contours on several cross-sections of type 2.

Figure 5 shows $S_{g,\Delta P}$ distributions of type 2. At $z = 0.5L$, it can be seen that the maximum of $S_{g,\Delta P}$ appeared at the center of the channel where the mixture of fluid is intense. However, at $z = L$, the value of $S_{g,\Delta P}$ near the channel wall where the fluid mixing of adjacent passages is intense is higher than the center of the channel. In addition, the velocity gradient increasing results in $S_{g,\Delta P}$ greater when Reynolds number increases.

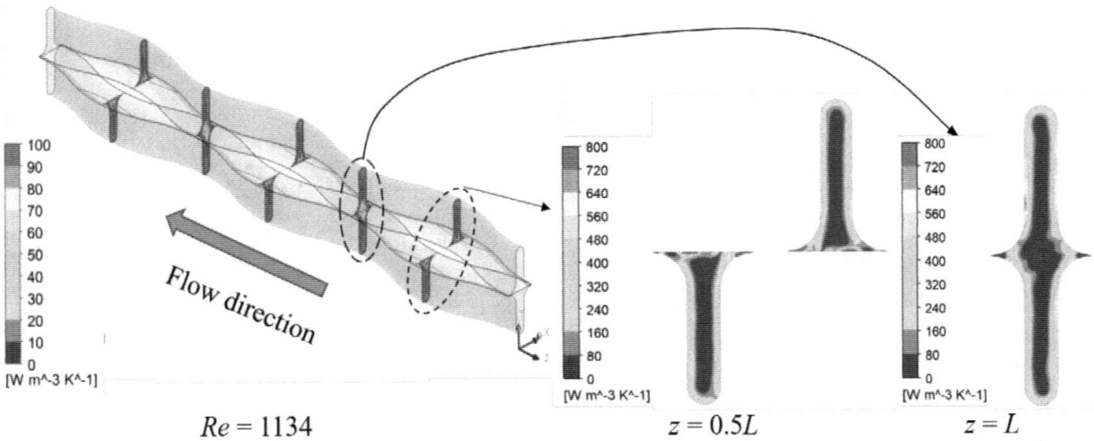

Figure 5. Frictional entropy generation contours on several cross-sections of type 2.

4.2 Effects of channel amplitude to pitch value on entropy generation

Figure 6 shows the influence of *A/P* values on $N_{g,\Delta T}$ and $N_{g,\Delta P}$ for different Reynolds number. As shown in figure 6(a), $N_{g,\Delta T}$ rising rate is fast when Reynolds number less than 700 and then flat. This is because the thermal boundary layer thickness is thinner and the temperature gradient increases with Reynolds number increases when Reynolds number less than 700. However, the thickness of thermal boundary layer has a little change with Reynolds number when it larger than 700. Besides, the greater *A/P* value is, the lower $N_{g,\Delta T}$ becomes. For example, $N_{g,\Delta T}$ has decreased by 32.9% at Re = 1134 with *A/P* value from 0.2333-0.8333. It is due to the greater channel amplitude results in the more drastic deflection of the fluid flow and fluid mixing which reduce the temperature gradient. Therefore, the thermal irreversibility of type 5 is the minimal. From figure 6(b), $N_{g,\Delta P}$ increases almost linearly with Reynolds number. In addition, the $N_{g,\Delta P}$ value of type 5 is greater when Reynolds number is larger than 700. This is because the greater amplitude of the channel results in more intense friction and impingement between fluid and channel wall. However, $N_{g,\Delta P}$ is much smaller than $N_{g,\Delta T}$ under same conditions.

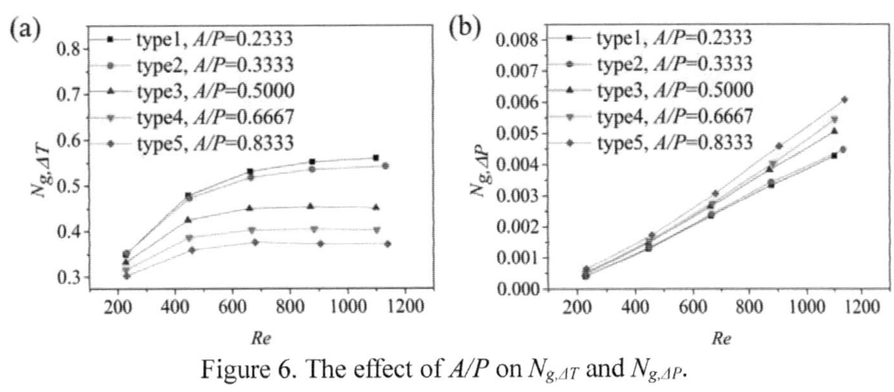

Figure 6. The effect of *A/P* on $N_{g,\Delta T}$ and $N_{g,\Delta P}$.

As shown in figure 7, N_g and $N_{g,\Delta T}$. have the same variation trend. The N_g value of type 5 is lowest under same conditions. Therefore, to minimize the irreversibility of heat transfer and fluid flow in cross-wavy channel, values of type 5 should be preferable considered.

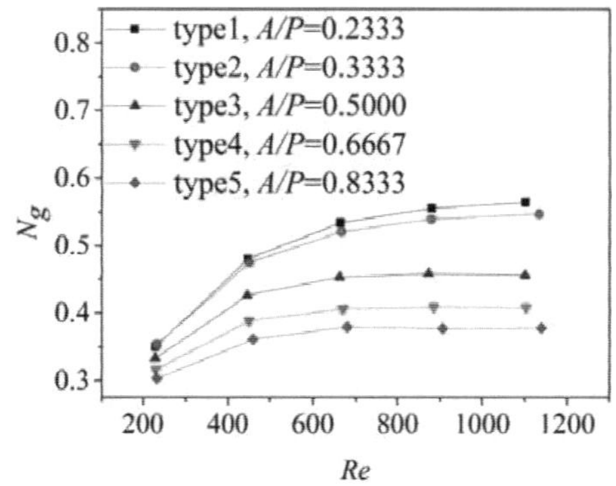

Figure 7. The effect of A/P on N_g.

Conclusion

In this paper, the influence of dimensionless parameter A/P and Reynolds number on $N_{g,\Delta T}$ and $N_{g,\Delta P}$ distributions at typical cross-sections are presented. The simulation results show $N_{g,\Delta T}$ rising rate is fast when Reynolds number less than 700 and then flat. In addition, it is smaller with increasing channel amplitude to pitch value. For example, thermal entropy generation has decreased by 32.9% at Re = 1134 with A/P value from 0.2333-0.8333. However, $N_{g,\Delta P}$ increases almost linearly with Reynolds number.

In a word, CW channels which have larger A/P value can reduce the total irreversibility to the least under the same flow conditions. And the lower Reynolds number also can reduce the total irreversibility.

Acknowledgment

This study was supported by the Key Research Program of the Chinese Academy of Sciences (ZDRW-CN-2017-2).

References

[1] McDonald, C.F. (2000) Low-cost compact primary surface recuperator concept for microturbines. Applied Thermal Engineering, 20: 471-497.

[2] Wang, Q.W., Liang, H.X., Luo, L.Q., Wang, J.W., Huang, Z.P., Feng, Z.P., Chen, Z.Q. (2005) Experimental Investigation on Heat Transfer and Pressure Drop in a Microtubine Recuperator With Cross-Wavy Primary Surface Channels. In: ASME Turbo Expo 2005: Power for Land, Sea, and Air. Reno. pp. 293-298.

[3] Xiao, G., Yang, T.F., Liu H.L., Ni, D., Ferrari, M., Li, M.C., Luo, Z.Y., Cen, K.F., Ni, M.J. (2017) Recuperators for micro gas turbines: A review. Applied Energy, 197: 83-99.

[4] Stasiek, J., Collins, M.W., Ciofalo, M., Chew, P.E. (1996) Investigation of flow and heat transfer in corrugated passages—I. Experimental results. International Journal of Heat and Mass Transfer, 39: 149-164.

[5] Ciofalo, M., Stasiek, J., Collins, M.W. (1996) Investigation of flow and heat transfer in corrugated passages—II. Numerical simulations. International Journal of Heat and Mass Transfer, 39: 165-192.

[6] Utriainen, E., Sundén, B. (2002) A numerical investigation of primary surface rounded cross wavy ducts. Heat and Mass Transfer, 38: 537-542.

[7] Du, L.X., Zeng, M., Wang, Q.W. (2011) A Simplified CFD Model With Multi-Periodic Boundary Conditions for Cross Wavy Channels.In: Asme Turbo Expo: Turbine Technical Conference & Exposition. Vancouver. pp. 863-869.

[8] Du, L.X., Ma, T., Zeng, M., Guo, Z.X., Wang, Q.W. (2014) Numerical Investigations on the Thermohydraulic Performance of Cross-Wavy Channels with Multi-Periodic Boundary Conditions. Numerical Heat Transfer, Part A: Applications, 65: 732-749.

[9] Zhang, L., Che, D. (2011) Turbulence Models for Fluid Flow and Heat Transfer Between Cross-Corrugated Plates. Numerical Heat Transfer, Part A: Applications, 60: 410-440.
[10] Tao, W.Q. (2001) Numerical heat transfer, Xi'an Jiaotong University Press, Xi'an.
[11] Bejan, A., Kestin, J. (1983) Entropy Generation Through Heat and Fluid Flow. Journal of Applied Mechanics, 50: 475.
[12] Bashi, M., Rashidi, S., Esfahani, J.A. (2017) Exergy analysis for a plate-fin triangular duct enhanced by a porous material. Applied Thermal Engineering, 110: 1448-1461.
[13] Ma, T., Du, X.L., Sun, N., Zeng, M., Sundén, B., Wang, Q.W. (2016) Experimental and numerical study on heat transfer and pressure drop performance of Cross-Wavy primary surface channel. Energy Conversion and Management, 125: 80-90.

Study on Agglomeration of Ultrafine Droplet Particles by Acoustic Air-jet Generators

Lintao Wu[1], Guangxue Zhang[1*], Yao Huihui[1], Zhenfang Ma[1], Zuohe Chi[1] and Shengjia Liu[2]

[1]Institute of Energy Engineering, China Jiliang University, Hangzhou, Zhejiang, 310018, China

[2]Hangzhou E-energy power technology Co., Ltd, Hangzhou, Zhejiang, 310018, China

*Corresponding author's e-mail: zhangguangxue@cjlu.edu.cn

Abstract. Acoustic agglomeration is a potential technology for fine particle pretreatment, but the high energy consumption restricts its industrial application. In order to develop a reliable sound source with low energy consumption, acoustic air-jet generators with frequencies ranging from 3 kHz to 8 kHz are adopted to study on the agglomeration effect of ultrafine droplet particles. The influences of frequency, flow rate and initial particle concentration on agglomeration efficiency are experimentally investigated. The results show that the acoustic frequency is a crucial factor for the agglomeration efficiency. The agglomeration efficiency of ultrafine droplet particles is higher than 55% at an optimal frequency of 6 kHz. The probability of collision between droplets increases with the increase of the initial mass concentration, which favors aerosol agglomeration. Moreover, it is shown that the sound pressure level increases with the increase of flow rate of compressed air.

1. Introduction

Generally, numerous industrial mist droplet aerosols could not be captured effectively, which induces some drawbacks in environmental pollution and waste of valuable substance. For example, the limestone-gypsum wet flue gas desulphurization technology is widely applied in coal-fired power plants for its unique advantages in low input costs, stable operation, and high desulfurization efficiency. However, it should be noted that the coal-fired gas is usually water-saturated through the wet desulfurization process. Commonly, the gas without proper treatment is considered to be an unneglectable reason for "gypsum rain" or acid rain, resulting in environmental pollution. The conventional particle filtering devices are inefficient for retaining particles smaller than 20 μm[1]. Moreover, the device will be clogged or corroded with the reason of water pressure, flue gas velocity, the cycle of backwashing, and so on. Therefore, we urgently need to find a reliable and effective technology to remove the droplet particles quickly.

Acoustic agglomeration is an aerosol preconditioning process in which intense sound waves promote relative motions and collisions of aerosol particles. It can significantly turn smaller aerosol particles into larger ones to enhance the performance of conventional filtering devices. A lot of theoretical works have been done on acoustic agglomeration by scholars all over the world. The process of acoustic agglomeration is complex but scholars think orthokinetic interaction[2] and acoustic wake effect[3, 4] play an important role in acoustic agglomeration. Orthokinetic interaction is based on the fact that different size particles enter the sound field at different rates, leading to the relative movement of them[5]. Small particles will stick to big particles after relative movement and collisions and turns into larger ones. The acoustic wake effect is based on the disturbance of particles toward the flow field which results in a low-pressure wake behind the particles. If the other particles travel in this acoustic wake, it moves towards to the leading one with accelerated motion in a reduced resistance flow field. As a consequence, the two particles converge within a short time and eventually collide.

Acoustic agglomeration technology is mainly used in flue gas pretreatment which can promote the dust removal efficiency of fine particles and reduce the emission of PM $_{2.5}$[6]. And this technology has been used in airport defogging to improve the visibility[7]. It is also applied to the recovery of sulfuric acid aerosol particles and sodium carbonate particles in sulfuric acid plants and paper mills[8]. Electrically driven transducer is widely applied in the experiments of acoustic agglomeration by scholars in domestic and overseas because of its convenient adjustment. Based on this, the experts got a lot of experimental results. Guangxue Zhang[6] studied the agglomeration experiment of coal-fired fly ash and investigated that the best agglomeration frequency of the fly ash particles was 1400-1600 Hz which was verified by simulation. Michael Volk[9] noticed the best agglomeration frequency of carbon aerosol particles is 3 kHz by exploring its behavior in the sound field. The frequency of 20 kHz is the better selection than 10 kHz for fluidized bed smoke[10]. Guangxue Zhang[11] performed an experimental investigation of acoustic agglomeration with the initial concentration and found that the acoustic agglomeration efficiency had a positive correlation with initial particle concentration. Yubo Kang[12] collected aerosol particles with a diameter of less than 100nm from combustion and studied the effect of acoustic agglomeration on it. And then, he obtained the same result as Guangxue Zhang. Shuangquan Hou [13] experimented to disperse the water mist with low frequency sound waves, and the result showed that the low frequency sound waves with high sound intensity can promote the agglomeration effect of sound waves on water mist. Ph. Capéran[14] studied ethylene glycol mist by two different frequencies of sound field, and found that the agglomeration rate is positively correlated with the square of acoustic amplitude. In addition, some scholars have added "additives" to the sound field in order to obtain better agglomeration efficiency. Jie Wang [15] added seed particles to the coal-fired fly ash aerosol, which promoted the agglomeration efficiency by 15.5%. Guangxue Zhang[16] significantly improved acoustic agglomeration of fine particles by droplet spray.

At present, sound energy in experimental study of acoustic agglomeration is converted from electromagnetic energy. Significant agglomeration efficiency obtained by the laboratory studies, but the low electro-acoustic conversion efficiency of transducer cannot be ignored. It requires a lot of energy, and the equipment is easy to be damaged in high temperature environment, which limits its industrial applications. Therefore, it's necessary to develop reliable high-power sound sources suitable for industrial environments. Comparatively speaking, the air sound conversion efficiency of acoustic air-jet generators can reach 80% [17] and it can adapt to the hostile environments which is an appropriate industrial sound source. There exist high-power acoustic air-jet generators at present. Such as the hooter, whose sound produced by the compressed air blowing the reed or the film vibrates, the siren cuts the airflow quickly by the rotating body to make the air vibrate and produce the sound, and the Hartmann whistle uses the impact of airflow on the periodic variation of the cavity. So the use of acoustic air-jet generators for acoustic agglomeration is almost the only feasible technical path to promote the technology to industrial application. However, there are no experimental reports on the use of acoustic air-jet generators for acoustic agglomeration up to now.

In this paper, the agglomeration experiment system of acoustic air-jet generators was built to study the influence of several key factors (acoustic frequency, flow rate of generators, and initial concentration of aerosol) on the agglomeration efficiency of ultra-fine droplets, which can provide reference for the industrial application of acoustic air-jet generators.

2. Experimental setup

2.1 Experimental systems and equipment

The experimental system is shown in Figure 1. The main body of the agglomeration chamber is a cylindrical structure made of organic glass with an inner diameter of 100 mm and a length of 700 mm. The small hole near the bottom serves as the inlet for compressed air. The agglomeration chamber has a high transmittance which allows most of the light to pass through. And the bottom is a rigid reflecting surface, which can effectively reflect the sound waves and promote the utilization efficiency.

The ultrafine droplets produced by the ultrasonic atomizer pass through the agglomeration chamber under the action of a vacuum pump. The air pump supplies the compressed air to the acoustic air-jet generators which emits a SPL (sound pressure level) of up to 135 dB. The ultrafine droplet particles agglomerate under the continuous action of the sound pressure of the intense SPL of the acoustic air-jet generators, and the measurement is performed after the working condition is stable.

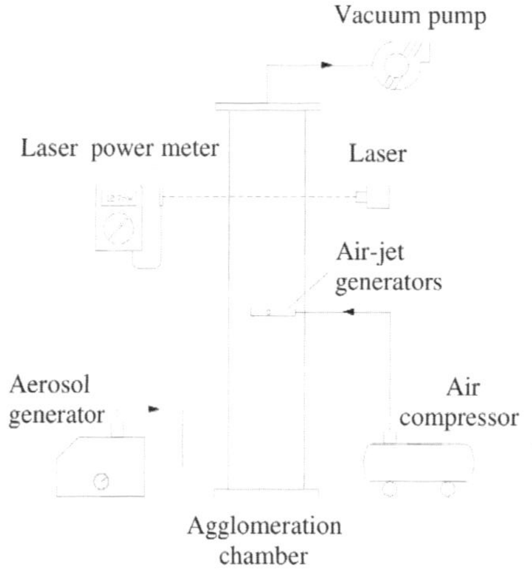

Figure 1. Schematic diagram of experimental device.

2.2 Features of the acoustic air-jet generators

The cylindrical acoustic air-jet generator is showed in Figure 2. The inlet of compressed air is on the top of acoustic air-jet generator, and an annular nozzle, formed by the round baffle and the wall, is located on the middle part. Moreover, the resonance chamber is located between the bottom and the annular nozzle and serves as the place to generate sound. The compressed air accelerates through the annular nozzle and rapidly enters the chamber which produces impact force. The center baffle added into the device connecting the round baffle with the bottom to enhance the support and improve the vocal efficiency.

The main frequency of the air-jet generators is determined by its cavity depth and diameter. Six different air-jet generators with frequencies ranging from 3 to 8 kHz designed according to the empirical formula (1). The SPL in the agglomeration room can reach 135 dB when the flow rate driven by the sound source exceeds a certain value, which can meet the requirements of experiment for high SPL.

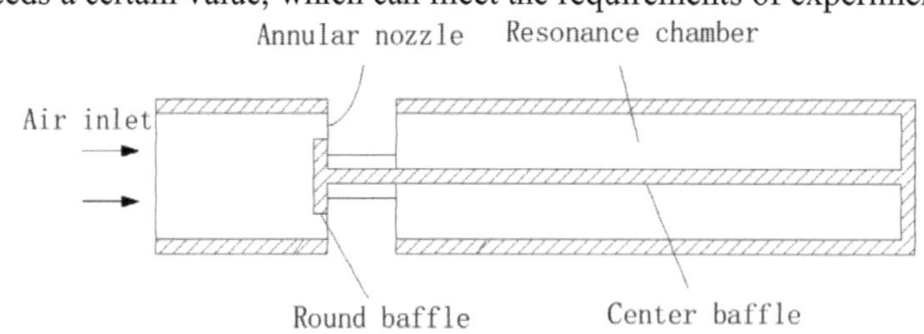

(a) The structure of the sound sourse

(b) The photo of the sound sourse

Figure 2. The acoustic air-jet generator

$$f = \frac{c}{4(L + 0.3d)} \tag{1}$$

where c is the speed of sound, L is the depth of the resonance chamber, and d is the diameter of the device.

The results of spectrum analysis of the air-jet generators are shown in Figure 3. It can be seen that the sound is strong at the design frequency. Taking Figure 3 (a) as an example, it can be seen that the SPL of the sound source exceeds 120 dB at a frequency of about 3 kHz, and there is also a distinct peak at the frequency of 6 kHz (twice as the fundamental frequency), but its SPL is only 80dB, which is far lower than the frequency of 3 kHz. In summary, the main SPL of the acoustic air-jet generator is concentrated on 3 kHz.

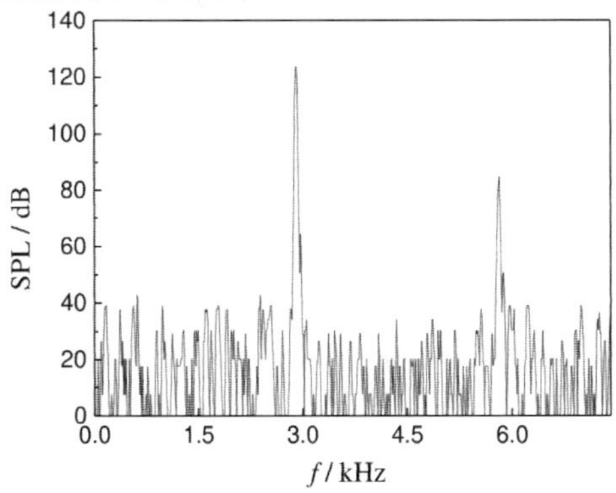

(a) 3 kHz

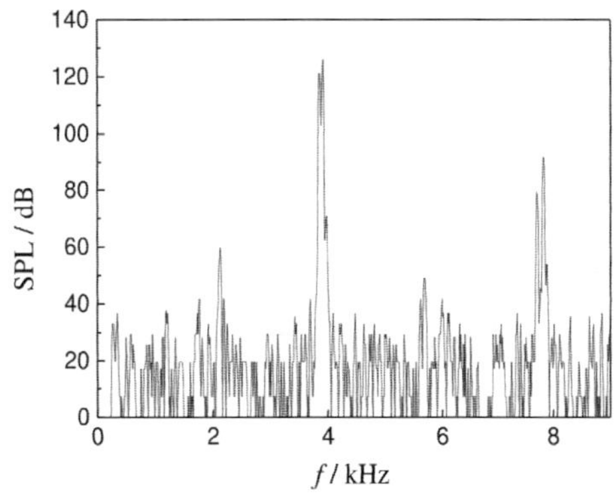

(b) 4 kHz

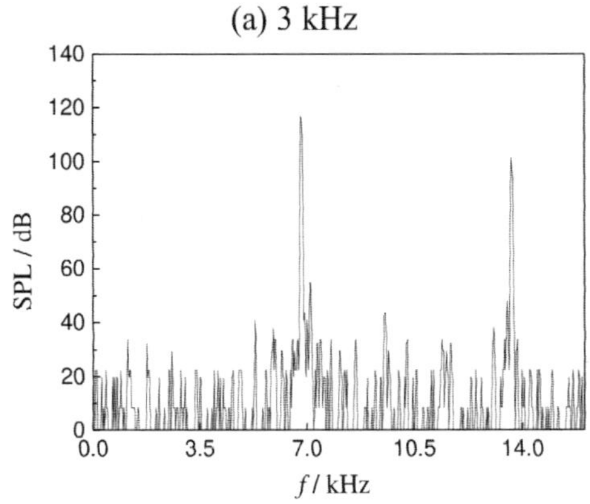

(c) 5 kHz

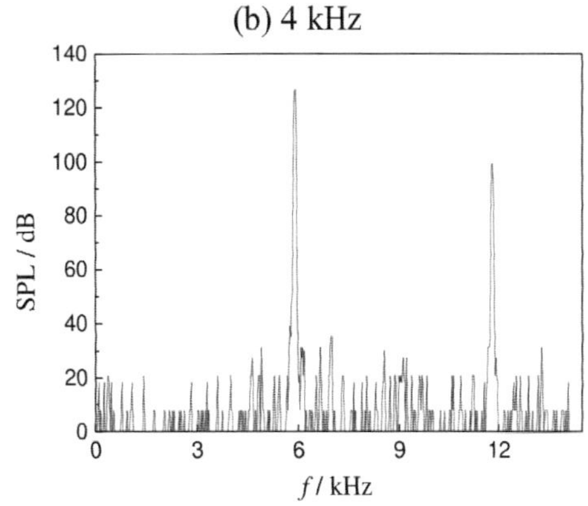

(d) 6 kHz

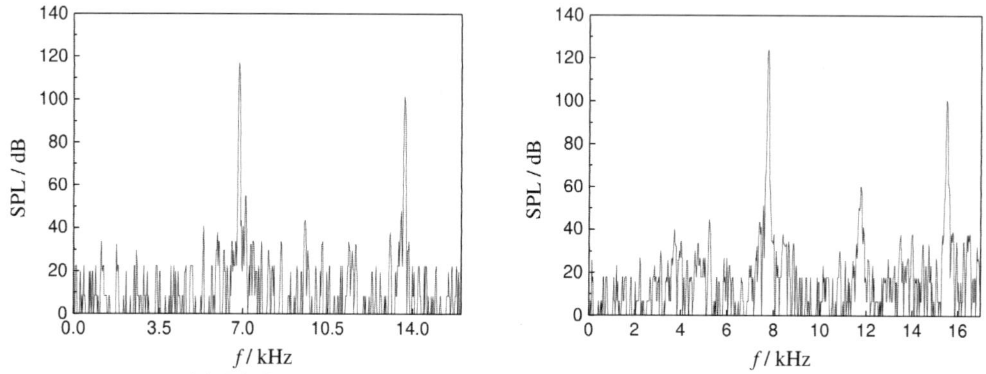

(e) 7 kHz (f) 8 kHz

Figure 3. Acoustic air-jet generators spectrum analysis.

2.3 Calculation of the agglomeration efficiency

The ultrafine droplet particle concentration is obtained by the light transmittance test method. The light is from the laser with a power of 60 mW and a wavelength of 650 nm, and the ultrafine droplet particle transmissivity[18] (T_a) was calculated based on the attenuation of the laser beam, given by

$$T_{\mathrm{a}} = \frac{I_{\mathrm{a}}}{I_0} \times 100\% \tag{2}$$

where I_a is the light intensity light intensity through the ultrafine droplet particle when the sound wave acted, and I_0 is incident light intensity.

In order to eliminate the effect of compressed air dilution, a comparative test was added. The light intensity of the mixed compressed air and ultrafine droplet particles was measured, and the new light transmittance T_0 was obtained. The calculation formula is as follows:

$$T_0 = \frac{I_{\mathrm{a}}'}{I_0} \times 100\% \tag{3}$$

where T_0 is the light transmittance of the ultrafine droplet particles in the agglomeration chamber with only compressed air acted, and I_{a}' is the light intensity when the only equal flow of compressed air flows through particles.

According to Lambert-beer's law, the relationship between the transmittance of aerosol particles and the volume fraction of the particles gained by Manoucheri[19], based on this, the agglomeration efficiency(η) was calculated by

$$\eta = 1 - \frac{V_{\mathrm{a}}}{V_0} = 1 - \frac{\ln(T_{\mathrm{a}})}{\ln(T_0)} \tag{4}$$

where V_a is the volume fraction of ultrafine droplet particles in the agglomeration chamber under the action of sound waves, and V_0 is the volume fraction of the ultrafine droplet particles in the agglomeration chamber when only the equal flow compressed air is passed.

3. Results and discussion

3.1 Tests of Acoustic air-jet generators and ultrafine droplet particle

The particle size distribution of the ultrafine droplets was measured by the FA-3 impactor (with a sampling range from 0.43 μm to 10 μm). The equipment has an aluminum alloy disc with a decreasing order of eight-stage orifice diameter and a filter plate. The measured particles are pumped into the device by the air pump, and the air velocity will increase as the size of the nozzle decreases. Different sizes of particles will impinge on the corresponding acquisition board with the action of inertia, and the submicron level particles are finally captured by the filter. Every stage of the acquisition board can be weighted to obtain the particle mass of each stage.

The mass data in10 min was extracted from the ultrafine droplets by the impactor at a sampling speed of 10 L/min, and the mass data was processed to obtain the particle size distribution map of the ultrafine droplets as shown in Figure 4. It can be seen from the distribution map that the particle size of the ultrafine droplet particles is between 1-10 μm, and the fine ultrafine droplets particles whose diameters are below 4 μm accounted for the main specific gravity.

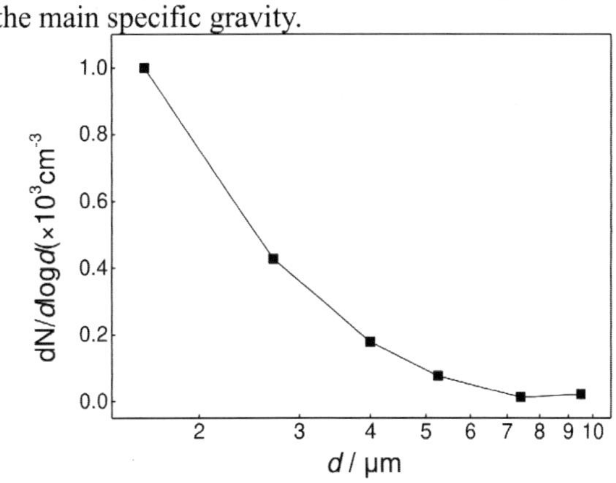

Figure 4. Particle size distribution of ultrafine droplet particles.

Figure 5 showes the relationship between the SPL and air flow or air pressure. It can be seen that the SPL of air-jet generators increases with the bigger air flow. And the SPL of six air-jet generators can reach 125 dB, some well-performing devices even get the high SPL of 130 dB. Different devices correspond to different frequencies of sound waves, and the SPL is changed by adjusting the air flow.

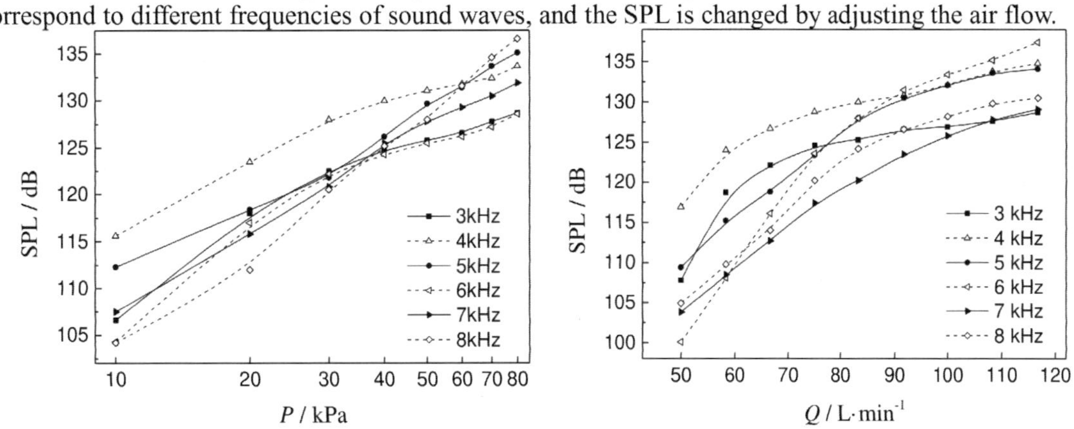

(a) The effect of air pressure on SPL. (b) The effect flow on SPL.

Figure 5. Effect of flow and pressure on SPL.

3.2 Influence of frequency sound on agglomeration effect

Figure 6 shows the agglomeration effect of an aerosol under the effect of frequency which is from 3 kHz to 8 kHz with the mass concentrations of 100 g/m3. Six types of air-jet generators were used in the acoustic agglomeration, and the experimental result shows that, comparing to the other frequencies, the air-jet generator of 6 kHz has the most significant effect on enhancing the agglomeration efficiency, even if at different flow rates. A reduction of 56% in total number concentration is gained by 6 kHz under the flow rate of 58 L/min, which is obviously greater than the elimination efficiency of 45% by 8 kHz.

As mentioned above, the acoustic agglomeration is mainly based on the orthokinetic interaction and the acoustic wake effect. The ultrafine droplet particles of different sizes correspond to their optimal frequencies under the agglomeration by the movement in the same direction, and the agglomeration efficiency will be reduced by the deviation from the optimal frequency. Moreover, the ultrafine droplet particles under the action of the sound field are also subjected to the entrainment of sound waves, which will regularly reciprocate under the action of the sound waves. And the entrainment effects of different particle sizes are different which causes a speed difference between the particles, eventually the collision of particles. Theoretically, there should be an optimal agglomeration frequency to achieve the optimal agglomeration effect. And the optimal frequency of aerosol agglomeration gained by equation (5).

$$f_{\mathrm{op}} = \frac{9\sqrt{3}\eta}{4\pi\rho_{\mathrm{p}}a^2} \tag{5}$$

where η is the gas viscosity coefficient, ρ_{p} is the density of the particles, and a is the radius of the particles.

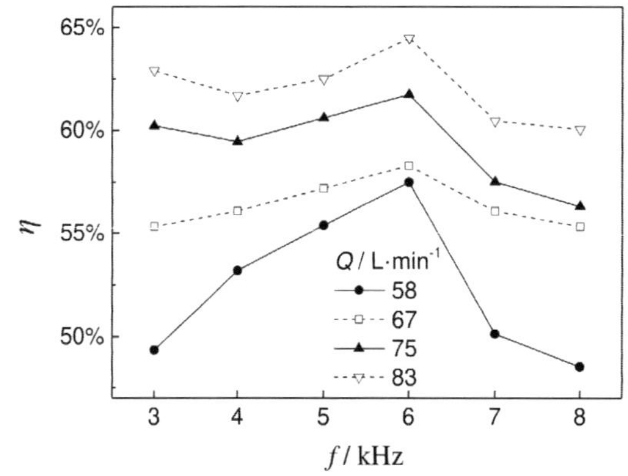

Figure 6. Effect of frequency on agglomeration efficiency.

3.3 Influence of initial concentration on agglomeration

The gas flow rate is fixed at 75 L/min, exploring the influence of initial concentration sound on agglomeration. The result shows that the acoustic agglomeration efficiency has a positive correlation with initial particle concentration. The initial particle concentration increased from 18 g/m3 to 64 g/m3, and the agglomeration efficiency of the 3 kHz increased from 25% to 52%. At the same time, we can find that it is obvious to improve the agglomeration effect by the increase of initial concentration for the low concentration particles

The initial concentration represents the number of particles in the unit volume. The initial concentration was positively correlated with the number of particles and negatively correlated with the relative distance between particles. And it improves the probability of collision between the ultrafine droplet particles under the action of sound waves. The aggregates will play the role of collecting core in the sound field because of their big volume and mass. It will easily collide with small particles and adsorb them on the surface, which promote the agglomeration efficiency[20].

In the numerical calculation of aerosol particles, the agglomeration kernel function is often used to describe the probability of collision between two particles, defined as the occurrence of unit time between two particles per unit volume and unit particle number density. The function can be expressed as

$$K_{ij} = \pi f(d_i + d_j)^2 \mu_{ij} \tag{6}$$

where K_{ij} is the agglomerated kernel function between particle i and particle j, f is the acoustic frequency, and d_i and d_j represent the diameters of the two particles.

The number of particle agglomeration per unit time can be expressed by equation (7):

$$dN = K_{ij} \times N'_i \times N'_j \tag{7}$$

where dN represents the number of agglomerations in two particles per unit time, the particle number concentration of particle i and particle j are represented by N'_i and N'_j.

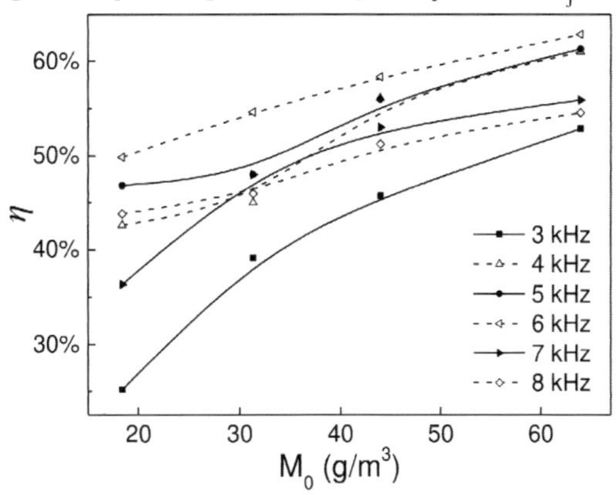

Figure 7. Effect of initial ultrafine droplet particle concentration on agglomeration efficiency.

3.4 Influence of the air flow rate on agglomeration

In this experiment, the air flow of the air-jet generators will dilute the concentration of the ultrafine droplet particles, and it also determines the SPL. Therefore, the influence of the gas flow rate on the agglomeration efficiency is important. Figure 8 shows the effect of air flow rate on agglomeration efficiency when acoustic waves are applied at different frequencies. It can be seen from the Figure 8 that at various acoustic frequencies, the acoustic agglomeration efficiency has a positive correlation with air flow rate. It is worth noting that at the condition of 3 kHz, with the increase of gas flow rate, the growth rate of agglomeration efficiency is significantly higher than that of other conditions. This is because the frequency spectrum is shifted to 6 kHz as shown in the Figure 3(a) with the flow rate increased, which significantly improves the agglomeration efficiency.

As can be seen from Figure 4(b), the SPL becomes larger as the flow rate increases and the gas medium in the agglomeration chamber also oscillates. The ultrafine droplet particles will oscillate accordingly because of the viscous force of the gas. Compared with small particles, the large particles whose inertial are large will be slightly affected by this. Relative motion occurs in the particles, and then they collide into the aggregates.

Agglomeration volume is the collision area of the large particles with the small particles in orthokinetic interaction, which is viewed as cylindrical region. The agglomeration volume is related to the diameter and amplitude of the particles size, specifically expressed as

$$V_{ij} = \frac{\pi}{4}(d_i + d_j)^2 \times \frac{u_0}{w} \mu_{ij} \tag{8}$$

where V_{ij} represents the volume of the agglomerate, d_i and d_j are the diameter of the two particles respectively, u_0 is the amplitude of the medium vibration, and w is the angular frequency, μ_0 indicating the relative carrying coefficient of particle i and particle j.

The sound agglomeration efficiency has a positive correlation with SPL, but the high SPL also represents the high intensity energy, and the energy consumption increases sharply with the increase of the SPL. Moreover, after the SPL reaches a certain level, the increase of the SPL slows down. The income and high energy consumption are not proportional. Therefore, the relationship between energy consumption and agglomeration efficiency should be coordinated to obtain the best economic benefits in industrial applications.

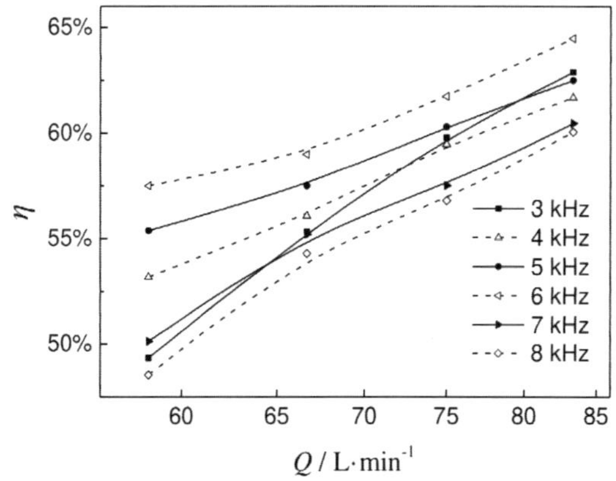

Figure 8. Effect of flow on agglomeration efficiency.

4. Conclusions

In this experiment, the agglomeration experimental system was built to verify the effect of air-jet generators on agglomeration of ultrafine droplets. The effects of acoustic frequency, initial concentration and compressed air flow rate on agglomeration efficiency were studied.

(1) In this experiment, the ultrafine droplets have the highest agglomeration efficiency at the frequency of 6 kHz, and the agglomeration efficiency of the other five frequencies is significantly lower, indicating there is a direct relationship between the agglomeration efficiency and frequency of the sound wave.

(2) In the range of the initial concentration of ultrafine droplets from 40 g/m3 to 140 g/m3, the collision efficiency between the particles increases with the increasing of initial concentration, and the agglomeration effect is significantly improved.

(3) Similar to sound pressure level, agglomeration is enhanced as the compressed air flow rate increases. Stronger relative action is gained by the increasing flow rate, and it also favors the vibration amplitude of particles, increases the collision probability between particles which cause the significant improvement in the sound agglomeration effect.

Acknowledgments

This study was supported by the National Natural Science Foundation of China (Grant No. 51876197 and 51408574).

References

[1] Sun Z C, Guo Y H, Xiao H P, J. (2010) Numercal Simulation and Experience Validation on Demising Characteristic of the Mist Eliminator for Jet Bubble Reactor Dusulfurization System. Proceedings of the CSEE. 8: 68-75.

[2] Zhang G X Liu J Z, Zhou J H, J. (2007) Retrospect and Prospect of Aerosol Acoustic Wave Agglomeration Technology. Thermal Power Generation, 36(11): 7-11.

[3] Zhang G X, Zhang L L, Wang J Q, J. (2018) A new model for the acoustic wake effect in aerosol acoustic agglomeration processes. Applied Mathematical Modelling. S0307904X18301434.

[4] HOFFMANN, T. L. An extended kernel for acoustic agglomeration simulation based on the acoustic wake effect [J]. Journal of Aerosol Science, 1997, 28(6): 919-36.

[5] Chen H T, Zhao B, Zhang Nu Xin, J. (2009) Experimental Study on the Enhancement of Removing PM2.5 in Coal-fired Fumes Under the Combined Effect of Acoustic Wave and Atomized Water Drop. Proceedings of the CSEE.2: 35-40.
[6] Zhang G X, Liu J Z, Zhou J H, J. (2009) A Theoretical Model and Experimental Verification on the Influence of Frequency on Acoustic Agglomeration of Coal-fired Fly Ash. Proceedings of the CSEE. 29(17): 97-102.
[7] Chou KH. J.(1980) Aerosol agglomeration and deposition in high-intensity acoustic fields.
[8] Denser HW, Neumann E. J. (2002) Industrial Sonic Agglomeration and Collection Systems. Indengchem. 41(11): 2439-42.
[9] Volk M, Moroz WJ, J. (1976) Sonic agglomeration of aerosol particles. Water Air & Soil Pollution. 5(3): 319-34.
[10] Gallego-Juárez JA, Sarabia RFD, Rodríguez-Corral G, J. (1999) Application of Acoustic Agglomeration to Reduce Fine Particle Emissions from Coal Combustion Plants. Environmental Science & Technology. 33(21): 3843-9.
[11] Zhang G X, Liu J Z, Zhou J H, et al, J. (2009) Acoustic agglomeration of coal-fired fly ash particles in low frequency sound fields. CIESC Journal. 60(4): 1001-6.
[12] Kang Y B, Zhu Y J et al, J. (2016) Influencing Factors of Acoustic Agglomeration of Ultradine Particles. JOURNAL OF SHANGHAI JIAO TONG UNIBERSITY. 50(4): 551-6.
[13] Hou S Q, Wu J, J. (2002) Experiments on acoustic dissipation of water fog at low frequency. Experiments and Measurements inFluid Mechanics. 16(4): 52-6.
[14] Capéran P, Somers J, Richter K, J. (1995) Acoustic agglomeration of a glycol fog aerosol: Influence of particle concentration and intensity of the sound field at two frequencies. Journal of Aerosol Science. 26(4): 595-612.
[15] Wang J, Zhang G X, Liu J Z, J. (2011) Effect of seed particles on acoustic agglomeration efficiency. CIESC Journal. 62(02): 355-61.
[16] Zhang G, Zhou TT, Zhang L, J . (2018) Improving acoustic agglomeration efficiency of coal-fired fly-ash particles by addition of liquid binders. Chemical Engineering Journal. 334(891-9.
[17] Ma D Z, J. (2004) The theoretical basis of modern acoustics. Science Press. Bei Jing.
[18] Fan W C, J. (1995) Concise tutorial on fire science. China University of Science and Technology Press. He Fei.
[19] Manoucheri M, Ezekoye OA, J. (1996) Polystyrene Soot Agglomeration Enhancement in an Ultrasonic Acoustic Field. Hazardous Waste & Hazardous Materials.13(1): 121-30.
[20] Hoffmann TL, Koopmann GH, J. (1994) A new technique for visualization of acoustic particle agglomeration. Review of Scientific Instruments. 65(5): 1527-36.

IOP Conf. Series: Materials Science and Engineering **721** (2020) 012027 doi:10.1088/1757-899X/721/1/012027

Heat Transfer Calculation of Organic Heat medium Heater and Analysis of maximum film temperature

Hao Hu[1*], lianqiang Liu[2], YanLiu[3], Liang Zhou[4] and YunzhouLiu[5]

[1,3,4,5]Thermal Engineering (Heater) Department, Beijing Aerospace Petrochemical Technology and Equipment Engineering Co. Ltd, Beijing, 100176, China

[2]Liquid and Rotating Machinery Division, Beijing Aerospace Petrochemical Technology and Equipment Engineering Co. Ltd, Beijing, 100176, China

[*]Corresponding author's e-mail:huh@calt11.cn

Abstract: Medium hot oil is applied to heat transmission when accurate temperature of process is required and process should be heated homogenously and slowly in petroleum industry. Organic heat medium heater is applied to process whose outlet temperature is below 400°C. Organic heat medium heater is listed as B grade, C grade and D grade as per Boiler Safety Technical Supervision Administration Regulation. Heat transfer calculation is mentioned about Organic heat medium heater and max. Film temperature calculation is also demonstrated in order to explain deterioration of hot oil.

1. Forword

Organic heat medium heaters are widely used in petroleum and chemical industries. In recent years, organic heat medium heater is also used in solar electric generation system, which is vigorously developed in China. The medium is heated to a target temperature in the heater then exchange heat with process in exchanger.

Heat mediums mainly divided into organic heat medium and inorganic heat medium. Inorganic salts can be heated above 400°C, such as molten salt. It is stable and safe that molten salt is heated to500°C.

Heat medium heater for heating inorganic salts is not inspected by the government, which is different from steam boilers and organic heat medium heaters. Organic heat medium, also known as hot oil, is generally used in situations below 400 °C. There are many kinds of hot oil, such as Dowtherm, YD, Syltherm and so on. Organic heat medium heaters are classified as B-grade, C-grade and D-grade boilers according to different loads in the scope of Boiler Safety Technical Supervision Administration Regulation.

2. Product Type

At present, there are two main type of organic heat medium heater: one is skid helical coil cylindrical type, the other is radiation-convection vertical cylindrical heater or vertical box type. The former is derived from Germany. This type of organic heat medium heater can be designed to 20 MW Generally. And it is assemble in shop before delivery and could be hoisted on site as whole with less site installation work. The heater arch is under positive pressure in operation. The latter comes from the fire heater type in API standard. The radiation tube in the heater is vertical or horizontal arrangement, and is located along the wall or in the middle. Horizontal finned tube of convection is used to enhance heat transfer. This kind of organic heat medium heater can be used for both large and small heat loads. API type organic heat medium heater has large site work, long installation period and relatively high cost. Large-scale organic heat medium heater must adopt API type. The heater arch is under negative pressure in operation. Figures 1 and 2 are outlines of two types of organic heat medium heater.

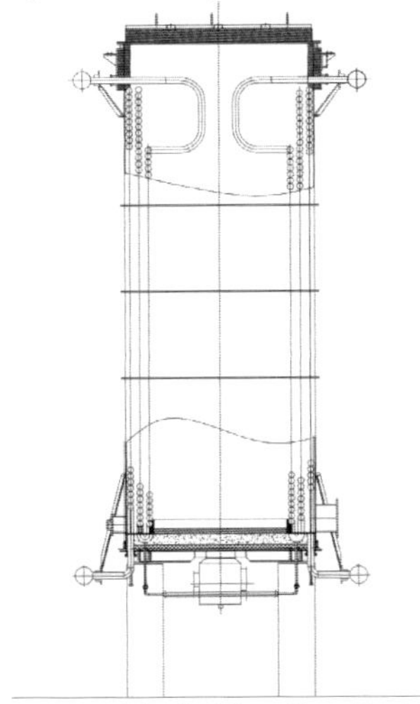

Figure 1-1. skid helical coil cylindrical type (Vertical)

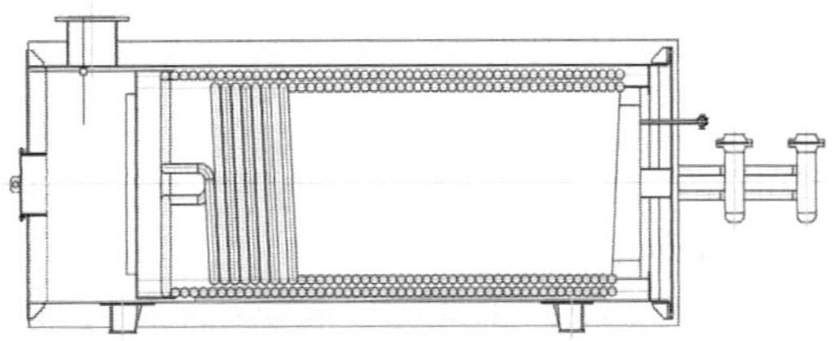

Figure 1-2. skid helical coil cylindrical type(horizontal)

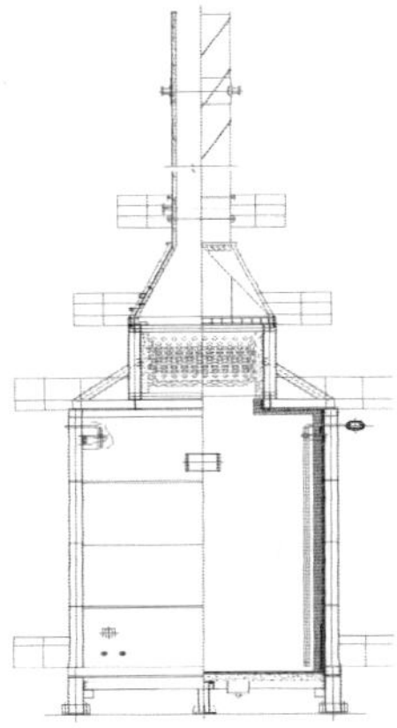

Figure 2-1.API vertical type(cylindrical)

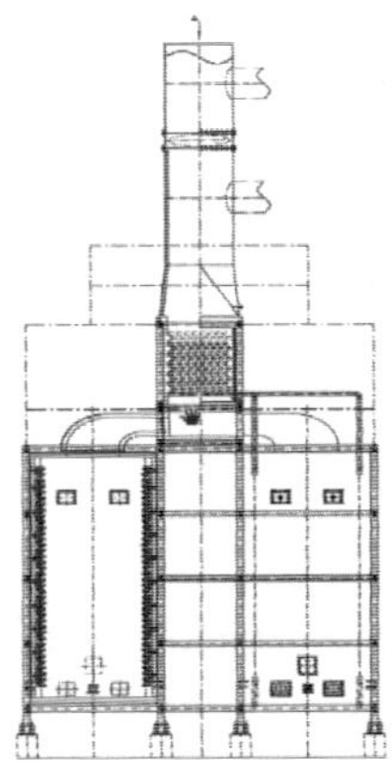

Figure 2-2.API vertical type(box)

3. Heat Transfer Calculation of Heater

The core component of the organic heat medium heater is the coil. The design of the heater coil is related to the thermal performance and safety of the organic heat medium heater. Organic heat medium heaters manufactured and operated in China must be inspected by government. Organic heat medium heaters exported usually require design and manufacture in accordance with ASME standards to obtain stamp. The design of coil and heat exchange surface in heater is determined by thermal calculation and hydraulic calculation. Checking calculation is usually used to calculate the heat transfer area. Firstly, the structure parameters and area are assumed preliminarily, and the heat transfer quantity is calculated. The heat transfer quantity and the required heat transfer quantity are found to be consistent (generally, the error is less than 2%). Checking calculation is the most commonly used calculation method [1]. Regardless of the type of heater, the heat transfer calculation block diagram in the heater is similar, mainly divided into radiation heat transfer and convection heat transfer.

The Robo-Evans module is used to calculate the radiant heat transfer in heater. The heat transfer rate equation of radiant chamber is [2]:

$$Q_R = 5.72\alpha A_{cp}F\left[\left(\frac{T_g}{100}\right)^4 - \left(\frac{T_w}{100}\right)^4\right] + h_{RC}A_R\left(T_g - T_w\right) \quad (1)$$

Q_R – Radiant quantity, W
α – Angular coefficient
A_{cp} – Cold flat area, m^2
A_R – Radiant area, m^2
F – Exchang factor
h_{RC} – Convective heat transfer coefficient in radiant chamber , W/ (m2 K)
T_g –Flue gas temperature in radiant chamber, K
T_w –Outer skin temperature of radiant chamber, K

Thermal equilibrium equation:

$$Q_R = BV_y\left(H_{max} - H_g\right) \quad (2)$$

Q_R – Radiant quantity, W
B – Fuel gas consumption, Nm3/s
V_y – Actual flue gas flow rate， Nm3/Nm3
H_{max} – Enthalpy of Flue Gas at Theoretical Flame Temperature , J/Nm3
H_g – Enthalpy of Flue Gas at Arch Temperature , J/Nm3

Skin temperature calculation formula:

$$t_w = \frac{1}{2}\left(t_j + t_c\right) + \left(\frac{1}{h_i} + R_i + \frac{\delta}{\lambda_s}\right)q_R\frac{d_0}{d_i} \quad (3)$$

t_j, t_c – Inlet and outlet temperature of process, °C
h_i – Convective heat transfer coefficient in tube, W/m2 K
R_i – Fouling resistance in tube, m2 K/W
δ – Thickness of tube, m
λ_s – Conductivity, W/m K
q_R – Flux of coil, W/m2
d_o – Outer diameter, m
d_i – Inner diameter, m

The calculation of hot oil film temperature in organic heat medium heater is also very important. Because the film temperature is higher than the allowable film temperature of hot oil during operation, it will cause the deterioration of hot oil and seriously affect the operation of the whole system. The calculated film temperature must be lower than the allowable liquid film temperature. The calculation of liquid film temperature is introduced separately.

The program diagram of radiation heat transfer in heater is as Figure 3.

About 50% to 80% of the total heat is radiant transfer heat, and the residual heat is transferred to hot oil through convective heat transfer. Because of the liquid phase in tube, the flow is in a turbulent state, and the flue gas velocity outside the tube is 10-20 m/s, which can also reach the turbulent state. However, the convective heat transfer coefficient in the tube is higher than that outside the tube. Even if the finned coil is used in API heater, the thermal resistance is subject to outside the tube. The heat transfer equation [3] is:

$$Q_c = KA\Delta t_m \quad (4)$$

Q_c – Convectionquantity, W
K – heat transfer coefficient, W/m2 K
A – Convection area, m2
Δt_m-Logarithmic temperature difference, °C

Thermal equilibrium equation:

$$Q_c = BV_y(H_{in} - H_{out}) \quad (5)$$

H_{in} − Inlet Enthalpy value of flue gas in convection section, J/Nm3
H_{out} − Outlet Enthalpy value of flue gas in convection section, J/Nm3

The program diagram of convection heat transfer in heater is as Figure 4.

The thermodynamic calculation process is a trial calculation process. The thermodynamic parameters printed by the calculation of radiation heat transfer and convection heat transfer need to be judged and analyzed by professional knowledge and codes, and unreasonable thermodynamic parameters can be judged. The thermodynamic parameters can be optimized through areal changes to meet the requirements of performance and safety.

Pressure drop calculation of medium is important in process calculation of organic heat medium heater. Pressure drop is divided into two categories: along-way loss h_f and local lossh_m. Pressure drop formula [4]:

$$P = h_f + h_m = \lambda \frac{l}{d} \frac{\rho v^2}{2} + \xi \frac{\rho v^2}{2} \quad (6)$$

P − Pressure drop, KPa
λ − Along − way loss coefficient
l − Length of coil, m
d − Inner diameter, m
ρ − Process density, Kg/m3
V − Velocity of inner tube, m/s
ξ − Local loss coefficient

Pressure drop in heater is an important basis for selecting circulating pump. The pressure drop of the medium in the furnace is large, the pump head needed is high, so the one-time investment is high, and the operation is not economical. The low pressure drop indicates that the flow rate of medium in the furnace tube is low, the heat transfer is not optimized, and the film temperature and skin temperature is high. Therefore, the pressure drop should consider both the economy of operation and optimization of thermodynamic parameters.

IOP Conf. Series: Materials Science and Engineering **721** (2020) 012027 doi:10.1088/1757-899X/721/1/012027

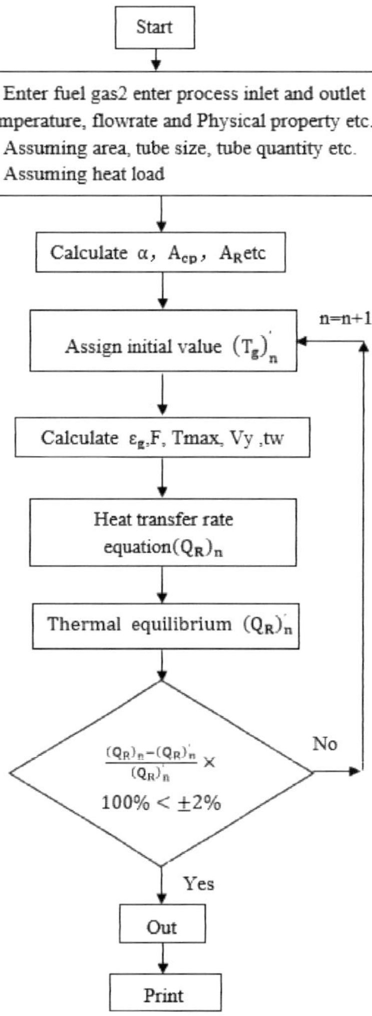

Figure 3. The Program Diagram of Radiation Heat Transfer

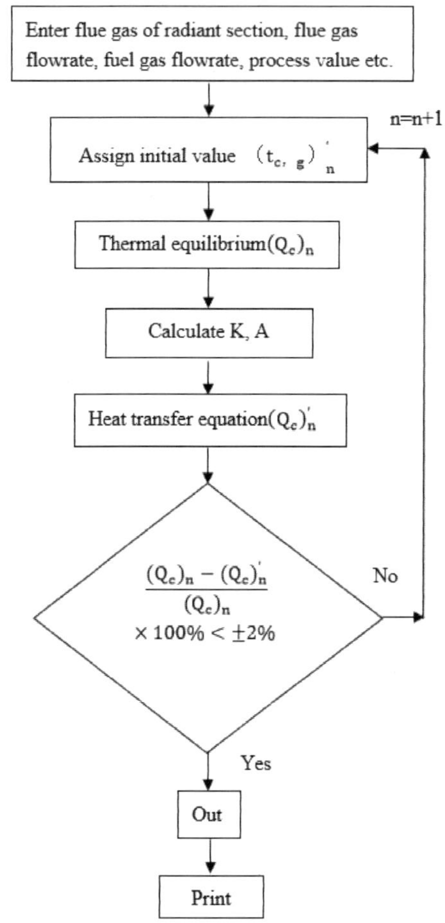

Figure 4. The Program Diagram of Convection Heat Transfer

4.Coking mechanism of organic medium

Water and inorganic salts are stable during heating process and are not easy to deteriorate, so there will be no deterioration of heat transfer caused by medium coking in the heater tube during operation. Organic medium belong to hydrocarbon organic compounds. When reaching certain temperature, free radicals R are generated at the weakness of molecular structure. RO2 leads to chain initiation, and then chain transmission, growth and termination. Free radicals are easy to decompose, leading to chain breakage and degradation, resulting in a certain amount of coking matrix with free radicals, while the combination of free radicals increases the molecular weight, gum formation and coking process. The coking rate is mainly related to the coking reaction temperature, i.e. the film temperature. The coking rate equation [5] is as follows:

$$R_c = C_1^n A e^{-E/RT} \quad (7)$$

R_c – Coking rate, $mg\ cm^{-2}h^{-1}$
C_1 – Coking Matrix Velocity, $mg\ cm^{-3}$
n – Reaction order

$A - $ Pre exponential factor$mg^{1-n}cm^{3n-2}h^{-1}$
$E - $ Reaction activation energy, KJ/mol
$R - $ General Gas Constants, KJ $mol^{-1}K^{-1}$
$T - $ temperature, K

For industrial heaters, Nu, Re and Pr affect film temperature. According to literature [6], the coking rate equation of industrial heaters as below:

$$R_c = C_2 d/Re^{0.8} \quad (8)$$

$C_2 - $ equation coefficient
$d - $ Inner diameter, mm
$Re - $ Reynolds number

The higher the temperature is, the higher the coking rate is; the smaller the Reynolds number is, that is, the smaller the flow rate is, the higher the coking rate is; the larger the diameter of the selected pipe is, the higher the coking rate is.

5. Film Temperature Calculation

The maximum allowable temperature of hot oil is determined by thermal stability test. The test data prove that the deterioration rate caused by thermal cracking will double when the working temperature rises by 10 ℃. However, the film temperature can not be determined by experimental method. It is determined by the chemical properties and thermal stability conditions of hot oil [7]. The calculation of film temperature only provides a condition for the safety of organic heat medium heater, and the calculation results are inconsistent with the actual film temperature. In the calculation, the spot film temperature where has the maximum heat flux or the minimum Reynolds number should not be greater than the maximum allowable liquid film temperature of heater.

Film temperature calculation as per Boiler Safety Technical Supervision Administration Regulation explanation Attachment 2:

$$t_{film} = t_{bulk} + \Delta t_{htf} \quad (9)$$

$t_{film} - $ Max film temperature, °C
$t_{bulk} - $ Bulk temperature °C

$$\Delta t_{htf} = \frac{q_{rmax}}{\alpha_{htf}} X \frac{d_{od}}{d_{id}} X \Phi_{whl} \quad (10)$$

$\Delta t_{htf} - $ Film temperature difference, °C
$q_{rmax} - $ Max flux, w/m2
$\alpha_{htf} - $ Convective heat transfer coefficient in tube, w/(m2 °C)
$d_{od} - $ Outer diameter, m
$d_{id} - $ Inner diameter, m
$\Phi_{whl} - $ Correction Coefficient Considering Tangential Radiation Heat Loss of Tube Wall,
$\Phi_{whl} \approx 0.9$

$$\alpha_{htf} = \Phi_{bd} X \frac{\lambda_{htf}}{d_{id}} X 0.12 X Re^{0.87} X Pr^{0.4} \quad (11)$$

$\Phi_{bd} - $ Coefficient Considering the Effect of Tube Bending and Heat Flow Direction,
$\Phi_{bd} \approx 0.8$
$\lambda_{htf} - $ Conductivity of bulk temperature, w/(m K)
$Re - $ Reynolds number
$Pr - $ Prandtl number
$q_{rmax} = \Phi_{saf} X \Phi_a X q_{fur}$
$\Phi_{saf} - $ safety factor, $\Phi_{saf} \approx 1.15$
Φ_a
$-$ Correction Coefficient Considering Combustion Air Temperature and Excess Air, as per Figure 5
$q_{fur} - $ Flux, w/m2

$$q_{fur} = \frac{d_{fire}}{D_{fur} X a_{fur} X \sigma_0 X (T_{f.1}^4 - T_{wal}^4)} \quad (12)$$

d_{fire} – Flame diameter, m
D_{fur} – Average diameter of combustion chamber, m
a_{fur} – Heater emissivity, as per Figure 6
σ_0 – Radiation Constant = $5.67X10^{-8} w/(m^2K^4)$
T_{f1} – flame Temperature, K
T_{wal} – Outer Skin Temperature, K

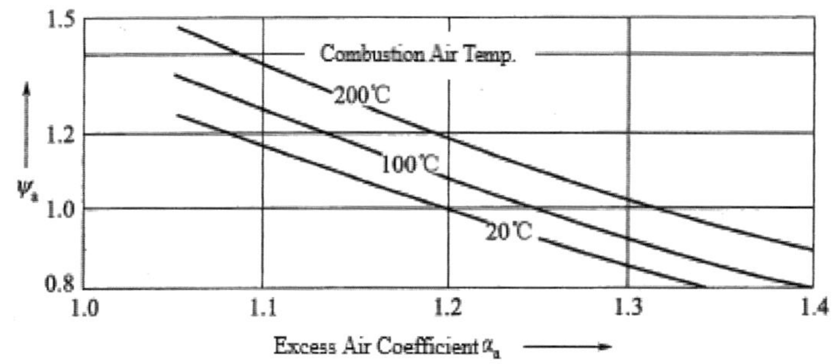

Figure 5.Correction Coefficient Considering Combustion Air Temperature and Excess Air

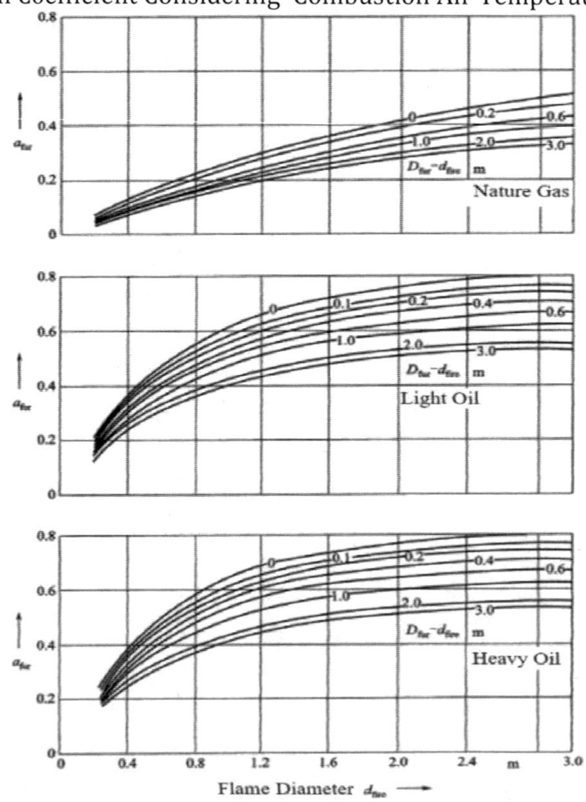

Figure 6. Heater Emissivity

6. Calculation Case

According to the actual project parameters, the relevant thermodynamic calculation parameters are given as below table 1:

Table 1. Comparison Thermal parameters of Two type Heaters

	API	skid helical coil cylindrical type
Hot Oil	Therminol VP-1	
Allowable bulk temperature℃	400	
Allowable film temperature℃	425	
Inlet temperature ℃	301	301
Outlet temperature℃	393	393
Heat load MW	18.5	18.5
Flue gas temperature of arch ℃	762	1127
Radiant flux KW/m2	26.6	60
Ratio radiant heat exchange	80%	60%
Discharge flue gas temperature℃	386	435
Direction of hot oil	Convection section-Radiant section	Inner Coil-Outer Coil
Velocity of Hot Oil m/s	2.8	Inner：3.46 outer：2.5
Inner diameter of tube mm	113.9	79
Calculated max. film temperature ℃	411	420

Because the ratio of radiation heat transfer is higher and the radiation heat transfer is stronger than convection heat transfer, the highest film temperature usually occurs in the radiation section. Convection heat transfer is relatively mild, generally there is no case of hot oil deterioration and coking. The average flux of the radiation section of the skid helical coil cylindrical type heater is high. Even if the low temperature inlet process is set in the radiation section, coking will inevitably occur. Therefore, this part of the Heater coil adopts high flow rate and small diameter to obtain lower film temperature.

7.Conclusion

The thermodynamic calculation process in heater needs repeated trial. The final thermodynamic parameters require the thermodynamic calculator to make rational judgment by using professional knowledge and experience.

The risk of deterioration of hot oil is much higher than that of inorganic heat medium and water. Therefore, the calculation of film temperature is particularly important. According to the properties of hot oil, the coil diameter, flow rate, pressure drop, heat transfer area and flux should be taken into account in the calculation of film temperature, so as to ensure that the calculated film temperature is below the allowable maximum film temperature and ensure the safe operation of the equipment.

References

[1] Feng, J.K. (2003) Principle and calculation of boiler (Rev.3). Science Press, Beijing.
[2] Qian, J.L. (2007) Tubular Heater (Rev.2). China Petrochemical Press, Beijing.
[3] Tao, W.Q. (1998) Heat transfer. Higher Education press, Beijing.
[4] Cai, Z.Q. (1999) Hydrodynamic Pumps and Fans (Rev.4). China Architecture & Building Press, Beijing.

[5] New sone, D.S., Leftin, H.P. (1980) Coking Rates in a Laboratory Pyrolysis Heater, Chem Eng Prog. 76(5):76.
[6] Wan, W.J., lai,X.J. (1995) An Investigation of Coke deposition in Steam Pyrolysis of Oil Fractions. Petrochemical Technology. 24(2):106.
[7] Guo, Y.L. (2013) Boiler Safety Technical Supervision Administration Regulation Explanation. Chemical Industry Press, Beijing.

IOP Conf. Series: Materials Science and Engineering **721** (2020) 012028 doi:10.1088/1757-899X/721/1/012028

Research of a blackbody cavity for effective integrated emissivity with finite volume method

Bo Liu[1*], Chong-yuan Wang[1], Jun-long Xu[1], Zhi-yuan Wang[2], Peng-fei Zhu[1]

[1]Shanghai Institute of Measurement and Testing Technology, Shanghai, 201203, China

[2]School of Energy and Power Engineering, University of Shanghai for Science and Technology, Shanghai, 200093, China

[*]Corresponding author's e-mail: liubo021@126.com

Abstract. Temperature is a critical parameter for energy efficiency measurement evaluation. High-performance blackbody radiation source is the key equipment for highly precision reproduction of radiation temperature. The effects of blackbody geometric parameters, cavity material emissivity, and temperature distribution on integrated effective emissivity $_{\mathrm{e}}$ and normalized spatial integrated effective emissivity variation were studied by finite volume method. The numerical simulation has a good agreement with the result computed by Monte-Carlo method. The results show that the isothermal cavity has a higher $_{\mathrm{e}}$ and a smaller than the non-isothermal. A smaller aperture diameter D_{a} and a higher material emissivity are benefit for a higher $_{\mathrm{e}}$ and a more uniform distribution on detector of $_{\mathrm{e}}$. The results provide references for the design and study on high-performance blackbody radiation source.

1. Introduction

Temperature is a critical parameter for energy efficiency measurement evaluation. Blackbody cavities are commonly used as standard radiation sources for the calibration of radiation thermometers, thermal imagers, and radiometers. Radiation reaching the detector from points at the inner surface of the cavities has two different components. One is due to direct emission from the point where the optical system is focused (depending on the local temperature and spectral-directional intrinsic emissivity ε). The other contribution is due to multiple reflections of the radiation originated at any other point inside the cavity, which finally reaches the detector by reflection in the focus point (depending on the reflectance: diffuse or specular) [1]. The effective emissivity ε_{e} which is central to blackbody designs accounts for both emitted and reflected radiation. In general, ε_{e} depends on the intrinsic surface emissivity, cavity geometry, and local temperature at the point of interest (and the cavity thermal gradient as a whole). However, a different reference temperature (instead of the local) can be used in the definition of ε_{e}, such as that of the bottom cavity.

Computational methods are needed to accurately predict ε_{e} since experimental determination of ε_{e} is limited to restricted geometries, spectral ranges, and cavity temperature distributions. The theoretical methods for calculating ε_{e} in blackbody cavities generally include the resolution (analytically or numerically) of complex integral equations [1].

The Monte-Carlo back ray tracing technique is nowadays the most common method used for solving radiative transfer problems. This statistical technique basically consists (when it is applied to the calculation of effective emissivity) in throwing a large number of photons from the detector toward

IOP Conf. Series: Materials Science and Engineering **721** (2020) 012028 doi:10.1088/1757-899X/721/1/012028

the point of interest inside the cavity and subsequently counting those exiting the cavity after multiple reflections [2]. The Monte-Carlo method, though effective, is time-consuming for large numbers of traced rays [3]. Other computational methods need to be developed and verified to assure their computational accuracy.

Due to advance in numerical simulation technology, Liu et al. [4] reported on integrated effective emissivity computations for nonisothermal nonaxisymmetric cavities using the finite volume method. The governing equations were solved by Ansys Fluent, with the results agreeing well with the Monte-Carlo method.

The axisymmetric cavities, including conical, cylindrical, and cylindro-conical cavities with and without grooved walls, are widely used and its ε_e have been calculated by many methods [5,6]. However, very few studies have calculated the ε_e by numerical simulation method. This paper studied the effect of geometric parameters, cavity material emissivity, and temperature distribution on ε_e. This study is useful to validate other method results for integrated effective emissivity computations.

2. Physical and numerical models

The cylinder cavity geometry configuration and temperature distribution are shown in Figure 1. D is the cylinder diameter, D_a is the cavity aperture diameter, H is the cavity length, ε is the cavity material emissivity. D was set to one throughout because of the scaling properties. The temperature along z axis was linear distributed. ΔT_r ($\Delta T_r = (T_a - T_b) / T_a$) is defined to describe the non-isothermal distribution. ΔT_r equals to zero for isothermal cavity. The material for the entire inner wall of the cavity was assumed to be uniform and grey so that the radiating cavity surfaces were spatially uniform and independent of the wavelength and to ensure that the surfaces exhibit diffused emissions and reflections.

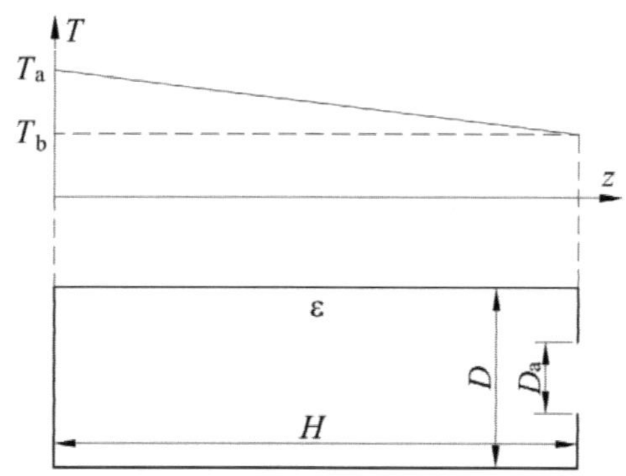

Figure 1. Cylinder cavity geometry and temperature distribution

The key factors influencing integrated effective emissivity of the cylinder cavity are summarized in Table 1. The factors range is selected based on blackbody cavities manufacturers. The effects of D_a, H, and are studied in Runs 1-2, 3-4, and 5-6, respectively. Isothermal and non-isothermal cavities are studied with $\Delta T_r = 0$ and 0.3 in Runs 1-6, respectively.

Table 1. Simulation conditions

Runs	D_a	H		ΔT_r / %
1	0.6 – 1.0	4	0.95	0
2	0.6 – 1.0	4	0.95	0.3
3	0.8	3 – 7	0.95	0
4	0.8	3 – 7	0.95	0.3
5	0.8	4	0.91-0.95	0
6	0.8	4	0.91-0.95	0.3

The local directional effective emissivity of a non-isothermal cavity in a non-refractive environment is

$$\varepsilon_{\mathrm{e,l}}(\xi,\boldsymbol{\omega})=\frac{L(T,\xi,\boldsymbol{\omega})}{L_{\mathrm{b}}(T_0)} \tag{1}$$

where L is the radiance emitted from a point ξ with a local temperature T on the cavity wall in the direction of $\boldsymbol{\omega}$; L_b is the radiance of a perfect blackbody at a reference temperature T_0 in the same direction. The bottom center temperature of the cavity was used as the reference temperature.
The integrated effective emissivity is defined as

$$\varepsilon_{\mathrm{e}}(D_{\mathrm{d}},H_{\mathrm{d}})=\frac{\Phi(D_{\mathrm{d}},H_{\mathrm{d}})}{\Phi_{\mathrm{b}}(T_0,D_{\mathrm{d}},H_{\mathrm{d}})} \tag{2}$$

where $\Phi(R_d,H_d)$ is the radiant flux falling onto the black detector that is irradiated by the radiating cavity surfaces, and $\Phi_b(T_0,R_d,H_d)$ is the radiant flux falling onto the same detector that is irradiated by a cavity with perfectly black walls. The integrated effective emissivity can be derived from Eq. (1) via integration over the appropriate areas and solid angles.

The outgoing flux from the radiating cavity surface consists of the flux that is intrinsically emitted and that reflected after incidence from the other radiating cavity surfaces. Therefore, the outgoing radiant flux of a radiating cavity surface element is expressed as

$$J_k=E_k+(1-\varepsilon_k)\sum_{j=1}^{N}F_{k,j}J_j \tag{3}$$

where E_k is the radiant flux emitted by a radiating surfaces element, ε_k is the emissivity of the radiating cavity surface, $F_{k,j}$ is the configuration factor for two surface elements, and N is the number of surface elements. The local radiant flux is calculated by iteratively solving Eq. (3). The radiant flux that falls onto the black detector irradiated by the radiating cavity surfaces is

$$\Phi(D_{\mathrm{d}},H_{\mathrm{d}})=\sum_{k=1}^{N}F_{k,\mathrm{d}}J_k \tag{4}$$

where $F_{k,\mathrm{d}}$ is the configuration factor for one surface element and the detector. The radiant flux falling onto the same detector that is irradiated by the cavity with perfectly black radiating surface is given by

$$\Phi_{\mathrm{b}}(T_0,D_{\mathrm{d}},H_{\mathrm{d}})=\sum_{k=1}^{N}F_{k,\mathrm{d}}J_{k,\mathrm{b}} \tag{5}$$

where $J_{k,b}$ is the outgoing radiant flux of a perfect black surface element. Thus, the integrated effective emissivity is calculated as

$$\varepsilon_{\mathrm{e}}(T_0,D_{\mathrm{d}},H_{\mathrm{d}})=\frac{\Phi(D_{\mathrm{d}},H_{\mathrm{d}})}{\Phi_{\mathrm{b}}(T_0,D_{\mathrm{d}},H_{\mathrm{d}})}=\frac{\sum_{k=1}^{N}F_{k,\mathrm{d}}J_k}{\sum_{k=1}^{N}F_{k,\mathrm{d}}J_{k,\mathrm{b}}} \tag{6}$$

The configuration factors were calculated using the surface-to-surface model in Ansys Fluent according to the geometry and the grid. The radiating cavity surface temperature distribution was defined using a UDF.

The detector does not affect the radiative transfer. Therefore, the detector emissivity was set to 1, and the temperature was set to 0 K.

3. Mesh independence test and model verification

The integrated effective emissivity of non-isothermal axisymmetric cylindrical cavities were calculated at D = 1, D_a = 0.8, H = 4, ε = 0.91, T_a = 500 K, and T_b = 498.5 K. A fictitious detector was set at the cavity aperture. In other words, the detector diameter equals to D_a and the distance from the detector to the aperture is zero. Figure 2 shows the computational results for various numbers of elements. As demonstrated, the difference is less with the increase of elements. Therefore, in order to save computational time, similar meshing method and grid size to the 2.3×10^5 elements were employed for the subsequent runs.

Comparison with Monte-Carlo results [7] had been made to verify the model setup in our previous study [8],[9]. The maximum relative difference is less than 0.005%, indicating that the present method is accurate.

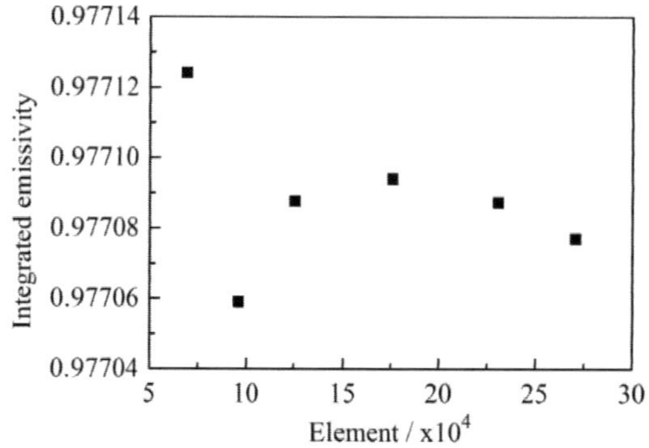

Figure 2. Effect of elements numbers on the integrated effective emissivity

4. Results and discussion

Figure 3 shows the effect of aperture diameter D_a on integrated effective emissivity $_e$ and normalized spatial integrated effective emissivity variation with H = 4 and = 0.95. is defined to describe the $_e$ distribution on detector, as $\eta = \varepsilon_e' / \varepsilon_{e\,mean}$, in which $\varepsilon_{e\,mean} = \int \varepsilon_e dA / \int dA$ and $\varepsilon_e'^2 = \int(\varepsilon_e - \varepsilon_{e\,mean})^2 dA / \int dA$. With the increase of D_a, the $_e$ decrease and the increase in both isothermal and non-isothermal conditions. A smaller D_a indicates that the cylinder cavity is closer to an ideal blackbody. An inverse trend is observed between $_e$ and . In other words, a higher $_e$ indicates its distribution is more uniform. The cavity bottom temperature is set as reference temperature for the calculation of $_e$. The isothermal cylinder cavity emits more radiation flux than the non-isothermal one. Therefore, the isothermal cavity has a higher $_e$ and a smaller than the non-isothermal.

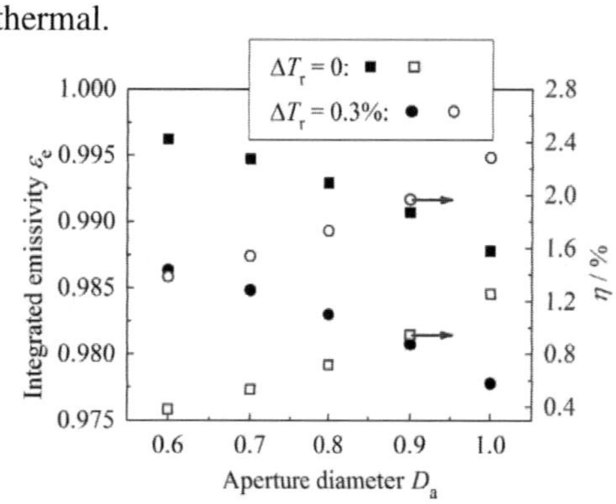

Figure 3. Effect of D_a on $_e$ and with H = 4 and = 0.95

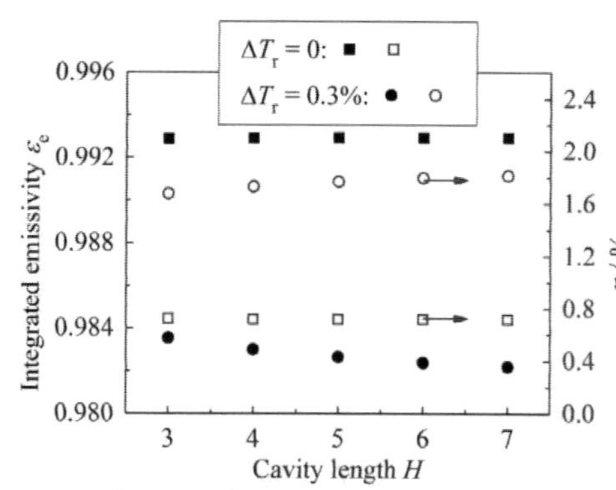

Figure 4. Effect of H on $_e$ and with D_a = 0.8 and = 0.95

Figure 4 shows the effect of cavity length H on $_e$ and with D_a = 0.8 and = 0.95. The influence of H is less significant in isothermal condition. The $_e$ decreases and the increases with the increase of H in non-isothermal conditions. For the blackbody source manufacturer, a method to improve cavity thermal homogeneity is more useful than extend the cavity length. Figure 5 shows the significant effect of cavity material emissivity on $_e$ and with D_a = 0.8 and H = 4. With the increase of , the $_e$ increases and the decreases in both conditions. A linear relationship is observed between the $_e$ and the in all the studied Runs, not shown in figure for brevity. Generally speaking, the blackbody source with a higher $_{\mathrm{e}}$ will have a uniform distribution of $_{\mathrm{e}}$ on the detector. This performance is benefit for thermometer calibration.

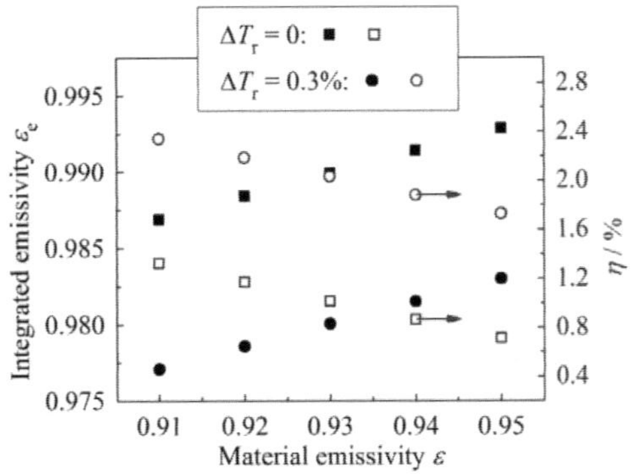

Figure 5. Effect of on $_{\mathrm{e}}$ and with $D_{\mathrm{a}} = 0.8$ and $H = 4$

5. Conclusions

High-performance blackbody radiation source is the key equipment for highly precision reproduction of radiation temperature. This paper studied the effect of geometric parameters, cavity material emissivity, and temperature distribution on integrated effective emissivity $_{\mathrm{e}}$ and normalized spatial integrated effective emissivity variation by finite volume method. The numerical simulation has a good agreement with the result computed by Monte-Carlo method. The results show that the isothermal cavity has a higher $_{\mathrm{e}}$ and a smaller than the non-isothermal. A smaller aperture diameter D_{a} and a higher material emissivity are benefit for a higher $_{\mathrm{e}}$ and a more uniform distribution on detector of $_{\mathrm{e}}$. The results provide references for the design and study on high-performance blackbody radiation source.

Acknowledgements

This work was supported by the Research Program of Shanghai Municipal Bureau of Quality and Technical Supervision (No. 2018-17).

References

[1] Bedford, R.E. (1988) Calculation of effective emissivities of cavity sources of thermal radiation. In: DeWitt, D.P. and Nutter, Gene D. (Eds.), Theory and practice of radiation thermometry. Wiley Inc., New York. pp. 653-772.
[2] Prokhorov, A.V. (1998) Monte Carlo method in optical radiometry. Metrologia, 35: 465-472.
[3] Liu, D. Duan, Y.Y. Yang, Z. (2013) Calculations of the average normal effective emissivity for nonaxisymmetric cavities using the modified finite volume method. Opt. Eng., 52: 039702.
[4] Liu, D. Duan, Y.Y. Yang, Z. (2013) Integrated effective emissivity computation for non-isothermal non-axisymmetric cavities. Chin. Opt. Lett., 11: 022001.
[5] Prokhorov, A. V. (2012) Effective emissivities of isothermal blackbody cavities calculated by the Monte Carlo method using the three-component bidirectional reflectance distribution function model. Appl. Opt., 51: 2322–2332.
[6] Prokhorov, A. V. (2012) Application of the three-component bidirectional reflectance distribution function model to Monte Carlo calculation of spectral effective emissivities of nonisothermal blackbody cavities. Appl. Opt., 51: 8003–8012.
[7] Prokhorov, A. V., Hanssen, L. M. (2004) Effective emissivity of a cylindrical cavity with an inclined bottom: I. Isothermal cavity. Metrologia, 41: 421-431.
[8] Liu, B., Zheng, W., Li, H.-Y., et al. (2018) Research of an isothermal blackbody cavity for effective integrated emissivity. In: 1st Chinese thermophysical properties conference, Hanzhong, China. TP2018050. (in Chinese)

[9] Liu B., Zheng W., Li H.-Y., et al. (2019) Study on integrated emissivity of an isothermal blackbody cavity with response surface methodology. Acta Metrologic Sinica, 40(4): 625-630. (in Chinese)

IOP Conf. Series: Materials Science and Engineering **721** (2020) 012029 doi:10.1088/1757-899X/721/1/012029

Simulation study of two phase flow in temperature and pressure reducing device

Lao Xingsheng[1*], Liu Yong[1], Dai Chunhui[1], Wujun[1] and Wang Wei[1]

[1]Science and Technology on Thermal Energy and Power Laboratory, Wuhan 2nd Ship Design and Research Institute, Hubei, 430205, P.R. China

[*]Corresponding author's e-mail: 460079620@qq.com

Abstract. In order to analyze the influence of water spray desuperheating measures on the pressure and flow velocity in the temperature and pressure reducing device, a small temperature and pressure reducing device is taken as the prototype, the grid is divided and the calculation model is established, and the flow field inside the device under different water spray conditions is simulated. The distribution of the vapor-condensate two-phase flow field in the temperature and pressure reducing device shows that the water injection measures have obvious effect of reducing the temperature, and have almost no effect on the two-phase flow pressure and flow rate fluctuation in the device.

1. Introduction

The temperature and pressure Reducing Device (PTRD) is usually employed to control the superheated steam temperature of the boiler outlet to the required range. The control of the internal temperature and pressure parameters of the PTRD is very important for the stable operation of the boiler system. Increasing the nozzle on the basis of the traditional orifice plate structure will increase the desuperheated steam speed and efficiency of the PTRD, but the water spray may affect the parameters such as temperature, pressure and speed inside the PTRD, and operational control of the boiler system is more difficult. The internal two-phase flow simulation calculation was carried out on the model of the PTRD with the size of about 350mm×200mm×250mm. The results show that the desuperheating effect brought by the water spray is significant while has little effect on the steam flow rate and pressure drop.

2. Model description

2.1Model parameters

The model is a PTRD with four-stage small hole throttle expansion decompression and water spray before the last stage. Where, the various levels of decompression throttle hole and related parameters are shown in Table 1 below.

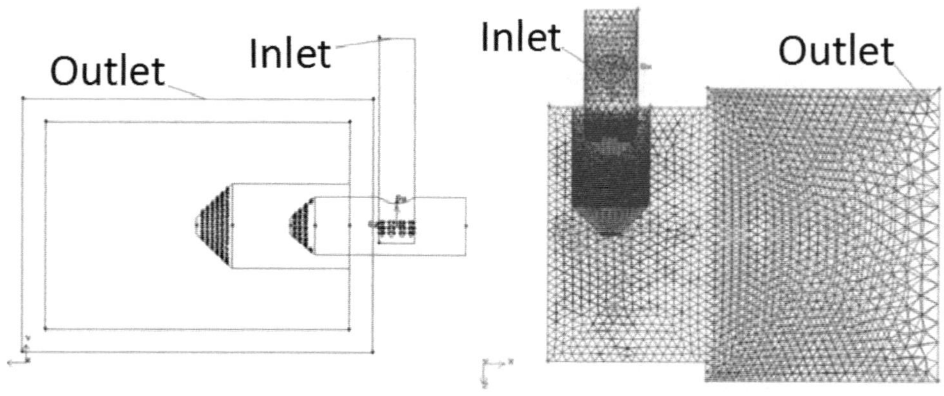

Figure 1. Model of PTRD

Table 1. Parameters of four-stage small hole throttle

Stage number / parameters	Stage 1	Stage 2	Stage 3	Stage 4
Number of holes	30	54	129	988
Aperture (mm)	5.40	5.40	5.00	5.00
Cross-sectional area (mm2)	687.10	1236.70	2532.90	19399.30
Hole center distance (mm)	12.60	8.00	8.00	8.00
Expansion pressure ratio	0.56	0.50	0.50	0.56

2.2Boundary conditions and assumptions

Boundary condition is set as following: (1) Inlet conditions: the use of pressure import boundary conditions, the direction, temperature, and pressure of the airflow are specified. (2) Outlet conditions: Pressure boundary conditions are used, the export back pressure is specified as atmospheric temperature and pressure, (3) Wall: insulation, no slip. The flow field of the two-phase flow of gas and liquid in the temperature and pressure reducer is assumed as following: (1) The flow field is steady, (2) The flow field medium is a two-phase flow of high-temperature high-pressure water vapor and water droplet particles. where, water vapor is regarded as the ideal fluid, (3) wall thickness is ignored.

3. Results and Discussions

3.1Temperature field distribution in PTRD

Figure 2(a) and figure 2(b) are the static temperature field distribution clouds inside the PTRD, in the case where the first three stages of the water spray device are not working and normal operation. From the comparison of figure 2(a) and figure 2(b), we can see whether the water spray has little effect on the first three stages of the temperature reduction and pressure reducer, and the temperature of the first three stages does not change much. This is because we assume that this is done under adiabatic conditions, so the temperature should be constant. When the water spray device sprays water, the temperature field at the end of the third stage begins to change, especially in the fully expanded area of the divergent water mist sprayed from the nozzle, and the temperature drops greatly. This is because the droplets absorb the heat of the high-temperature steam and then fully evaporate here. The temperature of the water vapor drops rapidly. From the speed vector below, we can see that the water vapor that comes in later is cooled after reaching here. Therefore, the temperature here is very obvious. This situation is exactly the opposite of the case of unsprayed water. Eventually the average temperature after spraying is lowered.

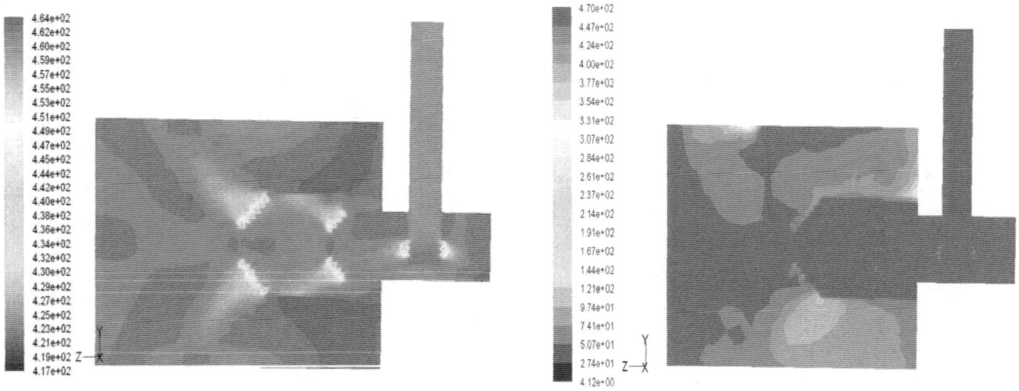

(a) without spray (b) spray

Figure 2. Static temperature field distribution clouds

Figure 3. shows the temperature distribution curve with different water spray amounts. From figure 3, we can see that the different water spray has almost no effect on the temperature of the first three stages. Only after the third-stage orifice plate, as the number of nozzles increases, that is, the amount of water spray increases, the steam temperature at the outlet of the temperature-reducing pressure reducer decreases. As the amount of water spray increases, the number of droplets that can be evaporated increases, and the amount of heat absorbed from the superheated steam increases. According to the law of conservation of energy, the average temperature of water vapor is also reduced, which is the purpose of reducing the temperature of water vapor.

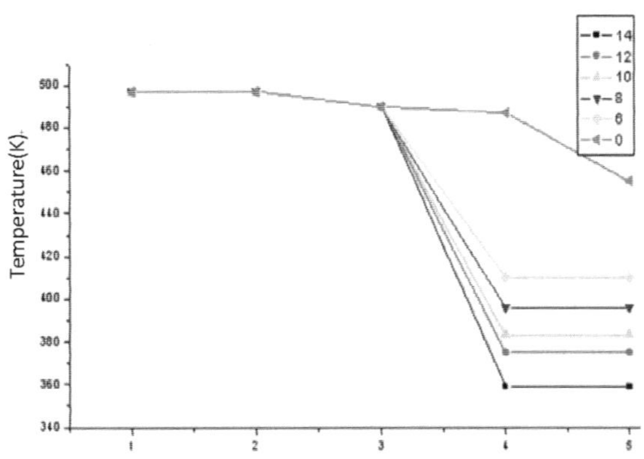

Figure 3. Temperature distribution at various stages with different nozzle numbers

3.2Internal Velocity Field in PTRD

Figure4(a) and figure4(b) are the velocity field distribution diagrams of the desuperheater and pressure reducer in the first three stages of the water spray device, and the table 5 shows the average speed distribution at all levels .From the comparison of figure 4(a) and figure 4(b)), we can see that the water spray has a certain disturbance effect on the velocity field inside the PTRD. However, as can be seen from table 2, the speed after the water spray treatment and the orifice plate of each stage without water spray treatment and the absolute speed of the outlet of the temperature reducer are not much different, indicating that the effect of the water spray on the flow rate of the steam is not Large, the impact is mainly concentrated on the velocity vector.

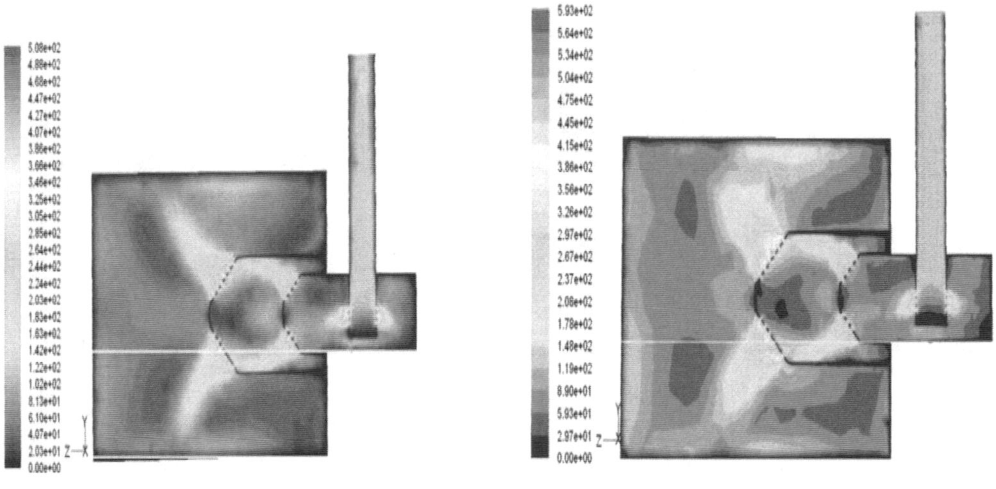

(a) without spray (b) spray

Figure 4. Flow field velocity distribution cloud map of PTRD

Table 2. Distribution of speed(m/s)

Nozzle number / Stage number	0	6	8	10	12	14
Inlet	132.6	132.8	132.8	132.8	132.8	132.8
Stage 2	72.3	74.4	74.6	74.6	74.0	74.2
Stage 3	43.4	44.6	44.7	44.7	44.4	44.5
Stage 4	36.8	37.5	37.9	37.8	37.0	37.3
Outlet	33.2	35.4	33.1	33.0	32.8	32.1

3.3Internal Pressure Field in PTRD

Figures 5(a) and figure 5(b) show pressure field distribution clouds inside the PTRD at the end of the third stage, respectively. From figure 5(a), figure 5(b) and table 3, we can conclude that since the water is sprayed in the third stage expansion chamber, the water spray has no effect on the pressure drop after the first two stages of the orifice plate. The pressure drop of the first two stages is mainly caused by the throttling of the orifice plate. After the third stage, the temperature of the steam is lowered due to the evaporation of the water mist, and a pressure drop is generated according to the ideal gas state equation. However, from the overall pressure drop, the pressure drop due to water spray is small, about an order of magnitude, has little effect on the pressure reduction effect of the entire steam discharge system.

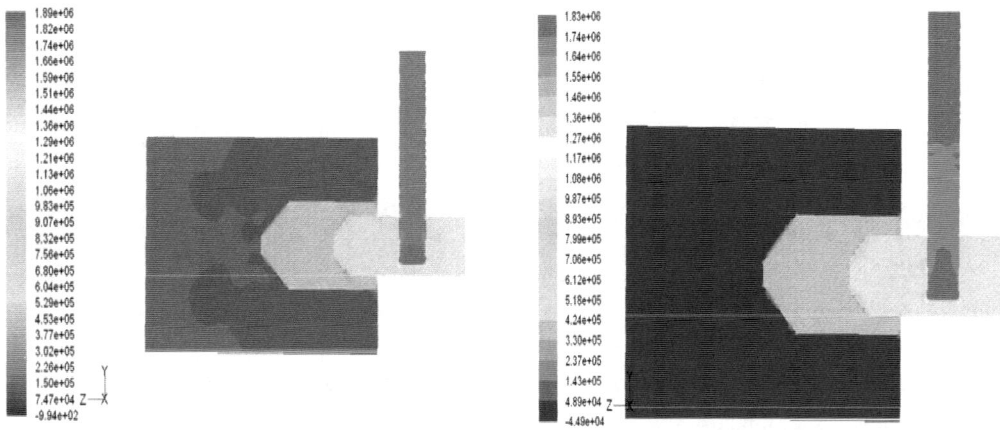

(a) without spray (b) spray

Figure 5. Pressure field distribution clouds inside the PTRDD

Table 3. Distribution of pressure (Pa)

Nozzle number / Stage number	0	6	8	10	12	14
Inlet	1740450	1740450	1740450	1740450	1740450	1740450
Stage 2	1042000	1033550	1033550	1033500	1033900	1033875
Stage 3	388000	378000	377500	377000	378000	377252
Stage 4	14850	2750	2400	2100	2500	2038
Outlet	0	0	0	0	0	0

4. conclusions

The model of the PTRD is established, and the internal flow field is numerically simulated. The DPM model was used to simulate the gas-liquid two-phase flow inside the steam exhaust system, and the distribution of pressure, velocity and temperature after the high temperature and high pressure steam passed through the orifice plates at various levels was determined. The results show that spray in PTRD has few effects on pressure and speed fluctuation inside, while the temperature reduction effect is significant.

References

[1] P. U. Akpan, J., W. F. Fuls. (2019) Application and limits of a constant effectiveness model for predicting the pressure of steam condensers at off-design loads and cooling fluid temperatures. Applied Thermal Engineering, 158: Article 113779.

[2] Roman Kwidzinsk. (2019) Experimental and theoretical investigations of two-phase flow in low pressure steam–water injector. International Journal of Heat and Mass Transfer, 144: Article 118618.

[3] Wiesław Zima. (2019) Simulation of steam superheater operation under conditions of pressure decrease. Energy, 172: 932-944

Effect of crack length on tip field of 304 stainless steel SCC

Xu Yang[1], Jianlong Zhang[1]

[1]Xi'an Special Equipment Inspection Institute, Xi'an 710065, China

Correspondence should be addressed to Jianlong Zhang; jlzhang916@hotmail.com

Abstract: Stress corrosion cracking (SCC) is one of the key issues affecting the long-term safe operation of high temperature pressure vessels and pipeline. In this paper, the effect of crack length on the SCC crack tip field of austenitic 304 stainless steel is studied by using a compact tensile specimen. Firstly, the finite element model of SCC crack tip creep is established by using ABAQUS, and the crack tip creep and creep to crack tip mechanics are analyzed. The influence of different crack lengths on crack tip creep and creep rate was studied. It can be obtained by numerical simulation analysis software. The high creep variable area is mainly concentrated in the near crack tip region. The crack length has a great influence on the SCC crack tip creep field. With the continuous expansion of the austenitic 304 stainless steel crack, crack tip creep variables and creep rates are also increasing. It provides a theoretical reference for the safety evaluation and analysis of high temperature pressure vessels and pipelines.

1.Introduction

Stress corrosion cracking (SCC) is a behavior of structural materials that generate cracks and then fail under the combined action of stress and corrosive environment. It is one of the main failure modes of austenitic stainless steel [1][2], and its existence will be a serious threat Safe operation and service life of pressure vessel piping equipment. Studies have shown that austenitic 304 stainless steel is used in high temperature and high pressure water for a long time. When SCC crack is generated and expanded under high stress conditions, the crack tip will creep. [3][4][5], and then the crack propagation have an important impact. Kassner M E et al. conducted a critical assessment of low temperature creep and irradiation creep in nuclear reactors. It is concluded that even under relatively low temperature and stress conditions, with or without radiation, metallic materials are subject to measurable creep deformation [6]. Wang guo zhen et al. studied the crack tipping effect of high temperature creep cracks and established a creep crack propagation life evaluation method for high temperature parts incorporating crack tip restraint effect [7]. Garud YS et al. applied low temperature creep and irradiation to nuclear reactors. The significance and evaluation of creep is discussed [8]. Kumar Y et al. evaluated the rate of creep crack propagation (CCG) and its micromechanical properties in welds [9].

The expansion process of stress corrosion cracking is similar to the expansion of fatigue cracks [10], including the two stages of initiation and expansion. The incubation period of the crack initiation phase lasts for a long time, and the expansion rate is slow, when the crack grows to the small crack stage. Under the action of the service load, the expansion rate will gradually increase.

In this paper, using the concept of "field" in the numerical simulation software ABAQUS, the creep strain field of crack-containing materials is established by the finite element model to obtain the micro-mechanical parameters of the crack tip, and the influence of the crack length a on the crack tip creep is analyzed. The effect of different crack propagation stages on the crack tip creep field of SCC was quantitatively studied.

2.Establishment of finite element model

2.1 Materials and geometric models

Most of the experiments on stress corrosion at laboratory and abroad use standard compact tensile (0.5T-CT) specimens for structural material selection and safety assessment [11] [12]. The model still uses compact tensile specimens. Figure 1 is a schematic view of the dimensions of a sample of a compact tensile specimen in accordance with ASTM-E399.

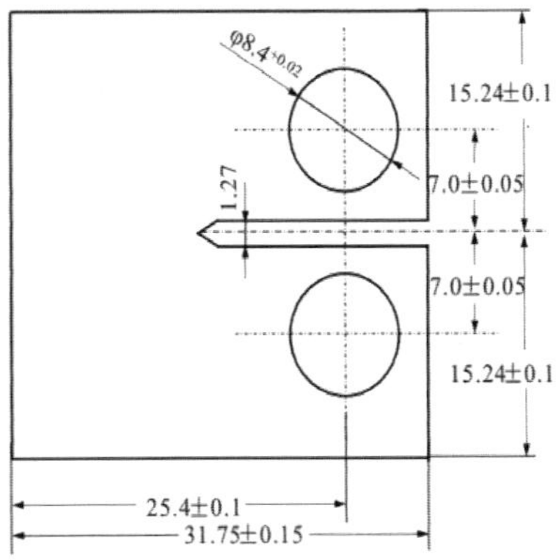

Fig.1 CT sample geometric model and size(mm)

The Ramberg-Osgood model is a stress-strain relationship model established by Ramberg and Osgood in 1943 without a significant yielding platform. The theoretical model is used to describe the stress-strain relationship of a material near its yield point, and is mostly used in solid mechanics [13]. [14], at the same time the model is also suitable for nonlinear metal materials, 304 austenitic stainless steel is a power hardening material, its mechanical properties meet the RO model, the expression is as in formula 1 [15]:

$$\frac{\varepsilon}{\varepsilon_0} = \frac{\sigma}{\sigma_0} + \alpha\left(\frac{\sigma}{\sigma_0}\right)^n \tag{1}$$

Where σ, ε are the actual stress and the actual strain, respectively, σ_0 is the yield limit of the material, ε_0 is the corresponding strain value when the stress reaches the yield limit, α is the hardening coefficient, and n is the hardening index. By fitting the true stress-strain curve of 304 stainless steel, it can be known that the 304 austenitic stainless steel material parameters commonly used in the primary circuit of nuclear power are shown in Table 1.

Tab.1 mechanical properties of 304SS Material

Materia l	Elastic modulus/MPa	Poisson' s ratio	Yield/MPa	Offset coefficient α	Hardenin g index *n*
304SS	204.55	0.33	366	15.01	2.38

2.2 Load selection and calculation

The stress intensity factor is usually used to characterize the stress field of the crack tip in the fracture mechanics sample [16]. Whether the crack propagation factor is measured by the stress toughness of the material exceeds the fracture toughness of the material, so this part simulates the external load on the specimen. Characterized by the crack tip stress intensity factor. The literature shows that the crack tip K is generally selected from 10~50MPam$^{1/2}$ [17] in the stress corrosion cracking test of stainless steel structural materials. The crack tip K applied in the finite element simulation calculation is set to 20MPam$^{1/2}$. The relationship between the crack stress intensity factor K and the applied load and crack length of the compact tensile specimen is shown in Table 2 [18], and the calculated load in the numerical

IOP Conf. Series: Materials Science and Engineering **721** (2020) 012030 doi:10.1088/1757-899X/721/1/012030

simulation is applied to the reference point coupled to the upper and lower force surface of the specimen. on.

Tab.2 Compact tensile specimen value

a/w	*f(a/w)*	a/w	*f(a/w)*	a/w	*f(a/w)*
0.450	8.34	0.485	9.23	0.520	10.29
0.455	8.46	0.490	9.37	0.525	10.45
0.460	8.58	0.495	9.51	0.530	10.63
0.465	8.70	0.500	9.66	0.535	10.80
0.470	8.83	0.505	9.81	0.540	10.98
0.475	8.96	0.510	9.96	0.545	11.17
0.480	9.09	0.515	10.12	0.550	11.36

$$K_{\mathrm{I}} = \frac{P}{B\sqrt{W}} f\left(\frac{a}{W}\right) \quad (2)$$

$$f\left(\frac{a}{W}\right) = \frac{\left(2+\frac{a}{W}\right)\left[0.886+4.64\frac{a}{W}-13.32\left(\frac{a}{W}\right)^2+14.72\left(\frac{a}{W}\right)^3-5.6\left(\frac{a}{W}\right)^4\right]}{\left(1-\frac{a}{W}\right)^{3/2}} \quad (3)$$

Where K_I is the stress intensity factor, P is the load, B is the specimen thickness, W is the sample size,a is the crack length, and the initial a/W=0.5 is obtained by looking up Table 2, which can be calculated by formula (2). At this time, the value of the load P is 470N.

2.3 Finite element mesh model

The finite element global mesh model and the refined mesh model at the crack tip are shown in Fig. 2(a), where the global model has a mesh number of 13516 and the unit type is CPE4R. In order to obtain a more detailed and accurate creep variable at the crack tip, the sub-model technique is used to analyze the mechanical state of the crack tip, and the mesh at the crack tip of the sub-model is refined, while maintaining the sub-model mesh type and the main model. Consistent, the number of grids is 15027, as shown in Figure 2(b).

In order to characterize the stress and strain of the crack tip and the change of creep in more detail, observation paths 1 and 2 are respectively arranged along the circumferential direction of the crack tip and the crack propagation direction, as shown in Fig. 2(c).

2.4 Setting of boundary conditions

In order to prevent the rigid body from rotating and moving in the X direction during the stretching process, the center point of the loading hole is created as a reference point, and the reference point is coupled with the upper and lower force receiving holes, and then the movement of the reference point in the X direction is restricted. And the rotation in the Z direction is only realized by releasing the movement in the Y direction.

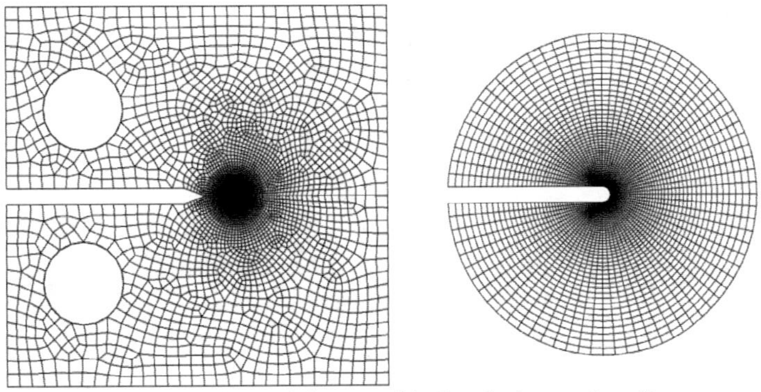

(a) Global mesh model (b) Crack tip mesh refinement model

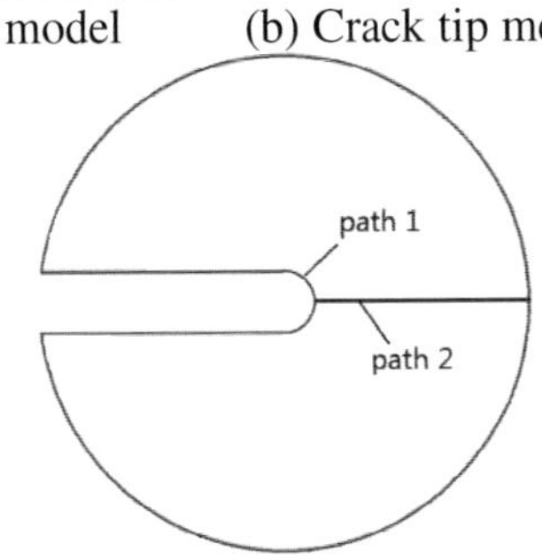

(c) Crack tip creep field observation path

Fig.2 Finite element mesh model and observation path of CT sample

2.5Creep numerical simulation process

The finite element analysis process of SCC creep is realized by two analysis steps: the first stage through Static Analysis, that is, the sample forms a stable crack tip stress strain field under the action of the crack tip stress intensity factor K; the second stage passes Creep analysis is used to obtain the influence of creep on the crack tip stress field under the time cumulative effect. It is achieved by setting the creep analysis step, and the creep equivalent loading time is t=500h. In the calculation process, the initial crack tip stress intensity factor K_0=20MPam$^{1/2}$ and the initial crack length a_0=2mm are guaranteed.

3. Calculation results and analysis

The crack length is set to 0.8, 0.9, 1.0, 1.1, and 1.2 times of the initial crack length a0, respectively. Considering the actual working conditions, the crack length a/W = 0.5 and the crack tip stress intensity factor is 20 MPa.m$^{0.5}$. As the applied load, the influence of different crack lengths on the crack tip creep field is compared, and the relationship between crack length change and crack tip creep and creep rate is obtained.

3.1Analysis of creep variables in crack tip regions

The distribution of the equivalent creep of the crack tip region is shown in Figure 3. The creep time is consistent and set to 500h. It can be seen from the figure that the high creep region is mainly concentrated in the near crack. In the tip region, at the same creep time, the area of the crack tip high creep variable increases with the crack propagation length, and the crack is in the relatively small creep region with the creep variable less than 0.03. The effect of the extended length on the creep variable is relatively small.

IOP Conf. Series: Materials Science and Engineering **721** (2020) 012030 doi:10.1088/1757-899X/721/1/012030

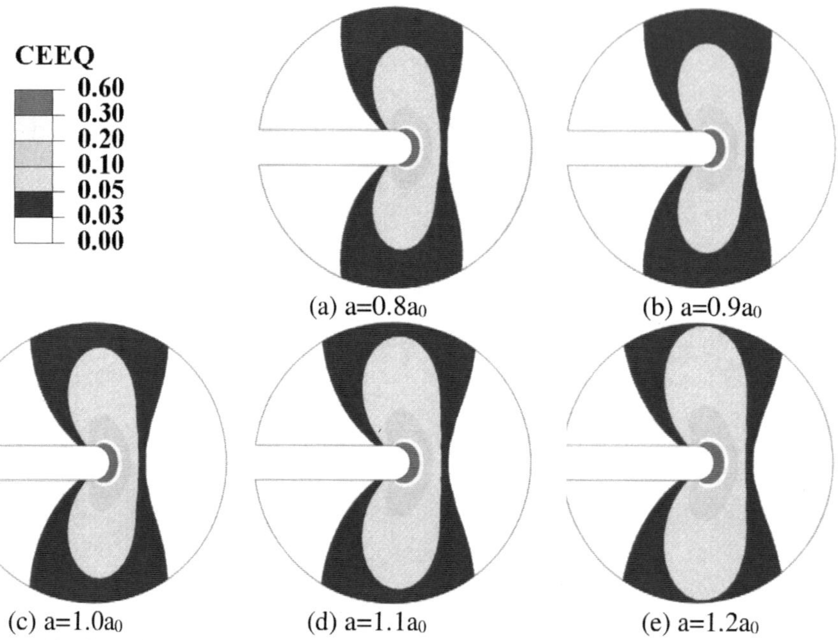

Fig.3 Distribution of crack tip creep variables in different crack propagation stages

3.2 Effect of crack propagation length on crack tip creep and creep rate

The creep curve after creeping along the crack tip path 2 and the path 1 under different crack propagation lengths a is shown in Figs. 4 and 5. It can be seen from Fig. 4 that the crack propagation length has little effect on the creep distribution on the horizontal path 2 of the crack tip, and is still bounded by the crack tip distance r=1μm. When the feature distance is smaller than the feature distance, the crack tip creep variable The sharp drop, after the characteristic distance, the creep variable value has little effect and the data remains stable. It can be seen from Fig. 5 that there is a positive correlation between the creep variable and the crack propagation length, and the crack length increases the same proportion. Within the range, the increase of the creep variable gradually becomes larger, that is, the increase of the creep variable is slightly larger than the increase of the crack propagation length.

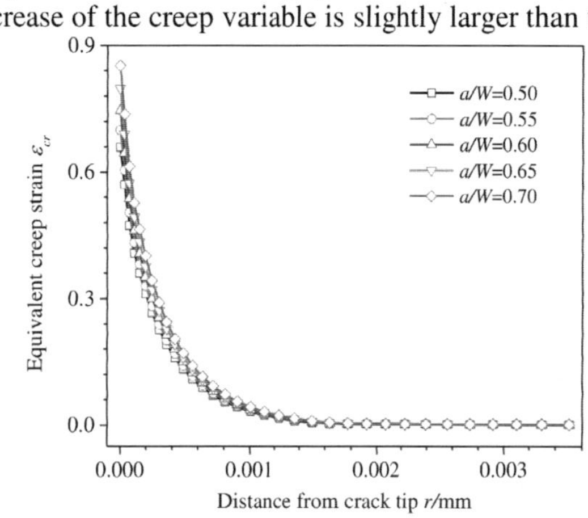

Fig.4 Effect of crack length on creep variable in expansion direction

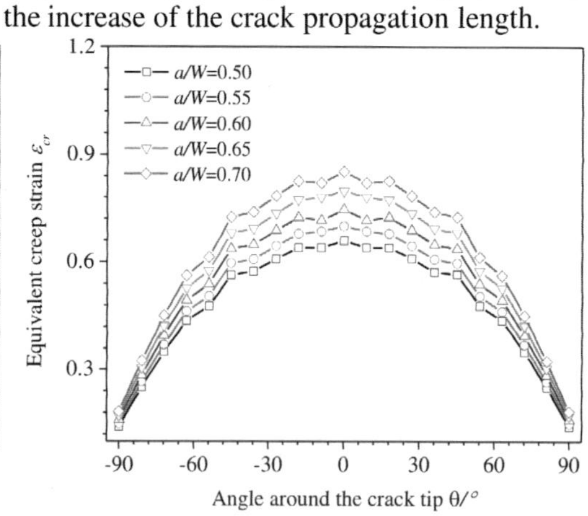

Fig.5 Effect of crack length on circumferential creep

Figure 6 shows the creep distribution curve of the crack tip under different crack propagation lengths. In the first stage of creep, the creep variable of the crack tip has little change under different crack propagation lengths. As the creep progresses, the crack can be seen. The tip creep rate increases with the increase of the crack propagation length. It can be more clearly seen from Fig. 7 that the creep rate of the crack tip increases with the increase of the crack propagation length in the second stage of creep, but the increase is larger. small.

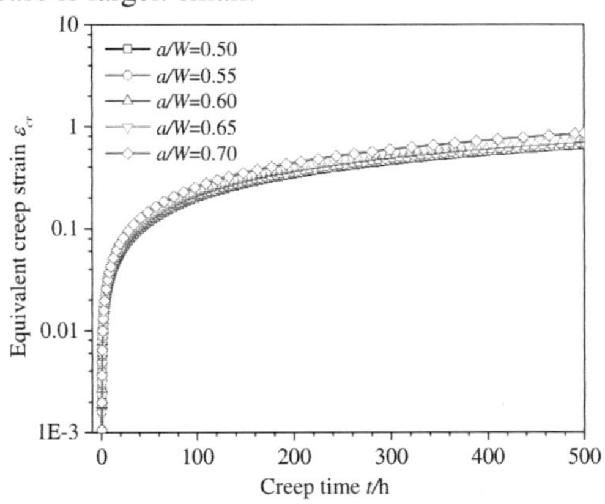

Fig.6 Analysis of Influence of Crack Length on Crack Tip Creep

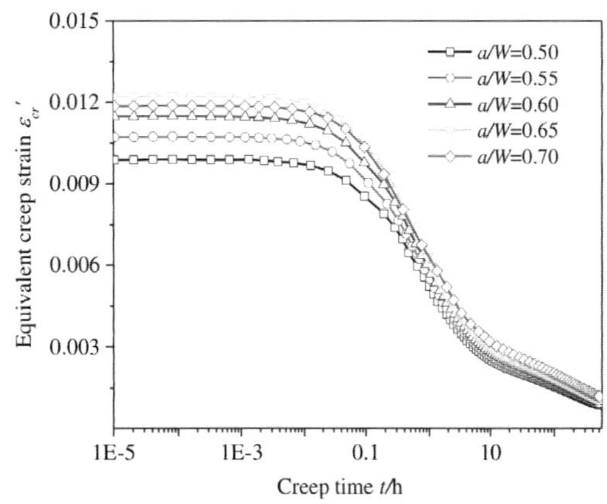

Fig.7 Analysis of the Influence of Crack Length on Crack Tip Creep Rate

4.conclusion

(1) At the initial stage of creep, the crack length is positively correlated with the crack tip creep. In the steady creep stage, the creep rate of the crack tip under different plastic parameters is reduced.

(2) The crack length of the material is positively correlated with the creep rate of the crack tip, and the increase of the crack length leads to the increase of the creep rate of the crack tip. As the creep progresses, the creep rate tends to the same level, and as the time continues to increase, the creep rate does not change much.

Reference

[1] Xue H, Sato Y, Shoji T. Quantitative estimation of the growth of envir666onmentally assisted cracks at flaws in light water reactor components[J]. Journal of Pressure Vessel Technology. 131:61-70.

[2] MA Cheng, PENG Qunjia, HAN Enhou, et al. Review of stress corrosion cracking of structural materials in nuclear power plants[J]. Journal of Chinese Society for Corrosion and Protection, 2014. 34(1): 37-45.

[3] Kassner M E, Smith K. Low temperature creep plasticity[J]. Journal of Materials Research & Technology, 2014, 3 (3): 280-288.

[4] Nie D F, Zhao J. An Approach to Estimate Room Temperature Creep Of Structural Steels[J]. Acta Metall Sin, 2011, 47(2): 142-147.

[5] Michael E. Kassner, Kamia S. Low temperature creep plasticity[J]. Joural of Materials Reaserch and Technology, 2014, 3(3):280-288.

[6] Kassner M E, Smith K. Low temperature creep plasticity[J]. Journal of Materials Research and Technology, 2014, 3(3): 280-288.

[7] Wang G Z, Xuan F Z, Tu S D. Creep crack-tip constraint effect in high temperature structures. Advances in Mechanics, 2017, 47: 201704.

[8] Garud Y S. Significance and Assessment of Low Temperature Creep and Irradiation Creep in Nuclear Reactor Applications [J]. Procedia Engineering, 2015, 130:1162-1176.

[9] Kumar Y, Venugopal S, Sasikala G, et al. Study of Creep Crack Growth Behaviour of a type 316(N) Stainless Steel Weld and its Mechanism[J]. Materials Science & Engineering A, 2018.

[10] Huang X G. Mechanism Study of Pit Evolution and Crack Propagation for Corrosion Fatigue[D].

Shanghai Jiao Tong University, 2013.
[11] Dong L, Peng Q, Han E H, et al. Microstructure and intergranular stress corrosion cracking susceptibility of a SA508-52M-316L dissimilar metal weld joint in primary water[J]. Journal of Materials Science & Technology, 2018, 34(8): 1281-1292.
[12] Dong L, Han E H, Peng Q, et al. Environmentally assisted crack growth in 308L stainless steel weld metal in simulated primary water[J]. Corrosion Science, 2017, 117:1-10.
[13] Byun T S, Hashimoto N, Farrell K. Temperature dependence of strain hardening and plastic instability behaviors in austenitic stainless steels[J]. Acta Materialia, 2004, 52(13):3889-3899.
[14] Ramberg W, Osgood W R. Description of stress-strain curves by three parameters[J]. Technical Report Archive & Image Library, 1943(902).
[15] Xue H, Li Z, Lu Z, et al. The effect of a single tensile overload on stress corrosion cracking growth of stainless steel in a light water reactor environment[J]. Nuclear Engineering and Design, 2011, 241(3):731-738.
[16] Ma J S. Finite Element Analysis of 3D Mixed-Mode Fracture Stress Intensity Factors[D]. Harbin Engineering University, 2008.
[17] Toivonen A. Stress corrosion crack growth rate measurement in high temperature water using small precracked bend specimens[J]. 2004.
[18] ASTM I. Standard test methods for plane-strain fracture toughness and strain energy release rate of plastic materials[J]. ASTM D5045-99, 2007.

IOP Conf. Series: Materials Science and Engineering **721** (2020) 012031 doi:10.1088/1757-899X/721/1/012031

Application of BP neural network based on genetic algorithm optimization in thermal conductivity of nanofluids

Jiang Wang and Yuling Zhai*

State Key Laboratory of Complex Nonferrous Metal Resources Clean Utilization, Kunming University of Science and Technology, Yunnan, 650093, China.

The National Natural Science Foundation of China (51806090)

*Corresponding author's e-mail: zhaiyuling00 @126.com

Abstract. In this paper, a hybrid model (BP-GA) including a back propagation network and a genetic algorithm is used to estimate the thermal conductivity of the nanofluids. The genetic algorithm is used to optimize the initial weight and threshold of BP neural network, so that the optimized BP neural network can better predict the function output. In this study, CuO-ZnO hybrid nanofluids with the mass fraction of 0%, 1%, 2%, and 3% were studied at temperatures of 25, 30, 35, 40, 45, 50, 55, and 60 °C, respectively. The thermal conductivity of the hybrid nanofluids base liquid is 20:80, 40:60, 50:50, 60:40, 80:20, and the mass ratio of CuO-ZnO is 50:50. The mass fraction of CuO-ZnO, the temperature and the mixing ratio of different base liquids were used as input parameters, and the thermal conductivity was the output parameter, forming a 3-input ANN neural network. The predicted thermal conductivity results of BP neural network and genetic algorithm optimized BP neural network (GA-BP) were compared with experimental data. The results illustrate that the BP neural network optimized by genetic algorithm exerts a positive influence on the accuracy and stability of the predicted output.

1. Introduction

Nanofluids have potential applications in many important areas, such as advanced heat transfer or cooling, energy harvesting, microelectronics, MEMS, microfluidics, transportation, medical, and numerous thermal management systems [1-7]. Due to the small size and large surface area of nanoparticles, various types of nanofluids have broad application prospects in engineering applications. Therefore, the main goal of the researchers is to estimate the transport parameters of nanofluids called thermal conductivity and viscosity. In the late 1950s, the perceptron network introduced artificial neural networks, which is a powerful statistical tool. In recent years, neural networks, fuzzy logic and genetic algorithms have been introduced as computational methods, therefore, neural network algorithms can be used to predict nanofluids thermal conductivity and viscosity. By reading the relevant literature [8,9], the neural network algorithm has higher prediction accuracy for thermal conductivity and viscosity.

Nanofluids have superior heat transfer performance than traditional liquids. One of the main reasons is that nanoparticles significantly increase the thermal conductivity of nanofluids. The thermal conductivity is the main parameter reflecting the heat transfer capacity of the medium and has important theoretical and application significance. So far, there is no molded fluid transient hot wire thermal conductivity tester at home and abroad. In addition, Kurt and Kayfeci [10] predicted the thermal conductivity of ethylene glycol-water-based nanofluids. Emmat Esfe et al.[11] designed an artificial neural network algorithm to estimate the thermal conductivity and dynamic viscosity of ferromagnetic nanofluids using 72 experimental data. They prepared two sets of relevant data to predict these thermophysical parameters and compared the experimental data with the relevant results. Zhao et al. [12] proposed two neural network models based on radial basis function (RBF) to predict the viscosity of Al_2O_3-water and CuO-water, evaluated the accuracy of the model by using to the root mean square error (RMSE), mean absolute percentage error (MAPE), the sum of squared errors (SSE), and coeffificient of determination (R^2). Hemmat Esfe et al. [13] proposed an artificial neural network to predict the viscosity of TiO_2-water nanofluids. They studied various structures of the network and applied a hidden layer and four neuron tansig activation functions to calculate the regression coefficients, the results used for viscosity estimation are compared with experimental data, and the maximum error is 0.5%. Ahmadloo and Azizi [14] used a 776 experiment to establish a 5-input artificial neural network model to estimate Al_2O_3-water, Al-water, TiO_2-water, Cu-water, CuO-water, ZrO_2-water, Al_2O_3-EG The thermal conductivity of Cu-EG, CuO-EG, $Mg(OH)_2$-EG, Al_2O_3-Oil, Al-Oil, CuO-Oil and Cu-Oil is evaluated by the average absolute error and correlation coefficient. Afrand et al. [15]used experimental data to estimate the relative viscosity of MWCNTs-SiO_2/AE40 nano-lubricants, and used 80 experiments to establish an artificial neural network with temperature and nanoparticle volume fraction as input variables, the number of hidden layer neurons, the transfer function, and the repetition of the training step size determine the optimal network. Esfe et al. [16-23] measured the thermal conductivity of other different types of nanofluids based on experimental methods. On this basis, the effectiveness of the multilayer perceptual feedforward network model in the prediction of thermal conductivity of nanofluids was verified. Hemmat and Esfe et al. [24] evaluated the accuracy of the artificial neural network model in predicting the thermal conductivity of Al_2O_3/EG-W. The volume fraction of 0~1.5% nanoparticle and the temperature range of 20~60°C are input data, and the measured thermal conductivity is the output data, the number of neurons with different transfer functions is evaluated to determine the optimal model of the number of neurons in the hidden layer. The results show that the artificial neural network structure has high simulation accuracy. The paper is arranged in the following structure. Section 1 argues the influence of base mixing ratio, temperature and mass fraction on thermal conductivity. Section 2 introduces the use of GA-BPNN and BPNN model to establish a thermal conductivity prediction model. Section 3 describes the prediction of the thermal conductivity between the GA-BPNN model and the BPNN model and the experimental data, and characterizes the influence of different factors (temperature, nanoparticle mass fraction, base-liquid mixing ratio) on the thermal conductivity. Based on the experimental results, it is found that the GA-BPNN model has higher prediction accuracy than the BPNN model.

2. Experiments

The specific preparation of the hybrid nanofluids, firstly, according to the mass fraction of (CuO-ZnO), calculate the corresponding mass of 40nm CuO and 30nm ZnO nanoparticles and the corresponding deionized water and ethylene glycol mass, secondly, weighed and mixed together in accordance with the ratio of ethylene glycol to deionized water. Finally, the mixture was stirred for 30 min with a magnetic stirrer and ultrasonically shaken for 45 min to form a uniformly dispersed nanofluids.

Preparation of nanofluids, used for CuO-ZnO/(ethylene glycol: deionized water) nanofluids with the equal mass fraction (1:1) nanofluids under different temperature and nanoparticle mass fraction and different basic mixing ratio under experimental evaluation. The nanoparticle mass fraction of CuO-ZnO was 0%, 1%, 2%,3%,.At temperatures of 25,30,35,40,45,50,55 and 60 ℃, ethylene glycol and deionized water at different mixing ratio (20:80, 40: 60, 50:50, 60:40, 80:20) with a Hot disk 2500 s thermal constant analyzer repeated four times, each sample taken four times the average thermal conductivity as the final measured values.

As shown in figure 1, the mass fraction of 0%, 1%, 2%, and 3%, depicts the temperature of 20,25,30,35,40,45,50,55 and 60 ℃, ethylene glycol and deionized water at different mixing ratio (20:80, 40:60, 50:50, 60:40, 80: 20) effect on the thermal conductivity, and through the experiment measuring the thermal conductivity of 3 d surface figure.

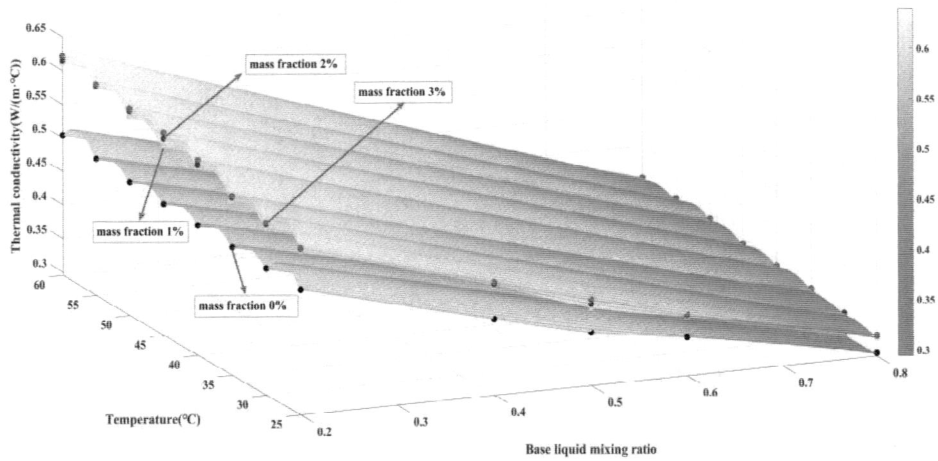

Figure 1. Three dimensional scatter surface of thermal conductivity under different mass fraction

It can be seen from Figure 1 that, when the temperature and nanoparticle mass fraction are constant, the ratio of ethylene glycol to deionized water (that is, the mass fraction of ethylene glycol to ethylene glycol and deionized water) increases. The decrease in thermal conductivity is chiefly due to the fact that the thermal conductivity of ethylene glycol is lower than that of deionized water.

When the temperature and the ratio of ethylene glycol to deionized water are constant, as the mass fraction increases from 0% to 1%, the thermal conductivity increase greatly. The primary reason is that the addition of nanoparticles is the main factor affecting the thermal conductivity. When the mass fraction increases from 1% to 3%, the thermal conductivity increases with the increase of the mass fraction. The cause is that the increase of the content of nanoparticles per unit volume leads to the increase of the heat exchange area, and the probability of collision between particles greatly improves the heat exchange efficiency.

When the mass fraction and the ratio of ethylene glycol to deionized water are constant, the thermal conductivity increases as the temperature rises. The cause for this is that the Brownian motion velocity increases, which contributes to the increase of the thermal conductivity of the nanofluids.

2.1 Creating an BP neural network model

The BP neural network is a multi-layer feedforward neural network. The main feature of this neural network is that the signal is transmitted forward and the error is transmitted backward. The BP neural network consists of three layers of input layer, hidden layer and the output layer, as shown in Figure 2. BP neural network usually uses sigmoid function (log-sigmoid or tan-sigmoid) as the transfer function. The so-called reverse propagation learning algorithm is that input information is passed from the input layer through the hidden layer to the output layer. If the output error is large, the error signal is returned along the original path for reverse learning. During the return process, the new output error is obtained by modifying the connection weight and threshold between the neurons until the error signal is minimum. As shown in Figure 1, the input values are the mass fraction W, temperature T, the ratio of ethylene glycol to deionized water R, and the output value is the thermal conductivity. Seen from the topology results graph, BP neural network is equivalent to a nonlinear function, mapping the input value to the output value.

R^2 and MSE are considered to be the two main parameters for determining the best mode of the neural network. In fact, when the error measurement standard reaches a predetermined amount, the training of the BP neural network ends. The error measurement criteria include mean square error (MSE) and Coefficient of determination (R-square). The values are defined as follows:

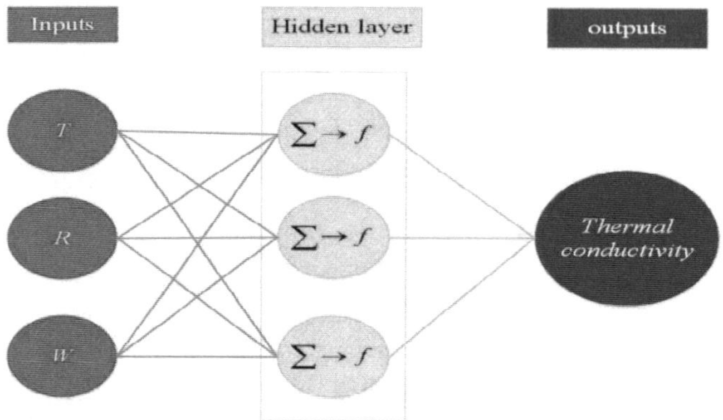

Figure 2. BP neural network extension structure

R^2 and MSE are considered to be the two main parameters for determining the best mode of the neural network. In fact, when the error measurement standard reaches a predetermined amount, the training of the BP neural network ends. The error measurement criteria include mean square error (MSE) and Coefficient of determination (R-square). The values are defined as follows:

$$R^2 = 1 - \frac{(\sum_{i=1}^{n}(T_i - P_i)^2)/n}{(\sum_{i=1}^{n}(T_i - \bar{T})^2)/n} \tag{1}$$

$$MSE = \frac{1}{n}\sum_{i=1}^{n}(T_i - P_i)^2 \tag{2}$$

In these equations, T_i is the actual value and P_i is the value of the model. $\bar{T}$ is the average experimental value of n sets of test data. MSE (Mean Squared Error) refers to the expected value of the square of the difference between the parameter estimate and the true value of the parameter. MSE can evaluate the degree of change of the data. The smaller the MSE value, the more accurate the prediction model describing the experimental data. R^2 is statistically used to represent the goodness of fit of the model. The larger the R-square is, the better the effect of our model is. The maximum value is 1.

2.2BP neural network prediction model construction

All the data of the 112 groups after the pretreatment were normalized with reference to the equation (3), and distributed between 0 and 1.

In the equation, X_{nor}: Normalized value. X: Actual input value. X_{max}: The maximum of the actual input value. X_{min}: The minimum of the actual input value.

$$X_{nor} = \frac{X - X_{min}}{X_{max} - X_{min}} \tag{3}$$

The normalized data is divided into the training group and the prediction group. The BP neural network is established by using the newff function in Matlab, and the predictable result of the newly established neural network model is normalized to obtain the final result. This paper uses a three-layer BP neural network structure of the input layer, the output layer and the hidden layer. After normalization, the three key parameters, in other words, temperature, nanoparticle mass fraction, the mixing ratio of different base liquids are used as the input layer, and the thermal conductivity is used as the output layer, so the number of neurons in the input layer and the output layer is 3, respectively, and 1. The choice of the number of hidden layer nodes is given by equation (4).

$$H = \sqrt{m+n} + L \tag{4}$$

In the equations, m: The number of nodes for the input layer.

n: The number of nodes for the output layer.

It can be known from equation (4) that the number of hidden layer nodes of the neural network is between 3 and 12. The new data is trained by using the newff function to build a BP neural network model. The BP neural network predictive model selects 70% of the data as training, 15% of which is used for verification, and the remaining 15% is used to test the accuracy of the predicted sample model.

2.3Genetic algorithm optimization BP neural network

Genetic algorithm (GA)-a parallel random search optimization method that simulates the natural genetic mechanism and biological evolution theory. Based on the biological evolution principle of "survival of the fittest, survival of the fittest", the coding series is formed by optimizing parameters. The selected fitness function is used to screen individuals through selection, crossover and mutation in the inheritance, so that individuals with good fitness values are retained, individuals with poor fitness are eliminated, and new groups inherit the information of the previous generation. Superior to the previous generation, repeated cycles until the conditions are met. If the optimal initial weights and thresholds obtained by genetic algorithm optimization can be given to the BP neural network and training is started, BP can be improved. The flow chart of the prediction accuracy and stability of the neural network model is as follows.

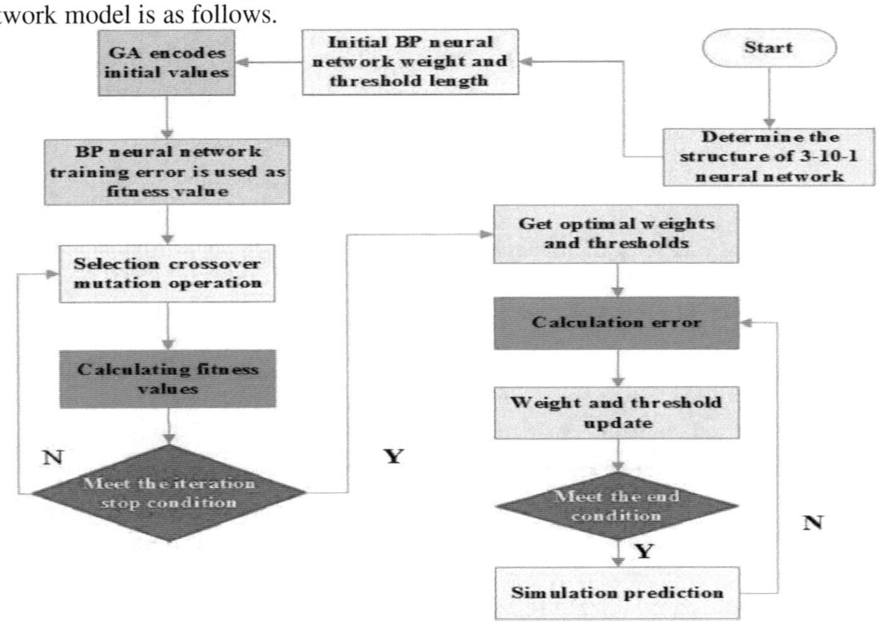

Figure 3. Flow chart of BP neural network algorithm optimized by genetic algorithm

The main optimization process is as follows:

Step 1: encode weights or thresholds and randomly generate connection weights or thresholds between a group of neurons.

Step 2: input the training sample, calculate its error, and define the sum of the absolute error as the fitness function.

Step 3: if the error is smaller, screen out the individuals with small sum of absolute error and directly pass them on to the next generation.

Step 4: evolve the current population and produce the next generation population through crossover, mutation and other operations.

Step 5: The initial weight or threshold of the BP neural network is continuously updated until the final condition is met, and the accuracy of the model is verified.

IOP Conf. Series: Materials Science and Engineering **721** (2020) 012031 doi:10.1088/1757-899X/721/1/012031

3. Results and Discussion

3.1BP neural network model prediction results

The predicted sample data were substituted into the BP neural network prediction model, and the predicted absolute and relative thermal conductivity errors were calculated according to equations (3) and (4), and then the number of neurons in different hidden layers and different transfer functions were calculated. MSE and R^2, as shown in table 1, BP neural network prediction model has the best thermal conductivity when the hidden layer is 10 neurons and tansig transfer function. BP neural network predicted the thermal conductivity.

Table 1. BP neural network results for all data

Candidate topology num	Number of hidden layers	Transfer Function	MSE	R-square
1	[3]	tansig	2.7720E-04	0.99763
2	[3]	logsig	4.3312E-04	0.99229
3	[4]	tansig	6.6516E-04	0.992806
4	[4]	logsig	3.0390E-03	0.98958
5	[5]	tansig	2.6150E-04	0.99775
6	[5]	logsig	8.3406E-04	0.99839
7	[6]	tansig	2.437E-03	0.99736
8	[6]	logsig	4.2303E-04	0.99832
9	[7]	tansig	3.5047E-04	0.99836
10	**[7]**	**logsig**	**1.1381E-04**	**0.99887**
11	[8]	tansig	1.4523E-04	0.99898
12	[8]	logsig	3.5191E-03	0.99899
13	[9]	tansig	7.8411E-04	0.99834
14	[9]	logsig	1.8542E-03	0.99716
15	[10]	tansig	8.9056E-04	0.99913
16	[10]	logsig	5.1475E-04	0.99793
17	[11]	tansig	4.0209E-03	0.99887
18	[11]	logsig	1.7804E-03	0.99841
19	[12]	tansig	3.9780E-04	0.99892
20	[12]	logsig	3.7639E-03	0.99663

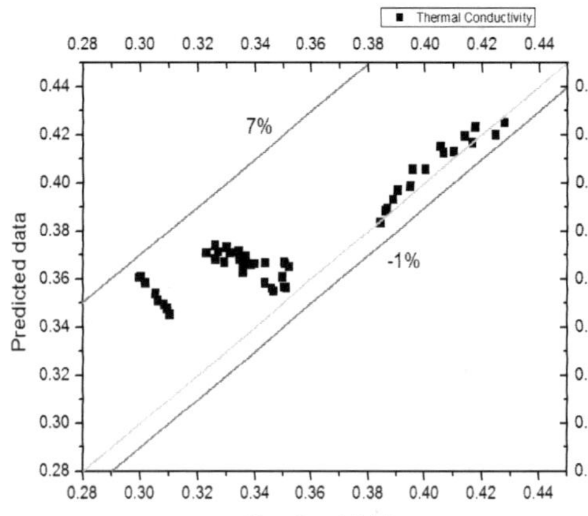

Figure 4. BP neural network predicts the ratio of thermal conductivity data to experimental data.

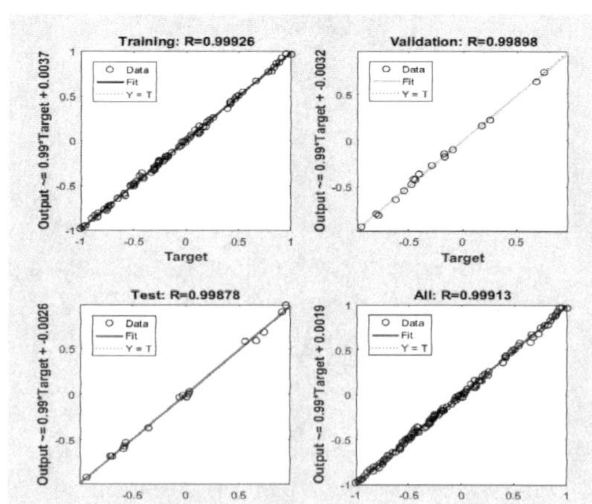

Figure 5. BP neural network predicts thermal conductivity regression.

There were 160 sets of data, 112 of which were used for training data. Of the 112 data sets, 70% was used for training, 15% for validation, 15% for testing, the remaining 48 data sets were used for new data, and the original training neural network was used for testing to ensure the stability of the model and prevent overfitting phenomenon. It can be seen from Table 1 that the number of neurons under the hidden layer is 10, and the transfer function is an s-type function, and the prediction performance is the best. From Figure 4, it can be seen that when the experimental data and the predicted data are equal, the line with the slope of 1 is detected on the graph, when the predicted data is larger than the experimental data, it will be revealed above the line, Forecast data is less than experimental data and will be reflected below this line. It can be seen from Figure 5 that the BP neural network algorithm predicts the regression coefficient of the thermal conductivity coefficient as 0.99913.

3.2Analysis of prediction results of BP neural network model optimized by genetic algorithm

In view of the disadvantage of BP neural network algorithm falling into local optimization when optimizing weights and thresholds, this study used the genetic algorithm to optimize the initial weights and thresholds of BP neural network, determined the optimal initial weights and thresholds, eliminated the randomness of BP neural network results, and improved the accuracy and stability of the model.

This study uses the genetic algorithms to perform 10 generations of evolution, with a crossover probability of 0.3 and a mutation probability of 0.1, to obtain the optimal individual fitness curve, as shown in Figure 6. It can be seen from the figure that the optimal individual fitness obtained is 1.3189, and the corresponding evolutionary algebra is 7, and the fitness does not change any more. Therefore, it can be determined that the optimal fitness of the individual has been reached at this time. The initial weights and thresholds obtained from the optimization results are given to the BP neural network for model training, and the prediction results of the multiple training are completely consistent.

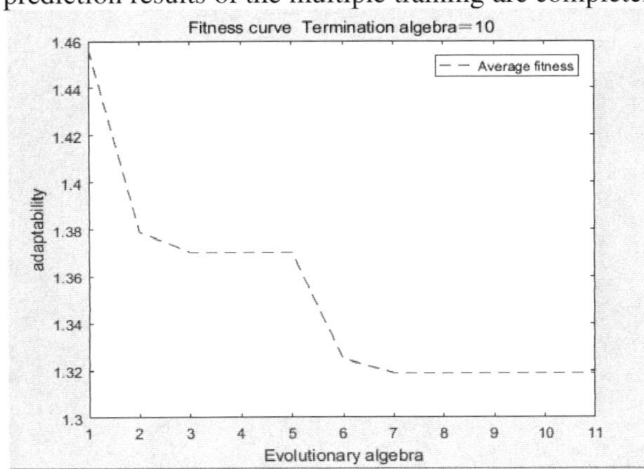

Figure 6. Genetic algorithm individual fitness optimization curve.

As shown in Table 2, the BP neural network optimized by the genetic algorithm has different transfer functions and the number of neurons is also different. For comparison with BP neural network, the selected transfer function is S type (tansig) and the number of hidden layer neurons is 10. The prediction model (GA-BP) obtained by BP neural network optimized by genetic algorithm is used to predict the thermal conductivity, and the prediction results are compared with the experimental data, as shown in Figure 7.It can be seen from the figure that the overall trend of the BP neural network model optimized by genetic algorithm is in good agreement with the experimental data, and there is basically no predicted value with large deviations, and the maximum error limit is 2%.

As shown in Figure 8, the maximum error of the relative error is 4%. Therefore, whether from the accuracy of prediction results or the stability of prediction, the BP neural network prediction model optimized by genetic algorithm can play a good predictive effect.

Table 2. BP neural network optimized by genetic algorithm results for all data

Candidate topology num	Number of hidden layers	Transfer Function	MSE	R-square
1	[3]	tansig	2.7184E-04	0.99861
2	[3]	logsig	4.0989E-05	0.99745
3	[4]	tansig	9.4888E-05	0.99889
4	[4]	logsig	3.8722E-04	0.99815
5	[5]	tansig	1.1465E-04	0.99864
6	[5]	logsig	2.7019E-04	0.99796
7	[6]	tansig	6.5392E-04	0.99902
8	[6]	logsig	3.1106E-04	0.99865
9	[7]	tansig	1.4937E-04	0.99927
10	**[7]**	**logsig**	**1.7827E-04**	**0.99849**
11	[8]	tansig	6.8901E-05	0.99917
12	[8]	logsig	7.7350E-04	0.99894
13	[9]	tansig	2.2214E-04	0.99927
14	[9]	logsig	8.1118E-05	0.99859
15	[10]	tansig	4.1656E-05	0.99947
16	[10]	logsig	6.4236E-05	0.99798
17	[11]	tansig	4.3754E-04	0.99941
18	[11]	logsig	5.9504E-04	0.99863
19	[12]	tansig	5.5396E-04	0.99919
20	[12]	logsig	2.3979E-04	0.99877

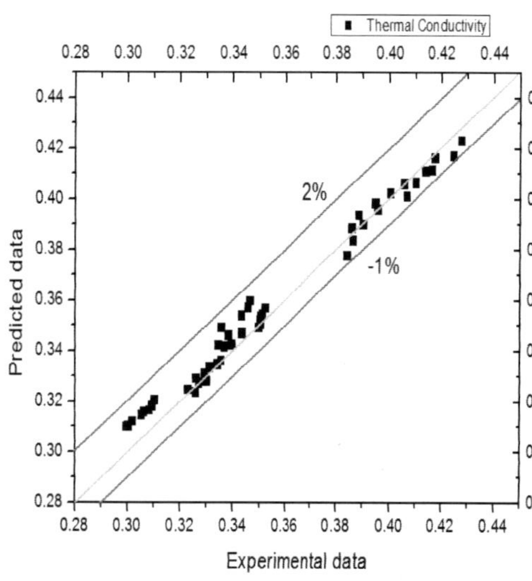

Figure 7. GA-BP neural network predicts the ratio of thermal conductivity data to experimental data

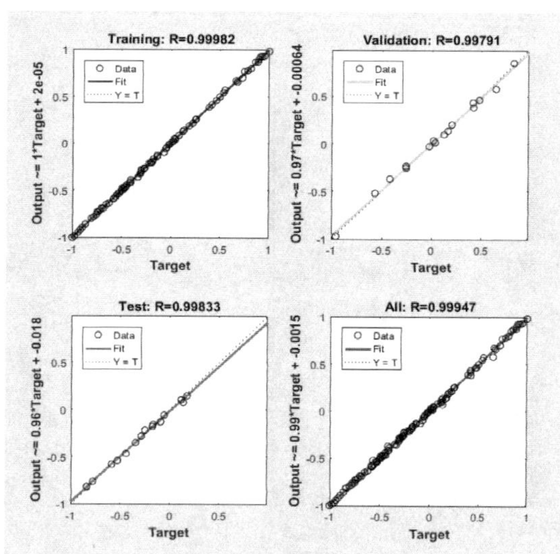

Figure 8. GA-BP neural network predicts thermal conductivity regression

4. Conclusions

Based on the experimental data in CuO-ZnO/(EG:DW) nanofluids with the same mass ratio (1:1), the CuO-ZnO mass fraction, temperature, and different base-liquid mixing ratios were used as input parameters, corresponding thermal conductivity is used as the output to apply the BP neural network optimized by genetic algorithm to construct the thermal conductivity prediction model. The results illustrate that the BP neural network optimized by genetic algorithm can establish a predictive model that can accurately predict the thermal conductivity, using 30% of the experimental data. Model verification, the predicted thermal conductivity maximum error limit is 2%, MSE is 0.000041656, and R2 is 0.99947. Compared with the unoptimized BP neural network, the genetic algorithm can not only solve the BP problem, but also improve the prediction accuracy. The neural network randomly selects the problem that the prediction result caused by the initial weight and threshold is unstable, and it can also give BP neural network the accuracy of the model by parallel random search optimal initial weight and threshold.

5. Appendices

Symbols	Units	meaning
R^2	1	Coefficient of determination
MSE	1	Mean Square Error Normalized value
X	1	Actual input value.
X_{nor}	1	Normalized value.
X_{max}	1	The maximum of the actual input value.
X_{min}	1	The minimum of the actual input value.
m	1	The number of nodes for the input layer.
n	1	The number of nodes for the output layer.
T_i	1	The actual value
P_i	1	The value of the model
$\overline{T}$	1	The average experimental value of n sets of test data.
T	℃	Temperature
R	1	ethylene glycol and deionized water at different mixing ratio
W	(W/(m·℃)	The mass fraction

Acknowledgments
I would like to thank my mentor Yuling Zhai for her careful guidance, my parents and family for their support, and all my brothers and sisters in the lab.

References

[1] Chein, R., Chuang, J. (2007) Experimental microchannel heat sink performance studies using nanofluids. Int. J. Therm. Sci., 46:57-66.
[2] Kleinstreuer, C., Feng, Y. (2011) Experimental and theoretical studies of nanofluid thermal conductivity enhancement: a review. Nanoscale Res. Lett., 6:1-13.
[3] Murshed, S.M.S., Leong, K.C., Yang, C. (2008) Thermophysical and electrokinetic properties of nanofluids-a critical review. Appl Therm Eng., 28:2109-2125.
[4] Wong, K.V., De Leon, O. (2010) Applications of nanofluids: current and future. Adv. Mech. Eng., 2:519659.
[5] Saidur, R., Leong, K.Y., Mohammad, H.A. (2011) A review on applications and challenges of nanofluids. Renew Sustain Energy Rev., 15:1646-1668.
[6] Harikrishnan, S., Roseline, A.A, Kalaiselvam, S. (2013) Preparation and Thermophysical Properties of Water-Glycerol Mixture-Based CuO Nanofluids as PCM for Cooling Applications. IEEE Trans. Nanotechnol., 12(4):629-635.
[7] Fang, Y.K., Osama, M., Rashmi, W., Shahbaz, K., Khalid, M., Mjalli, F.S., Farid, M.M. (2016) Synthesis and thermo-physical properties of deep eutectic solvent-based graphene nanofluids. Nanotechnology., 27(7):075702.
[8] Afrand, M., Toghraie, D., Sina, N. (2016) Experimental study on thermal conductivity of water-based Fe_3O_4 nanofluid: development of a new correlation and modeled by artificial neural network. Int. Commun. Heat Mass Transf., 75:262-269.

[9] Afrand, M., Najafabadi, K.N., Sina, N., Safaei, M.R., Kherbeet, A.S., Wongwises, S., Dahari, M. (2016) Prediction of dynamic viscosity of a hybrid nano-lubricant by an optimal artificial neural network. Int. Commun. Heat Mass Transf., 76:209-214.

[10] Kurt, H., Kayfeci, M. (2009) Prediction of thermal conductivity of ethylene glycol-water solutions by using artificial neural networks. App. Energy., 86(10):2244-2248.

[11] Esfe, M.H., Saedodin, S., Sina, N., Afrand, M., Rostami, S. (2015) Designing an artificial neural network to predict thermal conductivity and dynamic viscosity of ferromagnetic nanofluid, Int. Commun. Heat Mass Transf., 68:50-57.

[12] Zhao, N., Wen, X., Yang, J., Li, S., Wang, Z. (2015) Modeling and prediction of viscosity of water-based nanofluids by radial basis function neural networks, Powder Technol., 281:173-183.

[13] Esfe, M.H., Ahanga, M.R.H., Toghraie, D., Hajmohammad, M.H., Rostamian, H., Tourang, H., (2016) Designing artificial neural network on thermal conductivity of Al_2O_3/water-EG (60-40%) nanofluid using experimental data. J. Therm. Anal. Calorim., 126(2):1-7.

[14] Ahmadloo, E., Azizi, S. (2016) Prediction of thermal conductivity of various nanofluids using artificial neural network, Int. Commun. Heat Mass Transf., 74:69-75.

[15] Afrand, M., Najafabadi, K.N., Sina, N., Safaei, M.R., Kherbeet, A.S., Wongwises, S., Dahari, M. (2016) Prediction of dynamic viscosity of a hybrid nano-lubricant by an optimal artificial neural network, Int. Commun. Heat Mass Transf., 76:209-214.

[16] Esfe, M.H., Saedodin S., Naderi A., Karimipour A., Alirezaie A., Wongwises S., Goodarzi M., Dahari M. (2015) Modeling of thermal conductivity of ZnO-EG using experimental data and ANN methods. Int. Commun Heat Mass Transf., 63:35-40.

[17] Esfe, M.H., Wongwises, S., Naderi A., Asadi A., Safaei, R.M., Rostamian, H., Dahari M., Karimipour, A. (2015) Thermal conductivity of Cu/TiO_2-water/EG hybrid nanofluid: Experimental data and modeling using artificial neural network and correlation. Int Commun Heat Mass Transf., 66:100-104.

[18] Esfe M.H., Afrand, M., Wongwises, S., Naderi, A., Asadi, A., Rostami, S., Akbari M. (2015) Applications of feedforward multilayer perceptron artificial neural networks and empirical correlation for prediction of thermal conductivity of $Mg(OH)_2$/EG using experimental data. Int. Commun Heat Mass Transf., 67(4):46-50.

[19] Esfe, M.H., Saedodin, S., Sina, N., Afrand, M., Rostami, S. (2015) Designing an artificial neural network to predict thermal conductivity and dynamic viscosity of ferromagnetic nanofluid. Int.Commun. Heat Mass Transf., 2015, 68:50-57.

[20] Esfe, M.H., Rostamian, H., Afrand M, Karimipour A., Hassani, M. (2015) Modeling and estimation of thermal conductivity of MgO-water/EG (60:40) by artificial neural network and correlation. Int Commun. Heat Mass Transf., 63:35-40.

[21] Esfe, M.H., Tatar, A., Ahangar, M.R.H., Rostamian, H. (2018) A comparison of performance of several artificial intelligence methods for predicting the dynamic viscosity of TiO_2/SAE 50 nano-lubricant. Phys. E Low Dimens. Syst. Nanostruct., 96:85-93.

[22] Esfe, M.H., Naderi, A., Akbari, M., Afrand, M., Karimipour A. (2015) Evaluation of thermal conductivity of COOH-functionalized MWCNTs/water via temperature and solid volume fraction by using experimental data and ANN methods. J. Therm. Anal Calorim., 121(3):1273-1278.

[23] Esfe, M.H., Ahangar, M.R.H., Toghraie, D., Hajmohammad, M.H., Rostamian, H., Tourang H. (2016) Designing artificial neural network on thermal conductivity of Al_2O_3/water-EG (60-40%) nanofluid using experimental data. J. Therm. Anal Calorim., 126(2):1-7.

[24] Esfe, M.H., Esfandeh,S., Afrand, M. Rejvani, ,M., Rostamian, S.H. (2018) Experimental evaluation, new correlation proposing and ANN modeling of thermal properties of EG based hybrid nanofluid containing ZnO/DWCNT nanoparticles for internal combustion engines applications. Appl. Therm. Eng., 133:452-463.

The Range and Standards of Yang Dongfang Temporal Water Temperature Variation Angle Ⅰ. Model Calculation

Dongfang Yang[1, 2, a], Haixia Li[1], Haoyuan Ren[1], Sivakumar Manickam[3], Dong Yang[1]

[1]Accountancy School, Xijing University, Xi'an 710123, China;

[2]North China Sea Environmental Monitoring Center, SOA, Qingdao 266033, China;

[3]Department of Chemical and Environmental Engineering, Faculty of Science and Engineering, University of Nottingham Malaysia, Kuala Lumpur 43500, Malaysia

[a]dfyang_dfyang@126.com

Abstract: According to the investigation materials about water bodies of Jiaozhou Bay in May, August and October 1979, this paper studies the changing process of water temperature in the surface and bottom waters. The findings show that from May to August and then to October, water temperature in the waters from inside of bay mouth to the outside forms a peak line over the time. From May to August and then to October, water temperature changes from the lowest to the highest and then drops to a relatively higher level. Based on the definition and model of Yang Dongfang temporal water temperature variation angle, the results indicate that in the water bodies of inside, around and outside of bay mouth, the Yang Dongfang temporal water temperature variation angles in the surface layer range in 75.86-78.84°from May to August, and in -70.41--77.28°from August to October. While in the bottom layer, Yang Dongfang temporal water temperature variation angle ranges in 75.47-75.61°from May to August, and in -63.99--71.27°from August to October.

1. Introduction

Along with the rise of air temperature in spring, the quantity of heat heated by the sun is delivered to the seabed through water bodies. Similarly, along with the decline of air temperature in autumn, the quantity of cold cooled by the wind is delivered to the seabed through water bodies. Therefore, it is important to study the transmission process of heat in water bodies [1-10]. Based on the investigation materials about water temperature in Jiaozhou Bay in May, August and October 1979 and the definition and model of Yang Dongfang temporal water temperature variation angle, which provides a scientific basis for the research on vertical and horizontal transmission process in water bodies.

2. Investigation water fields, materials and methods

2.1 Natural environment in Jiaozhou Bay

Jiaozhou Bay is located in the south of Shandong Peninsula, ranging in 120°04′-120°23′E, 35°58′-36°18′N, bounded by Tuan Island and Xuejiao Island, and connecting to the Yellow Sea. With the area of 446km^2 and average depth of 7m, it is a typical semi-enclosed bay. There are more than 10 rivers flow to the sea, of which the Dagu River, the Yang River, and some rivers in the urban of Qingdao such as Haipo River, Licun River and Loushan River, are the rivers with large runoff and sediment concentration. These rivers are ephemeral streams with obviously seasonal hydrological characteristics [11, 12].

2.2 Materials and methods

The investigation materials about the water temperature in the water bodies of Jiaozhou Bay in May, August and October of 1979 applied in this study are offered by the North China Sea Environmental Monitoring Center of State Oceanic Administration. In May, August and October, setting three stations to take water samples from the surface and bottom respectively: H34, H35, H36 (as shown in figure 1). Based on the water depth to take samples: Taking from surface and bottom layer, where the depth > 10m; Taking from surface layer only, where the depth < 10m. This investigation method conforms to the national standards, and is recorded in *The Specification for Marine Monitoring (1991)* [13].

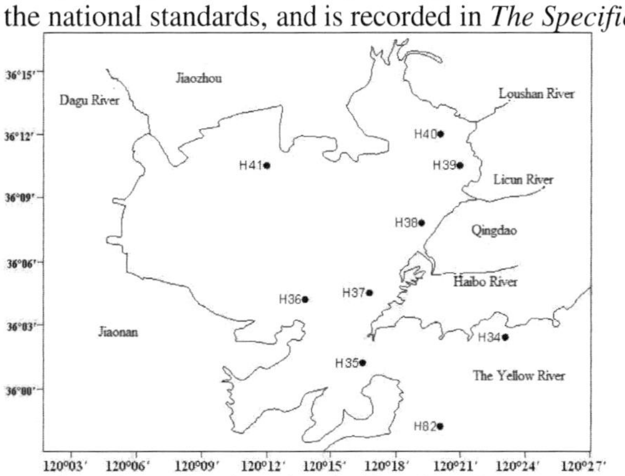

Fig.1 Investigation sites in Jiaozhou Bay

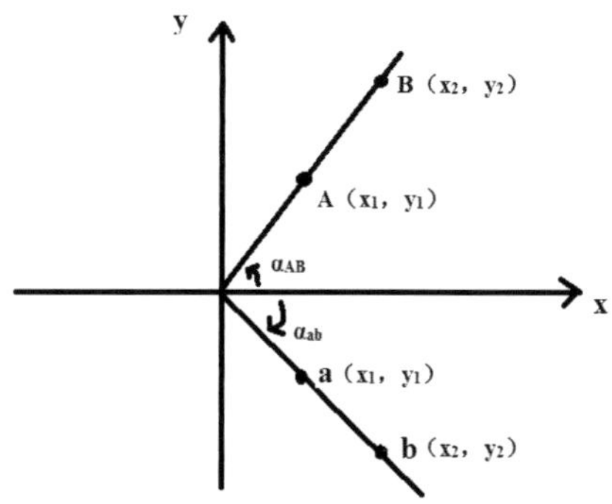

Fig.2 The model diagram of Dongfang Yang's temporal water temperature changing degree

3. Results and Discussion

3.1 The definition and model of Yang Dongfang temporal water temperature variation angle

Taking time as x-axis, water temperature as y-axis, forming an XOY plane. Space x changes from x_1 to x_2 and water temperature y changes from y_1 to y_2. Thus, on the plane XOY, the water temperature y changes along the straight line formed by the surface points A (x_1, y_1) and B (x_2, y_2), of which the slope is:

$$k_{AB}=(y_2-y_1)/|x_2-x_1| \tag{1}$$

The intersection angle made by the line and x-axis is:

$$\alpha_{AB}=\text{arctg } k_{AB} \tag{2}$$

It is called Yang Dongfang temporal water temperature variation angle, of which the range is $-90°<\alpha_{AB}<90°$.

3.2 The range and standards of Yang Dongfang temporal water temperature angle

Through the changing process of Yang Dongfang temporal water temperature variation angle (figure 2), this paper determines the changing degree of the water temperature varying with time and standards of variation degree.

When $0°<\alpha_{AB}<90°$, the water temperature rises with time. Meanwhile, the larger the Yang Dongfang temporal water temperature variation angle is in the interval (0°,90°), the faster the water temperature rises over time. The smaller the Yang Dongfang temporal water temperature variation angle is in the interval (0°,90°), the slower the water temperature rises over time.

When α_{AB} =45°, this angle is called standard rising angle of Yang Dongfang temporal water temperature. At this time, the water temperature rises in standard over time.

When $0°<\alpha_{AB}<30°$, this angle is called a slow rising angle of Yang Dongfang temporal water temperature. At this time, the water temperature rises slowly over time.

When $30°<\alpha_{AB}<45°$, this angle is called a slower rising angle of Yang Dongfang temporal water temperature. At this time, the water temperature rises slower over time.

When $45°<\alpha_{AB}<60°$, this angle is called a faster rising angle of Yang Dongfang temporal water temperature. At this time, the water temperature rises faster over time.

When $60°<\alpha_{AB}<90°$, this angle is called a fast rising angle of Yang Dongfang temporal water temperature. At this time, the water temperature rises fast over time.

When $-90°<\alpha_{AB}<0°$, the water temperature declines over time. In addition, the larger the Yang Dongfang temporal water temperature variation angle is in the interval (-90°, 0°), the slower the water temperature declines over time. The smaller the Yang Dongfang temporal water temperature variation angle is in the interval (0°,90°), the faster the water temperature declines over time.

When $\alpha_{AB}=-45°$, this angle is called standard drop angle of Yang Dongfang temporal water temperature. At this time, the water temperature declines in standard over time.

When $-30°<\alpha_{AB}<0°$, this angle is called slow drop angle of Yang Dongfang temporal water temperature. At this time, the water temperature declines very slowly over time.

When $-45°<\alpha_{AB}<-30°$, this angle is called slower drop angle of Yang Dongfang temporal water temperature. At this time, the water temperature declines slowly over time.

When $-60°<\alpha_{AB}<-45°$, this angle is called faster drop angle of Yang Dongfang temporal water temperature. At this time, the water temperature declines faster over time.

When $-90°<\alpha_{AB}<-60°$, this angle is called fast drop angle of Yang Dongfang temporal water temperature. At this time, the water temperature declines very fast over time.

When $\alpha_{AB}=0°$, the Yang Dongfang temporal water temperature variation angle is zero. At this time, the water temperature remains unchanged over time.

Thus, giving a quantitative description and standard on the change of water temperature varying over time when it ranges within $-90°<\alpha_{AB}<90°$.

3.3 The changes of water temperature in the surface layer

Station H36 locates inside waters of the bay mouth, station 35 in the waters of bay mouth, and H34 outside waters of bay mouth. In May, August and October, water temperature changes in both surface and bottom layers along with the changes of time (figure 1).

In May, the water temperature in the surface waters inside of bay mouth changes from the lowest figure 12.10℃, then rises gradually to the highest figure 27.12℃ in August. After that, it gradually declines to a relatively higher figure 18.25℃in October. Thus, the changing process of water temperature forms a peak line (figure 3). The waters temperature changes from lowest to highest then to a relatively high level in the period from May to August then to October.

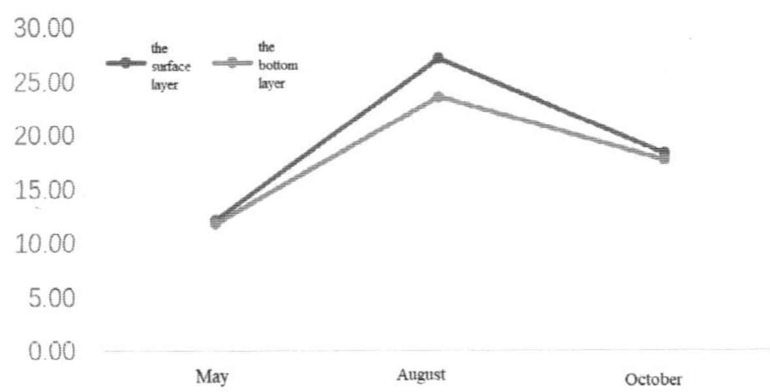

Fig.3 The change process of water temperature at the surface and bottom inside the bay mouth

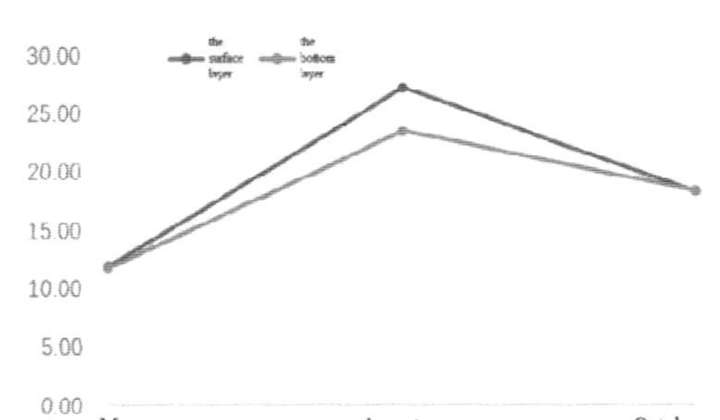

Fig.4 The change process of water temperature at the surface and bottom in the bay mouth

In the surface water bodies around bay mouth, the water temperature changes from the lowest figure 12.00℃in May, then rises gradually to the highest figure 27.21℃ in August. After that, it gradually declines to a relatively higher figure 18.38℃in October. Thus, the changing process of water temperature forms a peak line (figure 4). The waters temperature changes from lowest to highest then to a relatively high level in the period from May to August then to October.

In the surface water bodies outside of bay mouth, the water temperature changes from the lowest figure 12.10℃in May, then rises gradually to the highest figure 24.02℃ in August. After that, it gradually declines to a relatively higher figure 18.39℃in October. Thus, the changing process of water temperature forms a peak line (figure 5). The waters temperature changes from lowest to highest then to a relatively high level in the period from May to August then to October.

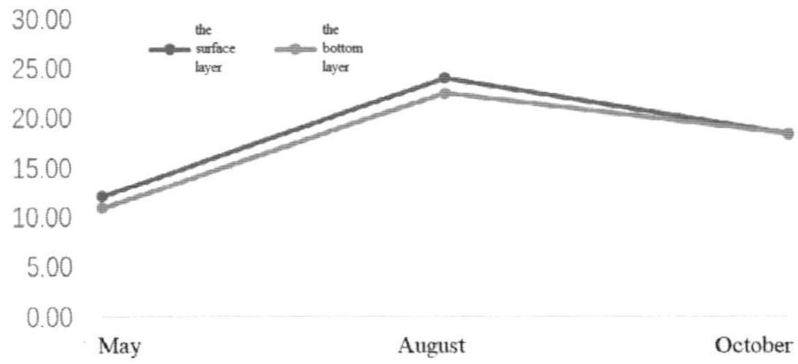

Fig.5 The change process of water temperature at the surface and bottom outside the bay mouth

3.4 The changes of water temperature in the bottom layer

Station H36 locates inside waters of the bay mouth, station 35 in the waters of bay mouth, and H34 outside waters of bay mouth. In May, August and October, water temperature changes in both surface and bottom layers along with the changes of time (figure 1).

In May, the water temperature in the bottom waters inside of bay mouth changes from the lowest figure 11.80℃, then rises gradually to the highest figure 23.52℃ in August. After that, it gradually declines to a relatively higher figure 17.62℃in October. Thus, the changing process of water temperature in the bottom layer forms a peak line (figure 3). The waters temperature changes from lowest to highest then to a relatively high level in the period from May to August then to October.

In the bottom water bodies around bay mouth, the water temperature changes from the lowest figure 11.80℃in May, then rises gradually to the highest figure 23.51℃ in August. After that, it gradually declines to a relatively higher figure 18.39℃in October. Thus, the changing process of water temperature forms a peak line (figure 4). The waters temperature changes from lowest to highest then to a relatively high level in the period from May to August then to October.

In the bottom water bodies outside of bay mouth, the water temperature changes from the lowest figure 10.90℃in May, then rises gradually to the highest figure 22.50℃ in August. After that, it gradually declines to a relatively higher figure 18.39℃in October. Thus, the changing process of water temperature forms a peak line (figure 5). The waters temperature changes from lowest to highest then to a relatively high level in the period from May to August then to October.

In the water bodies inside of bay mouth, around the bay mouth and outside of bay mouth, the water temperature in surface and bottom all forms a peak line from May to August and then to October. At the same time, the peak value in surface water bodies is greater than it in bottom water bodies. Meanwhile, the values of water temperature are almost same at the starting point of May and ending point of October.

3.5 Yang Dongfang temporal water temperature variation angle in the waters inside of bay mouth

Taking time as x-axis, and the water temperature in the waters inside of bay mouth as y-axis, the water temperature in the surface layer from May to August changes along with the line made by points A (5,12.10) and B (8,27.12). The slope of the line is k_{AB}=5.00 and the intersection angle made by the line and x-axis: α_{AB}= Yang Dongfang temporal water temperature variation angle =78.69°. From August to October, the water temperature changes along with the line made by points B (8,27.12) and C (10,18.25). The slope of the line is k_{BC}=-4.43 and the intersection angle made by the line and x-axis: α_{BC}= Yang Dongfang temporal water temperature variation angle = -77.28° (Tab. 1).

Through vertical transmission, water temperature passes through the surface water bodies to the bottom. Thus, the changes of water temperature in the bottom from May to August follow the line made by points a (5,11.80) and b (8,23.52), with the slope k_{ab}=3.90. The intersection angle angle made by the line and x-axis is α_{ab}=Yang Dongfang temporal water temperature variation angle = 75.61°. The changes of water temperature in the bottom from August to October follow the line made by points b (8,23.52) and c (10,17.62), with the slope k_{bc}=-2.95. The intersection angle angle made by the line and x-axis is α_{bc}=Yang Dongfang temporal water temperature variation angle = -71.27° (Tab. 1).

3.6 Yang Dongfang temporal water temperature variation angle in the waters around bay mouth

Taking time as x-axis, and the water temperature in the waters around bay mouth as y-axis, the water temperature in the surface layer from May to August changes along with the line made by points A (5,12.00) and B (8,27.21). The slope of the line is k_{AB}=5.07 and the intersection angle made by the line and x-axis: α_{AB}= Yang Dongfang temporal water temperature variation angle =78.84°. From August to October, the water temperature changes along with the line made by points B (8,27.21) and C (10,18.38). The slope of the line is k_{BC}=-4.41 and the intersection angle made by the line and x-axis: α_{BC}= Yang Dongfang temporal water temperature variation angle = -77.22° (Tab. 1).

Through vertical transmission, water temperature passes through the surface water bodies to the bottom. Thus, the changes of water temperature in the bottom from May to August follow the line made by points a (5,11.80) and b (8,23.51), with the slope k_{ab}=3.90. The intersection angle angle made by the line and x-axis is α_{ab}=Yang Dongfang temporal water temperature variation angle = 75.61°. The changes of water temperature in the bottom from August to October follow the line made by points b (8,23.51) and c (10,18.39), with the slope k_{bc}=-2.56. The intersection angle angle made by the line and x-axis is α_{bc}=Yang Dongfang temporal water temperature variation angle = -68.66° (Tab. 1).

3.7 Yang Dongfang temporal water temperature variation angle in the waters outside of bay mouth

Taking time as x-axis, and the water temperature in the waters outside of bay mouth as y-axis, the water temperature in the surface layer from May to August changes along with the line made by points A (5,12.10) and B (8,24.02). The slope of the line is k_{AB}=3.97 and the intersection angle made by the line and x-axis: α_{AB}= Yang Dongfang temporal water temperature variation angle =75.86°. From August to October, the water temperature changes along with the line made by points B (8,24.02) and C (10,18.39). The slope of the line is k_{BC}=-2.81 and the intersection angle made by the line and x-axis: α_{BC}= Yang Dongfang temporal water temperature variation angle = -70.41° (Tab. 1).

Through vertical transmission, water temperature passes through the surface water bodies to the bottom. Thus, the changes of water temperature in the bottom from May to August follow the line made by points a (5,10.90) and b (8,22.50), with the slope k_{ab}=3.86. The intersection angle angle made by the line and x-axis is α_{ab}=Yang Dongfang temporal water temperature variation angle = 75.47°. The changes of water temperature in the bottom from August to October follow the line made by points b (8,22.50) and c (10,18.39), with the slope k_{bc}=-2.05. The intersection angle angle made by the line and x-axis is α_{bc}=Yang Dongfang temporal water temperature variation angle = -63.99° (Tab. 1).

Table 1 Yang Dongfang temporal water temperature variation angle and degree

Water area	Water body	From May to August		Variation degree	From August to October		Variation degree
Waters inside of bay mouth	surface	k_{AB}=5.00	α_{AB}=78.69	fast rise angle	k_{BC}=-4.43	α_{BC}=-77.28	fast drop angle
	bottom	k_{ab}=3.90	α_{ab}=75.61	fast rise angle	k_{bc}=-2.95	α_{bc}=-71.27	fast drop angle
Waters around bay mouth	surface	k_{AB}=5.07	α_{AB}=78.84	fast rise angle	k_{BC}=-4.41	α_{BC}=-77.22	fast drop angle
	bottom	k_{ab}=3.90	α_{ab}=75.61	fast rise angle	k_{bc}=-2.56	α_{bc}=-68.66	fast drop angle
Waters	surface	k_{AB}=3.97	α_{AB}=75.86	fast rise	k_{BC}=-2.81	α_{BC}=-70.41	fast drop

outside of bay mouth				angle			angle
	bottom	k_{ab}=3.86	α_{ab}=75.47	fast rise angle	k_{bc}=-2.05	α_{bc}=-63.99	fast drop angle

4. Conclusion

In the waters inside of bay mouth to the around then to the outside, the water temperature in the surface and bottom waters forms a peak line from May to August then to October. Thus, from May to August then to October, the water temperature in the surface and bottom waters changes from the lowest to the highest then drops to a relatively high level.

From the waters inside of bay mouth to the around then to the outside, the water temperature in the surface and bottom layers both forms a peak line from May to August then to October. Meanwhile, the peak value in surface water bodies is greater than it in bottom water bodies. The values of water temperature are almost same at the starting point of May and ending point of October.

Taking time as x-axis, and the water temperature as y-axis, getting the calculation results based on the definition and model of Yang Dongfang temporal water temperature variation angle as follows:

In the surface waters inside of bay mouth, Yang Dongfang temporal water temperature variation angle is 78.69°from May to August, and -77.28°from August to October. In the bottom, Yang Dongfang temporal water temperature variation angle is 75.61°from May to August, and -71.27°from August to October.

In the surface waters around bay mouth, Yang Dongfang temporal water temperature variation angle is 78.84°from May to August, and -77.22°from August to October. In the bottom, Yang Dongfang temporal water temperature variation angle is 75.61°from May to August, and -68.66°from August to October.

In the surface waters outside of bay mouth, Yang Dongfang temporal water temperature variation angle is 75.86°from May to August, and -70.41°from August to October. In the bottom, Yang Dongfang temporal water temperature variation angle is 75.47°from May to August, and -63.99°from August to October.

From May to August, the water temperature in the surface waters inside, around and outside of bay mouth rises over time, and the corresponding Yang Dongfang temporal water temperature variation angle ranges in 75.86-78.84°. The water temperature in the bottom rises as well, of which the range of Yang Dongfang temporal water temperature variation angle is 75.47-75.61°.

From August to October, the water temperature in the surface waters inside, around and outside of bay mouth drops over time, and the corresponding Yang Dongfang temporal water temperature variation angle ranges in -70.41--77.28°. The water temperature in the bottom drops as well, of which the range of Yang Dongfang temporal water temperature variation angle is -63.99--71.27°.

References

[1] Teaching-Research Office in Marine Thermology, Shandong Institute of Oceanology, Oceanology [M]. Qingdao: Shandong Institute of Oceanology Press, 1980, 1-335.
[2] Ming Yang, Dongxing Xia, Dongqi Gu, et al. The evolution of geographical environment in coastal zone in Qingdao under the influence of global changes[J]. Advances in Marine Science, 2005, 23(3): 289-296
[3] Dongfang Yang, Zhenqing Miao. Gulf Ecology (1) [M]. Beijing: Maritime Press, 2010, 1-320.
[4] Dongfang Yang, Zhenqing Miao. Gulf Ecology (2) [M]. Beijing: Maritime Press, 2010, 1-330.
[5] Dongfang Yang, Zhenhui Gao, Peiyan Sun, et al.Silicon limitation on primary production and its destiny in Jiaozhou Bay, ChinaVI The ecological variation process of the phytoplankton [J]. Chin. J. Oceanol. Limnol. 2006, 24(2): 186-203.
[6] Dongfang Yang, Zhenhui Gao, Yingbin Yang, et al. Silicon limitation on primary production and its destiny in Jiaozhou Bay, ChinaVII The Complementary mechanism of the earth ecosystem [J]. Chin. J. Oceanol. Limnol. 2006, 24(4): 401-412.
[7] Dongfang Yang, Zhenhui Gao, Jie Qin, et al. Nutrient silicate complementary mechanism in earth

ecosystem. [J]. Advances in Marine Science, 2006, 24(4): 407-412.
[8] Dongfang Yang, Zhenhui Gao, Yuan Ma, et al. The impact of environment changes in Jiaozhou Bay on marine biological resources [J]. Marine Environmental Science, 2006, 25(4): 39-42.
[9] Dongfang Yang, Jianping Wu, Yanfeng Qu, et al. The complementary mechanism of air and water temperature in earth ecosystem [J]. Advances in Marine Science, 2007, 25(1): 117-122.
[10] Dongfang Yang, Zhenhui Gao, Wenfeng Li, et al. The influence of marine ecological changes on the climate and crop planting relationship research [J]. Journal of Modern Academic Research. 2007, 7: 104-109.
[11] Dongfang Yang, Yu Chen, Zhenhui Gao, et al. Silicon Limitation on primary production and its destiny in Jiaozhou Bay, China IV transect offshore the coast with estuaries [J]. Chin. J. Oceanol. Limnol. 2005, 23(1): 72-90.
[12] Dongfang Yang, Fan Wang, Zhenhui Gao, et al. The ecological phenomenon of floating algae in Jiaozhou Bay [J]. Marine Science, 2004, 28(6): 71-74.
[13] State Oceanic Administration. The specification for marine monitoring. Beijing: Maritime Press, 1991

IOP Conf. Series: Materials Science and Engineering **721** (2020) 012033 doi:10.1088/1757-899X/721/1/012033

The Range and Standards of Yang Dongfang Temporal Water Temperature Variation Angle II. Model Application

Dongfang Yang[1, 2, a], Haoyuan Ren[1], Dong Yang[1], Jing Fang[3], Haixia Li[1]

[1]Accountancy School, Xijing University, Xi'an 710123, China;

[2]North China Sea Environmental Monitoring Center, SOA, Qingdao 266033, China;

[3]College of Geography and Environmental Science, Tianjin Normal University, Tianjin 300387, China

[a]dfyang_dfyang@126.com

Abstract: According to the investigation materials about water bodies of Jiaozhou Bay in May, August and October 1979, this paper studied the changing process of water temperature in the surface and bottom waters. Based on the definition and model of Yang Dongfang temporal water temperature variation angle, the results indicated that in the water bodies of inside, around and outside of bay mouth, Viewing the waters inside of bay mouth, around the bay mouth and outside of bay mouth as one water body, the length of variation range of Yang Dongfang temporal water temperature variation angle is 2.98°when heating the surface of this water body, and 0.14°when heating the bottom. The length of variation range of this angle is 6.87°when cooling the surface and 7.28°when cooling the bottom, which displays the changing process of water temperature under the heating or cooling condition.

1. Introduction

Along with the rise of air temperature in spring, the quantity of heat heated by the sun is delivered to the seabed through water bodies. Similarly, along with the decline of air temperature in autumn, the quantity of cold cooled by the wind is delivered to the seabed through water bodies. Therefore, it is important to study the transmission process of heat in water bodies [1-10]. Based on the investigation materials about water temperature in Jiaozhou Bay in May, August and October 1979 and the definition and model of Yang Dongfang temporal water temperature variation angle, this paper exhibited the spatial and temporal changes of the source and transmission quantity of heat and cold energy in water bodies, which provided a scientific basis for the research on vertical and horizontal transmission process in water bodies.

2. Investigation water fields, materials and methods

2.1 Natural environment in Jiaozhou Bay

Jiaozhou Bay is located in the south of Shandong Peninsula, ranging in 120°04′-120°23′E, 35°58′-36°18′N, bounded by Tuan Island and Xuejiao Island, and connecting to the Yellow Sea. With the area of 446km^2 and average depth of 7m, it is a typical semi-enclosed bay. There are more than 10 rivers flow to the sea, of which the Dagu River, the Yang River, and some rivers in the urban of Qingdao such as Haipo River, Licun River and Loushan River, are the rivers with large runoff and sediment concentration. These rivers are ephemeral streams with obviously seasonal hydrological characteristics [11, 12].

2.2 Materials and methods

The investigation materials about the water temperature in the water bodies of Jiaozhou Bay in May, August and October of 1979 applied in this study are offered by the North China Sea Environmental Monitoring Center of State Oceanic Administration. In May, August and October, setting three stations to take water samples from the surface and bottom respectively: H34, H35, H36 (as shown in figure 1). Based on the water depth to take samples:

Taking from surface and bottom layer, where the depth > 10m;

Taking from surface layer only, where the depth < 10m.

This investigation method conforms to the national standards, and is recorded in *The Specification for Marine Monitoring (1991)* [13].

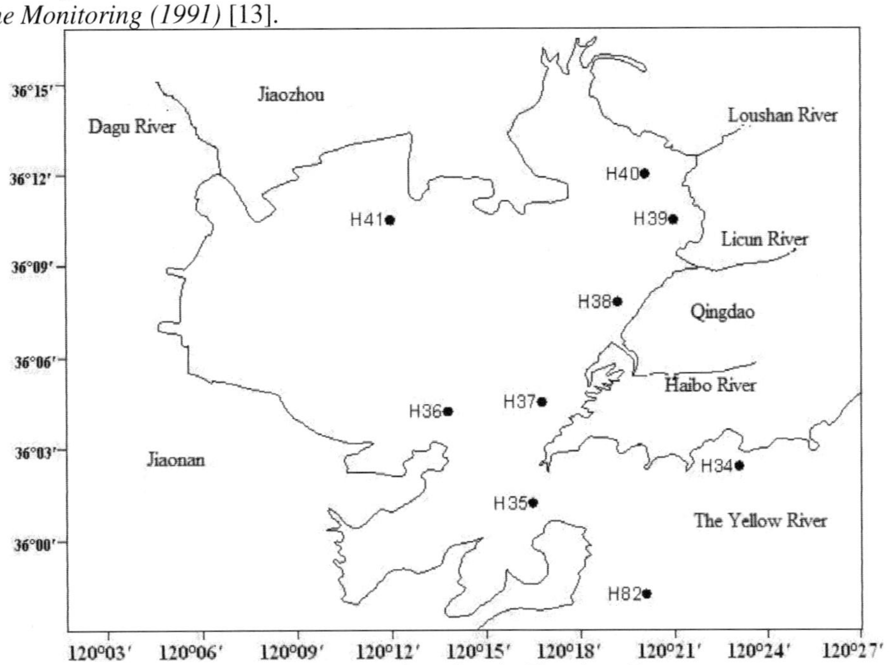

Fig.1 Investigation sites in Jiaozhou Bay

3. Results and Discussions

In May and August, it appears that the surface waters inside of the bay mouth transfer water heat to the marine waters around bay mouth of Jiaozhou Bay and that the waters outside of the bay mouth transfer water heat to the marine waters around bay mouth in October. Along with the changes of time, it is easily to quantify the rise or drop degree of water temperature on the basis of Yang Dongfang temporal water temperature variation angle proposed by the author, no matter the water temperature is rising or declining.

3.1 The rise and drop of water temperature in the waters inside of bay mouth

From May to August, water temperature in the surface rises and Yang Dongfang temporal water temperature angle is 78.69°. When $60° < \alpha_{AB} < 90°$, the angle is Yang Dongfang fast rise angle. At this time, the water temperature in the surface rises rapidly with the change of time. It indicates that the quantity of heat increases fast from May to August as the sun can heat the surface water directly.

From August to October, water temperature in the surface declines and Yang Dongfang temporal water temperature angle is -77.28°. When $-90° < \alpha_{AB} < -60°$, the angle is Yang Dongfang fast drop angle. At this time, the water temperature in the surface drops rapidly with the change of time. It indicates that the quantity of heat decreases fast from May to August under the cooling of wind.

From May to August, water temperature in the bottom rises and Yang Dongfang temporal water temperature angle is 75.61°. When $60°<\alpha_{AB}<90°$, the angle is Yang Dongfang fast rise angle. At this time, the water temperature in the bottom rises rapidly with the change of time. It indicates that the quantity of heat rises fast from May to August under the heating of sun.

From August to October, water temperature in the bottom declines and Yang Dongfang temporal water temperature angle is -71.27°. When $-90°<\alpha_{AB}<-60°$, the angle is Yang Dongfang fast drop angle. At this time, the water temperature in the bottom drops rapidly with the change of time. It indicates that the quantity of heat decreases fast from August to October under the cooling of wind.

3.2 The rise and drop of water temperature in the waters around bay mouth

From May to August, water temperature in the surface rises and Yang Dongfang temporal water temperature angle is 78.84°. When $60°<\alpha_{AB}<90°$, the angle is Yang Dongfang fast rise angle. At this time, the water temperature in the surface rises rapidly with the change of time. It indicates that the quantity of heat increases fast from May to August under the heating of sun.

From August to October, water temperature in the surface declines and Yang Dongfang temporal water temperature angle is -77.22°. When $-90°<\alpha_{AB}<-60°$, the angle is Yang Dongfang fast drop angle. At this time, the water temperature in the surface drops rapidly with the change of time. It indicates that the quantity of heat decreases fast from May to August under the cooling of wind.

From May to August, water temperature in the bottom rises and Yang Dongfang temporal water temperature angle is 75.61°. When $60°<\alpha_{AB}<90°$, the angle is Yang Dongfang fast rise angle. At this time, the water temperature in the bottom rises rapidly with the change of time. It indicates that the quantity of heat rises fast from May to August under the heating of sun.

From August to October, water temperature in the bottom declines and Yang Dongfang temporal water temperature angle is -68.66°. When $-90°<\alpha_{AB}<-60°$, the angle is Yang Dongfang fast drop angle. At this time, the water temperature in the bottom drops rapidly with the change of time. It indicates that the quantity of heat decreases fast from August to October under the cooling of wind.

3.3 The rise and drop of water temperature in the waters outside of bay mouth

From May to August, water temperature in the surface rises and Yang Dongfang temporal water temperature angle is 75.86°. When $60°<\alpha_{AB}<90°$, the angle is Yang Dongfang fast rise angle. At this time, the water temperature in the surface rises rapidly with the change of time. It indicates that the quantity of heat increases fast from May to August under the heating of sun.

From August to October, water temperature in the surface declines and Yang Dongfang temporal water temperature angle is -70.41°. When $-90°<\alpha_{AB}<-60°$, the angle is Yang Dongfang fast drop angle. At this time, the water temperature in the surface drops rapidly with the change of time. It indicates that the quantity of heat decreases fast from May to August under the cooling of wind.

From May to August, water temperature in the bottom rises and Yang Dongfang temporal water temperature angle is 75.47°. When $60°<\alpha_{AB}<90°$, the angle is Yang Dongfang fast rise angle. At this time, the water temperature in the bottom rises rapidly with the change of time. It indicates that the quantity of heat rises fast from May to August under the heating of sun.

From August to October, water temperature in the bottom declines and Yang Dongfang temporal water temperature angle is -63.99°. When $-90°<\alpha_{AB}<-60°$, the angle is Yang Dongfang fast drop angle. At this time, the water temperature in the bottom drops rapidly with the change of time. It indicates that the quantity of heat decreases fast from August to October under the cooling of wind.

3.4 The rise of water temperature from May to August

From May to August, the water temperature in the surface waters inside, around and outside of bay mouth rises and the corresponding Yang Dongfang temporal water temperature variation ranges in 75.86-78.84°, which belongs to Yang Dongfang fast rise angle. In this period, the water temperature in the surface rises fast over time. Meanwhile, the water temperature in the bottom also rises rapidly, and the Yang Dongfang temporal water temperature variation angle ranges within 75.47-75.61°, which is a

Yang Dongfang temporal water temperature variation angle too. The water temperature in the bottom rises fast as the time changes.

The range of Yang Dongfang temporal water temperature variation angle in the surface layer is greater than the range in the bottom, which indicates that the water temperature rises much faster in the surface than in the bottom. The range interval of the water temperature variation angle in the surface is 2.98°while 0.14°in the bottom, which indicates that the rising degree of water temperature in the surface layer is greater than it in the bottom layer. That is, the differences of rising degree of water temperature in the surface waters inside, around and outside of bay mouth is greater than it in the bottom waters. It reveals that the rising degree of water temperature is different in the surface waters when the sun heat different water areas at the same time. However, when the heat is delivered to the bottom waters through the water bodies, the water temperature rising degree is consistent in the bottom.

3.5 The drop of water temperature from August to October

From August to October, the water temperature in the surface waters inside, around and outside of bay mouth drops and the corresponding Yang Dongfang temporal water temperature variation ranges in -70.41--77.28°, which belongs to Yang Dongfang fast drop angle. In this period, the water temperature in the surface drops fast over time. Meanwhile, the water temperature in the bottom also drops rapidly, and the Yang Dongfang temporal water temperature variation angle ranges within -63.99--71.27°, which is a Yang Dongfang temporal water temperature variation angle too. The water temperature in the bottom drops fast as the time changes.

The range of Yang Dongfang temporal water temperature variation angle in the surface layer is greater than the range in the bottom, which indicates that the water temperature drops much faster in the surface than in the bottom. The range interval of the water temperature variation angle in the surface is 6.87°while 7.28°in the bottom, which indicates that the drop degree of water temperature in the surface layer is smaller than it in the bottom layer. That is, the differences of drop degree of water temperature in the surface waters inside, around and outside of bay mouth is smaller than it in the bottom waters. It reveals that the drop degree of water temperature is different in the surface waters when the wind cools different water areas at the same time. However, when the cold energy is delivered to the bottom waters through the water bodies, the changing degree of water temperature is much great in the bottom.

3.6 Water temperature variation process in water bodies

When heating or cooling the surface of water bodies, water temperature will rise or decline. Through the effect of vertical water bodies, the eddy turbulence in the vertical direction will result in the heat exchange between surface and bottom waters, which will lead to the rise or decline of the water temperature in the bottom.

When heating the surface of a water body, the variation interval of Yang Dongfang temporal water temperature variation angle is 2.98°in the surface layer, while it is 0.14°in the bottom. It shows that the changing degree of the rising of water temperature in the surface is greater than it in the bottom. In the surface, the sun heats various water bodies at the same time while the rising degree of water temperature is relatively small. In the bottom, heat is delivered to the bottom through water bodies so the rising degree of water temperature is much smaller than it in the surface.

When cooling the surface of a water body, the variation interval of Yang Dongfang temporal water temperature variation angle is 6.87° in the surface layer, while it is 7.28°in the bottom. It shows that the drop degree of water temperature in the surface is smaller than it in the bottom. In the surface, the wind cools various water bodies at the same time where the drop degree of water temperature is relatively large. In the bottom, the cooling energy is delivered to the bottom through water bodies where the drop of water temperature is much greater than it in the surface.

4. Conclusion

Taking time as x-axis, and the water temperature as y-axis, getting the calculation results based on the definition and model of Yang Dongfang temporal water temperature variation angle as follows:

Viewing the water bodies inside of bay mouth, around the bay mouth and outside of bay mouth as a whole, when heating the water, the range interval of the Yang Dongfang temporal water temperature variation angle in the surface is 2.98° while 0.14°in the bottom. When cooling the water, the range interval of the Yang Dongfang temporal water temperature variation angle in the surface is 6.87° while 7.28°in the bottom, which displays the changing process of water temperature in water bodies.

Acknowledgement

This research was sponsored by Doctoral Degree Construction Library of Guizhou Nationalities University and Research Projects of Guizhou Nationalities University ([2014]02), Research Projects of Guizhou Province Ministry of Education (KY [2014] 266), Research Projects of Guizhou Province Ministry of Science and Technology (LH [2014] 7376).

References

[1] Teaching-Research Office in Marine Thermology, Shandong Institute of Oceanology, Oceanology [M]. Qingdao: Shandong Institute of Oceanology Press, 1980, 1-335.

[2] Ming Yang, Dongxing Xia, Dongqi Gu, Zuo Xue, Aiping Feng. The evolution of geographical environment in coastal zone in Qingdao under the influence of global changes[J]. Advances in Marine Science, 2005, 23(3): 289-296

[3] Dongfang Yang, Zhenqing Miao. Gulf Ecology (Volume 1) [M]. Beijing: Maritime Press, 2010, 1-320.

[4] Dongfang Yang, Zhenqing Miao. Gulf Ecology (Volume 2) [M]. Beijing: Maritime Press, 2010, 1-330.

[5] Dongfang Yang, Zhenhui Gao, Peiyan Sun, Sheng Zhao, Youchi Zhang. Silicon limitation on primary production and its destiny in Jiaozhou Bay, ChinaVI The ecological variation process of the phytoplankton [J]. Chin. J. Oceanol. Limnol. 2006, 24(2): 186-203.

[6] Dongfang Yang, Zhenhui Gao, Yingbin Yang, Peiyan Sun, Xingping Wang. Silicon limitation on primary production and its destiny in Jiaozhou Bay, ChinaVII The Complementary mechanism of the earth ecosystem [J]. Chin. J. Oceanol. Limnol. 2006, 24(4): 401-412.

[7] Dongfang Yang, Zhenhui Gao, Jie Qin, Suxia Huo, Zhongqiang Li. Nutrient silicate complementary mechanism in earth ecosystem. [J]. Advances in Marine Science, 2006, 24(4): 407-412.

[8] Dongfang Yang, Zhenhui Gao, Yuan Ma, Peiyan Sun, Y Yingbinang. The impact of environment changes in Jiaozhou Bay on marine biological resources [J]. Marine Environmental Science, 2006, 25(4): 39-42.

[9] Dongfang Yang, Jianping Wu, Yanfeng Qu, Jun Hu, Yanrong Zhou. The complementary mechanism of air and water temperature in earth ecosystem [J]. Advances in Marine Science, 2007, 25(1): 117-122.

[10] Dongfang Yang, Zhenhui Gao, Wenfeng Li, Zijiang Yu, Changjiang Huang. The influence of marine ecological changes on the climate and crop planting relationship research [J]. Journal of Modern Academic Research. 2007, 7: 104-109.

[11] D F YANG, Y CHEN, Z H GAO, et al. Silicon Limitation on primary production and its destiny in Jiaozhou Bay, China IV transect offshore the coast with estuaries [J]. Chin. J. Oceanol. Limnol. 2005, 23(1): 72-90.

[12] Dongfang Yang, Fan Wang, Zhenhui Gao, et al. The ecological phenomenon of floating algae in Jiaozhou Bay [J]. Marine Science, 2004, 28(6): 71—74.

[13] State Oceanic Administration. The specification for marine monitoring. Beijing: Maritime Press, 1991

IOP Conf. Series: Materials Science and Engineering **721** (2020) 012034 doi:10.1088/1757-899X/721/1/012034

Thermal stress analysis of printed circuit heat exchanger based on thermal-structural coupling method

Jun Wu[1*], Shuo Yao[1], Zhenxing Zhao[1], Qi Xiao[1], Zhiwu Ke[1], Yuansheng Lin[1,2]

[1]Science and Technology on Thermal Energy and Power Laboratory, Wuhan Second Ship Design and Research Institute. Wuhan, Hubei Province, 430205, China

[2]School of Energy and Power Engineering, Xi'an Jiaotong University, Xi'an, Shaanxi Province, 710049, China

[*]Corresponding author's e-mail: wu_jun1986@126.com

Abstract. The supercritical carbon dioxide (S-CO_2) Brayton cycle has the advantages of high efficiency, compactness and simplicity. It is expected to become the main technical route for the future advanced ship nuclear power platform. The printed circuit heat exchanger (PCHE) with ultra-high pressure, ultra-compact and ultra-high efficiency typical characteristics is the ideal configuration of the S-CO_2 power generation system recuperator. The PCHE numerical analysis model is constructed and the thermal stress field of the PCHE core structure is analyzed by thermal-structural coupling method. The results show that the maximum thermal stress of the PCHE core structure is 117 MPa, which is less than the yield limit of the material and meets the requirement of strength safety. The thermal stress value of most regions in PCHE is less than 78.2MPa. Therefore, the current design scheme has a large safety margin. The PCHE compactness can be further improved by structural parameter optimization.

1. Introduction

Limited to narrow spaces and complex working environments, ship nuclear power platforms impose stringent design requirements for heat exchangers with high efficiency, compactness, high reliability and low vibration. The PCHE has a fine passage of 0.5 mm to 4 mm. Under the same exchanging heat quantity, the weight and volume of the PCHE are only 1/5 and 1/8 of the conventional shell-and-tube heat exchange r[1]. PCHE has significant advantages such as high pressure resistance, corrosion resistance, large temperature zone span, long design life and economy [2-3]. It is an advanced heat exchanger with great potential for future application in the ship nuclear power platform S-CO_2 power generation system.

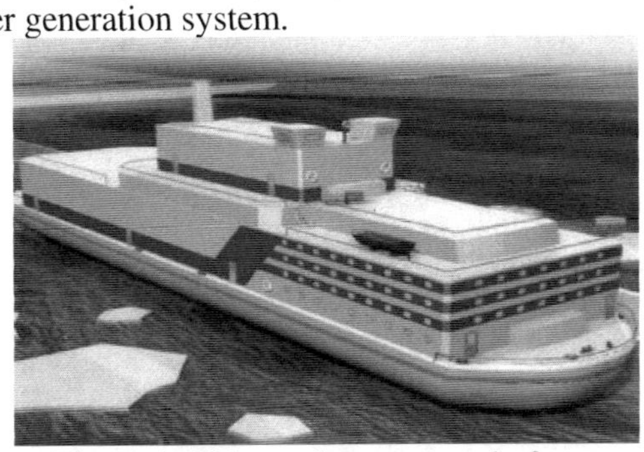

Figure 1. Ship nuclear power platform.

PCHE has been adopted in the recuperator and cooler of the SANDIA National Laboratory's 100-kilowatt S-CO_2 cycle test system. The recuperator power is 2.16 times that of the electric heater, which verifies the efficiency and compactness of PCHE [4]. The Korea Energy Research Institute has built Two S-CO_2 cycle test platform are established with the power rating of kilowatts and 10 kilowatt class respectively. The PCHE with a power rating of 600 kW is adopted to heat exchanger and cooler [5]. At present, the British company Heatric has mastered the PCHE manufacturing process. PCHE has been used in natural gas liquefaction/gasification plants for refrigeration cryogenic, petrochemical, onshore and offshore fixed platforms [6]. The PCHE design process is mainly composed of flow heat transfer design process and structural design process. Most of the current research focuses on flow heat transfer design, including PCHE flow heat transfer correlation [7], PCHE micro channel geometry design [8], PCHE flow rate distribution characteristics[9] and so on. Relatively conservative empirical values are adopted to the structural geometric parameters for the external solid part and the adjacent heat exchange channels of the PCHE to ensure structural strength safety requirements. The effect of the load caused by the fluid flow and heat transfer on the structural strength safety is not considered. It is thus difficult to quantitatively evaluate the structural safety of the PCHE design scheme. **X Q Zhang** perform the PCHE structural finite element numerical analysis by the single-section 3D solid channel model [10]. The effects of different working fluid pressure loads on the structural stress and deformation of the channel are studied. However, it is difficult to consider the coupling effect of structural deformation between multi-layer channels in single-section solid model. It is also difficult to reflect the influence of the actual structural boundary conditions of PCHE on the structural deformation field and stress field.

In this paper, the PCHE thermal-structural coupling numerical model of S-CO_2 Brayton cycle power generation system is established. The deformation and stress distribution of PCHE structure under the combined action of thermal load and pressure load are studied. The work of this paper provides technical support for the further development of PCHE structural design optimization and experimental verification research in the future.

2. PCHE scheme design for S-CO_2 recuperator

2.1 Design input

The S-CO_2 simple regenerative Brayton cycle system is adopted as the research background. According to the results of system heat balance analysis, the thermal parameters on the system key nodes are extracted as PCHE design inputs for the recuperator.

Table 1. Design input parameters of the PCHE.

Number	Variable	Unit	Value
1	Hot side/Cold side working fluid	—	S-CO_2
2	Hot side inlet pressure	MPa	7.96
3	Hot side inlet temperature	°C	440.08
4	Hot side outlet temperature	°C	95.32
5	Cold side inlet pressure	MPa	14.10
6	Cold side inlet temperature	°C	75.35
7	Cold side outlet temperature	°C	359.34
8	Hot side/Cold side flow rate	kg/s	0.015
9	Channel type	—	Semicircular straight channel

2.2Design result of the PCHE

According to the PCHE design input parameters, the temperature difference between the inlet and outlet of the S-CO_2 recuperator is large and the physical properties of the recuperator vary greatly.

However, the rationality of the heat transfer performance estimation using the traditional logarithmic mean temperature difference method is open to question. Therefore, the self-developed PCHE design program based on distributed calculation method is adopted in the PCHE design process in this paper. The PCHE design program performs heat transfer and resistance calculation based on the Gnielinski correlation. And the PCHE is calculated step by step by length. The physical properties of the hot side and cold side in each step range are updated according to the inlet temperature and pressure by the NIST REFPROP software. The temperature difference is calculated by the inlet hot side and cold side temperature in each step. Moreover, the local heat transfer coefficient, heat exchanging quantity, and pressure drop in each step are calculated. The temperature and pressure of the outlet in the step range are obtained as the inlet conditions for the next step calculation. Finally, the calculation is terminated when the heat exchanging quantity reaches the rated heat exchanging quantity. Compared with the traditional logarithmic mean temperature difference method, the distribution calculation method uses the temperature distribution along the path instead of the inlet and outlet conditions to calculate the temperature difference adopted in the heat exchange process is more reasonable.

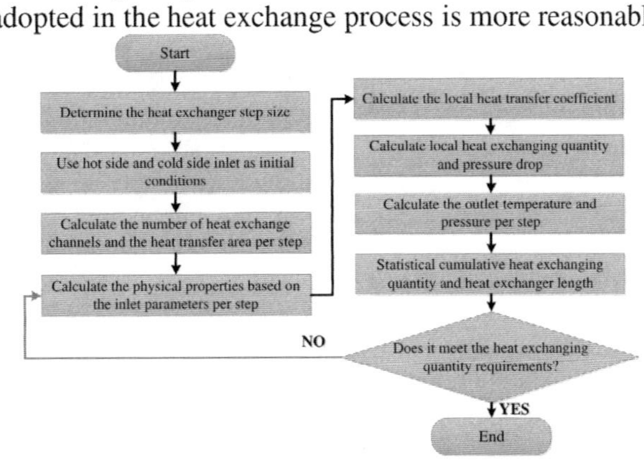

Figure 2. Design program calculation process of the PCHE.

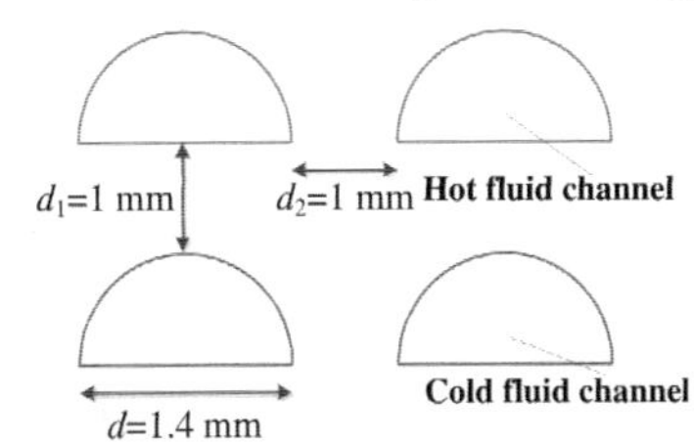

Figure 3. Micro heat exchange channel geometry size of the PCHE.

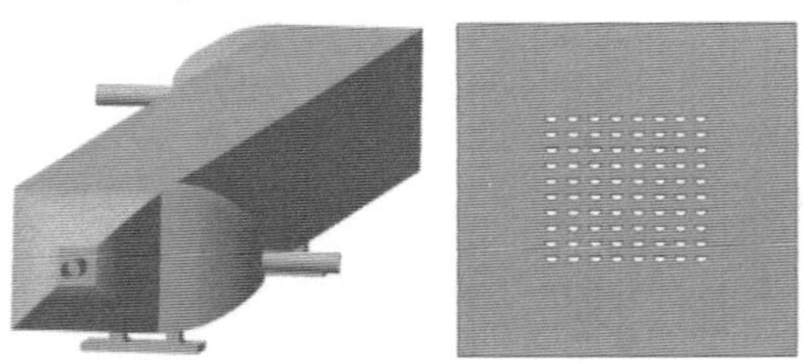

Figure 4. PCHE three-dimensional model and the micro channel cross section.

Table 2. Design results of the PCHE.

Number	Variable	Unit	Value
1	Channel diameter	mm	1.4
2	Heat exchange zone length	mm	713
3	Lateral spacing of adjacent holes	mm	1
4	Longitudinal spacing of adjacent holes	mm	1
5	Number of hot side and cold side plates	layer	10
6	External solid part thickness of the PCHE	mm	20
7	Number of horizontal micro channels	—	8

The S-CO_2 recuperator design result is obtained using the self-developed PCHE design program. Figure 4 shows the PCHE three-dimensional model and the fine channel section of the recuperator. Figure 5 and Figure 6 show the S-CO_2 temperature and heat transfer coefficient distribution along the PCHE channel. It can be seen that the heat transfer coefficient in the longitudinal direction of the hot side channel fluctuates slightly around 2000 W/(m^2•K). In contrast, the heat transfer coefficient of the cold side channel fluctuates greatly along the length direction. The heat transfer coefficient decreases from 3500 W/(m^2•K) at the inlet to 2200 W/(m^2•K) at the outlet. Since the physical property state at the inlet of the cold side channel is close to the critical point region, the drastically changing physical properties have a great influence on the heat transfer coefficient. Moreover, the total pressure drop on the hot side channel is 0.014 MPa, and the total pressure drop on the cold side channel is 0.008 MPa.

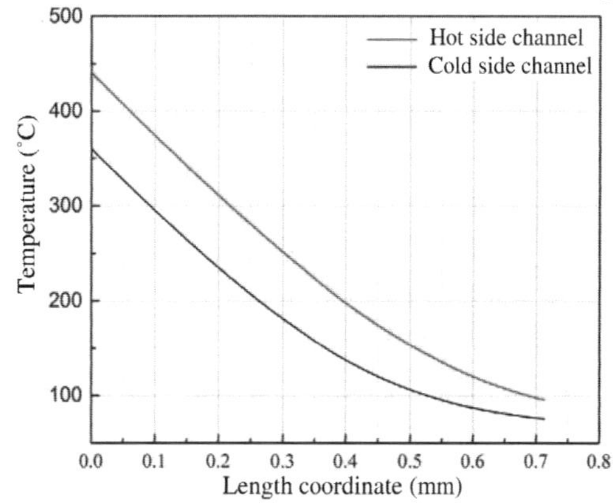

Figure 5. Temperature distribution of S-CO_2 working fluid in PCHE channel.

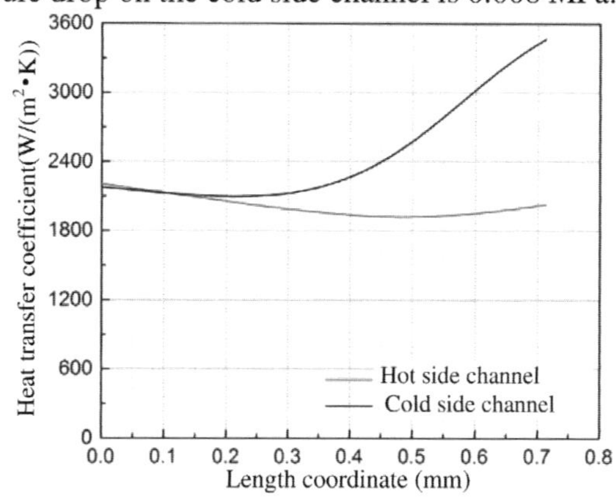

Figure 6. Heat transfer coefficient distribution of S-CO_2 working fluid in PCHE channel

3. Thermal stress analysis of PCHE

3.1Thermal-structural coupling analysis process of PCHE

Under the temperature load of S-CO_2 working fluid, the thermal expansion degree for each local structural region of PCHE is inconsistent. This results in the generation of thermal stress in the PCHE structure. Therefore, the thermal stress problem of the PCHE structure is derived from the interaction between two physical fields of heat and stress, which is a typical coupled field analysis problem. Based on the large-scale finite element analysis software ANSYS, combined with the parameterized programming language APDL, the sequential coupling method is applied to conduct the thermal-structural coupling analysis of PCHE. The analysis process of is shown in Figure 7.

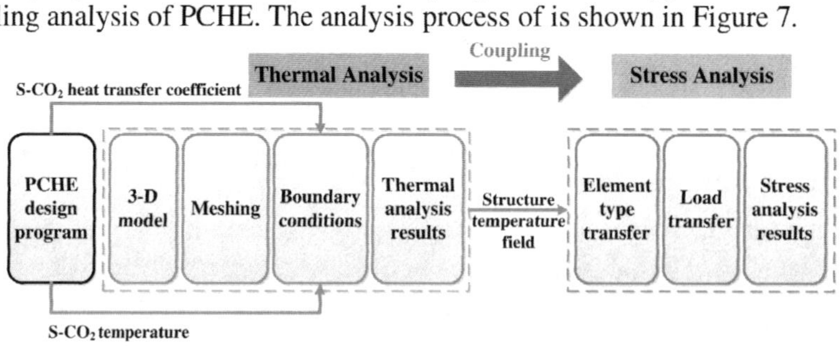

Figure 7. PCHE thermal-structural coupling numerical analysis process

3.2. 3-D computational geometry model and meshing

The structure is simplified according to the design result of PCHE. The core structure and support structure are extracted to construct the 3-D calculation model of PCHE thermal analysis.

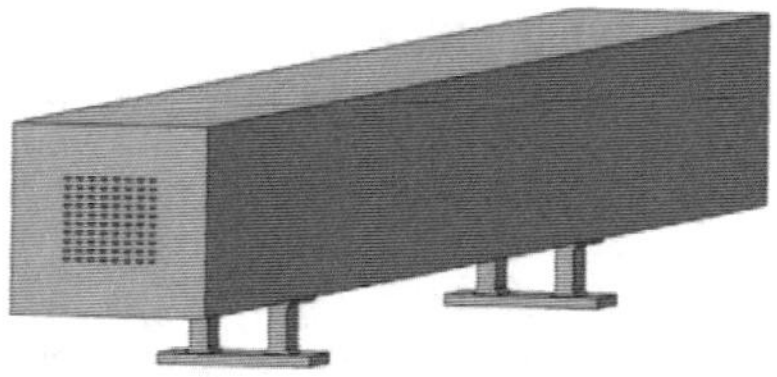

Figure 8. Construction of PCHE 3D Computing Model.

In this paper, the core heat exchange channel region is divided by relatively dense three-dimensional hexahedral solid elements. Other region is divided by relatively sparse three-dimensional hexahedral solid elements. The interfaces between the two regions use the coupling connection to transmit the degree of freedom information, which not only ensures the calculation accuracy but also the calculation efficiency. The total number of nodes of the calculation model is 1.68 million, and the total number of elements is 1.55 million.

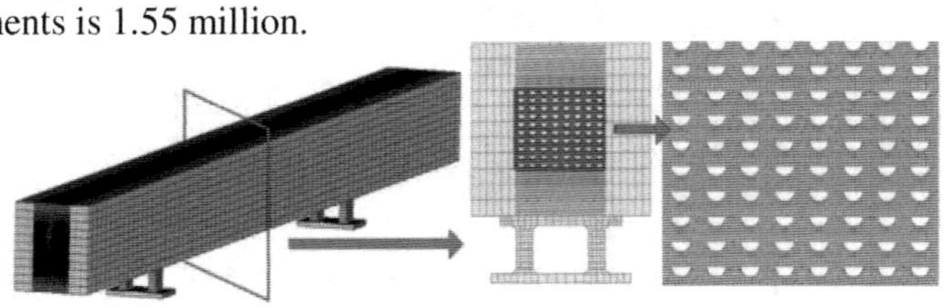

Figure 9. PCHE structure meshing.

3.3Thermal analysis

The PCHE material performance parameters are shown in Table 3. The material type of PCHE is austenitic stainless steel 06Cr17Ni12Mo2 (S31608).

Table 3. Material parameters of PCHE.

Number	Variable	Unit	Value
1	Material type	—	06Cr17Ni12Mo2（S31608）
2	Density	kg/m^3	8000
3	Elastic Modulus	GPa	169
4	Poisson's ratio	—	0.3
5	Linear expansion coefficient	°C^{-1}	18.5×10-6
6	Thermal Conductivity	W/(m• °C)	21.5
7	Yield Strength	MPa	121

According to the distribution result of S-CO_2 temperature and heat transfer coefficient in micro channel obtained by PCHE design scheme, corresponding boundary conditions are applied to the calculation model.

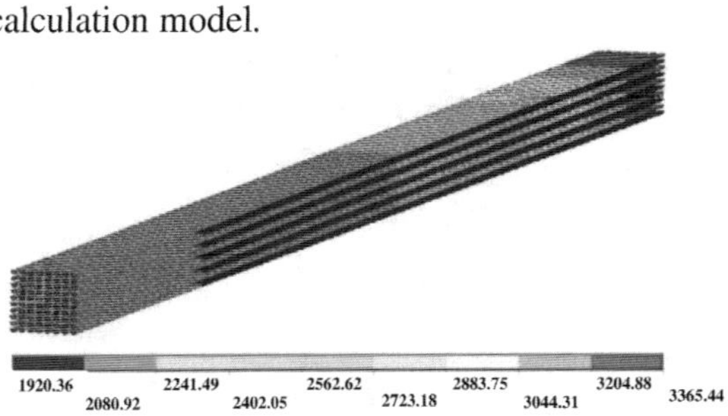

Figure 10. Heat transfer coefficient distribution in PCHE channels (W/(m^2•°C)).

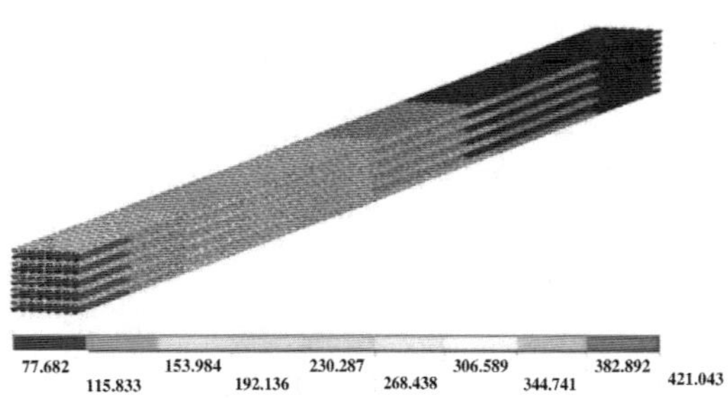

Figure 11. Temperature distribution in PCHE channels (°C).

1) Based on the heat transfer coefficient and the temperature distribution curve shown in figure 5 and figure 6, the heat transfer coefficient value and the temperature value at the local region of the micro channel are calculated by the linear difference method. This process is automatically conducted by APDL programming. Therefore, the convection boundary condition in the PCHE channel is applied.

2) The actual PCHE will be laid on the outside of the insulation layer. As a result, the adiabatic boundary condition is applied to the outer surface of the PCHE calculation model. That is, the heat flux density value is zero.

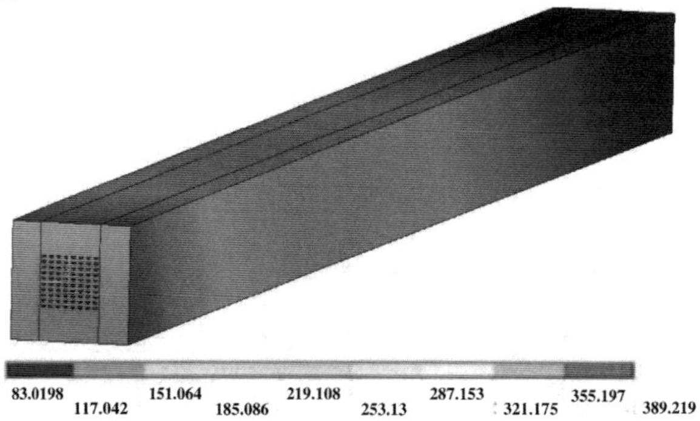

Figure 12. PCHE structure temperature field distribution (°C).

The temperature field distribution of the PCHE structure is obtained by steady state thermal analysis. Figure 12 shows that the PCHE structure temperature value gradually decreases along the X direction. The temperature distribution on the cross section of the same X coordinate is relatively uniform. The temperature field of the structure is basically consistent with the temperature distribution of the S-CO_2 working fluid in the micro channel. The maximum temperature and minimum temperature values of the PCHE structure are located on the inlet and outlet end faces of the hot channel respectively. The values are 389 °C and 83 °C respectively.

3.4Structural stress analysis

The temperature field distribution of the PCHE structure is obtained by thermal analysis above. On this basis, the structural stress analysis is carried out. The following boundary conditions are specifically applied.

1) First of all, the thermal analysis element is converted into the structural analysis element. The temperature field calculation result obtained by thermal analysis is read and loaded into the PCHE calculation model as a body load.

2) Secondly, degrees of freedom constraints in the XYZ and YZ directions are applied to the bottom nodes of the two support structures.

3) Finally, the pressure loads at the corresponding inlets are applied to the nodes of the cold and hot channel internal surfaces respectively. The values are 14.10 MPa and 7.96 MPa respectively.

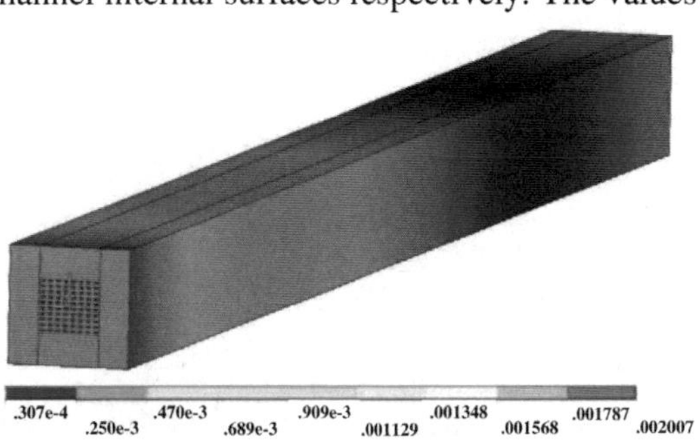

Figure 13. Total displacement distribution of the PCHE structure (m).

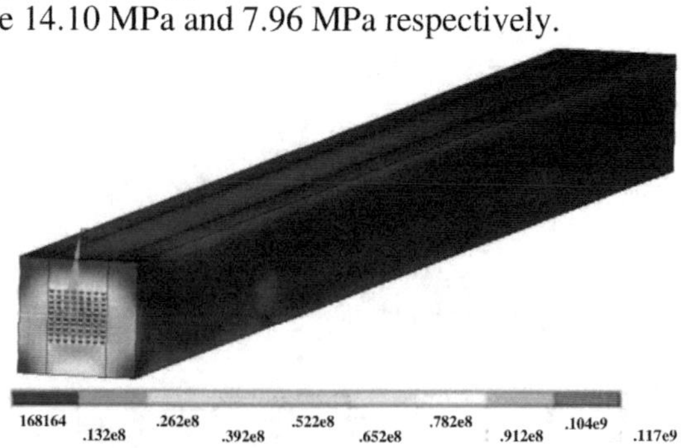

Figure 14. Von-Misses stress distribution of the PCHE structure (Pa).

The thermal deformation and thermal stress results of the PCHE structure under the combined action of temperature load and pressure load are obtained by structural static analysis. It can be seen from figure 13 that the maximum total displacement value of the PCHE structure is 2 mm, which is located at the inlet surface of the hot channel. It can be seen from figure 14 that the maximum equivalent stress of PCHE is 117 MPa, which is also located at the sharp corner of the first layer on the inlet surface of the hot side. As a result, the maximum equivalent stress of the PCHE structure is less than the material yield limit and meets the strength safety requirements. It can be observed that the equivalent stress value of the vast area in the PCHE structure is less than 78.2 MPa. This value is much smaller than the material yield limit. The margin of the structural size value of the current PCHE solid area is relatively large. Therefore, the structure size parameters can be further optimized to effectively improve the PCHE compactness.

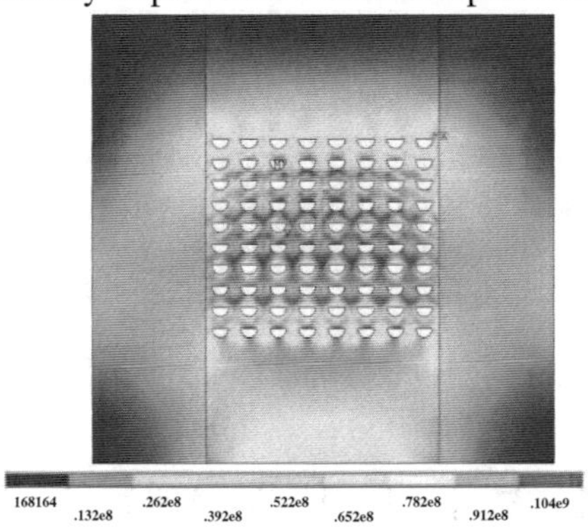

Figure 15. Von-Misses stress distribution of the hot side inlet cross section (Pa).

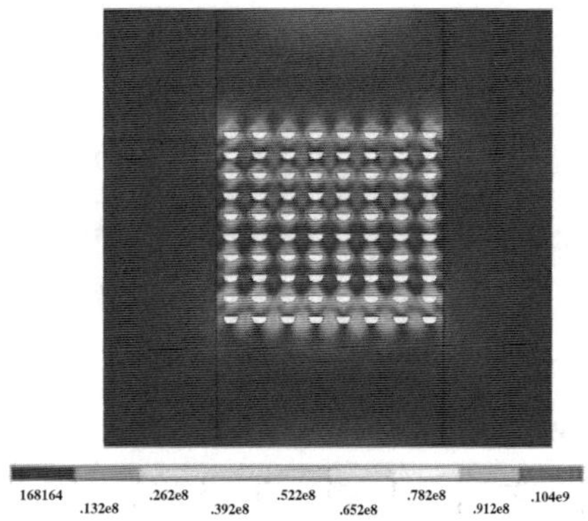

Figure 16. Von-Misses stress distribution of the cool side inlet cross section (Pa).

It can be seen from figure 15 and figure 16 that the stress distribution of the heat exchange channel near the central region in the cross section of the PCHE has a certain degree of similarity and periodicity. Under the influence of structural boundary conditions, the stress distribution of the heat exchange channel near the outer region is significantly different from the stress distribution of the heat exchange channel near the central region. However, the maximum stress value of the PCHE structure often appears at the location of the heat exchange channel near the outer region. The three-dimensional global structure model used in this paper can avoid the calculation error caused by the local structural model containing only partial periodic heat transfer channels. Therefore, a complete and reliable PCHE thermal stress calculation result is obtained.

4. Conclusions

In this paper, the S-CO_2 recuperator design scheme is obtained by the self-developed PCHE design program. Based on the ANSYS software combined with the parametric programming language APDL, the sequential coupling method is applied to perform the thermal-structural coupling numerical analysis of PCHE. The main conclusions are as follows:

1) The maximum total displacement and the maximum equivalent stress of the PCHE structure are 2 mm and 117 MPa respectively. The positions are both located at the inlet surface of the hot side channel. The maximum equivalent stress of the PCHE structure is less than the material yield limit and meets the strength safety requirement.

2) The equivalent stress of most regions in PCHE is much smaller than the yield limit of the material. The current PCHE structure design has a large margin. Therefore, the PCHE compactness can be further improved by the optimized design of size parameters.

3) The three-dimensional global structure model used in this paper can avoid the calculation error caused by the local structural model containing only partial periodic heat transfer channels. As a result, a complete and reliable PCHE thermal stress calculation result is obtained.

Acknowledgments
This research was financially supported by the Hubei Province Natural Science Foundation of China (NO. 2016CFA019, NO. 2017CFB325).

References

[1] Chang, W.S., Hee, C.N. (2017) Experimental study for pressure drop and flow instability of two-phase flow in the PCHE-type steam generator for SMRs. Nuclear Engineering and Design, 318: 109-118.
[2] Ma, T., Li L., Xu X.Y. (2015) Study on local thermal-hydraulic performance and optimization of zigzag-type printed circuit heat exchanger at high temperature. Energy Conversion and Management, 104:55-65.
[3] Bartel, N., Chen, M., Utgikar, V. P. (2015) Comparative analysis of compact heat exchangers for application as the intermediate heat exchanger for advanced nuclear reactors. Annals of Nuclear Energy, 81: 143-149.
[4] Conboy, T., Wright, S.A., Fuller, R. (2012) Performance characteristics of an operating supercritical CO_2 Brayton cycle. In: ASME Turbo Expo 2012. Copenhagen. pp. 415-420.
[5] Junhyun, C., Hyungki, S., Ho, S.R. (2016) Development of the supercritical carbon dioxide power cycle experimental loop in KIER. In: ASME Turbo Expo 2016. Seoul. pp. 460-467.
[6] Huang, Y.P., Wang, J.F. (2012) Applications of supercritical carbon dioxide in nuclear reactor system. Nuclear Power Engineering, 33(3): 21-27.
[7] Huang, Y.P., Liu S.H., Liu G.X. (2016) Evaluation and analysis of forced convection heat transfer correlations for supercritical carbon dioxide in tubes. Nuclear Power Engineering, 37(1): 28-33.
[8] Liu, S.H., Huang, Y.P., Lang, X.M. (2016) Geometry design and printed circuit heat exchanger based on flow and heat transfer correlation. Nuclear Power Engineering, 37(3): 106-109.
[9] Ke, H.B., Xiao, Q., Lin, Y.H. (2017) Research on flow rate assignment characteristic of printed circuit heat exchanger. In: Proceeding of Chinese Society of Engineering Thermophysics on Heat and Mass Transfer. Yangzhou. pp. 478-482.
[10] Zhang, X.Q., Shi, S.B., Richard, N.C. (2017) Review on mechanical design of printed circuit heat exchangers. In: ASME ICONE25. Shanghai. pp. 284-291.

IOP Conf. Series: Materials Science and Engineering **721** (2020) 012035 doi:10.1088/1757-899X/721/1/012035

Research and Analysis of Energy Conservation for organic heat medium heater

Yan Liu[1*], Yanlin Xi[2] and Zhaohui Liang[3]

[1]Thermal Engineering (Heater) Department, Beijing Aerospace Petrochemical Technology and Equipment Engineering Co.Ltd, Beijing, 100176,China

[2]Thermal Engineering (Heater) Department, Beijing Aerospace Petrochemical Technology and Equipment Engineering Co.Ltd, Beijing, 100176,China

[3]Thermal Engineering (Heater) Department, Beijing Aerospace Petrochemical Technology and Equipment Engineering Co.Ltd, Beijing, 100176,China

Corresponding author's e-mail: Liuy6@ calt11.cn

Abstract: Organic heat medium heater is defined as boiler of B grade,C grade and D grade as per heat load and is inspected by local government. Organic heat medium heater is widely used in finery factory and chemical plant, and a large amount of energy is consumed by heater. Increasing efficiency is important for energy conservation and saving operation cost. Method of saving energy is discussed and method how to operate safely and effectively is also discussed.

1. Introduction

Organic heat medium heaters are widely used in petrochemical industry. Even in the process, multiple organic heat medium heater operate in parallel, which is an important energy-consuming unit in factories. Generally, the thermal efficiency is used to measure the heater energy effective utilization. The parameter is an important index for evaluating fuel consumption, i.e. economy, design level and operation level. Organic heat medium heater belongs to the scope of government supervision, and there are strict assessment requirements for energy consumption. For organic heat medium heater that burns gas loads greater than 1.4MW, it is required that the exhaust temperature should be less than 170 °C and the efficiency should be greater than 90% [1]. If the fuel is natural gas, the efficiency requirement is 92%. Improving thermal efficiency not only meets the environmental needs of energy saving and emission reduction, but also can effectively reduce the operating costs of users and improve the operation economy of organic heat medium heater.

2. Theoretical Basis

There are two methods for calculating and measuring thermal efficiency: positive balance method and negative balance method, which are usually calculated and measured by negative balance method.

The Formula for Calculating the Thermal Efficiency of negative balance method [2]:

$$\eta = 100 - q_1 - q_2 - q_3 - q_4$$

q_1 – Heat loss of exhaust flue gas， %

q_2 – Heat Loss from Chemical Incomplete Combustion， %

q_3 – Heat Loss from Mechanical Incomplete Combustion， %

IOP Conf. Series: Materials Science and Engineering **721** (2020) 012035 doi:10.1088/1757-899X/721/1/012035

q_4 – Dissipation heat Loss, %

For oil and gas burns, $q_2 + q_3 = 0$. Heat efficiency can be improved by reducing heat loss of exhaust flue gas and dissipation heat loss Energy saving and emission reduction can start from these two aspects.

3. Energy Saving Measures

According to the conclusion drawn from the calculation formula of the negative balance, the energy saving measures of organic heat medium heater are also carried out from two aspects: reducing the heat loss of exhaust flue gas and reducing heat dissipation loss.

3.1 Energy saving is achieved by reducing heat loss of exhaust flue gas

There are two ways to reduce the heat loss of exhaust flue gas: first, to reduce the temperature of exhaust flue gas; second, to reduce the quantity of exhaust flue gas. There are several ways to reduce the exhaust temperature.

3.1.1Air preheater can be used to reduce the exhaust temperature.

The combustion air of the organic heat medium heater and the high-temperature flue gas at the outlet of the heater take part in the burning after heat exchange, and the flue gas temperature after heat exchange is lowered to the required temperature and discharged. This is the most common heat recovery method. The commonly used air preheaters are tubular type, heat pipe type air preheater and plate type air preheater. Theoretically speaking, the exhaust flue gas temperature can be reduced to tens of degrees. However, in practice, infinite reduction of exhaust flue gas temperature will lead to low temperature water dew point and acid dew point corrosion of equipment, and high combustion air temperature will lead to increase emission of nitrogen oxides. Therefore, it is not advisable to reduce the exhaust temperature indefinitely. Generally, the range of exhaust flue gas temperature is 110 ~ 170 C, and bypass is needed for combustion air to adjust the exhaust flue gas temperature according to the sulfur content of fuel during operation. In recent years, enameled tubes or ND steel have been widely used in the cold end of tubular air preheaters in order to solve dew point corrosion in low temperature section of equipment.

3.1.2The exhaust flue gas temperature can be reduced by using a flue gas waste heat boiler or a hot oil heat exchanger.

The high-temperature flue gas at the outlet of the heater exchanges heat with steam or low-temperature hot oil, and the flue gas is discharged after the temperature is lowered to the design temperature. If the flue gas can not reach the designed flue gas temperature after the waste heat boiler or the hot oil heat exchanger, a air preheater can be added to further reduce the flue gas temperature.

3.1.3 Clean heat exchanger tubes

Because of the serious fouling of heat exchanger tubes for a long time, the fouling of heat exchanger tubes should be cleaned up during the maintenance period to ensure the efficiency of heat exchange.

Generally speaking, under the same excess air, the thermal efficiency will decrease by about 1% [3] for every increase of the exhaust flue gas temperature of the heater by 15-20℃. This is an empirical estimation. The relationship between the exhaust temperature and the thermal efficiency needs to be determined by calculation. Taking a 2.5 million Kcal/h organic heat medium heater which has been operated in a project as an example, the excess air coefficient is 1.15 at this time. From the data collected in Table 2, it can be seen that the heat loss of exhaust flue gas increases by 0.6%~0.7% for every 13-14℃ increase of flue gas

IOP Conf. Series: Materials Science and Engineering **721** (2020) 012035 doi:10.1088/1757-899X/721/1/012035

Table 1. Fuel Gas Composition

Composition	V%
N2	1.19
CO2	0.62
CH4	94.34
C2H6	3.00
C3H8	0.85

Table 2. Comparing table of calculation results of exhaust temperature and heat loss

exhaust temperature℃	137	146	160	174.5	187	200
Heat Loss of Exhaust Flue gas	5.73	6.12%	6.79%	7.49%	8.09%	8.72%

Because the design calculation usually has a margin, and the field operation conditions often deviate from the design value. The following two tables are the summary tables of fuel conditions and energy efficiency test results measured in site.

Table 3. Fuel gas Composition in Site

Composition	V%
N2	1.717
CO2	0.547
CH4	93.151
C2H6	3.591
C3H8	0.634
Unsaturated hydrocarbons	0.36

Table 4. Summary of Energy Efficiency Test Results

Operation Load	250MMKcal/h
Excess air	1.14
Exhaust Temperature	137℃
Heat Loss of exhaust flue gas	5.56%
Dissipation Loss	2%
Efficiency	92.44%

Reducing the quantity of flue gas can also reduce the heat loss of exhausted flue gas. The direct way to reduce flue gas quantity is to reduce excess air, that is, to reduce excess air as much as possible under the condition of ensuring full combustion of fuel. This not only can achieve the purpose of energy saving, but also can effectively prevent the formation of nitrogen oxides. In other words, in order to achieve the same thermal efficiency, if the flue gas quantity is small, the exhaust temperature can be increased appropriately; if the excess air can not be effectively controlled and the flue gas quantity is increased, then the high thermal efficiency can only reduce the exhaust temperature, and the low exhaust temperature will cause the corrosion of the equipment. The control of excess air reflects the operation level of operators. Experienced on-site operators will adjust the air distribution according to the on-site residual oxygen analyzer. Some organic heat medium heater are designed with high-level automatic control system, which can automatically adjust the air distribution. Therefore, effective control of excess air is conducive to reducing the heat loss of exhaust flue gas. Table 5 is a comparison of energy efficiency test results of 1000 million Kcal/h organic heat medium heater running on site.

Table 5. Comparison of Energy Efficiency Test Results

Operation Heat Load	750 MMKcal/h	750 MMKcal/h
Excess air	1.15	1.71
Exhaust Temperature	148℃	110.5℃
Heat Loss of Exhaust Flue Gas	6.01%	6.02%
Dissipation Heat Loss	2%	2%
Efficiency	91.99%	91.98%

From the comparison data, it can be seen that reducing the excess air t can make the equipment run economically and safely. Therefore, site operators should also improve their own operation level, adjust excess air according to the site operation data, and improve the thermal efficiency of the equipment.

3.2 Energy saving by reducing heat loss

Organic heat medium heater generally reduces heat loss by setting thermal insulation layer. Generally, the calculation of insulation layer thickness must satisfy the requirements of no wind and ambient temperature of 27 ℃. The outer wall of radiation section, convection section and flue gas duct should not exceed 80 ℃, and the outer surface temperature of bottom should not exceed 90 ℃ [4]. Usually, the heat loss of fuel based on low calorific value ranges from 1.5% to 3%. The heat loss of dissipation is mainly caused by the convection heat transfer between the outer wall and the environment. The wind speed and temperature of the environment have a certain influence on the heat dissipation. When the wind speed is high, the convective heat transfer coefficient increases, but the temperature difference of heat transfer also decreases, so the total heat dissipation does not change much. Generally, the efficiency of newly built heaters is calculated according to the loss of heat dissipation of 2%. If the old heaters is reconstructed, air leakage and damaged insulation layer need to be repaired to reduce heat loss. The thermal insulation material should adopt the material with lower thermal conductivity as far as possible. If the material with lower thermal conductivity can not meet the requirement of heat temperature, composite lining material can be chosen to reduce the thermal conductivity of the material as far as possible.

4. Economic Analysis

The higher the thermal efficiency of the organic heat medium heater is, the more energy-saving it will be and the lower the operating cost will be. Moreover, the larger the load, the higher the thermal efficiency of the heater will bring more considerable benefits. For example, an organic heat medium heater with a design load of 10 million Kcal/h consumes natural gas with a design efficiency of 91%. The annual consumption of natural gas is 10720986Nm3, calculated by an annual operation of 8000 hours. If energy-saving renovation is carried out, the efficiency will be increased to 92%, and the annual consumption of natural gas will be 10604454 Nm3. The annual saving of natural gas is 116532 Nm3, and the price of natural gas is calculated at 4 yuan/Nm3, so the annual saving of operation cost is about 466000 yuan.

IOP Conf. Series: Materials Science and Engineering **721** (2020) 012035 doi:10.1088/1757-899X/721/1/012035

5. Conclusion

Reducing the exhaust temperature can improve the thermal efficiency. But the sulfur content in the fuel restricts the exhaust temperature. In order to ensure the safe and long-term operation of the equipment, the exhaust flue gas temperature must be reasonably calculated according to the sulfur content of the fuel to prevent acid dew point corrosion. In addition to considering dew point corrosion, too much sulfide is not allowed to be discharged into the atmosphere, so the fuel is best desulfurized.

Using advanced automatic control system control excess air and improve thermal efficiency. It is also very important to improve the operation level of site operators.

During the overhaul of the equipment, the dirt should be removed as far as possible and the damaged insulation should be repaired in time.

References:

[1] Guan,J.(2010)Supervision Administration Regulation on Energy Conservation Technology for Boiler. Xin Hua Publishing House, Beijing.
[2] Qian, J.L. (2007) Tubular Heater (Rev.2). China Petrochemical Press, Beijing.
[3] Li,Z.h. (1997) . Research on Methods to Reduce Flue Gas Temperature and Low Temperature Corrosion. Energy Conservation Technology, Harbin,15（3）：41~43
[4] Sun,Y. (2012)Fired Heaters for General Refinery Service. China petrochemical Press, Beijing.
[5] Guo, Y.L. (2013) Boiler Safety Technical Supervision Administration Regulation Explanation. Chemical Industry Press, Beijing.

IOP Conf. Series: Materials Science and Engineering **721** (2020) 012036 doi:10.1088/1757-899X/721/1/012036

Operation Performance and Economic Efficiency Analysis of Power Plant Boilers Ultra-low Emission Facilities

Hai Wang*

Center for Special Equipment Supervisal and Inspection in Jingdezhen City, Jiangxi, 333000, China

*E-Mail: wanghaijdz@qq.com

Abstract. In this paper, four units of a power plant are taken as the research object, and the air pollutant emission performance standards, equipment installed economic efficiency, operational economic efficiency and energy saving potential of four units using different technical methods for ultra-low emission transformation are analyzed. Take unit 4 as an example, the ultra-low emission operation cost is discussed. The results show that the operation performance, stability and reliability of some units of low-temperature economizer need to be improved; the K values of different ultra-low emission technology methods were between 7.6~12kW/MW. The installation of a low temperature economizer allows the wet desulfurization facility to have a greater water saving capacity.

1. Introduction

Coal, accounted for 64% China's total energy consumption in 2015, is a basic energy source in China in the next long period of time [1]. From the perspective of safety and economic efficiency, coal has the irreplaceable advantage compared with other energy sources. Using coal for thermal power generation and industrial boiler combustion is a major part of China's coal consumption. The pollution generated during coal combustion is the key issue in coal utilization, which is an important source of emissions affecting environmental quality, and ultra-low emission of coal-fired pollutants will play a positive role in reducing the total amount of pollutants discharged, improving the ecological environment and boosting the efficiency of coal clean utilization [2-4].

With the upgrading of boiler air pollutant control technology, especially the promotion of ultra-low emission technology, the technical selection of pollutant control and its economic analysis of construction and operation are still insufficient. Economic analysis of boiler air pollutant control technology is of great significance to the selection and evaluation of environmental protection process routes [5-6].

Based on the research on the operating parameters and energy consumption data of coal-fired power station boilers, this paper compares and analyzes the performance indicators and energy consumption data of different transformation technology methods, studies the economic efficiency of different process routes and the potential of energy saving and consumption reduction, and provide reference for the selection of process routes and improve the economics of system operation.

2. Operational evaluation of ultra-low emission environmental protection facilities

2.1 Performance compliance analysis

2.1.1 Analysis of the standard of dry electric dust removal

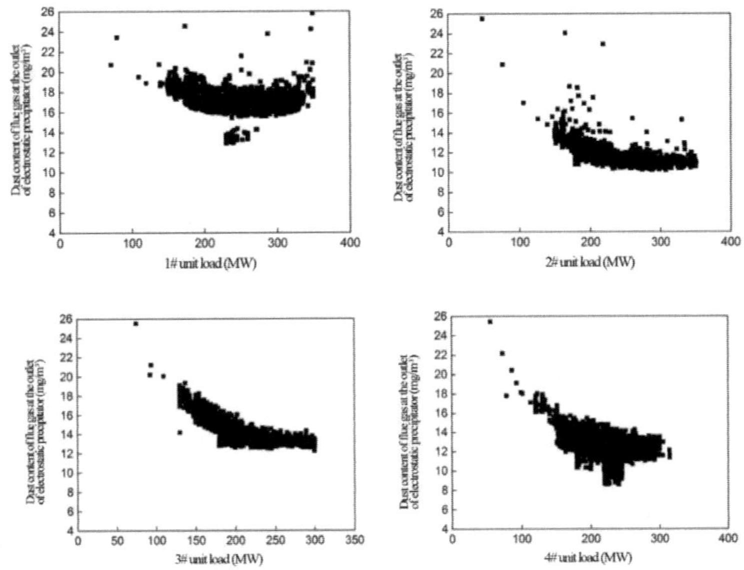

Fig. 1 Dust emission indicators of electrostatic precipitators in each unit

Figure 1 shows the dust concentration at the outlet of the electrostatic precipitator (desulfurization tower inlet) under different load conditions. After the high-frequency power supply is modified, the outlet concentration of the electrostatic precipitators of each unit is stable to reach the standard of < 20mg/Nm^3. The average concentration of smoke and dust at the outlet of the four units of electrostatic precipitators was 17.1/11.5/13.3/12.1mg/m^3, as shown in Figure 2.

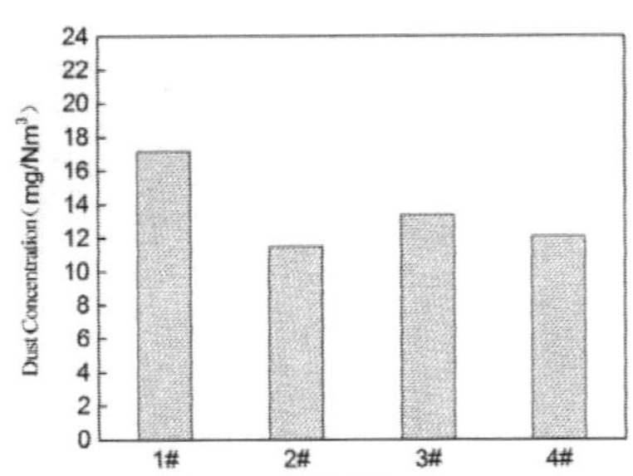

Fig. 2 The average value of the outlet smoke and dust concentration of the electrostatic precipitator after the ultra-low emission reform

Figure 3 shows the inlet flue temperature of the desulfurization tower under different loads after the low-temperature economizer is put into operation. After the economizer is put into operation, the flue temperature drops to 100 °C ~ 130 °C, and the system temperature of the 1# unit, 2# and 3# is about 100 °C, and unit 4 # system temperature is between 90 °C to 130 °C, which have the effect of reducing the temperature of the smoke. 1# unit, 2# system temperature is relatively stable, 3# unit, 4# system temperature fluctuation is large. Especially 4# unit, which performance stability needs to be improved.

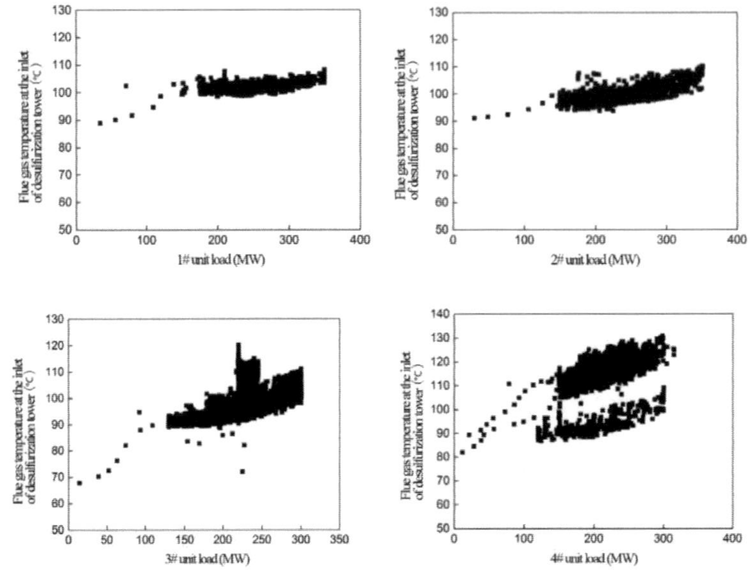

Fig. 3 Outlet temperature of electrostatic precipitator after ultra-low emission retrofit

2.1.2 SO_2 emission standard analysis

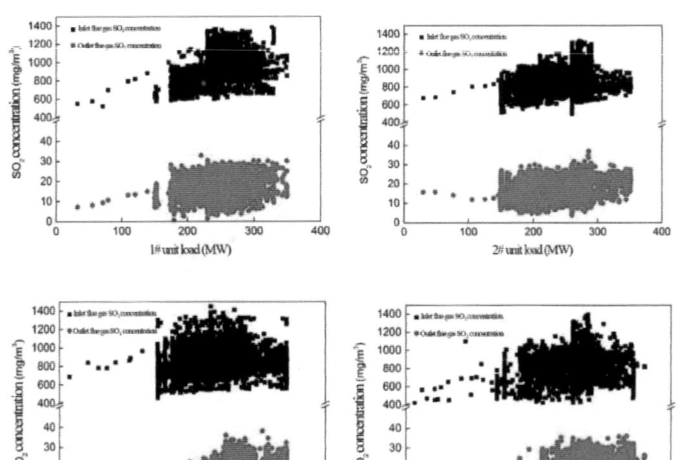

Fig. 4 SO_2 concentration at the inlet and outlet of each unit desulfurization tower

Figure 4 shows the SO_2 concentration at the inlet and outlet of four units. It can be seen from the figure that the SO_2 concentration at the outlet of the desulfurization tower is stable to reach the emission standard of 35mg/Nm^3. The above four technical routes can stably reach the technical requirements of ultra-low emission.

2.2 Analysis of Installed economic efficiency

2.2.1 Comparison of installed capacity of desulfurization systems with different process routes
Taking the 1# unit as an example, the power consumption of the desulfurization system is analyzed (without the pulping part), and the installed capacity of the desulfurization system is as shown in

Figure 5. From the figure, in the system installation, the circulating pump, the oxidation fan, the vacuum pump and the slurry disturbance pump. The power ratio is about 89%, which is the most important part of the energy consumption of the desulfurization system, and is also the main starting point for system economic analysis and energy saving.

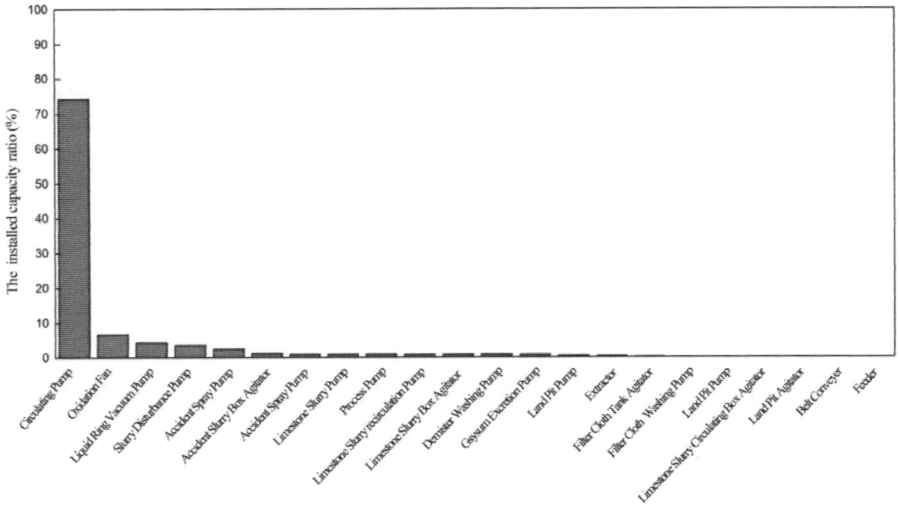

Fig. 5 Desulfurization system equipment installed capacity ratio

The desulfurization tower transformation plans of the first stage two units (1#, 2#) and the second stage two units (3#, 4#) are the same. The main equipment installation of the desulfurization system is shown in Figure 6. Comparing these two desulfurization system transformation plans, the installed capacity of the first stage (350MW) system is about 11% higher than that of the second stage, but the installed capacity of the first stage ultra-low emission technology route is about 35% higher than the second stage.

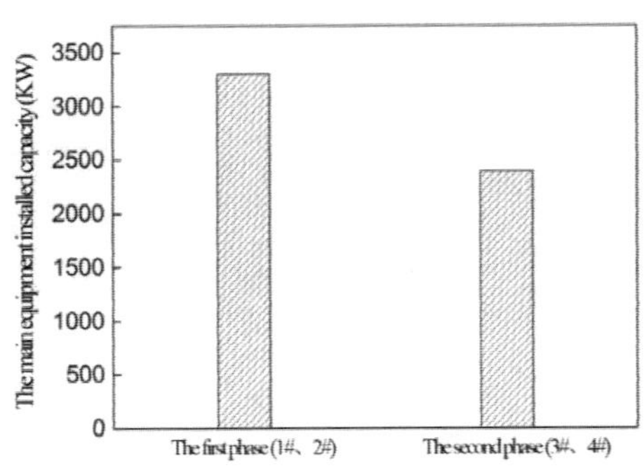

Fig.6 Main equipment installed capacity of different process routes

2.2.2 Comparison of installed capacity of different process wet dust collectors

The power consumption of the wet electrostatic precipitator is mainly the power consumption of the power supply. Its energy consumption accounts for more than 92% of the energy consumption of the wet dust removal system. Compared with the three units of wet electrostatic precipitators (as 3# unit is not equipped with wet electricity), the system power consumption is shown in Figure 7. As can be seen from the figure, the installed capacity of the 1# unit power supply is the smallest, the 2# unit has the installed capacity nearly double that of the 1# unit, and the 4# unit is between the two.

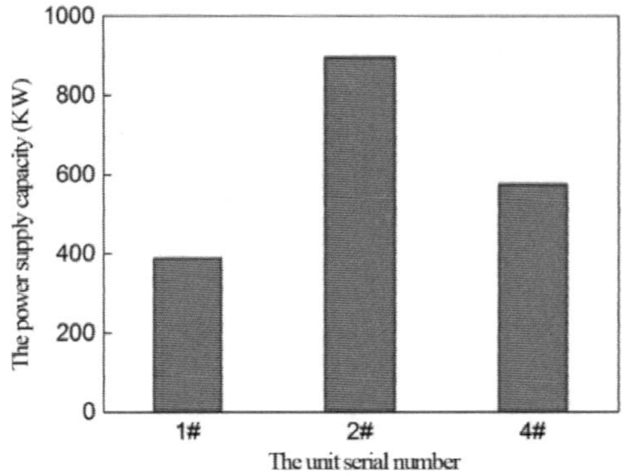

Fig. 7 Power supply capacity of different units of wet electrostatic precipitator
(1#, 2# are 350MW, 3#, 4# are 315MW)

2.2.3 Comparison of desulfurization and wet dust collector total installed capacity

The total installed capacity of the integrated desulfurization and wet electrostatic precipitator system, the total installed capacity of the four units is shown in Figure 8. As can be seen from the figure, since the 3# unit is not equipped with a wet electrostatic precipitator, its installed capacity is significantly lower than other units. Among the three units of wet electricity installed, the installed capacity of the 2# unit is the highest due to the high installed capacity of the wet electric power. Among these three units, the total installed capacity of the 4# unit is the lowest. The main reason is that 4# unit has one less spray layer than 1# unit, which makes the total installed capacity decrease obviously.

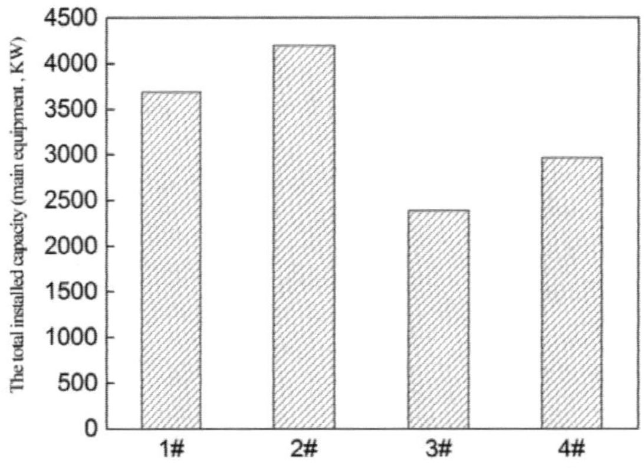

Fig. 8 Total installed capacity of main equipment of different units (kW)

Since the power generation capacity of the first and second stages is slightly different, the first stage is 350MW and the second stage is 315MW. Therefore, the unit installation K value can be used as the measurement index. The calculation formula is as follows:

$$K = \frac{Environmental\ protection\ installed\ capacity\ (Desulfurization + Wet\ dust\ collector)}{Power\ generation\ capacity}$$

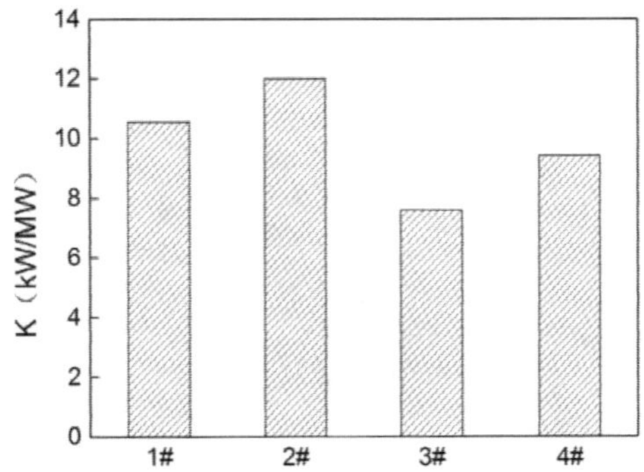

Fig. 9 Installed K value of different units

The K value of the unit installed in different units is shown in Figure 9. The K value of the unit installed in 4 units of ultra-low emission system is between 7.6~12kW/MW, of which the unit installed capacity in 2# unit is the largest, about 12kW/MW. Since 3 # unit does not install wet electricity system, its unit installed is smallest, about 7.5kW / MW.

2.3 Analysis of operational economic efficiency and energy saving potential of ultra-low emission facilities

The operation of the desulfurization facility is the final indicator for measuring the energy consumption of the facility. By analyzing the operational parameters of the facility, the economics of the technical route of the environmental protection facility is investigated. The operation of the desulfurization system is mainly related to the flue gas flow rate and the SO2 content in the flue gas, that is, the imported SO2 flow rate M:

$$M\ (kg\ /\ h) = \text{smoke volume}\ (Nm^3\ /\ h) \times SO_2\ \text{concentration}\ (kg\ /\ Nm^3)$$

2.3.1 Analysis of circulating pump operation economy and energy saving potential

The operating power of the first stage (1#, 2#) and the second stage (3#, 4#) circulating pumps are analyzed. The actual operation of the first stage is shown in Figure 10 (taking 1# as an example). The inlet SO_2 mass flow M mainly between 300kg / h to 1300kg / h. In most working conditions, the M value is between 500kg/h to 1200kg/h, three circulating pumps are turned on, and two or four circulating pumps are turned on under a few working conditions.

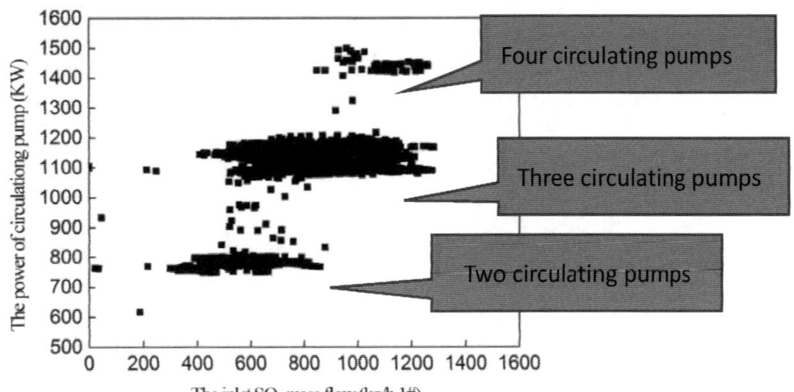

Fig.10 1# unit circulating pump power under different working conditions

The second stage (taking 3# as an example) shows the actual operating conditions as shown in Figure 11. The inlet SO_2 mass flow M is mainly between 200kg/h and 1100kg/h. Under most working conditions, the M value is between 300kg/h and 1000kg/h, two circulating pumps are turned on, and three circulating pumps are turned on under a few working conditions.

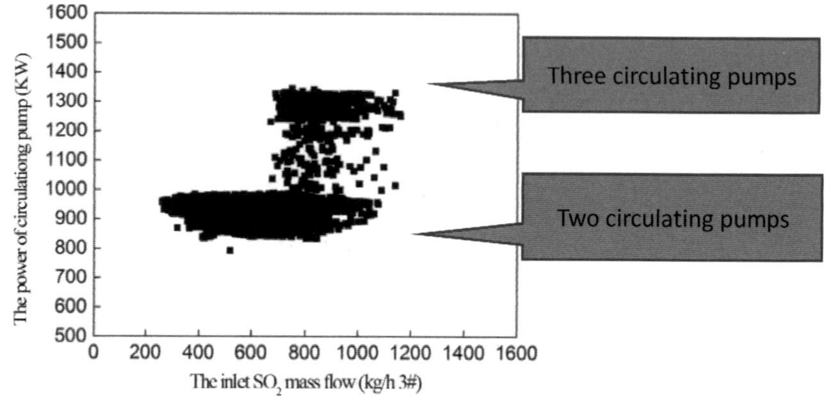

Fig.11 Circulating pump power under different working conditions of 1# unit

It can be clearly seen from Figure 10 and Figure 11 that there is no obvious SO_2 flow limit for the switching of the circulating pump. Taking 3# unit as an example, under the same inlet SO_2 mass flow rate M, there are a lot of operating condition points with 2 pumps and 3 pumps working, and the number is a huge amount. The same problem exists in the 1# unit. That is to say, the running of the circulating pump still has room for optimization. Under the condition that the 2 (or 3) pumps working can reach the standard, the new circulating pump should be no more put into operation, thus having the potential of energy saving.

In order to compare the energy consumption characteristics under different technical routes, the energy consumption of the two units under the same M value is analyzed, as shown in Figure 12:

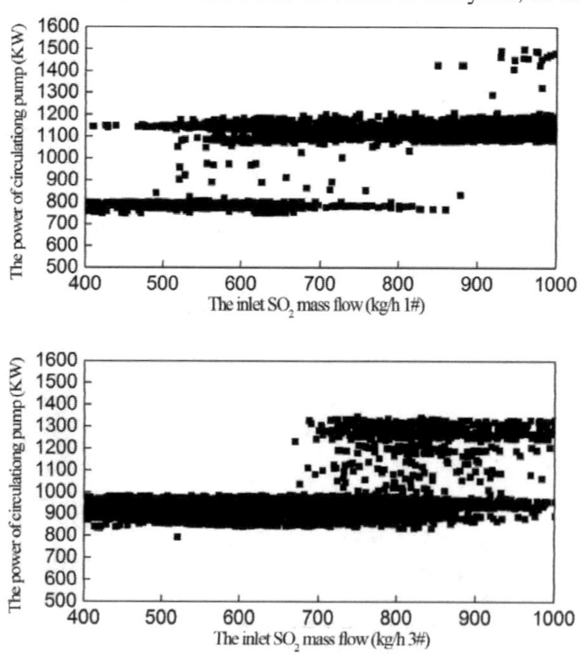

Fig. 12 Circulating pump operating power for different technical routes at the same M value

According to the different inlet SO_2 mass flow in Figure 12, the main points of the circulating pump energy consumption of each unit are summarized: from the operational point of view, the two process routes have advantages in certain intervals in different SO_2 flows. When the SO_2 flow rate is low (or higher), the 1# technical route with 4 circulating pumps is more advantageous because the adjustment is more flexible. When the M value is between 550mg/m^3 to 850mg/m^3, the technical route of the 3# unit has relative advantages.

2.3.2 Analysis of energy saving potential of oxidation fan

The concentration of imported SO_2 in the first and second stage desulfurization systems is 2000mg/Nm^3 and 1650 mg/Nm^3 respectively, while the maximum SO_2 concentration in actual operation is 1300mg/Nm^3 and 1100mg/Nm^3, respectively, which is only 65% and 67% of the design value. As the oxidation fan of the desulfurization system installed in the machine, which is second only to the circulating pump, it is currently operated at constant speed, that is, it maintains constant speed and constant power operation under all working conditions. Therefore, under the condition of ensuring sufficient oxidation, the oxidation fan has a huge potential for energy saving and consumption reducing.

2.4 Analysis of low temperature economizer for reducing desulfurization tower water consumption

In the process of wet desulfurization, a large amount of water is evaporated, which is an important part of the water consumption of coal-fired power plants. The low-temperature economizers installed in the units studied in this project will reduce the temperature of flue gas before desulfurization from 150 °C to 110 °C. In actual operation, the temperature of the desulfurization inlet flue gas is about 100 °C, which is beneficial to the improvement of the efficiency of the downstream electrostatic precipitator, reduce the flue gas temperature, the flue gas volume of the desulfurization tower and the flue gas temperature, which is beneficial to the improvement of the desulfurization efficiency and the reduction of the water evaporation, taking the 1# unit as an example (under full load), the evaporation water volume of the desulfurization tower at different inlet temperatures of the flue gas is shown in Figure 13.

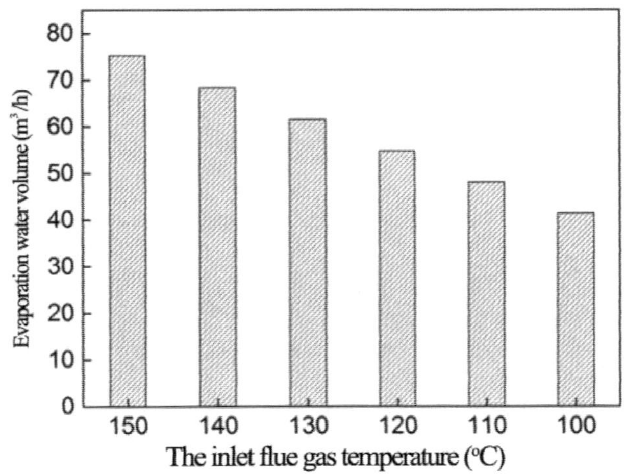

Fig. 13 Evaporation water volume of desulfurization system under different flue gas inlet temperature conditions under full load

After installing the low-temperature economizer in front of the electrostatic precipitator, the evaporation water volume of the full-load desulfurization tower is reduced from 75m^3/h to 40m^3/h, which is reduced by about 40%. Therefore, the use of front-end cooling method is an excellent water-saving measure.

3. Analysis of operating expenses of ultra-low emission facilities

Taking the 4# unit as an example, the installed capacity of the 4# unit ultra-low emission facility is about 3000kW (desulfurization + wet dust removal), and the newly installed capacity is about 1000kW (wet dust removal + fan power increase caused by new system resistance), annual operation time 4200h (China Electricity Council: the average utilization hours of firepower equipment in 2016). The electricity price is calculated according to the on-grid price, and the operating cost is increased at full load.

$$1000\text{kw/h} \times 4200\text{h} \times 0.045 \text{ yuan/degree} = 189000 \text{ yuan}$$

Which operating costs only increase less.

The power generation capacity in 2016 is shown in Figure 14. The average power is 224MW. According to the Notice on Relevant Issues Concerning the Implementation of the Support Policy for Ultra-Low Emission Electricity Price of Coal-Fired Power Plants, the price raise of the purchase of on-grid electricity is 0.01 yuan/kWh, and the annual operation time is 4200h. Annual subsidy fee

$$224000 \text{ degrees/h} \times 4200\text{h} \times 0.01 \text{ yuan/ kWh} = 9408{,}000 \text{ yuan}$$

The subsidy fee can better subsidize the construction cost of ultra-low emission facilities.

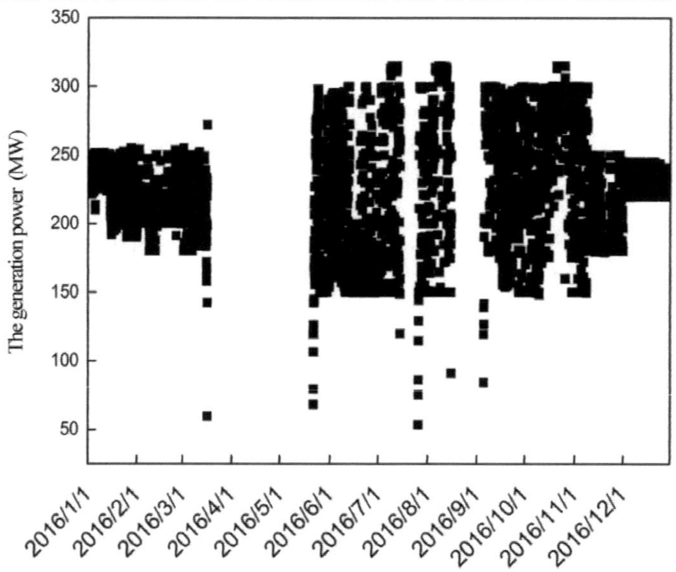

Fig. 14 Unit annual power generation power list

4. Conclusion

Based on the research on ultra-low emission retrofit technology and on-site investigation, the actual operating parameters and economics of the ultra-low emission reform project of coal-fired power plants were analyzed. The main conclusions are as follows:

(1) After installing the low-temperature economizer and electrostatic precipitator in front of the electrostatic precipitator, the original flue gas temperature is significantly reduced, and the electrostatic precipitator can stably reach the design index of 20mg/m^3. Performance, stability and reliability of some units operate at low temperature economizer still need to be improved.

(2) The process route adopted in the survey can stably reach the SO_2 ultra-low emission standard with an export concentration of 35mg/m^3, and the current technical level can stably meet the existing standard requirements.

(3) The total installed capacity of different ultra-low emission technology methods has a large difference. The unit installed K value of different units is between 7.6~12 kW/MW. Among them, the total installed economics without wet circuit line is the best. Dust removal technology can ensure that the dust is up to standard, try not to install a wet electrostatic precipitator.

(4) In the operation of ultra-low emission facilities, the economical operation of the system has a large optimization space; the operation time of the circulating pump still has optimization space, and the system energy consumption has great potential for saving; the first and second stages of desulfurization are extremely low. The emission of the two process routes has their own advantages in

different inlet SO_2 concentration ranges at different SO_2 flow rates.

(5) Since the actual operating parameters of the facility are much lower than the design parameters, the oxidizing fan operating at constant speed still has a large energy saving potential.

(6) The installation of the low temperature economizer enables the wet desulfurization facility to have a large water saving capacity.

Reference

[1] Wang Yongying. Research on current situation and countermeasures of coal-fired air pollutants control in China[J]. Coal Economic Research,2019,39(8):66-70.

[2] Yuan Jiahai, Xu Yan, Na Chunning. Clean and high efficient utilization status and outlook of coal and electric power[J].Coal Economic Research,2017,37(12):18-24.

[3] Yanfang Zheng. T Status of ultra-low emission technology in coal-fired power plant [J].Low Carbon World,2016(7):24-25.

[4] Yongchun Zhao, Siming Ma,J ianping Yang,et al,Status of ultra-low emission technology in coal-fired power plant[J].Jornal Of China Coal Society.2015,40(11):2629-2640.

[5] Zhixuan Wang. Considerations on Basic Issues of "Ultra-low Emissions"of Air Pollutants from Coal-fired Power Plants[J].Environment Impact Assessment.2015,37(4)： 14-17.

[6] Liming Cui, Lixia Meng, Hong Yuan. Ultra-low emission reform and its influence on the plant's electricity supply cost[J]. Thermal Power Generation.2017,46(6):119-124.

Study on Separation Efficiency of Cyclone Separator for 75t/h Circulating Fluidized Bed Boiler

Hai Wang*

Center for Special Equipment Supervisal and Inspection in Jingdezhen City, Jiangxi, 333000, China

*E-Mail: wanghaijdz@qq.com

Abstract: Taking a 75t/h circulating fluidized bed boiler cyclone separator in a boiler manufacturing plant as the research object, this study uses FLUENT fluid calculation software and the balanced orbit and dwell time model to analyze the influence of the particle trajectory tracking, exhausting pipe length and pipe diameter on the separation efficiency The calculation results show that when the relative diameter is equal to 0.3~0.5, the separation efficiency is already high. If the diameter of the core tube is reduced, the separation efficiency is not improved much. However, this will lead to a sharp increase in pressure loss; The increase of the powder concentration is markedly increased; the larger the particle size is, the higher the separation efficiency is. The separation efficiency is obviously improved after the particle size is larger than 30 μm. The separation efficiency is already high after the particle size is larger than 80 μm, and the increase of the particle size is not significant.

1. Introduction

In the development of circulating fluidized bed combustion technology, since the cyclone separator has no moving parts, the structure is relatively simple and the separation effect is good, and it is applied in engineering. Manufacturers and research institutes in all countries have a set of optimal design parameters, and the separation efficiency is high. Generally, the particle size distribution and operating conditions of the circulating fluidized bed can reach more than 99%. However, since the internal flow field of the cyclone is a three-dimensional turbulent rotation accompanied by gas-solid two phases, and the gas interacts with solid particles, solid particles and solid particles, the internal flow field is extremely complicated [1-5]. At present, through a certain empirical or semi-empirical formula, the combination of theory and practical experience in design is the design idea of the cyclone separator, which is used to obtain the structure and scale of the separator. However, the relationship between the insertion length of the central separator of the cyclone separator and the flow resistance is still insufficient, especially for the influence of the volute inlet and the slit inlet cyclone on the separation performance, which directly affects the influence. The accuracy of the cyclone design, and the traditional design method requires a lot of time and effort, and there are a lot of costs, and through numerical simulation, can shorten the research time and save research funding [6-8].

In this paper, the balance trajectory model and the residence time model are used to calculate the separation efficiency of the cyclone separator. The core tube is extended down to the lower end of the separator to form a CS cylindrical surface, and a balanced orbital model obtained by analyzing the resistance of the outward centrifugal force and the inward flowing airflow is established. Under the action of centrifugal force, the large particles move toward the wall of the cyclone separator and are replenished. The small particles escape and are carried into the exhaust pipe by the airflow. The reason is that the centrifugal force is proportional to the mass of the particles, that is, proportional to the resistance, that is, the skeletal force is proportional to the particle size, and the cyclone separator or cleavage particle size mainly means that the particle size is in equilibrium. A particle size is representative of a particle with a 50% probability of being captured, which is a measure of separator separation performance. The residence time model assumes that any position of the particles enters the cyclone at the inlet of the cyclone and the residence time is sufficient inside the cyclone to cause the particles to move radially to the wall and trapped at the bottom by the cyclone.

2. Physical model and meshing

2.1 Physical model

In this paper, numerical simulation is carried out for the cyclone separator of a 75t/h circulating fluidized bed boiler. One is to use the volute inlet according to the design drawings and the other is to use the tangential inlet. Figure 1 is a plan view of a design drawing of a volute inlet. 2 is a schematic view showing the structure and dimensions of a cyclone separator. The main geometric parameters are shown in Table 1.

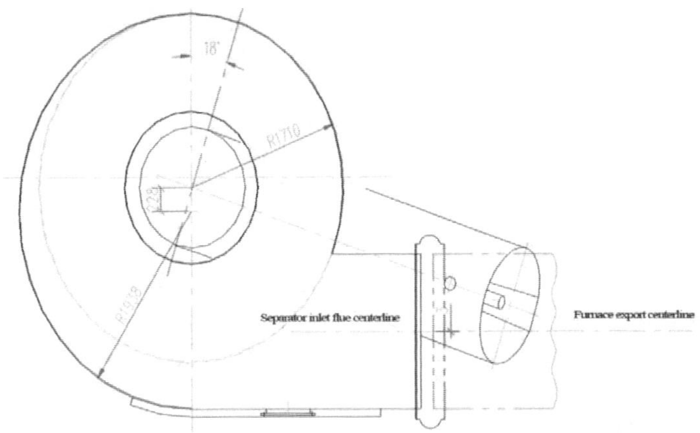

Figure 1 vortex shell inlet cyclone separator design plan

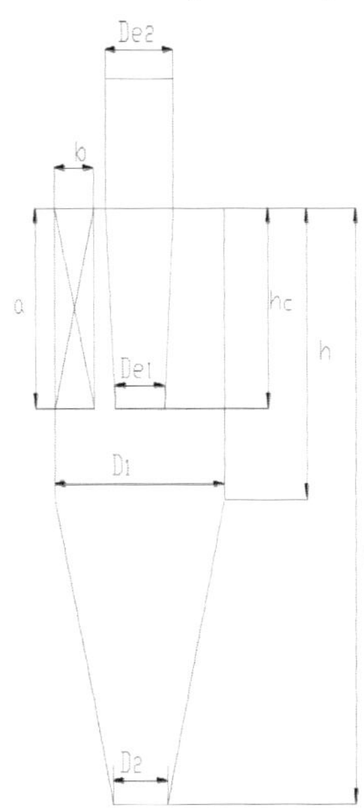

Figure 2 Schematic diagram of structure and size of cyclone separator

Table 1 volute inlet cyclone separator size (mm)

Separator inlet heiht a	Separator inlet width b	Volute diameter D_0	Separator diameter D_1	Exhaust tube outlet diameter D_{e2}
2600	1500	3865	3400	1500
Diameter of ash outlet of separator D_2	Exhaust pipe insertion depth h	Separator cylinder height h	Total height separator H	
1180	1100-3000	3600	6000	

2.2 Meshing

In this paper, FLUENT's own Gambit pre-processing program is used to establish a cyclone separator calculation model and generate the required grid for subsequent calculation and comparison. The cyclone separator meshing is shown in Figure 3 and Figure 4.

To carry out the numerical calculation of the flow field, it is first necessary to discretize the region to be calculated. The meshing of the region is to discretize the differential equation and lay the foundation for numerical solution. The way of meshing will directly affect the difficulty of discretization of the equation, affect the calculation speed and the required storage, and affect the convergence and accuracy of the numerical solution. In order to ensure the quality of the grid and improve the calculation accuracy, taking into account the sharpness of the connection between the intake pipe and the cylinder, the partition is used to generate the grid, and the whole model is divided into seven parts: the intake pipe, the annular space, the cylindrical separation space, Conical separation space, ash bucket portion, ash cone portion, and exhaust pipe. The intake pipe adopts a hexahedral unstructured grid, and the other parts adopt a hexahedral structured grid. The total number of calculated grids is controlled by the grid density along the height direction. In order to obtain a solution unrelated to meshing, a grid with a total number of 1 million is divided. The mesh model obtained by the division is shown in Fig. 3 and Fig. 4.

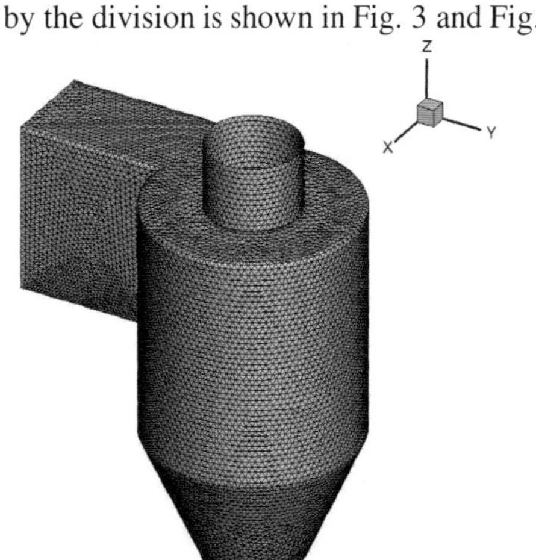

Figure 3 Schematic diagram of meshing of tangential inlet cyclone separator

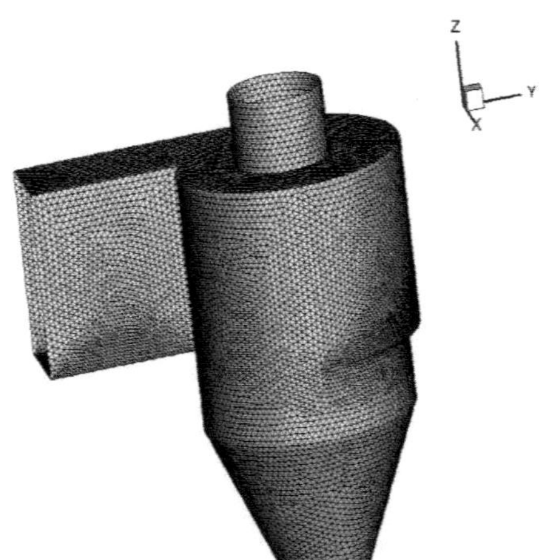

Figure 4 Schematic diagram of grid division of volute inlet cyclone separator

3. Mathematical models and calculation methods

3.1 Mathematical model

Adopt RSM turbulence.

3.2 Boundary conditions and calculation methods

(1) Entrance boundary

The inlet airflow is the air at normal temperature. According to the actual flue gas volume and the inlet cross-sectional area, the inlet velocity is calculated to be 18 m/s, the turbulence intensity is set to 10%, and the hydraulic diameter is 1.9.

(2) Exhaust port boundary

The exhaust port boundary is set to a pressure outlet, and the turbulence intensity is set to 10%, and the hydraulic diameter is 1.9.

(3) Particle trap boundary

Since there is almost no airflow flowing from the bottom, the airflow is set to zero.

(4) Solid wall boundary

No slip boundary wall, wall roughness defaults to 0.5.

The wall has a non-slip boundary with a default wall roughness of 0.5. The wall effect is the main source of vortices and turbulence, so the treatment of the near wall region has a significant impact on the accuracy of the numerical solution results. The turbulent diffusion of the laminar flow bottom layer is enhanced and the turbulent diffusion is relatively weakened in the region close to the solid wall surface, so that the turbulent transport equation acting on the high Reynolds number is not strictly effective. In this paper, the standard wall function method is used to treat the boundary turbulence to give the correct wall shear stress.

4. Results and analysis

4.1 particle trajectory tracking

In order to visually show the gas-solid separation process of the cyclone separator, it is necessary to track the movement trajectories of particles of various particle sizes by finding the principle of gas-particle two-phase separation in the separator. In this paper, the discrete phase model is used to analyze the trajectories of particles with different diameters and different incident positions in the cyclone separator.

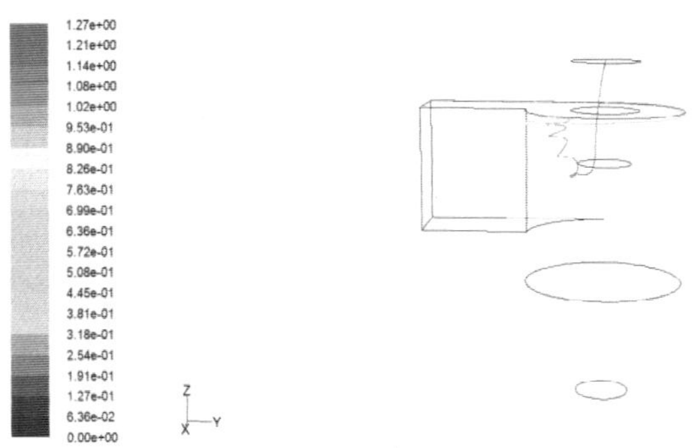

a) Particle size is 1μm

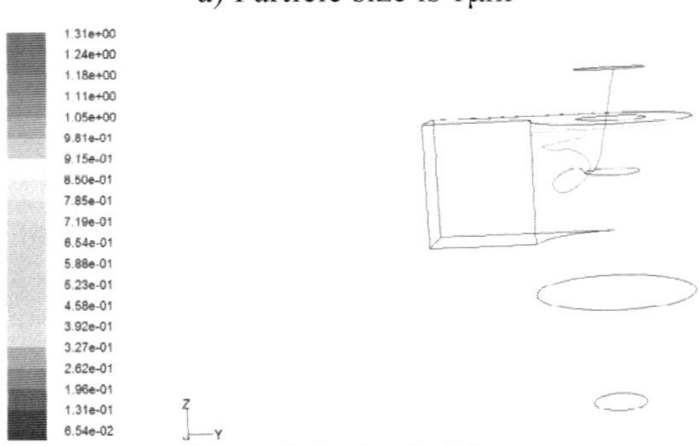

b) Particle size is 30μm

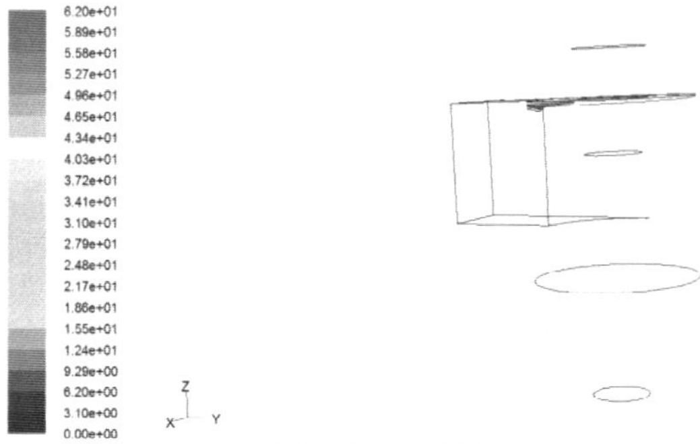

c) Particle size is 100μm

Figure 5: Particle tracking track at the upper left corner of the incident position (x = -1.7, y = 4, z = 2.6)

Fig. 5 is particles of three particle sizes, respectively 1 μm, 30 μm, 100 μm, and the incident position of the particles to be studied is the upper left corner of the inlet, see Fig. 5 (a) to (c). The movement of particles with particle sizes of 1 μm and 30 μm is similar, but it can be clearly seen that the 1 μm particle size is more complicated, that is, the motion trajectory is more unpredictable, but the particles of 1 μm and 30 μm particle size eventually escape. The 100 μm particle size is significantly different from the 1 μm and 30 μm particle size particles. It is a particle that can be trapped. After several rotations, it is free to fall to the bottom along the wall.

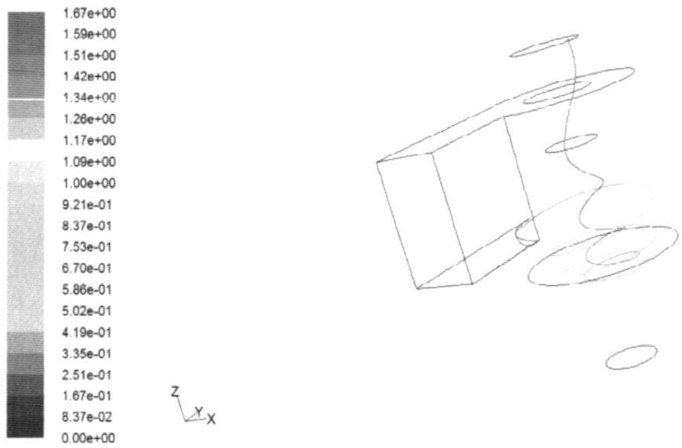

a) Particle size is 1μm

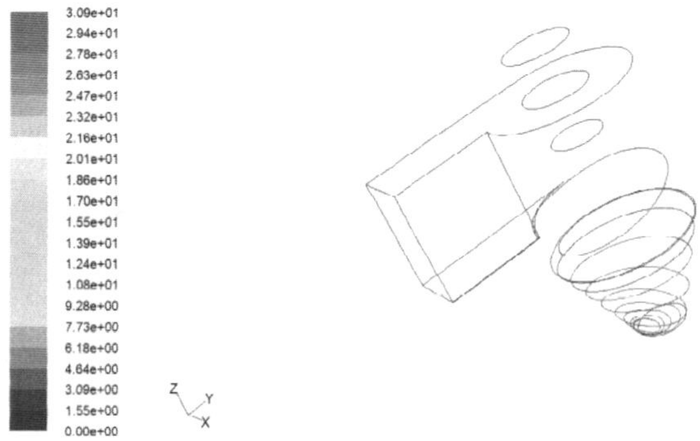

b) Particle size is 30μm

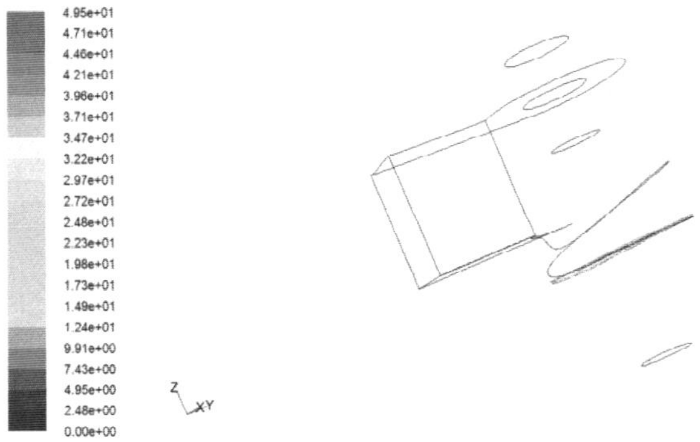

c) Particle size is 100μm

Figure 6: Particle tracking track at the lower right corner of the incident position (x = -0.2, y = -4, z = 0)

Fig. 6 shows particles having a particle diameter of 1 μm, 30 μm, and 100 μm, but the incident position is different from that of Fig. 6, and the incident position is the lower right corner. It can be seen from the figure that the incidence of the incident particle trajectory in the lower right corner is significantly different from that in the upper left corner. That is to say, the small particles are more affected by the turbulent pulsation, and the larger random particles are stronger, and the larger particles are less affected. Particles with a particle size of 1 μm are incident from the lower right corner. After the rotation, the final particles no longer collide with the inner surface of the wall and eventually escape. Particles with a particle size of 30 μm are incident from the upper left corner and eventually escape, but are incident from the lower right corner, collide with the wall, and are finally trapped by the lower end of the wall.

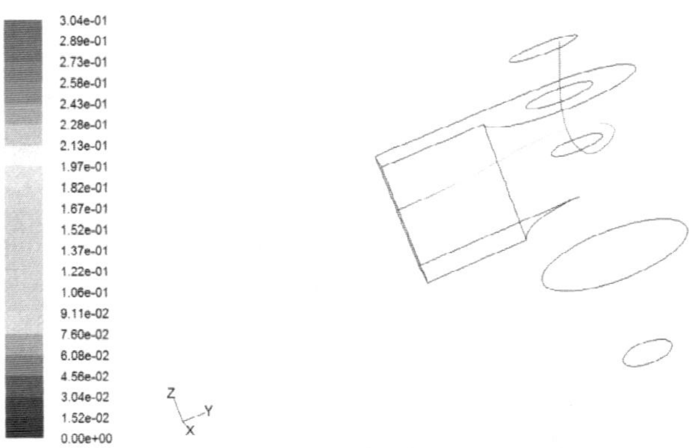

a) Particle size is 1μm

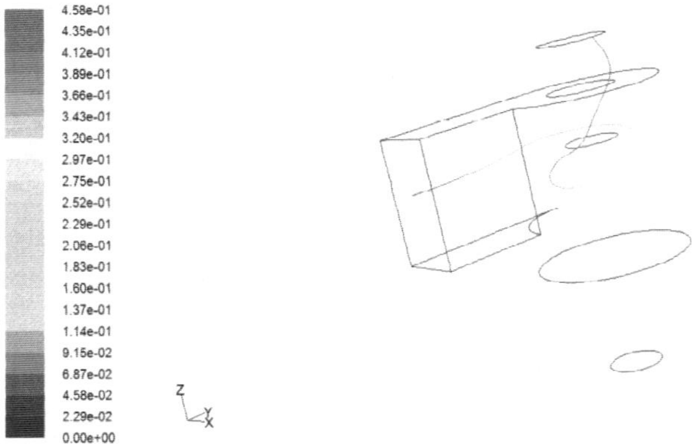

b) Particle size is 30μm

c) Particle size is 100μm

Figure 7: Particle tracking trajectory in the middle of the incident position (x = -0.95, y = -4, z = 1.5)

The incident particle diameter of Fig. 7 is also 1 μm, 30 μm, 100 μm, but the incident position is the middle of the entrance. As can be seen from the figure, the particles having a particle size of 1 μm and 30 μm formed a short circuit, and after the incident, the gas stream escaped from the exhaust pipe, and the particles having a particle diameter of 100 μm were trapped by the wall. It can be seen from Fig. 5 to Fig. 7 that small particles below 30 μm are not easily separated, and even if they are separated, the relationship with the incident position is large, and particles larger than 100 μm are easily trapped, and the relationship with the incident position of the particles is not Big.

In summary, most of the small particles incident from the upper left corner of the inlet will be carried away by the air flow, but there may be a small amount of particles staying near the top cover, forming a so-called "ash-up ring". The small particles injected from the entrance surface parallel to the exhaust pipe are likely to form a "short circuit", which seriously affects the separation efficiency of the cyclone. In addition to the 1 μm particles, the particles injected from the lower right corner of the entrance surface have a better separation effect. It can be seen from the numerical calculation results that the short circuit of the upper ash ring and the exhaust pipe has a great influence on the separation efficiency of the separator. Therefore, in the design process of the cyclone separator, attention should be paid to preventing or minimizing these two phenomena and improving the cyclone separator. Separation efficiency.

4.2 Influence of exhaust pipe length and pipe diameter on separation efficiency

The abscissa of Fig. 8 is the relative length of the center tube, that is, the ratio of the length of the center tube to the longitudinal length of the inlet, and the ordinate is the efficiency. It can be seen from the figure that as the length of the central tube increases, the separation efficiency increases. When the length of the central tube is about 0.4 to 0.5 times the longitudinal length of the inlet tube, the separation efficiency is the highest, and as the relative length of the central tube increases, The separation efficiency is reduced. Therefore, the influence of the excessive length of the central pipe on the separation efficiency is not conducive to the gas-solid separation. Because the insertion of the center tube is too shallow , it will cause the normal swirl core to bend or even break, resulting in unstable state, and it is easy to cause gas short circuit to reduce the separation efficiency, and the exhaust pipe inserted is too deep, which will shorten the exhaust pipe. The distance from the bottom of the cone increases the chance of secondary entrainment.

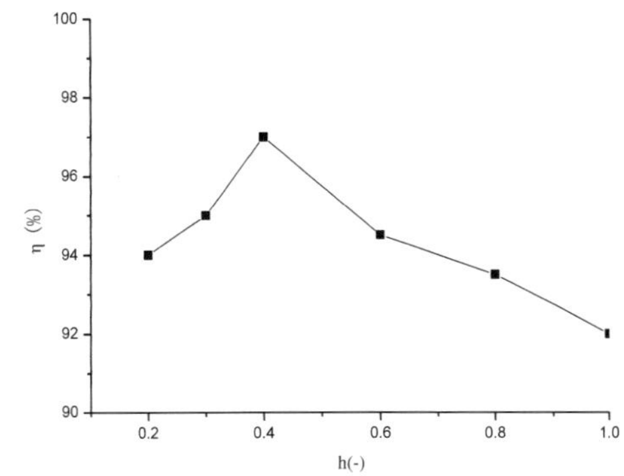

Figure 8 Effect of center tube length on separation efficiency

The abscissa of Fig. 9 is the relative pipe diameter, and the ratio of the diameter of the exhaust pipe to the diameter of the separator, and the ordinate is the separation efficiency of the cyclone. The size of the exhaust pipe affects both the separation efficiency and the pressure drop of the separator, and the two are contradictory. If the separation efficiency is to be increased, the diameter of the exhaust pipe should be reduced. To reduce the pressure drop of the separator, a larger exhaust pipe diameter is required. Therefore, in the design of the actual cyclone separator, it is necessary to comprehensively consider the separation efficiency and The relationship between pressure drop, as can be seen from the figure, when the relative diameter is equal to 0.3~0.5, the separation efficiency is already high. If the diameter of the core tube is reduced, the separation efficiency is not improved much. However, this will lead to pressure loss. increase rapidly. Therefore, it is suitable to take a relative pipe diameter of about 0.4.

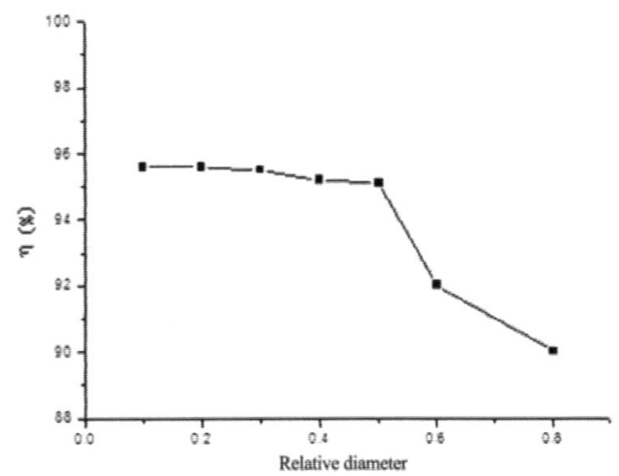

Figure 9 Effect of relative pipe diameter on separation efficiency

4.3 Effect of particle concentration on separation efficiency of cyclone separator

Figure 10 is a graph showing the variation of the separation efficiency of the tangential inlet cyclone with the powder concentration. In the figure, the abscissa is the solid powder concentration and the ordinate is the total separation efficiency. As can be seen from the figure, the total separation efficiency increases significantly as the powder concentration increases. When the concentration of the powder is 80g per cubic meter of gas, the increase of the total separation efficiency is obvious. The ratio of the concentration of the powder at the outlet of the cyclone core tube to the inlet concentration is only 1/3 of the low concentration. Therefore, although the amount of particles taken out from the cyclone vent is still increased as the powder concentration is increased, the proportion of the powder concentration in the inlet concentration is decreased.

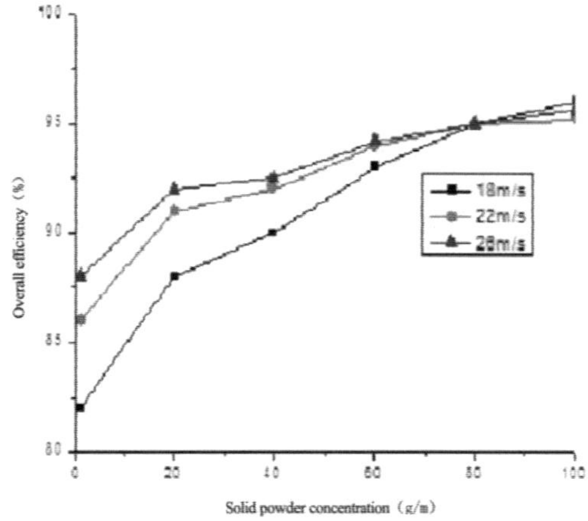

Figure 10 Relationship between separation efficiency of cyclone separator and solid particle concentration

4.4 Effect of Particle Size on Separation Efficiency of Cyclone Separator

Figure 11 is a plot of particle size versus separation efficiency, with the abscissa being the particle size and the ordinate being the separator efficiency. Due to the use and conditions of use of the separator, the overall efficiency of the cyclone separator is only meaningful for a specific particle group, and its performance as a separator is not versatile. It can be seen from Fig. 11 that the larger the particle size, the higher the separation efficiency, but the separation efficiency is obviously improved after the particle size is larger than 30 μm. The separation efficiency is already high after the particle size is larger than 80 μm, and the increase of the particle size is not significant.

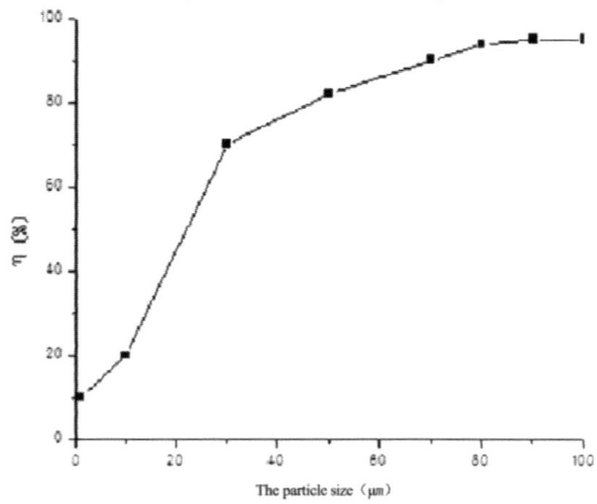

Figure 11 Effect of particle size on separation efficiency

5. Conclusions

(1) Most of the small particles incident from the upper left corner of the inlet will be carried away by the airflow, but a small amount of particles will stay near the top cover to form an "upper ash ring". The small particles injected from the entrance surface parallel to the exhaust pipe are likely to form a "short circuit", which seriously affects the separation efficiency of the cyclone. In addition to the 1 μm particles, the particles injected from the lower right corner of the entrance surface have a better separation effect. The short circuit of the upper ash ring and the exhaust pipe has a great influence on the separation efficiency of the separator.

(2) When the relative diameter is equal to 0.3~0.5, the separation efficiency is already high. If the core tube diameter is reduced, the separation efficiency is not improved much. However, this will

cause the pressure loss to rise sharply. A relative diameter of about 0.4 is suitable.

(3) The total separation efficiency increases significantly as the powder concentration increases. When the concentration of the powder is 80g per cubic meter of gas, the increase of the total separation efficiency is obvious. The ratio of the concentration of the powder at the outlet of the cyclone core tube to the inlet concentration is only 1/3 of the low concentration.

(4) The larger the particle size, the higher the separation efficiency, but the separation efficiency is obviously improved after the particle size is larger than 30μm. After the particle size is larger than 80μm, the separation efficiency is already high, and the increase of the particle size is not significant.

REFERENCES

[1] Qingyun Cao.Numerical Simulation of Gas-Solid Flow in Cyclone Separator of CFB Boiler[D].Beijing:China University of Petroleum Mast Degree Thesis,2008:10-14.
[2] Jianfei Song,Yaodong Wei,Mingxian Shi.Analysis of asymmetry of gas phase flow field in volute cyclone[J]..Journal of Chemical Industry and Engineering,2007,58(5):1091-1096.
[3] Chen J Y，Shi M X. A Universal Model to Calculate Cyclone Pressure Drop[J]. Powder Technology, 2007, 171:184-191.
[4] Gujun Wan,Yaodong Wei,Mingxian Shi.Numerical Simulation of Flow Field in a Cyclone Sepatator under High Temperatur[J]..The Chinese Joural of Process Engineering, 2007, 171：29-37.
[5] Shi L M， Bayless D J. Comparison of Boundary Conditions for Predicting the Collection Efficiency of Cyclones [J]. Powder Technology, 2007, 171：29-37.
[6] Hao Wang.The Study of Numerical Simulation on Two Phase Flow in Gas Cyclone Separator[D].Gansu：Lanzhou University of Technology Mast Degree Thesis. 2007：10-19.
[7] Hairui Yang,Lei Xue,Yuanxiong Guo,Junfu Lv,Guangxi Yue.Ignition Characteristic of Coal in CFB Boiler[J]..Journal of Combustion Science and Technology, 2005 (3):236-240.
[8] Speziale C G, Thangam S. Analysis of an RNG Based Turbulence Model for Separated Flows[J]. Internationnal Toural Engineering Science, 1992 30(2)：1379-1388.

IOP Conf. Series: Materials Science and Engineering **721** (2020) 012038 doi:10.1088/1757-899X/721/1/012038

A practical calibration method of the room type calorimeter for testing the energy efficiency of room air conditioner

HaiyunZhang[1*], ChenzheHang[1], ZhongjieWu[2] and DinghuaXu[1]

[1]National Institute of Metrology, Beijing, 100029, China

[2]Guangdong Institute of Metrology, Guangdong, 510405, China

[*]Corresponding author's e-mail:zhanghy@nim.ac.cn

Abstract. The room type calorimeter is generally used as the energy efficiency measurement standard equipment in the air conditioner industry. However, the calibration procedure is not investigated. In this study, based on the empty-load situation of calorimeter, the calibration model is proposed in order to restrain the error of heat balance caused by long-term use. Besides, two new parameters, which are background value of indoor-side and outdoor-side, are presented in accordance with the calibration model. Experiments are designed to obtain these two parameters and three heat leakage coefficients of partition wall, as well as the calibration procedure and application method of the parameters are illustrated. The corrected calorimeter measures ten different window type air conditioners in one year. The results show that the absolute value of heat balance is less than 2% in the cooling capacity range of 2000W~4500W and 1% in the range of 4500W~7000W.

1. Introduction

The room type calorimeter[1,3-4], as shown in figure 1, which provides approximate thermal insulators based on the conservation of energy and mass, converts the cooling capacity measurement into the measurements of electrical power, heat exchange of cooling coil, heat leakage of partition walls, heat transfer caused by evaporation and condensation water, et. al. Normally, the calorimeter has four test rooms, two of them are test chamber (shown as 2 and 10 in figure 1), and the other two are controlled-temperature air spaces (shown as 1 and 12 in figure 1). In order to decrease the heat leakages of partition wall, the air space can adjust the temperature to the test chamber's. Although calorimeter has more measurement parameters than the indoor air enthalpy test method apparatus [2], the calorimeter is used as the energy efficiency measurement standard equipment in the room air conditioner industry because of the higher accuracy [5-6].

Various type of sensors and instruments are installed in the calorimeter, and the measurement parameters include electrical power, temperature, pressure, flow and mass. Due to the long-term use of calorimeter, the measurement results of the sensors and instruments possibly drift, and this situation can lead to the unpredictable error of cooling capacity measurement. Therefore, periodic check, calibration and correction for the sensors and instruments are essential to the accuracy of calorimeter. For now, the calibration procedure of calorimeter focus on the sensors and instruments which can be calibrated in the field, such as the resistance temperature detectors, the pressure transmitters and the digital power meters. The calibration method, which regards the sensors, the data acquisition unit and the software as a integrated system, conducts the field calibration with the external sources. This part of procedure, which is proven by the practical calibration, is well applied in China air conditioner industry.

IOP Conf. Series: Materials Science and Engineering **721** (2020) 012038 doi:10.1088/1757-899X/721/1/012038

However, a part of sensors are inappropriate to the field calibration because of the field restrictions. For example, the thermocouples on the partition walls, which are dispersive in position, and surround which the space is narrow and small for the calibration equipments. Another example, the flow meter of cooling water, which need to disassemble because of the deficiency of calibration ports in the cooling water pipeline. For these sensors, the field calibration is usually not conducted, or the period of calibration is extended in practice. But this handling way can result in some issues, around which the zero drift of heat balance measurement is most apparent, to the accuracy of calorimeter.

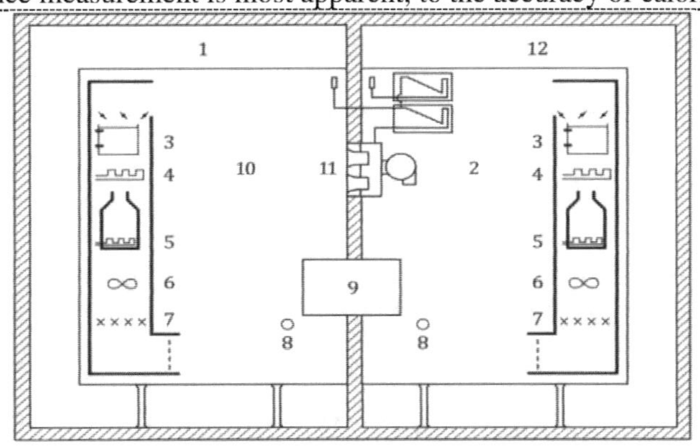

Figure 1. This is a diagram of room type calorimeter, of which 1 indoor-side air space, 2 outdoor-side test chamber, 3 cooling coil, 4 heating coil, 5 humidifier, 6 fan, 7 mixer, 8 air sampling tube, 9 equipment under test, 10 indoor-side test chamber, 11 pressure equalization device, 12 outdoor-side air space.

In this study, the calibration model, which converts the field calibration of thermocouples into the measurement of the coefficients of heat leakage, is proposed based on the empty-load situation of calorimeter. And the practical calibration procedure, which ensure the measurement accuracy of energy flow, improves the issue of heat balance zero drift. Finally, this procedure is applied to a calorimeter, and the heat balance measurement results verify its practicality.

2. The calibration model of calorimeter

The objective of calorimeter calibration is to guarantee the accuracy of heat flow measurement. The heat flow, which is related to the sensors unsuitable for filed calibration, is the heat leakage of partition wall. The heat leakage is the product of the temperature measurement results and the heat leakage coefficients. Under the condition that the temperature measurement sensors can not be calibrated, the accuracy of heat flow measurement also can be achieved through the adjustment of heat leakage coefficients.

Therefore, in order to obtain the coefficients of heat leakage, the calibration model of calorimeter is established based on the empty-load situation, under which the pressure equalization device and humidifier are closed. The energy flow chart of the empty-load is shown as figure 2.

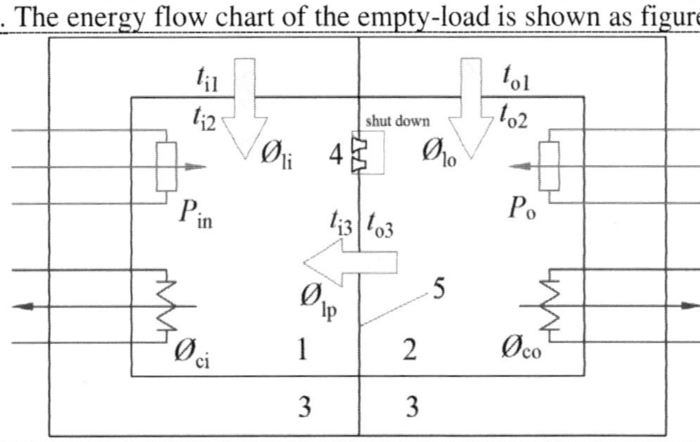

Figure 2. This is the integrated calibration model under empty load, of which 1 indoor-side test chamber, 2 outdoor-side test chamber, 3 controlled-temperature air space, 4 pressure equalization device, 5 the separating partition of test chambers.

From figure 2, it can be seen that only three kinds of energy flow are involved, they are the electrical power flowing in the test chambers (P_{in} and P_o), the heat removed by cooling coils (φ_{ci} and φ_{co}) and the heat leakage of the partition walls (φ_{li}, φ_{lo} and φ_{lp}). According to the conservation of energy, the energy that flowing in a test chamber equals to the energy flowing out. As a result, the indoor-side cooling capacity can be expressed as

$$\phi_{tci} = P_{in} - \phi_{ci} + \phi_{li} + \phi_{lp} \quad (1)$$

Similarly, outdoor-side heating capacity can be expressed as

$$\phi_{tco} = \phi_{co} - P_o - \phi_{lo} + \phi_{lp} \quad (2)$$

Bring $\phi_{li} = L_i \cdot (t_{i1} - t_{i2})$, $\phi_{lo} = L_o \cdot (t_{o1} - t_{o2})$ and $\phi_{lp} = L_p \cdot (t_{o3} - t_{i3})$ into Eq.(1) and Eq.(2), we have

$$\phi_{tci} = P_{in} - \phi_{ci} + L_i \cdot (t_{i1} - t_{i2}) + L_p \cdot (t_{o3} - t_{i3}) \quad (3)$$

$$\phi_{tco} = \phi_{co} - P_o - L_o \cdot (t_{o1} - t_{o2}) + L_p \cdot (t_{o3} - t_{i3}) \quad (4)$$

The Eq .(3) and Eq.(4) are the calibration model of calorimeter.

Moreover, in this model, the room type calorimeter is working under empty-load situation. In other words, there is no air conditioner under tested, which means indoor-side cooling capacity and outdoor-side heating capacity equal to zero. Then, the calibration model can be expressed as

$$P_{in} - \phi_{ci} + L_i \cdot (t_{i1} - t_{i2}) + L_p \cdot (t_{o3} - t_{i3}) = 0 \quad (5)$$

$$\phi_{co} - P_o - L_o \cdot (t_{o1} - t_{o2}) + L_p \cdot (t_{o3} - t_{i3}) = 0 \quad (6)$$

In Eq. (5) and Eq.(6), only three coefficients of heat leakage L_i, L_p and L_o are unknown, and the other parameters can be measured through the experiments. Then the coefficients of heat leakage can be obtained.

However, when these coefficients are brought back to Eq. (3) and Eq. (4) with the experiments data, the indoor-side cooling capacity φ_{tci} and outdoor-side heating capacity φ_{tco} not equal to zero. And φ_{tci} and φ_{tco} should be treated as the measurement background values of the calorimeter, which denoted by B_i and B_o.

To sum up, the system parameters include three coefficients of heat leakages and two background values.

3. The measurement method of system parameters

According to Eq. (5) and Eq. (6), the coefficients of heat leakage and background values can be calculated by designed experiments. The set values of ambient temperature are shown in table 1.

Table 1. Ambient temperature set values (dry bulb temperature, °C).

Experiment No.	Indoor-side test chamber	outdoor-side test chamber	Indoor-side air space	outdoor-side air space
1	20	40	20	40

2	20	20	40	40
3	27	35	27	35

From table 1, under the conditions of experiment 1, the heat mainly flows from the outdoor-side test chamber into the indoor-side test chamber through partition wall because of the obvious temperature difference between the two test chambers. On the contrary, under the conditions of experiment 2, the heats mainly flow from the air spaces into the test chambers through the surrounding walls because of the obvious temperature difference between the two sides of surrounding walls. The experiment 3, which is similar to T1 Standard cooling capacity rating conditions [5], aims to obtain the background values of empty-load situation.

The coefficients of heat leakage can be obtained from the experiment No. 1 and experiment No. 2. For the purpose of measurement accuracy, the temperature difference are extended to 20°C shown in table 1. In accordance with Eq. (5) and Eq. (6), we have

$$P_{\text{in-c1}} - \phi_{\text{ci-c1}} + L_{\text{i}} \cdot (t_{\text{i1-c1}} - t_{\text{i2-c1}}) + L_{\text{p}} \cdot (t_{\text{o3-c1}} - t_{\text{i3-c1}}) = 0 \tag{7}$$

$$\phi_{\text{co-c1}} - P_{\text{o-c1}} - L_{\text{o}} \cdot (t_{\text{o1-c1}} - t_{\text{o2-c1}}) + L_{\text{p}} \cdot (t_{\text{o3-c1}} - t_{\text{i3-c1}}) = 0 \tag{8}$$

$$P_{\text{in-c2}} - \phi_{\text{ci-c2}} + L_{\text{i}} \cdot (t_{\text{i1-c2}} - t_{\text{i2-c2}}) + L_{\text{p}} \cdot (t_{\text{o3-c2}} - t_{\text{i3-c2}}) = 0 \tag{9}$$

$$\phi_{\text{co-c2}} - P_{\text{o-c2}} - L_{\text{o}} \cdot (t_{\text{o1-c2}} - t_{\text{o2-c2}}) + L_{\text{p}} \cdot (t_{\text{o3-c2}} - t_{\text{i3-c2}}) = 0 \tag{10}$$

where the subscript -c1 denotes the results of experiment 1, and the subscript -c2 denotes the results of experiment 2.

In experiment 1, the temperature differences of surrounding walls are very small. Similarly, in experiment 2, the temperature difference of partition wall is also very small. Then, Eq. (7) to Eq. (10) can be simplified as

$$P_{\text{in-c1}} - \phi_{\text{ci-c1}} + L_{\text{p}} \cdot (t_{\text{o3-c1}} - t_{\text{i3-c1}}) = 0 \tag{11}$$

$$\phi_{\text{co-c1}} - P_{\text{o-c1}} + L_{\text{p}} \cdot (t_{\text{o3-c1}} - t_{\text{i3-c1}}) = 0 \tag{12}$$

$$P_{\text{in-c2}} - \phi_{\text{ci-c2}} + L_{\text{i}} \cdot (t_{\text{i1-c2}} - t_{\text{i2-c2}}) = 0 \tag{13}$$

$$\phi_{\text{co-c2}} - P_{\text{o-c2}} - L_{\text{o}} \cdot (t_{\text{o1-c2}} - t_{\text{o2-c2}}) = 0 \tag{14}$$

Further, the coefficients of heat leakage are expressed as

$$L_{\text{p}} = \frac{P_{\text{in-c1}} - \phi_{\text{ci-c1}} - P_{\text{o-c1}} + \phi_{\text{co-c1}}}{2 \cdot (t_{\text{i3-c1}} - t_{\text{o3-c1}})} \tag{15}$$

$$L_{\text{i}} = \frac{P_{\text{in-c2}} - \phi_{\text{ci-c2}}}{(t_{\text{i2-c2}} - t_{\text{i1-c2}})} \tag{16}$$

$$L_{\text{o}} = \frac{\phi_{\text{co-c2}} - P_{\text{o-c2}}}{(t_{\text{o1-c2}} - t_{\text{o2-c2}})} \tag{17}$$

The background values B_{i} and B_{o} can be measured through the experiment 3 with the coefficients which are obtained from experiment 1 and experiment 2. The background values can be expressed as

$$B_{\text{i}} = P_{\text{in-c3}} - \phi_{\text{ci-c3}} + L_{\text{i}} \cdot (t_{\text{i1-c3}} - t_{\text{i2-c3}}) + L_{\text{p}} \cdot (t_{\text{o3-c3}} - t_{\text{i3-c3}}) \tag{18}$$

$$B_{\text{o}} = \phi_{\text{co-c3}} - P_{\text{o-c3}} + L_{\text{p}} \cdot (t_{\text{o3-c3}} - t_{\text{i3-c3}}) - L_{\text{o}} \cdot (t_{\text{o1-c3}} - t_{\text{o2-c3}}) \tag{19}$$

where the subscript -c3 denotes the results of experiment 3.

4. Results and application

4.1The results of system parameters

In order to ensure the accuracy of measurement data, some preparations should be made before the measurements of the system parameters.

- The calibration and correction to the electrical power measurement system. The sensors and instruments in electrical power measurement system normally includes the digital power meters and the current transformers, and they all should be calibrated and corrected.
- The calibration and correction to the resistance temperature detectors which measure the inlet water temperature and the outlet water temperature of the cooling coil. We suggest the fitting correction for the purpose of the accuracy in the full range of temperature measurement.

Then, the pressure equalization device should be closed throughout the measurements of system parameters, and the ambient temperatures of the calorimeter should be set according to table 1. Besides, the mass flows of the cooling water for the cooling coils in indoor-side test chamber and outdoor-side test chamber are both set at 400 kg/h, and the temperature of the cooling water tank is set at 10 ℃. The measurement results are shown in tables 2, 3 and 4.

Table 2. Measurement results from experiment 1.

No.	P_{in}(W)	Φ_{ci}(W)	P_o(W)	Φ_{co}(W)	t_{i1}(℃)	t_{i2} (℃)	t_{i3} (℃)	t_{o1} (℃)	t_{o2} (℃)	t_{o3} (℃)
1	3549.9	3564.9	12977.2	12765.5	20.25	19.90	20.14	39.96	40.21	39.74
2	3556.4	3570.7	12988.7	12784.6	20.25	19.90	20.14	39.97	40.19	39.73
3	3547.6	3579.6	13018.0	12793.3	20.25	19.90	20.14	39.97	40.16	39.68
4	3748.8	3787.5	13373.1	13120.0	20.28	19.92	20.13	39.97	40.05	39.47
5	3755.1	3797.2	13253.5	13027.5	20.28	19.92	20.13	39.99	40.07	39.50
6	3765.2	3800.9	13163.5	12955.5	20.26	19.93	20.14	40.01	40.08	39.49
7	3792.9	3831.5	12369.8	12067.0	20.30	19.92	20.10	39.99	40.01	39.45
8	3793.7	3824.9	12311.3	12063.8	20.30	19.93	20.09	40.00	40.05	39.47
9	3784.3	3816.6	12280.8	12062.9	20.30	19.92	20.09	40.02	40.06	39.46
Average	3699.3	3730.4	12859.5	12626.7	20.27	19.92	20.12	39.99	40.10	39.55

Table 3. Measurement results from experiment 2.

No.	P_{in}(W)	Φ_{ci}(W)	P_o(W)	Φ_{co}(W)	t_{i1}(℃)	t_{i2} (℃)	t_{i3} (℃)	t_{o1} (℃)	t_{o2} (℃)	t_{o3} (℃)
1	3426.8	3760.5	3302.0	3777.9	39.68	20.19	19.90	39.62	20.44	19.95
2	3429.4	3755.1	3336.2	3761.4	39.69	20.19	19.87	39.90	20.42	19.92
3	3408.5	3744.0	3346.4	3746.4	39.69	20.19	19.87	39.61	20.39	19.90
Average	3421.6	3753.2	3328.2	3761.9	39.69	20.19	19.88	39.71	20.42	19.92

Table 4. Measurement results from experiment 3.

No.	P_{in}(W)	Φ_{ci}(W)	P_o(W)	Φ_{co}(W)	t_{i1}(℃)	t_{i2} (℃)	t_{i3} (℃)	t_{o1} (℃)	t_{o2} (℃)	t_{o3} (℃)
1	6540.4	6509.0	9996.7	9845.4	27.22	26.93	26.97	35.02	35.07	34.68
2	6533.7	6505.8	10017.8	9852.7	27.23	26.93	26.96	35.04	35.08	34.68
3	6511.8	6488.2	9990.1	9851.9	27.23	26.93	26.96	35.04	35.09	34.68
4	6528.3	6500.4	10001.1	9850.3	27.23	26.93	26.97	35.03	35.08	34.68
Average	6528.6	6500.9	10001.4	9850.1	27.23	26.93	26.97	35.03	35.08	34.68

Table 2 shows 9 groups of measurement results under the conditions of experiment 1. According to Eq. (15), the calculated results is in the range from 5.6 W/℃ to 8.8 W/℃, and the average value 6.8 W/℃ is used as the coefficient of heat leakage L_p.

Table 3 shows 3 groups of measurement results under the conditions of experiment 2. According to Eq.(16) and Eq.(17), the calculated results of L_i and L_o are in the range from 16.7 W/℃ to 17.2 W/℃ and from 20.8 W/℃ to 24.8 W/℃ separately, the average value 17.0 W/℃ and 22.5 W/℃ are used as the coefficients of heat leakage L_i and L_o.

4 groups of measurement results under the conditions of experiment 3 are shown in table 4. In accordance with Eq.(18) and Eq.(19), the measurement results of background value B_i is in the range from 81.2 W to 88.8 W, and the average value is 85.2 W. And the measurement results of background value B_o is in the range from -111.7 W to -84.6 W, and the average value is -97.8 W.

The results of system parameters are shown in table 5.

Table 5. Results of system parameters.

B_i (W)	B_o (W)	L_i (W/℃)	L_p (W/℃)	L_o (W/℃)
85.2	-97.8	17.0	6.8	22.5

4.2Application of the system parameters

The using conditions of the system parameters are determined based on the measurement situation of the calorimeter. Three coefficients of heat leakage should be applied to all of the measurement situation. The application of two background values is according to the working situation of the cooling coil in the test chamber.

In practice, the cooling coil in the indoor-side test chamber is possibly not working. For the basic situation, if the cooling coil does not remove heat from test chamber, the heating coil is not necessary. Under this basic situation, the energy flows of the indoor-side test chamber only include the heat leakages, which have been considered in the cooling capacity measurement. Therefore, the background value of indoor-side B_i equals to zero.

The application of the system parameters, especially the application of the background values, can improves the issues of heat balance zero drift. To verify the effect of the system parameters, the results in table 5 is applied to the calorimeter. And 10 window type air conditioners are measured using the calorimeter in one year, the arrangement photos are shown in figure 3.

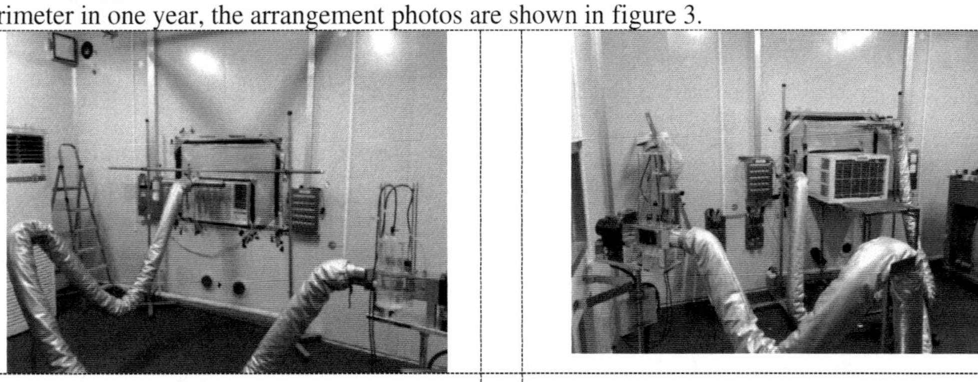

(a). indoor-side test chamber. **(b)** outdoor-side test chamber.

Figure 3. Arrangement photos of window type air conditioner in the calorimeter.

The heat balance results of the 10 measurements are shown in figure 4. The results show that the absolute value of heat balance is less than 2% in the cooling capacity range of 2000W~4500W and 1% in the range of 4500W~7000W.

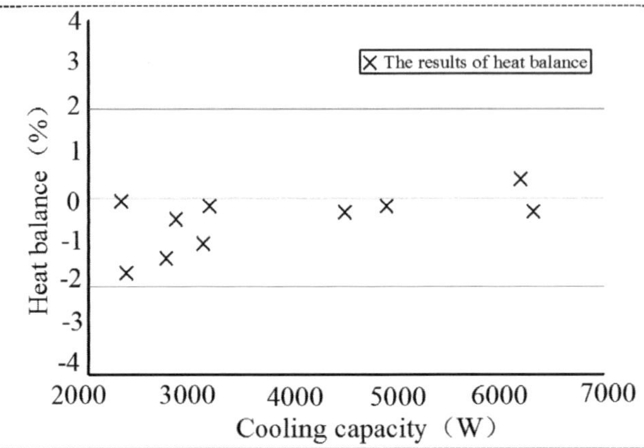

Figure 4. Heat balance results measured by room type calorimeter.

5. Conclusions

In this study, the calibration model of the room type calorimeter is established based on the empty-load situation. In accordance with the model, the system parameters, which include three coefficients of heat leakage and two background values of test chamber, are proposed to improve the issue of heat balance zero drift. And the calibration procedure of system parameters, which is given by the designed experiments, is applied to the calorimeter. Then the using conditions of the parameters are analyzed. Finally, the parameters are applied to the calorimeter which measures 10 different window type air conditioners in one year. The results show that the absolute value of heat balance is less than 2 % in the cooling capacity range of 2000W~4500W and 1% in the range of 4500W~7000W.

6. Appendices

Table 6. Nomenclature and symbols.

Φ_{tci}	indoor-side cooling capacity (W)	P_{in}	total electrical power consumed by indoor-side test chamber (W)
Φ_{tco}	outdoor-side heating capacity (W)	Φ_{ci}	cooling capacity of indoor-side cooling coil (W)
Φ_{li}	heat leakage of indoor-side surrounding wall (W)	Φ_{lp}	heat leakage of separating partition wall (W)
Φ_{co}	cooling capacity of outdoor-side cooling coil (W)	P_o	total electrical power consumed by outdoor-side test chamber (W)
Φ_{lo}	heat leakage of outdoor-side surrounding wall (W)	L_i	coefficient of heat leakage through indoor-side surrounding wall
B_i	Background value of indoor-side cooling capacity (W)	L_p	coefficient of heat leakage through partition wall
B_o	Background value of outdoor-side heating capacity (W)	L_o	coefficient of heat leakage through outdoor-side surrounding wall
t_{i1}	external surface average temperature of indoor-side test chamber surrounding wall (℃)	t_{o1}	external surface average temperature of outdoor-side test chamber surrounding wall (℃)
t_{i2}	Internal surface average temperature of indoor-side test chamber surrounding wall (℃)	t_{o2}	Internal surface average temperature of outdoor-side test chamber surrounding wall (℃)
t_{i3}	indoor-side average temperature of partition wall (℃)	t_{o3}	outdoor-side average temperature of partition wall (℃)

Acknowledgments
This work was funded by the National Key Research and Development Program of China (No. 2018YFF0212404).

References

[1] ANSI/ASHRAE 16-1983 (RA 2014). (2014) Method of Testing for Rating Room Air Conditioners and Packaged Terminal Air Conditioners. American Society of Heating, Air-Conditioning and Refrigeration Engineers, Inc, Atlanta.
[2] ANSI/ASHRAE 58-1986 (RA 2014). (2014) Method of Testing for Rating Room Air-Conditioner and Packaged Terminal Air-Conditioner Heating Capacity. American Society of Heating, Air-Conditioning and Refrigeration Engineers, Inc., Atlanta.
[3] GB/T 7725. (2004) Room air conditioners. China National Institute of Standardization, Beijing.
[4] ISO 5151. (2017) Non-ducted air conditioners and heat pumps - Testing and rating for performance. International Organization for Standardization, Geneva.
[5] Néstor, F.D. (2009) Methodology for uncertainty calculation of net total cooling effect estimation for rating room air conditioners and packaged terminal air conditioners. J. International Journal of Refrigeration., 32:1472–1477.
[6] Hang, C., Ma, G., Zhang, H., Xu, D., Xu, S. (2017) Cooling capacity source and its uncertainty analysis with water-enthalpy method. J. Chinese Journal of Scientific Instrument., 38:919–926.

IOP Conf. Series: Materials Science and Engineering **721** (2020) 012039 doi:10.1088/1757-899X/721/1/012039

Performance analysis of printed circuit heat exchanger for supercritical carbon dioxide and water

Huzhong Zhang[1, 2], **Keyong Cheng**[1, 2*], **Xiulan Huai**[1, 2*] **and Jiangfeng Guo**[1, 2]

[1] Institute of Engineering Thermophysics, Chinese Academy of Sciences, Beijing, 100190, China

[2] University of the Chinese Academy of Sciences, Beijing, 100049, China

*Corresponding author: chengkeyong@iet.cn (K. Cheng), hxl@iet.cn (X. Huai)

Abstract. Printed circuit heat exchanger could be used as pre-cooler in the supercritical carbon dioxide (S-CO_2) Brayton power cycle. Due to the drastic variations of S-CO_2 properties in pre-cooler, segmental design method is used to investigate the thermal-hydraulic performance of PCHE. When the heat duty is given, the local heat transfer entropy generation number has extreme value, which appears near the intersection point of the heat capacity flow rates of both sides. In addition, the local and total entropy generation mainly depend on the local and total heat transfer entropy generation, and the flow imbalance increases entropy generation, especially under conditions of the higher local heat capacity flow rate of water.

1. Introduction

In view of its high efficiency, compactness, and superior economy, the supercritical CO_2 (S-CO_2) Brayton cycle gains lots of attention in the waste heat recovery system, nuclear and concentrated solar [1, 2]. Heat exchanger in the S-CO_2 cycle has an important effect on system efficiency. The printed circuit heat exchanger (PCHE) is one of the leading heat exchanger candidates considering its resistance to extreme conditions, compactness, high effectiveness, and reliability. However, the drastic variations of S-CO_2 properties, especially near the pseudo-critical (PC) point, greatly affect the fluid flow and heat transfer characteristics. So it is quite necessary to study the PCHE performance with S-CO_2.

Lots of the existing papers focused on the channel structure optimization and thermal-hydraulic performance analysis based on the first law of thermodynamics. However, heat transfer is inherently irreversible according to the second law of thermodynamics, which is quite significant in the heat transfer analysis, and Bejan [3] proposed an entropy generation number to optimize the heat exchanger. Sarkar concluded the heat exchanger irreversibilities are significantly more compared with the other parts in the S-CO_2 cycle [4]. For the heat exchanger design, the traditional constant property based heat exchanger design method is inapplicable due to the S-CO_2 properties. Utamura et al. proposed the Generalized Mean Temperature Difference (GMTD) based on the logarithm mean temperature difference (LMTD) [5]. Guo et al. analyzed the low-temperature recuperator using segmental design method [6, 7]. There is little research on the pre-cooler entropy generation analysis in terms of heat exchanger design. In the present study, the water is used for cooling down the S-CO_2 in the pre-cooler device before flowing into compressor, and the pre-cooler is decomposed into a number of elementary units because of the rapid variations of S-CO_2 properties [8]. The local and whole performance of the pre-cooler is investigated from the viewpoint of heat exchanger design.

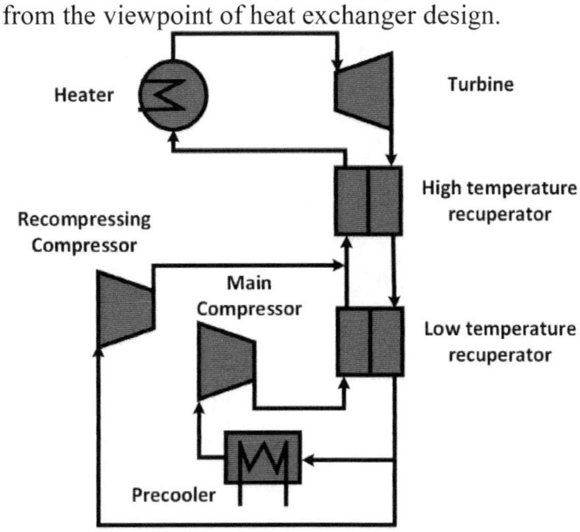

Figure 1. S-CO_2 recompressing cycle layout.

2. Theoretical analysis

The S-CO_2 recompressing cycle layout is shown in Figure 1. The fluid is split into two parts before entering the pre-cooler. One fraction of the fluid is cooled in the pre-cooler and the other is compressed by the recompressing compressor [1]. Notably, the splitting of flow may be changed [9], which could lead to changes in S-CO_2 heat capacity rates of the pre-cooler. The S-CO_2 properties change fiercely near the PC point as shown in Figure 2, so it is difficult for the whole heat exchanger design. The LMTD method is inapplicable in the whole heat exchanger design. To capture the influences of S-CO_2 thermo-physical properties, the pre-cooler is discretized into many elementary units whose properties are considered constant as shown in Figure 3 [7]. Each unit is considered as a black box which transforms two input temperatures to two output temperatures. These units are coupled to each other [8], the outlet temperature of hot side and the inlet temperature of cold side in one element is the inlet temperature of hot side and the outlet temperature of cold side in the next element.

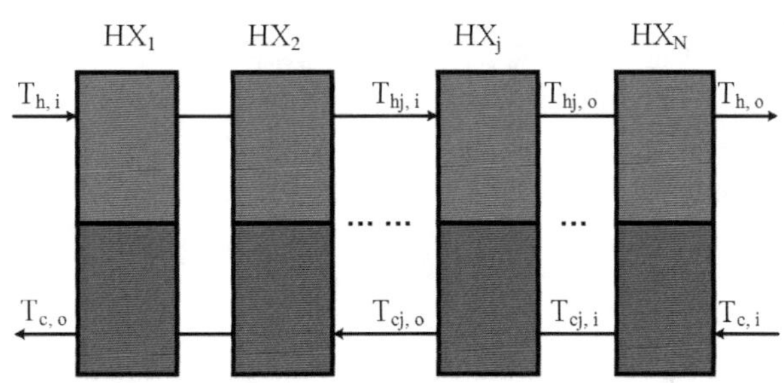

Figure 2. The schematic diagram of elementary units.

For the pre-cooler, straight semi-circular channel, single banking and counter-flow configuration are adopted. The design parameters are outlined in Table 1 and the fixed heat duty is considered. As the total heat duty is constant, the total heat transfer rate is divided equally. The heat transfer rate in elementary units is obtained as:

$$c_{p_{h,j}} m_h \left(T_{hj,i} - T_{hj,o}\right) = c_{p_{c,j}} m_c \left(T_{cj,o} - T_{cj,i}\right) = \frac{Q_{tot}}{N} = U_j A_j \Delta T_j \tag{1}$$

where c_p is specific heat; m is the mass flow rate; T is thermodynamic temperature; Q is the total heat duty; N is the elementary units number; U is heat transfer coefficient; A is heat transfer area; ΔT is logarithmic mean temperature difference; subscripts i and o represent inlet and outlet respectively; subscripts h and c indicate the hot and cold fluid respectively; j represents the local parameter.

Table 1. The initial design parameters.

Parameters	Hot side	Cold side
Inlet pressure (MPa)	8.25	0.55
Mass flow rate (kg/s)	0.40	0.86
Inlet temperature (K)	373.85	292.15
Diameter of channel (mm)	1.6	1.6
Number of plates	24	24
Number of channels in each layer	19	23
Number of elementary units	50	
Ideal gas constant (J/kg K)	189	

The hot working fluid is in turbulent flow regime. For the straight channel, the Gnielinski correlation [10] was recommended by Hesselgreaves [11]:

$$Nu_{h,j} = \frac{f_{h,j}/8 \left(Re_{h,j} - 1000\right) Pr_{h,j}}{1 + 12.7\sqrt{f_{h,j}/8}\left(Pr_{h,j}^{2/3} - 1\right)} \tag{2}$$

$$f_{h,j} = \left(0.79 lg\left(Re_{h,j}\right) - 1.64\right)^{-2} \tag{3}$$

where Nu is Nusselt number; Pr is Prandtl number; Re is Reynolds number; f is Darcy friction factor calculated from the Filonenko correlations.

The cold fluid is in the fully developed laminar flow state. Therefore, the semicircular channel Nusselt number and friction factor [11] is obtained as:

$$Nu_{c,j} = 4.089 \tag{4}$$

$$f_{c,j} = \frac{15.78}{Re_{c,j}} \times 4 \tag{5}$$

The S-CO_2 fluid is regarded as an ideal gas and water fluid is regarded as an incompressible liquid. The local entropy generation rate caused by heat transfer [12] is calculated as follows:

$$S_{gT,j} = m_h c_{p_{h,j}} ln\left(\frac{T_{hj,o}}{T_{hj,i}}\right) + m_c c_{p_{c,j}} ln\left(\frac{T_{cj,o}}{T_{cj,i}}\right) \tag{6}$$

The local entropy generation rate caused by pressure drop [13] is calculated as follows:

$$S_{gP,j} = -m_h R_g ln\left(\frac{P_{hj,o}}{P_{hj,i}}\right) + \frac{m_c \Delta P_{c,j}}{\rho_{c,j} T_{cj,i}} \tag{7}$$

Rg is the ideal gas constant for CO_2; P is the pressure; ΔP is the pressure drop; ρ is the density.
The total entropy generation rate in elementary unit is obtained:

$$S_{g,j} = S_{gT,j} + S_{gP,j} \tag{8}$$

The local dimensionless entropy generation rate is calculated as [11]:

$$N_{sT,j} = \frac{S_{gT,j} T_{hj,i}}{q_j} \tag{9}$$

$$N_{sP,j} = \frac{S_{gP,j} T_{hj,i}}{q_j} \tag{10}$$

3. Results and discussion

The initial design data are outlined in Table 1. The inlet temperature of hot side, the mass flow rate and the inlet temperature of cold side are constant. The total heat duty is 73.6 kW. Figure 3 illustrates the distributions of the local heat capacity flow rate and temperature. When the local heat flux accumulation is less than 30 kW, the heat capacity flow rates of hot side show little change. This is because the heat exchanger inlet is far from the PC point. The heat capacity flow rates of hot side have peak value from m_h = 0.30 kg/s to m_h = 0.40 kg/s. When the S-CO_2 flow rate is larger than 0.40 kg/s, the heat capacity flow rate is increased generally along the passage. The heat capacity flow rate of cold fluid remains almost unchanged. As the S-CO_2 flow rate is increased, the intersection point of the heat capacity flow rates of both sides moves to the right along the passage, and m_h = 0.30 kg/s has two intersection points. The temperature has curve variations as shown in Fig. 3(b), and the temperature change is greater along the passage as the S-CO_2 flow rate is decreased. When the S-CO_2 flow rate is larger than 0.30 kg/s, the heat capacity flow rate of hot side is greater than that of cold side and closes to the PC point at the S-CO_2 outlet. Therefore, the temperature is decreased slowly.

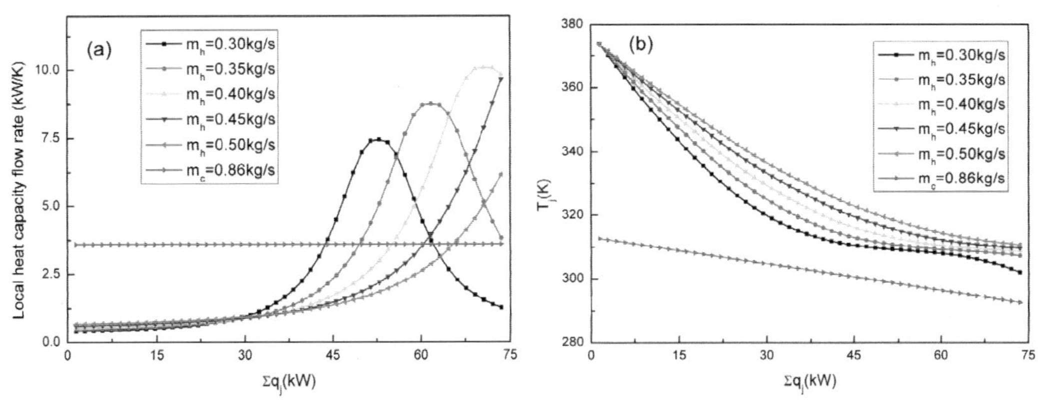

Figure 3. The relations of (a) local heat capacity flow rate and (b) temperature with the local heat flux accumulation.

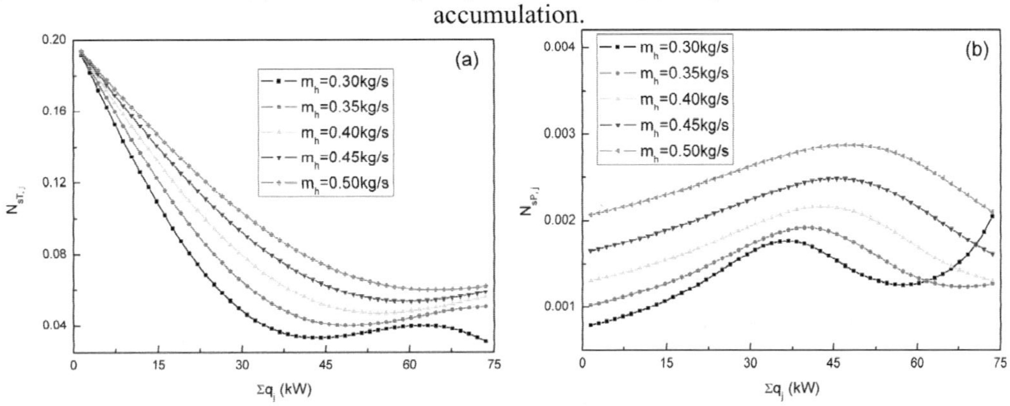

Figure 4. The relations of (a) $N_{sT,j}$ and (b) $N_{sP,j}$ with the local heat flux accumulation.

The distributions of the local entropy generation numbers caused by heat transfer and pressure drop are shown in Figure 4. $N_{sT,\,j}$ is generally decreased and then increased along the passage. The extremums of $N_{sT,j}$ appears near the intersection point of the heat capacity flow rates of both sides. As shown in Figure 4(b), $N_{sP,j}$ has the maximum value, which moves to the right along the passage with the increasing of the S-CO_2 flow rate.

To evaluate the whole heat exchanger performance, the total entropy generation numbers caused by heat transfer and pressure drop at different mass flow rates of hot fluid are illustrated in Figure 5. N_{sT} and N_{sP} are increased with the increasing of the S-CO_2 flow rate. However, N_{sT} has a larger magnitude than N_{sP}. It can be seen from Figure 4 that $N_{sT,j}$ also has a larger magnitude than the $N_{sP,j}$ along the passage. $N_{sT,j}$ and $N_{sP,j}$ have different changing tendencies, the trade-off between the irreversibilities of heat transfer and flow is not easy to obtain [6], reducing the fluid flow irreversibility may be not good for the heat transfer irreversibility, and vice versa [12].

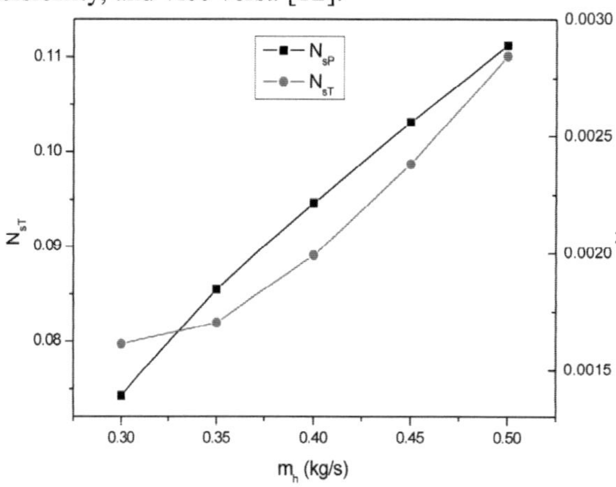

Figure 5. The relations of N_{sT} and N_{sP} with m_h.

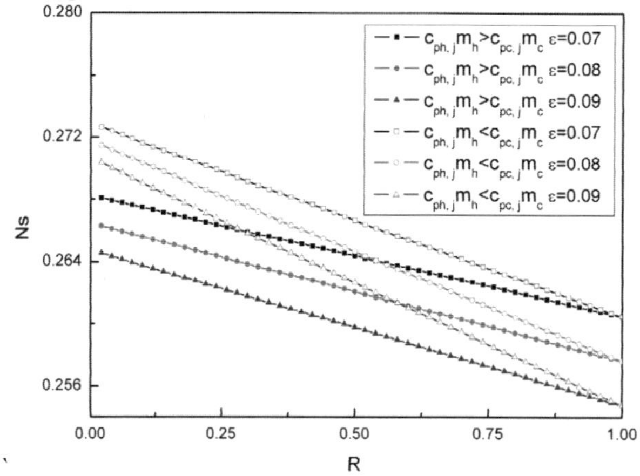

Figure 6. The relations of $N_{s,\,j}$ with the local heat capacity rate ratio.

It is concluded that the heat transfer entropy generation number is much greater than the frictional entropy generation number from Figure 4. Therefore, the contribution of pressure drop could be neglected. The entropy generation number of the elementary unit is calculated as follows [11]:

$$R_j = \frac{min\left(m_h c_{ph,j}, m_c c_{pc,j}\right)}{max\left(m_h c_{ph,j}, m_c c_{pc,j}\right)} \tag{11}$$

$$\varepsilon_j = \frac{q_j}{min\left(m_h c_{ph,j}, m_c c_{pc,j}\right)\left(T_{hj,i} - T_{cj,i}\right)} \tag{12}$$

$$N_{s,j} = \frac{S_{g,j} T_{hj,i}}{q_j} \quad \left(c_{p_{h,j}} m_h > c_{p_{c,j}} m_c\right)$$

$$= \frac{1}{\varepsilon_j \left(1 - \frac{1}{\left(T_{hj,i}/T_{cj,i}\right)}\right)} \left\{ log\left[1 + \varepsilon_j \left(T_{hj,i}/T_{cj,i} - 1\right)\right] + \frac{1}{R_j} log\left(1 - R_j \varepsilon_j \left(1 - \frac{1}{\left(T_{hj,i}/T_{cj,i}\right)}\right)\right) \right\} \tag{13}$$

$$N_{s,j} = \frac{S_{g,j} T_{hj,i}}{q_j} \quad \left(c_{p_{h,j}} m_h \le c_{p_{c,j}} m_c\right)$$

$$= \frac{1}{\varepsilon_j \left(1 - \frac{1}{\left(T_{hj,i}/T_{cj,i}\right)}\right)} \left\{ \frac{1}{R_j} log\left[1 + R_j \varepsilon_j \left(T_{hj,i}/T_{cj,i} - 1\right)\right] + log\left(1 - \varepsilon_j \left(1 - \frac{1}{\left(T_{hj,i}/T_{cj,i}\right)}\right)\right) \right\} \tag{14}$$

Equation (13) and (14) indicate the local entropy generation number is determined by the local effectiveness, the inlet temperature ratio and the heat capacity rate ratio. The relations of local entropy generation number with the local heat capacity rate ratio are illustrated in Figure 6 when the inlet temperature ratio is 1.28. The local entropy generation number is decreased as the effectiveness and heat capacity rate ratio are increased, it is evident imbalance increases entropy generation [11]. As the heat capacity rate of both sides is the same, the local entropy generation number reaches a minimum value. On the basis of the second law analysis, the higher the local heat capacity flow rate of hot fluid is, the better the thermodynamic performance of elementary unit is. In the conventional heat exchanger design, the ratio of heat capacity rate keeps the same in the whole heat exchanger because the working fluid properties are constant. However, due to the extreme changes of S-CO_2 properties, the whole PCHE is discretized into a number of elementary units. Many elementary units are in the unbalanced counter-flow case, which could increase the entropy generation. Notably, the local heat transfer entropy generation number is lower when the local heat capacity rate ratio is about 1 under the fixed heat duty conditions as shown in Figure 4(a). The changing trend of local entropy generation number and the heat transfer entropy generation number is similar because the heat transfer entropy generation is much greater than that caused by pressure drop. Therefore, more elementary units appear near the intersection point of the heat capacity flow rates of both sides, the total entropy generation would be less, such as the S-CO_2 flow rate is 0.30 kg/s under the constant heat duty condition.

4. Conclusions

The performance analysis of PCHE is conducted in the present work. The local heat transfer entropy generation number has extremums, which is near the intersection point between the heat capacity flow rates of both sides. The local and total heat transfer entropy generation numbers are much greater than the local and total frictional entropy generation numbers. Therefore, the entropy generation contributed by pressure drop could be neglected. In addition, flow imbalance caused by the rapid variations of S-CO_2 properties increases entropy generation, and the local thermodynamic performance is better when the S-CO_2 flow rate is higher based on second law analysis.

Acknowledgments
Our work was supported by the National Natural Science Foundation of China (No. 51606191), the National Key Research and Development Program-China (2017YFB0601803), and Key deployment project of Chinese Academy of Sciences (Y7220112H1).

References

[1] Dostal, V., Driscoll, M.J., Hejzlar, P., Wang, Y. (2004) Supercritical CO2 Cycle for Fast Gas-Cooled Reactors, 683-692.
[2] Ahn, Y., Lee, J., Kim, S.G., Lee, J.I., Cha, J.E., Lee, S.W. (2015) Design consideration of supercritical CO2 power cycle integral experiment loop. Energy, 86: 115-127.
[3] Bejan, A. (1977) The Concept of Irreversibility in Heat Exchanger Design: Counterflow Heat Exchangers for Gas-to-Gas Applications. Journal of Heat Transfer, 99: 374-380.
[4] Sarkar, J. (2009) Second law analysis of supercritical CO2 recompression Brayton cycle. Energy, 34: 1172-1178.
[5] Utamura, M., Nikitin, K., Kato, Y. (2008) A generalised mean temperature difference method for thermal design of heat exchangers. International Journal of Nuclear Energy Science and Technology, 4: 11-31.
[6] Guo, J.F., Huai, X.L. (2017) Performance Analysis of Printed Circuit Heat Exchanger for Supercritical Carbon Dioxide. J. Heat Transf.-Trans. ASME, 139, 9.
[7] Guo, J. (2016) Design analysis of supercritical carbon dioxide recuperator. Appl. Energy, 164: 21-27.
[8] Sekulić, D.P., Shah, R.K., Pignotti, A. (1999) A Review of Solution Methods for Determining Effectiveness-NTU Relationships for Heat Exchangers With Complex Flow Arrangements. Applied Mechanics Reviews, 52: 97-117.
[9] Ahn, Y., Bae, S.J., Kim, M., Cho, S.K., Baik, S., Lee, J.I., Cha, J.E. (2015) Review of supercritical CO2 power cycle technology and current status of research and development. Nucl. Eng. Technol., 47: 647-661.
[10] Incropera, F.P. (2006) Fundamentals of Heat and Mass Transfer, John Wiley & Sons, Inc.
[11] Hesselgreaves, J.E. (2001) Compact Heat Exchangers, Pergamon, Oxford, pp. 155-200.
[12] Bejan, A. (2016) Advanced Engineering Thermodynamics. Fourth Edition.
[13] Bejan, A. (1978) General criterion for rating heat-exchanger performance. Int. J. Heat Mass Transf., 21: 655-658.

IOP Conf. Series: Materials Science and Engineering **721** (2020) 012040 doi:10.1088/1757-899X/721/1/012040

Boiler Thermal Efficiency Test System Based on NB-IoT Technology

Zhigang Chen, Chongtao Xu, Shuai Gao

Tianjin Special Equipment Inspection Institute of Technology, Tianjin 300192, China

Corresponding author's e-mail: czghit@163.com

Abstract: This paper combines the wireless NB-IoT technology with the traditional boiler thermal efficiency test, and uses LabVIEW to develop a test analysis system. By installing the data export module and the transmission module together on the existing equipment to perform wireless and digital upgrade of the detection equipment, the purpose of improving the automation degree of the thermal efficiency test is achieved, thereby reducing the number of test staff on site. At the same time, it will avoid the dust and high temperature danger that the staff will encounter on site. The complex boiler thermal efficiency calculation formula is simplified according to the wireless, real-time and accurate needs of the boiler thermal efficiency test. A formula for calculating the enthalpy value that satisfies the requirements of online testing, a formula for the heat loss of the incomplete combustion of the gas, and a formula for fitting the ash combustibles are proposed.

1. Background

Boiler thermal efficiency test is one of the starting points and the most important content for carrying out boiler energy-saving renovation and improving boiler thermal efficiency. Efficient, accurate and comprehensive thermal efficiency testing is the technical and data foundation for the effective implementation of the boiler thermal efficiency improvement project. The boiler thermal efficiency test, especially the product qualification test and detailed test, has the problems of long test time, poor working environment of the test staff, and difficulty in real-time feedback of the thermal efficiency of the boiler, which affects test quality and productivity. The boiler thermal efficiency testing is not an online test conducted on a fixed number of days. The unstable test cycle lasting one to five days makes it impossible to establish a wired data transmission system, and real-time recording, calculation and analysis of integrated data cannot be realized. Since it is difficult to display the trend of data changes, this is not conducive to the control of the test process by the test staff on site. At present, the research on online test models related to boiler thermal efficiency at home and abroad focuses on power station boilers. Industrial boiler thermal efficiency testing is mostly off-line testing, which is difficult for on-site testing. In addition, there are few systematic studies on industrial boiler online test technology, especially the online thermal efficiency test system based on low-power IoT technology. [1-4] Therefore, for industrial boilers, it is of great practical value to develop a boiler thermal efficiency test system based on wireless transmission technology with universal, rapid and efficient performance.

2. Thermal efficiency test platform based on NB-IoT technology

The traditional thermal efficiency test is mainly based on single test, and the test parameters are numerous with poor correlations. The wired communication has high cost and complicated wiring, which can not fully meet the needs of industrial applications. In this paper, the existing thermal efficiency devices, including the flue gas analyzer, the ultrasonic flowmeter and the thermometer, are used to carry out the wireless upgrade of the detection equipment by adding the data exporting module and the NB-IoT transmission module. The wireless NB-IoT technology is applied to boiler thermal efficiency testing to achieve real-time interaction between testers, test equipment, and boilers under inspection.

2.1 Network architecture of the NB-IoT thermal efficiency test system

The NB-IoT technology in low-power wide-area IoT technology has the advantages of low power consumption, shielding against metal components, deep coverage, long battery life, no need for self-organizing networks, no repeaters, and high scalability. It is enabled to well avoid the problem that the host is difficult to receive data stably due to the interference of the boiler steel frame and the wall, realizing the wireless and digital stable transmission of the thermal efficiency test data. [5-6] The NB-IoT uses the license bands and can adopt three deployment modes: in-band, guard band, or independent carrier to coexist with the existing network. The communication module of NB-IoT technology connects multiple test equipment to form an efficient and stable data transmission wireless network. The wireless data acquisition terminal reads the data of various test equipment at the lower end through the connector. Through the NB-IoT wireless transmission module, the data is sent to the NB-IoT base station, and then uploaded to the management platform and directly forwarded to the host computer for analysis and calculation, as shown in Figure 1. The NB-IoT network includes an NB-IoT terminal, an NB-IoT base station, an NB-IoT packet core network, an NB-IoT connection management platform and an application end, as shown in Figure 2.

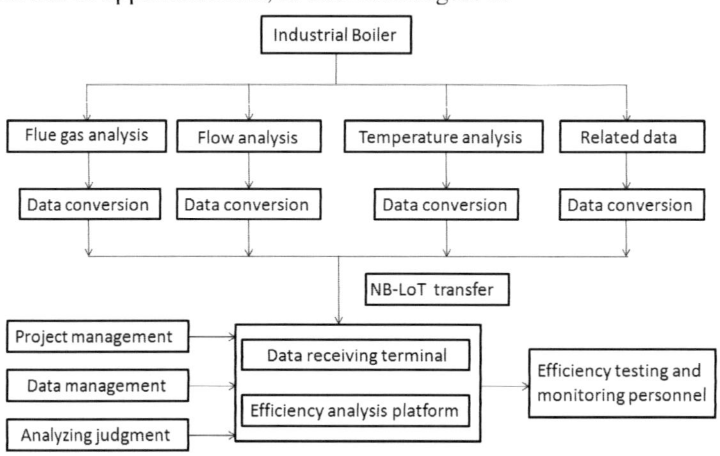

Fig.1 NB-IoT based thermal efficiency test system

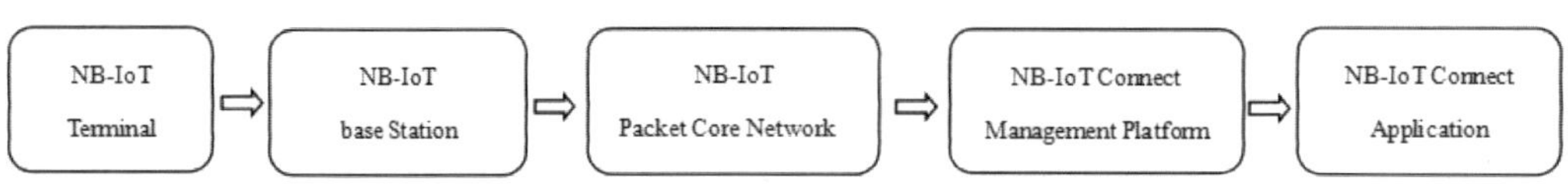

Fig.2 NB-IoT network composition

2.2 Founding of a calculation analysis system

Using the virtual instrument technology of the graphical programming language LabVIEW, the real-time data transmitted by the NB-IoT is read in order to develop a calculation analysis system. The system includes six module functions: test project management, test data management, result analysis and judgment, data self-learning, instrument and equipment management, and standard data. These modules can display values, averages and overall trend lines at each time point. The test module has a data manual resume function. The host calculates each data to obtain thermal efficiency, heat loss

IOP Conf. Series: Materials Science and Engineering **721** (2020) 012040 doi:10.1088/1757-899X/721/1/012040

values, instantaneous values of the output values, or an average of selected time periods. Besides, the rationality of key data is determined, alarm prompt values are set, and improvements are proposed.

3. Thermal efficiency calculation model analysis

The selection and improvement of the thermal efficiency calculation model is an important part of the boiler thermal efficiency test system design. This test system adopts three paths, including GB/T10180-2003 Thermal Performance Test *Code for Industrial Boilers*, TSG G0003-2010 *Energy Efficiency Test and Evaluation Rules for Industrial Boilers* and engineering experience formulas to achieve accurate and efficient calculation results of the test system. Formulas (1)-(3) are the ones for calculating the positive balance thermal efficiency, and formulas (4)-(9) are the ones for calculating the negative balance and the heat loss. Among them, the formula (3) corrects the problem that the positive balance thermal efficiency formula of the superheated steam boiler fails to calculate the difference in enthalpy between the superheated steam and the saturated steam in the GB/T10180-2003 *Thermal Performance Test Code for Industrial Boilers*. Formula (6) gives a simplified engineering calculation formula for the incomplete combustion heat loss of gas [7]. Formula (7) based on the *Energy Efficiency Test and Evaluation Rules for Industrial Boilers* further simplifies the engineering calculation formula for incomplete combustion heat loss. Based on the principle that accuracy and efficiency are equally important, the water and water vapor enthalpy is calculated with the least squares fitting formula.

Positive balance thermal efficiency:

For hot water boiler:

$$\eta_1 = \frac{G(h_{cs}-h_{js})}{BQ_r}\times 100\% \tag{1}$$

For saturated vapor boiler:

$$\eta_1 = \frac{D_{gs}(h_{bq}-h_{gs}-\frac{\gamma\omega}{100})-G_s\gamma}{BQ_r}\times 100\% \tag{2}$$

For superheated steam boiler:

$$\eta_1 = \frac{D_{gs}(h_{gq}-h_{gs})-G_s(h_{gq}-h_{gs}+\gamma)}{BQ_r}\times 100\% \tag{3}$$

G——the circulating water of the hot water boiler or the circulating oil of the hot oil carrier boiler(kg/h);

h_{cs}——Outlet enthalpy of the hot water boiler or the hot oil carrier boiler(kJ/kg);

h_{js}—— Intake enthalpy of the hot water boiler or the hot oil carrier boiler(kJ/kg);

h_{gq}—— Enthalpy of superheated steam(kJ/kg) (under test temperature and pressure);

D_{gs}——Feedwater flow of steam boiler(kg/h);

h_{bq}—— Enthalpy of saturated steam(kJ/kg);

h_{gs}——Feedwater enthalpy of steam boiler(kJ/kg);

r——Latent heat of vaporization(kJ/kg);

w——Steam moisture(%);

G_s——Boiler water sampling(kg/h);

B——Fuel consumption(kg/h);

Negative balance thermal efficiency:

$$\eta_2 = 100-(q_2+q_3+q_4+q_5+q_6) \tag{4}$$

Exhaust heat loss:

$$q_2 = (\mathrm{m}+\mathrm{n}\alpha_{\mathrm{py}})\left(\frac{t_{\mathrm{py}}-t_{\mathrm{lk}}}{100}\right)\left(1-\frac{q_4}{100}\right) \tag{5}$$

Gas incomplete combustion heat loss:

$$q_3 = 3.2a_{py}CO' \tag{6}$$

Solid incomplete combustion heat loss:

$$q_4 = \frac{328.66 \times A_{ar}}{Q_{net,v,ar}} \times \left(\alpha_{fh} \frac{C_{fh}}{100 - C_{fh}} + \alpha_{lz} \frac{C_{lz}}{100 - C_{lz}} \right) \tag{7}$$

Radiant heat loss:

$$q_5 = q_{5ed} \frac{Q_{ed}}{Q_{sc}} \tag{8}$$

Physical heat loss of cinder:

$$q_6 = \frac{Aar}{Q_r} \left[\frac{a_{lz}(ct)_{lz}}{100 - C_{lz}} \right] \tag{9}$$

Where, the calculation coefficient of m, n is selected:

α_{py}——Boiler air coefficient at the exhaust;

t_{py}——Exhaust gas temperature(℃);

t_{lk}——Cold air temperature into furnace(℃);

α_{fh}、α_{lm}、α_{lz}——Fly ash, coal leakage, slag ash content by weight of total ash in the furnace fuel;

c_{fh}、c_{lm}、c_{lz}——Fly ash, coal leakage, slag ash combustible content;

D_{ed}——Rated evaporation(kg/h);

q_{5ed}——Radiant heat loss at rated output(%);

D_{sc}——Output steam(kg/h);

$(ct)_{lz}$——Slag;

A_{ar}——Ash as received.

3.1 Correction of positive balance thermal efficiency calculation formula

Through theoretical analysis, it can be found that in the national standard formula, when the water supply amount deducting the output heat reference due to the sampling of the boiler water is higher than the effective water supply amount, the enthalpy difference between the superheated steam and the saturated steam is not considered, resulting in a calculated boiler thermal efficiency higher than the actual thermal efficiency. The formula (3) used in this paper is a formula for calculating the positive balance of a superheated steam boiler after the calculation formula (10) in the national standard is corrected.

$$\eta_1 = \frac{D_{gs}(h_{gq} - h_{gs}) - G_s\gamma}{BQ_r} \times 100\% \tag{10}$$

3.2 Processing of boiler output calculation parameters

Boiler output is the primary parameter for boiler thermal efficiency positive balance testing and characterizing the performance of the boiler. Calculations involve pressure, media flow, media temperature, and enthalpy. In this system, the medium flow, medium temperature and flue gas composition can be tested in real time by test equipment and transmitted to a thermal efficiency analysis platform. The boiler pressure is relatively stable, but the pressure gauge is connected to the pressure-bearing components, so it is not easy to disassemble on site. The system is processed by manual entry. The enthalpy value is obtained by the medium pressure and temperature fitting formula.

According to the type of working substance, the calculation of substance enthalpy value is divided into three parts, namely unsaturated water area, saturated steam area and superheated steam area. According to the water and water vapor property table data and the nature of the medium itself, the enthalpy values of the three areas are respectively fitted by least squares.

(1) Calculation of superheated steam enthalpy

The steam enthalpy near the boiler operating point can be obtained by look-up table method or by the IAPWS-IF97 fitting formula. However, in online applications in industrial sites, the look-up table method is too inefficient. In the meantime, the IAPWS-IF97 fitting formula is too complicated. Therefore, this paper selects the range of expected parameters of superheated steam, and uses the least squares method to perform quadratic fitting to obtain the calculation formula of superheated steam enthalpy under different temperatures and pressures.

The industrial boilers in this study are expected to have a pressure range of 1.6 MPa to 3.2 MPa and a temperature of 240° C to 360° C. This paper is dedicated to find a fitting formula that satisfies engineering applications in this area. After checking, the steam enthalpy value of the area is obtained, as shown in Table 1:

Table 1. Enthalpy table near the working point of industrial boiler kJ/kg

T/℃	P/MPa								
	1.6	1.8	2	2.2	2.4	2.6	2.8	3	3.2
240	2889.56	2880.34	2870.81	2860.94	2850.72	2840.12	2829.1	2817.62	2805.65
260	2938.39	2930.7	2922.8	2914.69	2906.37	2897.81	2889.01	2879.94	2870.6
280	2985.46	2978.89	2972.19	2965.35	2958.37	2951.24	2943.95	2936.5	2928.89
300	3031.32	3025.62	3019.83	3013.94	3007.96	3001.88	2995.69	2989.4	2983
320	3076.35	3071.34	3066.26	3061.12	3055.91	3050.63	3045.28	3039.85	3034.35
340	3120.82	3116.36	3111.86	3107.31	3102.71	3098.06	3093.36	3088.61	3083.81
360	3164.91	3160.91	3156.88	3152.81	3148.71	3144.57	3140.4	3136.18	3131.93

Since the enthalpy value is a binary function $h = f(T, P)$ of temperature and pressure, the T is regarded as the first independent variable and the least squares polynomial fitting is used to obtain the fitting result in Table 2:

Table 2. Results of fitting the temperature

P/Mpa	$h=aT^2+bT+c$
1.6	$h = -0.0011T^2 + 2.9739T + 2241.9$
1.8	$h = -0.0014T^2 + 3.1691T + 2200.6$
2	$h = -0.0017T^2 + 3.3796T + 2156.6$
2.2	$h = -0.002T^2 + 3.6086T + 2109.4$
2.4	$h = -0.0023T^2 + 3.8562T + 2058.9$
2.6	$h = -0.0027T^2 + 4.1251T + 2004.7$
2.8	$h = -0.0031T^2 + 4.4168T + 1946.6$
3	$h = -0.0035T^2 + 4.7373T + 1883.6$
3.2	$h = -0.004T^2 + 5.0874T + 1815.5$

Then P is regarded as the second independent variable and the least squares polynomial fitting is performed. Through the error comparison, the final determination is based on a sixth-order polynomial fit, which in Table 3 is:

Table 3. Results of fitting the pressure

a	$-0.002026P^6+0.028767P^5-0.16813P^4+0.517321P^3-0.883562P^2+0.79278P-0.29236$
b	$-0.047888P^6+0.69069P^5-4.0903P^4+12.773P^3-22P^2+20.651P-5.6984$
c	$6.7998P^6-98.818P^5+589.45P^4-1854.3P^3+3212P^2-3086.5P+3612$

Table 4. Comparison of the error between fitting point and true value under different pressure and temperature

P/ MPa	T=260℃			T=280℃			T=300℃			T=320℃		
	Fitting Value	Truth Value	Relative Error%	Fitting Value	Truth Value	Relative Error%	Fitting Value	Truth Value	Relative Error%	Fitting Value	Truth Value	Relative Error%
1.6	2941.80	2938.39	0.12	2989.546	2985.46	0.14	3036.416	3031.42	0.16	3082.415	3076.35	0.20
1.8	2931.598	2930.7	0.03	2980.094	2978.89	0.04	3027.487	3025.62	0.06	3073.777	3071.34	0.08
2	2922.056	2922.8	-0.03	2971.513	2972.19	-0.02	3019.626	3019.83	-0.01	3066.395	3066.26	0.00
2.2	2915.614	2914.69	0.03	2966.642	2965.35	0.04	3016.104	3013.94	0.07	3063.999	3061.12	0.09
2.4	2908.158	2906.37	0.06	2960.719	2958.37	0.08	3011.458	3007.96	0.12	3060.377	3055.91	0.15
2.6	2897.801	2897.81	0.00	2951.558	2951.24	0.01	3003.185	3001.88	0.04	3052.682	3050.63	0.07
2.8	2887.087	2889.01	-0.07	2942.12	2943.95	-0.06	2994.684	2995.69	-0.03	3044.779	3045.28	-0.02
3	2878.624	2879.94	-0.05	2935.446	2936.5	-0.04	2989.456	2989.4	0.00	3040.654	3039.85	0.03
3.2	2864.148	2870.6	-0.22	2921.974	2928.89	-0.24	2976.544	2983	-0.22	3027.858	3034.35	-0.21

Table 5. Comparison of the error between the fitted and true values of non-fit points at different pressures and temperatures

P/ MPa	T=270℃			T=290℃			T=310℃		
	Fitting Value	Truth Value	Relative Error %	Fitting Value	Truth Value	Relative Error %	Fitting Value	Truth Value	Relative Error %
1.7	2961.634	2958.58	0.10	3009.364	3005.47	0.13	3056.118	3051.26	0.16
2.5	2930.108	2928.86	0.04	2982.312	2980.14	0.07	3032.547	3029.29	0.11
2.7	2892.231	2893.44	-0.04	2946.574	2947.61	-0.04	2998.618	2998.8	-0.01

According to the statistical results of Table 4 and Table 5, the absolute value of the relative error is between 0 and 0.24%, which is acceptable in engineering.

(2) Calculation of saturated steam enthalpy

The saturated steam enthalpy is relatively simple compared to the superheated steam enthalpy. The medium pressure has a one-to-one correspondence with temperature and enthalpy, so it is easy to get a fitting formula (11).

$$h=4.0091P^3 - 29.694P^2 + 79.347P + 2727.7 \tag{11}$$

Table 6 shows the error between the fitting point and the true value of the saturated steam enthalpy at different pressures:

Table 6. Comparison of the error between fitting point and true value under different pressures

Enthalpy /Pressure MPa	1	1.2	1.4	1.5	1.6	1.7	1.8	2	2.2
Fitting Value kJ/kg	2781.36	2787.08	2791.58	2793.4	2795.06	2796.47	2797.69	2799.69	2801.23
Truth Value kJ/kg	2781.32	2787.11	2791.56	2793.40	2795.02	2796.43	2797.68	2799.71	2801.19
Relative Error %	0.00129	-0.00094	0.0007	0.00117	0.00131	0.00121	0.00039	-0.00069	0.00126

The maximum relative error between the fitted value and the true value is only 0.0013%, which shows high accuracy of the fitting formula.

(3) Calculation of unsaturated water enthalpye

Similar to the case of superheated steam, there are multiple temperatures corresponding to different enthalpy values at the same pressure. However, the enthalpy here is not sensitive to pressure changes but is sensitive to temperature. This paper selects the maximum and minimum ultimate pressures that can be encountered in the test and the average pressure between them (0.1MPa, 2.0MPa, 1.0MPa), and lists the corresponding water enthalpy values at common temperatures. By comparison, it can be found that the maximum relative error of enthalpy due to pressure changes is only 0.28%. In fact, the pressure difference between the inlet and outlet of the boiler water is generally about 0.2-0.5 MPa, and the above error should be reduced by about 0.14%, which is acceptable in engineering. The formula (11) for fitting the enthalpy value based on 1.0 MPa is:

$$h = 0.000006t^3-0.00084t^2+4.2327t-0.0631 \tag{11}$$

Table 7. Errors between fitting points and true values at different temperatures

Enthalpy/ Temperature ℃	50	70	90	110	130	150
Fitting Value kJ/kg	210.0969	293.8249	377.7209	462.0249	546.9769	632.8169
Truth Value kJ/kg	210.1	293.8	377.7	461.9	546.8	632.5
Relative Error %	-0.00148	0.008475	0.005533	0.02704	0.032352	0.050103

The maximum relative error between the fitted value and the true value is only 0.05%. It can be seen that the fitting of unsaturated water also has higher accuracy.

3.3 Gas incomplete combustion heat loss q_3

In the actual test, the unburned combustible gas mainly refers to the CO content. TSG G0003-2010 Energy Efficiency Test and Evaluation Rules for Industrial Boilers only gives the table value of the incomplete combustion heat loss of the gas based on the CO content of the exhaust gas. However, there is a large error between look-up table and calculation, and the maximum difference is nearly 30 times. The main influencing factors of q3 are the carbon monoxide content CO' and the excess air ratio α_{py} at the exhaust. Although natural gas is dangerous to sample but its composition is relatively stable, the typical natural gas composition in Tianjin is selected as the reference fuel component. Simplified calculation formula (12) based on theoretical formula is obtained:

$$q_3 = (3.36a_{py} - 0.34)\ CO' \tag{12}$$

In order to verify the accuracy of the derived formula and the results when the fuel composition changes greatly, this paper uses the data of the thermal efficiency test of some gas and coal-fired boilers in our college since 2015 (number 1 ~ 6 are the gas boilers, number 7 ~ 8 are coal-fired boilers, number 9 to 10 are biomass boilers, and number 11 to 12 are oil-fired boilers). At the same time, the calculation of q3 in the simple test under the operating conditions of TSG G0003-2010 is compared. The results are shown in Table 8.

Table 8. Comparison of engineering formula, standard formula and simple test method q_3 data

Serial Number	a_{py}	CO' (%)	Type Testing q_3 (%)	Simple Testing q_3 (%)	Error of Simple Testing	Engineerin g Formula q_3 (%)	Error of Engineerin g formula
1	1.15	0	0	0.2	0.2	0	0
2	1.13	0.01	0.035	0.2	0.165	0.034568	-0.00043
3	1.15	0.03	0.11	0.2	0.09	0.10572	-0.00428
4	1.09	0.005	0.017	0.2	0.183	0.016612	-0.00039
5	1.29	0.062	0.25	0.5	0.25	0.247653	-0.00235
6	1.16	0.081	0.29	0.5	0.21	0.288166	-0.00183
7	1.47	0.0054	0.023	0.2	0.177	0.024836	0.001836

8	1.38	0.0004	0.0019	0.2	0.1981	0.001719	-0.00018
9	1.64	0.0143	0.07	0.2	0.13	0.073937	0.003937
10	1.65	0.0016	0.0078	0.2	0.1922	0.008326	0.000526
11	1.05	0.0039	0.012	0.2	0.188	0.012433	0.000433

It can be seen from the calculation results that the results of the derivation formula are very close to the test results and the error rate is slight for different fuels. The gas boiler is repaired by retaining two significant figures, and the error between the calculation result and the final test is 0.0043%. The maximum error for coal-fired, biomass-burning, and oil-fired boilers does not exceed 0.0039%. The calculation accuracy is one lever higher than the calculation method of the simple test under the operating conditions of TSG G0003-2010.

3.4 Solid incomplete combustion heat loss q_4

In the simplified calculation formula of the solid incomplete combustion heat loss q_4 , the ash and the net calorific value as received basis of the fuel are basically stabilized to a known amount, and the ash and combustible content need to be determined experimentally. Standard calculation of the ash content share is performed after the test is completed, so it is impossible to provide real-time data of the test process. From the experience of the previous customary thermal efficiency test, the data relates to the slag weighing, the slag moisture content and the fly ash combustible content, the calculation accuracy is not high and the process is cumbersome. Therefore, the data in this paper is selected based on the empirical table of engineering application in TSG G0003-2010 *Energy Efficiency Test and Evaluation Rules for Industrial Boilers*, and the values are selected according to Table 9. [8]

Table 9. Weight percentage of fly ash and slag ash content in total ash content of furnace fuel (%)

Combustion Mode	Type of Coal	
	Fly Ash (α_{fh})	Slag (α_{lz})
Reciprocating Grate	20~10	75~85
Chain Grate	20~10	75~85
Stoker Grate	30~20	65~75
Fluidized Bed	50~40	50~60
Pulverized Coal Fired Boiler	90~80	10~20
Coal Water Slurry	80~70	20~30

Note: When selecting, $\alpha_{fh}+\alpha_{lz}=100$ should be satisfied.

For fuel and gas boilers, q_4 is 0

For technical and economic reasons, it is difficult to achieve accurate on-line measurement of slag and fly ash combustibles contents. In the case where the hardware detection device cannot complete the measurement, a statistically-based soft measurement scheme can be used to solve the problem. The author has previously studied the application of the above calculation method for q_4 in engineering, and derived the simplified fitting formula, which will not be elaborated here. [9]

4. Conclusion

In this paper, NB-IoT technology is applied to boiler thermal efficiency test, and wireless transformation based on existing thermal efficiency test equipment is realized. By establishing a thermal efficiency test software platform, real-time interaction between testers, test equipment, and boilers under inspection can be realized to form an efficient and stable data transmission wireless network.

The simplified boiler thermal efficiency calculation formula, the enthalpy value calculation formula and the ash combustible fitting formula are obtained by theoretical derivation and experimental verification. This not only takes into account the wireless, real-time and accurate calculation of boiler thermal efficiency, but also can be used for on-line monitoring of boiler thermal efficiency. However, due to the single subject and few experimental samples, it needs to be supplemented and revised in the subsequent actual study.

Author: Zhigang Chen (1983 ~), male, postgraduate, senior engineer, research interests: boiler safety, energy conservation, environmental protection, 13821201831

Source: Science and technology project of General Administration of quality supervision, inspection and Quarantine of the people's Republic of China: Research on comprehensive test system of boiler thermal efficiency based on wireless network.

References

[1] Zhao Hui, et al. Technical Analysis of Remote On -line Monitoring System for Industrial Boiler[J]. Energy Conservation Technology,2018,36(1):22-25.

[2] Zheng Ning, et al. A Survey of Low-Power Wide-Area Network Technology [J]. Information and Communications Technologies,2017 (01) :47-52.

[3] GB/T 10180—2003, Thermal performance test code for industrial boilers[S].

[4] Xu Chongtao, et al. Tianjin Industrial Boiler Statistical Analysis of Energy Efficiency and Countermeasures of Energy Saving [J].Industrial Boiler,2016(4) : 42-45.

[5] Xing Yulong, et al. The Deployment Strategy of Narrow-Band Internet of Things [J]. Information and Communications Technologies,2017.2(15):37-41.

[6] Liu Hui, et al. NB-loT Technology and Application Charactics of loT Product [J]. Micro controllers & Embedded Systems,2018(12):11-14.

[7] Zhang Xiaoming, et al. Improvement Suggestions on the Calculation of the Incomplete Combustion Loss of Gas in TSG G0003—2010: Energy Efficiency Testing and Evaluation Rule on Industrial Boiler[J]. China Special Equipment Safety,2015,(11):6-10.

[8] TSG G0003-2010, Energy Efficiency Test and Evaluation Regulation for Industrial Boiler[S].

[9] Zhang Xu, et al. Statistical Analysis of the Correlations between q4, the Carbon Content of the Slag and the Fly Ash with the Operating Parameters of the Coal-Fired Boiler[J].Energy Conservation Technology,2017,35(205) : 424-428.

IOP Conf. Series: Materials Science and Engineering **721** (2020) 012041 doi:10.1088/1757-899X/721/1/012041

Numerical simulations of the combustion characteristics for a 660MW tangentially fired boiler with different closing-to-wall air velocities

Kui Liang, Guangkui Liu* and Feng Liu

Boiler Department, China Special Equipment Inspection and Research Institute, Beijing, 100029, China

*Corresponding author's e-mail: liuguangkui9527@163.com

Abstract. To solve the problem of high temperature corrosion of water-cooled wall of power station boiler, this paper proposes a new type technology of closing-to-wall air. Taking a 660MW tangentially fired boiler as the research object, numerical simulations were conducted to study the influences of closing-to-wall air velocity on the combustion characteristics in the boiler. The results show that O_2 concentration in the flowing region is significantly increased, CO concentration is significantly reduced, and the oxidizing atmosphere in the near-wall region is enhanced. As the closing-to-wall air velocity increases, the rigidity of the airflow is gradually stronger, and the penetration depth of the airflow increases, so that the coverage area of the airflow is increased. The influence of the closing-to-wall air velocity on the combustion effect of the boiler is negligible.

1. Introduction

About 80% of China's utility boilers burning lean coal suffer from the problem of the high temperature corrosion of water-cooled wall with different degrees. In recent years, low NOx combustion technologies were generally adopt on the utility boilers to control the amount of NOx generated during the combustion process by reducing the oxygen atmosphere in the main combustion zone, but at the same time, the reducing atmosphere aggravate the high temperature corrosion of the water-cooled wall.

Closing-to-wall air technology is an effective means to alleviate high temperature corrosion of water-cooled wall in the furnace [1-2]. In this paper, a new type of closing-to-wall air technology suitable for tangentially fired boiler is proposed. Taking a 660MW tangentially fired boiler as a research object, the influence of the closing-to-wall air velocity on the combustion characteristics in the furnace, especially near the water-cooled wall, was studied.

2. The new type of Closing-to-wall air technology

The new type of closing-to-wall air proposed in this paper is shown in Figure 1. The closing-to-wall air nozzle is placed at the position where the water-cooled wall is prone to high-temperature corrosion. Some water-cooled wall tubes are bent toward the inside of the furnace. The gap between the inner wall and the furnace wall forms a double of the closing-to-wall air nozzles. The closing-to-wall air box is arranged outside the furnace wall, completely covering the bended tubes, and is connected with the secondary air box through the connecting pipe. The amount of the closing-to-wall air is controlled by the adjusting valve on the connecting pipe.

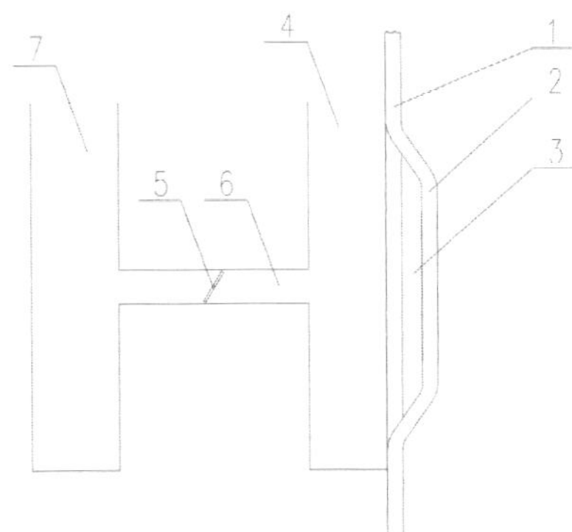

1. Furnace wall, 2. Inner wall, 3. Closing-to-wall air nozzle, 4. Closing-to-wall air box, 5. Adjusting valve, 6. Connecting pipe, 7. Secondary air box

Figure 1. The diagrammatic sketch of the new type of closing-to-wall air.

3. Object and research method

3.1. Research object

The research object in this paper is a 660MW wall-type tangentially fired boiler. The section size of the furnace is 20.402 meters (width) × 20.072 meters (depth), and the overall height of the furnace is 68 meters. A group of burners are arranged on each side of the furnace wall. The burner is arranged in two stages, with a total of 6 layers of nozzles. A four-layer over-fire air is placed at the four corners of the furnace above the burners. The design fuel of the boiler is lignite.

3.2. Research method

The numerical simulation is performed on a computer equipped with the Fluent 6.3.26 software. The selection of the calculation model is shown in Table 1. The velocity inlet and the pressure outlet are selected as the inlet and outlet boundary conditions respectively. A "no-slip" boundary condition was employed for fluid along the wall. A mesh independence test is conducted. An initial mesh with about 3036812 nodes was first created in the computational domain. The number of mesh nodes was then increased to 3561648, 4052674, 4634352 and 5034782. Mesh independence was checked by comparing the O_2 concentration in the flue gas at the furnace outlet. As shown in Figure 2, the finest mesh (5034782 nodes) and medium-density mesh (4032674 nodes) yield similar results. For this reason, we chose a mesh density of 4634352 nodes and have applied it to all further work reported in this study.

Table 1. Mathematical model for numerical simulation.

Project	Model
Turbulence model	Can be realized k-ε model [3]
Gas-solid model	Lagrangian particle trajectory model [4]
Radiative heat transfer model	P-1 model [5]
Volatile pyrolysis	Dual competition model [6]
Volatile combustion	Mixing fraction PDF method [7]
Coke combustion	Power - diffusion model [7]

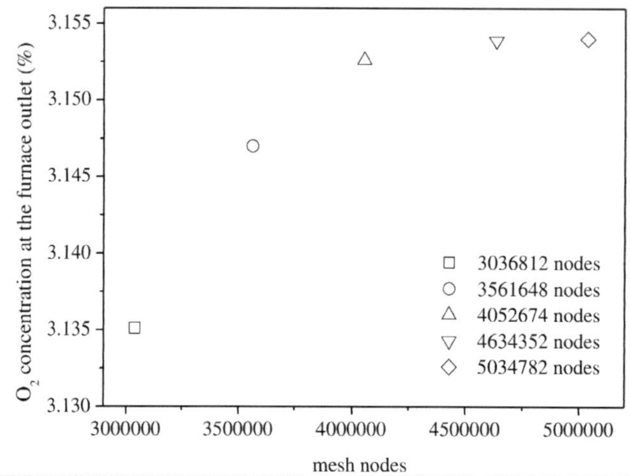

Figure 2. Mesh-independence test based on O_2 concentration at the furnace outlet.

3.3. Arrangement of conditions

Four cases are arranged with different closing-to-wall air velocity: 10m/s, 15m/s, 20m/s and 25m/s. The remaining parameters of the four cases are the same. The number of the closing-to-wall air nozzle is 20, and the closing-to-wall air ratio accounts for 2% of the total air supplied into the furnace. The distribution of pulverized coal particle size follows the Rosin-Rammler rule, the average particle diameter is 50 microns.

4. Results and discussion

Figure 3 shows the temperature field near the wall under different closing-to-wall air velocities. It can be seen that the flue gas temperature in the flowing region is significantly reduced, which is basically less than 1000K. In the region that the closing-to-wall air uncovered, the flue gas temperature is generally in the range of 1300K to 1800K. With the increase of the closing-to-wall air velocity, the area of the low temperature region is gradually enlarged.

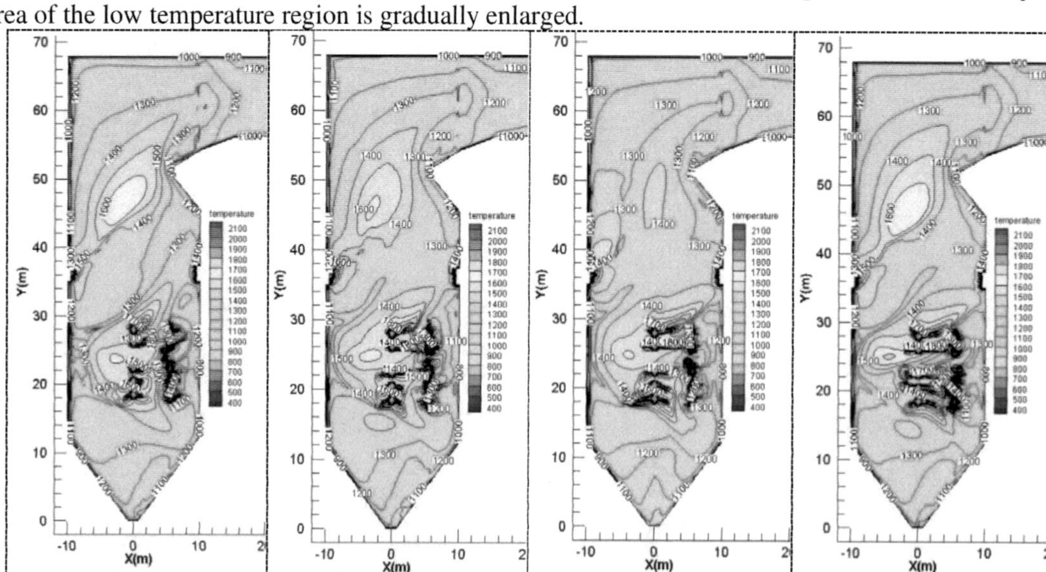

(a) 10 m/s (b) 15 m/s (c) 20 m/s (d) 25 m/s

Figure 3. Temperature field near the wall under different closing-to-wall air velocities.

Figure 4 and 5 show the O_2 and CO concentration field in the near wall area under different closing-to-wall air velocities. The O_2 concentration in the flowing region is significantly increased to be greater than 2%, while the CO concentration was reduced to be less than 1%. Because of the poor diffusivity of airflow, in the region upper and lower the closing-to-wall air nozzle, the O_2 concentration is still less than 2% and the CO concentration is higher than 1%. Due to the influence of the main airflow in the furnace, the penetration depth of the closing-to-wall air jet on the left and right sides is not the same, one in the same direction as the main airflow in the furnace has greater penetration depth. As the closing-to wall air velocity increases, the rigidity of the airflow is gradually stronger, and the penetration depth of the airflow increases, the coverage area of the airflow is increased, so that the area with oxidizing atmosphere near the water-cooled wall also increases accordingly.

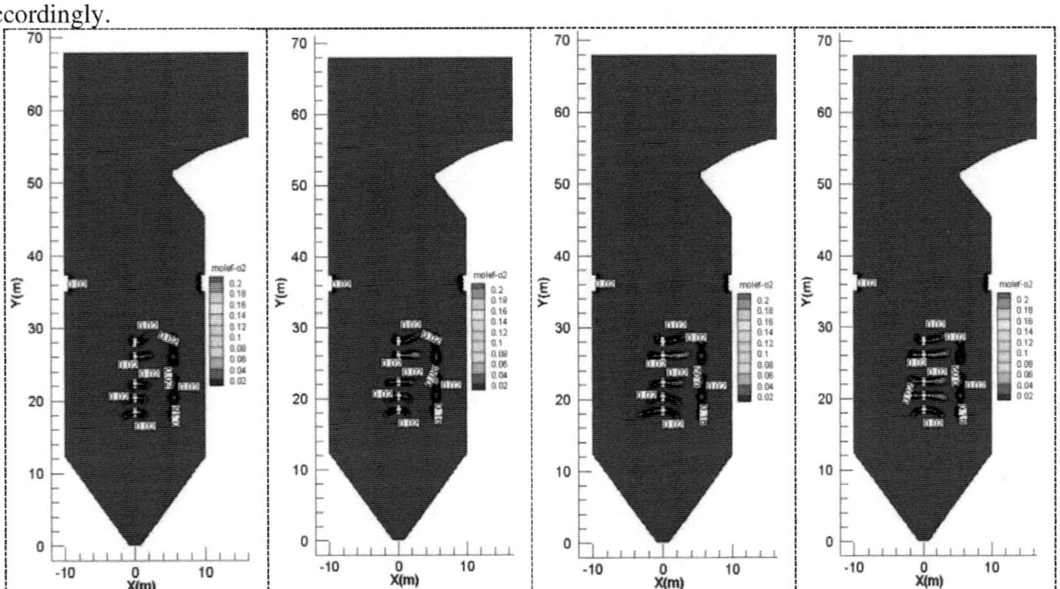

(a) 10 m/s | (b) 15 m/s | (c) 20 m/s | (d) 25 m/s

Figure 4. O_2 concentration near the wall under different closing-to-wall air velocities.

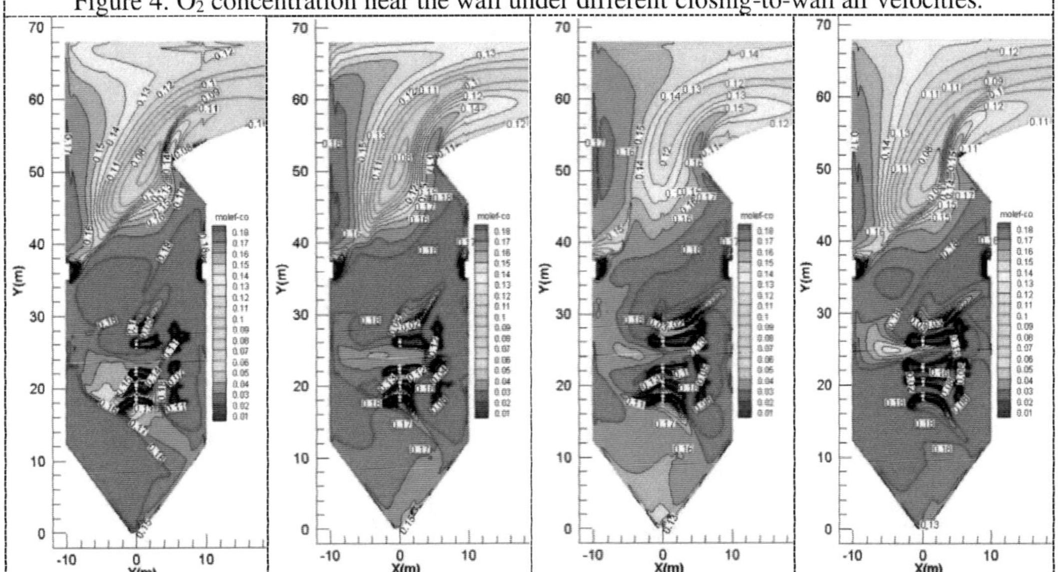

(a) 10 m/s | (b) 15 m/s | (c) 20 m/s | (d) 25 m/s

Figure 5. CO concentration near the wall under different closing-to-wall air velocities.

Figure 6 shows the temperature field in the central cross-section of the upper layer of closing-to-wall air nozzles. The temperature distribution on the cross-section is similar for the cases with different closing-to-wall air velocities. The airflows from the burners are quickly mixed with the high temperature flue gas, forming a relatively complete tangential circle in the furnace center. The flame in the furnace is not skewed, not attached to the wall, and the fullness is good. The heat load distributes uniformly near the four water-cooled walls. As the closing-to wall air velocity increases, the rigidity of the airflow is gradually stronger, and the penetration depth of the airflow increases, the area with lower temperature near the wall also increases.

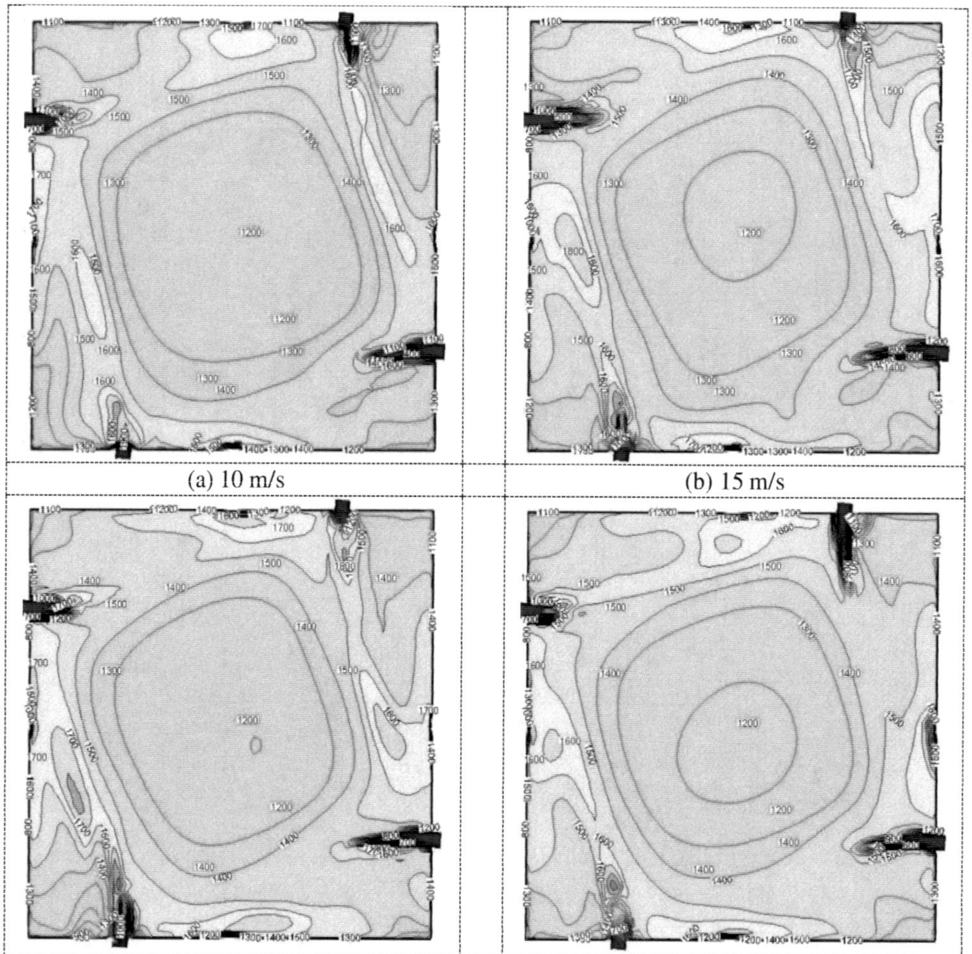

(a) 10 m/s (b) 15 m/s

(c) 20 m/s (d) 25 m/s

Figure 6. Temperature field in the central cross-section of closing-to-wall air nozzles.

Table 2 shows the average value of the flue gas temperature and component concentration at the furnace outlet. The flue gas temperature and component concentration are basically the same under different working conditions, indicating that the change of the closing-to-wall air velocity does not affect the combustion effect in the furnace. The combustibles content in the fly ash and flue gas at the exit of the furnace is not much different, indicating the influence of the closing-to wall air velocity on the combustion effect of the boiler is negligible.

Table 2. Simulation results of the parameters at the furnace outlets.

Closing-to-wall air velocity (m/s)	Gas temperature (K)	O_2 content (%)	CO content (%)	Combustibles content in the fly ash (%)
10	1299	2.89	0.18	6.82
15	1295	2.83	0.18	6.76
20	1297	2.86	0.17	6.79
25	1291	2.79	0.18	6.78

5. Conclusions

Taking a 660MW tangentially fired boiler as the research object, numerical simulations were conducted to study the influences of closing-to wall air velocity on the combustion characteristics in the boiler. The following conclusions can be drawn:

(1) The new type of closing-to-wall air proposed in this paper has good penetrability, can flow along the water wall to a distant location, and protects the water-cooled walls in the flowing region.

(2) The closing-to-wall air make the oxidizing atmosphere near the water-cooled wall more strengthen. The O_2 concentration in the flowing region is significantly increased to be greater than 2%, while the CO concentration was reduced to be less than 1%, which Can effectively alleviate high temperature corrosion of water wall.

(3) As the closing-to wall air velocity increases, the rigidity of the airflow is gradually stronger, and the penetration depth of the airflow increases, so that the coverage area of the airflow is increased.

(4) The influence of the closing-to wall air velocity on the combustion effect of the boiler is negligible.

Acknowledgments

This work was sponsored by National Key R&D Program of China (Grant No. 2016YFC0801904).

References

[1] Zhang Z.X., Cheng D.N., Wang Y.G., Zhao Q.X. (2011) Structure Design and Optimizing Simulation of a new type closing-to-wall air device. Journal of Chinese Society of Power Engineering, 2: 79-84, 102.

[2] Yao L., Chen T.J., Liu J.M. (2015) Influence of Closing-to-wall Air on Combustion Process in 660MW Opposed Firing Boiler. Journal of southeast university (Natural Science Edition), 45: 85-90.

[3] Shih T.H., Liou W.W., Shabbir A.A. (1995) New k-ε Eddy Viscosity Model for High Reynolds Number Turbulent Flows-model Development and Validation. Computers Fluids, 24: 227-238.

[4] Gosman A.D., Loannides E. (1981) Aspects of Computer Simulation of Liquid-fuelled Combustors. AIAA 19th Aerospace Science Meeting. St. Louis, pp: 323-333.

[5] Cheng P. (1964) Two-dimensional Radiation Gas Flow by a Moment Method. AIAA Journal, 2: 1662-1664.

[6] Smoot L.D., Smith P.J. (1989) Coal Combustion and Gasification. Plenum Press, New York.

[7] Zhou, L.X. (1993) Theory and Numerical Modeling of Turbulent Gas-Particle Flows and Combustion. CRC Press Inc, Florida.

IOP Conf. Series: Materials Science and Engineering **721** (2020) 012042 doi:10.1088/1757-899X/721/1/012042

Numerical simulation of combustion and emission of blast furnace gas boiler

Xuhui Ma[1,2], Songsong Zhang[2*], Hongling Yu[1], Guoli Qi[2]

[1]School of energy and power engineering, University of Shanghai for Science and Technology, Shanghai, 200093, China

[2]China Special Equipment Inspection And Research Institute, Beijing, 100029, China

[*]Corresponding author's e-mail: mxh1020@qq.com

Abstract. Taking an 110t/h full-fired blast furnace gas boiler as the research object, the numerical simulation method is used to calculate the combustion process in the furnace. The influence of flow field on combustion is analyzed, and the influence of air preheating temperature and oxygen concentration on combustion and NOx emission is discussed, which provides a theoretical basis for optimal design and reasonable operation of blast furnace gas boiler. The results indicate that with the increase of air preheating temperature and oxygen concentration, the CO consumption rate is increased, the in-furnace maximum temperature and average temperature are increased to different extents, the temperature distribution is basically unchanged, and the NOx concentration at the furnace outlet is also increased. The NOx concentration multiplies as the oxygen concentration increases.

1. Introduction

Blast furnace gas is a by-product gas of the steel industry, and is usually emitted into atmosphere by most middle and small steel enterprises. It can cause serious environmental pollution and waste of resources. How to make efficient utilize of blast furnace gas is the key to achieving energy saving and emission reduction [1-3].

Blast furnace gas as a low calorific value gas, the main components are CO, N_2 and CO_2, and the calorific value is generally 3000~3800KJ/m^3. Since blast furnace gas contains a large amount of inert gas and has low calorific value, it is difficult to ignite and unstable combustion in practical application. In response to these problems, the current stable combustion technologies are mainly: mixing high calorific value fuel, preheated air and gas, oxygen-enriched combustion and swirling combustion, etc. These stabilizing measures can achieve the purpose of optimizing the ignition conditions, increasing the flame temperature and optimizing the temperature field distribution, but also affecting NOx emissions [4]. It is of great significance to study the effect of stable combustion measures on NOx emission, and to consider the stable combustion and low nitrogen emission of blast furnace gas, so as to optimize the operation of blast furnace gas boiler and the efficient utilization of blast furnace gas.

In this paper, the 110t/h pure-burning blast furnace gas boiler is taken as the research object, meanwhile using CFD software Fluent for simulating the combustion process in furnace under different air preheating temperature and oxygen enriched concentration. The flow field, temperature field and concentration field distribution in the furnace are analyzed, which provides a theoretical basis for the optimal design of blast furnace gas boiler and the rational utilization of blast furnace gas.

2. Physical and mathematical models

2.1Physical model

The structure of the boiler and burner is shown in Figure 1. The length and width of the furnace are 6140mm and the height is 21300mm. Boiler combustion system adopt tangential firing combustion method, as well as burner adopts the swirl burners which obtain 2 layers each with four ones. The fuel is blast furnace gas with calorific value of 3338.09KJ/Nm^3.

Burner adopts unstructured tetrahedron grid, structured hexahedral grid is adopts in other regions. Also zoning meshing method is adopted to obtain high quality calculation mesh. In order to simulate the field of the burner export area accurately, the grid around the burner area should be encrypted. The total number of grid is 4.5 million, as shown in Figure 2.

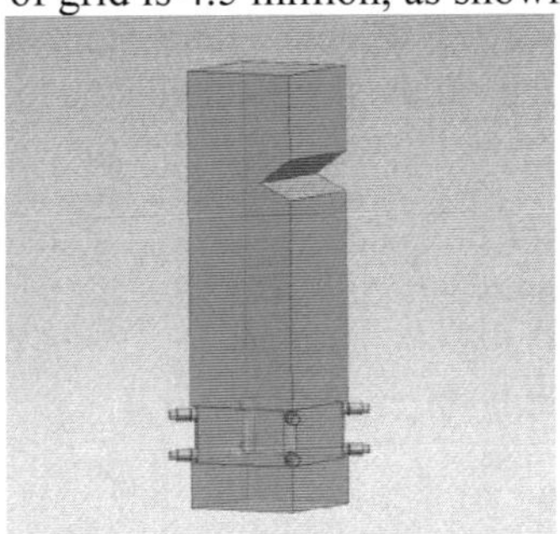

Figure 1. Furnace configuration.

Figure 2. Furnace grid.

2.2Mathematical model

For combustion, it is necessary to follow the three conservation equations of momentum, mass, and energy, and also follow the chemical composition equilibrium equation. Considering that the reliability of the model and the feasibility of engineering application, "k-epsilon model" is adopted as the model of Gas-phase Turbulent Flow, meanwhile the P-1 is adopted as the model of radiative heat transfer [5-6]. Combustion model uses the "Finite-Rate/Eddy-Dissipation" model. Model solves the coupling of pressure-speed by using standard wall function and SIMPLE algorithm [7-9].

The simulation calculation of NOx is completed by post-processing after calculating the combustion field. Since there are few nitrogen compounds in the gas, the Fuel NOx is negligible, so only Thermal NOx and Prompt NOx are considered [10]. The gas and air import boundary conditions are set to the speed inlet, and the furnace outlet is set to the pressure outlet boundary. Wall boundary conditions are thermal insulation.

3.Simulation results analysis

3.1Velocity field

Figure 3 shows the velocity distribution diagram of the cross section of the lower burner (z=2.3m), which reflects flow field characteristics of the tangential firing combustion method. After ejecting from the burner, the four streams of gas converge into a tangential circle at the center of the furnace. The flow velocity inside the tangential circle is low, thus forming a vacuum area. Figure 4 shows the velocity distribution of section at the different height. As the height increases, the flow velocity in the furnace gradually decreases. Due to the influence of the updraft, the swirling intensity of the flow is weakened. The vacuum zone in the center of the furnace gradually disappears, and the flow diffuses toward the center.

The jets interact with each other and produce a strong swirl. The heat, mass and momentum exchange of turbulent is strong, which can accelerate the fuel burning velocity after fire. The swirling flame sweeps through most of the furnace volume, and the entire furnace forms a large cyclone combustion chamber, which makes the combustion more complete and the furnace thermal load more uniform.

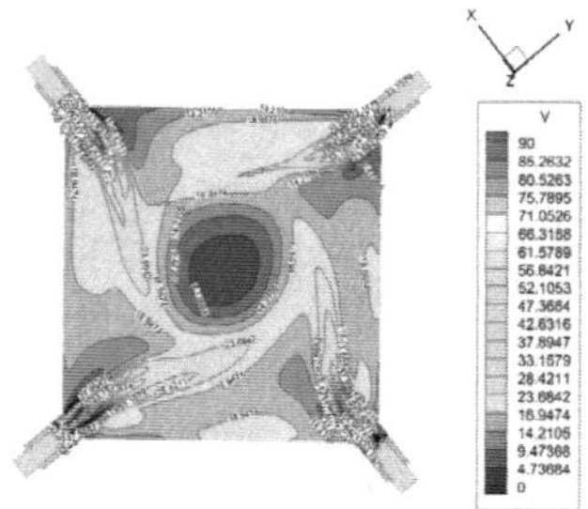

Figure 3. The velocity distribution diagram of the lower burner.

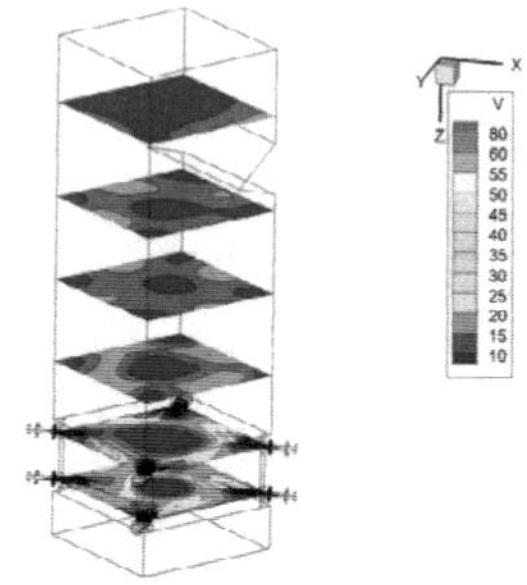

Figure 4. The velocity distribution of section at the different height.

3.2Effect of air preheating temperature on combustion process

By preheating the air, the initial temperature field of combustion can be optimized. The different preheating temperature is different, and the heat introduced into the furnace is different, which will affect the combustion efficiency and burning velocity, as well as the NOx production.

Figure 5 is the temperature nephogram of the burner region under different preheating air temperatures. As can be seen from the figure, the temperature at the outlet of the burner rises rapidly, beginning violent burning and releasing a large amount of heat. As the preheating temperature increases, the temperature variation gradient of the burner exit becomes larger, the combustion is more rapid, the high temperature zone in the furnace increases, and the temperature distribution tends to be uniform.

It can be seen from Figure 6 and Figure 7 that as the air preheating temperature increases, the maximum temperature in the furnace and the average temperature at different heights are significantly improved. The temperature rise in the furnace is most obvious when the air preheating temperature is raised from 373K to 423K. When the air preheating temperature reaches 473K, the maximum temperature in the furnace is increased by 63K. The temperature rises and then decreases along the height of the furnace.

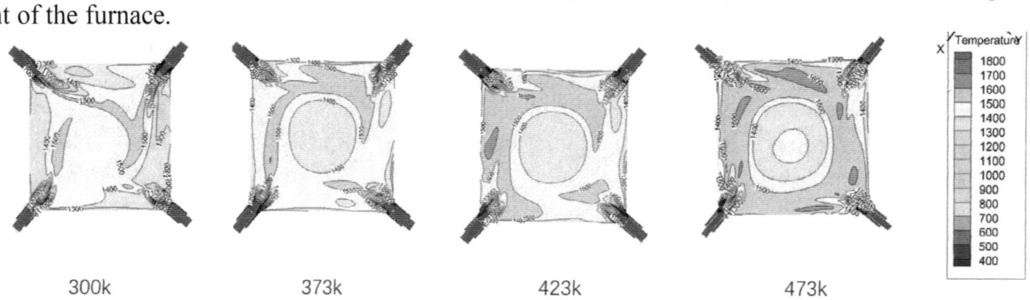

Figure 5. The temperature nephogram of the burner region under different preheating air temperatures.

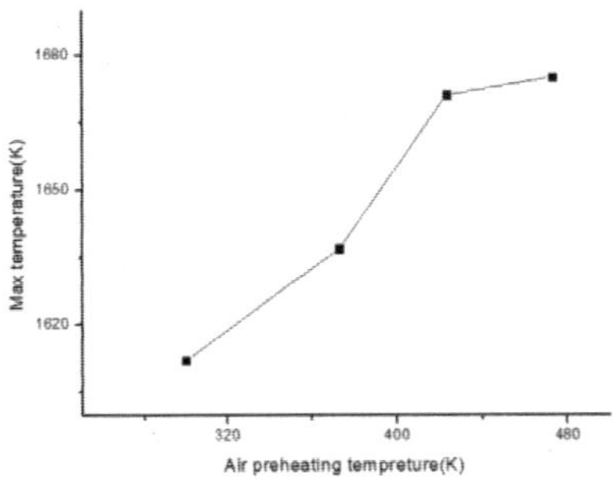

Figure 6. Variation of maximum temperature at different air preheating temperature.

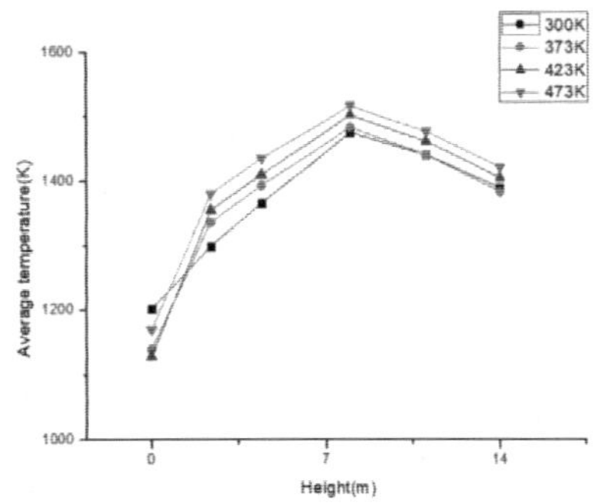

Figure 7. Variation of average temperature at different height with preheating temperature.

In order to investigate the effect of air preheating temperature on combustion rate and burnout rate, mass fraction of CO at different distance from burner outlet was calculated. As shown in Figure 8, the CO concentration is basically consistent at different air preheating temperature. At the outlet of burner, the reaction starts violently and the CO concentration decreases rapidly. The CO is almost completely consumed at 2m away from the burner outlet, and the reaction is relatively complete. When the air preheating temperature is raised from 300K to 373K, the reaction rate at the outlet of the burner will increase significantly. As the temperature continues to rise, the magnitude of this change tends to flat.

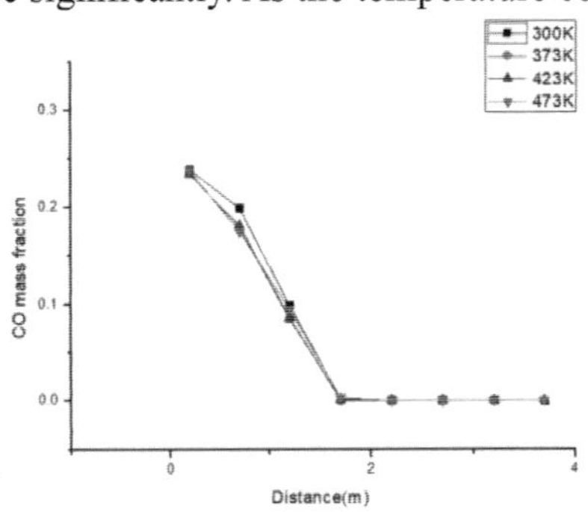

Figure 8. Mass fraction of CO at different distance from burner outlet.

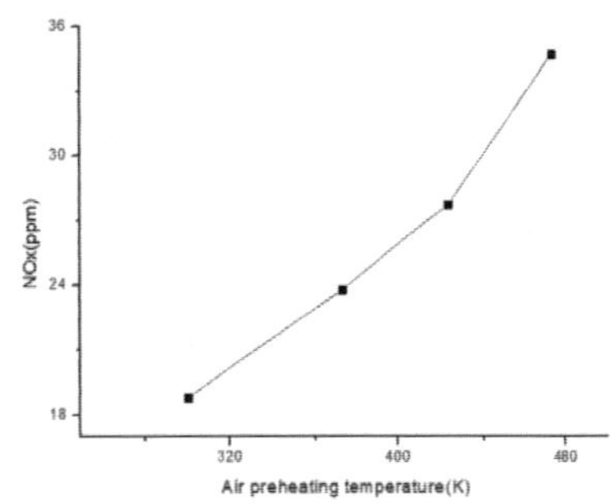

Figure 9. Variation of NOx emission at different preheating temperature.

The NOx generated by combustion of blast furnace gas boiler is mainly thermal NOx, and the generation rate can be expressed by the following equation. It can be seen that the main control factors of thermal NOx are temperature and oxygen concentration.

$$\frac{d[NO]}{dt}=6\times10^{16}[O_2]^{0.5}[N_2]T^{-0.5}e^{-69090/T} \tag{1}$$

Figure 9 shows the change of NOx concentration at the outlet of the furnace with different air preheating temperatures. As the air temperature increases, the temperature in the furnace rises, resulting in increasing in NOx concentration. When the air preheating temperature increased from 423K to 473K, the NOx concentration increased the most. The concentration of NOx at 473K preheating temperature was about 85% higher than that at 300K.

3.3Effect of oxygen concentration on combustion process

Oxygen-enriched combustion uses combustion-supporting air with higher oxygen content than normal air. This mode of combustion can speed up the reaction rate, reduce the amount of flue gas and thus reduce the heat loss from exhaust and improve the thermal efficiency. It has positive effect on the stable combustion of blast furnace gas.

Numerical simulations were carried out using three different conditions: oxygen concentration of 21%, 25%, and 29%, respectively. The effect of oxygen enriched concentration on the combustion process of blast furnace gas was analyzed. Figure 10 is the temperature nephogram of the section of the burner under three working conditions. The blast furnace gas is quickly burned from the corners to form a high temperature combustion region, and a relatively low temperature region appears in the center of the furnace. As the oxygen concentration increases, the high temperature region increases significantly, the temperature distribution becomes more uniform, and the combustion is more rapid. Figure 11 and Figure 12 show that the temperature in the furnace is improved after the oxygen-enriched combustion is carried out. The maximum temperature at 21% and 29% oxygen concentrations were 1688K and 1729K, respectively. When the oxygen concentration is increased from 21% to 25%, the average temperature in the furnace will increase significantly. When the oxygen concentration increases to 29%, the average temperature does not change significantly.

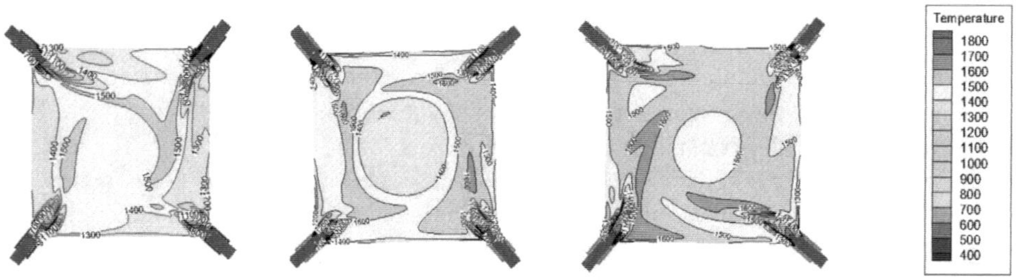

Figure 10. The temperature nephogram of the section of the burner at different oxygen concentration.

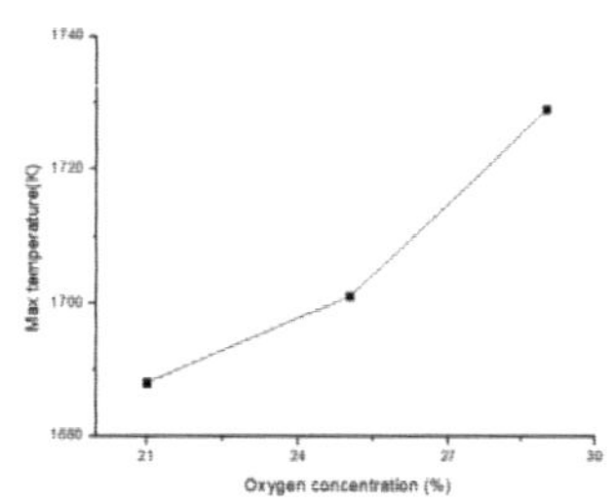

Figure 11. Variation of maximum temperature at different oxygen concentration.

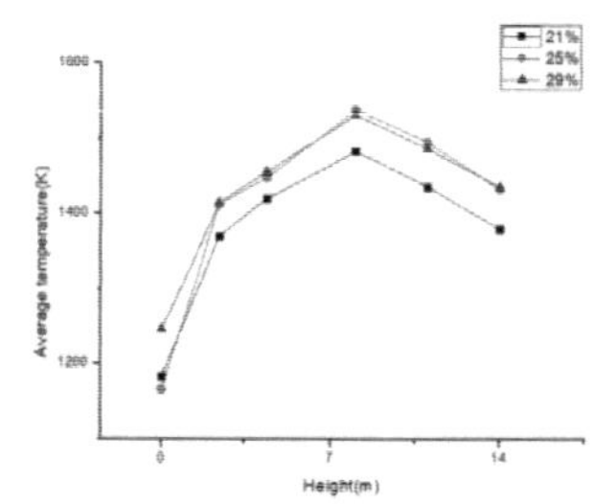

Figure 12. Variation of average temperature at different height with oxygen concentration.

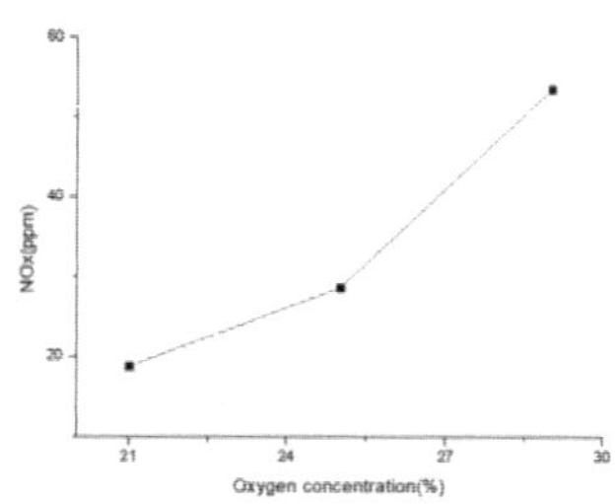

Figure 13. Variation of NOx emission at different oxygen concentration

Figure 13 shows the relation between the oxygen concentration in air and NOx concentration at the outlet of the furnace. It can be seen from the figure that the variation rule of NOx is basically the same as the maximum temperature. As the oxygen concentration increase, the amount of NOx is increasing, and the increment is large. When the oxygen content in the air is 21%, the NOx concentration at furnace outlet is the lowest; when the oxygen content is 25%, the NOx emission is 1.53 times higher than that of normal air combustion; when the oxygen content is 29%, the NOx emission reaches the normal air combustion 2.85 times. This indicates that as the oxygen content in the air increases, the NOx emissions multiply.

According to the analysis, on the one hand, oxygen-enriched combustion can reduce fuel consumption and the amount of flue gas, so that NOx emissions can be reduced to a certain extent. On the other hand, according to equation (1), the thermal NOx formation rate is exponentially related to temperature, which is proportional to the 0.5^{th} power of the oxygen concentration. Due to the increase of oxygen concentration and temperature in oxygen-enriched combustion furnace, the amount of thermal NOx generated is greatly increased. Overall, after oxygen-enriched combustion, the NOx emission is multiplied.

4.Conclusions

The numerical simulation method was used to study the blast furnace gas combustion status under different air preheating temperature and oxygen concentration, and obtained the velocity field and temperature field, concentration field, and draw the following conclusions:

1) The boiler adopts swirling burners and tangential firing combustion method. The gas flow converges into a tangential circle, form a moving vortex from bottom to top, and enhance the disturbance in the furnace. Four flames ignite each other during combustion, which is beneficial to stable and efficient combustion of blast furnace gas.

2) As the air preheating temperature increases, the combustion process of blast furnace gas can be improved. When the preheating temperature reaches 473K, the maximum temperature in the furnace increases 63K, and the high temperature region is obviously increase and the temperature distribution is more uniform. When the preheating temperature is raised from 300K to 473K, the NOx emissions from furnace outlet increased from 18.8ppm to 34.7ppm. The increase of the preheating temperature has little effect on NOx emission, and the combustion air should be fully preheated when the blast furnace gas is burned.

3) The increase of oxygen content in the air will also have a positive effect on the combustion of blast furnace gas, so that the highest temperature and the average temperature at different heights in the furnace are improved. Meanwhile, the increase of oxygen concentration has a great impact on NOx formation, the oxygen concentration is increased from 21% to 29%, and the NOx concentration at the furnace outlet is increased by 2.85 times. Therefore, the oxygen concentration of blast furnace gas during the oxygen-enriched combustion should not be too high. Considering comprehensively, oxygen concentration in combustion air should not exceed 25%.

Acknowledgement

This work was financially supported by the project supported by "National Key R&D Program of China"（2017YFF0209806）

References

[1] Worrell, E., Price, L., Martin, N. (2001) Energy efficiency and carbon dioxide emissions reduction opportunities in the US iron and steel sector. Energy, 26(5): 513–536.

[2] Shen, Y. S., Guo, B. Y., Yu, A. B., Zulli, P. (2009) A three-dimensional numerical study of the combustion of coal blends in blast furnace. Fuel, 88(2): 255–263.

[3] Matino, I., Dettori, S., Colla, V., Weber, V., Salame, S. (2019) Two innovative modelling approaches in order to forecast consumption of blast furnace gas by hot blast stoves. Energy Procedia, 158: 4043–4048.

[4] WANG, A., CAI, J., XIE, G. (2009) Numerical Simulation of Combustion Characteristics in High Temperature Air Combustion Furnace. Journal of Iron and Steel Research, International, 16(2): 6–10.

[5] Zeng, L., Chen, Q., & Zhao, X. (2011). Numerical Simulation of Combustion Processes in an Ultra-Supercritical Boiler. 2011 Asia-Pacific Power and Energy Engineering Conference. Wuhan.

[6] Mannella, C. A., Ribeiro, A., Cognon, B., & D'Arcangelis, D. (1992). Structure of paracrystalline arrays on outer membranes of rat-liver and rat-heart mitochondria. Journal of Structural Biology, 108(3):227–237.

[7] KOUTMOS, P., & MCGUIRK, J. (1991). Isothermal modeling of gas turbine combustors -Computational study. Journal of Propulsion and Power, 7(6): 1064–1071.

[8] Wu, K.-K., Chang, Y.-C., Chen, C.-H., Chen, Y.-D. (2010) High-efficiency combustion of natural gas with 21–30% oxygen-enriched air. Fuel, 89(9): 2455–2462.

[9] Wen, W., Jin, S., Huang, S., Fang, S., Zhang, X., Ao, S., Shu, L. (2011). Numerical simulation of ultra-low calorific value gas combustion. 2011 International Conference on Electronics, Communications and Control (ICECC). Ningbo. PP. 4411-4414.

[10] Park, S., Choi, G. M., Tanahashi, M. (2019). Demonstration of a gas turbine combustion-tuning method and sensitivity analysis of the combustion-tuning parameters with regard to NOx emissions. Fuel, 239: 1134–1142.

IOP Conf. Series: Materials Science and Engineering **721** (2020) 012044 doi:10.1088/1757-899X/721/1/012044

Study on Combustion Characteristics and Kinetics Analysis of Zhundong Lignite

Guanyu Wang [1], Xianchun Li [2,3*], Fanrui Meng [3,4] and Huanran Wang [2,3]

[1]China Boiler and Boiler Water Treatment Association, Beijing, 100029, China

[2]School of Chemical Engineering, University of Science and Technology Liaoning, Anshan, Liaoning, 114051, China

[3]Research Center of Advanced Coal & Coking Technology and Efficient Utilization of Coal Resources, Liaoning Province, Anshan, Liaoning, 114051, China

[4]School of Materials and Metallurgy, University of Science and Technology Liaoning, Anshan, Liaoning, 114051, China

[*]Corresponding author's e-mail: askd1972@163.com, guywan@163.com

Abstract. The high moisture and low calorific value of Zhundong lignite restricts its long-distance transportation and large-scale application. In this paper, the elemental composition changes and combustion characteristics of the upgraded lignite were studied, and then the combustion reaction kinetics of the upgraded lignite was analyzed by the Flynn-Wall-Ozawa (FWO) method. The results show that with the increase of the upgrading temperature, the content of volatiles and water in the coal decrease, which leads to an increase in the proportion of carbon, nitrogen and sulfur, and a decrease in the content of hydrogen and oxygen. The thermogravimetric curve of the upgraded lignite increases between 573K and 673K as the heating rate increases due to the influence of thermal buoyancy and inhalation gain. The burnout index Hj obviously increases with the increase of the heating rate, and the burnout condition of the coal is obviously improved. The apparent activation energy obtained at the same conversion rate and decreases with the increase of the upgrading temperature, which proves that the increase of the upgrading temperature reduces the apparent activation energy of the upgraded lignite.

1. Introduction

Fossil fuel will still play an important role in the future power generation industry [1]. Zhundong lignite is a kind of lignite with high moisture content, low calorific value and high alkali metal content, which is produced in Xinjiang, China [2]. Due to the inconvenience of transportation, low quality, slagging and other problems, Zhundong lignite is greatly restricted in the utilization process [3]. However, the low price of lignite makes it have unique advantages in the energy market. Therefore, how to make full use of lignite has important economic and social value.

Upgrading technology is a common pretreatment technology of lignite. Pre-drying and quality improvement technology include evaporation drying technology [4] and non-evaporation drying technology. Methods of evaporation dehydration include mixing grinding and drying, fluidized bed drying, pressurized steam drying, microwave drying and solar drying [6]. Non-evaporation drying methods include Reynard process, K-fuel process, hydrothermal dehydration, lignite technology upgrading, mechanical hot extrusion, etc. [6]. Among them, the technology of pre-drying and upgrading lignite with flue gas contain more waste heat produced by combustion in power plant (i.e. the waste gas that we usually think cannot be used) has a good application prospect. Zhu Xin et al. [7] studied the application and thermodynamic performance of pre-drying technology in wood fuel power plant. Xiaoqu Han et al. [8] have studied the recovery and kinetic analysis of water after pre-drying lignite, but have little knowledge of the change of combustion properties of coal during pre-drying. Hongyu Zhao [9] et al. Studied the drying and dynamic characteristics of lignite in different atmospheres, but did not explore the change of combustion characteristics of lignite in the presence of different oxygen or carbon dioxide. Thermogravimetric analysis has been widely used in the kinetic analysis process (such as co-pyrolysis of biomass and coal in different atmospheres [10], CO pyrolysis of coal and semi coke [11]). Most of the work on the thermogravimetry of pre-dried coal is focused on the change of physical and chemical properties of lignite samples before and after pre-drying. However, there is very little information about the influence of different upgrading conditions on the combustion characteristics of lignite. Combined with the theory of thermal analysis dynamics, the combustion characteristics and combustion dynamics parameters of the samples are obtained by analyzing the combustion thermogravimetry curve, so as to evaluate the combustion properties of the improved lignite.

In this paper, Zhundong lignite was improved by simulated flue gas of different temperatures, and the thermogravimetric combustion experiment was carried out by synchronous thermal analyzer, and the combustion kinetics was analyzed by FWO method.

2. Experiment

2.1. Quality improvement system and samples

The research object of this paper is Zhundong lignite from Xinjiang. Use crusher to crush Zhundong block raw coal, and then use vibrating screen machine to screen it into 0.075-0.125mm particle size for standby. See Table 1 for industrial analysis and element analysis results of coal sample.

Table 1. The proximate and ultimate analysis of ZD lignite

Proximate analysis (%)				Ultimate analysis (%)			
M_{ad}	V_{ad}	A_{ad}	FC_{ad}	C_d	H_d	O_d	N_d
6.33	51.64	29.51	12.52	75.64	4.09	14.00	0.79

The screened Zhundong raw coal is improved by using the experimental system shown in Figure 1. Take 3g ± 0.1mg of raw coal and put it into the quartz reactor, and inject 300ml / min nitrogen into the quartz reactor for evacuation. The crucible resistance furnace was heated to 423K, 623K and 773K respectively, and the constant temperature was 30min. Adjust the mass flow meter, the air flow into the quartz reactor is 300ml / min, N_2 (95%) / O_2 (5%) atmosphere (marked as II atmosphere in this experiment). Then put the quartz reactor into the furnace for 30min, take out the quartz reactor, and put the sample into the plastic bag for standby after the nitrogen of 300ml / min is introduced to cool the sample to room temperature. The samples were labelled as II-423K, II-623K and II-773K respectively.

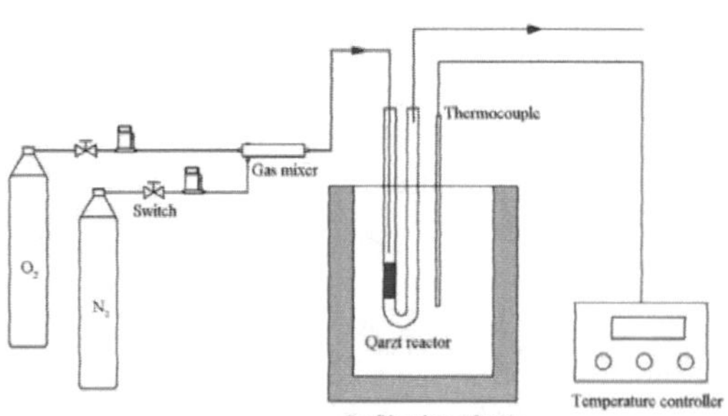

Figure 1. The lignite upgrade experiment rig

2.2. Thermogravimetric experiment

The combustion characteristics of the improved lignite under different conditions were analysed by the synchronous thermal analyzer. The non-isothermal thermogravimetry is to make the sample combust under the condition of linear temperature rising. The mass of the sample decreases with the increase of temperature. The combustion reaction characteristics of the sample are characterized by the change of conversion rate. The experimental equipment is STA449F5 type atmospheric synchronous thermal analyzer produced by NETZSCH.

Preheat the Al_2O_3 crucible to 1000K, then make it cool to room temperature and take out the crucible. Weigh 10 ± 0.005mg of sample put in the crucible with air flow of 100ml / min, set heating temperature range of 30 °C ~ 1000 °C, and heating rate of 10K / min, 20K / min and 40K / min respectively.

The sample is marked on the basis of the rising temperature rate. For example, the 10K / min sample of II-423K is marked with II-423K-10.

2.3. Data processing

The ignition temperature reflects the initial temperature of combustion reaction. In this paper, TG-DTG tangent method is used to determine the ignition temperature. The TG and DTG curves of coal combustion reaction are obtained from the thermogravimetry experiment. The weight loss peak point of the DTG curve is used as the vertical line, the intersection point of the vertical line and the TG curve is used as the tangent line of the TG curve, and the temperature corresponding to the intersection point of the tangent line and the weight loss starting parallel line is the ignition temperature T_i.

The burnout index [12] H_j is the characteristic parameter to judge the burnout performance of pulverized coal. The formula of H_j is

$$H_j=\frac{(\frac{dw}{dt})_{max}}{T_i \cdot T_{max} \cdot \frac{\Delta T_h}{\Delta T}} \tag{1}$$

In this paper, multiple scanning rate method is used, that is, TGA curves obtained at different heating rates are used to analyze combustion dynamics. Because the data at the same conversion rate on multiple TGA curves are usually used in calculation, it is also called equal conversion method [13]. The outstanding advantage of using the equal conversion method to estimate the activation energy is that the activation energy value can be calculated directly without considering the reaction mechanism function, so the error caused by the wrong choice of the reaction mechanism function can be effectively avoided.

The reaction rate of combustion, defined as the change of unit time, is calculated according to the following formula:

$$\frac{d\alpha}{dt}=k(T)f(\alpha) \tag{2}$$

$$\alpha=\frac{m_i-m_\alpha}{m_i-m_f}\times 100\% \tag{3}$$

Where α is the conversion rate, t is the time, T(K) is the absolute temperature, and k(T) is the rate constant about the temperature. m_0, m_α and m_f are the initial mass, the current mass and the residual mass after upgrading, respectively.

According to Arrhenius equation, the expression of k (T) is

$$k(T)=A\exp(-\frac{E}{RT}) \tag{4}$$

Where a is the frequency factor (1 / min), E is the activation energy, in kJ / mol, R is the molar gas constant, and the value is 8.314 $J/mol^{-1} \cdot K^{-1}$.

The combustion activation energy of the samples was calculated by Flynn Wall Ozawa (FWO) method [14]. The equation of Flynn wall Ozawa (FWO) method can be expressed as follows:

$$\ln(\beta)=\ln[\frac{0.0084AE_\alpha}{RG(\alpha)}]-1.5016\frac{E_\alpha}{RT} \tag{5}$$

At the same conversion rate α, G (α) is a fixed value. From the equation, it can be seen that ln (β) has a linear relationship with 1 / T. The apparent activation energy value Eα under the conversion rate α can be calculated by the slope of the straight line.

3. Results and discussion

3.1. Element analysis

The elemental analysis of different lignite is carried out by using the elemental vario macro cube constant element analyzer, and the data is listed in Table 2. It can be seen from the data in the table that with the increase of upgrading temperature, the contents of carbon, nitrogen and sulfur elements increase, while the contents of hydrogen and oxygen elements decrease. This is because with the increase of the temperature of upgrading, the amount of volatile matter and water in the coal decrease. At this time, due to the low oxygen content, the amount of carbon combustion loss is relatively small, which leads to the increase of the proportion of carbon, nitrogen and sulfur. Because the change of the content of elements in the improved lignite samples will have an important impact on its combustion properties, especially the increase of the proportion of combustible elements such as carbon will increase the calorific value of the fuel, and the proportion of different elements will also affect the combustion difficulty of the improved lignite.

Table 2. Ultimate analysis of upgraded lignite

	II-423K	II-623K	II-773K
C_d (%)	58.06	59.43	64.35
H_d (%)	4.016	3.162	2.12
S_d (%)	0.88	1.01	1.03
N_d (%)	0.194	0.242	0.319
O_d (%)	36.85	36.156	32.181

3.2. Thermogravimetric Characteristics

3.2.1. TG and DTG curves of lignite combustion at 10 K / min

Figure2 shows the TG and DTG curves of the improved lignite during combustion at different temperatures, in which the solid line is the TG curve and the dotted line is the DTG curve. It can be seen from Figure2 that the thermogravimetry curve of lignite with different upgrading temperatures has little difference at the initial stage of reaction, but it is arranged from low to high in the order of 437K, 623K to 773K from 573K to 673K. This is due to the different rate of combustion and heat release of the samples, resulting in the different magnitude of the heated buoyancy. The samples with high heating buoyancy are the samples with 773K upgrading temperature, and the TG curve of the samples has a large upward shift. At the same time, in the process of upgrading lignite, water and volatile matter will be lost, resulting in a relatively developed pore structure. In the process of combustion, a small amount of oxygen absorption or nitrogen absorption will increase weight, which will also cause the TG curve to shift upward [15]. With the increase of the temperature of upgrading, the moisture and a large amount of volatile matter released from lignite will increase, the coke pore will be more developed and more gas will be absorbed. The thermogravimetric curve shows that the maximum weight loss rate is reached between 673K and 773K, indicating that the combustion reaction is the most intense at this time. The coincidence degree of TG and DTG curves after 773K is high, which shows that the combustion process has little difference.

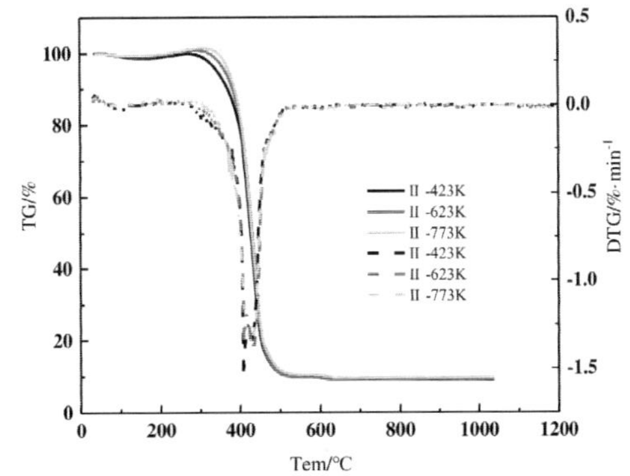

Figure 2. TG and DTG curves of upgraded ZD lignite under different temperature

3.2.2. TG and DTG curves of the same lignite under different heating rates

As shown in Figure 3, the slope of TG curve decreases with the increase of heating rate between 673K and 873K, and the peak point of DTG curve increases with the increase of heating rate. This is because with the increase of the heating rate, the temperature difference for inside and outside of the sample particles increases, resulting in the thermal lag phenomenon. The internal temperature of the particles is relatively low, and the reaction rate is relatively smaller than outside, resulting in the delay of volatilization analysis, and the temperature corresponding to the maximum weight loss rate point increases [16]. It can be seen clearly in the figure that the reaction starting temperature of the sample with lower heating rate is lower, because the sample has relatively sufficient time to absorb oxygen and be heat up under the condition, and then starts the combustion reaction. The peak area of DTG curve increases with the increase of heating rate. The peak area of DTG curve corresponds to the mass loss of sample during combustion, so it means that when the weight loss is reached during the reaction, the sample with higher heating rate will have lower temperature.

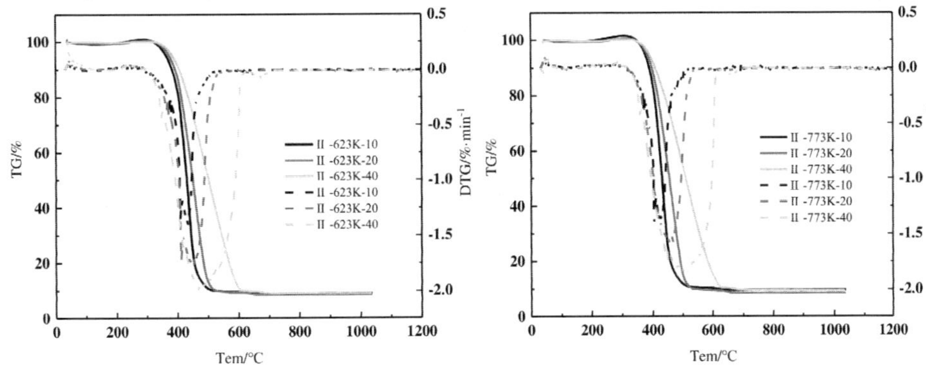

(a) II-623K Upgraded Lignite (b) II-773K Upgraded Lignite

Figure 3. TG and DTG curves of upgraded ZD lignite at different heating rates

3.3. Combustion characteristic parameters

The combustion characteristic parameters of raw coal and improved lignite with different heating rates are shown in Table 3. The combustion characteristics of the same sample with different heating rates are obviously different. The maximum weight loss rate temperature T_{max} increases with the increase of heating rate, which is also due to the effect of thermal lag on sample combustion. The burnout index HJ increases obviously with the increase of heating rate, indicating that the burnout of coal increases obviously with the increase of heating rate. But the Δ t of lignite increased from 423K, 623K to 773K at different temperatures. This is because the higher the temperature of lignite is, the more complete the non-combustible material such as water is, the more the proportion of the remaining coal combustible material is. This is also consistent with the results of element analysis that the higher the temperature of lignite is, the higher the content of carbon, nitrogen and sulfur is, and the lower the content of hydrogen and oxygen is.

Table 3. Combustion characteristic parameter of upgraded lignite

Sample No.	T_i	T_{max}	ΔT	$H_j(\times 10^{-6})$
RC-10	384.4	405.9	36.6	0.90
RC-20	397.4	444.4	77	1.53
RC-40	385.2	455	180	3.43
II-423K-10	382.5	407.9	41.9	0.96
II-423K-20	391.2	437.4	89	2.01
II-423K-40	389.9	459.9	180	3.47
II-623K-10	394.7	430.3	49.4	0.90
II-623K-20	395.7	453.1	86	1.77
II-623K-40	394.9	466.3	195	3.60
II-773K-10	394.2	429.9	50	0.85
II-773K-20	397.2	456.9	100	1.55
II-773K-40	400.6	499.4	215	3.22

3.4. Kinetic parameters

The FWO method is used to fit the combustion activation energy of raw coal and improved lignite samples, and the fitting image is shown in Figure 4.

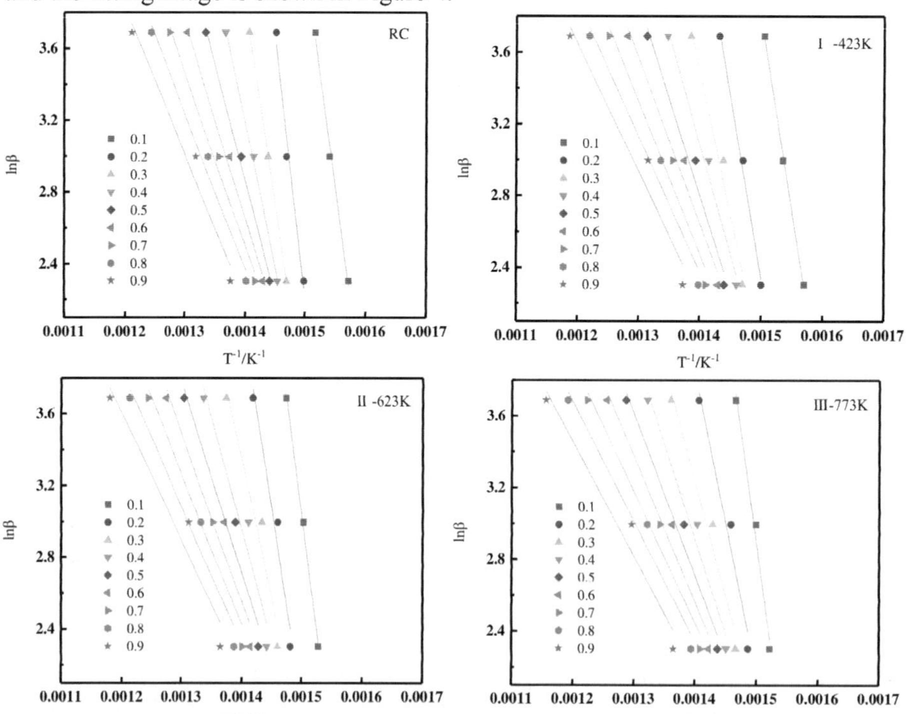

Figure 4. Fitting images of sample combustion activation energy by FWO method

According to the data fitting in Figure4, the activation energy value is calculated by FWO method, which is arranged in Table 4.

As shown in Figure 4, the fitting effect is better at the lower conversion rate, while the straight line fitted at the higher conversion rate deviates from the data greatly.

Except that the value of $\alpha = 0.1$ of RC sample is smaller than that of $\alpha = 0.2$, the activation energy of other samples decreases from 0.1 to 0.9. With the increase of conversion, the activation energy of the sample decreased obviously. The higher the apparent activation energy is, the lower the activation energy is and the easier the combustion reaction is.

The apparent activation energy values of the samples with the same conversion rate at different upgrading temperatures decrease with the increase of the upgrading temperature, which proves that the increase of the upgrading temperature reduces the apparent activation energy of the improved lignite and enhances the reactivity of the improved lignite.

Table 4. The activation energies obtained from FWO method for pyrolysis of biomass samples

α	RC	II-423K	II-623K	II-773K
0.1	197.82	237.28	202.96	255.43
0.2	224.78	190.25	165.67	161.16
0.3	173.24	140.46	123.04	113.8
0.4	128.16	113.7	100.23	93.58
0.5	104.04	94.2	84.96	77.27
0.6	87.32	81.53	74.15	68.31
0.7	75.77	71.57	65.88	61.26
0.8	67.9	64.61	59.56	55.33
0.9	63.92	60.56	55.57	51.95

4. Conclusion

With the increase of upgrading temperature, the amount of volatile matter and water in the coal decrease, so the proportion of carbon, nitrogen and sulfur increases, and the content of hydrogen and oxygen decrease.

The thermogravimetric curve of the improved lignite increases with the increase of heating rate due to the influence of thermal buoyancy and suction between 573K and 673K. The slope of TG curve decreases with the increase of heating rate between 673K and 873K, and the peak point of DTG curve increases with the increase of heating rate.

The maximum weight loss rate temperature T_{max} and the burnout index H_j increases with the increase of heating rate. Indicating that the burnout of coal increases with the increase of heating rate.

The value of apparent activation energy of lignite samples with the same conversion rate under different upgrading temperature decreases with the increase of upgrading temperature, which proves that the increase of upgrading temperature reduces the apparent activation energy of lignite and enhances the reactivity of lignite.

References

[1] Davies K., Malik A., Li J., et al. (2017) A meta-study on the feasibility of the implementation of new clean coal technologies to existing coal-fired power plants in an effort to decrease carbon emissions. PAM Review: Energy Science & Technology, 4: 30-45.

[2] Li J., Zhu M., Zhang Z., et al. (2016) Characterisation of ash deposits on a probe at different temperatures during combustion of a Zhundong lignite in a drop tube furnace. Fuel processing technology, 144: 155-163.

[3] Li J., Zhu M., Zhang Z., et al. (2016) The mineralogy, morphology and sintering characteristics of ash deposits on a probe at different temperatures during combustion of blends of Zhundong lignite and a bituminous coal in a drop tube furnace. Fuel Processing Technology, 149: 176-186.

[4] Han X., Karellas S., Liu M., et al. (2017) Integration of Organic Rankine Cycle with Lignite Flue

Gas Pre-drying for Waste Heat and Water Recovery from Dryer Exhaust Gas: Thermodynamic and Economic Analysis. Energy Procedia, 105: 1614-1621.

[5] Pusat S., Erdem H.H. (2017) Drying characteristics of coarse low-rank-coal particles in a fixed-bed dryer. International Journal of Coal Preparation and Utilization, 37(6): 303-313.

[6] Rong L.K., Song B., Yin W.Z., et al. (2017) Drying behaviors of low-rank coal under negative pressure: Kinetics and model. Drying technology, 35(2): 173-181.

[7] Xu C., Li X., Xu G., et al. (2018) Energy, exergy and economic analyses of a novel solar-lignite hybrid power generation process using lignite pre-drying. Energy Conversion and Management, 170: 19-33.

[8] Yan J., Bai Z., Hao P., et al. (2017) Comparative study of low-temperature pyrolysis and solvent treatment on upgrading and hydro-liquefaction of brown coal. Fuel, 199:598-605.

[9] Zhu X., Wang C., Wang L., et al. (2018) Thermodynamic and economic analysis on a two-stage predrying lignite-fueled power plant. Drying Technology, 2018: 1-12.

[10] Han X., Yan J., Karellas S., et al. (2017) Water extraction from high moisture lignite by means of efficient integration of waste heat and water recovery technologies with flue gas pre-drying system. Applied Thermal Engineering, 110: 442-456.

[11] Han X., Karellas S., Liu M., et al. (2017) Integration of Organic Rankine Cycle with Lignite Flue Gas Pre-drying for Waste Heat and Water Recovery from Dryer Exhaust Gas: Thermodynamic and Economic Analysis. Energy Procedia, 105: 1614-1621.

[12] Liu Y., et al. (2018) Thermal analysis on combustion characteristics of predried dyeing sludge. Applied Thermal Engineering, 140: 158-165.

[13] Li Q., Zhao C., Chen X., et al. (2009) Comparison of pulverized coal combustion in air and in O_2/CO_2 mixtures by thermo-gravimetric analysis. Journal of Analytical and Applied Pyrolysis, 85(1-2): 521-528.

[14] Ma B.G., Li X.G., Xu L., et al. (2006) Investigation on catalyzed combustion of high ash coal by thermogravimetric analysis. Thermochimica Acta, 445(1): 19-22.

[15] Hu R.Z., Gao S.L., et al. (2008) Thermal Analysis Kinetics. Science Press, Beijing.

[16] Flynn J.H., Wall L.A. (1966) A quick, direct method for the determination of activation energy from thermogravimetric data. Journal of Polymer Science Part C: Polymer Letters, 4(5): 323-328.

IOP Conf. Series: Materials Science and Engineering **721** (2020) 012045 doi:10.1088/1757-899X/721/1/012045

Prediction and Analysis of Bamboo heating value Near Infrared Spectroscopy Based on Competitive Adaptive Weighted Sampling Algorithm

Peiwei Jin[1], Yan Yao[1*], Qian Qiu[1], Huijun Liu[1], Xiaoyu Liang[1], Jinhui Cai[1]

[1]college of metrology & measurement engineering, China Jiliang university, Zhejiang, 310018, China

[*]Corresponding author's e-mail: yaoyan@cjlu.edu.cn

Abstract. China is one of the largest bamboo producers in the world. Bamboo has the characteristics of large biomass and fast growth, which is considered to be an ideal biomass energy source. In order to satisfy the needs of modern biomass conversion processes, there is an urgent need to develop a method for rapidly determining the heating value of bamboo. In this paper, the spectral data of 80 bamboo samples were obtained by Near-Infrared Spectroscopy (NIRS), and then the heating value of bamboo was measured by conventional methods. Through the analysis of bamboo sample data, the Partial Least Squares Regression (PLSR) algorithm, and the Competitive Adaptive Weighted Sampling and PLSR combination (CARS-PLSR) algorithm were used to establish the relationship between spectral data and heating value. The results show that the number of characteristic wavelengths obtained by the CARS algorithm is 24, while the wavelength of the original medium-long wavelength band is 1320, which reduces the number of wavelengths by 98.18%, and greatly reduces the data size. In addition, the selected feature wavelengths can better reflect the key information of the bamboo spectrum, which have a better model effect than the unscreened model. In this study, the heating value index (R2c, R2p, RPD) of the CARS-PLSR model was 0.97, 0.98, and 6.22, respectively. The bamboo heat generated PLSR model was established by using the characteristic wavelength selected by CARS algorithm, which greatly simplified the complexity of the model and improved the predictive ability of the model.

1. Introduction

Energy shortage has become a worldwide problem. With the rapid development of China's economy and the continuous improvement of people's living standards, energy issues have become increasingly prominent [1]. As the world's fourth largest energy source, biomass energy is located behind the three fossil energy sources of coal, oil and natural gas [2], and is one of the most promising renewable energy sources [3]. At present, biomass provides only 15%~20% of global energy demand, but it is more important for developing countries, especially in rural areas [4].

Born in China, bamboo is a forestry resource with Chinese characteristics, which has valuable features such as large biomass, cluster growth, short material production period and easy breeding [5] and great advantages as a biomass energy development and utilization. China is one of the countries with the widest distribution of bamboo in the world, mainly distributed in the south, and concentrated in Fujian, Zhejiang, Jiangxi, Anhui, and Sichuan, Guangxi, Yunnan and some other provinces in the western China [1]. According to the results of the Eighth Forest Resources Survey, the existing artificially operated bamboo forests in China are over 6 million hectares and are growing at an annual rate of over 150,000 hectares [6]. In China, as bamboo's planting area is large and still in rapid growth, it can be used as an ideal source of biomass. In recent decades, as the forest area on the earth's surface has been decreasing year by year but the area of bamboo forest has been expanding, and combined with that bamboo is an valuable biomass resources, it can be foreseeable that bamboo will play an increasingly important role in China's sustainable development strategy.

The heating value is an indicator of the energy content of the reaction fuel [7-8], which refers to the amount of heat generated per unit mass of fuel when it is completely burned. The heating value can be further divided into high heating value and low heating value. The high heating value refers to the heating value, which generated by the unit mass of fuel when it is completely burned, minus the heat generated by nitric acid and the heating value obtained after the sulfuric acid is corrected. The lower heating value is the heating value obtained by subtracting the heat of vaporization of water from the higher heating value. The heating value directly affects the availability of biomass energy. At present, there are two direct measurement methods and approximate analysis methods for fuel heating value [9]. The direct measurement method is based on the oxygen bomb method and is directly measured by a calorimeter. Although this method is relatively simple and inexpensive, it is a time-consuming and tedious work [10]; the approximation analysis uses a correlation between the composition and the heating value to obtain a simple prediction model to predict the heating value. The main approximation analysis methods include three methods of chemical components, elemental components and industrial component analysis. Although this method avoids dependence on lengthy experimental techniques [11], the data for mathematical models need to build relying on expensive experimental equipment, require professional technicians for experimental operation, and consume lots of labor and financial resources. Due to the widespread use of biomass materials, the traditional analysis speed has not met the needs of biomass thermal conversion process, it is necessary to explore a new rapid heating value measurement method.

NIRS is considered as a fast, non-destructive, non-polluting analytical method that can be used to evaluate the properties of biomass [12]. As a faster laboratory alternative, NIRS has been successfully applied to the heating value detection of a variety of biomass energy sources. Huang [13] and others used NIRS to establish a heating value prediction model of straw, and combined with Local algorithm to explore the model optimization, which confirmed that NIRS can realize the determination of straw heating value. Xiong [14] and others based on NIRS to establish a prediction model for the high heating value of five crop straw species. The results show that NIRS can provide basic model support for the rapid detection of high-heating value of crop straw. Sheng [15] and others found that the rapid detection of biomass material characteristics is of great significance for the production of high-quality compression-molded fuel. Therefore, the NIRS technique was used to establish the heating value prediction model of pine, fir and cotton stalks. The evaluation indexes R2 and RPD are 0.87 and 2.73, respectively. The experimental results show that the NIRS technology can completely replace the traditional industrial analysis method to achieve rapid determination of heating value. In addition to the widespread use of crop waste, bamboo has been actively used in the production of fuel as a material widely used in Asia. Thailand is one of the major bamboo producing countries in the Asia-Pacific Bamboo Region. It consists of 13 genera and 60 species, accounting for 5% of the world's total. Jetsada Posom [16-17] et al. used Fourier transform NIRS instead of the traditional bomb calorimetry to determine the heating value of Thailand bamboo, and used PLSR algorithm to establish predictive model of different positions of 80 bamboo buds. The model can better predict the characteristics of bamboo samples, thus improving the efficiency of bamboo as a biomass energy conversion process.

It can be seen that the NIRS technology is expected to achieve rapid detection of heating value. However, there are more than 500 kinds of bamboo species distributed in China, and the chemical composition of different varieties is also different, resulting in different information carried by the near-infrared spectrum, and the bamboo heating value prediction model that Jetsada Posom studied only includes few bamboo species and is not applicable to China. In this paper, NIRS is used to obtain the spectral information of bamboo samples. PLSR algorithm and combination algorithm of CARS and PLSR are used to establish a bamboo heating value prediction model for long spectral bands and characteristic bands to achieve rapid acquisition of bamboo heating value.

2. Materials and Method

2.1Sample preparation

The bamboo samples were collected from the China Bamboo Expo Park in Anji County, Zhejiang Province. 40 bamboo varieties including Guizhu, Danzhu and Leizhu were collected. A total of 80 experimental samples including 40 bamboo leaves and 40 bamboo branches were sampled from the different growth parts of bamboo. The collected fresh bamboo samples were placed in an outdoor environment at a temperature of 37°C to be air-dried to dryness. After the sample is separated from the external moisture, it is placed in a 65°C electric heating blast drying oven (Hangzhou Blue Sky Laboratory Instrument Factory, DHG-9070A) for 6 hours, and then pulverized and dried by a pulverizer (Xulang, XL-10B). The sample is placed in a sealed plastic bag through a 60 mesh screen (0.25 mm), placed in a sealed container, stored in a desiccator at room temperature (25°C), and coated with Vaseline at the dryer ground. In this way, the sample is stored until the experiment is completed.

2.2heating value determination method

The commonly used method for direct determination of heating value is the oxygen bomb method. The oxygen bomb calorimeter (Shanghai Changji Geological Instrument, HRY-1B) is the main equipment for heating value determination. The experiment was carried out in accordance with the national standard GB/T 213-2008 coal heating value determination method [18]. Before the experiment, the calorimeter was first calibrated with benzoic acid, and the calibration experiment was performed at least 5 times until the relative error of the 5 results was less than 0.2%. The experiment was carried out by wrapping 0.8 g ± 0.1 g of the experimental sample with a known quality of rubbing paper. The weighing required during the experiment was carried out by using a METTLER TOLEDO electronic analytical balance. Each sample is made in three parallels and averaged. The measured result is the heating value of the cartridge of the sample. In practical applications, it needs to be converted into high heating value and low heating value. The heating value of the cartridge minus the heat of nitric acid formation and the heat of sulfuric acid is the high heating value. The value obtained by subtracting the heat of dehydration from the high heating value is the low heating value. The sulfur content and hydrogen content required to calculate the high and low heating values were determined using an EA3000 elemental analyzer (EURO, Italy).When the S content is less than 0.1%, it is negligible. The sample moisture was measured in a blast drying oven according to the national standard GB/T 28731-2012 [19]. The formula for calculating the high and low heating values is shown as follows:

$$HHV = Q_b - 94.1S - 0.0012Q_b \tag{1}$$

$$LHV = HHV - 206H - 23M \tag{2}$$

Among them, HHV is the high heating value of the sample, Qb is the bomb heating value of the sample, S is the sulfur content of the sample, 94.1 is the correction value of each 1% sulfur in the sample, 0.0012 is the heat of nitric acid formation, and LHV is the low heating value of the sample, H is the hydrogen content of the sample, and M is the moisture content of the sample.

2.3Near infrared spectroscopy

Powdered samples usually use diffuse reflection analysis to obtain chemical information. The light signals obtained by diffuse reflection analysis mainly come from the shallow layer of the sample. The short-wave near-infrared penetrating ability is strong, the reflected back light is weak, and the short-wave diffuse reflected light is not loaded with sufficient chemical information for measurement. Therefore, when using the diffuse reflection analysis method, the near-infrared short-wave region should not be applied to chemical analysis. Therefore, the medium-long wave (1100nm~2500nm) is selected for NIRS quantitative analysis. Savitzky-Golay Convolution Smoothing (SG) is used to smooth the spectral data to eliminate noise, and combined with Multi-Scatter Correction (MSC) to eliminate the influence of sample particle distribution and particle size.

3. Results and discussion

3.1High and low heating value measurements

The high and low heating value and related component determination results of 80 samples are shown in Table 1. The moisture content of the sample ranged from 4.49% to 8.97%, the average value was 5.83%, and the standard deviation was 0.92. The hydrogen content ranged from 5.26% to 6.34%, the average value is 5.84%, and the standard deviation is 0.21. The sulfur element content ranges from 0% to 0.68%, the average value is 0.14%, and the standard deviation is 0.11. The high heating value range is 16077.38J/g~19127.72 J /g, the average value is 18050.26 J/g, and the standard deviation is 670.73. The low heating value range is 16064.72 J/g~19113.79 J/g, the average value is 18036.88 J/g, and the standard deviation is 670.45. The variation range of sample moisture content is relatively wide, and the standard deviation is large, indicating that the sample source is wide, which can represent most of the bamboo energy utilization research; the average content of S is only 0.14%, which is much smaller than 1% of the standard sulfur content of coal in China, and can be reasonably developed and utilized without affecting the ecological environment and is an environmentally friendly renewable energy source. Its high and low heating value are similar with large range span, which has a high representativeness.

Table 1. Statistical results of bamboo heating value and related components

Characteristic	Bamboo		
	Range	Average value	Standard deviation
Moisture (%)	4.49~8.97	5.83	0.92
H (%)	5.26~6.34	5.84	0.21
S (%)	0~0.68	0.14	0.11
High heating value (J/g)	16077.38~19127.72	18050.26	670.73
Low heating value (J/g)	16064.72~19113.79	18036.88	670.45

3.2Near infrared spectroscopy acquisition results

Figure 1 is a plot of the original medium-long-wave NIRS data for 80 samples, which uses PCA combined with Mahalanobis distance to eliminate sample outliers. The wavelengths covered in the figure range from 1100 nm to 2500 nm, and the ordinate is the sample concentration corresponding to each wavelength. The chemical constituents of bamboo are mainly cellulose, hemicellulose and lignin. Cellulose's structure is mainly composed of C-H, N-H and O-H bonds. It can be seen from the figure that the absorption peaks of the hydrogen-containing groups of the sample mainly appear near 1450 nm, 1720 nm, 1940 nm, 2150 nm, and 2350 nm. These absorption peaks reflect the constituent groups of the main active components of the sample, and the O-H double-frequency absorption band is around 1450 nm. The vicinity of 1720 nm is the C-H double frequency absorption band; the vicinity of 1940 nm is the absorption band of O-H and H2O; the vicinity of 2150 nm is the combined absorption band of N-H; and the vicinity of 2350 nm is the combined absorption band of C-H. The NIRS results of the collected samples accord with the chemical composition of the bamboo.

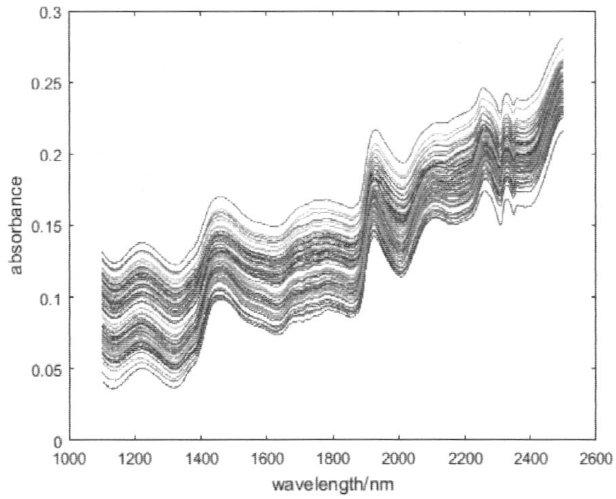

Figure 1. Long wave near infrared spectra of bamboo

3.3Model evaluation index

The criteria for judging the pros and cons of the model include the pros and cons of the model itself and its predictive ability. This paper uses the following criteria to evaluate the model:

(1)Decisive factor

$$R^2 = 1 - \sum_{i=1}^{n} (y_i - \hat{y}_i)^2 / \sum_{i=1}^{n} (y_i - \bar{y})^2 \quad (3)$$

R2 represents the degree of fit between the measured and predicted values of the model. The closer R2 is to 1, the better the model effect. yi is the measured value of the i-th sample, $\hat{y}_i$ is the predicted value of the i-th sample, $\bar{y}$ is the average value of all samples, and n is the number of samples.

(2)Root Mean Square Error (RMSE)

$$RMSE = \sqrt{\frac{1}{n} \sum_{i=1}^{n} (\hat{y}_i - y_i)^2} \quad (4)$$

RMSE represents the root mean square error of the training set. When nonlinearly fitting, the smaller the RMSE, the better the result.

(3) Root Mean Square Error Of Prediction (RMSEP)

$$RMSEP = \sqrt{\frac{1}{N} \sum_{i=1}^{N} (y_i - \hat{y}_i)^2} \quad (5)$$

RMSEP is an important indicator to predict the prediction accuracy of the prediction set. The smaller the RMSEP value, the closer the predicted value is to the measured value.

(4) Root Mean Square Error of Cross Validation (RMSECV)

$$RMSECV = \sqrt{\frac{1}{N-1} \sum_{i=1}^{N} (y_i - \hat{y}_i)^2} \quad (6)$$

RMSECV is an important indicator of the feasibility of modeling methods and the ability of model prediction.

(5) Standard Deviation (SD) of prediction set

$$SD = \sqrt{\frac{1}{N-1} \sum_{i=1}^{N} (\hat{y}_i - \bar{y})^2} \quad (7)$$

SD represents the degree of deviation between the predicted value of the prediction set and the mean.

(6) Relative Percent Deviation

$$RPD = SD / RMSEP \tag{8}$$

RPD is the relative analysis error of the prediction set and is an important evaluation parameter for evaluating the resolving power of the model. When RPD>3, the model has good predictive ability and can be used for practical application; when 2.5<RPD<3, the model is feasible with good prediction effect; when 2<RPD<2.5, the model prediction effect is not good and can be used for approximate analysis.

3.4Medium and long waveband PLSR modeling

Using SPXY partitioning, 80 samples were divided into training and prediction sets in a ratio of 3:1. The PLSR algorithm is used to establish a heating value prediction model for the long-wavelength of the sample. The RMSECV value and the model RPD value obtained by leaving the cross-validation method are used to confirm the number of hidden variables of the model. Figure 2 and Figure 3 show the relationship between the number of hidden variables and RMSECV and RPD in the PLSR modeling process. When the number of latent variables is 15, the RMSECV of the thermal value reaches the minimum and the RPD reaches the maximum.

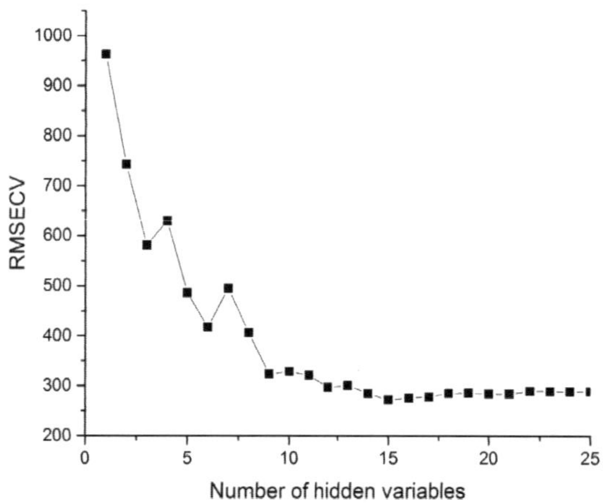

Figure 2. erelationship between the number of hidden variables and RMSECV

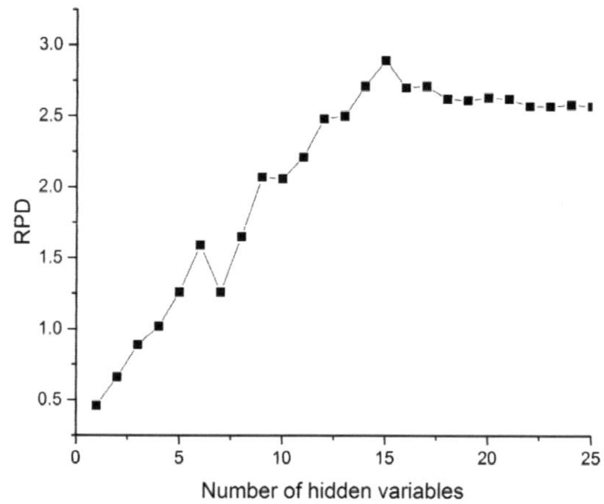

Figure 3. erelationship between the number of hidden variables and RPD

3.5Characteristic wavelength screening

Reducing the model wavelength by wavelength selection can better explain the relationship between the digitized spectrum and the characteristics to be studied [20]. The CARS characteristic wavelength screening algorithm is based on the principle of survival of the fittest in Darwin's theory of evolution, with the advantages of faster calculation speed and fewer characteristic wavelength need to be screen, and have been widely used in the screening of Near-Infrared characteristic variables. The characteristic wavelength screening is performed by the CARS algorithm, the number of MSCs is set to 100, and the screening process and results are shown in Figure 4. Figure 4 shows the number of variables involved in modeling during the CARS algorithm screening process, the RMSE of the adaptive weighted sampling model, and the regression coefficient of each variable as a function of the number of MSCs. From the figure, the number of MCS is 62. When the RMSE reaches a minimum value, the number of characteristic wavelengths screened is 24.

IOP Conf. Series: Materials Science and Engineering **721** (2020) 012045 doi:10.1088/1757-899X/721/1/012045

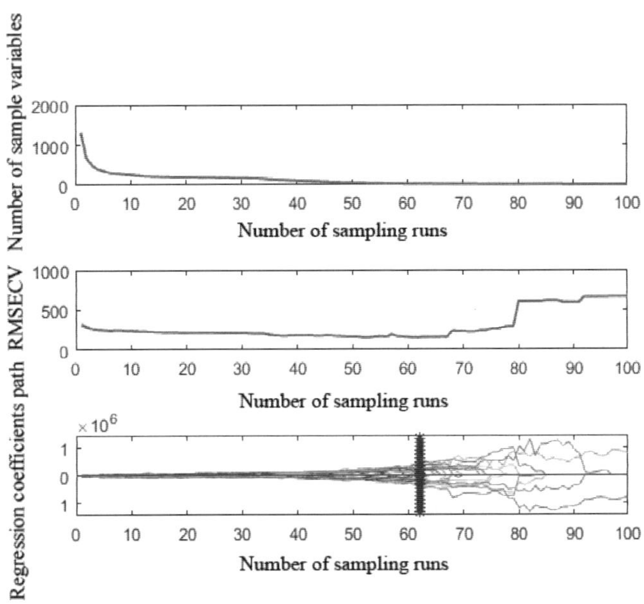

Figure 4. CARS Algorithm screening process

3.6CARS-PLSR modeling

Using SPXY partitioning, 80 samples were divided into training and prediction sets in a ratio of 3:1. The PLSR algorithm is used to establish the heating value prediction model for the characteristic wavelengths selected by CARS. Figures 5 and 6 show the relationship between the number of hidden variables and RMSECV and RPD in the CARS-PLSR modeling process respectively. When the number of latent variables is 13, the RMSECV of the heating value is minimized while the RPD is maximized.

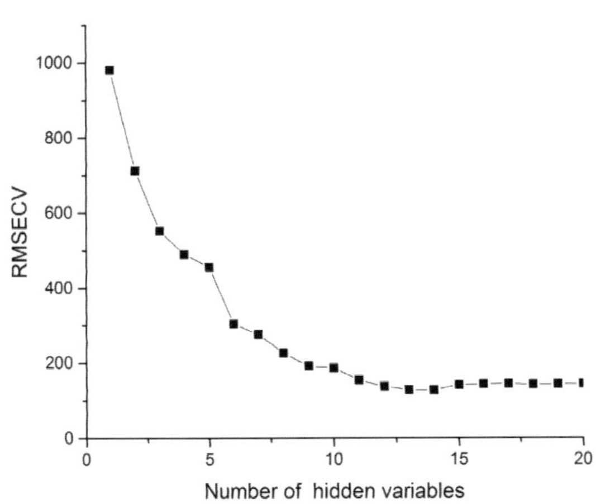

Figure 5. erelationship between the number of hidden variables and RMSECV

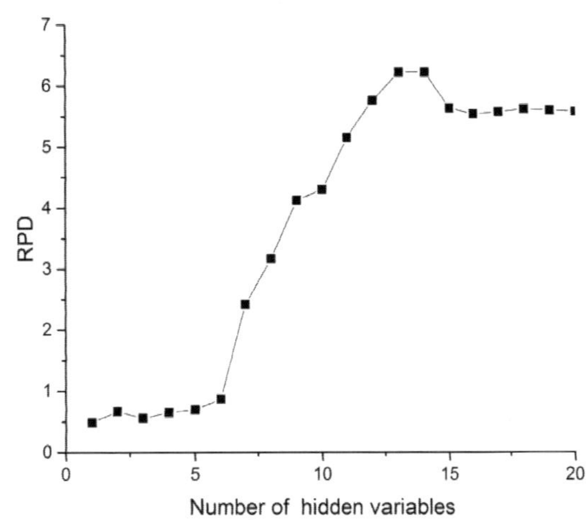

Figure 6. erelationship between the number of hidden variables and RPD

3.7PLSR and CASR-PLSR modeling results

Table 2 shows the results of PLSR modeling in the medium and long-band PLSR modeling and the CARS characteristic wavelength screening. The results of the high and low heating value models are similar. Taking HHV as an example, using PLSR, the thermal index training set model index R2c of the 1320 wavelengths in the long-wavelength of the sample reaches 0.98, the prediction set model index R2p reaches 0.90, and the RPD value is greater than 2.5, which is 2.89. High predictive ability, the model is feasible. The number of characteristic wavelengths obtained by the wavelength screening of CARS algorithm is 24, which is greatly reduced compared with the mid-long band spectrum, which is only 1.8% of the medium-long spectrum. The prediction ability of the model is greatly improved, and the prediction set R2p is 0.98, which is increased. 0.08, RMSEP is 124.37, which is 53.10% lower than that of medium-length spectrum. The RMSE after wavelength screening is increased by 12.98%, and the RPD is increased from 2.89 to 6.22, which is far beyond the practical application of models with RPD value of 3. The heating value prediction model established by using CARS algorithm combined with PLSR is better than the PLSR model without wavelength screening, which can be used in practical applications. Figure 7 and Figure 8 are the results of the high and low heating value prediction models built by the two methods.

Table 2. Bamboo heating value modeling prediction results

Modeling method	Expected target	Number of hidden variables	Number of wavelengths	R^2_c	R^2_p	RMSEP	RMSECV	RMSE	RPD
PLSR	HHV	15	1320	0.98	0.90	265.19	272.08	89.87	2.89
	LHV	15	1320	0.98	0.90	265.18	272.07	89.86	2.88
CARS-PLSR	HHV	13	24	0.97	0.98	124.37	127.60	103.26	6.22
	LHV	13	24	0.97	0.98	124.39	127.62	103.25	6.22

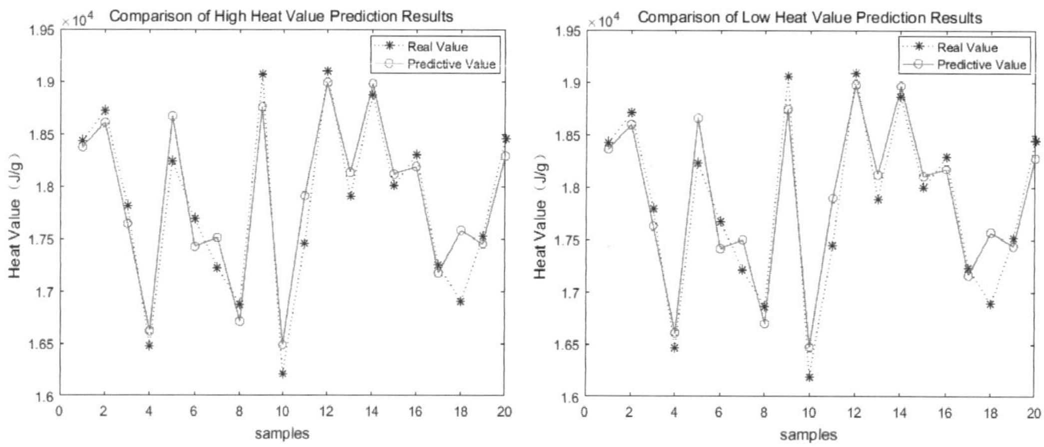

(a) HHV model (b) LHV model

Figure 7. PLSR model prediction diagram

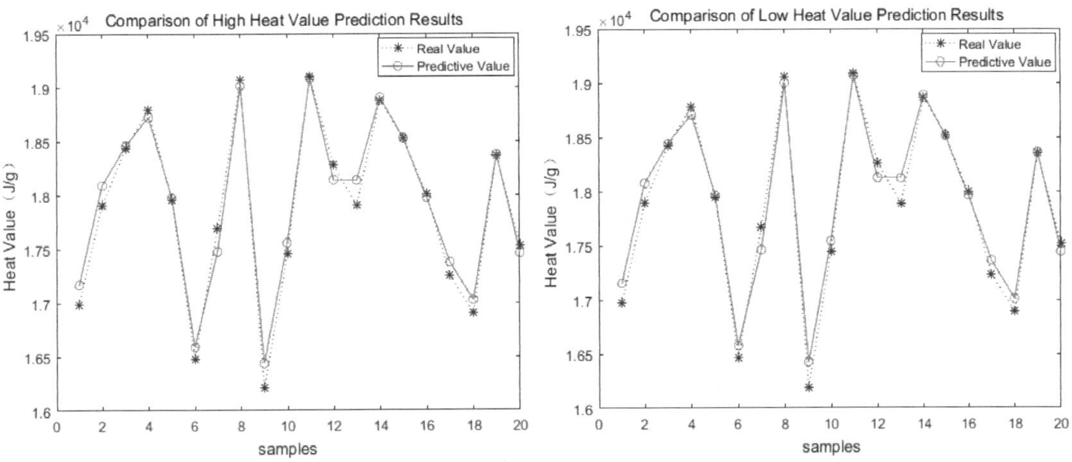

(a) HHV model (b) LHV model

Figure 8. CARS-PLSR Model prediction diagram

4. Conclusion

In this paper, the spectrum of bamboo in Zhejiang Province was determined by NIRS. The rapid prediction models of bamboo heating value were established by using PLSR and CARS-PLSR modeling methods. Based on Darwin's "survival of the fittest" principle, the CARS algorithm selects a combination of key wavelengths in an efficient and competitive way, which reduces redundant information. The 24 characteristic wavelengths obtained by CARS screening were involved in the PLSR modeling. The model evaluation indexes R2c, R2p and RPD were 0.97, 0.98 and 6.22, respectively. Compared with the PLSR modeled by the long-wavelength 1320 wavelengths in the sample, the data scale was greatly simplified, and the model effect has been greatly improved, which is almost a perfect model. The results show that wavelength screening is very necessary. Selecting several characteristic wavelengths with chemical significance has better prediction effect than using continuous wavelength bands. The rapid prediction model of bamboo heating value established by CARS-PLSR is outstanding which can provide reference for the research and development of heating value rapid testing equipment, and has certain application value.

Acknowledgments

Authors wishing to acknowledge assistance or encouragement from colleagues, special work by technical staff or financial support from organizations should do so in an unnumbered Acknowledgments section immediately following the last numbered section of the paper.

This work was funded by the Development and Application of High Throughput Organic Elemental Analyzer (National Key R & D Program 2018YFF01014002) and the title of the project is the application development and method verification of organic matter element analysis in biomass fuel.

References

[1] Zhou, B.Z., Fu, M.Y., Yang, X.S., et al. (2006) Energy bamboo plant resources and its development potential in China. World Forestry Research., 19(6): 49-52.

[2] Liu H.X., Feng Y.M. (2015) The current situation and future development trend of biomass energy in the world. World Agriculture., 05: 117-120.

[3] China electricity council. (2016) Biomass energy: the most potential renewable energy. http://www.cec.org.cn/xinwenpingxi/2016-02-25/149454.html.

[4] Vargas-Moreno J.M., Callejón-Ferre A.J., Pérez-Alonso J., et al. (2012) A review of the mathematical models for predicting the heating value of biomass materials. Renewable and Sustainable Energy Reviews., 16(5): 3065-3083.

[5] Chen B.K., Yang Y.M., Zhang G.X., et al. (2007) Study on the cultivation technology and comprehensive utilization of large clumped bamboo. Journal of West China Forestry Science., 36(2): 1-9.
[6] Lin M.Y. (2019) Present situation of bamboo development and utilization in China. China Southern Agricultural Machinery., 50(02): 105.
[7] Erol M., Haykiri-Acma H., Kucukbayrak S. (2010) heating value estimation of biomass from their proximate analyses data. Renew Energy., 35: 170-173.
[8] Sheng C., Azevedo J.L.T. (2005) Estimating the higher heating value of biomass fuels from basic analysis data. Biomass Bioenerg., 28: 499-507.
[9] Posom J., Sirisomboon P. (2017) Evaluation of the higher heating value, volatile matter, fixed carbon and ash content of ground bamboo using near infrared spectroscopy. Journal of Near Infrared Spectroscopy., 25(5): 301-310.
[10] Hosseinpour S., Aghbashlo M., Tabatabaei M. (2018) iomass higher heating value (HHV) modeling on the basis of proximate analysis using iterative network-based fuzzy partial least squares coupled with principle component analysis (PCA-INFPLS). Fuel., 222: 1-10.
[11] Kathiravale S., Yunus M.N.M., Sopian K., et al. (2003) Modeling the heating value of Municipal Solid Waste. Fuel., 82(9): 1119-1125.
[12] Lestander T.A., Johnsson B., Grothage M. (2009) NIR techniques create added values for the pellet and biofuel industry. Bioresource Technology., 100(4): 1589-1594.
[13] Huang C.J., Han L.J., Liu X., et al. (2009) Fast analysis of straw moisture and heating value by near infrared spectroscopy based on LOCAL algorithm. Journal of Infrared and Millimeter Wave., 28(3): 301-306.
[14] Xiong X.Q., Qian S.P., Sheng K.C., et al. (2017) High heating value of crop straw was predicted based on industrial analysis/elemental analysis and visual-near-infrared spectroscopy. Spectroscopy and Spectral Analysis., (05): 301-306.
[15] Sheng K.C., Sheng Y.Y., Yang H.Q., et al. (2012) Rapid determination of components and heating value of biomass based on spectroscopic technique. 32(10): 2805-2809.
[16] Posom J., Sirisomboon P. (2017) Evaluation of lower heating value and elemental composition of bamboo using near infrared spectroscopy. Energy., 121: 147-158.
[17] Posom J., Sirisomboon P. (2017) Evaluation of the higher heating value, volatile matter, fixed carbon and ash content of ground bamboo using near infrared spectroscopy. Journal of Near Infrared Spectroscopy., 25(5): 301-310.
[18] Standardization Administration of the People's Republic of China. (2008) GB/T 213-2008 Methods for determination of heating value of coal. Standards Press of China, Bei Jing.
[19] Standardization Administration of the People's Republic of China. (2012) GB/T 28731- 2012 Industrial analysis methods for solid biomass fuels. Standards Press of China, Bei Jing.
[20] Li H., Liang., Xu Q., et al. (2009) Key wavelengths screening using competitive adaptive reweighted sampling method for multivariate calibration. Analytica Chimica Acta., 648(1): 77-84.

IOP Conf. Series: Materials Science and Engineering **721** (2020) 012046 doi:10.1088/1757-899X/721/1/012046

Analysis of Hydrogen Corrosion of Water Wall Tube of a Power Plant Boiler

Tian Chengchuan, Jiang Chong

Huadian Electric Power Research Institute Co., Ltd.

Address: NO. 78-1, Fumin south street, Hunnan new district, Shenyang city, Liaoning, province, 110180

Chengchuan Tian's e-mail: cc_tian@163.com

Abstract: During the operation of a power station boiler with the rated evaporation of 1025 t/h, its water wall tube, which is made of SA-210C, bursts continuously. Based on the fracture analysis, material inspection, metallographic examination, energy spectrum analysis and mechanical property analysis of the tube sample, this paper confirms that the cause of the water wall tube failure is hydrogen corrosion, which is accelerated by the copper impurity.

1. Introduction

The unplanned shutdown of units by the boiler in recent years caused is mainly due to the leakage of the heating surface. The water wall is located in the worst environment, making it the most possible to suffer from that problem [1]. Therefore, the effective analysis of the water wall leakage accident can provide guidance for the reliable operation of boiler.

The HG-1025/17.5-YM36 boiler of a power plant has a specification of Φ 63.5×7mm and is made of SA-210C. The internal medium of the water wall tube is a steam-water mixture at a temperature of 366 °C and under a pressure of 20.21 MPa. During its operation, leakage spots are generated on the fire side of the water wall tube with uneven corrosion pits.

2. Leakage spot test

2.1 Spot

The leakage spot on the water wall tube is on the fire side. The outer wall of the tube where the leakage spot locates is partially peeled off, and the outer wall of the fire side is dotted with brown marks. The inner wall of the leakage spot is shown in Figure 1. The inner walls of the back side and the fire side are evenly covered with a layer of sediment, while the inner wall of the fire side presents ulcer morphology. It can be seen that the color of the leakage spot is reddish brown and partially dark gray. A crack can also be clearly observed, as indicated by the red arrow. In addition, many water flow wash marks can be observed, as indicated by the yellow arrow. Figure 2 is a secondary electron mosaic of the inner wall of the leak spot at a lower magnification when the sample is observed under a scanning electron microscope. The energy spectrum analysis is performed in the 1~5 regions near the crack and the results are shown in Table 1.

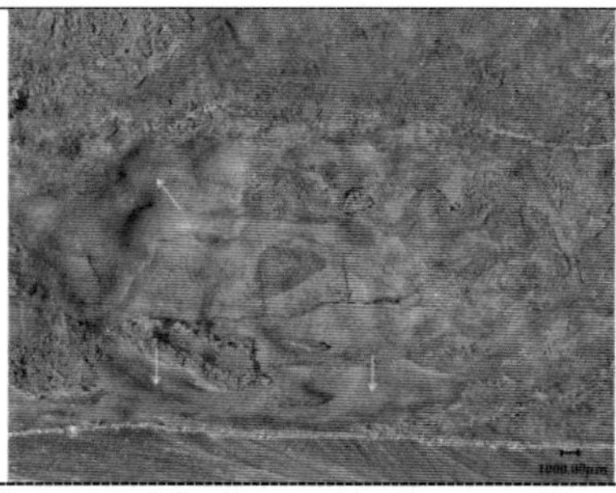

Figure 1. Macroscopic picture of inner wall of the leakage spot

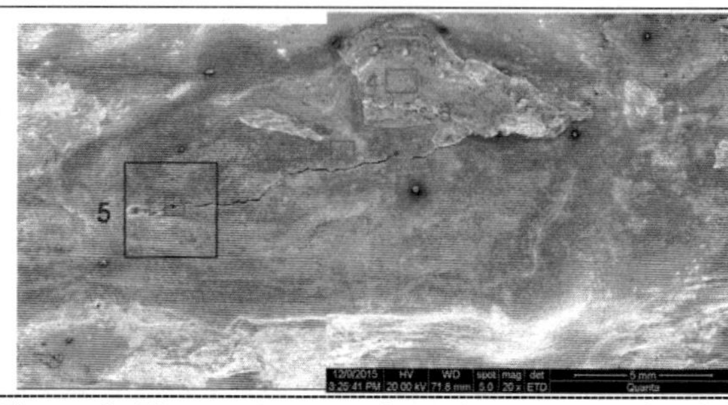

Figure 2. Electronic secondary electron mosaic of the inner wall of leakage spot

Table 1. Energy spectrum analysis results of leakage spot (wt, %)

Element	1	2	3	4	5	6
C	13.82	9.71	13.88	15.6	14.06	13.74
Si	-	0.42	-	0.41	0.42	0.38
Mn	1.38	0.62	-	1.81	-	1.74
Fe	13.88	81.19	3.66	46.14	71.75	52.16
Cu	50.88	0.37	80.13	13.16	0.85	0.66
O	17.13	7.31	2.34	18.41	11.89	28.33
Mg	0.67	-	-	0.66	-	0.74
Al	0.60	0.38	-	1.06	0.43	0.65
Cr	1.44	-	-	1.7	0.36	0.19
Ca	0.21	-	-	0.77	0.25	0.72

According to Table 1, the oxygen content of each spot is relatively high and the copper content of spots 1 and 3 in Figure 2 is the highest. Spot 1 is located inside the crack, with a copper content of 50.88%, while spot 3 has higher a copper content of 80.13%, and the microscopic morphology is granular. According to the results of the energy spectrum analysis, the local copper content at the leakage spot is high, and there may be copper impurities.

2.2 Fracture observation

Two dual fractures can be obtained after opening the leakage spot along the crack. The minimum wall thickness of the fracture is 3.7 mm, while the wall thickness of the water wall tube is 7 mm, indicating the existence of significant corrosion reduction. Figure 3 is what can be observed after putting fracture 1 under a scanning electron microscope. Figure 4 is the secondary electron mosaic at a lower magnification. This paper observes in detail all parts on the fracture, and conducts energy spectrum analysis of the A~E regions. The results are listed in Table 2.

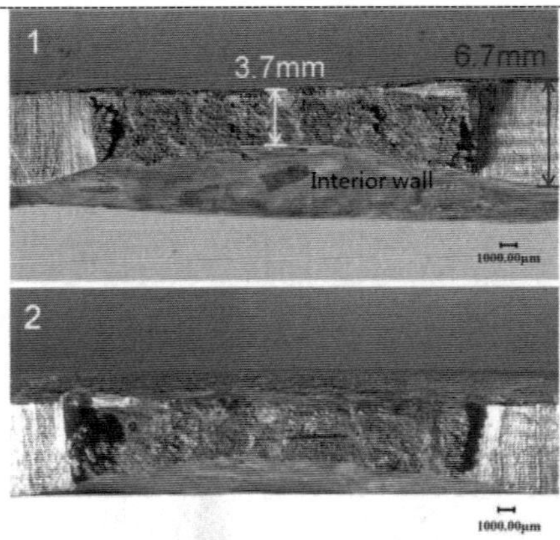

Figure 3. Macroscopic view of the fracture

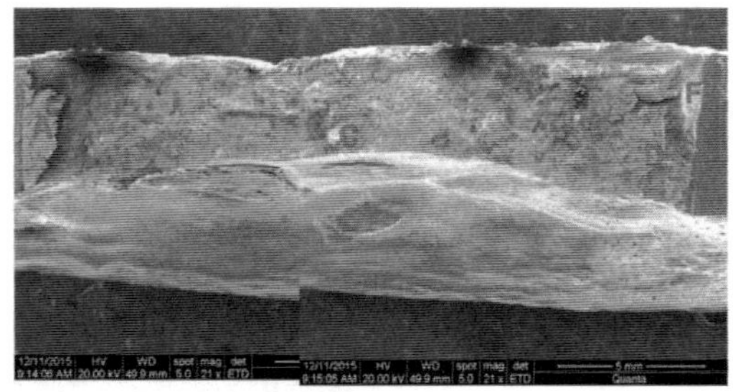

Figure 4. Electronic secondary electron mosaic of the fracture

Table 2. Energy spectrum analysis results of different parts of the fracture (wt, %)

	C	Si	Mn	Fe	Cu	O	Al
A	2.67	-	0.96	89.00	-	7.38	-
B	5.65	-	-	59.64	19.75	17.95	-
C	2.52	0.33	-	81.83	0.33	14.45	0.55
D	2.91	0.37	0.83	89.03	-	6.85	-
E	6.58	-	0.39	4.46	82.22	6.35	-

The fracture morphology is basically intergranular fracture. There are also oxidized particles, and the oxygen content is high. Element distribution result indicates the local aggregation of copper. Figure 19 shows the morphology of the spot C, where there are traces of water scouring and the oxygen content is high. Spot D is near the edge of the fracture and has an intergranular morphology. Spot E has morphology of granular, with a copper content as high as 82.22%, which should be copper impurity.

3. Test analysis

3.1 Metallographic observation

The metallographic structure of the fracture sample is shown in Figures 5 and 6. In Figure 5, the bright yellow copper impurity and a layer of gray-black iron oxide covering the fracture can be observed. Figure 6 is the metallographic structure of the inner wall near the fracture, which is basically ferrite with a large number of small cracks. It can be seen that all the cracks are along the grain boundary, and a few obvious ones are indicated by the red arrows.

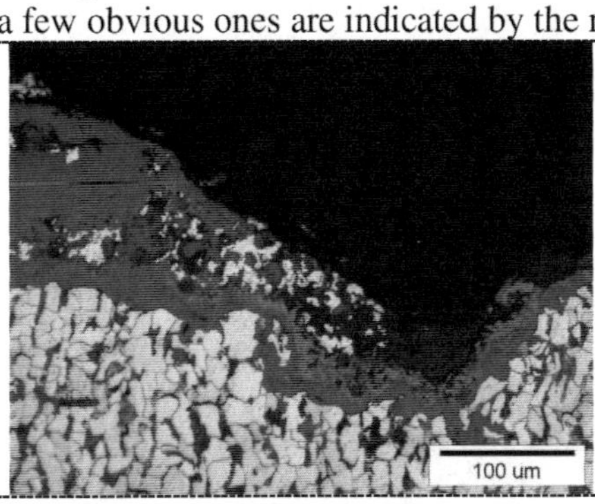

Figure 5. Metallographic phase of the inner wall near the fracture

Figure 6. Micro-crack in the metallographic structure

According to the above observation, the tissue away from the fracture is ferrite and pearlite; that on the inner wall near the fracture is mostly ferrite, and that on the outer wall is ferrite and pearlite. This indicates the existence of decarburization in the inner wall of the fracture. This paper puts the metallographic sample of Figure 5 under a scanning electron microscope to obtain the back-reflection electron image, which is shown Figure 7. It can be seen that there are a large number of small cracks near the inner wall, and the number of crack increases as it gets closer to the fracture. A boundary is formed between the inner and outer walls, as shown by the red dotted line. Divide the wall into A and B. Part A is close to the inner wall, with a lot of small cracks; Part B is close to the outer wall with almost no crack.

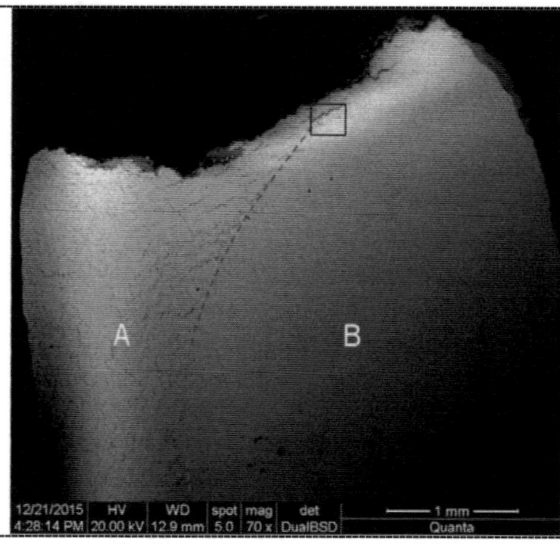

Figure 7. Back reflection electron image of the fracture

Based on the metallographic observation, it is found that the inner wall tissue at the fracture is decarburized, with a large number of small intergranular cracks.

3.2 Chemical component analysis

This paper takes a 30×30 mm sample on the back side of the water wall, and sands its surface for the chemical component analysis using direct-reading spectrometer. Through the comparison of the results in Table 3 with the ASME SA-210/SA-210M standard, it is found that all elements, except copper, are in compliance with the standard requirements [2].

Table 3. Analysis results of main chemical components

	C	Si	Mn	Cu	P	S	Fe
Measured value	0.22	0.23	0.83	0.11	0.012	0.008	remain
ASME A-210C	≤ 0.35	≥0.1	0.29~1.06	-	≤ 0.035	≤ 0.035	remain

3.3 Tensile properties

Take and prepare three tensile specimens of the intact areas of both the fire side wall and the back-side wall of the tube for tensile test. The measured tensile properties are listed in Table 4. All properties meet the standard requirements, and the strength of the fire side is slightly lower than that of the back side. The microscopic topography of the tensile fracture is shown in Figures 36 and 37, which belongs to the dimple morphology.

Table 4. Mechanical properties of the material

Tested area	**No.**	**Yield strength (MPa)**	**Tensile strength (MPa)**	**Elongation (%)**
Back side	1#	332	568	32.5
	2#	350	569	32.5
	3#	348	566	31.0
	Average	343	568	32.0
Fire side	1#	329	562	32.0
	2#	317	552	33.0
	3#	333	562	32.5
	Average	326	559	32.5
Standard specified value		**≥275**	**≥485**	**≥30**

4. Result analysis

4.1 Result analysis

According to macroscopic observation, the leakage spot is on the fire side. The thickness of the inner wall of the fracture is significantly reduced, indicating that the leakage starts at the inner side of the fire side. Microscopic observation shows that most of the fracture surface is heavily oxidized and the slightly oxidized area is intergranular fracture, whose artificial fracture belongs to the dimple morphology. The energy spectrum analysis indicates the existence of copper impurity in the fracture. Metallographic examination shows: 1) The tissue of inner wall of the leakage spot is basically ferrite, and that of outer wall is ferrite and pearlite, indicating the existence of decarburization of the inner wall [3]. 2) There are many small intergranular cracks on the inner wall. 3) Divide the wall into A and B. Part A is close to the inner wall, with a lot of small cracks; Part B is close to the outer wall with almost no crack. Moreover, based on the trend of the boundary line between A and B (red dotted line in Figure 7), the number of crack increases as it gets closer to the fracture, indicating that more cracks are located in the thicker part. That phenomenon shows that the fracture is formed through the connection of intergranular cracks at high temperature and under high pressure.

The above results show that the failure starts at the inner wall of the fire side with the characteristic of decarburization at the fracture and a large number of intergranular cracks on the inner wall.

4.2 Hydrogen corrosion

Due to the complicated working environment, the water wall pipe of power station boilers often suffers from damages and failures which are caused by acid corrosion, high temperature corrosion, fatigue, wear, hydrogen corrosion, overheating, etc. [4]

Hydrogen corrosion is a form of high temperature damage caused by hydrogen damage, which happens to carbon steel or low alloy steel at a high temperature and under high pressure environment containing hydrogen for a long time. The source of hydrogen may be the hydrogen produced by steam corrosion, or the hydrogen released by the under-deposit acidic corrosion due to the low pH of the water and the concentration of acidic substances into the sediment [5]. Regardless of the source, if the hydrogen cannot be quickly taken away by water or steam, the atomic hydrogen will permeate into the interior of the metal to react with cementite at a certain temperature and pressure (for the carbon steel, the temperature shall be greater than 250 °C and the hydrogen partial pressure shall be greater than 2 MPa).

$Fe3C+4H \rightarrow 3Fe+CH4\uparrow$

On the one hand, this reaction decomposes cementite, leading to surface decarburization or internal decarburization. On the other hand, the generated CH4 molecules will accumulate on the impurity or grain boundaries to form CH4 bubbles, which will grow as CH4 continues to aggregate and finally form hydrogen corrosion cracks along the grain boundaries.

4.3 Comprehensive analysis

The water wall tube is made of high-quality carbon steel, which has the possibility of hydrogen corrosion. According to the analysis results, the failure is characterized by the decarburization at the fracture and a large number of intergranular cracks on the inner wall. These characteristics are completely consistent with those of hydrogen corrosion. Therefore, the leakage is estimated to be caused by hydrogen corrosion. Further, the wall gets thinner as it gets closer to the inner wall, which is also caused by hydrogen corrosion.

In addition, the material of the water wall tube is SC-210C, which should not contain copper according to the standard. However, the test result shows that it contains 0.11% copper. SEM and metallographic observation all prove the existence of copper impurity located at the fracture, which will make the internal structure of the material not loose and discontinuous. When hydrogen corrosion occurs, copper impurity will facilitate the accumulation of methane gas, making the tube more likely to fail.

5. Conclusion and recommendations

The direct cause of the water wall burst is hydrogen corrosion, and the indirect cause is the presence of copper impurity in the water wall that accelerates the corrosion. Therefore, it is necessary to strictly check the tube material during the installation process of power station boilers so as to avoid unplanned shutdown caused by unqualified materials. Secondly, it is important to ensure the water quality by implementing water quality inspection and chemical instrument calibration during the operation of the power station boiler. Thirdly, it is necessary to strengthen the anti-wear and explosion-proof inspection of the heating surface so as to eliminate potential safety hazards in time before accidents.

References

[1] Gao Jinsong, invalid mechanism and prevention actions research at boiler heating pipe surface[D], Nanchang University, 2007
[2] ASTM Handbook, Vol.13 Corrosion, P311-388.
[3] Li Bing, Hydrogen damage at water-cooling wall pipe in heat power station[J], power technology, 1980(12:30-36)
[4] Cui Lun, Sun Kai, Chen Junfeng, Hydrogen corrode at water-cooling wall pipe in power station[J], Jilin Power, 2004(5:40-43)
[5] Zheng Xiaohong, invalid mechanism research and calculated working life [D], Zhejiang university, 2002

IOP Conf. Series: Materials Science and Engineering **721** (2020) 012047 doi:10.1088/1757-899X/721/1/012047

An Energy Efficiency Indicator and Grading of Shell-and-Tube Heat Exchangers

Kelei Chen[1], Guoli Qi[2], Xueming Liu[3], Jianfei Zhang[4] and Zhiguo Qu[1*]

[1, 4] MOE Key Laboratory of Thermo-Fluid Science and Engineering, Xi'an Jiaotong University, Xi'an, Shaanxi, 710049, China

[2, 3]China Special Equipment Inspection and Research Institute, Beijing, 100029, China

[*]Corresponding author's e-mail: zgqu@mail.xjtu.edu.cn

Abstract. The shell-and-tube heat exchanger is the most commonly used heat exchange equipment in the industry. Whether shell-and-tube heat exchangers are energy-saving is of great significance for the efficient using of energy. Therefore, the corresponding evaluation system and index are needed for evaluating the heat transfer process in the shell-and-tube heat exchanger. In this paper , based on Kern method, the energy efficiency evaluation index of shell-and-tube heat exchanger suitable for oil-water working medium is derived. And the relevant database is used to obtain the overall energy efficiency distribution of the industry, which is subject to the normal distribution, finally according to the overall energy efficiency distribution, the energy efficiency of the shell-and-tube heat exchanger is graded.

1. Introduction

Heat exchangers are widely used in petrochemical, energy power, urban heating, mechanical light industry and other fields, accounts for 30%-40% of the total equipment investment usually. In some thermal power plants, they can even account for more than 70% of the total investment of the entire power plant[1]. Therefore, research on energy efficiency and energy-saving technologies in heat exchangers is of great significance for energy conservation and emission reduction.

The shell-and-tube heat exchanger is a partition wall heat exchanger which heat exchange surface is a heat exchange tube bundle enclosed in a shell. This kind of heat exchanger has the advantages of simple structure, easy manufacture, low cost, wide flow section and easy cleaning, but at the same time, the heat transfer coefficient is low relatively and the floor space is large. Shell-and-tube heat exchangers can be manufactured from a variety of structural materials (metal materials mainly) and can be operated under high temperature and high pressure conditions. They are the most widely used heat exchanger types and they play a vital role in industrial systems around the world nowadays.

There have been many studies on energy efficiency indicators for heat transfer processes. The heat transferred by the power of the unit fluid was proposed as the evaluation standard, that is, Q/N as the evaluation index[2]. Based on the work of the predecessors, R.L.Webb proposed a relatively complete set of PEC criteria[3]. Bejan proposed the number of entropy production units based on the second law of thermodynamics as an evaluation index, and demonstrated that Q/N could not fully reflect the rationality of energy utilization through model calculation[4].

There are also some researches on the energy efficiency index of heat exchangers. Zhang Yanfeng et al. theoretically deduced the energy efficiency index of plate heat exchangers, proposed a new energy efficiency index EEI, and obtained a normal distribution map of energy efficiency

IOP Conf. Series: Materials Science and Engineering **721** (2020) 012047 doi:10.1088/1757-899X/721/1/012047

distribution[5]. However, there is no established energy efficiency level and related standards for shell-and-tube heat exchangers at home and abroad. There is no universally recognized energy efficiency evaluation method or indicator in academia and related industries. The commonly used indicators can only be used from one aspect of energy utilization and lack of rigorous theoretical derivation, it is difficult to compare shell-and-tube heat exchangers of different structures.

In this paper, from the theoretical point of view, the specific form of the energy efficiency index of the shell-and-tube heat exchanger applicable to the oil-water working medium is derived based on the Kern method, and the overall energy efficiency distribution of the industry is obtained by combining the performance data of the shell-and-tube heat exchanger.

2. Energy efficiency indicator of shell-and-tube heat exchanger

2.1Theoretical derivation

Kern method for designing shell-and-tube heat exchangers was proposed in the 1950s[6]. The method focuses on the flow and heat transfer on the shell side, and the calculation error for the shell side convective heat transfer coefficient is small. The calculation of the drop is slightly not accurate compare with the experiment data. However, the qualitative analysis of the Kern method is accurate and basically covers the main structural parameters of the shell-and-tube heat exchanger (tube spacing, shell side inner diameter, number of baffles), the equation structure is easy to deduct and simplify, so the energy efficiency index of the shell-and-tube heat exchanger can be derived based on this method.

In shell-and-tube heat exchangers, since the thermal resistance of the tube side and the shell side is usually on the order of 10^{-5} magnitude, much less than the thermal resistance of the working medium during heat transfer, it is ignored in the derivation. The total heat transfer coefficient of the heat exchanger can be expressed by equation (1):

$$\frac{1}{k}=\frac{1}{h_i}\frac{d_o}{d_i}+\frac{d_o}{2\lambda_w}\ln\left(\frac{d_o}{d_i}\right)+\frac{1}{h_o} \tag{1}$$

Where k is the total heat transfer coefficient of the heat exchanger, $\mathrm{W\cdot m^{-2}\cdot K^{-1}}$; h_i , h_o are the convective heat transfer coefficients of the tube side and the shell side, respectively, $\mathrm{W\cdot m^{-2}\cdot K^{-1}}$; d_i , d_o are the inner and outer diameters of the internal heat transfer tube, m; λ_w is the thermal conductivity of the tube wall, $\mathrm{W\cdot m^{-1}\cdot K^{-1}}$.

If the convective heat transfer coefficient of the shell side of the shell-and-tube heat exchanger is much smaller than the convective heat transfer coefficient of the tube side, and the heat conduction resistance caused by the wall thickness is ignored, the main thermal resistance of the heat transfer process is reflected in shell side. It can be approximated:

$$k \approx h_o \tag{2}$$

Therefore, if the convective heat transfer coefficient of the shell side working fluid is much smaller than the convective heat transfer coefficient of the pipe side working fluid (such as cooling oil and water), the heat transfer of the shell side working fluid can be used to approximate the shell-and-tube heat exchanger's overall heat transfer situation.

For the shell side of shell-and-tube heat exchangers, the experimental correlation of Nusselt number and pressure drop is:

$$Nu_{shell}=\frac{h_o D_e}{k}=0.36\,\mathrm{Re}_{shell}^{0.55}\,\mathrm{Pr}^{1/3}\left(\frac{\mu}{\mu_w}\right)^{0.14} \tag{3}$$

$$\Delta P_{shell}=\frac{4f_{shell}G_s^2 D_s(N_b+1)}{2\rho D_e\left(\frac{\mu}{\mu_w}\right)^{0.14}} \tag{4}$$

$$\mathrm{Re}_{shell}=\frac{G_s D_e}{\mu} \tag{5}$$

$$f_{shell} = 0.25e^{0.576-0.19\ln \mathrm{Re}_{shell}} \tag{6}$$

Where Nu_{shell} is Nusselt number; Re_{shell} is Reynolds number for shell side flow; Pr is Prandtl number; ΔP_{shell} is pressure drop of shell side, Pa; μ , μ_w are the dynamic viscosity coefficients of hot fluid and cold fluid respectively. $\mathrm{N \cdot s/m^2}$; f_{shell} is Shell side friction factor; G_s is mass flow under the minimum flow area of the shell side, $\mathrm{kg/m^2 \cdot s}$; D_s is inner diameter of the shell, m; N_b is the number of baffles; ρ is shell side working medium density, $\mathrm{kg/m^3}$; The specific calculation method is as shown in the formula (7) (8):

$$D_e = \frac{4(\frac{P_t}{2} \times 0.86P_t - \frac{1}{2}\frac{\pi d_o^2}{4})}{\frac{\pi d_o}{2}} \text{ (For square pitch)} \tag{7}$$

$$D_e = \frac{4(P_t^2 - \frac{\pi d_o^2}{4})}{\pi d_o} \text{ (For triangular pitch)} \tag{8}$$

Where P_t is the pipe center spacing, m.

The total heat transfer coefficient represents the benefit of the heat exchanger operation, which reflects the heat transfer characteristics of the heat exchanger as a whole; the pressure drop represents the energy consumption during the operation of the heat exchanger, which reflects the flow characteristics of the heat exchanger during operation. The physical meanings of the two are different, and the total amount of heat transfer and the pumping power, the heat loss and the helium loss of the flow, and the entropy increase are about one order of magnitude, so in order to reflect the k and Δp 's importance in energy efficiency, we introduce $k/\Delta p^n$, combining equations (3)-(8) we can get:

$$\frac{h_o}{\Delta p^n} = \frac{0.36\frac{\lambda}{D_e}(\frac{D_e u_o \rho}{\mu})^{0.55}\mathrm{Pr}^{1/3}}{(0.5\mathrm{e}^{0.576}\rho^{0.81}(N_b+1)\mu^{0.19}\mathrm{u}_o^{1.81}D_e^{-1.19n})^n} \tag{9}$$

Simplify:

$$\frac{h_o}{\Delta p^n} = \frac{0.36}{0.8895^n}\lambda(N_b+1)^{-n}\mathrm{Pr}^{1/3}\mu^{-0.55-0.19n}\rho^{0.55-0.81n}D_e^{1.19n-0.45}u_o^{0.55-1.81n} \tag{10}$$

Where u_o is the shell side working fluid flow rate, $\mathrm{m/s}$.

If the energy efficiency index is weakly correlated or not correlated with the working fluid flow rate, thereby reducing or even eliminating the influence of the working fluid flow rate on the energy efficiency index, the speed index term 0.55-0.81n needs to be near zero, so the value of n is:

$$\mathrm{n} \approx 0.304 \tag{11}$$

Substituting into (10), the energy efficiency indicator can be written as:

$$\frac{h_o}{\Delta p^{0.304}} = \frac{0.373}{(N_b+1)^{0.304}D_e^{0.088}}\frac{\lambda\rho^{0.3038}\mathrm{Pr}^{1/3}}{\mu^{0.608}} \tag{12}$$

2.2Influencing factor

As can be seen from Equation 15, the energy efficiency indicator can be divided into two parts. The first part represents the influence of the shell-and-tube heat exchanger structure on energy efficiency indicators, structurally related parameters include the number of baffles and the equivalent diameter of the shell side, reflects the design and production level of shell-and-tube heat exchangers; The second part represents the influence of fluid thermal properties on energy efficiency indicators, including density, dynamic viscosity, Prandtl number and thermal conductivity. The energy efficiency index will be affected by the type of fluid, even if the fluid type is constant, the change of fluid temperature will affect the thermal properties of the fluid, which in turn affects the energy efficiency index.

2.3Shell-and-tube heat exchanger flow heat transfer performance database

After collecting and collating the articles on the shell-and-tube heat exchanger experiments in each database, a total of 68 liquid-liquid phase-change shell-and-tube heat exchangers were collected. The structure mainly includes a spiral baffle plate and a conventional bow plate, and the size of the heat exchanger is mostly a small heat exchanger for testing, and the detailed data for the industrial large-sized shell-and-tube heat exchanger is still to be perfected. For the shell-and-tube heat exchangers with oil-water working fluids, there are 38 groups, some of which lack geometric parameters or flow parameters. The size of the heat exchanger ranges from 0.5m to 5m, and some of the heat exchangers use enhanced heat transfer means such as new high-efficiency heat transfer tubes.

2.4Distribution of energy efficiency indicators of shell-and-tube heat exchangers

Taking the energy efficiency index of the shell-and-tube heat exchanger with oil-water as the working fluid in the database as a sample, the overall energy efficiency distribution of the shell-and-tube heat exchanger industry can be estimated by mathematical statistics. The energy efficiency indexes of 38 sets of shell-and-tube heat exchangers with oil-water as working fluid under standard working conditions were calculated, and the distribution histogram was obtained as shown in Fig. 1.The probability density distribution of the overall energy efficiency index of the shell-and-tube heat exchanger satisfies the normal distribution function $f(x)=N(38.4,7.97^2)$.

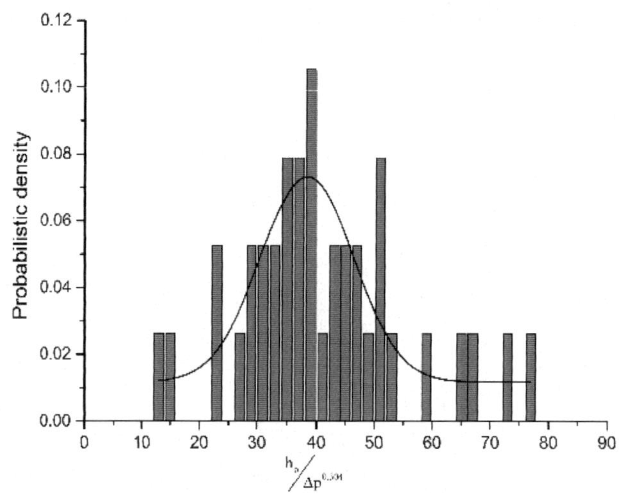

Figure 1: Total energy efficiency indicator histogram and fitted normal distribution curve

According to the probability energy function of the overall energy efficiency index of the shell-and-tube heat exchanger $f(x)=N(38.4,7.97^2)$, the energy efficiency interval can be divided. For example, the energy efficiency of shell-and-tube heat exchangers is divided into three energy efficiency intervals: low energy efficiency, medium energy efficiency and high energy efficiency, and the proportions are 30%, 50%, 20%, respectively. When the energy efficiency index of the heat exchanger is less than 32.2 and the heat exchanger is at a low energy efficiency level, the heat exchanger should be phased out or updated; when the heat exchanger energy efficiency index is between 32.2 and 64.2, the heat exchanger is at a medium energy efficiency level, then the heat exchanger can continue to be used; when the energy efficiency index of the heat exchanger is greater than 64.2, the heat exchanger is at a high energy efficiency level, and the heat exchanger should be vigorously promoted.

3. Conclusion

From the theoretical point of view, based on the Kern formula, the energy efficiency index $h_o/\Delta p^{0.304}$ of the traditional bow-baffle shell-and-tube heat exchanger based on oil-water working fluid is proposed. The indicator can be divided into two parts, which respectively reflect the influence of the structural parameters of the shell-and-tube heat exchanger and the thermal properties of the running medium on the flow and heat transfer of the heat exchanger. Combined with the flow and heat transfer performance database of shell-and-tube heat exchanger, the overall energy efficiency distribution of shell-and-tube heat exchanger $f(x) = N(38.4, 7.97^2)$ is obtained. According to this, the quantitative energy efficiency evaluation and energy efficiency level division of the shell-and-tube heat exchanger are completed.

References

[1] Shi M, Wang Z. (2014) Heat exchanger principle and design. Southeast University Publishing, Nanjing.
[2] Gao X, Feng X. (2007) Analysis of shell side heat transfer evaluation method for shell-and-tube heat exchanger. Journal of North China Electric Power University (Natural Science Edition). 34(2):95-97.
[3] Webb, R L. (1981) Performance evaluation criteria for use of enhanced heat transfer surfaces in heat exchanger design. International Journal of Heat & Mass Transfer. 24(4):715-726.
[4] Bejan, A. (1978) General criterion for rating heat-exchanger performance. International Journal of Heat and Mass Transfer. 21:655~658.
[5] Zhang Y, Jiang C, Shou B, (2018) A Quantitative Energy Efficiency Evaluation and Grading of Plate Heat Exchangers. Energy. 142.
[6] Kuppan, T. (2014) Heat exchanger design handbook. CRC Publishing, Ohio.

IOP Conf. Series: Materials Science and Engineering **721** (2020) 012048 doi:10.1088/1757-899X/721/1/012048

Theoretical Analysis of Acid Dew Point Temperature and Vapour Condensation Characteristics of the Flue Gas

Ning Wang[1], Xuemin Liu[2], Jian Guan[2], Guoli Qi[2], Jianfei Zhang[1], Zhiguo Qu[1,*]

[1]MOE Key Laboratory of Thermo-Fluid Science and Engineering, School of Energy and Power Engineering, Xi'an Jiaotong University, Xi'an, Shaanxi 710049, PR China

[2]China Special Equipment Inspection and Research Institute, Beijing 100029, PR China

[*]Corresponding author's e-mail: zgqu@mail.xjtu.edu.cn

Abstract. Based on the gas-liquid phase equilibrium and multi-component diffusion theory, an analytical model for predicting acid dew point temperature and vapour condensation characteristics of flue gas was developed. The effects of components content, ash particle radius, wall temperature, coal-burned type and excess air ratio have been investigated. The results indicated that the acid dew point temperature and vapour condensation rate will increase with both water and sulfuric acid vapour contents increase. Vapours are not prone to condense on the ash particles when the particle radius is less than the corresponding critical condensing radius. The vapour condensation characteristics can be divided into two different stages with the wall temperature varies. The particle radius has a significant influence on the vapour condensation characteristics in the higher temperature stage. The acid dew point temperatures present significant differences for different coal-burned types, and the water vapour condensation rate is more sensitive to excess air ratio compared with the acid condensation rate.

1. Introduction

During the last several years, the waste heat recovery has received extensive attention in achieving energy highly efficient utilization and pollution emission reduction of the industrial systems. As an important industrial waste heat resources, the different temperature and component flue gas is basically utilized by the heat exchangers, such as air preheaters and economizers. However, the severe challenge of the flue gas utilization is the presence of SO_2 and ash particles at the coal-burned boilers. The sulfuric acid vapour increases the ADP (acid dew point), which significantly causes the heat exchanger corrosion, resulting in ash blocking, combustion deterioration and operating unsafety of the system [1]. Previous studies have shown that the low temperature corrosion is not only affected by the ADP, but also related to the acid vapour condensation characteristics [2].

Several scholars [3-4] have corrected several empirical or semi-empirical formulas by the experimental data combined with thermodynamic theory to predict the ADP. In order to improve the accuracy and application range of the model, more in-depth researches were conducted recently. ZareNezhad and Aminian [5] adopted a three-layer feedforward artificial neural network intelligent algorithm to establish an accurate ADP model of flue gas under multi-parameter variables, which can be applied to low temperature corrosion prevention and waste heat recovery. Wang et al [6] developed a numerical model to compare the ADP distributions of H-type, dimple, vortex generators and composite fin structures in economizers of the boiler, the effect of ash particle size was also discussed.

With respect to the prediction of vapour condensation characteristics, Yin et al [7] established a water vapour laminar film condensation with a small amount of oxygen (0%-3%) in a three-dimensional microchannel by the VOF (volume of fluid) method. It is worth noting most of the current studies considered the water vapour condensation characteristics, but there are only few researches focused on multi-component vapour condensation in flue gas. Wilson [8] revised the mathematical model of the component partial pressure on the gas-liquid interface of sulfuric acid solution, and proposed the calculation method of the sulfuric acid and water vapour condensation rate. Jin et al. [9] numerically explored the acid condensation rate in the chimney based on the above method. The result can be implemented to predict the life and design anti-corrosion configuration of the chimney.

In this paper, an analytical prediction model is developed to investigate the various parameter effect on the ADP and vapour condensation characteristics, such as flue gas component content, ash particle radius, wall temperature, coal-burned type and excess air coefficient. This work can provide guidance for the prevention of low temperature corrosion on the heat transfer surface in the flue gas.

2. Methodology

2.1 Mathematical model

In this section, the gas-liquid phase equilibrium and multi-component diffusion theory are introduced to establish the analysed model. The ADP and vapour condensation characteristics on the heat transfer wall and ash particles are compared respectively. The following assumptions are made: Thermal resistance of the condensate film is neglected; Flue gas should be considered as ideal mixture gas; Coal should be considered completely burned and the ash particles in the flue gas are spherical.

The flue gas composition after complete combustion can be determined by the combustible elements in the fuel. The required theoretical air amount and SO_2 volume can be expressed as:

$$V^0=0.0889(C_{ar}+0.375S_{ar})+0.265H_{ar}-0.0333O_{ar}, \quad V_{SO_2}=0.7S_{ar}/100$$

Only little SO_2 is oxidized to SO_3 in the flue gas, the SO_3 is further reacted with H_2O to form sulfuric acid vapour. The conversion percent of SO_3 is depended on the flue gas temperature (Figure 1).

CO_2, N_2 and H_2O volume after complete combustion:

$$V_{CO_2}=1.866C_{ar}/100, \quad V_{N_2}=0.79V^0+0.8N_{ar}/100, \quad V^0{}_{H_2O}=11.1H_{ar}/100+1.24M_{ar}/100+0.016V^0$$

The actual flue gas volume can be calculated as:

$$V_g=V_g^0+(\alpha-1)V^0+0.0161(\alpha-1)V^0$$

where C_{ar}, S_{ar}, H_{ar}, O_{ar} and N_{ar} denote the content of carbon, sulfur, hydrogen, oxygen and nitrogen.

The component partial pressure on the gas-liquid interface of sulfuric acid solution can be obtained based on the gas-liquid phase equilibrium theory [8]:

$$\text{In}p_i=a_iA+b_iB+c_iC+\Delta H_i^{v}D+\Delta S_i^{v}E+\overline{C}_{pi}^{1}F+\overline{L}_i^{1}G+\alpha_iH+\text{In}\overline{a}_i$$

where the A-H is the coefficient, which is only related to the temperature; i represents different components; a_i, b_i, c_i, ΔH_i^v, ΔS_i^v are pure component thermodynamic parameters; the others are component dependent parameters. These can be found by reference [8].

Part of the sulfuric acid solution will be decomposed into water and SO_3 vapour. It is necessary to correct the partial pressure of each component by the apparent fugacity coefficient:

$$p_{io}=p_i/\phi_{io}, \quad \phi_{ao}=p_w/[p_w+K_0+K_1(p_w)^2], \quad \phi_{wo}=1/[1+K_1p_a]$$

where ϕ_{ao} and ϕ_{wo} are the apparent fugacity correction coefficients of sulfuric acid and water vapor; K_0 and K_1 are reaction equilibrium constants.

When the sulfuric acid vapour condenses on the heat transfer surface, the vapour needs to pass through the other vapours. The vapour mass transfer rate can be expressed as [10]:

$$m_i = -\rho D_{i,\mathrm{eff}} \nabla y_i, \quad D_{i,\mathrm{eff}} = (1-x_i) / \sum_{\substack{j=1 \\ j\neq i}}^{n} \frac{x_j}{D_{ij}}, \quad D_{ij} = \frac{0.0104T^{1.75}(1/M_i + 1/M_j)^{0.5}}{P(v_i^{1/3} + v_j^{1/3})^2}$$

The relationship between the two condensation rates is as below:

$$X_\mathrm{a} = R_\mathrm{a} / \left(R_\mathrm{a} + R_\mathrm{w}\right)$$

where X_a is the sulfuric acid solution molar concentration; R_a and R_w are the mass transfer rates of sulfuric acid and water vapor.

The surface curvature of the ash particles increases the component partial pressure at the gas-liquid phase interface of the sulfuric acid solution, which can be corrected as the below [10]:

$$p_{\mathrm{a,i}} = p_{\mathrm{a,i}}^{*} \exp(2\sigma M / \rho R T r_\mathrm{p})$$

Based on the above mathematical model, the ADP and vapour condensate characteristics of flue gas under different parameters can be determined by the iterative method (Figure 2).

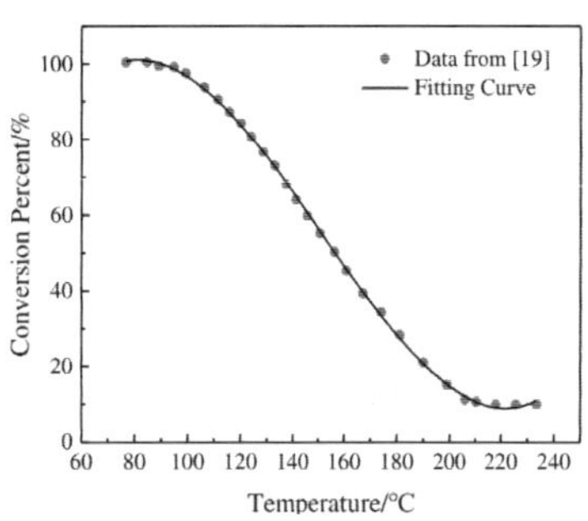

Figure 1. Sulfuric acid vapor conversion percent

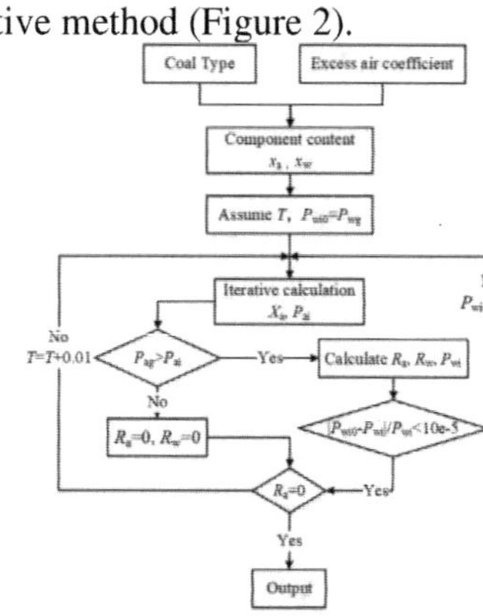

Figure 2. Iterative flow chart

2.2Model validation

During the iterative process, the accurate prediction of the vapour partial pressure on the gas-liquid interface of the sulfuric acid solution is strongly determined to the accuracy of the whole model. Therefore, it is necessary to verify the water and sulfuric acid vapour partial pressure. The water and sulfuric acid vapour partial pressure on the gas-liquid interface at different temperatures and solution concentration were compared with the experimental data [8], as shown in Figure 3. It can be seen that the experimental data and the calculation results show good consistency.

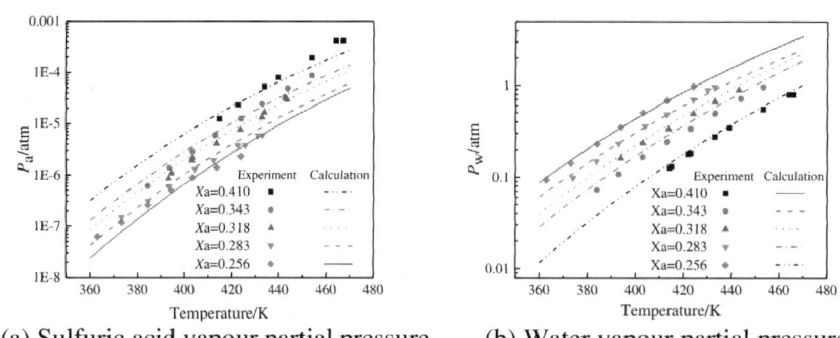

(a) Sulfuric acid vapour partial pressure (b) Water vapour partial pressure

Figure 3. Model validation of vapour partial pressure

3. Results and Discussion

3.1Effect of vapour contents

The vapour condensation characteristics of the flue gas are mainly influenced by the water and sulfuric acid vapour contents. Generally, the water and sulfuric acid vapour contents in the flue gas of coal-burned boilers are 5-50 ppm and 5%-16%, respectively. Figure 4 presents the effect of vapour contents on ADP of the wall and ash particles (0.1 μm). It can be seen that as the water or sulfuric acid vapour contents increase, the ADP gradually increases, while the corresponding increasing rate gradually decreases, which means the higher vapour contents have a less effect on the ADP. When the sulfuric acid vapour content increases from 5 ppm to 50 ppm, the ADP of the heat transfer wall and the ash particles can be increased by 18.85% and 18.89%. Additionally, when the water vapour content increases from 5% to 16%, both ADPs are increased by 7.36% and 7.61%, indicating that the ADP is more susceptible to the sulfuric acid vapour content. Moreover, the ADP of the ash particles is always lower than the ADP of heat transfer wall under the same conditions, it means the surface of the ash particles is more difficult to condense, which is caused by the increase of the component partial pressure due to the curvature of the particles.

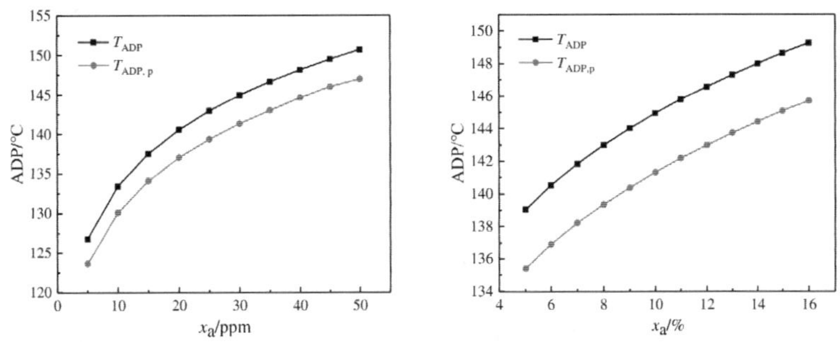

(a) Sulfuric vapour content (b) Water vapour content

Figure 4. ADP variations with the vapor contents

Figure 5 shows the effect of vapour contents on condensation characteristics. It can be found that the water and sulfuric acid vapour condensation rates increase linearly with the sulfuric acid vapour increases under the constant water vapour content. The sulfuric acid vapour condensation rate increases more quickly. Remarkably, the vapour content has little effect on the sulfuric solution concentration. As shown in Figure 5(b), when the sulfuric acid vapour content in the flue gas is constant, both the two vapour condensation rates show an increasing trend. The influence of water vapour content in the flue gas on the condensation rate is relatively weaker compared with the sulfuric acid vapour. Moreover, the condensation rate on the ash particles is smaller than the heat transfer wall under the same conditions. However, the vapour condensation rate on the particle surface gradually approaches the value of the heat transfer wall with the increase of water vapour content, indicating that the effect of ash particles on the condensation rate is weaker in the case of high water vapour content. On the other hand, unlike the influence of the above-mentioned sulfuric acid vapour, an increasing water vapour content causes a sharp drop in the sulfuric acid solution concentration.

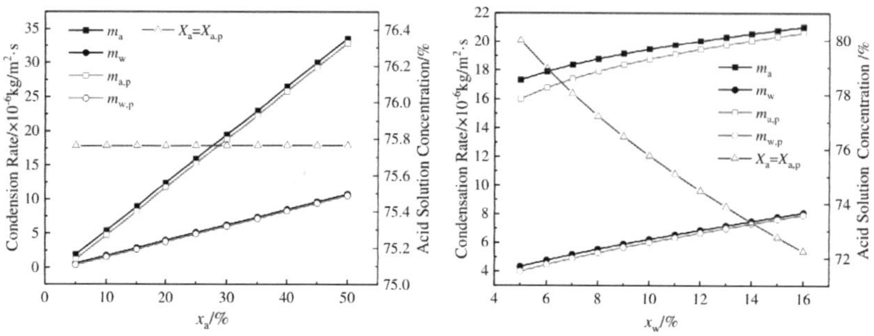

(a) Sulfuric vapour content (b) Water vapour content

Figure 5. Condensation rate variations with the vapour contents

3.2 Effect of heat transfer wall temperature

The heat transfer wall temperature is the main factor affecting the vapour condensation characteristics under the constant vapour contents. Figure 6 shows the vapour condensation rate variation curves with the heat transfer wall temperature. The curves can be divided into two different stages. When the wall temperature is varied from 50 °C to 110 °C (Stage 1), the water vapour condensation rate drops sharply and then begins to decrease slowly, while the sulfuric acid vapour condensation rates are increased linearly by 31.66% in this stage. As the wall temperature increases (Stage 2), the water vapour condensation rates continue to decrease quickly and start to separate, while the corresponding sulfuric acid vapour condensation rate begins to decrease sharply until the condensation disappears.

Additionally, as the wall temperature increases, an increasing trend is presented for the sulfuric acid solution concentration. It is also worth noting that the high temperature wall has a significant influence on the vapour condensation characteristics of the fly ash particles. At the first stage, when the wall temperature is relatively low, the vapour condensation rate of the different fly ash particle radius under the same input parameters is almost constant. As the wall temperature continues to rise, the curves of the vapour condensation rate of the different fly ash particles radius begin to separate. The smaller of the fly ash particles radius, the lower the corresponding condensation rate at the same temperature, the lower the corresponding ADP.

3.3Effect of coal-burned type and excess air ratio

The composition contents of the flue gas are mainly determined by the coal-burned type and excess air coefficient. In this section, five common coal-burned types are considered, which are divided into 10 specific coal-burned groups. Figure 7 shows that the ADP variations for different coal types, the temperature difference of the ADP corresponding to the wall surface and the fly ash particles for all coal-burned types is nearly constant, which are about 3.6 °C. The ADP of the different coal-burned type flue gas are significantly different, the maximum temperature difference is about 30 °C.

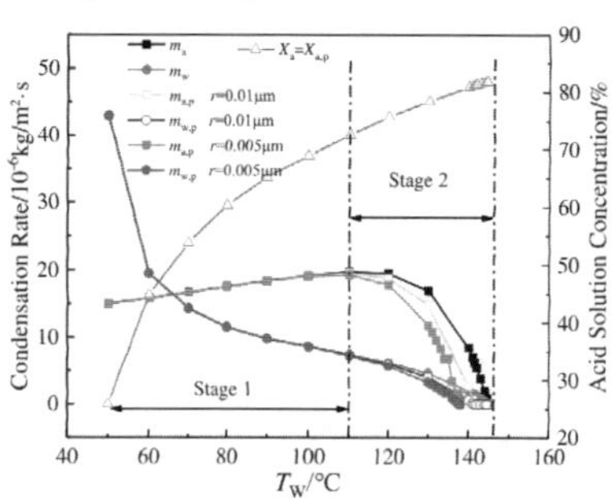

Figure 6. Effect of wall temperature on condensation characteristics

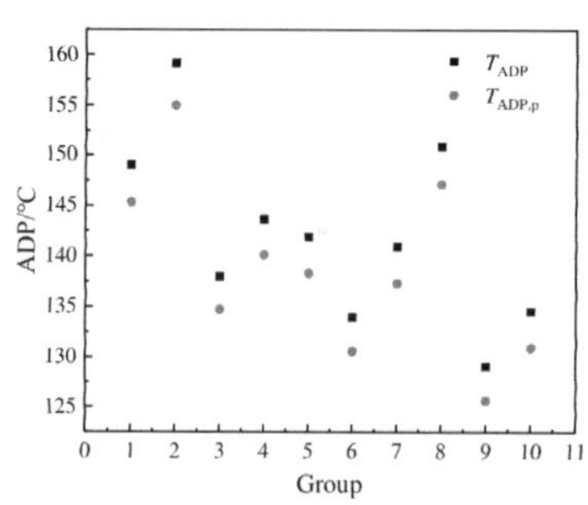

Figure 7. Effect of coal-burned types on ADP

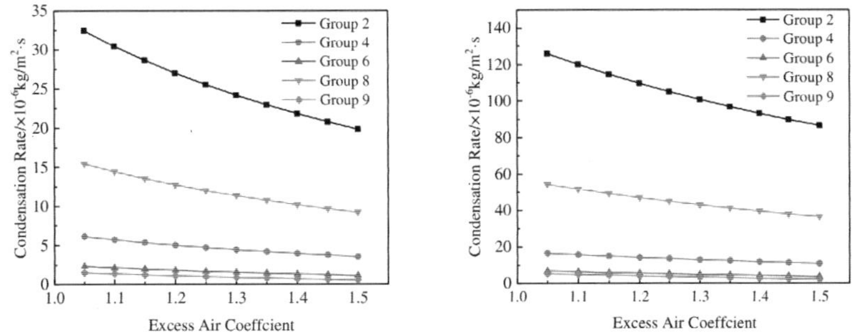

(a) Water vapour (b) Sulfuric acid vapour

Figure 8. Effect of excess air ratio on vapour condensation characteristics

Figure 8 shows the effect of excess air ratio on the vapour condensation rate. As the excess air ratio increases from 1.05 to 1.5, the water and sulfuric acid vapour condensation rate decreases correspondingly, and the decreasing speed gradually becomes slower. Moreover, the water vapour condensation rate variation under the same coal-burned type is larger than the sulfuric acid vapour. For example, the sulfuric acid vapour condensation rate of the group 2 is reduced by 32.61%, but the corresponding water vapour condensation rate is decreased by 40.23%, indicating that the water vapour condensation rate is more sensitive to the changes of the excess air ratio.

4. Conclusions

Based on the gas-liquid phase equilibrium and multi-component diffusion theory, a theoretical analysed model about ADP and vapour condensation characteristics of flue gas is developed. The effects of different parameters are discussed. The main conclusions are as follows:

(1) The ADP of the flue gas gradually increases with the water or sulfuric acid vapour content increases, and the ADP is more susceptible to the sulfuric acid vapour content. The ADP of the ash particles is always lower than the ADP of the heat transfer wall under the same conditions, it means the surface of the ash particles is more difficult to condense.

(2) The vapour condensation characteristics can be divided into two different stages according to the wall temperature variation, the high wall temperature has an obvious influence on the vapour condensation rate of the fly ash particles.

(3) The ADP corresponds to different coal-burned types coal are different completely, the water vapour condensation rate is more sensitive to the excess air coefficient than the sulfuric acid vapour.

Acknowledgement

This work was financially supported by National Key R&D Program (No. 2017YFF0209802)

References

[1] Xiang, B., Tang, B., Wu, Y., Yang, H., Zhang, M., Lu, J. (2016) Predicting acid dew point with a semi-empirical model. Appl. Therm. Eng., 106: 992-1001.

[2] He, Y., Tang, S., Wang, F., Zhao, Y., Tao, W. (2016) Gas-side fouling, erosion and corrosion of heat exchanger for middle and low temperature flue gas waste heat recovery. Chinese Science Bulletin, 61(17): 1858.

[3] Muller, P. (1959) A contribution to the problem of the action of sulfuric acid on the dew point temperature of flue gases. Chem. Eng. Technol., 31: 345-350.

[4] Tang, Z., Jin B., Sun K., Zhong, Z. (2005) The contrast and evaluation of the calculation formulas of acid dew point temperature of the flue gas after the WFGD. Journal of Power Engineering,25.

[5] Zarenezhad, B., Aminian, A. (2011) Accurate prediction of the dew points of acidic combustion gases by using an artificial neural network model. Energy convers. manage., 52(2): 911-916.

[6] Wang, Y., Tang, G. (2016) Prediction of sulfuric acid dew point temperature on heat transfer fin surface. Appl. Therm. Eng., 98: 492-501.
[7] Yin, Z., Wen, J., Wu, Y., Wang, Q., Zeng, M. (2016) Effect of non-condensable gas on laminar film condensation of steam in horizontal minichannels with different cross-sectional shapes. Int. Commun. in Heat Mass Transfer, 70: 127-131.
[8] Wilson, R.W. (1990) Condensation of sulfuric acid and water mixtures in laminar flow. Ph.D thesis.
[9] Jin, B., Tang, Z., Sun, K., Zhong, Z. (2006) Numerical prediction of the deposit rate of condensed sulfuric acid from the flue gas of power plant chimney. Proceedings of the CSEE, 26(9).
[10] Han, H., He, Y.L, Tao, W.Q. (2013) A numerical study of the deposition characteristics of sulfuric acid vapor on heat exchanger surfaces. Chem. Eng. Sci., 101: 620-630.

IOP Conf. Series: Materials Science and Engineering **721** (2020) 012049 doi:10.1088/1757-899X/721/1/012049

Optimization design method of PRHRS air heat exchanger based on natural circulation characteristics

Yandong Wang, Yongdong Chen*, Xiaohong Wu,Zhiqiang Yu, Pei Cheng

Hefei General Machinery Research Institute Co., Ltd., Hefei, Anhui, 230031, China

*Corresponding author's e-mail: chenyongdong@hgmri.com

Abstract: Passive residual heat removal system (PRHRS) was the key equipment to ensure the inherent safety of the fourth generation of nuclear reactors, which can effectively reduce the probability of core meltdown in the event of an accident. A multi-physics coupling model of sodium cooled fast reactor PRHRS was established in this paper, which based on the natural circulation characteristics of PRHRS. The model was analytically calculated according to the flow balance and energy balance, and finally the model was dynamically balanced. The influence of the tube spacing of the serpentine finned tube and the opening area of the air door on the thermal behavior of the system was analyzed through PRHRS model, and the optimal design method of the air heat exchanger (AHX) of PRHRS was mastered. The research results provide a theoretical basis for the rational design of the air heat exchanger and the safe operation of the sodium cooled fast reactor PRHRS.

1. Introduction

The safety and reliability of nuclear power systems have been received more and more attention after the nuclear accident in Fukushima, Japan [1-3]. At present, the cooling methods of the passive residual removal system（PRHRS）mainly include water cooling and air cooling. Compared with the water cooling method, the method of using air as a cooling medium is not restricted by geographical location, and is convenient for nuclear power facilities in water shortage areas.

Because of these advantages, research on air-cooled PRHRS air heat exchanger has gradually deepened in recent years. Hie-Chan Kang et al. [4-5] found that the pressure drop and heat transfer coefficient of the air heat exchanger did not change much when tube banks were arranged in the rectangular pattern and the parallelogram pattern according to the study of Korea advanced liquid Metal reactor PRHRS. However, the pressure drop and heat transfer coefficient of the staggered tube banks were higher than the parallelogram tube banks by 80% and 40% respectively. V.Vinod et al. [6-7] conducted a forced ventilation test on the sodium-air heat exchanger in the Indian prototype fast breeder reactor (PFBR). The results showed that the heat transfer coefficient of the sodium-air heat exchanger is mainly determined by the air side heat resistance. Li Jing et al [8] used CFD software to calculate the natural circulation flow in the air heat exchanger of China experimental fast reactor, and gave the air velocity distribution and temperature distribution in the air heat exchanger under various working conditions. Finally, the change law of the heat removal capacity of the air heat exchanger with the outdoor temperature and the sodium tube temperature was summarized.

In summary, the experimental and numerical methods for PRHRS air heat exchangers of the third generation and the fourth generation reactor are mostly used, and there are few analytical models to be studied, especially the analytical modeling for the fourth generation reactor. Therefore, a multi-physics coupling model of sodium cooled fast reactor PRHRS is established in this paper, which is based on the characteristics of natural circulation. The general thermal equation of the sodium cooled fast reactor PRHRS is obtained by formula derivation, and the calculation software is used to analyze and calculate the system flow and obtain the relevant thermal parameters. Finally, the influence of the variation of the tube spacing of serpentine fin tube and the opening area of air door on the heat transfer and flow of the PRHRS air heat exchanger is explored, and the optimal design method of the PRHRS air heat exchanger is mastered.

2. Heat transfer and flow model of PRHRS

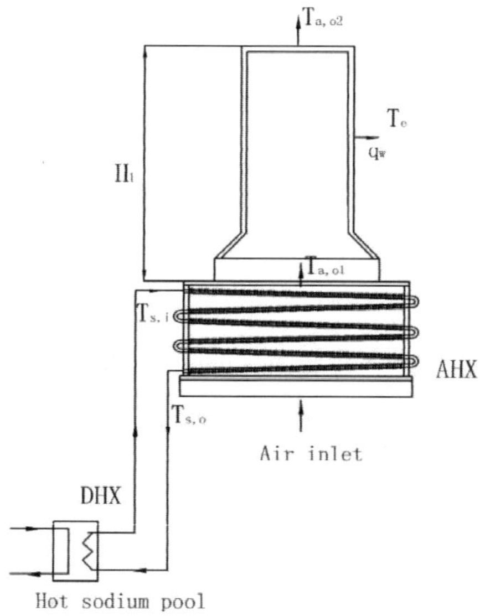

Figure 1. Schematic diagram of PRHRS model.

Figure 1 is a schematic diagram of the model of the sodium cooled fast reactor PRHRS. In the model, temperature field, density field and pressure field interact with each other and reach dynamic equilibrium. The system mainly consists of the air heat exchanger (AHX), the decay heat exchanger (DHX), and the chimney and the air door. The system includes of three natural circulation sections, the DHX in the hot sodium pool and reactor core constitute a natural circulation of the first loop, then the DHX and the AHX constitute the natural circulation of the intermediate loop, and finally the AHX in the chimney and the external environment constitute a natural circulation of air cooling.

The total heat transfer amount of the PRHRS air heat exchanger is calculated as

$$Q_o = \phi K n_s N_p L_A A_o \Delta T_m \tag{1}$$

Where

$$K = \left[\frac{\beta}{h_s} + r_i \beta + \frac{A_o}{2\pi\lambda_w} \ln\left(\frac{d_o}{d_i}\right) + \frac{r_o}{\eta} + \frac{1}{h_a \eta} \right]^{-1} \tag{2}$$

In formula (1), K is the total heat transfer coefficient of the AHX, $W/m^2.K$; n_s is the number of tube; N_P is the number of tube passes; L_A is the length of the tube per pass; A_o is the outer surface area of the fin tube base tube per unit length, m^2/m; Φ is the logarithmic mean temperature difference correction coefficient; ΔT_m is the logarithmic mean temperature difference, °C.

In formula (2), h_s and h_a are the sodium side and air side heat transfer coefficients, respectively, $W/m^2.K$; r_i and r_o are the sodium side and air side fouling thermal resistance, $m^2.K/W$; d_i and d_o are inner and outer diameters of finned tube base tube, respectively, m; β is the finned ratio; η is the total efficiency of the finned tube wall surface; λ_w is the thermal conductivity of the tube material, $W/m.K$.

In the chimney channel, the buoyancy force of the height of H1 chimney on the air is given as follow

$$F_a = \int_0^{H_1} (\rho_e - \rho_{a,om}) g dh \tag{3}$$

In formula (3), $\rho_e$is the air density at ambient temperature, kg/m^3; $\rho_{a,om}$ is the air density at average value between the outlet temperature of the AHX and the outlet temperature of the chimney, kg/m^3.

The total pressure drop on the air side is given

$$\Delta P_a = f_a N_p N_r \frac{(G_{\max})^2}{2\rho_{a,m}} + f_c \frac{(m_a/S_c)^2}{2\rho_{a,om}} \frac{H_1}{D_i} + \xi_c N_w \frac{(m_a/S_c)^2}{2\rho_{a,om}} + \varepsilon_c \frac{(m_a/S_c)^2}{2\rho_{a,om}} \tag{4}$$

In formula (4), G_{max} is the maximum mass flow rate among the bundles, $kg/m^2.s$; m_a is the air mass flow rate, kg/s; S_c is the cross-sectional area of the chimney channel, m^2; H_1 is the height of the chimney, m; f_a is the frictional resistance coefficient among the bundles; f_c、 ξ_c and ε_c are the friction coefficient of the chimney, the local resistance coefficient and the resistance coefficient of the inlet and outlet, respectively.

The heat loss of the pipeline can be neglected because the heat loss of the pipeline and the environment is small, so the buoyancy force of sodium side is given as follow

$$F_s = \int_0^{H_2} (\rho_{s,o} - \rho_{s,i}) g dh \tag{5}$$

In formula (5), $\rho_{s,i}$ is the sodium density of the DHX outlet, kg/m^3; $\rho_{s,o}$ is the sodium density of the AHX outlet, kg/m^3; H_2 is thermal center height difference between the AHX and the DHX, m.

The resistance loss inside the pipe is given as follow

$$\Delta P_s = \sum_{i=1}^{n} \left(f_{s,i} \cdot \frac{\rho_{s,i} v_{s,i}^2}{2} \cdot \frac{L_{s,i}}{d_{s,i}} + \xi_{s,i} N_{s,i} \cdot \frac{\rho_{s,i} v_{s,i}^2}{2} \right) \tag{6}$$

In formula (6), $f_{s,i}$ is the friction resistance coefficient inside the pipe, $L_{s,i}$ is the length of each pipe, m; $N_{s,i}$ is the number of pipe bends; $\xi_{s,i}$ is the local resistance coefficient in the pipe.

Combined with formula (3) and formula (4), it can be found that there are many factors affecting the ventilation mass flow rate of PRHRS air heat exchanger, which is related to the ambient temperature, chimney geometry, finned tube geometry and heat loss.

3. Flow chart of PRHRS

Figure 2 is the cyclic flow chart of the sodium cooled fast reactor PRHRS. The flow is mainly divided into four calculation procedures, including two flow balance calculation modules and one energy balance calculation module, and each module must be performed a balance determination.

(1) The air side buoyancy force F_a is calculated according to the equation (3), and the air side total pressure drop ΔP_a is calculated according to the equation (4). Judging the relative error between the two, the air side flow balance is determined if the relative error is within 1%, and the system energy balance calculation is entered. Otherwise, the air side outlet temperature is adjusted.

(2) The sodium side buoyancy force F_s is calculated according to the equation (5), and total resistance loss ΔP_s is calculated according to the equation (6). Judging the relative error between the two, the sodium side flow balance is determined if the relative error is within 1%, and the system energy balance calculation is entered. Otherwise, the sodium side inlet and outlet temperature is adjusted.

(3) The total heat transfer amount Q_o is calculated by the formula (1), and the relative error between the Qo and the designed heat load Q is judged. The system energy balance is determined and the thermal parameters are outputed if the relative error between both is within 1%. Otherwise, the inlet and outlet temperatures on both sides of the flow balance module are adjusted.

(4) After the system flow balance and energy balance are determined, the physical characteristic parameters of the fin tube and the structural parameters of the related equipment are adjusted, which is to obtain the optimized design result of AHX.

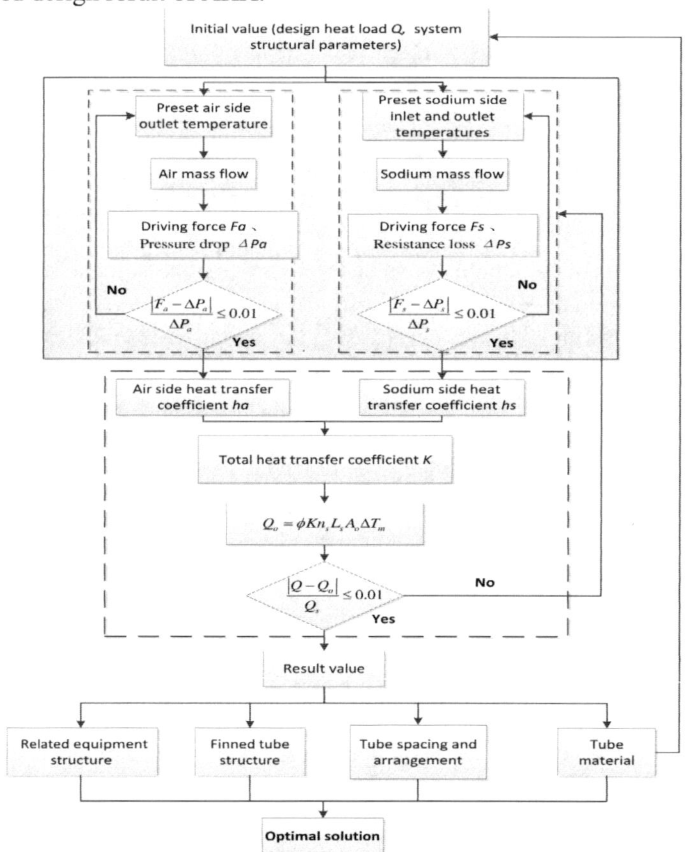

Figure 2. Flow chart of the sodium cooled fast reactor PRHRS.

4. Calculation results and analysis

Table 1. Design heat load and main structural parameters of PRHRS.

Design Parameters		Design Parameters	
Design heat load [MW]	2	Number of tubes per row	22
ID of tube [mm]	32.9	Tube arrangement [°]	30
OD of tube [mm]	38.1	Ambient temperature [℃]	40
Transverse tube spacing [mm]	85	Chimney inner diameter [m]	1
Fin height [mm]	13	Chimney height [m]	30
Fin thickness [mm]	1.22	Thermal center height [m]	41.5
Fin pitch [mm]	5.1059	Tube material	316L

According to the experimental model from literatures [6-7]. Table 1 lists the design heat load of the sodium cooled reactor PRHRS, the main structural parameters of the finned tube and the related equipment.

As shown in Figure 3a, with the increase of tube spacing, the air side outlet temperature of the AHX decreases, while the sodium inlet and outlet temperature increase. This is because the increase of tubes spacing leads to a reduction in the total heat exchange area of the fin tubes while increases the inlet air flow. As shown in Figure 3b, the total heat transfer coefficient and pressure drop of the AHX decrease with the increase of tube spacing, and the decreasing trend is slowing down. This is because that although the cross-sectional area between tube bundles increases with the increase of tube spacing, the inlet air flow rate increases correspondingly, so the speed rate of the air flow becomes slower. When the tube spacing increases from 70mm to 90mm, the total heat transfer coefficient reduces by 13.64%, the heat exchanger pressure drop reduces by 90.75%, so the influence of the tube spacing on the pressure drop is greater. Therefore, the reduction of the pipe spacing increases the heat transfer coefficient and reduces the size of the tube bundle, while the increase of the pressure drop requires a higher height of the chimney, so the tube bundle size and the height of the chimney should be comprehensively considered in practical engineering designs.

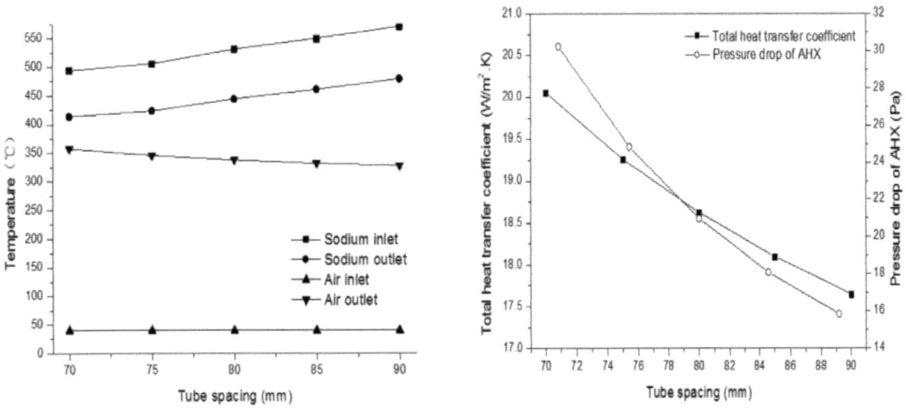

(a) (b)

Figure 3. The AHX thermal parameters under different tube spacing.

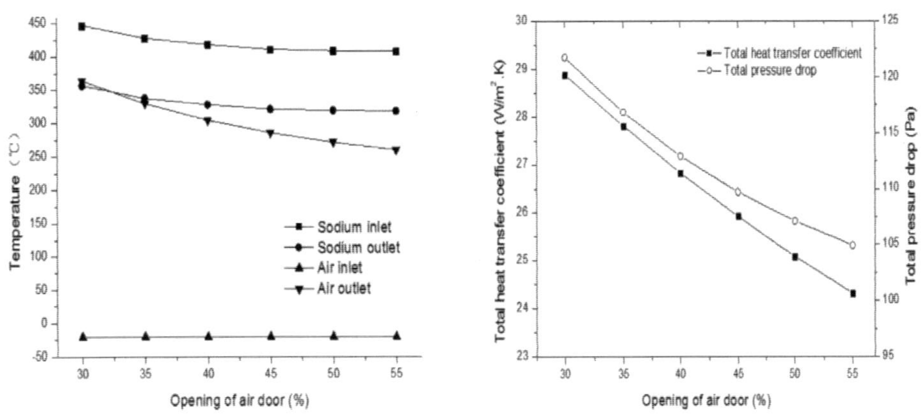

(a) (b)

Figure 4. The AHX thermal parameters under different opening area of air door.

Figure 4a and Figure 4b show the thermal parameters of AHX as a function of the ambient temperature of -20 °C. It can be seen that when the opening area reduces to about 30%, the outlet temperature of sodium can increase by about 20 °C. At the same time, the opening area of air door reduces from 60% to 30%, the total heat transfer coefficient increases by 25.45%, and the total pressure drop (the sum of the AHX and the chimney) increases by 24.46%. Therefore, the method of reducing the opening area can effectively increase the operating temperature of the equipment and the heat transfer effect of the system when the PRHRS encounters extremely low temperature environmental conditions

5. Conclusions

The multi-physics coupling model of the sodium cooled reactor PRHRS was established, and the calculation method of the system circulation process was given. The influence of the variation of the tube spacing of fin tube and the opening area of air door on the thermal behavior of the PRHRS was obtained. The following conclusions were reached in this study.

(1) With the increase of the tube spacing, the inlet and outlet temperatures of the sodium side were increased, the total heat transfer coefficient and pressure drop of AHX were decreased and the trend was slowing down. The influence of the tube spacing on the pressure drop is much greater than the heat transfer coefficient. Therefore, the tube bundle size and the height of the chimney should be comprehensively considered in practical engineering designs.

(2) When the PRHRS encountered extremely low temperature environmental conditions, the method of reducing the opening area of air door could effectively improve the operating temperature of the AHX and the heat transfer effect of the system.

Acknowledgments

This work was supported by Anhui Provincial Natural Science Foundation (No.: 1708085QE122), National key research and development program(2017YFF0209802).

References

[1] Wang Q, Chen X, Yi-Chong X. (2013) Accident like the Fukushima unlikely in a country with effective nuclear regulation: Literature review and proposed guidelines. Renewable and Sustainable Energy Reviews, 17:126-146.

[2] Bae K H , Kim H C , Chang M H , et al. (2001) Safety evaluation of the inherent and passive safety features of the SMART design. Annals of Nuclear Energy, 28(4):333-349.

[3] Sato T, Kojima Y. (2007) Variations of a passive safety containment for a BWR with active and passive safety systems. Nuclear Engineering & Design, 237(1):74-86.

[4] Kang H C , Eoh J H , Cha J E , et al. (2013) Numerical study on pressure drop and heat transfer for designing sodium-to-air heat exchanger tube banks on advanced sodium-cooled fast reactor[J]. Nuclear Engineering and Design, 254(none):5-15.

[5] Kang H C, Jang H S. (2011) Natural Convection Correlations of Circular Finned Tube Heat Exchanger Asme-jsme-ksme Joint Fluids Engineering Conference. American Society of Mechanical Engineers.

[6] Vinod V , Pathak S P , Paunikar V D , et al. (2013) Experimental evaluation of sodium to air heat exchanger performance. Annals of Nuclear Energy, 58(8):6-11.

[7] Vinod V, Suresh Kumar V A, Noushad I B, et al. (2008) Performance assessment of sodium to air finned heat exchanger for FBR.14th International Conference on Nuclear Engineering. American Society of Mechanical Engineers Digital Collection, 115-118.

[8] Li Jing. (2007) The thermal hydraulic characteristics of the air heat exchanger in China experimental fast reactor[D]. China Institute of Atomic Energy.

Numerical simulation research of influencing factors of fin temperature in finned tube heat exchanger

Zhiqiang Yu, Yongdong Chen*, Xiaohong Wu, Yandong Wang, Pei Cheng and Gaige Yu

Hefei General Machinery Research Institute Ltd, Hefei, Anhui, 230031, China

*Corresponding author's e-mail: chenyongdong@hgmri.com

Abstract. In the using process of the finned tube, the temperature on the top of the windward face of the fin in the first and second row of finned tube is the highest temperature of finned tube. It restricts and determines the high temperature conditions and application fields of finned tube heat exchanger. This paper establishes the research model of the first row and second row finned tube, to carry out the numerical simulation calculation of the fluid flow and heat transfer of high temperature air flowing through the finned tube and exchanging heat with cooled liquid water, and to research the influence of different tube transverse spacing, fin spacing and other factors on the fluid flow and heat transfer characteristics of finned tube and the highest fin temperature. The air outlet temperature, heat exchange capacity, Nu number, pressure drop and maximum fin temperature are obtained in the conditions of different tube transverse spacing and fin spacing, and the variation characteristic, mechanism and law of temperature field and velocity vector for the finned tube are also gained.

1. Introduction

The energy crisis has caused a huge impact on the world's economic development and scientific research, which has greatly promoted the research and development of enhancement heat transfer technology objectively. Finned tube heat exchanger is the earliest and most successful achievement in the process of strengthening tubular heat transfer, and it is widely used in power, chemical industry, petrochemical industry, air conditioning engineering and refrigeration engineering [1].

In recent years, domestic and foreign scholars have carried out many studies on finned tube heat exchanger [2-5]. Romero-Mendez et al. [6] revealed that fin spacing had different influence trends on fluid flow and heat transfer; for certain constraints, fin spacing had an optimal value. Glazar et al. [7] compared and analysed the heat transfer efficiency and pressure drop of microchannel heat exchangers with square and rectangular sections. Ammar et al. [8] experimentally studied the pressure drop characteristics of R134a in microfinned microchannel and smooth microchannel. Al-Neama et al. [9] studied the heat transfer performance of the microchannel with a hydraulic diameter of 1.5mm and a "V" shaped fin. Sarafraz et al. [10] experimentally studied the heat transfer performance of multi-walled carbon nanotubes/water nanofluids, and found that increasing the volume fraction of nanoparticles could enhance the heat transfer effect.

In the process of using the finned tube, the temperature on the top of the windward face of the fin in the first and second row of finned tube is the highest temperature of finned tube. It restricts and determines the high temperature conditions and application fields of finned tube heat exchanger. This paper establishes the research model of the first row and second row finned tube, to carry out the numerical simulation calculation of the fluid flow and heat transfer of high temperature air flowing through the finned tube and exchanging heat with cooled liquid water, and to research the influence of different tube transverse spacing, fin spacing and other factors on the fluid flow and heat transfer characteristics of finned tube and the highest fin temperature. The air outlet temperature, heat exchange capacity, Nu number, pressure drop and maximum fin temperature are obtained in the conditions of different tube transverse spacing and fin spacing, and the variation characteristic, mechanism and law of temperature field and velocity vector for the finned tube are also gained. This work can provide theoretical guidance and engineering reference for the extending application of finned tube heat exchanger under special and extreme conditions.

2. Research model and related setting

2.1Research model

In order to simplify the model, the heat transfer and flow field distribution of the front two rows tubes are studied under high temperature air. Calculation models 1: tube transverse spacing 56mm, 60mm, 61mm, 62mm, 63mm, 64mm, 68mm, 74mm, and equilateral triangle arrangement (30°). Take the tube transverse spacing 74mm as an example, as shown in figure 1. Calculation models 2: fin spacing 2.0mm, 2.5mm, 3.0mm, 3.5mm, 6.0mm. Take fin spacing 6.0mm as an example, as shown in figure 2.

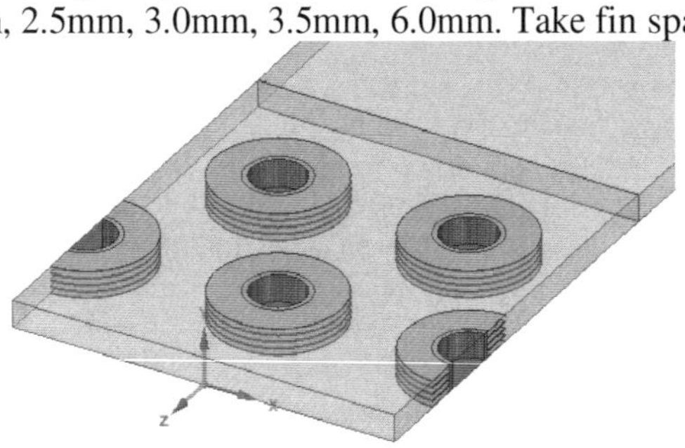

Figure 1. Model of tube transverse spacing 74mm.

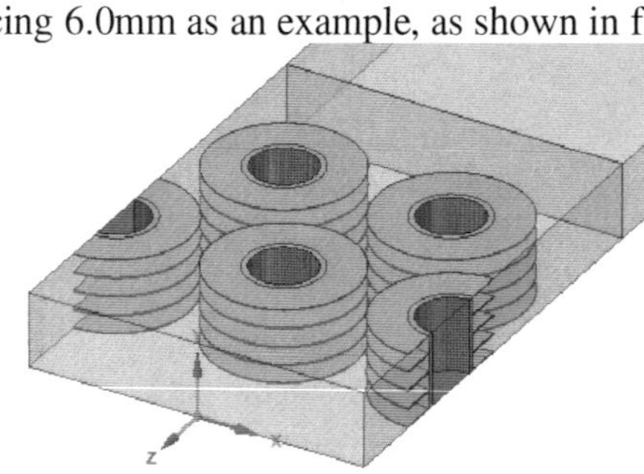

Figure 2. Model of fin spacing 6.0mm.

2.2Related setting

The related settings of fin spacing 6.0mm is taken as an example. Model: length, width and height are 112mm, 648.5mm and 24mm respectively. Round straight tube, the tube Φ25×2mm, tube transverse spacing is 56 mm, equilateral triangle arrangement (30°). The annular fin is 12mm high, 0.5mm thick and 6.0mm apart. The height of 4 fin spacing is calculated here. Extending 500mm for outlet to avoid backflow. The grid is unstructured. Taking the fin spacing 6.0mm as an example, the number of nodes is about 710,000 and the number of grids is about 1.37 million, as shown in figure 3.

Figure 3. Calculation grid of the model of fin spacing 6.0mm.

The hot fluid is air; its temperature is 924.75K (651.6°C) and its inlet velocity is 8m/s. The cold fluid is liquid water; its temperature is 332.15K (59°C) and its inlet velocity is 1m/s. The tube and fin are made of stainless steel. Finite volume method is adopted for numerical calculation, second-order upwind format is adopted for discretization of each equation, SIMPLE algorithm is used for pressure-velocity coupling, and SST k-ω turbulence model is selected [11-12].

3. Calculation results and analysis

3.1Influence of tube transverse spacing

In order to obtain the influence of different tube transverse spacing on the fluid flow and heat transfer characteristics, numerical calculations are carried out on the models with the transverse spacing 56mm, 60mm, 61mm, 62mm, 63mm, 64mm, 68mm and 74mm respectively. For transverse spacing 56mm and 74mm, the temperature fields in the calculated area are shown in figures 4a to 4b, and the velocity vector diagrams are shown in figures 5a to 5b. It can be found that when the tube spacing is 56mm, the air is disturbed with the largest level, and the vortex formed behind the finned tube is also the largest one. As the spacing of finned tube increases, the air velocity decreases at the narrowest point, which weakens the heat exchange between tube bundle and air. Inside the tube bundle, the air temperature distribution is the most uniform with 56mm tube spacing. With the increase of tube spacing, the influence between adjacent finned tubes on the same tube row position is gradually weakened, and the heat transfer area between hot and cold air is also increased along the air flow direction.

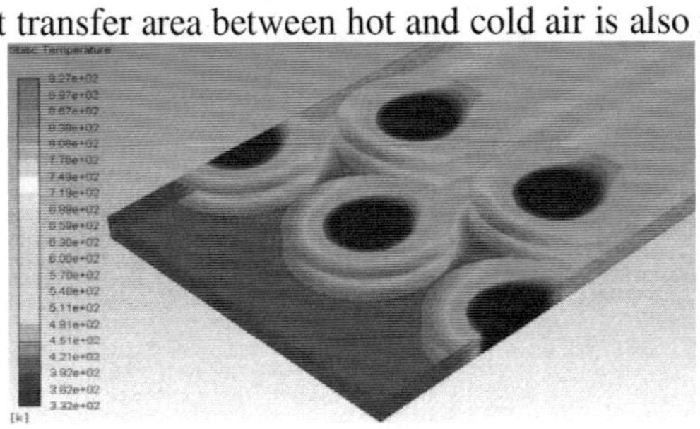

Figure 4a. Temperature field for S_{tube}=56mm.

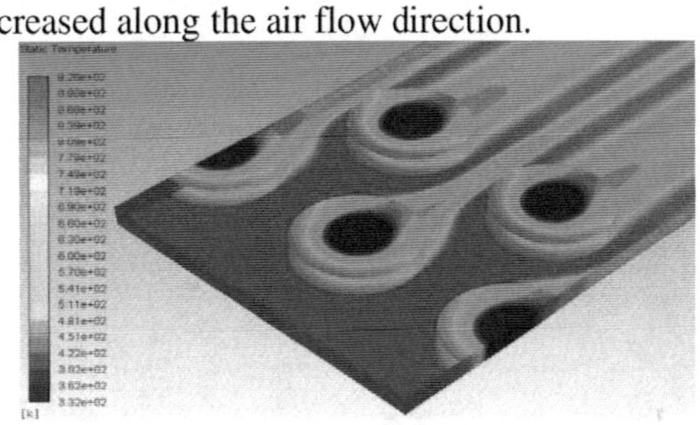

Figure 4b. Temperature field for S_{tube}=74mm.

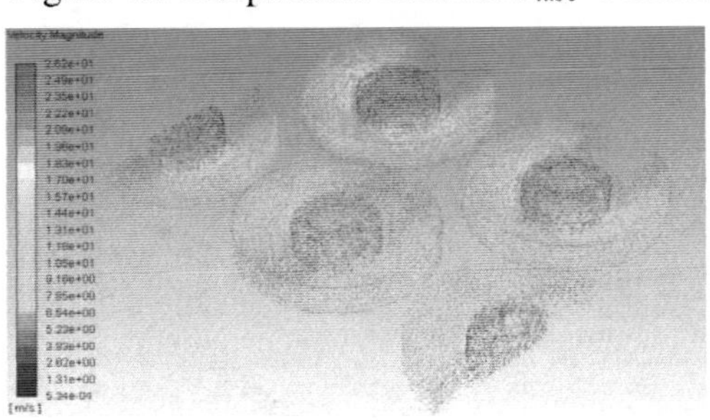

Figure 5a. Velocity vector for S_{tube}=56mm.

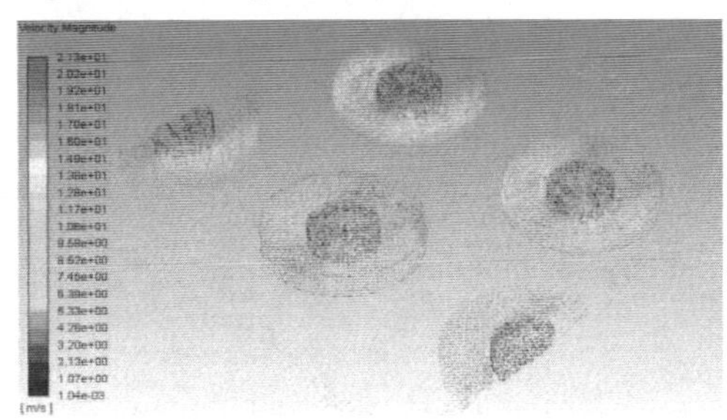

Figure 5b. Velocity vector for S_{tube}=74mm.

The influences of different tube spacing on some thermodynamic parameters are shown in figures 6a to 6e. As the tube spacing increases, the Nusselt number decreases, the total heat transfer power decreases, and the Nusselt number decreases slowly. This is because when the tube spacing increases, the flow cross-sectional area between the transverse bundles increases, and the air velocity decreases slower and slower at the same inlet air mass flow rate, so the heat transfer coefficient decreases slowly. In addition, with the increase of tube spacing, the pressure drop decreases gradually. When the tube spacing increased from 54mm to 74mm, the Nusselt number decreases by 21.43% and the pressure drop decreases by 66.67%. Tube spacing has a greater effect on the pressure drop. Therefore, the decrease of tube spacing will reduce the size of tube bundle, while the increase of pressure drop will require a higher height of duct. Therefore, the size of tube bundle and the height of duct should be considered comprehensively in practical engineering.

At the same time, it can be found that the maximum temperature of the fin in the first row of finned tube decreases gradually, and the maximum temperature of the fin in the second row of finned tube increases first and then decreases. This because with the increase of the tube spacing, the heat load received by the second row of tubes is more and more large, and the heat transfer capacity is gradually slowing down. Therefore, there is an optimal tube spacing, and the heat exchange between hot air outside the tube and cooling water inside the tube is the most sufficient.

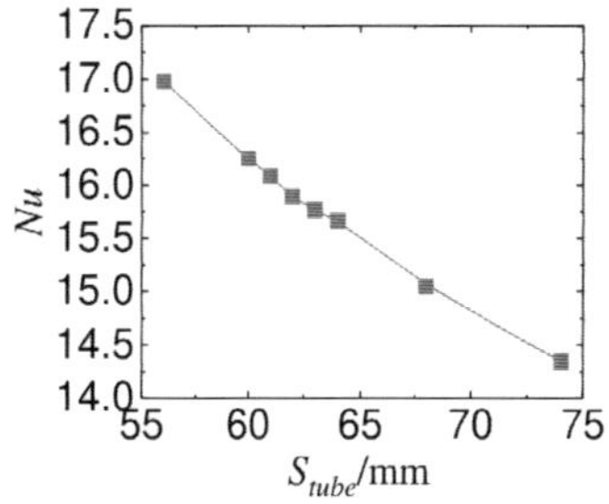

Figure 6a. Tube spacing influence on *Nu*.

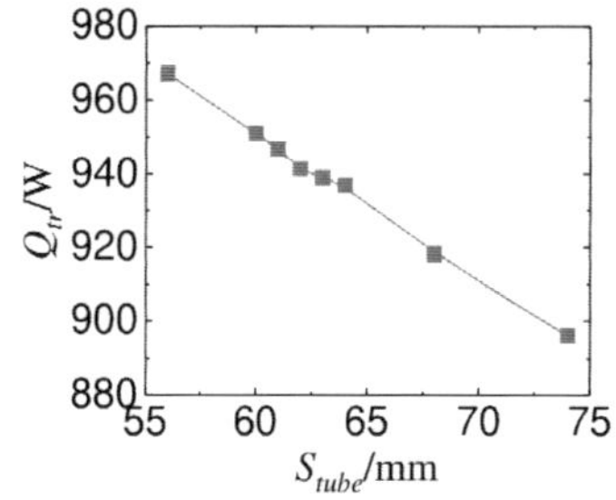

Figure 6b. Tube spacing influence on Q_{tr}.

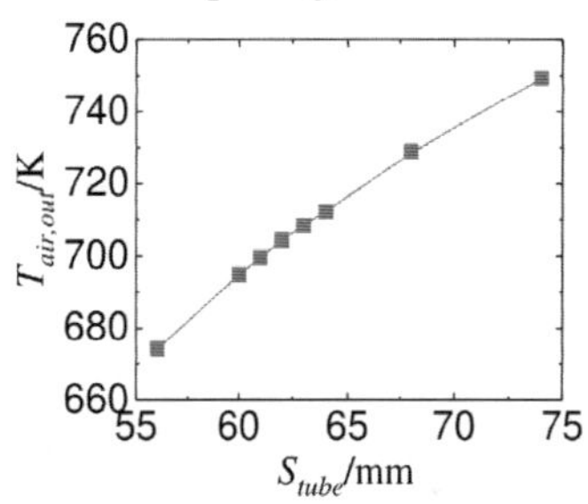

Figure 6c. Tube spacing influence on $T_{air,out}$.

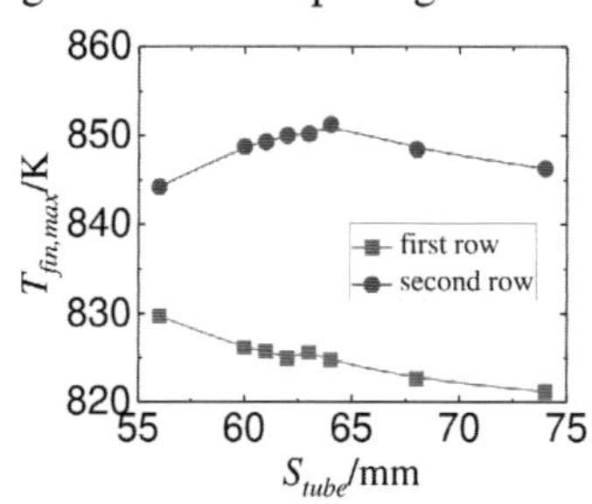

Figure 6d. Tube spacing influence on $T_{fin,max}$.

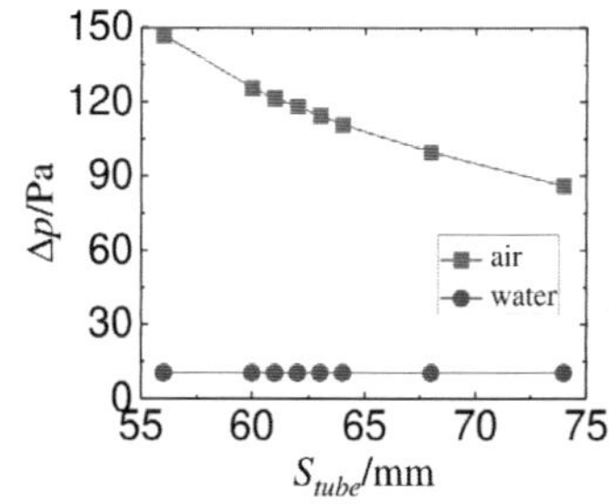

Figure 6e. Tube spacing influence on Δp.

3.2Influence of fin spacing

Here, the tube bundle models with fin spacing 2.0mm, 2.5mm, 3.5mm and 6.0mm are respectively calculated. For fin spacing 2mm and 6mm, the temperature fields in the calculated area are shown in figures 7a to 7b, the velocity vector diagrams are shown in figures 8a to 8b, and figures 9a to 9d are the diagrams of the influences of different fin spacing on some thermodynamic parameters.

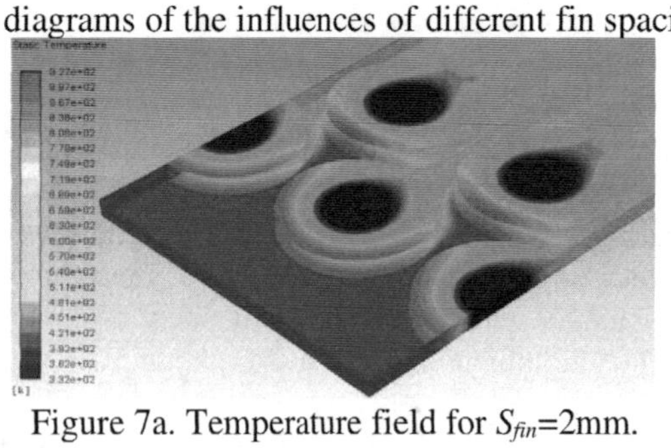

Figure 7a. Temperature field for S_{fin}=2mm.

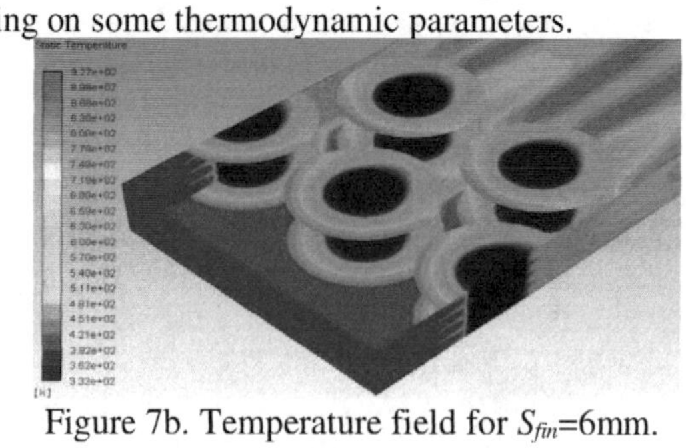

Figure 7b. Temperature field for S_{fin}=6mm.

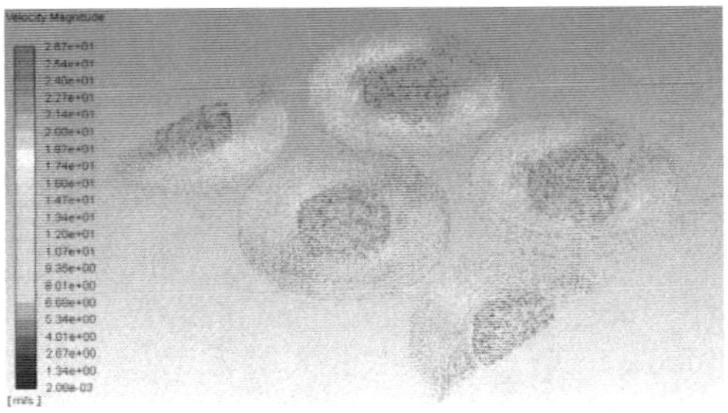

Figure 8a. Velocity vector for S_{fin}=2mm.

Figure 8b. Velocity vector for S_{fin}=6mm.

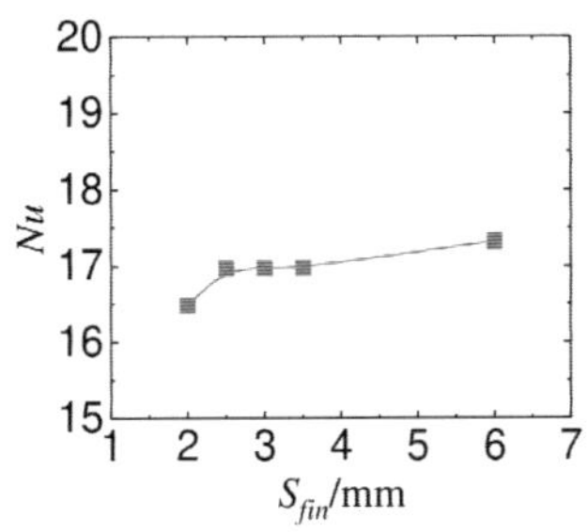

Figure 9a. Fin spacing influence on *Nu*.

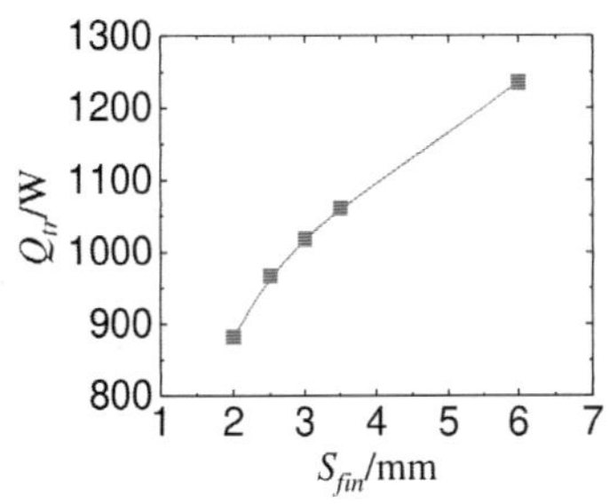

Figure 9b. Fin spacing influence on Q_{tr}.

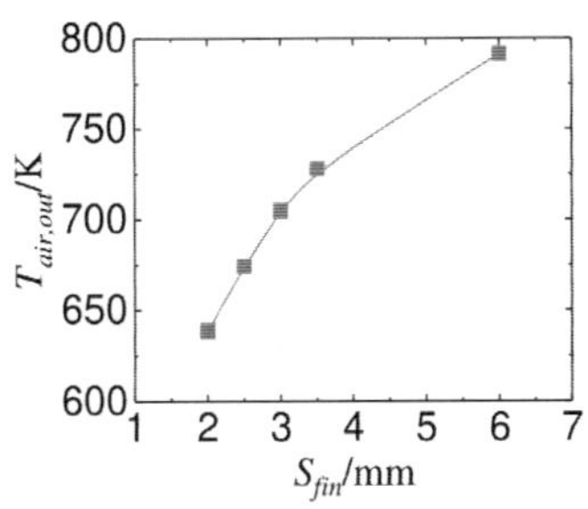

Figure 9c. Fin spacing influence on $T_{air,out}$.

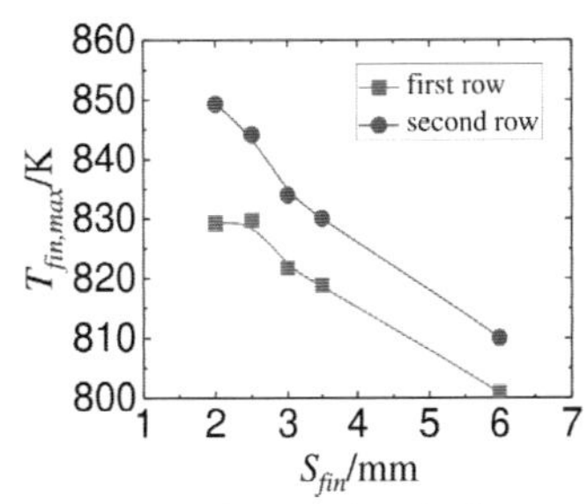

Figure 9d. Fin spacing influence on $T_{fin,max}$.

Based on the above results, when the fin spacing increases, the heat exchange between the air outside the tube and the cooling water inside the tube is more complete. When the fin spacing reaches 6mm, the thermal boundary layer of the adjacent fins has been separated, and the middle region is filled with high-temperature air whose temperature is about equal to the air inlet temperature, then full heat exchange is achieved. After the thermal boundary layer is completely separated, the influence of fin spacing on heat transfer power is mainly achieved by increasing the surface area of the base tube between fins, so the increase of the heat transfer power in the calculation area gradually becomes slow. As the fin spacing increases, the air flow velocity decreases. When the fin spacing increases, the fin disturbance intensity is enhanced, which can reduce the thickness of boundary layer and enhance the heat transfer. Nusselt number respectively increases by 3.66%, 0.59%, 0.58%, 1.75%, so the influence of fin spacing on heat transfer coefficient is decreasing.

In addition, with the increase of fin spacing, fin ratio of fin tube decreases, which give rise to the decrease of heat exchange area of finned tube per unit length. In order to obtain the same heat exchange effect, the required equipment size increases, so that the whole machine volume increases. Therefore, for extreme conditions with high air velocity, the influence of spacing on heat transfer coefficient and heat transfer area should be considered comprehensively, so as to select the fin spacing that is most conducive to heat transfer.

4. Conclusions

In this paper, the heat transfer characteristics of finned tube bundles are researched by numerical simulation method, the calculation model of finned tube bundles is established and the influence of tube transverse spacing and fin spacing on heat transfer characteristics of finned tube bundles is analysed. The research results show that:

1) For the arrangement of fork and row, the reduction of tube spacing will reduce the size of the tube bundle and increase the heat transfer power of the tube bundle, but the air pressure drop will increase in a higher proportion. Therefore, the size of tube bundle and the height of air duct should be considered comprehensively in practical engineering application.

2) The average Nusselt number of fin bundle increases with the increase of fin spacing, and the increase rate gradually decreases in this process. Because the increase of fin spacing will reduce the total surface area of finned tube per unit length. So, the increase of fin spacing and the size of equipment should be considered comprehensively in practical engineering applications.

Acknowledgments

This work was supported by Anhui Provincial Key Research and Development Program (201904a05020050) and Anhui Provincial Natural Science Foundation (1708085QE122).

References

[1] Sheikholeslami M, Gorji-Bandpy M, Ganji D.D. (2015) Review of heat transfer enhancement methods: Focus on passive methods using swirl flow devices. Renewable and Sustainable Energy Reviews, 49: 444-469.

[2] Chien N.B., Choi K.I., Oh J.T., et al. (2018) An experimental investigation of flow boiling heat transfer coefficient and pressure drop of R410A in various minichannel multiport tubes. International Journal of Heat and Mass Transfer, 127: 675-686.

[3] Zheng L, Zhang D, Xie Y, et al. (2016) Thermal performance of dimpled/protruded circular and annular microchannel tube heat sink. Journal of the Taiwan Institute of Chemical Engineers, 60: 342-351.

[4] Meyer C.J., Kroger D.G. (1998) Plenum flow losses in forced air-cooled exchangers. Applied Thermal Engineering. 1998, 18(9-10): 875-893.

[5] Wongwises S, Chokeman Y. (2005) Effect of fin pitch and number of tube rows on the air side performance of herringbone wavy fin and tube heat exchangers. Energy Conversion and Management. 46(13-14): 2216-2231.

[6] Romero-Mendez R, Sen M, Yang K.T., et al. (2000) Effect of fin spacing on convection in a plate fin and tube heat exchanger. International Journal of Heat Mass Transfer, 43(1): 39-51.

[7] Glazar V, Frankovic B, Trp A. (2015) Experimental and numerical study of the compact heat exchanger with different microchannel shapes. International Journal of Refrigeration, 51: 144-153.

[8] Ammar S.M., Abbas N, Abbas S, et al. (2019) Experimental investigation of condensation pressure drop of R134a in smooth and grooved multiport flat tubes of automotive heat exchanger. International Journal of Heat and Mass Transfer, 130: 1087-1095.

[9] Al-Neama A.F., Khatir Z, Kapur N, et al. (2018) An experimental and numerical investigation of chevron fin structures in serpentine minichannel heat sinks. International Journal of Heat and Mass Transfer, 120: 1213-1228.

[10] Sarafraz M.M., Hormozi F. (2016) Heat transfer, pressure drop and fouling studies of multi-walled carbon nanotube nano-fluids inside a plate heat exchanger. Experimental Thermal and Fluid Science, 2016, 72: 1-11.

[11] Tao W.Q. (2001) Numerical Heat Transfer, 2nd edition. Xi'an Jiaotong University Press, Xi'an.

[12] Yang S.M., Tao W.Q. (2006) Heat transfer, 4th edition. Higher Education Press, Beijing.

IOP Conf. Series: Materials Science and Engineering **721** (2020) 012051 doi:10.1088/1757-899X/721/1/012051

Research on Low Temperature Condensation of Flue Gas from Gas Industrial Boiler

Yu Ji-Ming[1], Li Juan[2]*, Guan Jian[1]

[1]China Special Equipment Inspection and Research Institute, 100029 Beijing, China

[2]China Energy Conservation and Environmental Protection Consulting Co., Ltd. 100089 Beijing, China

*Corresponding author's e-mail: 47215786@qq.com

Abstract. The paper analyzed the heat transfer in the condensing section of condensing gas industrial boiler. As to the operation of boiler, the field test of the flue gas condensing section was carried out where the field test data was compared with the theoretical values to verify the calculation formula of the convective exothermic coefficient of the condensing section in the condensing boiler's spiral rib.

1. Introduction

The energy available in the flue gas generated by the combustion of gas fuel is divided into two parts: the obvious heat of flue gas and the latent heat of water vapor in flue gas. The difference between condensing boiler and traditional boiler is that the latent heat of water vapor in flue gas is effectively utilized. For natural gas in northern Shaanxi, about 70% of the latent heat of water vapor in the flue gas can be utilized if the exhaust smoke temperature of the boiler is reduced to 40°C [2].

The higher temperature of smoke extraction in traditional boiler disables the latent heat of water vapor in the flue gas. Based on the assumption of "dry flue gas", only the sensible heat of flue gas and steam is calculated in the thermodynamic calculation of boiler. As the latent heat of water vapor in the flue gas of condensing boiler is released and used, the calculation of condensing section of condensing boiler is different from that of traditional boiler.

2. Analysis of Moisture Condensation of Flue Gas

The flue gas of condensing boiler refers to the condensation of water vapor in a large amount of non-condensable mixed gases. According to Dalton's principle of partial pressure, the total pressure of flue gas refers to the sum of the partial pressure of steam in flue gas and the partial pressure of non-condensable gas. When water vapor mixed with a large amount of non-condensable gases encounters the heating surface below the temperature of its acid dew point, the water vapor molecules condense at the condensation core, where the partial pressure of water vapor decreases while the partial pressure of non-condensable gas increases. After condensation of water vapor, the liquid film is surrounded by non-condensable gas to form a layer of non-condensable gas film. Water vapor molecules outside the film need to pass through it to reach the surface of the liquid film to condense. Different from the columnar condensation of pure water vapor, the condensation of steam in condensing boiler flue gas has an impact on non-condensable gas, which is worthy of analysis [3].

The columnar condensation of steam in flue gas can be divided into three steps:

IOP Conf. Series: Materials Science and Engineering **721** (2020) 012051 doi:10.1088/1757-899X/721/1/012051

2.1. The water vapor starts to condense in the condensation core to form tiny liquid droplets, which gradually grow up to form condensation chamber on the wall's surface. The initial formation of condensation is the so-called nucleation process.

2.2. The condensation droplets grow further to form a thin liquid film on the whole condensation surface. With the continuous thickening of the liquid film, it reaches the critical thickness and splits into small droplets, a process called liquid film growth.

2.3. The small liquid droplets formed by liquid film splitting are constantly polymerized and fall off from the heating surface of the boiler when they reach the critical size. The is the aggregation of bead condensation. During liquid droplet polymerization, the speed of droplet polymerization is faster with a minimum amount of heat transfer, which can be regarded as an adiabatic process [3].

3. Analysis of Convection Heat Transfer in Condensing Section of Boiler Flue Gas

Through the analysis of the whole condensation process of steam in the flue gas, it can be found that when the temperature of steam in the flue gas drops to the acid dew point, the latent heat of vaporization is released and condensed into liquid state. As for the heat transfer in the condensing section of the boiler, it is different from the heat transfer in the condensing part of the flue gas: the temperature of the flue gas is above the acid dew point, and the heat transfer process of the two is consistent, both are sensible heat of non-condensable gas in flue gas and heat exchange between sensible heat and boiler heated medium. The heat exchange of flue bas in the traditional boiler is terminated and exhausted from the system before its temperature drops to the acid dew point. When the temperature drops to the acid dew point, the condensation heat transfer consists of sensible heat from non-condensable gas and heat exchange between latent water vapour and heated medium in the boiler, with the location of heat transfer being flexible. As the flowing flue gas encountering the heated surface in the process of cooling, latent heat is released from water vapor condensation. This is inseparable from the position of acid dew point, and the condensation of water vapor in the flue gas is a continuous process.

The heat transfer of boiler can be divided into radiation heat transfer and convection heat transfer, where heat transfer in condensing section of boiler is categorized as convection heat transfer. Coefficient of convective heat transfer represents the intensity of convective heat transfer, directly related to fluid property, flow status and the arrangement of heated tubes. Based on the structure of the boiler, the thermodynamic calculation of the convective heat transfer in heated surface is carried out through the following formula.

3.1. Convective exothermic coefficient of cross - flush straight tube bundles

$$a_d = 0.2CsCz\frac{\lambda}{d}Re^{0.65}Pr^{0.33} \quad Kw/(m^2\ ℃) \tag{1}$$

where: a_d Convection exothermic coefficient Kw/(m^2 ℃)

Cs Correction coefficient of geometric arrangement of tube bundle
Cz Correction factor of the number of pipe rows in the direction of flue gas travel
λ Average temperature of flue gas thermal conductivity Kw/(m2 ℃)
Re Reynolds criterion number
Pr Prandtl number
D Equivalent diameter m

3.2. Convection exothermic coefficient of staggered cross-flow tube bundles

$$a_d = CsCz\frac{\lambda}{d}Re^{0.6}Pr^{0.33} \quad Kw/(m^2\ ℃) \tag{2}$$

3.3. Convection exothermic coefficient of longitudinal scour heating surface

$$a_d = 0.023CtC1\frac{\lambda}{d}Re^{0.8}Pr^{0.4} \quad Kw/(m^2\ °C) \qquad (3)$$

Where

C_t Correction coefficient, Ct=1 when the flue gas is cooled

C_l Relative length correction coefficient, the inlet of the round pipe is straight only when l/d < 50. It is adopted when there is no rounded edge.

3.4. Convective exothermic coefficient of cross flow convection with transverse strip bundle or circular fin in line arrangement

$$a_d = 0.015CzCs\frac{\lambda}{S_{p6}}\left(\frac{d}{S_{p6}}\right)^{-0.54}\left(\frac{h_{p6}}{S_{p6}}\right)^{-0.14}\left(\frac{\omega S_{p6}}{\nu}\right)^{0.72} \quad Kw/(m^2\ °C) \qquad (4)$$

where: S_{p6} Fin pitch mm

h_{p6} Fin height mm

ω Flue gas flow ratem/s

ν Kinematic viscosity of flue gas m^2/s

With SZS type gas condensing boiler as an example, the convection bundle between the upper and lower boiler barrel is built by cross-flow light tube bundles in line arrangement. Calculated by Formula (1), the economizer adopts cross-flow spiral finned tube. When Calculated by Formula (4), the structure of the condenser is also cross-flow spiral finned tube. Considering the heat from condensation of flue gas, the results would be below the normal range if calculated by Formula (4).

For condensing boiler with natural gas as the fuel, as the composition of natural gas is relatively stable with a low content of sulfur, the acid dew point of flue gas does not generally go higher than 130 °C. When the temperature of inlet water into the condenser goes over 40 °C, the inlet flue gas temperature below160 °C and outlet smoke temperature above 80 °C, we combined the coefficient formula of convective heat transfer for traditional cross - flush tube bundles with transverse strip or circular fin in line arrangement with the field test data of different furnaces to form a formula suitable for calculation within the following temperature range.

$$a_d = 0.015CzCs\frac{\frac{1.6}{(t_p/100)^{2.4}}\lambda}{S_{p6}}\left(\frac{d}{S_{p6}}\right)^{-0.54}\left(\frac{h_{p6}}{S_{p6}}\right)^{-0.14}\left(\frac{\omega S_{p6}}{\nu}\right)^{0.72} \quad Kw/(m^2\ °C) \qquad (5)$$

t_p Mean flue gas temperature °C

Table 1 Summary of Theoretical Calculation and Field Data

Furnace	Type	Inlet smoke temperature of condenser (°C)	Inlet water temperature of condenser (°C)	Outlet smoke temperature of condenser (°C)	Outlet water temperature of condenser (°C)	Heat transfer coefficient	Difference between Smoke Temperature (°C)
SZS58-1.6 /130/70-Q	Formula (4)	152	45	89	62	40.5	6
	Field Data	152	45	83	64	/	/
	Formula (5)	152	45	81	65	51.8	2
SZS70-1.6 /130/70-Q	Formula (4)	159	42	90	61	41.2	10
	Field Data	159	42	80	65	/	/

	Formula (5)	159	42	81	64	52.6	1
SZS46-1.6 /130/70-Q	Formula (4)	155	50	93	67	41.2	9
	Field Data	155	50	84	72	/	/
	Formula (5)	155	50	85	70	52.6	1
WNS14 -1.6/130/7 0-Q	Formula (4)	147	45	88	61	40	8
	Field Data	147	45	80	63	/	/
	Formula (5)	147	45	80	64	51	0

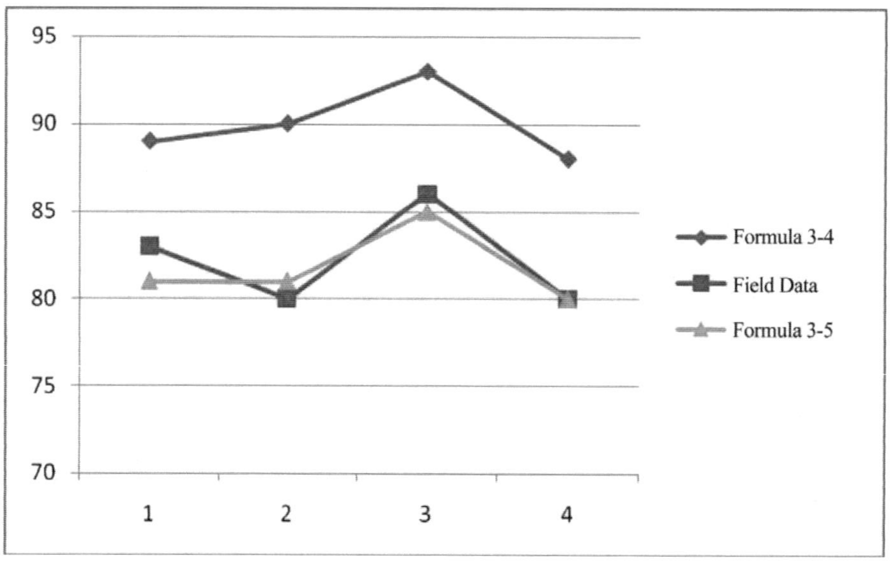

Figure 1: Schematic Diagram of Comparison between Two Formulas and Calculations

Table 1 gives some typical field data of gas condensing hot water boiler condenser. Figure 1 shows that when Formula (5) is used for calculating convective exothermic coefficient in the condenser section of condensing boiler, the result is quite close to the field data. When calculated by traditional Formula (4), there is a greater difference between the exhaust gas temperatures, an inaccurate reflection of the actual heat transfer performance of the boiler. Because the exhaust gas temperature is lower than 80°C, the amount of condensed water will increase greatly, as indicated by the results by Formula (5), which is greatly different from the actual value.

4. Conclusion

Based on differences betwee condensing boiler and traditional boiler in heat transfer in condensing section, Formula (6) is given based on convective heat transfer coefficient of spiral finned tube of traditional boiler and numerous field calculations.

$$a_d = 0.015CzCs\frac{\frac{1.6}{(t_p/100)^{2.4}}\lambda}{S_{p6}}\left(\frac{d}{S_{p6}}\right)^{-0.54}\left(\frac{h_{p6}}{S_{p6}}\right)^{-0.14}\left(\frac{\omega S_{p6}}{\nu}\right)^{0.72} \quad \text{Kw/(m}^2\ ^\circ\text{C)} \qquad (6)$$

When the inlet water temperature of the condenser is above 40°C, the inlet temperature of the flue gas is less than 160°C, and the outlet smoke temperature is more than 80°C, the calculated value of the formula is highly consistent with the field data (see Figure 1)

IOP Conf. Series: Materials Science and Engineering **721** (2020) 012051 doi:10.1088/1757-899X/721/1/012051

Acknowledgments
Thanks to the National Key Research and Development Program (2017YFF0209800)

References
[1] Che D (2002) Condensing boiler and its system Mechanical Industry Press Beijing
[2] Che D, Kang Y, Liu W, Fang S (2001) Analysis of ultimate thermal efficiency and exhaust heat loss of natural gas boiler Energy research and information 4 217-224
[3] Sun J, Li X, Jia L, (2000) Analysis of heat transfer process of flue gas condensation Journal of Beijing Institute of Construction Engineering 4 23-26
[4] Osakable M. Jshida K. Yagi k. Ohmasa K. (2001) Condensation Heating Transfer on Tubes in Actual Flue Gas[J]. Heat Transfer Asian Research., 30(2):139-151
[5] Jeff Ferenc (2007) Conventional and Condensing Boilers: Industry experts answer commonly asked boiler questions [C] HPAC Engineering 79(6)
[6] Richard N. Carbon Robert P. Wilson (2003). Field Measurements of the Performance of A Condensing Residential Boiler [C] Sympsium on Condensing Heating Exchanges Proceedings.1-2
[7] Dong L (2010) Discussion on calculation method of heat balance of condensing boiler industrial boiler 4 15-16
[8] 1976 Standard method for thermal calculation of boiler unit1973 Mechanical Industry Press Beijing
[9] Zang B, Chen X, (2006) Energy saving condensing heating boiler Energy research and utilization 3 2-4
[10] Cao Y, Ai X, Guo Q, Pan s, Fu Z, Wang S, (2000) Experimental study on convective heat transfer of flue gas with water vapor condensation 6 729-733
[11] Li X, Deng B, Tao W (2005) Three dimensional numerical simulation of finned tube bundle shell heat exchanger Journal of Engineering Thermophysics 2 316-318
[12] Sun J, Li X, Jia L (2001) Research and application of separated condensing natural gas boiler Energy saving technology 2 12

IOP Conf. Series: Materials Science and Engineering **721** (2020) 012052 doi:10.1088/1757-899X/721/1/012052

Numerical Study of Enhanced Heat Transfer of Microchannel Heat Sink with Nanofluids

Yan Hou, Jialin Yang* and Wenqi Zhang

20th Research Institute of Electronic Technology Corporation, Xi'an, China

*Corresponding author's e-mail: 17791290017@163.com

Abstract: The electronic devices on the circuit board of the navigation device have the defects of high power and difficulty in heat dissipation, thereby restricting their performance. In view of the above defects, this paper proposes a miniature water-cooled radiator with a trapezoidal cross-section spoiler arranged at equal intervals. The numerical simulation of the three-dimensional laminar flow and heat transfer is carried out to study the performance of the micro water-cooled heat sink. The effects of Reynolds numbers and the types of nanofluids on the heat transfer performance of microchannel heat sink are also investigated. The results show that when the Reynolds number ranges from 100 to 500, the existence of the spoiler column makes the heat dissipation effect of the microchannel heat sink significantly improved, and the friction coefficient decreases with the increment of the Reynolds number. With the increasing of the volume fraction in the nano-particles, the heat transfer performance of microchannel heat sink is also improved. The heat transfer performance of SiC nanofluids is better than that of TiO_2 nanofluids under the same flow rate.

1. Introduction

With the rapid development of microelectronics technology, highly integrated miniaturized and high-power devices have become the main development trend, and the integration of devices is becoming higher and higher, resulting in the local heat flux density of devices increasing dramatically [1-3]. Studies have shown that chip-level heat flux can be as high as $100W/cm^2$ and for component level can reach up to $25W/cm^2$. The heat dissipation of high-performance modules and power devices of high-power navigation equipment, such as the satellites and radars, seriously affects the reliability of equipment and restricts the development of high technology. Therefore, there is a very high requirement for thermal design, and the problems of high heat generation and dissipation are urgently needed to be solved.

The microchannel heat sink has the advantages of compact structure, such as the weight and high heat exchange area, and also has a better heat dissipation performance than macroscopic heat dissipation method [4-7]. Despite of the significant advances in the microchannel heat sink system technology; more efficient coolants are needed to further increase cooling efficiency in some cases. By dispersing the nano-sized particles with a diameter about 1-100 nm into the base liquid, the heat transfer performance of the new heat conducting fluid can be improved obviously, and the heat conductivity is higher than that of the conventional liquid, thus it is also called as nanofluids [8-10]. As it can be uniformly distributed in the base solution and has a high thermal conductivity, it has been drawn a lot of attention by the researchers. However, the thermal conductivity properties of nanofluid can be determined by varied factors, such as solid particle (shape and size), the kinds of base solution, annexing agent, temperature and pondus hydrogenii values [11-14]. Especially for the SiC nanofluids and TiO_2,

IOP Conf. Series: Materials Science and Engineering **721** (2020) 012052 doi:10.1088/1757-899X/721/1/012052

which have a high thermal conductivity capacity [15-18]. However, there are no literatures reporting the enhance thermal by combing microchannel with nanofluids to cool the electronic devices.

Based on the above research, in this paper, the microchannel and nanofluid strengthening technology are combined to dissipate heat from a typical circuit boards in the navigation electronic devices. A new type of micro channel heat dissipating device for trapezoidal cross section spoiler is proposed and its heat transfer characteristics are numerically simulated. The average nussel number and average fanning friction coefficient of microchannel heat sink under different Reynolds numbers are studied. The effects of SiC nanofluid and TiO_2 nanofluid on the microchannel heat sink are compared in details.

2. Mathematical-physical model

2.1Description of physical problems

The microchannel heat sink designed in this paper is based on a channel matrix made of copper, and water and nanofluids are cooling fluids. The specific physical model is shown in Fig. 1. As the spoiler column is arranged symmetrically in the channel, a unit in the microchannel heat sink is chosen for simulation study. Fig. 2 shows a detailed schematic diagram of calculation simulation region. The exact geometries of microchannel heat sink are shown in Fig. 3 and Table 1, respectively.

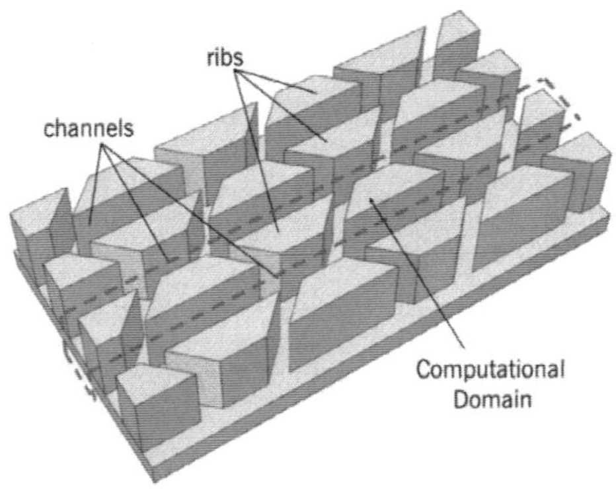

Fig.1 Schematic diagram of microchannel heat sink

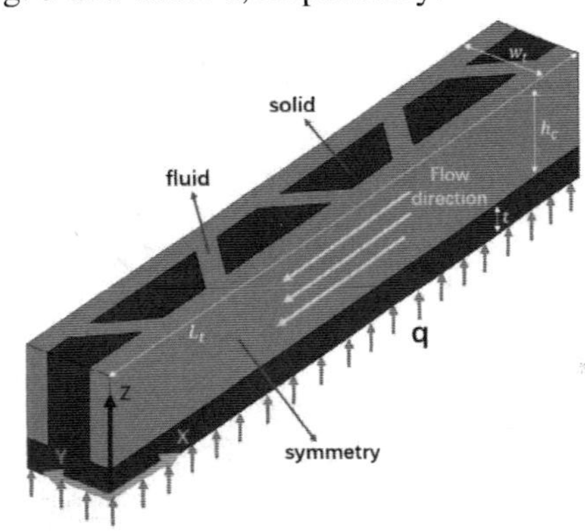

Fig. 2 Schematic diagram of simulation region

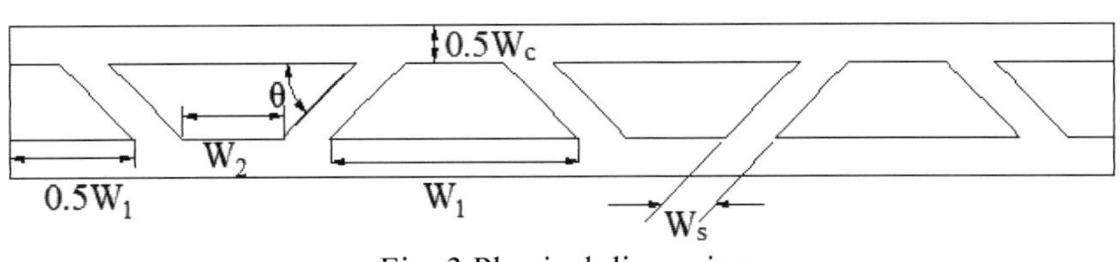

Fig. 3 Physical dimension

Table 1 Detail values of physical dimension

Parameters	W_s	h_c	t	L_t	W_c	W_1	θ	W_t
Values	0.1	0.3	0.1	2.25	0.15	0.5	45	0.3

2.2Mathematical modelling

Steady state and incompressible flow are adopted in this paper. Heat sink surface is smooth; the effects of radiation heat transfer and surface tension are ignored and the deposition of nanoparticles in the nanofluid solution is not considered. The thermophysical properties of the liquid have a great influence on the flow characteristics of the microchannel heat sink. All physical parameters are fixed values without internal heat source. Under the above assumptions, the continuity equation, momentum conservation equation and energy conservation equation of the fluid can be expressed as follows:

IOP Conf. Series: Materials Science and Engineering **721** (2020) 012052 doi:10.1088/1757-899X/721/1/012052

The continuity equation can be expressed as:

$$\nabla \cdot (\vec{v}) = 0 \tag{1}$$

The momentum conservation equation can be written as:

$$\nabla \cdot (\rho \vec{v} \vec{v}) = \nabla P + \nabla \cdot (\mu \nabla \vec{v}) \tag{2}$$

The energy conservation equation can be written as:

$$\begin{aligned} &\nabla \cdot (\rho \vec{v} c_p T) = \nabla \cdot (k \nabla T) \quad (\text{liquid}) \\ &\nabla \cdot (k \nabla T) = 0 \quad (\text{solid}) \end{aligned} \tag{3}$$

1) Boundary condition

Inlet:

$$\begin{aligned} &x = -2.25, \ u = u_f, \ v = w = 0, \\ &For\ fluid: T_f = T_{in} = 300\text{K}, \\ &For\ solid: -k_s \left(\frac{\partial T_s}{\partial x} \right) = 0 \end{aligned} \tag{4}$$

Outlet:

$$\begin{aligned} &x = 0 \quad P_f = P_{out} = 1\text{atm}, \\ &For\ fluid: -k_f \left(\frac{\partial T_f}{\partial x} \right) = 0, \\ &For\ solid: -k_s \left(\frac{\partial T_s}{\partial x} \right) = 0 \end{aligned} \tag{5}$$

Top:

$$\begin{aligned} &z = 0.4mm \qquad u = v = w = 0 \\ &-k_s \left(\frac{\partial T_s}{\partial y} \right) = 0 \end{aligned} \tag{6}$$

Convection heat transfer contact surface can be expressed as:

$$\begin{aligned} &fluid\,/\,solid \ \ surface: \ u = v = w = 0, \\ &-k_s \left(\frac{\partial T_s}{\partial n} \right) = -k_f \left(\frac{\partial T_f}{\partial n} \right) \end{aligned} \tag{7}$$

Periodic boundary:

$$\begin{aligned} &At \ \ y = 0mm \ \ symmetry \\ &At \ \ y = 0.3mm \ \ symmetry \end{aligned} \tag{8}$$

Bottom:

$$\begin{aligned} &z = 0mm \\ &-k_s \left(\frac{\partial T_s}{\partial y} \right) = q = 100W\,/\,\text{cm}^2 \end{aligned} \tag{9}$$

2) Parameter definition

The characteristic length of the channel is defined as:

$$D_h = \frac{2W_c h_c}{W_c + h_c} \quad (10)$$

The expression of Reynolds number is:

$$\mathrm{Re} = \frac{\rho u_m D_h}{\mu} \quad (11)$$

where ρ is fluid density, u_m is average fluid velocity, μ is hydrodynamic viscosity coefficient, W_c is channel width and h_c is channel height.

The density of nanofluids is defined as:

$$\rho_{nf} = (1-\varphi)\rho_b + \varphi\rho_p \quad (12)$$

where φ is the volume fraction of nanoparticles in solution, ρ_b is the density of base solution, ρ_p is the density of nanoparticles.

The dynamic viscosity of nanofluids is defined as:

$$\mu_{nf} = \frac{1}{(1-\varphi)^{2.5}}\mu_f \quad (13)$$

where μ_f is the dynamic viscosity of the base liquid.

The specific heat capacity of the nanofluid is defined as:

$$C_{pnf} = \frac{(1-\varphi)\rho_b C_{pb} + \varphi\rho_p C_{pp}}{(1-\varphi)\rho_b + \varphi\rho_p} \quad (14)$$

where C_{pb} is the specific heat capacity of the base solution, and C_{pp} is the specific heat capacity of the nano particles.

The thermal conductivity of nanofluids is defined as:

$$K_{nf} = K_b\left[\frac{K_p + (n-1)K_b - \varphi(n-1)(K_b - K_p)}{K_p + (n-1)K_b + \varphi(K_b - K_p)}\right] \quad (15)$$

where n can be expressed as:

$$n = \frac{3}{\psi} \quad (16)$$

where K_b is the thermal conductivity of the base solution, K_p is the thermal conductivity of the nanoparticles, n is the shape factor of the nanoparticles, and ψ represents sphericity. For spherical nanoparticles, the usual value ψ is 1. The physical properties of other parameters are shown in Table 2 and Table 3.

Table 2 Properties of SiC and TiO_2

Thermophysical properties	H_2O	SiC	TiO_2
Density, ρ [kg m^{-3}]	*994.2*	3160	4157
Specific heat cacacity, C*p* [J kg^{-1} K^{-1}]	4178	675	710
Thermal conductivity, k [W m^{-1} K^{-1}]	0.625	490	8.4
Viscosity, μ [N *s* m^{-3}]	7.25×10^{-4}	……	……

Table 3 physical parameters under the different percent of SiC and TiO_2 nanofluids

Properties	SiC-H_2O			TiO_2-H_2O		
	1%	5%	9%	1%	5%	9%
ρ_{nf} [kg m^{-3}]	1020	1100	1190	1030	1150	1280
$\mu_{nf} \times 10^{-4}$ [N s m^{-3}]	7.43	8.23	9.17	7.43	8.23	9.17
$C_{pnf} \times 10^{-4}$ [J kg^{-1} K^{-1}]	4070	3670	3340	4040	3550	3161
k_{nf} [W m^{-1} K^{-1}]	0.644	0.723	0.809	0.67	0.703	0.771

The average fanning friction coefficient is defined as:

$$\bar{f} = \frac{\Delta p D_h}{2\rho L_t u_m^2} \tag{17}$$

where L_t is the total length of the microchannel, Δp is the pressure drop along the length. $\Delta p = \bar{p}_{in} - p_{out}$, where $\bar{p}_{in}$ is weighted average pressure of import quality.

The average heat transfer coefficient is defined as:

$$\bar{h} = \frac{qA_w}{A_{con}\left(\bar{T}_w - \bar{T}_f\right)} \tag{18}$$

The average nussel number is defined as:

$$\bar{N}u = \frac{\bar{h}D_h}{k_f} \tag{19}$$

where q is the heating source at the bottom surface, A_w is the area of the heating source at the bottom, A_{con} is the coupling area between solid and liquid, $\bar{T}_w$ is the weighted average temperature at the bottom surface, $\bar{T}_f$ is the weighted average temperature of the fluid mass, and k_f is the thermal conductivity of the fluid.

3. Independence check of grid

The continuity equation adopts the Standard format, and the SIMPLE algorithm is used to deal with the coupling problem of pressure and velocity. The momentum equation adopts the upwind format of second-order accuracy during the simulation process.

In order to ensure the quality of flow field, the domain decomposition method is often used for practical devices with complex geometric shapes. Due to the complex flow field in the trapezoidal spoiler region, the microchannel model is divided into two regions, the inlet and outlet region and the spoiler region. The two regions are divided into grids, which can ensure the accuracy of calculation and save time. In this paper, ICEM software is used to grid the 3D model. For the near wall surface and the spoiler column area, due to the large velocity and pressure gradient, a dense grid is adopted. The meshing results are shown in Figs. 4 (a) and (b).

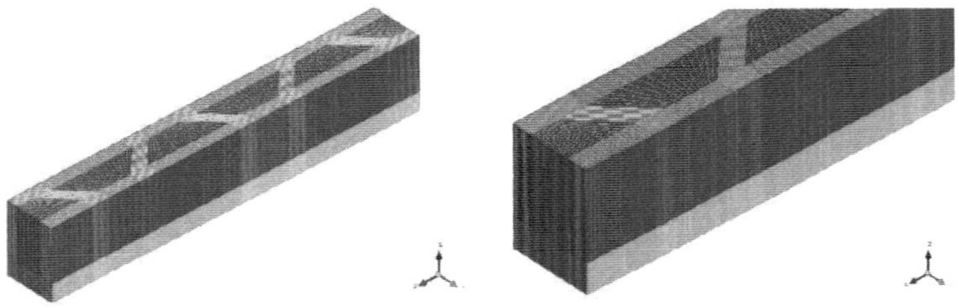

(a) Three-dimensional grid diagram (b) Three-dimensional mesh boundary encryption diagram

Fig.4 Three-dimensional grid diagram model

When the Reynolds number is 200, the grid numbers are 12124, 89736, 271648 and 486472, respectively, and the average nussel number is compared, as shown in Fig. 5. The model with 271648 grid number is selected for the factor analysis and calculation in consideration of the accuracy of the result and calculation resource.

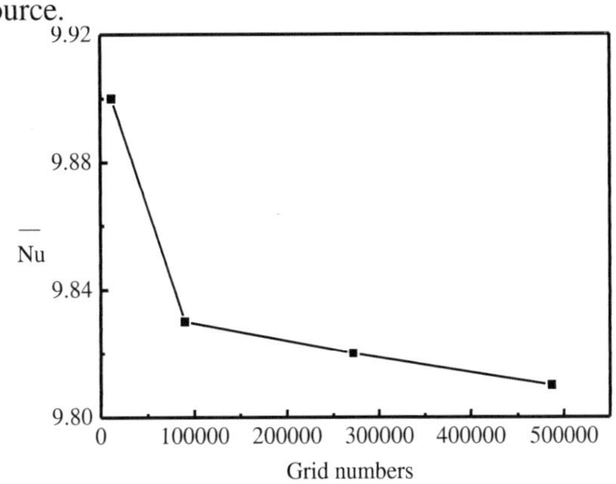

Fig. 5 Grid independence validation

4. Results and discussion

4.1The effect of Reynolds number

With the increase of Reynolds number, the convective heat transfer capacity of the micro-channel increases, thus the average nussel number increases and the average surface temperature decreases as shown in Fig. 6 and Fig. 7 (a). This is because the secondary flow is generated in the micro-channel radiator by introducing an inclined channel, which leads to the destruction of the velocity boundary layer, thus reducing the thickness of the average thermal boundary layer and achieving the purpose of strengthening heat dissipation. But with the increase of Reynolds number, the increase of average nussel number slows down, and the decrease of surface temperature also slows down. With the increase of Reynolds number, fanning friction coefficient of the channel shows a downward trend as shown in Figs. 6 and 7(b). Therefore, with the increase of Reynolds number, heat transfer is further enhanced and system work loss is reduced.

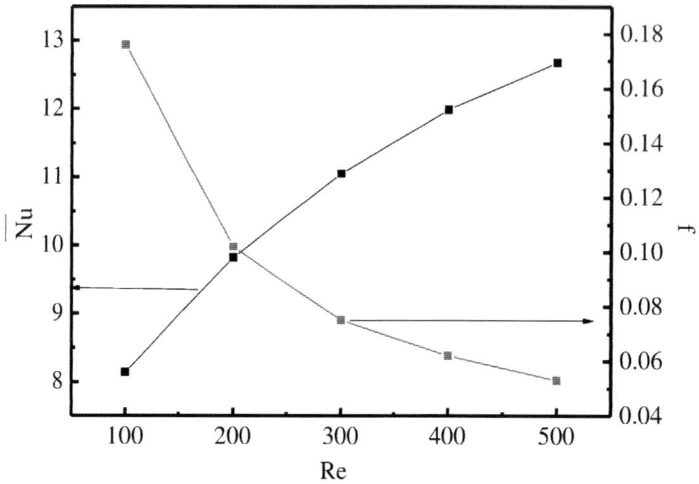

Fig. 6 Diagram of the change of the average nussel number and fanning friction coefficient under different Reynolds numbers

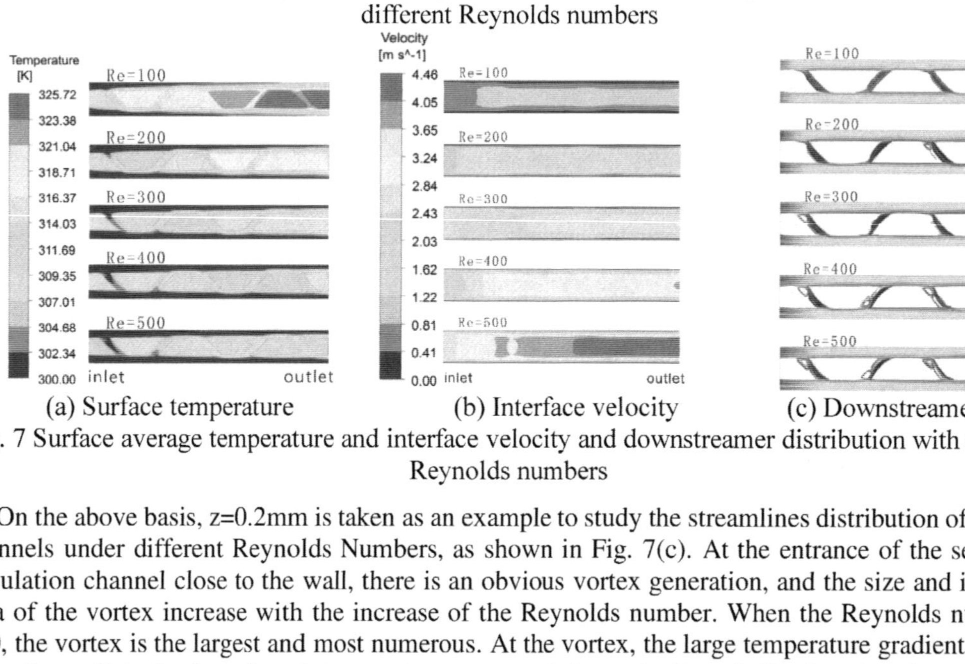

(a) Surface temperature (b) Interface velocity (c) Downstreamer

Fig. 7 Surface average temperature and interface velocity and downstreamer distribution with different Reynolds numbers

On the above basis, z=0.2mm is taken as an example to study the streamlines distribution of inclined channels under different Reynolds Numbers, as shown in Fig. 7(c). At the entrance of the secondary circulation channel close to the wall, there is an obvious vortex generation, and the size and influence area of the vortex increase with the increase of the Reynolds number. When the Reynolds number is 500, the vortex is the largest and most numerous. At the vortex, the large temperature gradient changes from the wall to the boundary between the vortex and the main flow, indicating that the vortex can enhance heat transfer.

4.2The effects of nanofluids on heat transfer

On this basis, nanofluids are introduced to further enhance heat transfer through the microchannel heat sink. As shown in Fig. 8 and Fig. 9, both nanofluids have an enhanced effect on heat transfer of the microchannel heat sink, and it can be seen that with the increase in the concentration of nanofluids, the maximum temperature on the surface of the radiator gradually decreases, indicating that the heat transfer ability is gradually enhanced. And as the flow rate of the liquid increases, the temperature drop gradually slows down. Compared Fig. 8 with Fig. 9, when two kinds of nanofluids with the same concentration are selected for heat transfer comparison, the heat transfer capacity of the two kinds of nanofluids increases with the increase of refrigerant flow rate, which is mainly due to the enhanced convective heat transfer capacity when the flow rate increases. At the same flow rate, the maximum temperature drops to 2.5 K. At the same time, it can be found that the heat transfer effect of SiC nanofluid is better than that of TiO_2 nanofluid, which is mainly due to the fact that the thermal conductivity of SiC nanofluid is better than that of TiO_2 nanofluid when the content of nanofluid is 9%.

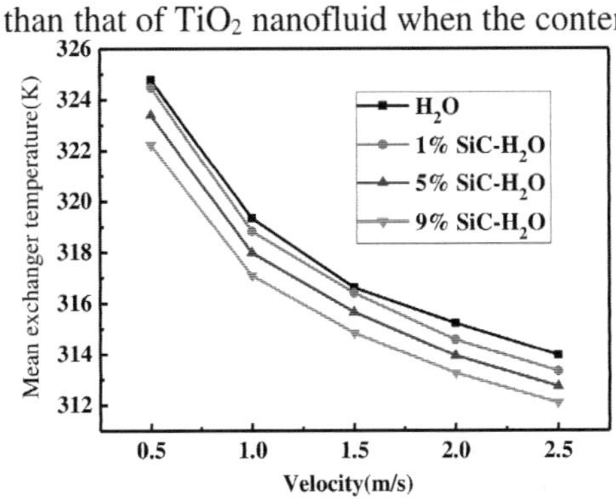

Fig. 8 Heat transfer effects of nanofluids containing different proportions of SiC

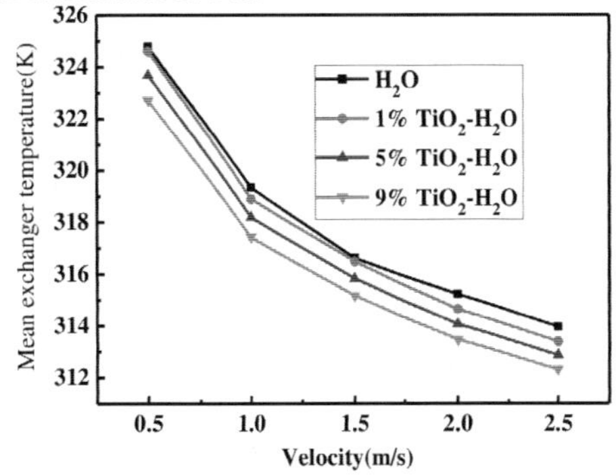

Fig. 9 Heat transfer effects of nanofluids containing different proportions of TiO_2

5. Conclusions

In this paper, the numerical study of three-dimensional laminar flow and heat transfer of a miniature water-cooling radiator with trapezoidal section and evenly spaced spoiler columns designed for the circuit board of navigation equipment is carried out. The results show that:

(1) In the range of 100-500 Reynolds number, the larger the Reynolds number is, the smaller the friction resistance is, and the decreasing trend gradually becomes gentle. For the micro-channel radiator studied in this paper, the Nu value increases with the increase of Reynolds number.

(2) As for the microchannel radiator studied in this paper, secondary flows is generated due to the trapezoidal section structure of the spoiler column, which destroys the velocity boundary layer, reduces the thickness of the thermal boundary layer, and improves the heat transfer capacity of the microchannel.

(3) Both SiC nanofluids and TiO_2 nanofluids have enhanced heat transfer of the radiator, and with the increase of flow rate, the enhanced heat transfer effect gradually slows down, and the former has better heat transfer effect than the latter.

References

[1] Honda H, Wei J J. (2004) Enhanced boiling heat transfer from electronic components by use of surface microstructures. Experimental Thermal and Fluid Science, 28:159-169.

[2] Sparrow E M, Vemuri S B, Kadle D S. (1983) Enhanced and local heat transfer, pressure drop, and flow visualization for arrays of block-like electronic components. International Journal of Heat and Mass Transfer, 26(5):689-699.

[3] Mandal S K, Sen D. (2016) A brief review on mixed convection heat transfer in channel flow with vortex generator for electronic chip cooling. International Journal of Engineering Research & Applications.

[4] Chai L, Xia GD, Wang HS. (2016) Numerical study of laminar flow and heat transfer in microchannel heat sink with offset ribs on sidewalls. Applied Thermal Engineering 92 32-41.

[5] Xu JL, Gan YH, Zhang DC, Li XH. (2005) Microscale heat transfer enhancement using thermal boundary layer redeveloping concept. International Journal of Heat and Mass Transfer 48: 1662-1674.

[6] Hayder A. (2014) Numerical study of jet impingement cooling on a smooth curve surface. Malaysia: University Tun Hussein Onn Malaysia.

[7] Kuppusamy NR, Saidur R, Ghazali N, Mohammed H. (2014) Numerical study of thermal enhancement in micro channel heat sink with secondary flow. International Journal of Heat and Mass Transfer 78: 216-223.

[8] Lee S P, Choi S, Li S, Eastman JA. (1999) Measuring thermal conductivity of fluids containing oxide nanoparticles. Journal of Heat Transfer, 121(2):280-289.
[9] Buongiorno J. (2006) Convective Transport in Nanofluids. Journal of Heat Transfer, 128(3):240-250.
[10] Keblinski P, Phillpot S R, Choi S U S, AEastman J. (2002) Mechanisms of heat flow in suspensions of nano-sized particles (nanofluids). International Journal of Heat and Mass Transfer, 45(4):855-863.
[11] Gupta M, Singh V, Kumar R, Kumar R, Said Z. (2017) A review on thermophysical properties of nanofluids and heat transfer applications. Renewable & Sustainable Energy Reviews, 74(2):638-670.
[12] Yu W, France D M. (2008) Review and comparison of nanofluid thermal conductivity and heat transfer enhancements. Heat Transfer Engineering, 29(5):432-460.
[13] Devendiran D K, Amirtham V A. (2016) A review on preparation, characterization, properties and applications of nanofluids. Renewable & Sustainable Energy Reviews, 60:21-40.
[14] Vanaki S M, Ganesan P, Mohammed H A. (2016) Numerical study of convective heat transfer of nanofluids: A review. Renewable & Sustainable Energy Reviews, 54:1212-1239.
[15] Lee S W, Park S D, Kang S, Bang IC, Kim JH. (2011) Investigation of viscosity and thermal conductivity of SiC nanofluids for heat transfer applications. International Journal of Heat and Mass Transfer, 54:433-438.
[16] Timofeeva E V, Smith D S, Yu W, France DM, Singh D, Routbort JL. (2010) Particle size and interfacial effects on thermo-physical and heat transfer characteristics of water-based alpha-SiC nanofluids. Nanotechnology, 21(21):215703.
[17] Batmunkh M, Tanshen M R, Nine M J, Myekhlai M, Choi H, Chung H, Jeong H. (2014) Thermal Conductivity of TiO2 Nanoparticles Based Aqueous Nanofluids with an Addition of a Modified Silver Particle. Industrial & Engineering Chemistry Research, 53(20):8445-8451.
[18] Sen S, Govindarajan V, Pelliccione C J, Wang J; Timofeeva EV, Miller D J. (2015) Surface modification approach to TiO_2 nanofluids with high particle concentration, low viscosity, and electrochemical activity. ACS Applied Materials & Interfaces, 7(37):20538.

IOP Conf. Series: Materials Science and Engineering **721** (2020) 012053 doi:10.1088/1757-899X/721/1/012053

Energy conservation certification for coal layer-combustion industrial boiler

Xuemin Liu[1*], Jiming Yu[1], Songsong Zhang[1] and Lihua Zhao[2]

[1]China Special Equipment Inspection and Research Institute, Beijing, 100029, China

[2]China Electronics Standardization Institute, Beijing, 100007, China

[*]Corresponding author's e-mail: liuxuemin@csei.org.cn

Abstract. The effect of boiler load on main operation parameters and the boiler efficiency were investigated, and the thermal efficiency distribution of nearly 300 coal-fired boilers was further analyzed. The results show that the exit gas temperature decreases as the boiler load decreases, while the unburned carbon contents in ash and slag, and CO content in flue gas usually increase. Correspondingly, the loss due to exit gas decreases, and the loss due to unburned solids usually increases. The loss due to surface radiation and convection increases. Taking into account changes of all the losses, the boiler efficiency decreases at lower loads. With reference to TSG G0002, Technical Specification for Energy Conservation Certification of Coal Layer-Combustion Industrial Boiler under off-Design Conditions has been drafted and the evaluating values of energy conservation for products have been proposed, including boiler efficiency under different test loads and CO concentration in flue gas. The target efficiency values vary according to the boiler capacity and fuel type. The CO concentration value is a limit that can be achieved by most boilers.

1. Introduction

There are more than 400,000 industrial boilers in China, most of which are coal-fired boilers. Industrial boilers are widely used for steam production in various industries and building heating in northern regions. According to different combustion modes, coal-fired industrial boilers can be further divided into layer combustion boilers, fluidized bed boilers and pulverized coal boilers, among which the layer-combustion boiler is the most widely used type.

In recent years, with the improvement of design and manufacturing level of industrial boilers, and the elimination of small-capacity coal-fired boilers year by year, the overall energy efficiency level of coal-fired industrial boilers has been increased. In 2016, the Supervision Administration Regulation on Energy Conservation Technology for Boiler (TSG G0002-2010) stipulates that the thermal efficiency target values of layer-combustion boilers with different capacities and different coal types under the rated condition are all higher than 85%, and the limit values higher than 80%[1]. However, the actual efficiency during daily operation is much lower[2]. And it is a very common phenomenon that industrial boilers operate under non-rated conditions. Therefore, improving the energy efficiency level of coal-fired industrial boilers under variable working conditions is of great significance for promoting boiler energy conservation and emission reduction. **B X Zhang** proposed an optimization design of the control system for low load operation of coal-fired chain furnace, and determined the boiler parameters suitable for low load operation[3]. The economic operation under low load can be achieved by coal layer thickness and grate speed optimization[4]. M Tontu conducted performance analysis of a coal-fired steam power plant with varying loads[5].

The Chinese regulations and standards have specified the energy efficiency indexes of industrial boilers under rated conditions and industrial boiler systems[6]. The energy efficiency index in Japan specifies the excess air coefficient for a certain load range. The EU code considers the weighting efficiency of different loads and the US Energy Star examines the annual fuel utilization efficiency. In general, the thermal efficiency indexes for industrial boilers under variable working conditions have not yet been seen in the literature.

The effect of boiler load on exit gas temperature, CO content in flue gas and unburned carbon loss was investigated by field tests in this work. The thermal efficiency distribution of nearly 300 coal-fired boilers was analyzed. Based on the above analysis, with reference to TSG G0002, the energy conservation certification technical specification for coal layer-combustion industrial boiler under off-design conditions was developed.

2. Principles and methods of boiler efficiency test

2.1Efficiency calculation

The basis of thermal efficiency calculation is the first law of thermodynamics, namely the law of conservation of energy. It can be stated as

$$QrF\left(\text{Input from fuel}\right)+QrB\left(\text{Credits}\right)=QrO\left(\text{Output}\right)+QrL\left(\text{Losses}\right) \tag{1}$$

According to GB/T 10180, the boiler efficiency is calculated as equation (2), which is a typical formula of gross efficiency.

$$\eta=100\times\frac{QrO}{QrF+QrB}\left(\%\right) \tag{2}$$

Combining equation (1) and equation (2), equation (3) can be obtained,

$$\eta=\left[100\text{-}100\times\frac{QrL}{QrF+QrB}\right]\left(\%\right) \tag{3}$$

which can be further expressed as:

$$\eta=\left[100\text{-}\left(q_2+q_3+q_4+q_5+q_6+q_7\right)\right]\left(\%\right) \tag{4}$$

where

q_2 -- loss due to flue gas, %;

q_3 -- loss due to unburned gases, %;

q_4 -- loss due to unburned solids, %;

q_5 -- loss due to surface radiation and convection, %;

q_6 -- loss due to sensible heat of residue, %;

q_7 -- loss due to sulfur capture, %。

2.2Testing methods

During the test, coal and slag were sampled, exhaust gas composition was analyzed and main operating parameters were recorded. Coal sampling timing should be advanced properly according to the time for fuel from the sampling point to the furnace to ensure that the sample is representative. Exhaust gas was sampled at the gas duct outlets of the air heaters. The content of each flue gas component was measured by using a multi-component flue gas analyzer.

Main operating parameters were recorded once every 5 or 15 minutes. The averages of the following parameters during the entire test period were taken: outlet water temperature and pressure, feed water flow, temperature and pressure, air temperature and exhaust gas temperature, etc.

3. Field test results

In order to study the impact of load changes on boiler operation, hot water boilers with capacities of 58MW and 29MW were tested in Changchun, Mudanjiang and some other cities. Exit gas temperature is one of the most important factors affecting boiler efficiency, which determines the sensible heat loss due to flue gas together with the flue gas flow. The change in flue gas temperature with boiler load is shown in figure 1. It can be seen that the exit gas temperature decreases as the boiler load decreases, mainly caused by decrease in furnace temperature.

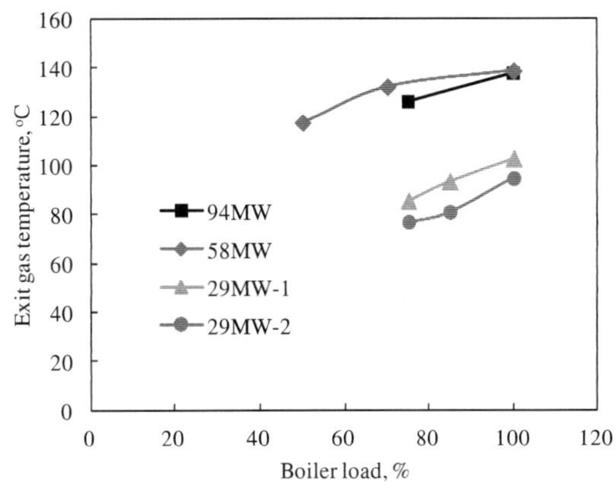

Figure 1. Change in flue gas temperature with boiler load.

The unburned carbon content reflects the combustion status of the boiler. The change in unburned carbon content with boiler load is shown in figure 2. The furnace temperature is lower under lower boiler load. The combustion intensity decreases, usually leading to higher carbon content in ash or slag. But there are some other factors affecting combustion. For instance, in many cases, the excess air ratio will be increased properly at lower load to ensure coal particles burning-out. Therefore, the change trend is usually complicated. Another indicator that reflects the combustion status is CO content in flue gas. Generally, CO content increases at lower boiler load. But increasing the excess air ratio has a diluting effect, leading to decrease in CO content.

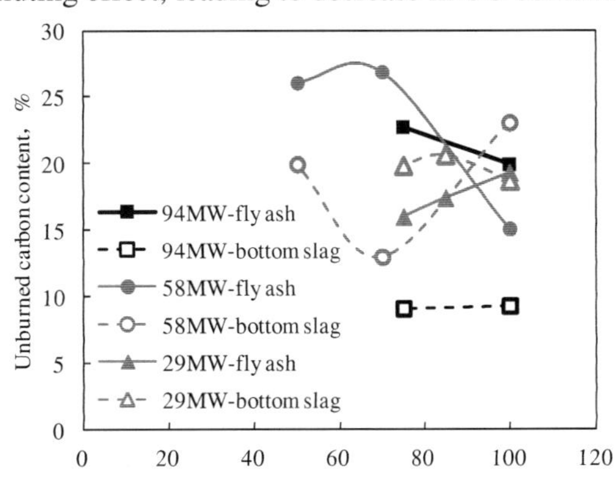

Figure 2. Change in unburned carbon.

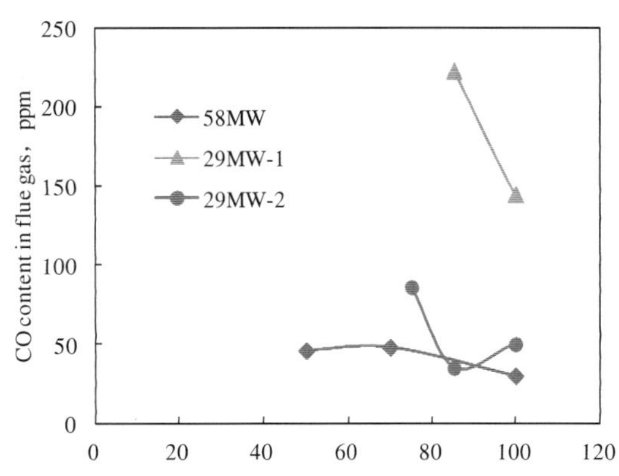

Figure 3. Change in CO content.

The efficiency test and calculation results of a 58 MW boiler in Changchun are shown in table 1 and figure 4. It can be clearly seen that the boiler efficiency decreases as the load decreases. The loss due to exit gas can be calculated using equation (5). It means that the loss due to exit gas will decrease if the excess air ratio remains constant. K_{q4} is the correction factor, which can be calculated by equation (6). As the boiler load decreases, the loss due to unburned solids usually increases, as shown in figure 4, leading to decrease in K_{q4}.

$$q_2 = \frac{K_{q4}}{Q_{in}}\left(h_{ds} - h_{ca}\right)\times 100 \tag{5}$$

where Q_{in} is the energy entering the system, h_{ds} is the exit gas enthalpy, and h_{ca} is the cold air enthalpy.

$$K_{q4} = \frac{100 - q_4}{100} \tag{6}$$

The loss due to surface radiation and convection increases as the boiler load decreases. The temperature change on the outer surface under different loads is relatively small, which means that the heat exchange due to surface radiation and convection almost remains constant. The input energy decreases at lower load, so the ratio of heat exchange due to surface radiation and convection to input energy increases. Taking into account changes of all the losses, the boiler efficiency decreases at lower loads.

Table 1. Test results of a 58 MW boiler in Changchun

Item	**Unit**	**Case 100%**	**Case 70%**	**Case 50%**
Boiler output	MW	56.57	41.04	29.39
Boiler efficiency	%	84.72	83.52	80.01
Exit gas temperature	°C	138.70	132.18	117.50
Excess air ratio	/	1.43	1.81	1.52
Loss due to flue gas	%	7.99	7.47	5.54
Loss due to unburned gases	%	0.013	0.027	0.021
Loss due to unburned solids	%	6.07	7.35	12.16
Loss due to surface radiation and convection	%	0.8	1.14	1.6
Loss due to sensible heat of residue	%	0.41	0.49	0.67

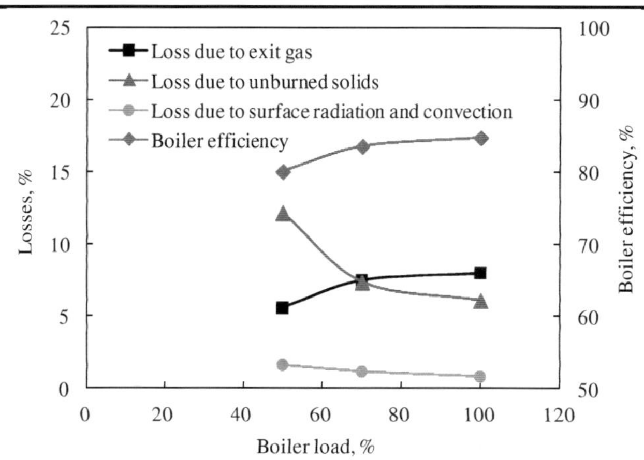

Figure 4. Boiler efficiency of a 58 MW boiler in Changchun.

4. Platform data analysis

A large number of boiler efficiency test data have been accumulated in "Industrial boiler efficiency test data calculation and management platform". The boiler efficiency distribution, exit gas temperature, excess air ratio and CO concentration of different coal type boilers under different test loads were analyzed. The results show that with the decrease in test load, the boiler efficiency shows a downward trend, the exit gas temperature decreases and the excess air ratio increases. The thermal efficiency distributions of coal-fired industrial boilers with different capacities are shown in figure 5 and figure 6. The statistical results of CO concentration are shown in figure 7. It can be seen that the distribution is dispersed, but most of the data are below 200ppm.

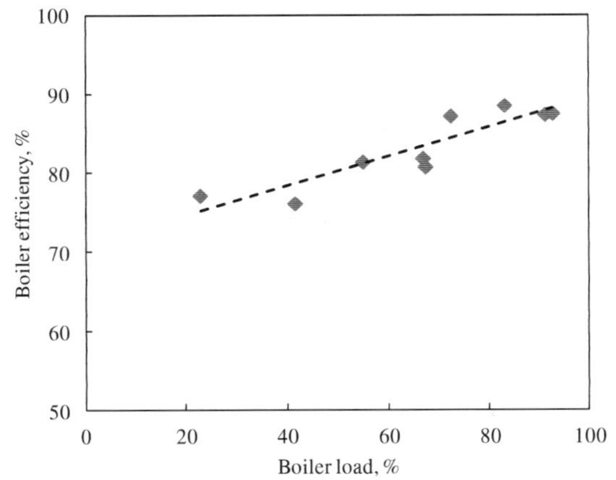

Figure 5. Efficiency distribution (Rated capacity>20t/h)

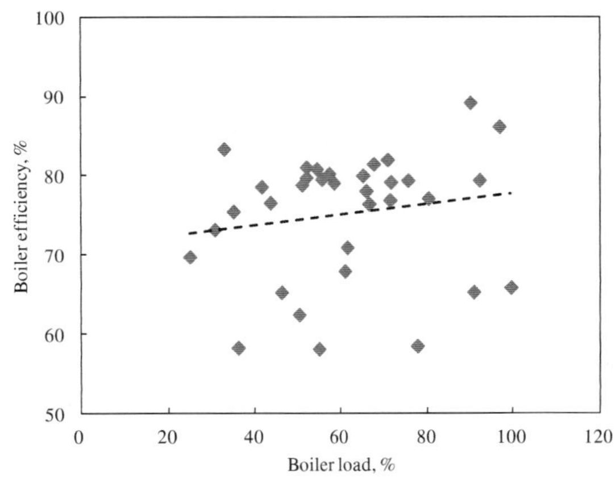

Figure 6. Efficiency distribution (≤20t/h)

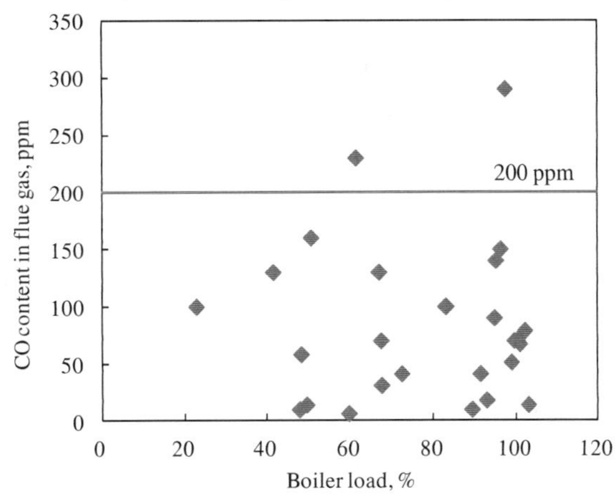

Figure 7. CO concentration in flue gas

5. Evaluating values of energy conservation

Based on the field test results and platform data statistics, Technical Specification for Energy Conservation Certification of Coal Layer-Combustion Industrial Boiler under off-Design Conditions has been drafted and the evaluating values of energy conservation for products have been proposed, as shown in table 2. The evaluating values of energy conservation include boiler efficiency and CO concentration. The boiler efficiency is the most intuitive indicator of energy conservation level. The target values vary according to the boiler capacity and fuel type. Three typical loads are chosen as the test loads, namely, 100%, 70% and 50%. The target values under the 100% load are consistent with TSG G0002. The CO concentration can well reflect the combustion situation of the boiler. On the one hand, reducing the CO concentration will help the boiler to improve the thermal efficiency. On the other hand, the CO index will help to avoid the phenomenon that the fuel efficiency is sacrificed to achieve pollutant emission reduction. The CO concentration limit value is a necessary condition to achieve energy conservation, which is the limit that can be achieved by most boilers.

Table 2. Evaluating values of energy conservation

Item	Fuel type		Lower heating value $Q_{net.v.ar}$ (KJ/kg)	Rated capacity D≤20 t/h or Q≤14 MW			Rated capacity D>20 t/h or Q>14 MW		
				Load			Load		
				100%	70%	50%	100%	70%	50%
Boiler efficiency (%)	Bituminous coal	II	$17700 \le Q_{net.v.ar} \le 21000$	85	83	80	86	84	81
		III	$Q_{net.v.ar} > 21000$	87	84	82	89	86	84
	Lignite		$Q_{net.v.ar} \ge 11500$	85	83	80	87	85	82
CO concentration in flue gas (ppm)				200					

6. Conclusions

The effect of boiler load on main operation parameters and the boiler efficiency were investigated. The exit gas temperature decreases as the boiler load decreases, while the unburned carbon contents in ash and slag, and CO content in flue gas usually increase. Correspondingly, the loss due to exit gas decreases, and the loss due to unburned solids usually increases. The loss due to surface radiation and convection increases at lower load. Taking into account changes of all the losses, the boiler efficiency decreases at lower loads. The thermal efficiency distribution of nearly 300 coal-fired boilers was further analyzed. With reference to TSG G0002, the evaluating values of energy conservation for products were proposed, including boiler efficiency under different test loads and CO concentration in flue gas. The target efficiency values vary according to the boiler capacity and fuel type. The CO concentration limit value is a necessary condition to achieve energy conservation, which is the limit that can be achieved by most boilers.

Acknowledgments

This study was supported by the National Key R&D Program of China (No. 2017YFF0211603).

References

[1] General Administration of Quality Supervision, Inspection and Quarantine of P.R.C. (2010) Supervision administration regulation on energy conservation technology for boiler, TSG G0002-2010.
[2] Qiu, Z.Y., Xiong, W.D., Zhao, H. (2017) Analysis on overall situation and energy efficiency status of industrial boilers used in Hangzhou. Industrial Boiler, 4: 42-46.
[3] Zhang, B.X. (2014) Energy optimization industrial coal-fired chain boiler under low load operation. Tianjin University.
[4] Wang, Y.L., Xiao, Y.T., Cui, Q., et al. (2015) Combustion economy analysis of low load operation optimization in chain grate stoker boiler. Industrial Boiler, 4: 42-45.
[5] Tontu, M., Bilgili, M., Sahin, B. (2018) Performance analysis of an industrial steam power plant with varying loads. International Journal of Exergy, 27(2): 231-250.
[6] National Energy Administration. (2017) Energy efficiency index and grading of industrial boiler system, NB/T 47061-2017.

Study on combustion and nitrogen oxide emissions of gas boiler

Yanyan Ji[1, 2], Songsong Zhang[2*], Kejian Wang[1*], Guoli Qi[2]

[1] School of mechanical engineering, Beijing University of Chemical Technology, Beijing, 100029, China

[2]China Special Equipment Inspection and Research Institute, Beijing, 100029, China

[*]Corresponding author's e-mail: zhangsong0536@163.com,wangkj@mail.buct.edu.cn

Abstract. In order to reduce the emission of nitrogen oxides in gas-fired boilers, the effects of excess air ratio and flue gas recirculation on nitrogen oxide emissions during natural gas combustion are studied by numerical calculation. By adding flue gas recirculation, the generation of nitrogen oxides is significantly reduced. When the flue gas recycling rate is 20%, the emission of nitrogen oxides is minimized to about 40 mg/m^3.

1. Introduction

The impact of NO_x on the environment is receiving more and more attention. The emission standards for boiler air pollutants issued by Beijing in 2015 [1]: The NO_x emission limit of new boilers before March 31, 2017 is 80mg/m^3 (standard status), and the new boiler NO_x emission limit after April 1, 2017 is 30mg/m^3. From this point of view, it is necessary to effectively control the nitrogen oxides produced during the combustion process to achieve low nitrogen combustion. Natural gas low-nitrogen combustion technology mainly focuses on how to reduce combustion temperature and reduce thermal NO_x formation. Its main technologies include staged combustion (air classification, fuel classification), lean-burn premixed combustion, flue gas recirculation, flameless combustion, etc. [2].

As a commonly used method for reducing nitrogen oxides, flue gas recirculation is widely used in industry and is divided into internal flue gas recirculation and external flue gas recirculation. Internal flue gas recirculation refers to the structural design of the burner and the furnace (including the entrainment of high-speed jets, swirling or bluffing of the bluff body), so that the flue gas is recirculated and returned to the combustion zone in the furnace to participate. The flue gas external circulation is to spray a part of the lower temperature of the boiler tail into the furnace to reduce the oxygen content in the furnace and make the natural gas stay in an anoxic combustion state, thereby reducing the combustion temperature and reducing the formation of nitrogen oxides.

The external flue gas recycling technology extracts a part of the flue gas at the tail of the flue, and uses the circulating fan to send the flue gas to the furnace inlet for re-combustion without the need to modify the burner, which is relatively economical [3]. Xie Zhengwu [4] discussed the impact of flue gas recirculation on boiler thermal calculation. Zhang Liqin [5] and others have shown that under oxygen-rich conditions, flue gas recirculation makes NO_x content in flue gas increase after pulverized coal combustion, and NO_x conversion rate decreases significantly. Zhao Zengwu [6] and Hu Manyin [7] studied the effect of flue gas recirculation on NO_x generation by numerical simulation. It was found that flue gas recirculation effectively reduced the oxygen content in the combustion-supporting gas and reduced the overall furnace and flame. The temperature in the local high temperature zone effectively suppresses the generation of thermal NO_x and achieves low nitrogen combustion.

2. Model and numerical simulation method

2.1 Physical model

The combustion mode of the gas boiler studied in this paper is diffusion type combustion, and the burner and furnace are properly simplified. The final model is shown in Figure 1. The furnace is simplified to a cylinder with a diameter of 800 mm and a length of 2000 mm. In the pre-processing software ICEM provided by Fluent, the grid is divided into a structured grid and an unstructured grid. The structured grid is dominated by hexahedrons, and the results of calculations are easy to converge. The non-structured grids are mainly tetrahedral. The advantage is that meshing is easy, and very complex models can be quickly divided. So in engineering practice it can be widely used on the application. Since the actual burner model is studied, the structure is more complicated, and unstructured meshing is used. By setting the maximum grid value, some important faces are locally encrypted, and finally a grid of about 3 million is obtained, and the quality is above 0.35, which can meet the requirements of simulation.

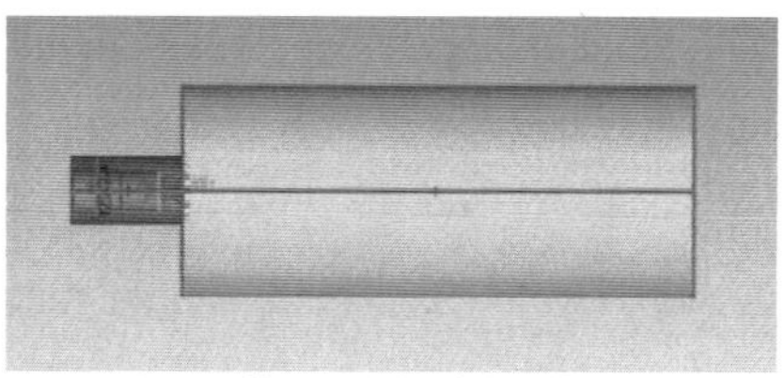

Figure 1. Overall model of the furnace.

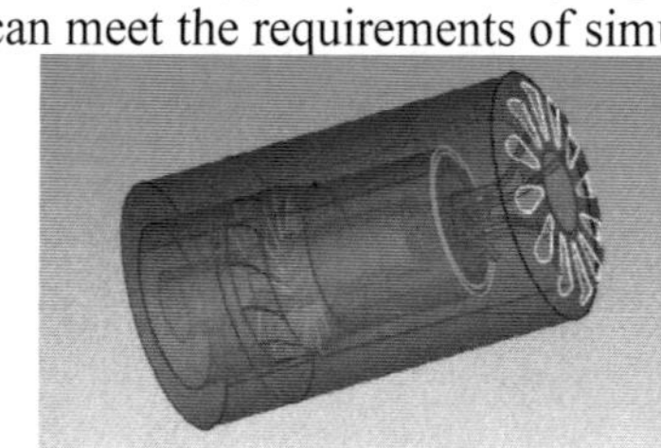

Figure 2. Three-dimensional structure of the burner.

2.2 Mathematical model

Gas and air are a complicated process in the process of combustion, with flow and heat transfer and accompanying chemical reactions. This process includes turbulence, radiant heat transfer, and component transport. In the simulation of natural gas combustion, the standard k-ε model, P1 radiation model and finite rate model are used. Finally, when calculating the boiler temperature field, the nitrogen oxides use a pollutant model. The air and gas inlets adopt a speed inlet, the walls of the furnace and the burner are set at a constant temperature, and the outlet of the furnace is a pressure outlet.

2.3 Simulation conditions

The conditions simulated in this paper include smokeless gas recirculation and flue gas recirculation. In the case of smokeless gas recirculation, the effects of different excess air ratios (1.0-1.3) on furnace temperature and nitrogen oxide formation are simulated; when the flue gas recirculation is simulated, the excess air ratio is set to 1.1. The effects of different flue gas recirculation rates (0%-20%) on furnace temperature and nitrogen oxide formation are simulated.

The composition of natural gas is shown in Table 1 below. The theoretical air volume of natural gas of this component is 9.71 m^3/m^3, and the low calorific value is 34417 KJ/m^3.

Table 1. Beijing natural gas composition table.

Species	CH_4	C_2H_6	C_3H_8	C_4H_{10}	C_5H_{12}	N_2	H_2O
Percentage	93.8	3.36	0.57	0.21	0.06	0.52	1.48

3. Simulation results analysis

3.1 Influence of different excess air ratio on NOx emissions

The excess air ratio has a certain influence on the temperature and oxygen concentration in the furnace. From the below cloud Figure 3, the maximum temperature of the z=0 section under each working condition can be seen, and the average temperature of the z=0 section can also be integrated, and the temperature distribution size and shape can also be seen. Because of the diffusion combustion, natural gas and air are mixed while burning, and will burn at the exit of the furnace. The temperature of the flame is sequentially lowered from the inside to the outside, and the flame is mainly concentrated in the middle of the furnace. It can be seen from the cloud diagram that as the excess air coefficient increases, the temperature will decrease to a certain extent, and the high temperature region will also shrink. When the excess air coefficient is 1.3, the performance is more obvious, the length of the flame becomes shorter, and the red high temperature portion also significantly reduced.

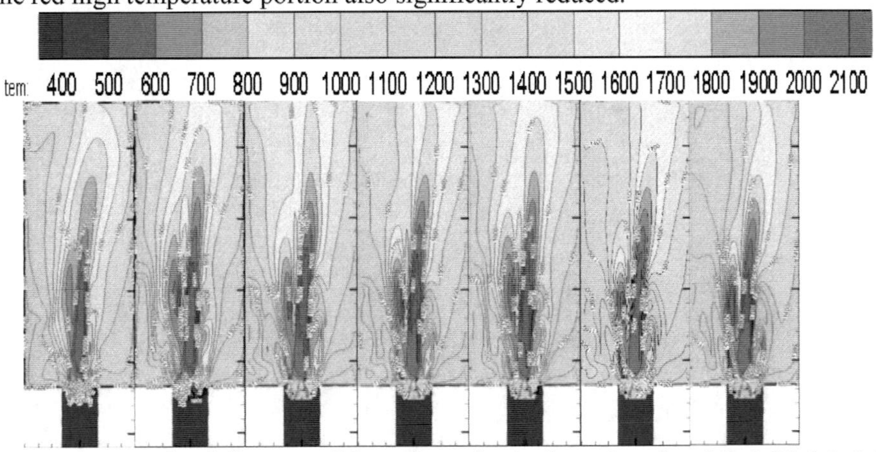

Figure 3. The temperature cloud diagram of the z=0 section in the order of α=1.0, 1.05, 1.1, 1.15, 1.2, 1.25, 1.3.

The data of the highest temperature and the average temperature of the section can be extracted from the cloud Figure 3 for analysis, as shown in Figure 4. It can be seen from Figure 4 that the abscissa is the excess air coefficient, the ordinate is the highest temperature and the average temperature of the section, and the maximum temperature first increases and then decreases with the increase of the excess air coefficient. When the excess air coefficient is 1.05, z =0 The temperature of the section is the highest, which is 2178K. At this time, the ratio of natural gas to oxygen in the furnace is close to 1, and complete combustion can be performed, and the heat release is the most complete. When the average temperature of the z=0 section exceeds 1.15, the average temperature of the section decreases significantly. At this time, it can be seen from the cloud Figure 3 that the flame becomes shorter and the high temperature region is correspondingly reduced. When the excess air ratio is 1.3, the average temperature of the z=0 section is the lowest, which is nearly 90K lower than the highest temperature.

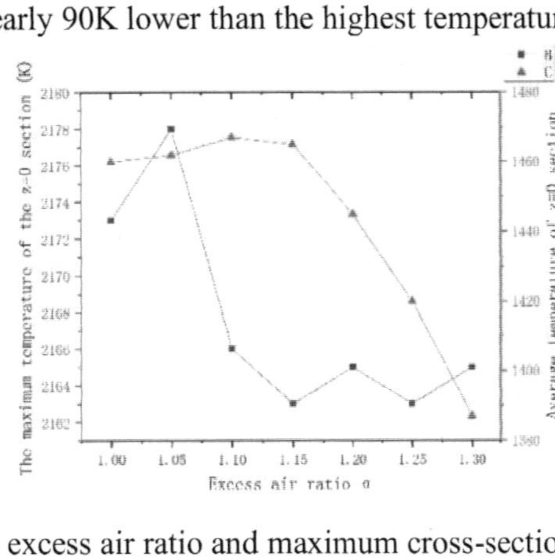

Figure 4. Relationship between excess air ratio and maximum cross-section temperature and average temperature.

In the process of combustion, the amount of oxygen will have an adverse effect on the combustion. If the oxygen is too small, the chemical heat of the fuel will not be fully exerted. However, if the oxygen is too big, the volume of the flue gas of the furnace will increase, and the temperature of the furnace will decrease. As a result, the thermal efficiency of the heating device is reduced [8]. It is found in the cloud Figure 5 below that there is a significant residual oxygen in the furnace when the excess air ratio exceeds 1.15. In Figure 4, the average temperature of the section shows a significant decrease. It can be seen that as the excess air ratio increases, the residual amount of oxygen in the section of the furnace is continuously increased. The increased oxygen cannot be consumed and the heat of the furnace is absorbed. Further discharge into flue gas will increase the smoke loss [9], and excessive excess air coefficient will reduce the temperature of the furnace, and will also increase the energy consumption of the fan, resulting in a certain waste.

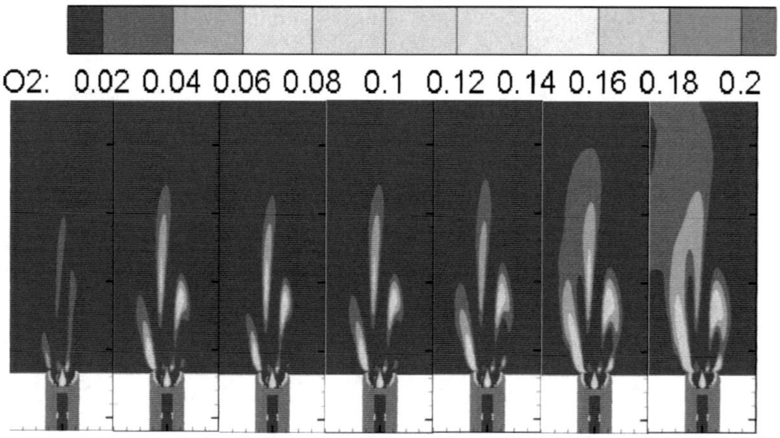

Figure 5. The oxygen component cloud diagram of the z=0 section in the order of α=1.0, 1.05, 1.1, 1.15, 1.2, 1.25, 1.3.

The excess air ratio has a large effect on the average temperature of the z=0 section, while the temperature type NO_x is sensitive to temperature changes. At temperatures above 1500°C, the rate of formation of temperature-type nitrogen oxides becomes very fast. The effect of excess air ratio on nitrogen oxides is shown in Figure 6. The area in which nitrogen oxides are formed is mainly concentrated in the middle and rear portions of the furnace, and the position at the exit of the burner is small. When the natural gas is injected into the furnace, it first reacts with oxygen, so the concentration of nitrogen oxides at the inlet of the furnace is very low, and in the middle of the furnace, the temperature is very high, reaching 1900°C, at this time, nitrogen oxidation and of hearth is generated very quickly.

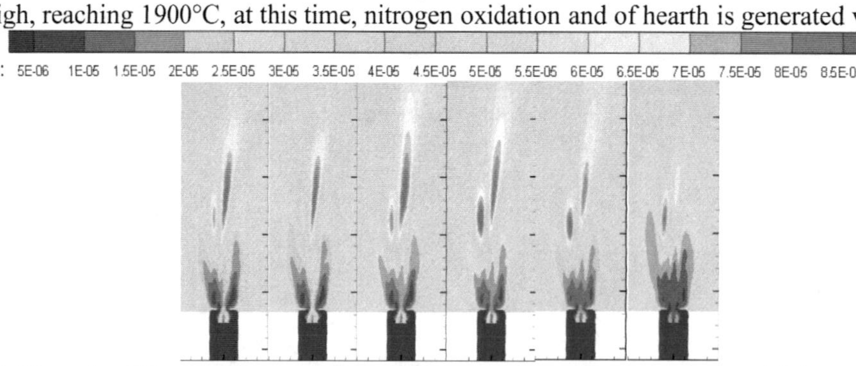

Figure 6. Show the NOx cloud map of the z=0 cross section with the excess coefficients of 1.05, 1.1, 1.15, 1.2, 1.25, and 1.3.

Figure 7 shows the relationship between the average temperature of the z=0 section and the concentration of nitrogen oxides at the outlet of the furnace and the excess air ratio. It can be seen that the law of the variation curve of the concentration curve of nitrogen oxides at the outlet of the furnace and the average temperature of the section is basically. As the excess air coefficient increases, the first slowly changes rapidly. This is mainly because the major factor affecting nitrogen oxides is temperature. When the average temperature of the section decreases, the high temperature zone is shrinking, and the formation of nitrogen oxides naturally decreases. Finally, when the excess air ratio is 1.3, the outlet is at the exit. The lowest concentration of nitrogen oxides should be the best, and the concentration of nitrogen oxides at the outlet is 50 ppm.

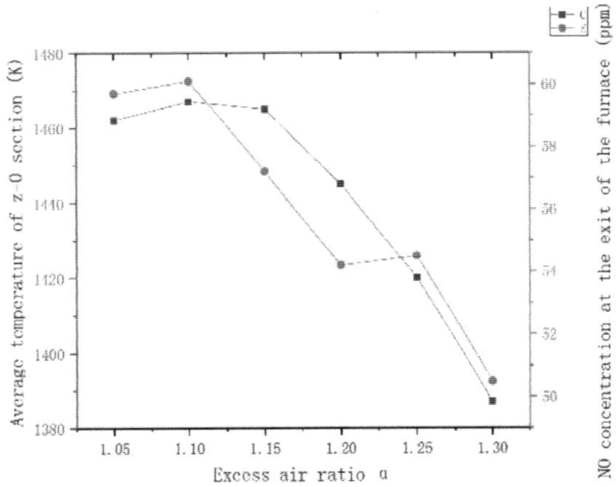

Figure 7. Relationship between excess air ratio and average cross-sectional temperature and nitrogen oxide concentration at the exit of the furnace.

3.2 Influence of different flue gas recirculation rates on NOx emission

Figure 8 shows the cloud map distribution of z=0 cross section when different flue gas ratios are added. After the flue gas recirculation is added, the furnace temperature is about 200K lower than that without flue gas recirculation, and the temperature drops significantly. This is because the reflux flue gas is an inert gas. When mixed with air, it will reduce the oxygen content in the furnace, thereby reducing the temperature of the furnace and reducing the formation of thermal nitrogen oxides.

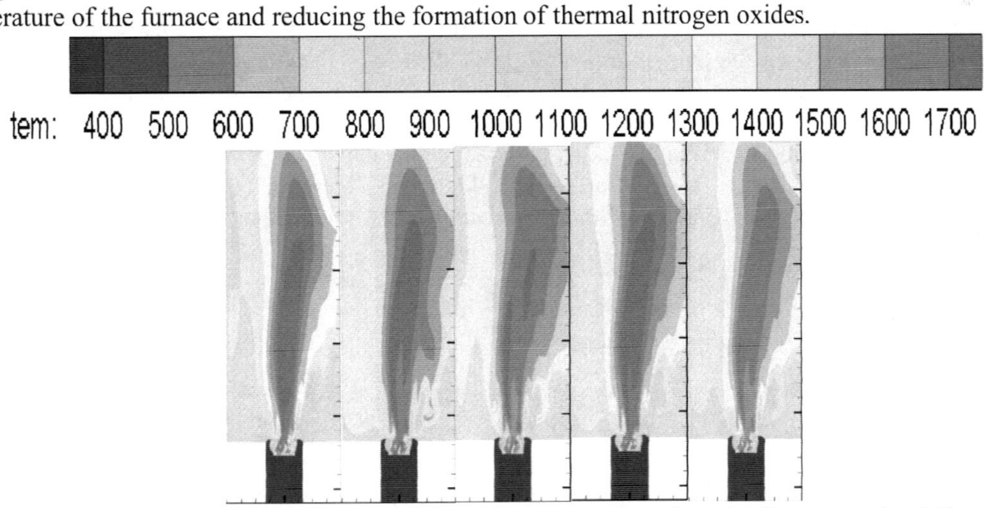

Figure 8. The temperature cloud diagram of the z=0 cross section when the flue gas recirculation rate is 0%, 5%, 10%, 15%, and 20%.

As the flue gas recirculation rate increases, the maximum temperature of the furnace is continuously decreasing, and the color of the high temperature region is becoming shallower and the range is also decreasing. As shown in Figure 9, after the addition of flue gas recirculation, the average temperature of the section is reduced from 1480K to 1350K, and the degree of decline in each stage is different. In the interval of 0%-5%, the average temperature of the section is the most. As you can see, nitrogen oxides also fall the fastest in this range, from 140mg/m^3 to 80mg/m^3. It can be seen that the formation of nitrogen oxides is closely related to temperature. Eventually, when the flue gas recirculation rate reached 20%, the nitrogen oxide emissions reached a minimum of 40 mg/m^3.

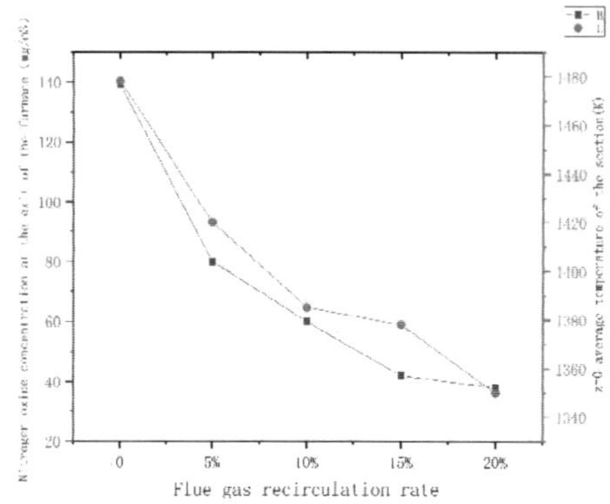

Figure 9. Relationship between flue gas recirculation rate and NOx and average temperature of section.

Under the operating conditions of the actual gas boiler, when the flue gas recirculation rate exceeds 20%, the flame temperature in the furnace is detected to change drastically, and the CO volume fraction increases sharply. At this time, the combustion becomes unstable at this time. The oxygen concentration inside is too low, causing the natural gas to not fully burn [10]. In order to ensure the smooth operation of the boiler and reduce nitrogen oxides, the flue gas recycling rate should be kept within 20%.

4. Conclusions

The simulation of the diffusion gas boiler with smokeless gas recirculation is carried out, and it is found that the flue gas recirculation can effectively reduce the furnace temperature and nitrogen oxide emissions, and has the following rules:

(1) With the increase of excess air coefficient, the maximum temperature of the furnace first rises and then decreases. When the air volume is too large, there will be excess oxygen remaining in the furnace, which will take away part of the heat, causing the average temperature of the furnace to decrease. Therefore, the formation of nitrogen oxides will also decrease.

(2) Adding flue gas recirculation can effectively reduce the maximum temperature of the furnace, so that the formation of nitrogen oxides will be less. When the excess air ratio is 1.1 and the flue gas circulation rate is 20%, the formation of nitrogen oxides is the lowest. It is about 40mg/m^3, which is about 70% lower.

Acknowledgement

This work was financially supported by the project supported by "National Key R&D Program of China" (2017YFF0209806)

References

[1] Beijing Municipal Environmental Protection Bureau. DB11/139—2015 Boiler Air Pollutant Emission Standard [S]. Beijing Municipal Environmental Protection Bureau, 2015.

[2] Song Shaopeng, Zhuo Jiankun, Li Na et al. Mechanism of fuel grading and flue gas recirculation on low-nitrogen combustion characteristics of natural gas[J]. Proceedings of the CSEE, 2016(24):258-267.

[3] Wang Jiegang, Huang Yabin, Wei Bo, et al. Calculation of thermal parameters of flue gas recirculation of steam injection boiler based on iteration and its influence on combustion[J]. Petroleum and Petrochemical Energy Conservation and Emission Reduction, 2019(3).

[4] Xie Zhengwu, Liang Haijie, Che Defu. The Role of Flue Gas Recirculation and Its Influence on Boiler Thermal Calculation[J]. Energy Research and Information, 1999(3): 42-48.

[5] Zhang Liqin, Song Wei, Wu Ning, et al. Emission Characteristics of Polluted Oxygen-Enriched

Combustion Pollutants[J]. Proceedings of the CSEE, 2009, 29(29): 35-40.
[6] Zhao Zengwu, Cang Daqiang, Li Baowei, et al. Effect of flue gas recirculation on high temperature air flat flame combustion and its NOx emission characteristics [C]// Proceedings of the 2007 Annual Meeting of China Society of Mechanical Engineering. 2007.
[7] Hu Manyin, Qiao Huan, Du Xin, et al. Numerical Simulation of Influence of Flue Gas Recirculation on Nitrogen Oxide Formation in Furnace[J]. Journal of North China Electric Power University, 2007, 34(6): 77-82.
[8] Tongji University...[etc.]. Gas Combustion and Application [M]. China Building Industry Press, 2011.
[9] Wang Xuedong, Yan Tao, Cheng Lin et al. Experimental study on the effects of boiler combustion adjustment on NOx emissions and boiler efficiency[J]. Journal of Power Engineering, 2008(1).
[10] Zeng Qiang, Liu Hanzhou, Yan Liang. Effect of Flue Gas Recirculation on NOx Emission Characteristics of Natural Gas Non-Premixed Combustion[J]. Journal of Combustion Science and Technology, 2018, v.24; No.128(04):85- 91.

IOP Conf. Series: Materials Science and Engineering **721** (2020) 012055 doi:10.1088/1757-899X/721/1/012055

Numerical Simulation Research on Agglomeration between Coal-fired Fly Ash Fine Particulate and Atomized Droplets

Yiquan Guo[1*], Junying Zhang[2]

[1]China Ship Development and Design Center, Wuhan, Hubei, 430064, P.R. China

[2]School of Energy and Power Engineering, Huazhong University of Science and Technology, Wuhan, Hubei, 430074, P.R. China

[*]Corresponding author's e-mail: quattrogyq@163.com

Abstract. In this paper, the interaction process between atomized droplets and particles after spraying chemical agglomeration solution was numerically simulated, the force model of particles in agglomeration solution was established, and the effects of density, viscosity, flow rate of agglomeration solution, and diameter of atomized droplets and other factors on the velocity, quality, diameter and number of particles after agglomeration were analyzed. The simulation results showed that the speed, mass, diameter and number of particles after agglomeration were greatly affected by the physical and chemical parameters of the chemical agglomeration solution. Increasing the density and viscosity of the solution could increase the variation trend of each characteristic quantity of particles. As the flow rate of agglomeration solution and diameter of atomized droplets became larger, the collision probability between the solution and particles got higher as a result. Since the particles gradually agglomerated, their mass and diameter increased. As their speed gradually decreased, the probability of collision and agglomeration with chemical agglomeration solution decreased as well.

1. Introduction

China ranks first in the world in annual coal production and consumption. Fine particulate matter (PM2.5) emitted from coal combustion can enter the lungs due to its small particle size, causing various respiratory diseases and even lung cancer[1]. Existing dust removal technologies, including electrostatic dust removal and bag dust removal technologies, have a removal efficiency of 99.9% for coarse particles, but a lower removal efficiency for fine particles. A large number of fine particles cannot be captured and removed by dust collectors due to their small sizes[2]. In view of this situation, chemical agglomeration technology[3] was adopted. A spray gun was installed in the flue in front of the dust collector to inject chemical agglomeration agent into the flue through compressed air and fully mixed with flue gas. Under the action of agglomeration solution, fine particles were agglomerated into particles with larger size, and then captured by dust removal equipment.

Agglomeration phenomenon among particles has always been the focus of research in colloid science, materials science, physics and microbiology. Exploring the mechanism of particle collision agglomeration has always been the goal pursued by most scientific researchers. By clarifying the mechanism of particle agglomeration, we can better master the chemical agglomeration technology, thus applying it to the engineering and technical fields of air pollution control, sewage treatment, etc.

In the process of studying the agglomeration mechanism of fine particles, researchers have established various particle agglomeration models according to their own experimental research needs. Research by Schaefer and Mathiesen[4]shows that, according to the size and viscosity of agglomerated agent droplets, the agglomeration mechanism can be divided into two types, namely distribution mode and immersion mode, as shown in Fig. 1. When the droplet size of the agglomeration solution was smaller than the particle size, or the two sizes were similar, it was mainly the distribution mode. In this mode, the droplets of agglomeration solution were uniformly wrapped on the surface of the particles, and then the droplets and the droplets were agglomerated together under the action of a liquid bridge to form agglomerates. When the droplet size of agglomeration solution was much larger than the particle size, it was mainly in immersion mode. In this mode, the particles adhered to the surface of the agglomeration solution droplets. Generally speaking, in the process of agglomeration, both modes existed. When the size of agglomeration agent drops was small, the viscosity was low and the speed was slow, the distribution mode dominated. When the size of agglomeration agent drops was large, the viscosity was high and the speed was fast, the immersion mode dominated. Goldszal and Bousquet[5] summarized the four stages of the combination of atomized droplets and particles according to the addition or moisture content of the solution when the agglomeration solution interacted with the particulate matter, as shown in Fig. 2.

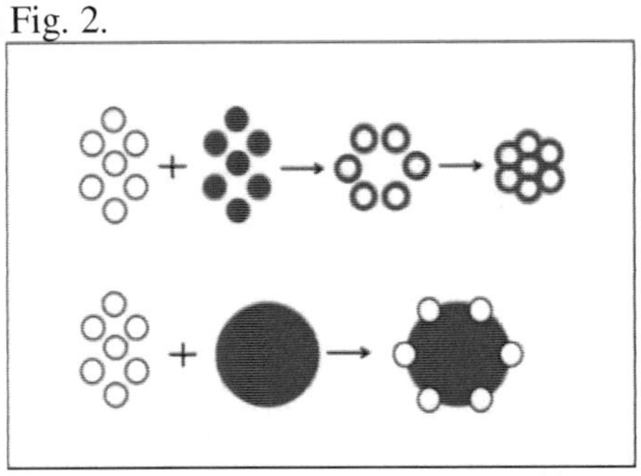

Figure 1. Mechanism of distribution agglomeration and immersion agglomeration

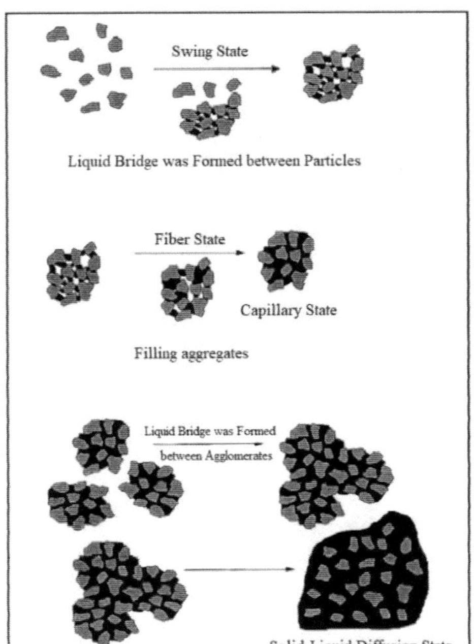

Figure 2. State diagram of droplet and particle agglomeration process

2. Theoretical analysis of spray process of chemical agglomeration solutions

2.1. Collision model of agglomeration solutions atomized droplets with particles

A collision model between spraying droplets and particles was established. The effects of various factors on the agglomeration effect of particles after the agglomeration solutions were sprayed were simulated. After spraying the chemical agglomeration solution, atomized liquid drops were formed. They then collided with the fly ash fine particles and combined together to form larger particles[6, 7]. The model assumed the following:

- The particle size was characterized by the average particle size of fine particles of coal-fired fly ash and atomized droplets of chemical agglomerating agent;
- Assuming that the fly ash particles were agglomerated together after colliding and no longer separated. The shape of atomized droplets and particles before and after the agglomeration process were all regular spheres.
- Assuming that the fly ash particles had uniform distribution of particle mass and density. The total mass and volume of the system including atomized droplets and particles were conserved before and after agglomeration;
- Assuming that the particles were continuously dynamic during the air flow process. The movement of gas molecule was ignored.
- Assuming that tension of the agglomeration solution remained unchanged.

2.2. Force analysis of particles in solutions

After the fly ash particles were dissolved in the agglomeration solution, colloidal suspension was formed. In solution, the particles were affected by various forces. According to the force type classification, including volume force, such as gravity and electrostatic force; area force, including buoyancy and drag force. The trajectory and heat transfer of particles in solution were calculated by various equilibrium forces acting on particles by droplets and heat transfer caused by convection/radiation, respectively[8]. In Cartesian coordinate system, the force balance equation of particles in the X-axis direction was:

$$\frac{dv_p}{dt} = F_D(v_l - v_p) + \frac{g_x(\rho_p - \rho_l)}{\rho_p} + F_x \tag{1}$$

Among them, $F_D(v_l - v_p)$ was the drag force, F_x was the additional force, including thermophoretic force F_{th}, brownian force F_b, Saffman lift F_s, Van der Waals force F_v, electrostatic force F_e, liquid bridge force F_{lb} and Basset force. Besides, the drag force was:

$$F_D = C_D \frac{\pi d_p^2}{4} \frac{\rho(v_l - v_p)^2}{2} \tag{2}$$

C_D was the drag coefficient, which was related to Re. d_p was the particle diameter, v_l was the atomized droplet velocity, v_p was the droplets velocity, and Re was:

$$Re \equiv \frac{\rho d_p |v_p - v_l|}{\mu} \tag{3}$$

According to the research of Haider and Levenspiel[9], the drag coefficient C_D was:

$$C_D = \frac{24}{Re}\left(1 + b_1 Re^{b_2}\right) + \frac{b_3 Re}{b_4 + Re} \tag{4}$$

3. Analysis of Factors Affecting Particle Aggregation Effect

During the time period of t to t+Δt, the volume increase of atomized droplets of agglomeration solution caused by adhering particles was expressed by the volume of particles contained in the volume swept by the droplets as follows:

$$(v_l - v_p)\frac{\pi}{4} d_a^2 n_0 \frac{\pi}{6} d_p^3 dt \tag{5}$$

According to droplet evaporation drying theory, the volume reduction due to evaporation can be expressed as:

$$\frac{2\pi d_a K(1 + 0.276 Re^{0.5} Sc^{0.33})}{\rho_l \gamma} \Delta T_m dt \tag{6}$$

So, the formula of particle velocity change with time was:

$$\frac{dv_a}{dt} = \frac{18\mu(v_p - v_l)}{\rho_l d_a^2} - \frac{(v_l - v_p)^2 \frac{\pi}{4} d_a^2 n_0 m_p}{m_a + (v_l - v_p)\frac{\pi}{4} d_a^2 n_0 m_p} \quad (7)$$

The formula for the change of particle mass m_a with time was

$$\frac{dm_a}{dt} = \left[(v_l - v_p)\frac{\pi}{4} d_a^2 n_0 m_p\right] - \left[\frac{2\pi d_a K(1 + 0.276 Re^{0.5} Sc^{0.33})}{\gamma} \Delta T_m\right] \quad (8)$$

The initial parameters of chemical agglomeration solution droplets and fly ash particles were set as follows:

Table 1. Initial Parameters of Numerical Simulation of Chemical Agglomeration.

Item	Unit	Value
Smoke Temperature	K	423/373
Initial Velocity of Particles	$m \cdot s^{-1}$	5
Initial Diameter of Particles	μm	0.1
Initial Mass of Particles	μg	0.001×10^{-3}
Initial Number of Particles	-	10000
Particle Viscosity	$kg \cdot m^{-1} \cdot s^{-1}$	2.4×10^{-5}
Initial Velocity of Agglomeration Solution Droplets	$m \cdot s^{-1}$	12
Initial Diameter of Agglomeration Solution Droplets	μm	20/15/10
Density of Agglomeration Solution Droplets	$kg \cdot m^{-3}$	2400/1800/1200
Viscosity of Agglomeration Solution Droplets	mPa·s	260/200/130

3.1. Effect of density of chemical agglomeration solution on agglomeration effect

This section simulated the effect of spraying three kinds of chemical agglomeration solutions with different densities on the characteristic quantities of particles after agglomeration. The solution densities were 2400 $kg \cdot m^{-3}$, 1800 $kg \cdot m^{-3}$ and 1200 $kg \cdot m^{-3}$, respectively.

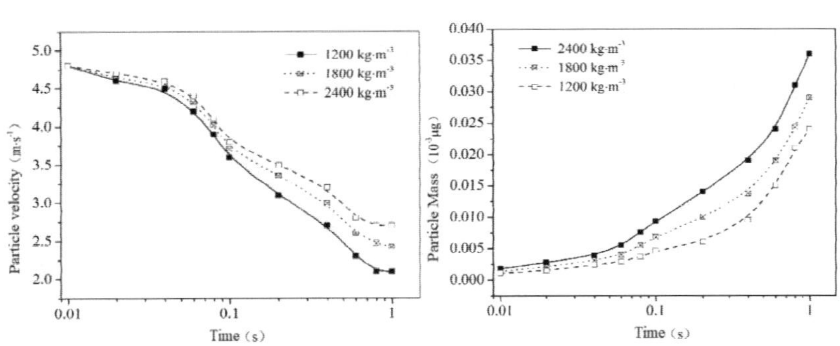

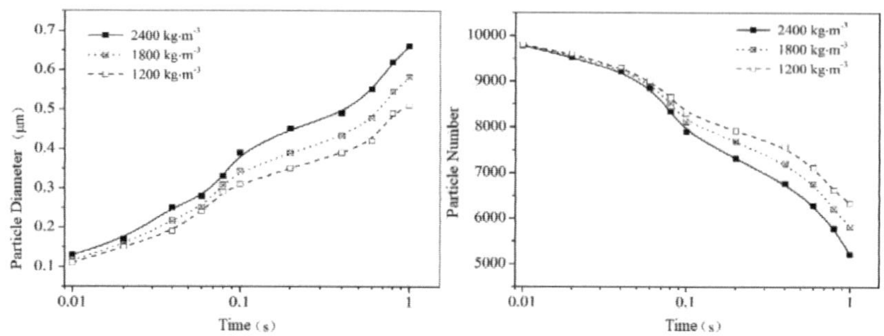

Figure 3. Effect of density of chemical agglomeration solution on characteristic quantity of particles after agglomeration

As can be seen from Fig. 3, after the chemical agglomeration solution was injected, the speed of fly ash particles gradually decreased. When the time closed to 1s, the particle velocity tended to be stable. It showed that particles and droplets of the chemical agglomeration solution aggregated together to form polymers, and the speed reached a stable value. In addition, under the same conditions, the higher the density of the agglomeration solution, the faster the particle velocity decreased. In the same time, the particle mass increased faster after injecting the agglomeration solution with higher density. Aggregates formed by aggregation of agglomeration solutions and particles were easier to adsorb surrounding particles as their mass increased. Under the same conditions, the effect of spraying agglomeration solution with higher density was better. The diameter of the formed polymer was increased from 0.1 μm to more than 0.6 μm, and the effect was better than that of the agglomeration solution with lower density (0.5 μm). After spraying chemical agglomeration solution, the particle number decreased significantly. Therefore, in order to improve the agglomeration efficiency, the density of agglomeration solution should be appropriately increased.

3.2. Effect of viscosity of chemical agglomeration solution on agglomeration effect

There were three kinds of chemical agglomeration solutions with different viscosities. The viscosities were 260 mPa·s, 200 mPa·s and 130 mPa·s, respectively. The effect of the additive on the characteristic quantities of the agglomerated particles was investigated.

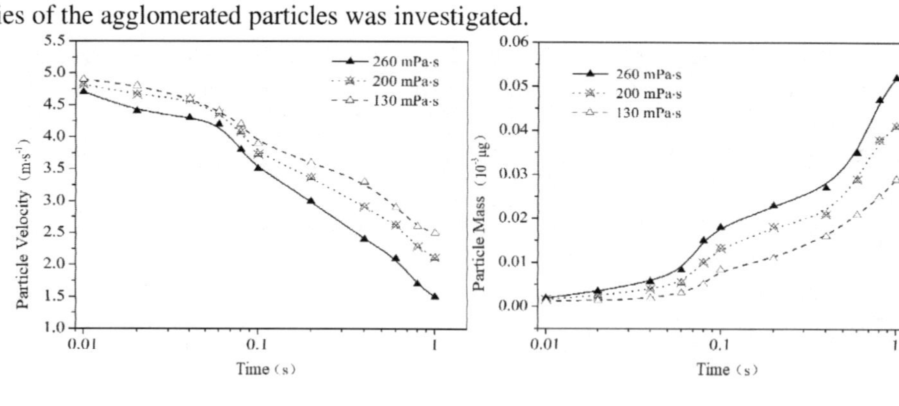

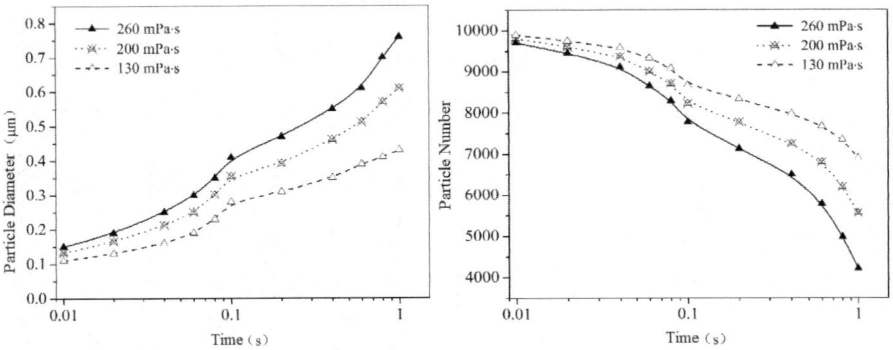

Figure 4. Effect of viscosity of chemical agglomeration solution on characteristic quantity of particles after agglomeration

As shown in Fig. 4, solution with higher viscosity also have greater resistance to particles. Therefore, the particles were easier to be captured by the agglomeration solution with higher viscosity and the agglomeration efficiency was higher under the same conditions. Agglomeration solution with higher viscosity were easier to agglomerate with particles, thus increasing the mass of particles faster. Injecting agglomeration solution with higher viscosity increased the particle diameter more obviously. The initial particle size was 0.1 μm, the particle size increased to more than 0.7 μm 1 s later after the agglomeration solution with a viscosity of 260 mPa·s was injected. However, after the agglomeration solution with a viscosity of 130 mPa·s was injected, the particle size only increased to about 0.4 μm. Solution with higher viscosity had better capture effect on particles. Therefore, the particles were easier to aggregate and grew up, and the diameter change was more obvious.。 The initial particle number was about 10,000. After the agglomeration solution with a viscosity of 260 mPa·s was injected, the particle number decreased to about 4,000, which decreased about 60%.

3.3. Effect of flow rate of chemical agglomeration solution on agglomeration effect

The influence of spraying three kinds of chemical agglomeration solutions with different flow rates on each characteristic quantity after agglomeration was simulated. The flow rates were 0.05 kg·s^{-1}, 0.04 kg·s^{-1}, and 0.025 kg·s^{-1}.

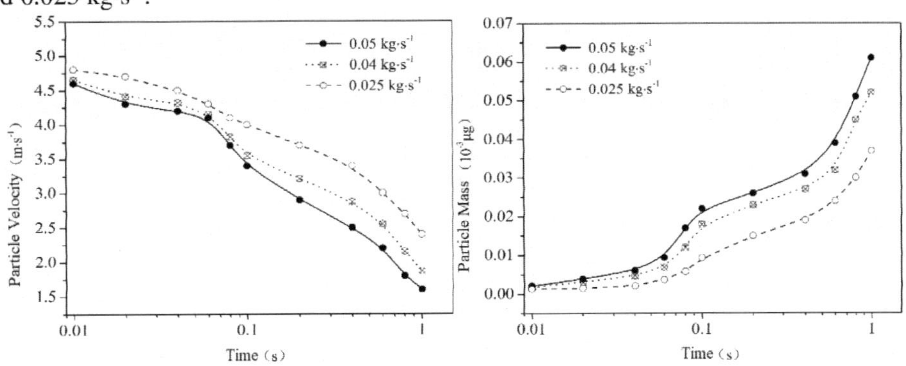

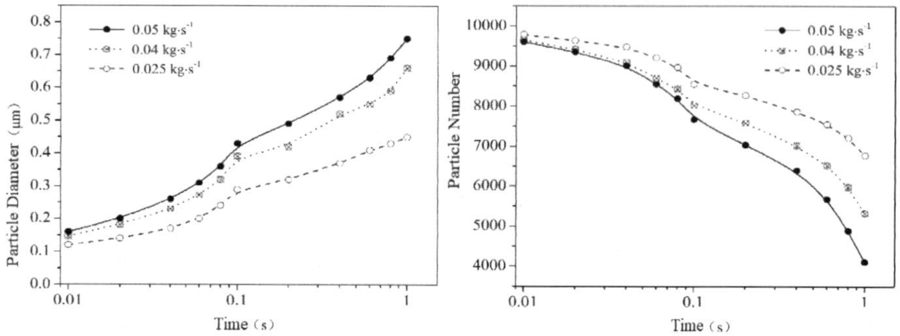

Figure 5. Effect of flow rate of chemical agglomeration solution on characteristic quantity of particles after agglomeration

As can be seen from Fig. 5, the larger flow rate, the larger number of agglomeration solution droplets in the same time. As the mass was larger, the probability of collision between atomized droplets and particles was relatively increased. The diameter of particles increased faster when the agglomeration solution with larger flow rate was injected during the same time. As well as the diameter of agglomerates formed was larger. Therefore, in order to improve the agglomeration efficiency, the flow rate of agglomeration solution can be appropriately increased.

3.4. Effect of atomized droplet diameter on agglomeration effect

The influence of spraying three kinds of chemical agglomeration solutions with different diameters of atomized droplets on the characteristic quantities of particles after agglomeration was simulated. The droplet atomization sizes were 20 μm, 15 μm and 10 μm, respectively.

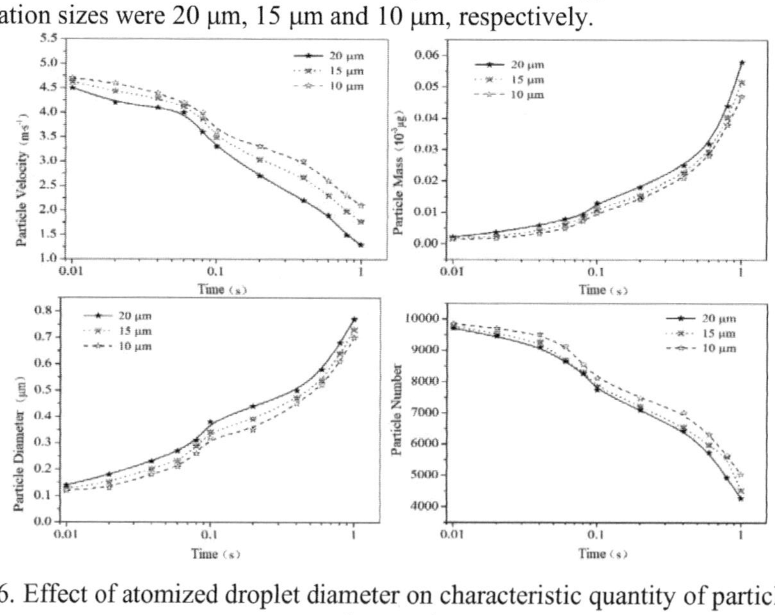

Figure 6. Effect of atomized droplet diameter on characteristic quantity of particles after agglomeration

As shown in Fig. 6, as the droplet diameter increased, the specific surface area of the droplet also increased. So was the collision probability between droplets and particles. Particles were easier to combine with droplets to form aggregates. Therefore, the particle speed drop faster. Under the same conditions, the mass and diameter of particles increased faster after spraying solution with larger atomized droplet diameter. Increasing the atomized droplet diameter could increase effective contact area with the particles. Therefore, properly increasing the atomized droplet diameter can effectively improve the agglomeration efficiency of fine particles.

4. Conclusions

In this paper, the interaction process between atomized droplets and particles after spraying chemical agglomeration solution was numerically simulated. The stress model of particulate matter in agglomeration solution solution was established. The effects of agglomeration solution density, viscosity, flow rate and atomized droplet diameter on velocity, mass, diameter and number of particles after agglomeration were analyzed. The main conclusions are:

The simulation results show that the speed, mass, diameter and number of particles after agglomeration were greatly affected by the physical and chemical parameters of the chemical agglomeration solution. Increasing density and viscosity of the chemical agglomeration solution properly can increase the variation trend of each characteristic quantity of the particles. Within a certain range, the larger the flow rate of agglomeration sloution and the larger the diameter of atomized droplets, the higher the collision probability with particles and the better the agglomeration effect.

References

[1] Pui D Y H, Chen S C, Zuo Z. $PM_{2.5}$ in China: Measurements, sources, visibility and health effects, and mitigation[J]. Particuology, 2014, 13: 1-26.

[2] Xiong Guilong, Li Shuiqing, et al. Development of advanced electrostatic precipitation technologies for reducing PM2.5 emissions from coal-fired power plants. [J]Proceedings of the CSEE, 2015, 35(9): 2217-2223

[3] Guo YQ, Zhang JY, Zhao YC, Wang SL, Jiang C, Zheng CG. Chemical agglomeration of fine particles in coal combustion flue gas: Experimental evaluation. Fuel, 2017, 203: 557-569.

[4] Schæfer T, Mathiesen C. Melt pelletization in a high shear mixer. IX. Effects of binder particle size. International journal of pharmaceutics, 1996, 139(1-2): 139-148.

[5] Goldszal A, Bousquet J. Wet agglomeration of powders: from physics toward process optimization. Powder Technology, 2001, 117(3): 221-231.

[6] Lian G, Thornton C, Adams M J. Discrete particle simulation of agglomerate impact coalescence. Chemical Engineering Science, 1998, 53(19): 3381-3391.

[7] Hetsroni G. Handbook of multiphase systems. 1982.

[8] Jones M R, McCarthy A, Booth A. Characteristics of the ultrafine component of fly ash. Fuel, 2006, 85(16): 2250-2259.

[9] Haider A, Levenspiel O. Drag coefficient and terminal velocity of spherical and nonspherical particles. Powder technology, 1989, 58(1): 63-70.

IOP Conf. Series: Materials Science and Engineering **721** (2020) 012056 doi:10.1088/1757-899X/721/1/012056

Design and Research of High-accuracy and Intelligence Test Device for Gas Water Heater Energy Efficiency

Zhongjie Wu[1*],Dinghua Xu[2],Haiyun Zhang[2], Chenzhe Hang[2]and Junhong Zhou[1]

[1] Guangdong Institute of Metrology, Guangdong, 510405, China;

[2] National Institute of Metrology, Beijing, 100013, China

[*]Corresponding author's e-mail:13751895093@163.com

Abstract. The research subject aims to solve the defects of slow test speed and inefficient for gas water heater energy efficiency. It not only applies modern sensor technology, high reliability data acquisition card, solenoid valve control technology and stable industrial computer, but also adopts LabVIEW language, PID control technology and flow stability judgment algorithms to develop an efficient, accurate and intelligent test system for gas water heater energy efficiency. The test results shows that the measurement repeatability of the test device is superior to the artificial test device and common intelligent test device, and the assessed measurement uncertainty meets the requirements of the national supervision and inspection technical specifications.

1. Introduction

Energy efficiency labeling system is an important mandatory energy conservation management measures in China. With the in-depth implementation of the energy efficiency labeling system of gas water heaters, market regulators gradually increased the supervision and inspection of the energy efficiency labeling, which impels higher requirements for the energy efficiency testing devices to manufacturers and authoritative testing institutions, and accelerated the improvement and upgrading of the testing devices for energy efficiency [1]. Now, the traditional artificial test devices is widely used in the energy efficiency test in China, which simply assembling all kinds of testing sensors to a device, and calculate all parameters manually. This practice not only has a large workload, a long test period and inefficiency, but also bring human factors into measurement, which affects the accuracy and reliability of the measurement results. It will increase the judge risk of the government in energy efficiency supervision and inspection.

In recent years, with the rapid development of science and technology, the application of intelligent measurement and control technology in the measurement field is getting more mature. The traditional means relying on artificial measurement has been replaced gradually by efficient and intelligent measurement technology[2]. The development of high-precision intelligent test device for gas water heater energy efficiency, will solves the defects of slow test speed and inefficient for traditional device. It applies modern sensor technology, high reliability data acquisition card, solenoid valve control technology and stable industrial computer. It uses LabVIEW language, precise and reliable PID control technology and flow stability judgment algorithms to develop an efficient, accurate and intelligent test system for gas water heater energy efficiency, which can meet the test requirements of national standard GB6932 and national supervision and inspection specifications JJF1261.9[3-4].

IOP Conf. Series: Materials Science and Engineering **721** (2020) 012056 doi:10.1088/1757-899X/721/1/012056

Through verification tests, the measurement repeatability and measurement uncertainty of the device is further evaluated.

2. General design scheme and main research contents

2.1 General design scheme

The research and development of high-precision intelligent test device for gas water heater energy efficiency, including the general scheme design, hardware circuit design, software development, system assembly and debugging of the test system [5-6]. As shown in Figure 1, the test device is mainly composed of industrial computer, data collector, PID controller, solenoid valve, gas and water pressure regulating valve, gas flow meter, water flow meter, gas pressure gauge, gas and water thermometer, etc., By collecting the parameters of gas temperature, gas pressure, gas flow, water flow, inlet and outlet water temperature, the industrial computer automatic integrated control according to the test methods, and then export results based on the specific algorithm.

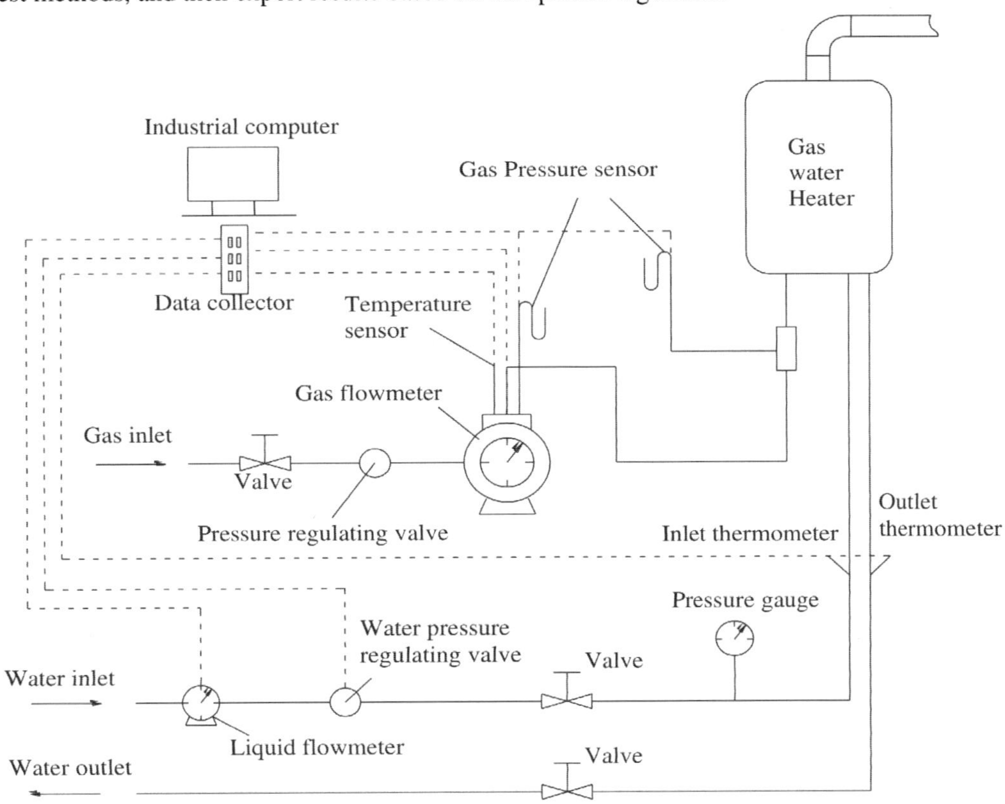

Figure 1. Structure of intelligent test device for gas water heater energy efficiency

2.2 Hardware design

As shown in Figure 2, the system adopts industrial computer, data collector, PID controller and various sensors as the core hardware.

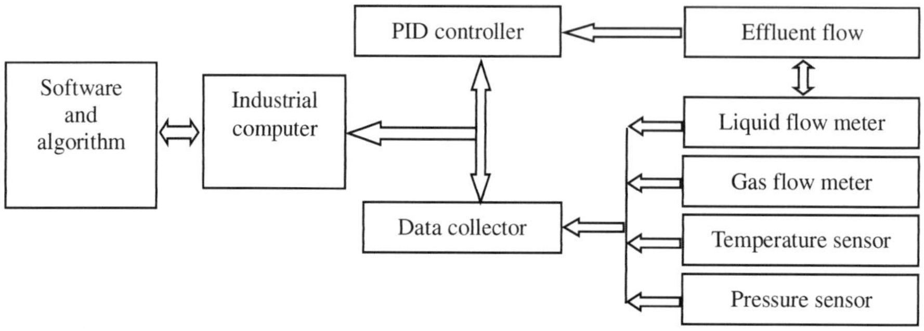

Figure 2. Hardware structure of test system

The data collector chooses to Agilent 34980A, which has the characteristics of high resolution, fast response and high precision. It can collect the data of all sensors quickly and accurately, and realize the best matching of measurement and collection. The function of PID controller is to precisely control the water flow. The system adopts EUROTHERM 2404 high-precision PID controllers, which have the functions of self-tuning and PID gain meter. The current signal of flow meter is input to the controller, and the output of valve position is controlled by relay to the optimal control of flow.

The sensor is an important part of the device, the accuracy of the sensor is the key to ensure the accuracy of measurement. All kinds of sensors adopts high-quality brands at home and abroad, and the measurement technical specifications are superior to the mainstream products in the market. The technical specifications of the main sensors are shown in Table 1.

Table 1. Main sensor specifications

Sensor name	Range	Accuracy	Resolution
Temperature sensor	0～100 °C	± 0.1 °C	0.05 °C
Water Pressure sensor	0～5 MPa	± 0.2%FS	1 kPa
Gas Pressure sensor	0～20 kPa	± 0.5%FS	1 Pa
Liquid flowmeter	0～20 L/min	± 0.5%	0.001 L/min
Gas flowmeter	0～5000 L/h	± 1.0%	50 mL

2.3 Software design

According to the industrial computer program, the test system software is designed to realize the functions of visual interface, data acquisition, serial communication, process control, parameter setting, data storage, etc. The main design includes: (1) Based on the consideration of high efficiency, convenience and friendly human-computer interface, the software is developed by LabVIEW language, which not only has a powerful visual component library, but also has the characteristics of flexible language and high execution efficiency. (2) The automatic control and measuring method of the test system applies the computer centralized and distributed measurement and control. The subordinate controller collects and displays the data, while the superior master computer analyzes and processes the data collected by the subordinate controller, and sends the set value instruction and data collection instruction to the subordinate controller. (3) Considering the subsequent data storage function, the built-in database can complete the data storage for test system, can directly access the specific data sources, and easy to operate the database with the provided database components.(4) The test system has a friendly human-computer interface, provides real-time graphic processing and intuitive simulation process control panel, which can monitor, record and store in real time.

3. Key Technologies of the device

3.1 Flow control

PID control is adopted for water heater outlet flow, which can monitor and control the operation status

IOP Conf. Series: Materials Science and Engineering **721** (2020) 012056 doi:10.1088/1757-899X/721/1/012056

of the system in real time. As shown in Figure 3, the controller adjusts the deviation between the measured value and the set value from the flow meter by proportion, differentiation and integration. If the actual flow does not meet the preset flow, the actuator will adjust according to the deviation until meet the preset requirements, so as to achieve the purpose of fast and accurate flow control.

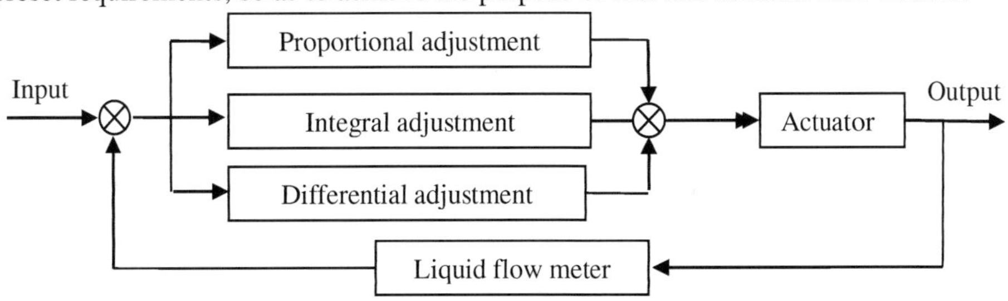

Figure 3. Structure diagram of the PID controller

The PID controllers have the functions of self-tuning and PID gain meter, which can automatically adjust the parameters to ensure better dynamic and static response of controlled flow, so as to meet the requirements of accurate control.

3.2 Test stability judgment

The objective influencing factors of accurate detection mainly come from outlet flow and temperature, while the stability of water temperature depends on the test water heater, not the device. Therefore, the stability control of flow is particularly critical, which directly affects the accuracy of test results. In order to ensure the stable operation in the test process, the system needs to judge the flow stability, ensure the flow to meet specific requirements in the stable phase. The algorithm of flow stability judgment, using the range value to evaluate the flow dispersion. Figure 4. Is the flowchart of stability judgment in stability stage.

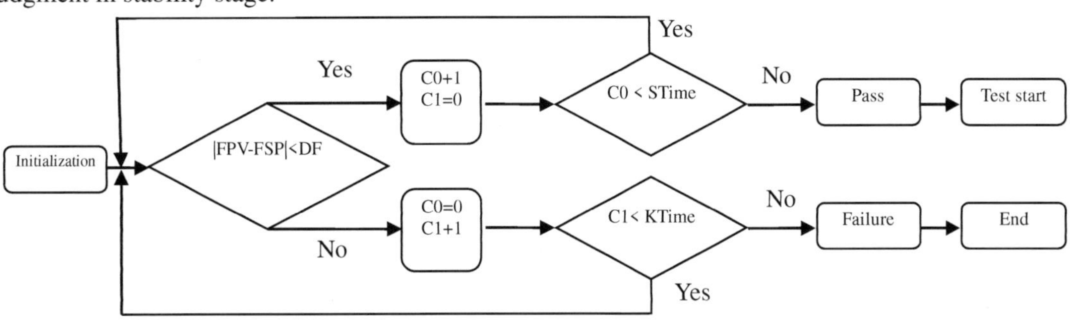

Figure 4. Flow chart of stability judgment in stability stage

The specific process for stability judgment is as follows: (1) The system should initialize the settings when entering the stable phase. Initialization set to: flow setting value is FSP, flow current value is FPV, stability deviation is DF, stable timer C0 set to zero, unstable timer C1 set to zero, stable timer setting is STime, unstable timer setting is KTime; (2) After initialization, collect the current flow value, calculate whether the deviation between the collected value and the set value meets the requirements. If the deviation meets the requirements, the stable timer C0 starts to accumulate until the stability judgment is successful, and then enter the test phase. Otherwise, the stability timer is set to zero, and the unstable timer C1 is accumulated to enter the unstable cycle.

4. Technical capability evaluation of the device

4.1 Measurement repeatability of the device

The measurement repeatability is the key technical index to evaluate the stability of the test device [7]. In order to evaluate the measurement repeatability of the device, choose a gas water heater with stable performance as the measurement sample, then the thermal efficiency is measured ten times continuously, the results were 94.5%, 94.5%, 94.4%, 94.8%, 94.4%, 94.4%, 94.6%, 94.5% and 94.3%.

According to the evaluation principle of uncertainty in JJF 1059.1-2012 Evaluation and Expression of Uncertainty in Measurement [8], the measurement repeatability of thermal efficiency is calculated by Bessel Formula, and the calculation formula is as follows:

$$\phi(\eta)=\sqrt{\frac{\sum_{i=1}^{n}(\eta_i-\bar{\eta})^2}{n-1}} \tag{1}$$

Where: $\phi(\eta)$ is the repeatability of thermal efficiency measurement; η_i is the thermal efficiency value obtained from the i-th measurement; $\bar{\eta}$ is the average value of thermal efficiency obtained from the n-th measurement; n is the number of independent measurement.

Substitute the measurement results into the above formula, the repeatability of thermal efficiency measurement is 0.14%. According to the same test method, the repeatability is measured by traditional artificial test device and common intelligent test device, and the repeatability of ten consecutive measurements is 0.65% and 0.42%. It can be seen that the repeatability of the developed high-precision test device is much better than the artificial and the common intelligent test device.

4.2 Measurement uncertainty

Ensure the measurement accuracy of the device and continuous control of the measurement process, using uncertainty of measurement results as evaluation index. The evaluation of measurement uncertainty according to JJF 1059.1-2012 Evaluation and Expression of Uncertainty in Measurement and JJF1261.9 Rules of metrology testing for energy efficiency label of domestic gas instantaneous water heater and gas fired heating and hot water combi-boilers, the thermal efficiency is measured ten times and then the uncertainty is evaluated. Refer to the evaluation method in Appendix A of JJF1261.9, the measurement uncertainty components of thermal efficiency are shown in Table 2.

Table 2. Subscale of uncertainty assessment

Source of uncertainty	Measured value	Relative standard uncertainty	Distribution	Uncertainty component
Repeatability	0.137 %	0.137%	Normal	0.137%
Water volume	12.255 kg	0.289%	Rectangle	0.289%
Temperature difference	40.0 K	0.204%	Rectangle	0.204%
Gas temperature	297.8 K	0.019%	Rectangle	0.019%
Gas flow	0.07 m^3	0.289%	Rectangle	0.289%
Calorific value of gas	34.04 MJ/m^3	0.577%	Rectangle	0.577%
Pressure correction value	99.009 kPa	0.058%	Rectangle	0.058%

According to the composition principle of uncertainty in JJF 1059.1-2012 Evaluation and Expression of Uncertainty in Measurement, the relative standard uncertainty is 0.75%. When the confidence probability is 95%, the coverage factor k=2, and the relative expanded uncertainty of thermal efficiency is 1.5%. The results meet the requirement of relative expanded uncertainty of thermal efficiency (1.6%) in national supervision and inspection technical specification (JJF1261.9).

5. Conclusion

The development of high-precision intelligent test device for gas water heater energy efficiency will greatly improve the technical level of energy efficiency test in gas water heater industry, provide accurate and reliable measurement data for authoritative testing institutions, and reduce the judge risk of government supervision and inspection. In conclusion, the research of high-precision intelligent test device for gas water heater energy efficiency can be summarized as follows:

(1) Based on modern sensor technology, high reliability data acquisition card, solenoid valve

control and stable industrial computer and other technical measures, develop an intelligent device which can meet the requirements of rapid and accurate test for gas water heater energy efficiency.

(2) The measurement technical specifications of sensors used in the device are superior to the technical level of mainstream products on the market; through test results analysis, the measurement repeatability of the device is superior to the artificial test device and common intelligent test device, and the measurement uncertainty of the device meets the requirements of the national supervision and inspection technical specifications, which will further reduce the judgment risk of supervision and inspection.

(3) The test system is developed with LabVIEW language, and adopts PID control technology and the algorithm of stability judgment to ensure more accurate test results.

Acknowledgments

This work is funded by the National Key Research and Development Program of China (No. 2018YFF0212404).

References

[1] Cao, X.B., Zhang, Z.Y. (2017) Discussion on the implementation of China's energy efficiency labeling system. Standard & Quality of Light Industry., 03:67-69.
[2] Hu, H.M., Liu, B. (2017) Analysis on the Development and Application of Modern Measurement and Control Technology. Mechanics-electronics Applied Technique., 23 :76-83.
[3] GB6932. (2015) Domestic gas instantaneous water heater. Standards Press of China, Beijing.
[4] JJF1261.9. (2013) Rules of metrology testing for energy efficiency label of domestic gas instantaneous water heater and gas fired heating and hot water combi-boilers. Standards Press of China, Beijing.
[5] Xi, J. (2007) On-line detection of a gas fired water heater. Mechanical and Electrical Equipment., 09:14-17.
[6] Gai, H.T., Su, X.G. (2011) Design of Detection System for Gas Water Heater Based on PCI1711L. Jiangsu Aviation., 01 :19-21.
[7] Jia, S.H., Li, J.J. (2010) Application of Measurement System Repeatability and Reproducibility in Product Quality Management. Journal of Dalian Jiaotong University., 5:96-100.
[8] JJF 1059.1. (2012) Evaluation and Expression of Uncertainty in Measurement. Standards Press of China, Beijing.

IOP Conf. Series: Materials Science and Engineering **721** (2020) 012057 doi:10.1088/1757-899X/721/1/012057

Design-Based Thermal Efficiency Correction Method for Hot Water Boiler

Wang Lin[1*], Liu Chen[1], Jin Bing[1], Gao Yue[2]

[1]Boiler business department, Liaoning Academy of safety Sciences, Shenyang, Liaoning province, 110004, China

[2]Chemical analysis center, Liaoning ecological environment monitoring center, Shenyang, Liaoning province, 110161, China

*Corresponding author's e-mail: wangchao0019@163.com

Abstract: In the boiler design process, the thermal efficiency of the boiler operating under specified parameters is given in accordance with the design specifications. However, under the actual operating conditions, the specified parameters in the design book cannot be met at the same time. Especially for the hot water boiler, the coal quality component and the deviation of low calorific power, the feed water temperature and the ambient temperature will directly affect the operating efficiency of the boiler. In the current boiler energy efficiency test standards, the test conditions are very demanding. It is required to be the design coal type and it also gives fluctuation ranges for several parameters that affect the boiler operation efficiency. For non-design conditions, the boiler operation condition test method can only be executed referring to the standard. However, the measured efficiency is highly variable. This paper aims to analyze and study several parameters affecting boiler efficiency based on the perspective of boiler design, and give corresponding efficiency correction method to provide reference for the correction of energy efficiency test results under non-design conditions.

1.Introduction

"Hot water boiler" refers to a boiler that uses various fuels to convert chemical energy into heat energy and outputs it in the form of hot water. This type of boiler is used as a heating boiler mainly in the "three north" areas, namely northeast, north and northwest. According to incomplete statistics, there were about 2,000 hot water boilers in Liaoning Province at the end of 2018. In recent years, with the adjustment of industrial structure and environmental protection requirements, hot water boilers have developed towards large capacity and high parameters. From 2015, Liaoning Province eliminated a total of 2,000 small coal-fired boilers, most of which were small-scale regional heating boilers. From 2016, 10/h and below coal-fired small boilers were eliminated, and new coal-fired boilers of 20t/h and below were prohibited. In particular, at the end of 2017, 10 ministries and commissions such as the National Development and Reform Commission and the National Energy Administration issued the "*Northern District Winter Clean Heating Plan (2017-2021)*", clearly stated that "in 2021, all urban areas will realize clean heating, all coal-fired boilers below 35 steam tons be dismantled; the clean heating rate of the county and urban-rural junctions reached over 80%, and the coal-fired boilers below 20 steam tons be dismantled, and the clean heating rate in rural areas reach over 60%". At the same time, amount requirement of ultra-low emission was proposed for the unremoved boilers (under the condition of 6% oxygen content, the emission concentration of soot, sulfur dioxide and nitrogen oxides should be not higher than 10, 35, 50 mg/ m^3 respectively). Under the double impact of industry and environmental protection requirements, the average single-unit capacity of the hot water boiler in Liaoning Province has been increased from the previous 8.7MW to 25.6MW, and only one or several large heating boiler rooms in the urban built-up area which have a heating radius of 50-100km. For some townships and industrial and mining enterprises with relatively low population and low heat demand, some small hot water boilers are reserved.

Thermal efficiency is the main indicator[1] of boiler thermal performance. It is directly related to the operating cost of hot water boilers. For heating units, fuel cost accounts for more than 70% of the total operating cost. To this end, heating companies generally use the method of monitoring the main thermal parameters of the boiler to indirectly calculate the boiler operating efficiency and achieve the assessment of the boiler room management level and the furnacemen's capacity.

The *Test Procedure for Thermal Performance of Industrial Boilers* (GB/T10180-2017) gives the test and conversion method for boiler thermal load and operating thermal efficiency under full load operation conditions. In the actual operation of the boiler, it is necessary to calculate the downstream heat consumption according to factors such as the ambient temperature and the heating area to determine the load of the operation, so that the current standard has great limitations. In the initial design of the boiler, the radiation in the furnace and the convection heating surface are designed according to the rated load of the boiler. When the boiler load drops, due to the innate design, the heating surface of the boiler will not automatically increase or decrease according to the operating load. In this condition, the design thermal efficiency of the boiler will change accordingly. It is obviously biased to evaluate the boiler energy efficiency level under non-rated operating conditions based on the thermal efficiency index under full load conditions. Therefore, it is necessary to propose a boiler energy efficiency test method suitable for different working conditions and different operating parameters, and the corresponding calculation method, that the natural defects in the existing standards can be solved scientifically an objectively.

2.Calculation Method and Model Introduction

This method attempts to follow the QXL46-1.25-70/115-AII boiler designed by a boiler plant to ensure that the boiler structure, heating area, medium and flue gas flow rate are constant, and only changes several important index parameters affecting the thermal efficiency of the boiler to obtain the influence curve of the value change on the boiler's thermal efficiency, and give the correction coefficient correcting back to the original design parameters.

Refer to the correction method on net power of ASME PTC46-1996 for the overall performance test standard of power plants. Considering the above factors have a direct impact on energy efficiency, the multiplication correction method[2] is adopted for this purpose. The calculation formula is as follows:

$$\eta_c = \eta \prod_{i=0}^{4} \alpha_i \tag{1}$$

Of which: η_c - conversion thermal efficiency, %;

η - test thermal efficiency, %;

α_1 - correction of fuel calorific value;

α_2 - correction of ambient temperature;

α_3- correction of feed water temperature deviation;

α_4 - correction of output heat.

3.Calculation Results and Discussion

3.1Correction of Low Calorific Value of Fuel

The input heat has a great influence on the thermal efficiency of the boiler. The input heat[3] of the general hot water boiler mainly includes: fuel heat, physical sensible heat of fuel, external heating fuel or air heat, self-use steam heat and fuel oil atomizing steam heat. Among them, fuel heat is the mainstay, accounting for more than 90% of the total input heat. In the traditional boiler thermal efficiency calculation method, fuel heat is the product of the fuel low heat value and the fuel amount. For oil and gas boilers, the composition's consistency is high and the calorific value is stable, but the calorific value is related to the ratio of hydrogen content. The higher the proportion of hydrogen, the higher the calorific value. Compared with gas and liquid fuels, solid fuel components have poor consistency and unstable calorific value. Taking coal as an example, the main elements in coal that react with oxygen and release heat are carbon, hydrogen and sulfur. Hydrogen and sulfur are mainly concentrated in the volatile matter of coal, while the low calorific value of coal needs to deduct the heat released by the combustion of some hydrogen and all sulfur. The hydrogen content in coal is small, so it can be simply considered that the low calorific value of coal depends mainly on the carbon content. After the fuels of same quality are sent into the boiler, the coal with high carbon content generates greater amount of heat, which results in a higher furnace heat volume. According to the Stephen Boltzmann formula, it can be calculated that it has larger amount of heat exchange with the radiation heating surface in the furnace. The heat exchange of the medium is enhanced, thereby the operating efficiency of the boiler is improved.

After the coal particles enter the boiler, they are heated in the furnace. After the water is evaporated and volatilized, the porous medium structure will be formed. This will facilitate the penetration of oxygen into the core of the coal particles and react with the carbon elements. The heat generated by the reaction softens the coal particles. Meanwhile, the carbon dioxide gas is generated and overflows, and the carbon particles are broken to complete the condensation reaction[4]. After the coal particles break, the particle size decreases and the specific surface area increases, which is beneficial to the carbon particles to further react with oxygen, and the carbon particles are burned out. When the coal ash contains an alkali metal element compound, it will be melted and will be wrapped on the surface of the particle after being heated, which will hinder the carbon and oxygen from further reacting, resulting in excessive carbon content of the slag or fly ash, and the combustion efficiency is lowered. From the industrial analysis point of coal, the sum of the components is 100%, the higher the ash content, and the lower the carbon content, then the smaller the calorific value will be.

The low calorific value of the original designed coal of the boiler is 17693kJ/kg. Calculate the boiler thermal efficiency corresponding to several typical coal types in the second type bituminous coal interval. The efficiency curve shown in Figure 1 is obtained. The abscissa is the dimensionless quantity. It is the ratio of the low calorific value of fuel coal to the low calorific value of the designed coal.

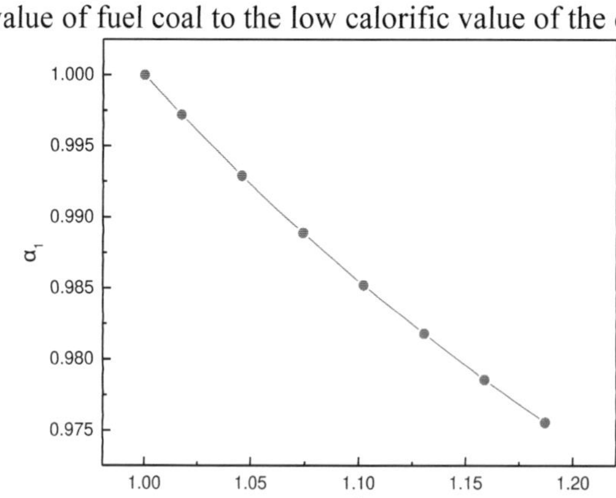

Figure1. Correction of Boiler Heat Efficiency to Boiler Thermal Efficiency

Fit the curve and get the formula (ii):

$$\alpha_1 = -7\times10^{-6}x + 1.1301 \tag{2}$$

3.2Correction of Ambient Temperature

The ambient temperature only affects the heat loss and the supply air temperature of the boiler. In the case of the same type of coal, the furnace temperature is basically the same, and after the heat is absorbed by the heating surface, part of the heat is transferred to the furnace wall, so that the furnace temperature is basically the same. Because of the combustion in the furnace, the internal heat source is formed, the temperature of the furnace wall must be higher than the ambient temperature. When the ambient temperature is high, the temperature difference between the furnace wall and the surrounding environment is small, From the convective heat transfer formula (iii), when the coefficient is constant, the smaller the temperature difference is, the smaller the heat exchange amount is, that is, the smaller the heat loss is, the greater the thermal efficiency of the boiler is.

$$q = h \times \Delta t \tag{3}$$

Of which: q - heat flux density, W / m^2;

h - convective heat transfer coefficient, $W / (m^2 \cdot ℃)$;

Δt - temperature difference, °C.

The inlet of the hot water boiler's air blower is generally set in the boiler room, and the supply air temperature is equivalent to the ambient temperature. When the ambient temperature is low, a large amount of heat is generated in the furnace due to the combustion of coal particles. The low-temperature air will absorb a large amount of heat and finally form a high-temperature flue gas, and the heating surface of the furnace will complete the entire heat exchange process; and for the high temperature environment, the air absorbs less heat, and then the combustion efficiency is increased. In order to improve boiler design efficiency, some boiler manufacturers generally set a preheater at the tail of the boiler and use high-temperature flue gas to heat the boiler air supply. This is the same principle.

Using the same method, the effect of ambient temperature change on boiler thermal efficiency is shown in *Figure 2*.

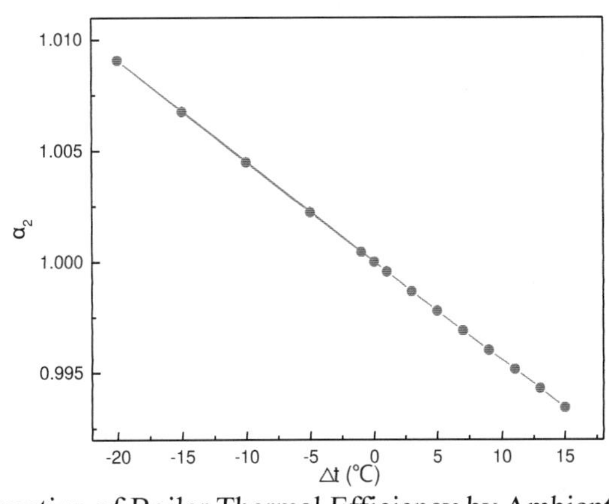

Figure2. Correction of Boiler Thermal Efficiency by Ambient Temperature

Fit the curve to get the formula (iv):

$$\alpha_2 = -0.0004x + 1 \tag{4}$$

3.3Correction of Feed Water Temperature Deviation

The *Parameters for Hot Water Boiler* (GB/T 3166-2004) gives four series of rated inlet and outlet water temperatures of 46 MW boiler, which are 115/70, 130/70, 150/90 and 180/110. For the heating pipe network, when the pressure in the local area is lower than the partial pressure of the saturated water, the saturated water will be vaporized to generate steam in a large amount, resulting in uneven distribution of heat in the pipe network, the heat balance of the pipe network cannot be achieved. There will be too much stream generated and will form a pipe network water hammer, seriously affecting the safety of the pipe network. Therefore, the current standards have been seriously divorced from reality and cannot serve as technical guidance. At present, the 46MW boiler operated by the heating company generally adopts the method of low temperature supply, that is, under the premise of ensuring the temperature difference, the effluent's temperature is lowered below the boiling point of water to ensure the safety of the pipe network.

Considering the heat exchange between the working medium side and the flue gas side, it can be known that more heat exchange amount because of larger temperature difference. Therefore, under the premise of ensuring that the heat capacity of the hot water boiler is constant, that is, the temperature difference and the flow rate are the same, refer to the operating parameters of the hot water boiler, calculate the typical boiler feed water temperature downward based on 70 °C, and calculate the boiler operating efficiency, then the *Figure 3* is obtained. The abscissa is the dimensionless amount, which is the ratio of the feed water temperature to 70 °C.

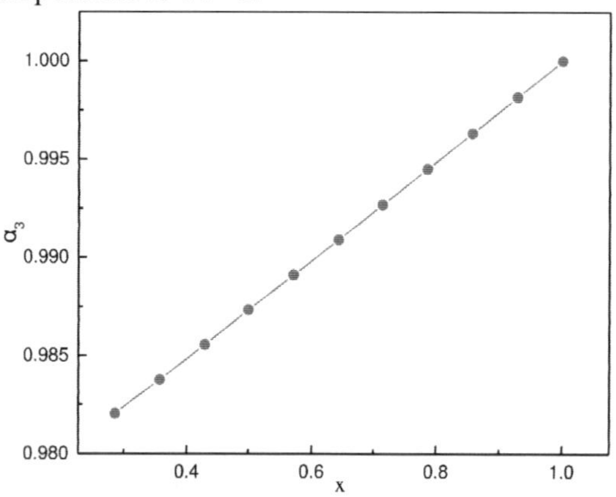

Figure3. Boiler Efficiency Correction Caused by Water Temperature Deviation

Fit and get the formula (V):

$$\alpha_3 = 0.0251x + 0.9748 \tag{5}$$

3.4Correction of Heat Output of Hot Water Boiler (Including Boiler Feed Water Temperature Difference and Flow Rate)

Figure4 shows the efficiency indexes of the boiler under different inlet and outlet water temperature and circulating water volume under the rated load. From the figure, it can be clearly seen that the efficiency of large flow rate and small temperature difference design method is larger than the large temperature difference and small flow design method.

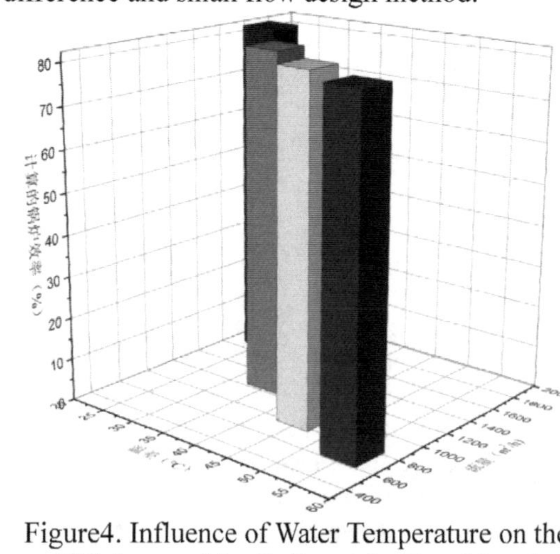

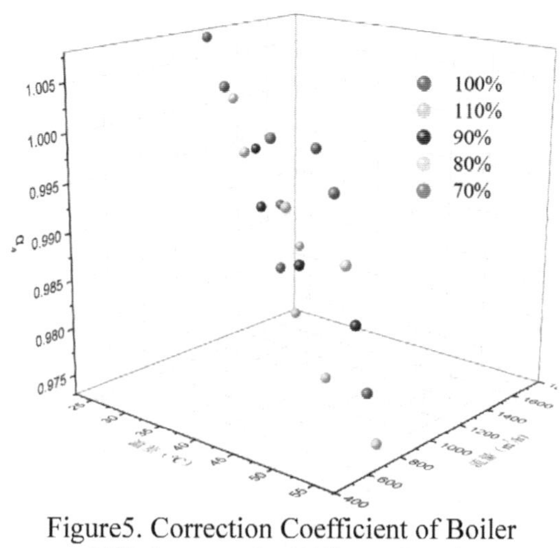

Figure4. Influence of Water Temperature on the Efficiency of the Boiler under Rated Load

Figure5. Correction Coefficient of Boiler Efficiency under Different Loads

The change in the heat output of the hot water boiler is characterized by the change of the temperature difference and flow rate of the boiler inlet and outlet water. In the design document, what given is the thermal efficiency under the rated load of the boiler and few boiler design documents can give the efficiency under non-rated load. Generally speaking, under low load, the heat output of the boiler is reduced, which determines that the heat input of the boiler is reduced, accordingly, the furnace heat load is reduced, the heat exchange effect is deteriorated, and the boiler operation efficiency is lowered. When the boiler operating load is too high, the furnace heat load is higher than the designed value, and it is limited by the arrangement and area of the heating surface, more heat is discharged out of the furnace with the flue gas without exchanging heat with the heating surface, resulting in an increase in boiler exhaust loss, and the boiler efficiency is reduced. In the design process, since the heat loss of the boiler is the selected empirical data, while in actual operation, the value increases with the decrease of the load rate, so the corresponding value calculated under the low load according to the design is relatively high. However, it can be seen from Figure 5 that without considering the heat loss, when the boiler load rate reduces, the boiler efficiency increases.

4.Conclusion

Through the analysis of several main factors affecting boiler operating efficiency, the following conclusions are obtained:

(1) When the boiler is operated in a non-involved state, the boiler design efficiency will change greatly, and it can be corrected by multiplication factor. The correction coefficient can be obtained by data fitting.

(2) The analysis of several main factors affecting boiler operating efficiency has indirectly obtained that under the premise of other conditions being unchanged, the better the calorific value of fire coal, the higher the boiler efficiency; the lower the feed water temperature, the higher the thermal efficiency of the boiler; The higher the ambient temperature, the higher the efficiency of the boiler; the thermal efficiency of the boiler increases with the decrease of the load rate without considering the heat loss.

(3) From the perspective of operation, the boiler with high temperature difference and large flow rate has high efficiency.

REFERENCES

[1] Chen X, Chen T(1991). Boiler principle. Mechanical Industry Press,Xian.
[2] Wang X(2003). Characteristic and scope of AMSE PTC6 and PTC46. Power Engineering, 1:30-34
[3] Fan Y, Lin A(2006). Discussion on calculation method of heat input of CFB boiler. Northeast Electric Power Technology,27(2):7-8
[4] Mostaghimi P , Armstrong R T , Gerami A , et al(2017). Cleat-scale characterisation of coal: An overview[J]. Journal of Natural Gas Science and Engineering, 39:143-160.

IOP Conf. Series: Materials Science and Engineering **721** (2020) 012058 doi:10.1088/1757-899X/721/1/012058

Numerical Study on Temperature Uniformity of Regenerative Heating Furnace

Hongtao Li[1], Yashan Zhao[1], Yunguang Ji[1*], Shuqi Xue[1] and Yazhi Liu[2]

[1]College of Mechanical Engineering, Hebei University of Science and Technology, Shijiazhuang, Hebei, 050018, China.

[2]Norendar International Ltd., Shijiazhuang, Hebei, 050011, China.

*Corresponding author's e-mail: lht9542@163.com

Abstract. Using CFD simulation technology, the influence of three kinds of operation modes (air change on the same side mode, cross air change mode and staged air change mode) on the temperature uniformity of regenerative heating furnace was studied. The results show that: 1) in the same side air exchange mode, the high temperature area is easy to form in the middle of the furnace, which makes the local temperature of the billet surface high; 2) in the cross air exchange mode, the high temperature area appears in the middle of the furnace, the heat exchange in the middle of the billet is strong, and the heating speed is fast; 3) in the staged air exchange mode, the high temperature area moves to both ends, and the temperature at both ends of the billet is higher than that in the middle. Therefore, in order to keep the temperature of billet constant, the best operation mode of regenerative heating furnace is alternating operation of cross air change mode and staged air change mode.

1. Introduction

The application of high temperature air combustion technology (HTAC) in the field of steel rolling heating furnace not only has better energy-saving effect, but also because the burners of the furnace are arranged on both sides of the furnace, and the alternative reversing makes the temperature in the width direction of the furnace more uniform, which is conducive to improving the uniformity of billet heating [1]. In order to get the best operation of regenerative heating furnace, scholars at home and abroad have done a lot of research. For example, Tang et al.[2] used the numerical calculation software FLUENT to study the gas temperature distribution in the heating furnace, and by optimizing the combustion mode of the burner, the injection speed and the injection angle to achieve a more uniform temperature distribution in the furnace. Ou et al.[3] studied the influence of different structure of reversing combustion on the entrainment and combustion of multi burner jet in the heating furnace. Qiu et al.[4] established a coupled heat transfer model of the furnace and the aluminum for a representative rectangular regenerative aluminum melting furnace. Based on the above research, this paper proposes a method to determine the temperature uniformity in the furnace through the temperature distribution on the billet surface, and simulates the influence of different operation modes on the temperature uniformity of the furnace.

2. Numerical simulation

2.1Physical Model

The research object is a regenerative forging furnace with a production capacity of 80 T/h. The furnace size is 4800mm×3850mm×1573mm, the billet specification is Φ550×1500, double row charging, the two sides of the regenerative burner are symmetrically arranged, the total number of burners is 10, and the reversing time is 30s. The unstructured tetrahedron and hexahedron hybrid mesh is used in the simulation, and the mesh is encrypted locally. The number of mesh nodes is 1093798. The main dimensions and section diagram are shown in Figure 1.

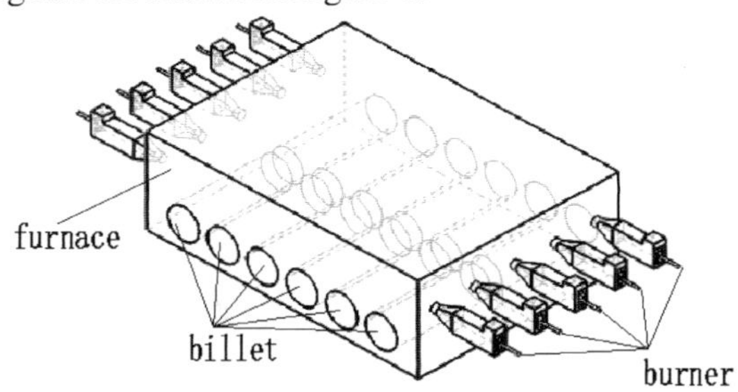

Figure 1. Layout of regenerative heating furnace and billet

In actual operation, the heating furnace is a nonlinear and complex time-varying system with many interference factors. To facilitate the calculation, the following assumptions are made for the model:

(1) The internal sealing of the heating furnace is good, there is no escape phenomenon, and there is no heat loss of the furnace wall.

(2) It is assumed that only the furnace gas is coupled to the billet for heat exchange.

(3) The influence of iron oxide scale on billet is not considered for the time being.

(4) It is assumed that the air temperature after the preheating of the regenerator can reach 1073K.

(5) It is assumed that the burner can be commutated at the same time when commutating.

2.2Mathematical formulations

2.2.1Flow and energy equations

The basic mathematical equations such as continuity, momentum, energy, turbulent kinetic energy, for gas flow and heat transfer are given by [5]:

Continuity equation:

$$\frac{\partial \rho}{\partial t}+\frac{\partial}{\partial x_i}(\rho u_i)=0 \tag{1}$$

Where: ρ is the fluid density, t is the time the fluid flows, and u_i is the velocity.

Momentum equation:

$$\frac{\partial}{\partial t}(\rho u_i)+\frac{\partial}{\partial x}(\rho u_i u_j)=F-\frac{\partial P}{\partial x_i}+\frac{\partial \tau_{ij}}{\partial x_j} \tag{2}$$

Where: p is the static pressure of the control body, F is the volume force, μ is the dynamic viscosity, τ_{ij} is the stress tensor.

Stress tensor is given by:

$$\tau_{ij}=[\mu(\frac{\partial u_i}{\partial x_i}+\frac{\partial u_j}{\partial x_i})]-\frac{2}{3}\mu\frac{\partial u_l}{\partial x_l}\delta_{ij} \tag{3}$$

Whereδij is the molecular viscosity

Energy equation:

$$\frac{\partial}{\partial t}(\rho E)+\frac{\partial}{\partial x_i}(u_i(\rho E+P))=\frac{\partial}{\partial x_i}(k_{eff}\frac{\partial T}{\partial x_i})+S_h \tag{4}$$

Where: k_{eff} is the heat conduction coefficient; T is the thermodynamic temperature; E is the energy, and S_h is user-defined.

2.2.2Turbulent combustion model

The k-ε model is generally applied to solve the turbulence, which shows a good agreement with the experimental observations. The k equation and the ε equation are expressed as follows:

Turbulent kinetic energy equation k:

$$\frac{\partial(\rho k)}{\partial \mathrm{t}}+\frac{\partial}{\partial x_i}(\rho u_i k)=[(\mu+\frac{\mu_t}{\sigma_k})\frac{\partial k}{\partial x_j})-\rho\varepsilon+G_k+G_b+S_k \tag{5}$$

Turbulent dissipation equation ε:

$$\frac{\partial(\rho\varepsilon)}{\partial \mathrm{t}}+\frac{\partial}{\partial x_i}(\rho u_i\varepsilon)=\frac{\partial}{\partial x_j}[(\mu+\frac{\mu_t}{\sigma_\varepsilon})\frac{\partial\varepsilon}{\partial x_j})]-\rho C_2\frac{\varepsilon^2}{k}+C_1\frac{\varepsilon}{k}(G_k+C_3G_b)+S \tag{6}$$

C_1, C_2, C_3 are constant, S_k and S are user-defined, Prandtl number is average turbulent flow, andμ_t is turbulent viscosity. The G_k formula is as follows:

$$G_k=-\rho\overline{u_i'u_j'}\frac{\partial T}{\partial x_i} \tag{7}$$

The G_b formula is as follows:

$$G_b=-\frac{1}{\rho}(\frac{\partial\rho}{\partial T})_P g_i\frac{\mu_t}{\mathrm{Pr}_t}\frac{\partial T}{\partial x_i} \tag{8}$$

Where: g_i is gravitational vector, P is fluid density and T is thermodynamic temperature.

2.3Burner operation mode

In order to analyze the influence of heating mode on the uniformity of temperature field in the furnace, three typical operating conditions was studied, which are air change on the same side mode, cross air change mode and staged air change mode, and the nozzle reversing cycle is 60s. The layout of specific nozzle operation mode is shown in Figure 2.

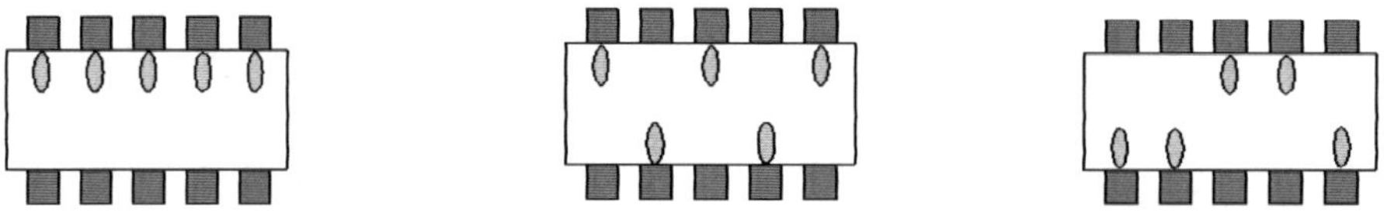

Air change on the same side mode Cross air change mode Staged air change mode

Figure 2. Schematic diagram of burner operation mode

2.4Boundary and initial conditions

The heating furnace uses natural gas as fuel, the calorific value is 8300 × 4.18 kJ/m^3, and the oxidant is air. The inlet boundary is velocity inlet, the velocity flow of fuel inlet is 36m^3/h, the inlet pressure is 6kPa, the excess air coefficient is 1.13, and the air inlet pressure is 2kPa. The outlet boundary is the pressure outlet and the value is 0. The billet is treated according to the boundary condition of heat conduction solid and there is no heat source inside. The surface blackness and furnace wall blackness of billet are taken as 0.8. The thermal conductivity of 7800kg / m3 manganese steel is calculated according to the thermal conductivity solid, and the thermal conductivity and specific heat are set

IOP Conf. Series: Materials Science and Engineering **721** (2020) 012058 doi:10.1088/1757-899X/721/1/012058

according to the reference [6]. During initialization, the air preheating temperature is 1073k, and the initial billet temperature is 1073k.

3. Results and analysis

3.1Uniformity analysis of furnace

Fig. 3 is a cloud chart of three operation modes at the time of single commutation cycle. It is divided into three groups: temperature field in the furnace, velocity field in the furnace and temperature field on the outer surface of billet. There are three operation modes corresponding to each group from left to right, which are air exchange mode on the same side, cross air exchange mode and staged air exchange mode. It can be seen from Fig. 3 (a) and (b) that at the time of single reversing cycle, the peak value of flame temperature in the same side air exchange mode is low, because the jet direction is in the same direction, the generated air flow is orderly, and there are less vortices between the air flows, forming a low-pressure area in the furnace interior, and the flame air flow is inclined to the inner part, so that a large vortices are generated in the furnace center, resulting in the length of the intermediate flame size. On the contrary, the jet direction of cross air exchange mode is staggered, and the jet direction between adjacent jets is opposite, and there will be vortex between each two jets. After that, the original

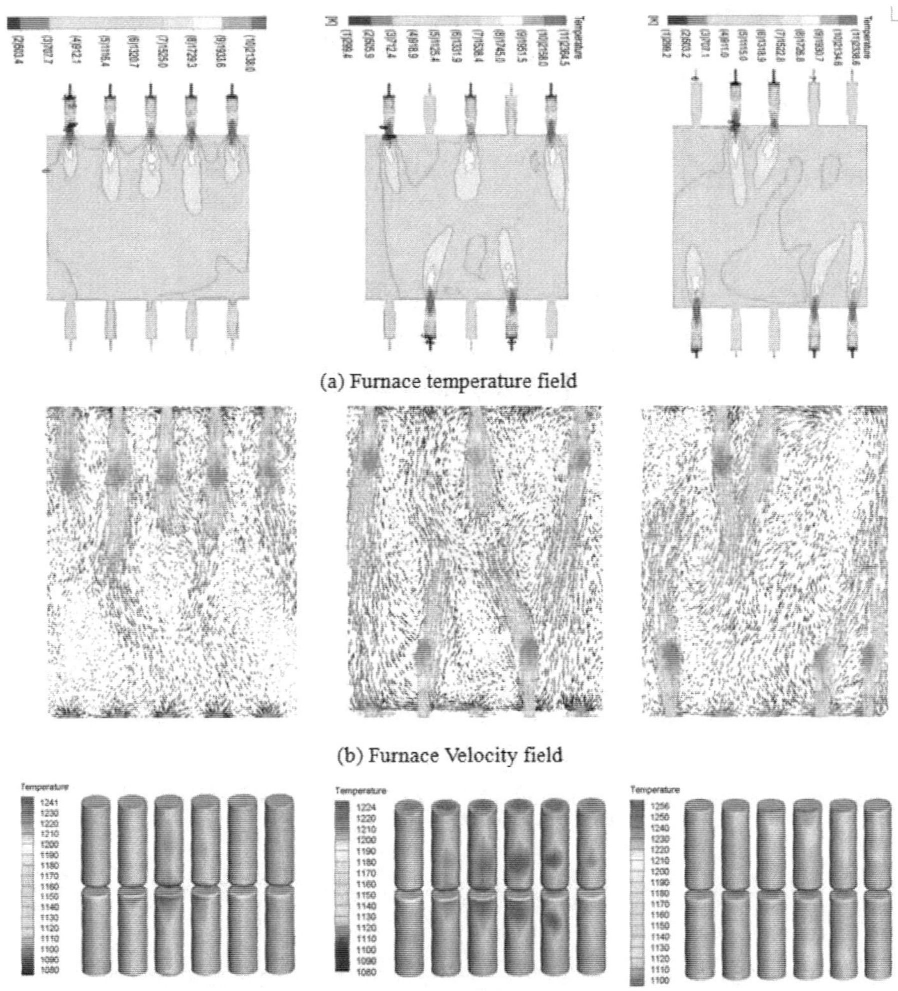

(a) Furnace temperature field

(b) Furnace Velocity field

(C) Billet temperature field

Figure 3. Cloud chart of three operation modes in single commutation period vortex will be broken up, resulting in vortex in the opposite direction. The vortex increases the degree of disorder in the furnace, making the combustion more complete, the peak temperature higher, the range of high temperature area larger and mainly located in the low-pressure area in the middle of the furnace. There are both the same direction jet and the reverse jet in the staged air exchange mode. The multi direction jet and the multi direction jet interact with each other, resulting in irregular size vortex. The air flow deviates seriously, and the high temperature area deviates to both ends.

According to the analysis of furnace uniformity based on the temperature field on the outer surface of billet, it can be seen from Fig. 3 (c) that a stable and concentrated high-temperature area is formed in the furnace central area of the same side air exchange mode, and the local temperature of billet in the furnace central area is high; the high-temperature area of the cross air exchange mode is enlarged, and the billet is heated and concentrated along the depth of furnace; the air flow of the staged air exchange mode is severely skewed, and the high-temperature area is inclined to both ends.

3.2Uniformity analysis of billet heating

In order to study the uniformity of billet heating, a coordinate system is established based on the axial center position of furnace and the end face of the third row of billet. The specific location of the origin is shown in Figure 4 (a). Fig. 4 (b) and (c) show the distribution of the axial surface temperature of billet and the axial surface temperature of furnace in three heating modes respectively. It can be seen from Fig. 4 that in the same side air exchange mode, the axial surface temperature of billet changes greatly, the central heat load of furnace is high, and the uniformity of heating is poor; in the cross air exchange mode, the surface temperature of billet along the axial direction of furnace is high and even, but the axial surface temperature of billet changes obviously. On the side deviating from the axial direction of furnace, the surface temperature of billet is low; in the graded air exchange mode, the axial steel along the furnace is low. The surface temperature of billet is high and even, but the temperature is low. The axial surface temperature of billet tends to increase away from the axial side of furnace.

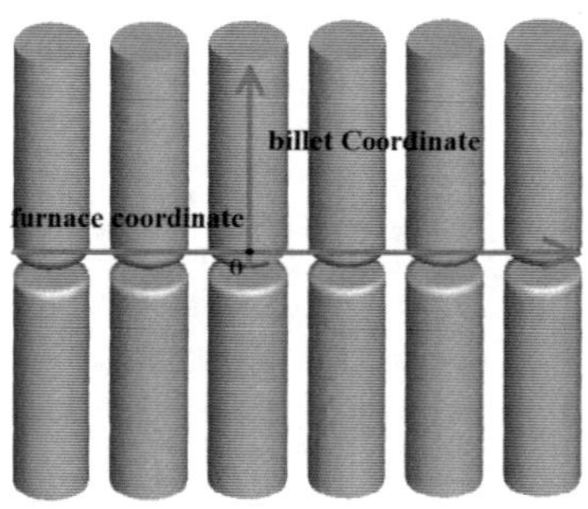

(a) Coordinate system diagram

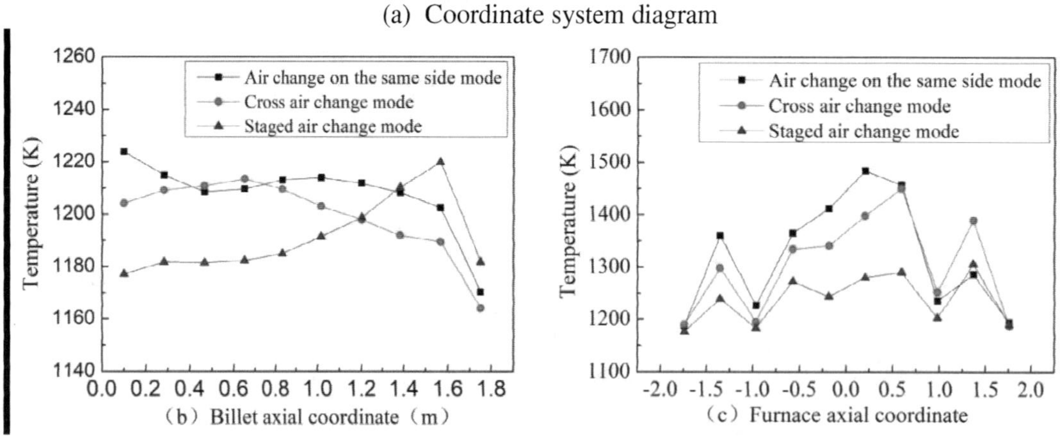

Figure 4. Surface temperature distribution of billet

4. Conclusion

The thermal information in the heating furnace which is difficult to get by common experimental methods can be obtained by numerical simulation, which provides reference for improving the design and operation of HTAC heating furnace and realizing the optimal control of heating furnace. Through numerical simulation, the following conclusions are obtained.

1) At the time of single reversing cycle, the peak value of flame temperature in the same side air exchange mode is lower than that of cross air change mode and staged air change mode.

2) In the same side air exchange mode, the axial surface temperature of billet changes greatly, the central heat load of furnace is high, and the uniformity of heating is poor; in the cross air exchange mode, the surface temperature of billet along the axial direction of furnace is high and even, but the axial surface temperature of billet changes obviously.

3) In order to keep the temperature of billet constant, the best operation mode of regenerative heating furnace is alternating operation of cross air change mode and staged air change mode.

Acknowledgement

The authors acknowledge the financial support from Key R&D Program Projects in Hebei Province (No. 18274006D), Science and Technology and Development Plan Project of Shijiazhuang (No. 181010213A).

References

[1] Geng, X., Niu, H.H., Chen, J. (2010) Design and Application of Combustion Optimal Control System for Rolling Heating Furnace. In: 2018 meeting of metallurgical branch of China Metrology Association. Ma Anshan. pp. 91-93.

[2] Tang, X.Q. (2013) Optimum Study on Temperature Uniformity of Heating Furnace. Northeastern University.

[3] Ou, J.P., Ma, A.C., Zhan,S,H. (2007) Numerical Simulation of Influence of Reversing Mode on Thermal Process of Regenerative Heating Furnace. Journal of Wuhan University of Technology., 04:68-71.

[4] Qiu, L., Feng, Y., Chen, Z. (2018) Numerical simulation and optimization of the melting process for the regenerative aluminum melting furnace. Applied Thermal Engineering., 145: 315-327.

[5] FLUENT 6.3 User's Guide, Fluent Inc., Lebanon, USA, 2006.

[6] Cui, J.L., Zhu, T. Numerical Simulation Study on Heat Transfer Process and Oxidation Burning of Billet in Experimental Furnace. Energy Saving Technology., 31(02):99-103+107.

IOP Conf. Series: Materials Science and Engineering **721** (2020) 012059 doi:10.1088/1757-899X/721/1/012059

Economic Benefit Evaluation of Waste Heat Recovery in Coal-fired CHP System Based on Exergoeconomic Analysis

Zaihai Yu[1*], Jintong Gao[2]

[1]Liaoning Academy of Safety Science, Shenyang, Liaoning, 110004, China

[2]Institute of Refrigeration and Cryogenics, Shanghai Jiao Tong University, Shanghai, 200240, China

[*]Corresponding author's e-mail: yzh25998@126.com

Abstract: As one of the main energy-saving measures and an important part of the power industry, CHP (combined heat and power) has achieved rapid development in recent years, which have the advantages of low heat loss, flexible operation mode, and cascade utilization of energy compared with heating boilers and traditional power plants. However, there is a lack of reasonable evaluation of CHP systems, due to the difference in energy quality of heat and electricity. It is urgent to establish considerable and reasonable evaluation indexes of CHP systems. In addition, in the actual operation of CHP plant, a large amount of waste heat is released into the environment, and there is still room for further improvement in the waste heat utilization. In this paper, an exergoeconomic analysis model of CHP system with circulating fluidized bed boiler is established, and the economic benefits of waste heat recovery in CHP systems are studied. After the application of waste heat recovery, the system exergy efficiency increased from 32.94% to 34.14%. The unit exergoeconomic cost of electricity and process steam did not change much, but the unit exergoeconomic cost of heating decreased by 17.1%, which indicates the economic benefits of waste heat recovery.

1. Introduction

With the in-depth development of haze control in China, CHP has gradually become one of the important measures to replace coal-fired boilers and solve the problem of loose coal pollution, and has been vigorously promoted by all sides. Due to the coal-based energy structure in China, the development trend of coal-fired CHP (combined heat and power) plant was accelerated in recent years, and the unit capacity has been increasing, which will be the main way of central heating in northern China in the future. Although the thermal efficiency of CHP is significantly improved compared to traditional power plants, in the operation of CHP plants, a large amount of waste heat is released into the environment, and there is still room for further improvement in the overall energy efficiency. The waste heat resources generated during the operation of CHP plants mainly include flue gases, boiler bottom ash, and exhaust steam, etc. The flue gases released from the boiler is generally between 120 °C - 150 °C. When the gas temperature is increased by 10 °C, the increased heat loss is close to 1%. Due to the ash content of coal and the insufficient combustion, there is a large amount of unused waste heat in the high temperature ash discharged from the boiler with high-grade temperature. For example, circulating fluidized bed boiler generally burns the lignite with poor quality and high ash content, and discharges large amount of ash. The hot ash discharged from the bottom of the boiler is usually around 950 °C. In addition, the heat of exhaust steam from the steam turbine in power plant accounts for about 35% - 60% [1] of the total input energy. This part of heat is finally discharged into the environment through the cooling tower.

The grade of the waste heat carried by the circulating cooling water of the cooling tower is low, but the total amount is large, and it can be extracted by heat pumps to preheat the return water and reduce steam extraction.

Due to the difference in energy quality of heat and electricity, the reasonable evaluation of CHP systems has become a difficult point, and it is urgent to establish a reasonable evaluation index of CHP systems. In addition, the improvement of overall exergy efficiency through waste heat recovery is based on the increase of investment, and it is necessary to consider both energy conversion efficiency and investment cost. Therefore, this paper conducts a comprehensive analysis of waste heat recovery of coal-fired CHP system based on exergoeconomic method, and provides a theoretical basis for the energy-saving reconstruction of power plants.

Table 1. Waste heat resources.

Resources	Initial temperature	Final temperature	Specific heat capacity	Mass flow rate	Heat flow
	°C	°C	kJ/ (kg K)	t/h	MW
Bottom ash	890	150	1.0421	40.37	8.65
Flue gas	137	97	0.8556	1335.86	12.70
Cooling water	37.27	28	-	12000	129.13

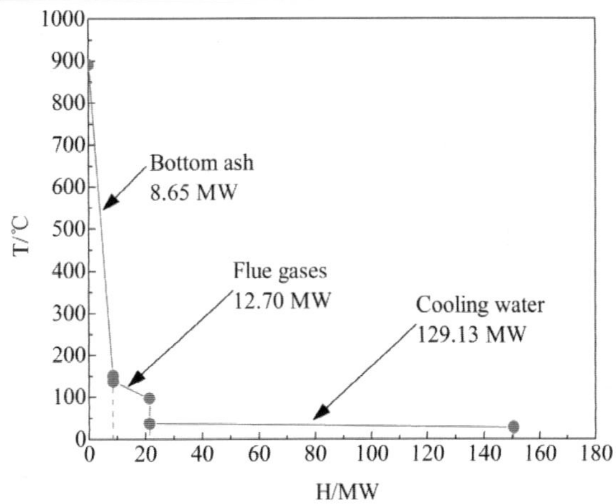

Figure 1. Waste heat profile for the CHP system.

2. Waste Heat Resources and Potential of CHP systems

The summary of waste heat resources of the case plant is shown in Table 1. The heat flows of the three waste heat resources are 8.65 MW (boiler bottom ash), 12.7 MW (boiler flue gas) and 129.13 MW (circulating cooling water), respectively. In this case, the released flue gases are generally below 140 °C. Except the corrosion caused by the acidic compounds in the flue gases, the cold source temperature must be below the flue gas dew point temperature, so that the latent heat of the flue gas can be recovered. When the moisture content of the flue gas is high, the dew point temperature is very low (for example, when the humidity is 0.11 kg/kg, the dew point temperature is only 57.5 ° C), and the initial temperature of the cold source (such as heating return water) is generally above 50 ° C. In this case, there is little space for heat recovery. The waste heat profile [2, 3] is obtained by plotting the data in Table 1. As shown in Figure 1, the horizontal axis represents the heat flows and the vertical axis represents the corresponding temperature range. It can be seen from the figure that the temperature of the boiler bottom ash is the highest among the three waste heat resources. Generally, the discharged ash of the circulating fluidized bed boiler can reach about 900 °C, which can be cooled to below 150 °C by the ash cooler. The outlet temperature of circulating cooling water from condenser is about 38 ° C, then it enters the cooling tower and is cooled to 27 ° C. About 129 MW of low temperature waste heat is released into the environment in this process, which is the largest among the three waste heat resources.

3. Waste Heat Recovery System Models of CHP Systems

The structure of the waste heat recovery in CHP system is shown in Figure 2, which uses rolling-cylinder ash cooler and absorption heat pump to recover the waste heat of the boiler bottom ash and exhaust steam, respectively It is composed of 24 equipments: Boiler, high pressure cylinder (HPT), intermediate pressure cylinder (IPT), low pressure cylinder (LPT), boiler feed pump turbine (BFPT), condensate pump (CONP), gland condenser heater (GC), low pressure heater (LPH), deaerator (DEA), feed water pump (FWP), high pressure heater (HPH), steam user, absorption heat pump (AHP), steam to water heat exchanger (SWHE), rolling-cylinder ash cooler (RAC), generator (GEN), condenser (CON) and two dissipative units (Stack). The case plant has a 1025 t/h circulating fluidized bed boiler and a supporting 300 MW unit. The circulating fluidized bed boiler has a large amount of ash discharge with high temperature. The ash discharge at high temperature is not easy to handle and transport. Therefore, it is necessary to use the ash cooler to cool the hot ash, and the waste heat of the bottom ash can be recovered to increase the temperature of the feed water. In addition, the heating extraction temperature is relatively high (246.7 ° C, 0.39 MPa), while the temperature of the return water from heating network is only 56.2 ° C. There is a serious energy level mismatch, and the exergy loss is great during the heat exchange process, so it is necessary to preheat the return water to reduce the heat transfer temperature difference of the heat exchanger. A part of the heating steam is extracted to drive the absorption heat pump, and the heat of the circulating cooling water can be recovered to produce hot water at about 86 ° C. In this case, the purpose of preheating the return water and reducing the heat transfer temperature difference of the heat exchanger is achieved.

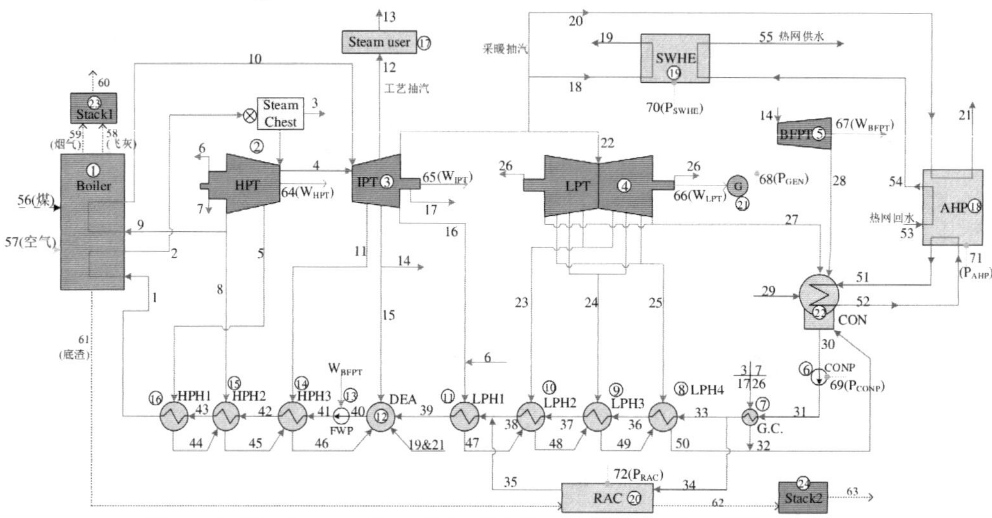

Figure 2. The integration of the RAC and AHP in the CHP system.

3.1 Rolling-cylinder Ash Cooler (RAC)

Four sets of GTL15C-20*L rolling-cylinder ash coolers are used, which are composed of inner cylinder and outer cylinder. The single bottom ash processing capacity is 25 t/h, with a 15 kW motor. After the bottom ash of the boiler enters the ash cooler from the inlet, the spiral blade inside the cylinder is pushed towards the outlet. The bottom ash exchanges heat with the boiler feed water arranged in the counter-flow in the outer cylinder during the moving process, thereby achieving the purpose of reducing the ash temperature and recovering the waste heat.

3.2 Absorption Heat Pump (AHP)

The circulating cooling water in the case plant has low temperature, large flow, and carries a large amount of low-grade waste heat, which can be recovered by the AHP to preheat return water In this case, the steam extraction from the turbine can be reduced. This part of steam can continue to work in the steam turbine to increase the system power generation and improve the overall energy efficiency [4, 5].

The basic principle of the single-effect absorption heat pump is shown in Figure 3. It mainly includes five heat exchange units (generator, absorber, condenser, evaporator and solution heat exchanger, solution pump and throttle valve). In the case plant, a part of the heating steam is used as the heat source of the absorption heat pump, which absorbs the waste heat of circulating cooling water. The pressures of the evaporator and the condenser were 3.4 kPa and 53.4 kPa. The heat load of the generator is 99.09 MW, and the generation temperature is 141.8 °C. The LiBr solution (4) is heated in the generator to generate refrigerant vapor (8). The concentration of the remaining solution (7) is increased to 62.54%. After exchanging heat with the diluted solution (3), it enters the absorber to absorb the refrigerant vapor (1) from the evaporator, and the concentration is reduced to 58.61% with heat output. The released heat is used for the heating of the return water from heating network. The condensed water (9) is depressurized and then enters the evaporator. It is heated by the circulating cooling water to a saturated steam state (1), and then enters the absorber to complete the cycle. After being heated in the absorber and condenser, the temperature of the return water is raised from 56.2 °C to 86.16 °C, and then entered into the heat exchanger to be heated to the required temperature. The total investment cost of the heat pump system in the case plant is 78.2 million RMB.

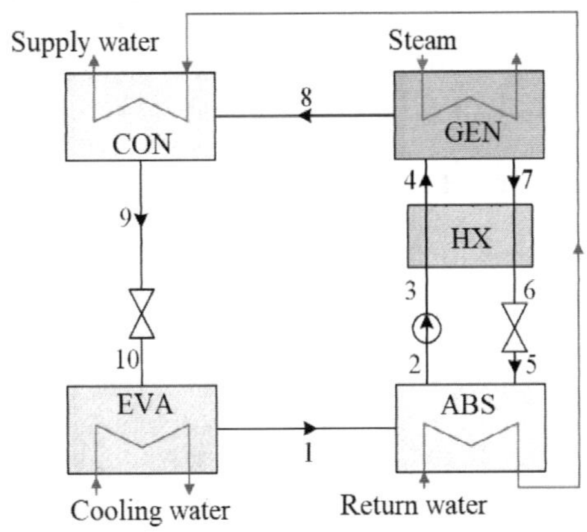

Figure 3. The schematic diagram of single-effect absorption heat pump.

4. Exergoeconomic Analysis

4.1 Exergy Cost Formation

In order to allocate the waste heat cost to the final product, it is necessary to determine the source of the waste heat. After the fuel products of each equipment and the production structure of the system are determined, the external resources consumed in the production process and the cost of the waste generated are distributed to the final product according to the production structure. According to the literature [6-9], the exergy cost of the equipment product is equal to the exergy cost of the fuel plus the allocated waste cost, as shown in equation (1).

$$C_{P,i} = C_{F,i} + C_{R,i} \tag{1}$$

Where, $C_{P,i}$, $C_{F,i}$, and $C_{R,i}$ represent the product cost, fuel cost, and allocated waste cost of equipment i. The fuel cost can be divided into two parts, one is from the resources outside the system boundary, and the other is the product of other equipment (equipment j), which is represented by $C_{e,i}$ and $C_{j,i}$ in equation (2).

$$C_{F,i} = C_{e,i} + \sum_j C_{j,i} = C_{e,i} + \sum_j \frac{E_{j,i}}{P_j} C_{P,j} = C_{e,i} + \sum_j \kappa_{j,i} C_{P,j} \tag{2}$$

The allocation coefficient of the waste cost is defined as $\psi_{i,r}$, which means the percentage of waste cost generated by the dissipative equipment r in the total cost of equipment r, and the waste cost of equipment i is:

$$C_{R,i} = \sum_r C_{r,i} = \sum_r \psi_{i,r} C_{P,r} \tag{3}$$

Regarding the determination of the waste cost's allocation coefficient $\psi_{i,r}$, the method in Ref. [7] is used in this paper. Then, the equations (2) and (3) are substituted into the equation (1) to obtain the exergy cost balance equation of the device *i*.

$$C_{P,i} - \sum_j \kappa_{j,i} C_{P,j} - \sum_r \psi_{i,r} C_{P,r} = C_{e,i} \quad (4)$$

4.2 Exergy Cost Calculation in Matrix Mode

Applying the exergy balance equation from the previous section to each device in the system yields a series of linear equations that can be represented in matrix form, as shown in equations (5) and (6),

$$(\mathrm{U_D} - \langle \mathrm{KP} \rangle - \langle \mathrm{KR} \rangle \mathrm{C_P}) = \mathrm{C_e} \quad (5)$$

$$\mathrm{C_P} = (\mathrm{U_D} - \langle \mathrm{KP} \rangle)^{-1} (\mathrm{C_e} + \langle \mathrm{KR} \rangle \mathrm{C_P}) \quad (6)$$

where $\mathbf{U_D}$ is a unit matrix, and both $\langle \mathbf{KP} \rangle$ and $\langle \mathbf{KR} \rangle$ are square matrices, the elements of which are the product cost allocation coefficient $\kappa_{j,i}$ and the waste cost allocation coefficient $\psi_{i,r}$. Then, the operator $\mathbf{C_R} = \langle \mathbf{KR} \rangle \mathbf{C_P}$ and $\langle \mathbf{P}^* | = (\mathbf{U_D} - \langle \mathbf{KR} \rangle)^{-1}$ are defined to simplify the equation and get the equation (7).

$$\mathrm{C_P} = \langle \mathrm{P}^* | (\mathrm{C_e} + \mathrm{C_R}) = \mathrm{C_{P,e}} + \mathrm{C_{P,R}} \quad (7)$$

It can be seen from the above formula that the product exergy cost in the system is divided into two parts: external resource exergy cost ($\mathbf{C_{P,e}}$)and waste cost ($\mathbf{C_{P,R}}$). Before establishing the economic analysis model, it is necessary to obtain the investment cost of the main equipment according to the cost estimation method [10], and to annualize it, that is, to equalize the annual service life of the asset to the annual amount, as shown in equation (8) and (9), where Z_i is the annualized cost of equipment *i*; *CRF* is the capital recovery coefficient, calculated by equipment life *N* and interest rate *IR*; φ is the system maintenance factor; *num* is the annual operation hours.

$$Z_i = IC_i \cdot CRF \cdot \varphi / (N \cdot num) \quad (8)$$

$$CRF = \frac{IR(1+IR)^N}{(1+IR)^N - 1} \quad (9)$$

The equipment annualized cost vector $\mathbf{Z}$ and the resource consumption cost vector $\mathbf{C_e^*}$ are brought into equation (7), and the exergoeconomic cost balance equation is shown in equation (10), where $\mathbf{C_P^*}$ is a vector, representing the exergoeconomic cost of each system equipment product; $\mathbf{C_e^*}$ represents the unit cost of external resources; $\mathbf{C_R^*}$ represents the unit cost unit of waste. In this paper, the flue gases treatment cost is included in the economic analysis category, which is included in equation (11)[11,12].

$$\mathbf{C_P^*} = \langle \mathbf{P}^* | (\mathbf{C_e^*} + \mathbf{C_R^*} + \mathbf{Z}) = \mathbf{C_{P,e}^*} + \mathbf{C_{P,R}^*} + \mathbf{C_{P,Z}^*} \quad (10)$$

$$\mathbf{C_P^*} = \langle \mathbf{P}^* | (\mathbf{C_e^*} + \mathbf{C_R^*} + \mathbf{Z} + \mathbf{G}) = \mathbf{C_{P,e}^*} + \mathbf{C_{P,R}^*} + \mathbf{C_{P,Z}^*} + \mathbf{C_{P,G}^*} \quad (11)$$

5. Exergoeconomic Analysis Results

The exergoeconomic analysis results of the case plant with and without waste heat recovery are shown in Table 2. The second column in the table is the equipment investment cost (Z_i), and the unit exergy cost of fuel and product are respectively represented by $c^*_{F,i}$ and $c^*_{P,i}$. f_i and r_i represent the exergoeconomic factors and relative cost differences[13]. It can be seen from the data that the relative cost difference of the boiler is the largest and has the greatest potential for improvement. In the case of no waste heat recovery, the relative cost difference r_i of the steam-water heat exchanger (SWHE) is 1.0871, which is only second to the boiler's 1.4986. In the waste heat recovery case, the relative cost difference of SWHE is only 0.6774. In the case of that the investment cost is unchanged, this shows that the cost increase caused by the irreversibility inside the equipment is reduced, that is, the cost of the equipment to produce the same product (the irreversibility in the production process) is reduced. The reason is that the return water is preheated by the absorption heat pump and then enters the heat exchanger In this case, the heat exchange temperature difference and exergy dissipation are reduced, and the performance of the heat exchanger is significantly improved. In addition, after the application of the ash cooler, the performance of LPH4 is also improved.

Table 2. Exergoeconomic analysis results.

	Z_i	Base case				Recovery case			
		$c^*_{F,i}$	$c^*_{P,i}$	f_i	r_i	$c^*_{F,i}$	$c^*_{P,i}$	f_i	r_i
	RMB /h	RMB /kWh	RMB /kWh			RMB /kWh	RMB /kWh		
Boiler	242.22	0.0746	0.1862	0.0067	1.4986	0.0746	0.1802	0.0067	1.4245
HPT	60.85	0.1967	0.2304	0.0202	0.1708	0.1928	0.2258	0.0206	0.1709
IPT	84.47	0.1967	0.2159	0.0379	0.0982	0.1928	0.2113	0.0387	0.0983
LPT	17.20	0.1967	0.3255	0.0069	0.6552	0.1928	0.3202	0.0044	0.6634
BFT	9.74	0.1967	0.2436	0.0284	0.2387	0.1928	0.2363	0.0305	0.2273
CONP	0.17	0.2337	0.3142	0.0070	0.3436	0.2337	0.3129	0.0054	0.3410
GC	0.34	0.1967	0.3030	0.0075	0.5398	0.1928	0.2918	0.0065	0.5142
LPH4	0.91	0.1967	0.2462	0.0485	0.2528	0.1928	0.2179	0.0748	0.1319
LPH3	0.91	0.1967	0.2528	0.0242	0.2847	0.1928	0.2542	0.0204	0.3198
LPH2	0.91	0.1967	0.2654	0.0109	0.3473	0.1928	0.2588	0.0101	0.3433
LPH1	0.91	0.1967	0.2872	0.0016	0.4580	0.1928	0.2740	0.0015	0.4229
DEA	9.28	0.1967	0.2205	0.0171	0.1210	0.1928	0.2152	0.0173	0.1168
FWP	27.78	0.2436	0.3228	0.0649	0.3235	0.2363	0.3090	0.0676	0.3062
HPH3	1.14	0.1967	0.2377	0.0032	0.2085	0.1928	0.2297	0.0033	0.1927
HPH2	1.14	0.1967	0.2410	0.0016	0.2246	0.1928	0.2330	0.0016	0.2094
HPH1	1.14	0.1967	0.2548	0.0013	0.2961	0.1928	0.2469	0.0013	0.2816
User	0	0.1967	0.1967	0	0	0.1928	0.1928	0	0
AHP	60.90	-	-	-	-	0.2403	0.3486	0.0258	0.4507
SWHE	0	0.1967	0.4106	0	1.0871	0.1928	0.3235	0	0.6774
RAC	1.35	-	-	-	-	0.1809	0.8992	0.0017	3.9667
GEN	36.99	0.2311	0.2337	0.0666	0.0108	0.2311	0.2337	0.0637	0.0108
CON	16.07	0.1961	0.4549	0.0200	1.3164	0.1921	0.6734	0.0107	2.5000
Stack1	0	0.1862	0.1862	-	-	0.1802	0.1802	-	-
Stack2	0	-	-	-	-	0.8992	0.8992	-	-

The comparison ofexergy efficiency and unit exergoeconomic cost is shown in Figure 4. The heat output of the CHP plant includes process steam and hot water. The steam turbine extraction is directly supplied to the process user. The hot water is heated by the steam extracted from the steam turbine. In addition, in the waste heat recovery case, the absorption heat pump also provides a part of the heating load, so the average cost of the hot water is calculated by the exergy value, as shown in equation (12).

$$c^*_{hw} = (c^*_{P,AHP} \cdot E_{AHP} + c^*_{P,SWHE} \cdot E_{SWHE})/(E_{AHP} + E_{SWHE}) \quad (12)$$

After the application of waste heat recovery, the overall exergy efficiency of the system increased from 32.94% to 34.14%. The process steam cost decreased slightly from 0.20 RMB/kWh fell to 0.19 RMB/kWh, and the cost of hot water decreased greatly, from 0.41 RMB/kWh to 0.34 RMB/kWh. The economic model includes fuel cost, waste cost, flue gas treatment cost, and operation cost including waste heat recovery equipment. Therefore, the economic benefits of waste heat recovery can be indicated by the unit exergoeconomic cost comparison.

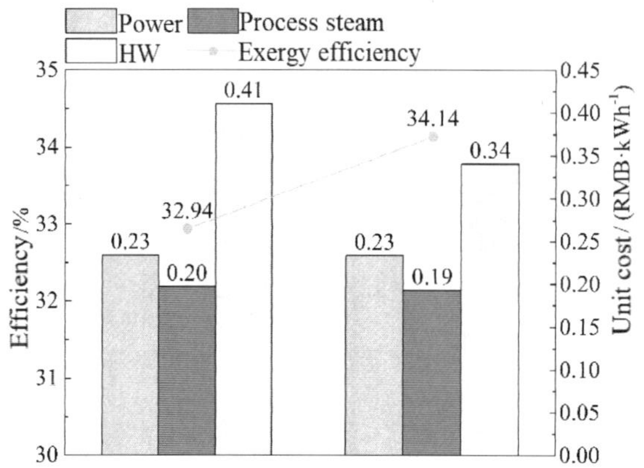

Figure 4. The comparison of the exergy efficiency and product cost in the two cases.

6. Conclusion

In this paper, models of waste heat recovery based on rolling-cylinder ash cooler and absorption heat pump are established, and the economic benefits of waste heat recovery are studied through exergoeconomic analysis. Among them, the rolling-cylinder ash cooler is used to recover the waste heat of the bottom ash. The absorption heat pump is driven by the steam turbine extraction steam to recover the waste heat of the circulating cooling water and preheat the return water. The analyzing result of the base case (no waste heat recovery) and the waste heat recovery case (applying the rolling-cylinder ash cooler and the absorption heat pump) shows that the system exergy efficiency is increased from 32.94% to 34.14%, and the performance of the heat network heat exchanger has been significantly improved. The unit cost of hot water has dropped from 0.41 RMB/kWh to 0.34 RMB/kWh, of which a decrease of 17.1%. As indicated by the analysis result, further optimization of heat integration is the improvement direction of the CHP system.

References

[1] Rattner A.S., Garimella S. (2011) Energy harvesting, reuse and upgrade to reduce primary energy usage in the USA. Energy, 36(10): 6172–83.

[2] Linnhoff B., Hindmarsh E. (1983) The pinch design method for heat exchanger networks. Chemical Engineering Science, 38(5): 745-763.

[3] Dhole V.R., Linnhoff B. (1993) Total site targets for fuel co-generation, emissions, and cooling. Computers & Chemical Engineering, 17(1): S101–S109.

[4] Hu T.L., Xie X.U., Jiang Y. (2017) Simulation research on a variable-lift absorption cycle and its application in waste heat recovery of combined heat and power system. Energy, 140: 912-921.

[5] Xu Z.Y., Mao H.C., Liu D.S., et al. (2018) Waste heat recovery of power plant with large scale serial absorption heat pumps. Energy, 165: 1097-1105.

[6] Torres C., Valero A., Rangel V., et al. (2008) On the cost formation process of the residues. Energy, 33(2): 144-15219.

[7] Agudelo A., Valero A., Torres C. (2012) Allocation of waste cost in thermoeconomic analysis. Energy, 45(1): 634-643.

[8] Luo X., Hu J., Zhao J., et al. (2014) Improved exergoeconomic analysis of a retrofitted natural gas-based cogeneration system. Energy, 72(1): 459–75.

[9] Torres C., Valero A., Perez E. (2007) Guidelines to developing software for thermos-economic analysis of energy systems. Part II: application to thermoeconomic deagnosis. the 20th International Conference on Efficiency, Cost, Optimization, Simulation and Environmental Impact of Energy Systems (ECOS).

[10] Ghaebi H., Saidi M.H., Ahmadi P. (2012) Exergoeconomic optimization of a trigeneration system for heating, cooling and power production purpose based on TRR method and using

evolutionary algorithm. Applied Thermal Engineering, 36: 113-125.

[11] Gao J.T., Zhang Q., Wang X.Z., et al. (2018) Exergy and exergoeconomic analyses with modeling for CO_2 allocation of coal-fired CHP plants. Energy, 152: (562-575).

[12] Zhang Q., Gao J.T., Wang Y., et al. (2018) Exergy-based analysis combined with LCA for waste heat recovery in coal-fired CHP plants. Energy, 169: (247-262).

[13] Boyaghchi F.A., Asgari S. (2017) A comparative study on exergetic, exergoeconomic and exergoenvironmental assessments of two internal auto-cascade refrigeration cycles. Applied Thermal Engineering, 122(25): 723-737.

IOP Conf. Series: Materials Science and Engineering **721** (2020) 012060 doi:10.1088/1757-899X/721/1/012060

Study on Calculation of Flue Gas Temperature at the Exit of Heating Furnace of Oil and Gas Field

Liu Chen [1*], Wang Lin [1] and Jin Bing[1]

[1] Boiler business department, Liaoning Academy of safety Sciences, Shenyang, Liaoning province, 110004, China

[*]Corresponding author's e-mail: 15802482964@163.com

Abstract. It is of significance to ascertain the flue gas temperature at the exit of heating furnace, which lays the foundation of thermodynamic calculation of boiler and its design. This study analyses some normal calculation formulas of the flue gas temperature at the exit of heating furnace of oil and gas field, and makes comparisons of different calculation methods via examples. Appropriate parameters are presented, so as to give guide to the design and calculation of heating furnace of oil and gas field.

1. Introduction

In petroleum industry, heating furnaces are widely used in heating of crude oil, natural gas and well products in the oil-gas gathering and transportation systems, so as to realize the processes of transportation, sedimentation, separation, dehydration and initial processing. With the adjustment of energy consumption policy and the enhancement of environmental protection, the heating furnace of oil and gas field has been developed rapidly in recent years. However, there is no unified standard for the thermodynamic calculation of the furnace in the gas oil field.

The thermodynamic calculation of boiler is important, of which the crucial one is the heat transfer calculation of furnace radiation. To ascertain the flue gas temperature at the exit of heating furnace is the main task. This study gives some calculation formulas and makes comparisons via examples.

2. Common Calculation Methods

2.1Union of Soviet Socialist of Thermal Standard method for Boiler Units

As for heat calculation for boiler, most designers followed the joint standards drawn up by the former Soviet Union's Thermal Research Institute and the Central Boiler Steam Turbine Research Institute, which had a solid foundation in both theory and experimental verification, with certain authority. The calculation formula for the flue temperature of the furnace in the Soviet version in the 1973th "Thermal Standard Method of Boiler Units" [1, 2]:

$$\theta_{11}'' = T_{11} / [\, M\,(\delta_0 \varphi_{pj} F_1 \alpha_1 T^3_{11} / \phi Bj\, V_{cpj})^{0.6} + 1\,] - 273 \tag{1}$$

Where: T_{11} is theoretical combustion temperature, K;

δ_0 is blackbody radiation constant, 5.67×10^{-11} KW/(M^2K^4);

F_1 is radiation heating area of furnace, m^2;

φ_{pj} is average thermal efficiency coefficient in furnace;

α_1 is systematic emissivity of furnace;
ϕ is heat preservation coefficient;
B_j is calculation of combustion consumption, kg/s;
V_{cpj} is average specific heat capacity of combustion products, kJ/kg·K;
M is fudge factor

This formula is based on the heat transfer model of chamber combustion furnace, applying the "similarity theory" amid the process. However, special attention should be paid to the physical meaning and calculation of the following parameters when applying this formula to horizontal internal combustion gas boiler.

(1) Average thermal efficiency coefficient φ_{pj}

φ_{pj} is the ratio of heat absorbed by the heating surface to the heat projected on the furnace wall. Specifically, the heat transfer calculation of furnace radiation is:

$$\varphi_{pj}=\varepsilon\times x \quad (2)$$

Where: x is fractional cold of furnace;
ξ is slagging factor, gas fuel, ξ=0.65

(2) Systematic blackness of furnace α_1

The radiation heat transfer between the flame and the furnace can be regarded as the radiation heat transfer between the closed systems of the cavity and inner convex object. It can be illustrated the relation by α_1 as follows:

$$\alpha_1=\alpha_{hy} / [\alpha_{hy}+\vartheta_{pj}(1-\alpha_{hy})] \quad (3)$$

$$\alpha_1 = m\alpha_{fg}+(1-m)\ \alpha_g \quad (4)$$

Where: α_{hy} is flame blackness;
α_{fg} is luminous flame blackness (carbon black particle);
α_g is non-luminous flame blackness (Triatomic gas);
m is the proportion of luminous flame, depends on heat load of furnace

For gas fuels, when $q_v \leq 400kw/m^3$, m = 0.1, and when $q_v \geq 1200kw/m^3$, m = 0.6, linear interpolation method can be used to determine the relationship between the two.

(3) Fudge factor M

M is the key coefficient to ascertain the flue gas temperature at the exit of heating furnace. It is a semi-empirical conversion coefficient in the thermodynamic calculation of the furnace, which fully takes into account the structure of the furnace, the arrangement of the burner and the fuel composition. The radiation heat transfer of furnace has the characteristics of uniform radial direction and one-dimensional linear distribution along the furnace, so the expression of coefficient m adopts linear function relation:

$$M=a-bX_d \quad (X_d = L_{flame}/L_{furnace}) \quad (5)$$

Where: a and b is coefficient 1.01 and 0.49, respectively.

$$M=1.01-0.49X_d \quad (6)$$

After the correction of this coefficient, the outlet smoke temperature calculated by this method is consistent with the actual one. Therefore, this paper uses the above formula as the standard of calculation to investigate the accuracy of other formulas.

2.2The Calculation Method in Theoretical Analysis and Calculation of Heat Transfer in Boiler Tank of Shell Oil and Gas Boiler

Bu Yinkun puts forward the calculation formula 7 [3] in Theoretical Analysis and Calculation of Heat Transfer in Boiler Tank of Shell Oil and Gas Boiler. This formula is derived on the basis of thermal radiation theory, which reflects the quartic temperature difference relationship of radiation heat exchange, which is reliable.

$$Q^{cr}_{1.2}=5.67\times10^{-11}\times A(\varepsilon^{ty}_{1}T_{y}^{4}-\varepsilon^{tw}_{1}T_{w}^{4})/(1+0.042\times\varepsilon^{tw}_{1}) \quad (7)$$

Where: $Q^{cr}_{1.2}$transfers heat via radiation heat transfer object 1 to object, KW;

A is the internal surface area of bellows furnace, m^2;

ε^{ty}_1 is object 1, the blackness of flue gas at the average thermodynamic temperature in the furnace tank;

T_y is average thermodynamic temperature of flue gas with high temperature in furnace K;

ε^{tw}_1 is object 1, the blackness of smoke on the heating surface of furnace at thermodynamic temperature on the inner surface of soot;

T_w is thermodynamic temperature of smoke and soot on heating surface of furnace tank, K.

2.3Calculation Method of Oil- and Gas-Fired Boiler

Zhao Qinxin and Hui Shi'en give a simplified calculation formula8 [4] in their book of *Oil- and Gas-Fired Boiler*. The theoretical basis is still the basic equation of radiation heat transfer, which reflects the quartic temperature difference relationship of radiation heat exchange. The parameters are few and easy to obtain.

$$Q=C\times H_f\times[(\frac{T_{hy}}{100})^4-(\frac{T_b}{100})^4]/Bj \quad (8)$$

Where: Q is radiation heat exchange, kJ/kg, kJ/m^3;

C is radiation heat transfer coefficient, kW/ (m^2·K), C = 11～15 is recommended in the book;

H_f is effective radiation heating area, m^2 ;

T_{hy} is average temperature of the flame in the furnace, K;

T_b is temperature of inner surface wall of furnace, K;

Bj is calculated fuel consumption, kg/s, m^3/s.

It is crucial to calculate T_{hy}, the average temperature of the flame and T_b, the temperature of inner surface wall of furnace. Generally speaking, T_b can be calculated by the formula of $T_b=T_{gb}+90$, while T_{hy} can be calculated by the following methods:

(1) Arithmetic interpolation method

$$T_{hy}=T''_{l}+\frac{1}{4}(T_{ll}-T''_{l})\ K \quad (9)$$

Where: T''_l is outlet temperature of furnace assumed in design and calculation;

T_{ll} is theoretical combustion temperature determined by incoming heat

(2) Geometric mean method

$$T_{hy}=0.9T^{(1-n)}_{ll}\ T''^{n}_{l}\ \ K \quad (10)$$

For oil- and gas-fired boiler, n=0.5

(3) Parametric method

$$T_{hy}=\tau T_L''K \quad (11)$$

Where: parameter $\tau=[0.44\times\theta^4+0.14(\theta^3+\theta^2+\theta+1)]^{0.25}$; $\theta=1+\frac{1}{3}(\frac{T_{11}}{T''_1}-1)$

(4) Geometric parameter method

$$T_{hy}=T''_1(\tau(1\text{-}x_{max}))^{0.25}\ K \tag{12}$$

Where: parameter $\tau=\dfrac{1}{[(\frac{T''_1}{T_{11}})^3+(\frac{T''_1}{T_{11}})^2+(\frac{T''_1}{T_{11}})]}$

Taking a 350 KW heating furnace as an example, the average flue gas temperature is calculated according to the above four methods. The results of it are as follows: For arithmetic interpolation method and geometric mean method, T_{hy} are 1290.05 k and 1329.04 k, respectively; that of parametric method and geometric parameter method are 1270.29 k and 940.2548 k, respectively. The calculation error of T_{hy} via the above three methods is small, expect for geometric parameter method. The arithmetic interpolation method is used in the calculation of this study.

3. Calculation Examples by Various Calculation Methods

3.1Union of Soviet Socialist of Thermal Standard method for Boiler Units

The calculation results of the flue gas temperature at the exit of heating furnace with 350 KW of oil and gas field is showed in Table 1.

Table 1. The calculation of the flue gas temperature at the exit of heating furnace of 350 KW oil and gas field.

Order number	Name	Symbol	Unit	Formula or data source	Result
1	flue gas temperature at the exit of heating furnace	θ''_1	°C	$\theta''_1=\dfrac{T_{ll}}{M(\frac{\sigma\psi a_1 A_b T^3_{ll}}{\varphi BjV_{cpj}})^{0.6}+1}-273$	704.93
2	Calculation error	Δ	°C	<100°C meets the requirements	-0.00073

By applies the above methods, the flue gas temperature at the exit of heating furnace of other series of heating furnace can be calculated as follows: the temperature of 350 KW and 820 KW heating furnace is 705 °C and 70 °C, respectively; that of 1,250 KW and 1,600 KW furnace at 715 °C and 710 °C, respectively; that of 2,800 KW and 4,000 KW at 720 °C and 880 °C, respectively.

As the actual temperature of flue gas at the exit of heating furnace is close to the above calculated one, the above data are used as the standard for the calculation of following formula.

3.2The Calculation Method in Theoretical Analysis and Calculation of Heat Transfer in Boiler Tank of Shell Oil and Gas Boiler

The calculation results of the flue gas temperature at the exit of heating furnace with different power of oil and gas field is showed in Table 2.

Table 2. The calculation data based on the formula in Theoretical Analysis and Calculation of Heat Transfer in Boiler Tank of Shell Oil and Gas Boiler

Bu Yinkun *Theoretical Analysis and Calculation of Heat Transfer in Boiler Tank of Shell Oil and Gas Boiler* Formula $Q=5.67\times10^{-11}\times A(\varepsilon^{ty}{}_1 T_y{}^4 - \varepsilon^{tw}{}_1 T_w{}^4)/(1+0.042\times\varepsilon^{tw}{}_1)$

Order number	Name	Symbol	Unit	Calculation Result					
				350 KW	820 KW	1250 KW	1600 KW	2800 KW	4000 KW
1	Flue gas heat transfer	Qcr	KJ/Nm^3	24505	28461	23702	24911	28152	24812
2	Calculation error	Δ	%	-4.98	-21.06	-2.90	-4.41	-17.59	-18.40

3.3The Calculation Method in Oil- and Gas-Fired Boiler
The calculation results of the flue gas temperature at the exit of heating furnace with different power of oil and gas field is showed in Table 3.

Table 3. The calculation data based on the formula in Oil- and Gas-Fired Boiler

Zhao Qinxin and Hui Shi'en *Oil- and Gas-Fired Boiler* Formula $Q=C\times H_f\times[(\frac{T_{hy}}{100})^4-(\frac{T_b}{100})^4]/Bj$

Order number	Name	Symbol	Unit	Calculation Result					
				350 KW	820 KW	1250 KW	1600 KW	2800 KW	4000 KW
1	Flue gas heat transfer	Qcr	KJ/Nm^3	23334	23429	22935	23860	23979	20937
2	Calculation error	Δ	%	0.034	0.345	0.426	-0.00227	-0.160	0.0924

Though the above formula, the calculation error may be large, if it is applied the formula C=11-17, which proves the recommended C value of this formula is not suitable to calculate the flue gas temperature at the exit of heating furnace of oil and gas field. This paper analyzed C value should adopt different power of various kinds of furnaces in order to reduce error.

Parameter C of 350 KW and 820 KW heating furnace are 5.1 and 8.9, respectively; that of 1,250 KW and 1,600 KW at 10.4 and 10.7, respectively; that of 2,800 KW and 4,000 KW at 11.5 and 8.8. The relationship between the thermal power of the heating furnace and the parameter C is drawn as a line, as shown in figure 1.

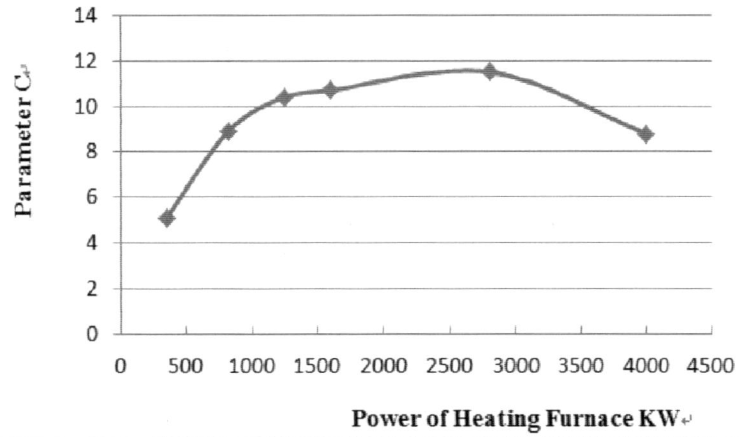

Figure 1. Relation Curves of thermal power of heating furnace and Parameter C

When the flue gas temperature at the exit of heating furnace is calculated by this formula, the recommended value of parameter C under the corresponding power can be found out from the fitting curves in figure 1, so that the calculation error can be reduced.

4. Analysis, Comparison and Conclusion

The flue gas temperature at the exit of heating furnace calculated by the method in formula 1 is in good agreement with the actual one. Although it is not reflected the difference of radiation heat exchange of quartic temperature difference, indeed, it is the combination of simultaneous heat exchange equation and heat balance equation. More importantly, all the factors affecting radiation heat exchange in this formula are expressed accurately. φ_{pj}, the effective coefficient of radiation heating surface, presents the structural factors of furnace and combustion chamber heating surface (guard fuel belt, dry back, wet back, etc.), especially the furnace blackness, which is calculated from the thickness of effective radiation layer reflected by furnace structure, until the positive pressure of furnace, luminous and non-luminous flame blackness, and then combined with the factors that affect the blackness. Coefficient M is a semi-empirical conversion coefficient in the calculation of furnace thermodynamic, which fully considers the furnace structure, burner arrangement, fuel composition and so on. Therefore, the flue gas temperature at the exit of heating furnace calculated by this formula can be in good agreement with the actual one, which is more accurate. And this formula is widely used in the industry of boiler with certain authority.

The heat transfer error calculated in formula 7 is based on formula 1. The heat transfer error of 350 KW and 820 KW heating furnace is 4.98% and 21.06%, respectively, that of 1,250KW and 1600 KW at 2.9% and 4.41%, respectively; that of 2,800 KW and 4,000 KW at 17.59% and 18.40%. Through the calculation, it can be seen that the design or checking calculation by this method is in accordance with the reality in most cases. The error of flue gas temperature is generally less than 100 °C, and that of heat transfer is generally not more than 20%. Moreover, this method is simple, with few independent variables, fast operated and sufficient theoretical basis.

Formula 8 can only roughly determine the coefficient C, which is selected appropriately according to different conditions of radiation heat exchange. In fact, the variation range of this value is very large, which is difficult to select exactly. Therefore, the calculation is certainly not very accurate. When the flue gas temperature at the exit of heating furnace is calculated by this formula, the recommended C value under the corresponding power can be found out from figure 1, so that the calculation error is very small. The application of this formula is simple, with few parameters and relatively sufficient theoretical basis, which is a good method for calculating the flue gas temperature at the exit of heating furnace.

In conclusion, one of the three formulas can be selected in the calculation of heat transfer of heating furnace of gas and oil field, according to the specific situation. In fact, the theoretical basis of these three formulas is derived from the radiation heat transfer equation of flame and furnace wall, and the furnace heat balance equation, with same basic theory. Only formula 1 has more empirical components and required parameters. Formula 7 has a solid theoretical foundation, with fewer parameters to make it easy to calculate, but in some cases the error is larger than that of other methods. Formula 8 also has a solid theoretical foundation, with clear concept, few parameters, with is easy to obtain and more convenient to calculate. However, it is necessary to accumulate more basic data and make the corresponding c-value fitting curve, in order to obtain more accurate data which is suitable for the corresponding furnace design.

References

[1] Zhang W, Chai X, Li S. (2001) Determination of outlet smoke temperature and influencing factors of horizontal internal combustion fuel (gas) boiler [J]. Industrial Boiler, 4:2-6.
[2] Yang J. (2004) Calculation formula of flue gas temperature in furnace outlet of horizontal internal combustion fuel gas boiler [J]. Industrial Boiler, 2:21-22.
[3] Bu Y. (2001) Theoretical Analysis and Calculation of Heat Transfer in Boiler Tank of Shell Oil and Gas Boiler [J]. Industrial Boiler, 5:10-13.
[4] Zhao Q, Hui S. (2000) Oil- and Gas-Fired Boiler [M]. Xi'an Jiaotong University Press. Xi'an.347-350.

IOP Conf. Series: Materials Science and Engineering **721** (2020) 012061 doi:10.1088/1757-899X/721/1/012061

Failure Analysis of SA-213TP347H High-temperature Reheater of 600 MW Supercritical Boiler

Jia'an Wang, Jun He

China Huadian Electric Power Research Institute Co., Ltd., Hangzhou, China

*Corresponding author's e-mail: jiaan-wang@chder.com

Abstract. Failure analysis of SA-213TP347H steel tube for high-temperature reheater of 600MW supercritical boiler was studied by means of macroscopic analysis, chemical composition, mechanical properties, microstructure, scanning electron microscopy and energy spectrum analysis. The results show that the chemical composition, hardness and strength of the tube samples meet the limit of ASTM A213/A213M standard. The microstructure of tube-burst sample is austenite, there are some precipitates containing Cr and Nb element at the grain boundaries. The test results for the spare part sample near elbow show that the hardness exceeds the limit of SA-213TP347H steel, there are a lot of slip lines and some non-metallic inclusions containing S in the microstructure. It is concluded that solid solution treatment missing in its manufacturing is the main reasons for the fracture, which makes Cr precipitate on the grain boundary and form the Cr-depleted zone, which is easy to cause intergranular stress corrosion and lead to the tube burst.

1. Introduction

SA-213TP347H austenitic stainless steel not only has good intergranular corrosion resistance, high-temperature oxidation resistance, creep performance, but also has good welding and thermal processing performance, which has been widely used in the high-temperature heating surface tubes of ultra-supercritical or supercritical large-capacity boiler [1]. However, in the production of SA-213TP347H steel, some domestic manufacturers did not carry out solid solution heat treatment after cold processing in strict accordance with the standard requirements, or even did not carry out solid solution treatment at all, which caused a great impact on the service performance of TP347H steel, especially the performance of elbow.

On June 22, 2018, the high-temperature reheater of a power plant leaked. On July 12, the leakage of the high-temperature reheater tube happened again. After the leakage, the relevant personnel of the Electric Power Academy rushed to the scene for the first time, and preliminarily judged that the first explosion port was the sixth of the 40nd screen tube from the left of the furnace, and the burst tube was the inside of the U-shaped elbow of the high-temperature reheater. The burst tube material was SA-213TP347H, and the specification was Φ 51 mm x 4.5 mm. The burst tube sample (No. 1) and the spare part (No. 2) were cut off on site, and the tube was analyzed by optical microscope (OM), scanning electron microscopy (SEM) and energy dispersive spectroscopy (EDS), etc., so as to find out the reason why the high-temperature reheater frequently leaked.

2. Experiment

2.1 Macroscopic morphology analysis
The explosion position of the high-temperature reheater is located on the arc side of the elbow, and the explosion is extended along the axial direction of the tube. It's found that the high-temperature reheater elbow is not well formed, and the elbow is triangular in section, forming an indentation during the pressing process. There are excessive defects in the elbow. The surface of the spare part is bright, and it has not been subjected to solution treatment after cold working, and it is magnetically measured by a magnetic measuring instrument. At the same time, the ellipticity of the elbow is measured, and the maximum ellipticity is 17%, which seriously exceed the standard requirements.

2.2 Chemical composition analysis
Table 1 show the chemical composition of sample No.1 and No.2 tubes. The results show that the main composition of the material conforms to the chemical composition requirements of SA-213TP347H steel in ASTM A213/A213M.

Table 1. Chemical analysis results of samples (mass fraction, %)

Chemical composition	C	Mn	S	Si	Cr	Ni	Nb
No.1	0.07	1.5	0.004	0.5	17.81	10.88	0.778
No.2	0.08	1.5	0.002	0.5	17.81	10.25	0.919
Standard value	0.04-0.10	≤2.00	≤0.030	≤1.00	17.0-19.0	9.0-13.0	0.8-1.10

2.3Mechanical performance analysis
Table 2 shows the hardness test results of the explosion position (No. 1-1), the distance from the explosion position 400 mm (No. 1-2), and the spare part straight section (No. 2-1), the inner arc surface on the elbow of the spare part (No. 2-2) and the outer arc surface (No. 2-3), each position is measured 3 times. The results show the hardness is consistent with the requirements of ASTM A213/A213M. The hardness of the explosion position is low, and the hardness of the spare part is obviously increased. It may be due to incomplete or no solution treatment of the spare part, and other microstructures existed in the microscopic phase.

Table 3 shows the mechanical tensile test on the fire side (No. L1) and the backfire side (No. L2) of the burst tube, and the spare part (No. L3 and No. L4). The tensile properties meet the requirements of the standard specification.

Table 2. The hardness results of samples (HRB)

Sample No.	①	②	③	Average value	Standard value
No.1-1	58.3	63.0	62.6	61.3	
No.1-2	69.0	72.5	73.0	71.5	
No.2-1	83.2	80.0	81.4	81.5	≤90
No.2-2	85.7	85.0	86.0	90.6	
No.2-3	83.3	83.8	82.0	83.0	

Table 3. The tensile test results of samples (MPa)

Sample No.	Sampling locations	$R_{p0.2}$/MPa	R_m /MPa	A / %
		≥205	≥515	≥35
No.1	L1	284	605	47
	L2	279	600	53
No.2	L3	284	655	50
	L4	274	655	48

2.4 Microstructure analysis

Figure 1 shows the microstructure of the sample on the burst tube, and Figure 2 shows the microsturcture of the sample at the elbow of the spare part. The microstructure of the tube is austenite and grain boundary carbide, and the grain size is 8 grade. No twins and slip lines are observed. The microstructure at the elbow of the spare part is austenite and a small amount of carbide distributed along the grain boundary, and the twins and the slip line can be clearly observed.

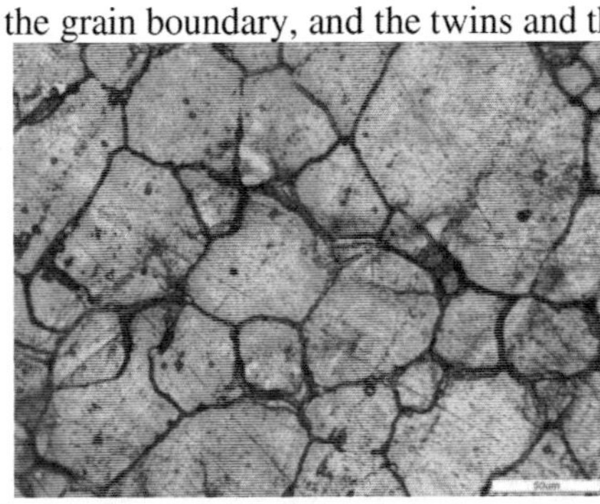

Figure 1. Microstructure of the burst tube

Figure 2. Microstructure of the spare part

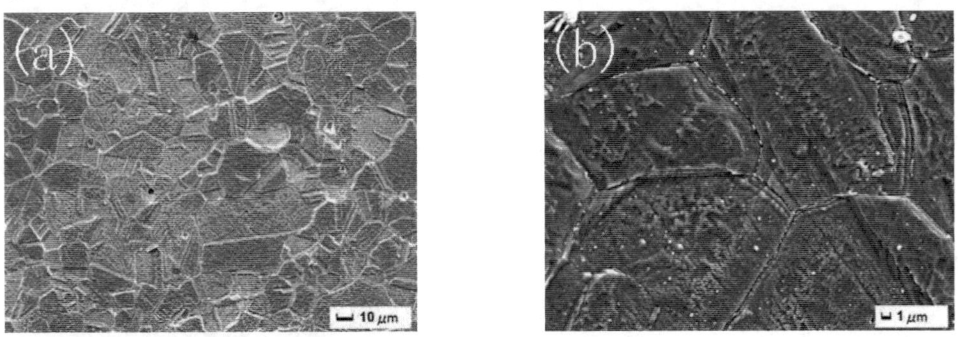

Figure 3. SEM of the sample No.1-1. (a) low-magnification; (b) high-magnification

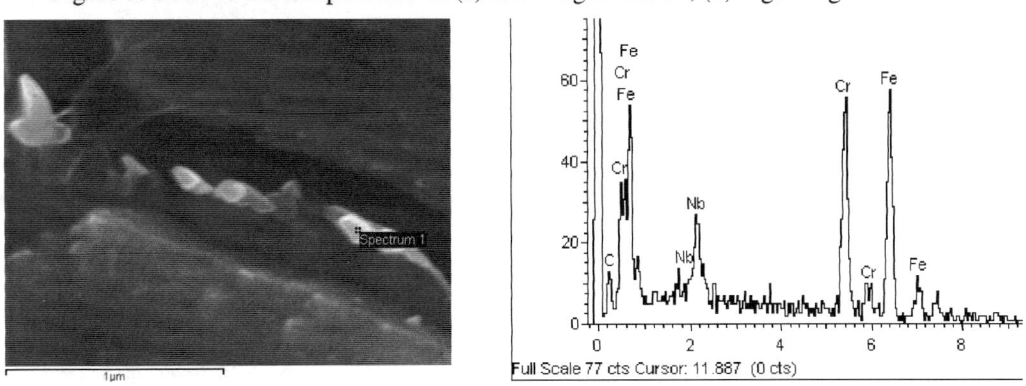

Figure 4. EDS of the precipitates. (b) EDS of the precipitates on grain boundary of (a)

2.5SEM and EDS analysis

The microstructure is further analyzed by SEM. Figure 3 shows the microstructure of the No. 1-1 sample with low-magnification and high-magnification. It can be seen that the precipitates are aggregated into a linear shape at the grain boundary. Figure 4 shows EDS of the precipitates. The results of EDS show that the precipitates are rich in Cr and Nb, which aggregate along the grain boundary.

3. Results and Discussion

The results of chemical composition analysis and mechanical property test at room temperature on the sample of the burst tube and the spare part show that the sample conforms to the requirements of ASTM A213/A213M on SA-213TP347H, which indicating that the chemical composition is normal. The operation time of the boiler is short, and no oxide layer is found on the inner and outer wall of the explosion position, and there are no signs such as bulging, bulging, etc., indicating that it is not caused by overheating [2]. There are excessive defects in the elbow, and the maximum ellipticity is 17%, which seriously exceeds the standard requirements. According to the requirements, the solution treatment should be carried out after the cold working of the elbow is completed. The solution treatment can reduce the residual stress and dislocation generated by the processing. The literature [3] points out that the residual stress and the dislocations accelerate the precipitation of chromium in the grain boundary, the chromium boundary is sharply depleted, and the deformation-induced martensite potential is low, which seriously weakens the resistance of the grain boundary. As well, this high-temperature reheater elbow has not been solution treated after processing. Therefore, the unsolvent treatment after cold working is an important cause of high hardness on the sample of the spare part and the presence of more slip lines [4].

Scanning electron microscopy and energy spectrum analysis show that there are precipitates of oxides and carbide particles with Cr and Nb at the grain boundary of the burst tube. Precipitates containing Cr are precipitated at the grain boundaries, which resulting in the formation of a chromium-depleted region at the grain boundaries, which deteriorates the corrosion resistance of the grain boundaries. The formation and development of chromium-depleted regions is controlled by the diffusion of Cr [5]. The large number of dislocations and vacancies generated during the cold processing of the elbows greatly promote the diffusion of elements. And the effect of deformation is accelerated sensitization for 18Cr-9Ni steel with carbon content of 0.04% or more. It is controlled by the diffusion of Cr and further accelerates the sensitization. It can be seen that the unsolvent treatment of the elbow severely reduces the corrosion performance of the grain boundary [6].

In addition, the flue gas, coal ash contains chloride and excessive sulfide, sulfate, etc., may affect the solubility of oxides in the vicinity of the crack tip, thereby increasing the corrosion rate [7] The literature [8-9] shows that the main reason for the failure of austenitic stainless steel tubes is that sulfur element promotes grain boundary stress corrosion cracking. It can be seen that inclusions such as S can adversely affect the performance of the tube.

4. Conclusion

Based on the above analysis, one of the reasons for the frequent bursting of the high-temperature reheater is that due to the influence of the manufacturing quality, the defects in the bending process and the elliptic deformation exceed the standard, resulting in a decrease in the mechanical strength of the tube. Since the high-temperature reheater tube is not solution treated, the dislocations and residual stresses at the elbow increase, which accelerate the precipitation of the Cr element carbide at the grain boundary, and the grain boundary is sharply depleted in Cr, resulting in a decrease in grain boundary corrosion resistance. With the operation of the unit, the SA-213TP347H without solution treatment of the high-temperature superheater is frequently bursting under the combined action of steam pressure, thermal stress, residual stress and corrosive medium.

References

[1] Li, Z.L., Wang, J.A., Xiao, M.Y. (2015) Failure analysis of SA-213TP347H steel reherater tube. Heat Treatment of Metals., 40: 191-194.
[2] Ma, Q., Liang, P., Yang, S.E., et al. (2009) Investigation of high-temperature steam oxidation for TP347H steel. Transactions of Materials and Heat Treatment., 30: 172-176.
[3] Ma, G.B., Li, X.H., Xie, A.M. (2008) Short-time cracking analysis of TP347H tubes of high temperature heat delivery surfaces of boilers. Engineering Journal of Wuhan University., 40: 80-84.

[4] Tao, J., Yao, Z.J., Xue, F. (2006) Fundamentals of Material Science. Chemical Industry Press, BeiJing.
[5] Wang, J.A., Li, J.P., Li, J. et al. (2015) Failure analysis of tube-burst of SA-213TP347H high-temperature reheater of 600 MW supercritical unit. Transactions of Materials & Heat Treatment., 36:122-127.
[6] Zhang, D.K. (1982) Local Corrosion of Stainless Steel. Science Press, BeiJing.
[7] Long, H.G., Chen, H.D., Wan, K.Y. (2008) Failure analysis of high temperature super-heater bend for boiler. Corrosion & Protection, 29: 157-159.
[8] Natesan, K., Kraus, C. (1998) Corrosion performance of structural alloys in oxygen/ sulfur/ chlorine-containing environments. In: 12th Fossil Energy Materials. Knoxville. pp. 18-30.
[9] Natesan, K., Baxter, D.J. Oxygen-sulfhr corrosion of metals in mixed-gas atmospheres. In: 3rd Corrosion-Erosion Wear of Materials at Elevated Temperatures. Houston. pp. 67-86.

IOP Conf. Series: Materials Science and Engineering **721** (2020) 012062 doi:10.1088/1757-899X/721/1/012062

Analysis of influence of peak regulation on life of once-through boiler

En Ren Liu[1], Yu Lan Zhao[2] Qing Yan[3] Yuan Yuan Li[4] Ying Kun Han[5] Xiang Rong Meng[6]

Shandong Electric Power Research Institute, Jinan250002, China

e-mail:zaihuadianzhuce@163.com

Abstract. In this paper, a 670 MW once-through boiler of a power plant is taken as the analysis object. The method of ANSYS numerical analysis is used to calculate the stress of the once-through boiler in different starting modes and different peaking depths. We calculate the life loss of different operating conditions according to the stress value, which provides reasonable suggestions for the life management of the once-through boiler during peak shaving, which can reduce the impact of peak shaving on boiler life.

1. Introduction

In the future, with the continuous expansion of the scale of new energy power generation, thermal power will play a major role in peaking the power grid. When coal-fired units participate in peak shaving, the load often changes rapidly and the unit starts and stops frequently, which will adversely affect the life of the boiler.

The steam-water separator is one of the main pressure-bearing components in the once-through boiler, and is the hub of the boiler steam-water system. If the steam-water separator is damaged during operation, the repair is extremely difficult. Therefore, the study of the effect of peak regulation on boiler life can be based on the study of steam-water separators.

2. Research object

The research object of this paper is a 670MW ultra-supercritical unit of a power plant with the boiler type SG-1978/29.3-M6012. The dimensions of the steam separator and material properties are shown in Tables 1 and 2.

Tables 1. Steam-water separator structure size

Part name	Pipe diameter
Cylinder	φ610mm×75mm
Soda water introduction tube	φ170×35
Steam extraction tube	φ334×55
Water extraction tube	φ324×5

Tables 2. Physical properties of WB36 steel

Steel temperature (°C)	Thermal Conductivity (W/m·°C)	Specific heat capacity (KJ/(kg·K))	Elastic Modulus (GPa)	density (kg/m^3)
20	38	0.46	211	7850
100	41	0.5	206	—
200	42	0.5	200	—
300	41	0.54	192	—
400	39	0.54	184	—
500	38	0.63	175	—

Due to the limited computing ability of the computer, the steam-water separator are simplified:

(1) The steam-water separator is a three-dimensional solid structure with bilateral symmetry. It is ok to take 1/2 structure for calculation.

(2) This paper mainly studies the position of the dangerous point and the law of stress change during the peak load and start-up process of the low load operation of steam-water separator, The maximum stress is near the tube introduction tube, so in order to save the calculation time, the calculation model is as shown in Figure 1 and 2.

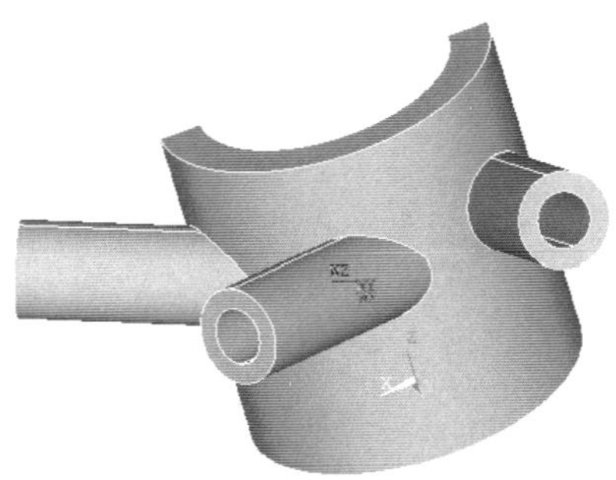

Figure 1. Internal structure

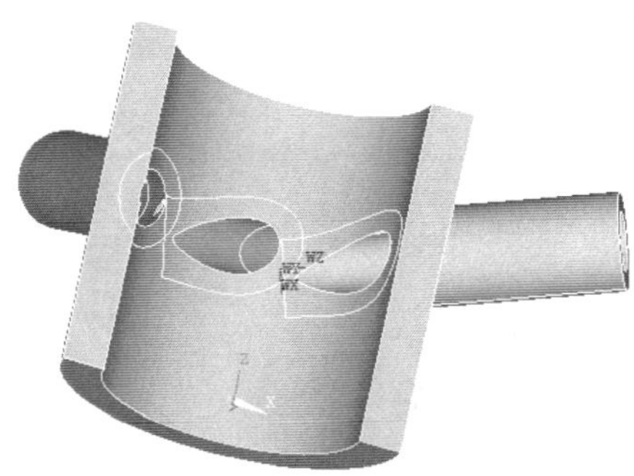

Figure 2. External structure

3. Stress analysis of steam-water separator in peak-adjusting operation

There are mechanical stresses and thermal stresses in the steam-water separator during start-up and peak-shaving. The alternating stress is the main cause of fatigue damage of metal materials. Therefore, it is necessary to analyze the stress distribution during the startup and peaking process of the steam separator to analyze the life loss. This paper mainly simulates the cold start, warm start, hot start and different peak shaving conditions.

3.1 Cold start stress analysis

Figure 3 shows the thermal stress field distribution of the thermal stress at the maximum moment. It can be seen that the thermal stress at the intersection of the inlet pipe and the inner wall of the cylinder is the largest. This is mainly because the thermal stress gradient here is large and the structure is discontinuous, so that the thermal stress changes drastically, and stress concentration occurs

The mechanical stress is mainly caused by the working fluid pressure inside the steam-water separator. Therefore, the maximum mechanical stress appears at the end of the start-up period. At this time, the mechanical stress distribution of the steam-water separator is shown in Figure 4. It can be seen from the figure that the stress gradients of the three steam-water inlet pipes of the steam-water separator are relatively large, and the upper mechanical stress of the intersecting line between the inner wall of the cylinder and the inner wall of the inlet pipe has a maximum value.

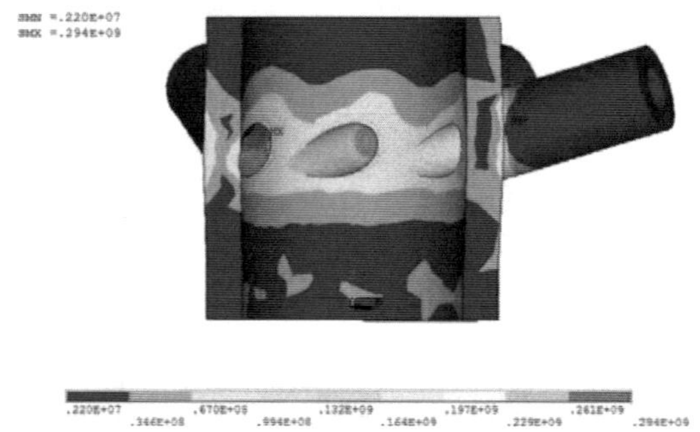

Figure 3. Cold start thermal stress distribution cloud map

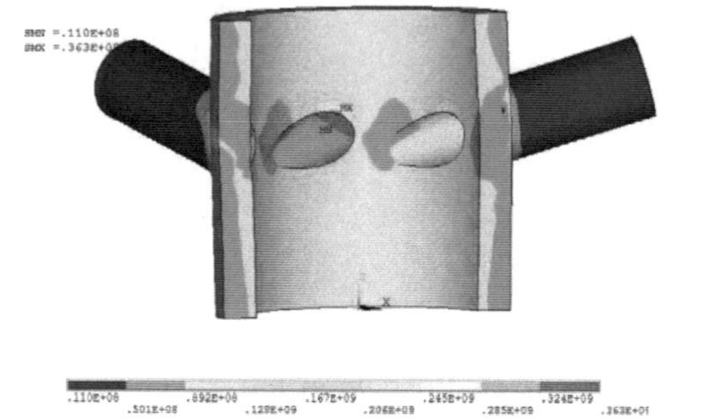

Figure 4. Cold start mechanical stress distribution cloud map

The total stress during the cold start process is the result of the coupling of mechanical and thermal stresses. Figure 5 shows the total stress distribution cloud at the start of 600s and the start of startup. The maximum stress point within 600s before the start of the initial stage is located at node 746 (maximum thermal stress point) on the intersection of the inner wall of the cylinder and the soda inlet tube.

The maximum stress point after 600s in the middle and late stages of the start is located at the node 901 (maximum mechanical stress point) on the intersection line between the inner wall of the cylinder and the soda water introduction tube. Therefore, the maximum stress value of the total stress at the different moments of starting is always on the line connecting the cylinder and the soda water inlet pipe, but the specific position is uncertain.

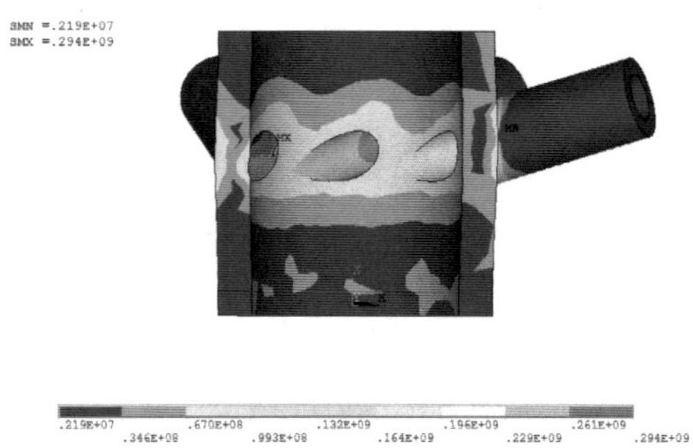

Figure 5. Total stress distribution at t=600s

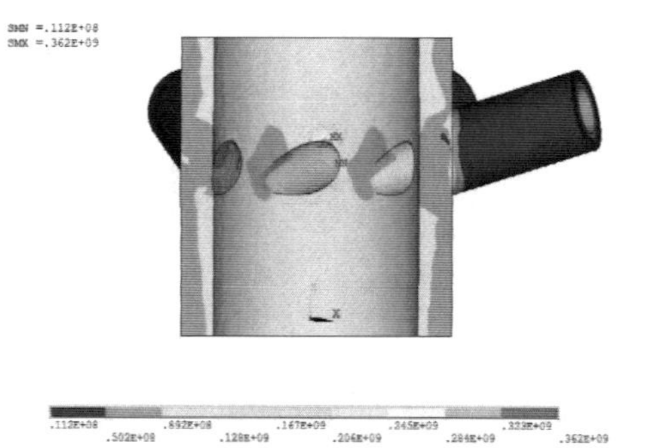

Figure 6. Total stress distribution cloud map at t=25800s

Figure 7 is a graph showing the relationship between the two maximum stress points (node 746 and node 901) over time during the cold start process. It can be seen from the figure that both curves are reduced first and then gradually increased. This is because in the initial stage of startup, the internal pressure of the working medium is small, the mechanical stress is small, and the temperature change rate is relatively large, resulting in a relatively large thermal stress.

The total stress mainly depends on the thermal stress and the total stress value reaches the secondary peak at this time. In the middle of the start-up, the thermal stress is small, and the total stress mainly depends on the mechanical stress, so the total stress remains basically unchanged.

During the late stage of the lifting process, the mechanical stress increases with the increase of the working medium pressure, and the thermal stress increases due to the increase of the working temperature. However, the thermal stress is less than the mechanical stress, and the total stress mainly depends on the mechanical pressure. At the end of the start-up, as the boiler full-load working fluid pressure reaches a maximum, the thermal stress tends to zero, the mechanical stress reaches a maximum, and the total stress of the steam-water separator also reaches a maximum.

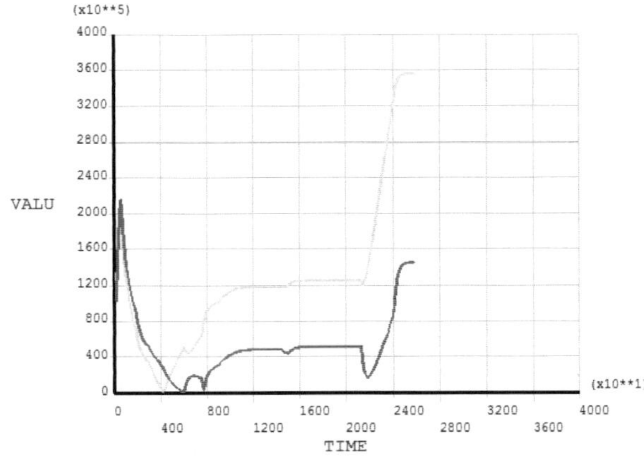

Figure 7. Total stress changes at nodes 746 and 901 under cold start

Since the node 901 is numerically larger, this point is taken as the maximum stress point, and the value of the point is taken for calculation in the life analysis.

3.2 Warm start stress analysis

Figure 8 shows the stress distribution cloud diagram when the total stress is maximum. The maximum stress value during the warm state startup is always located in the intersection line between the inner wall of the cylinder and the soda water introduction tube. This is different from the cold start, where the position of the maximum point of total stress during the cold start is uncertain. During warm start, the maximum total stress occurs at the end of the start, when the mechanical stress reaches a maximum.

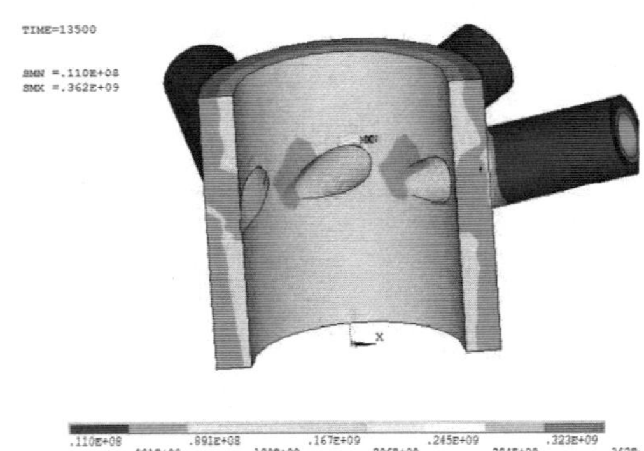

Figure 8. Total stress distribution cloud map under warm start

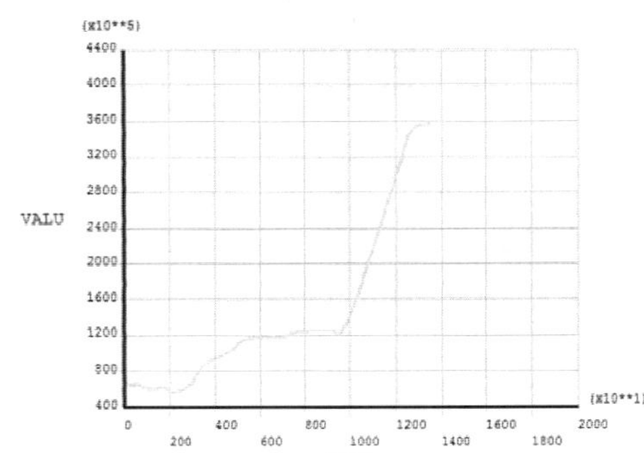

Figure 9. Regularity of total stress change under warm start

3.3 Hot start stress analysis

It can be seen from Figure10 that the dangerous point position of the total stress in the hot start process is consistent with the warm start. At the beginning of the hot start, the total stress changes are relatively flat. This is because the initial thermal stress value is much smaller than the total stress value. The total stress is mainly determined by the mechanical stress value. Therefore, the change in the total pressure is relatively flat. In the middle and late stages of the start-up, the change of the total pressure is consistent with the warm start and the cold start.

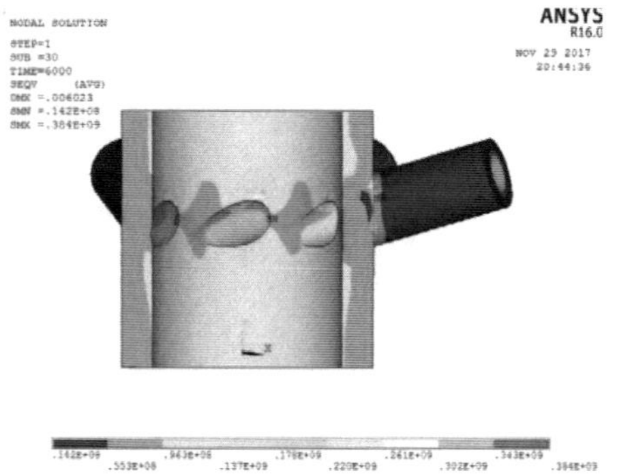

Figure 10. Total stress distribution cloud map under hot start

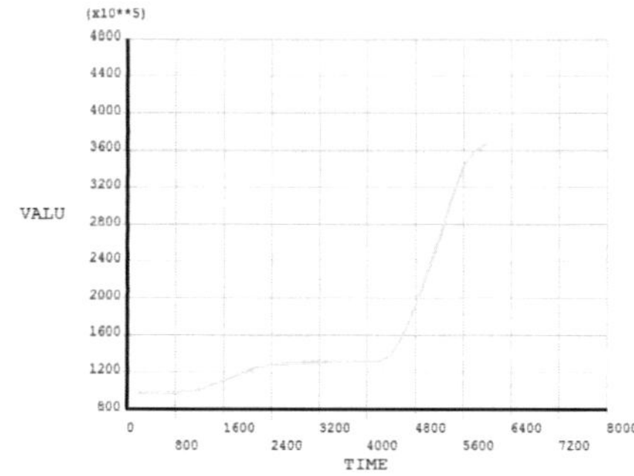

Figure 11. Regularity of total stress change under hot start

3.4 Deep peaking stress analysis

When the grid is at a low point, the deep peak-shaving operation prevents the unit from starting and stopping frequently. During the start-stop process, the pressure-bearing equipment needs to withstand severe temperature changes and pressure changes.

In order to compare and analyze the influence of different peak shaving depths on the life of the steam separator, in the text, we select three variations from 234MW-670MW, 268MW-670MW and 302-670MW, and the stress during the low load peaking operation is calculated separately.

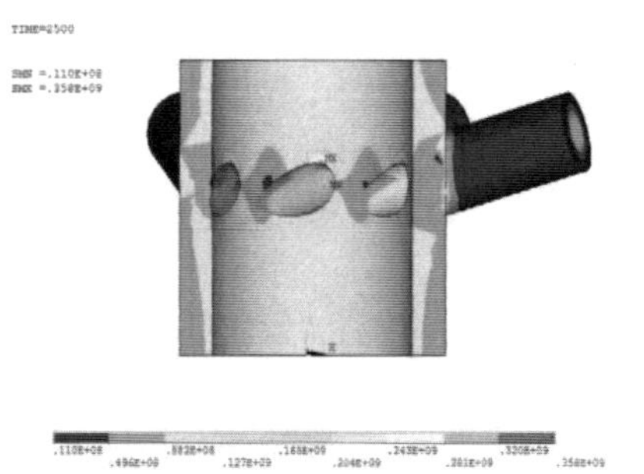

Figure 12. Total stress distribution cloud map during peak load operation

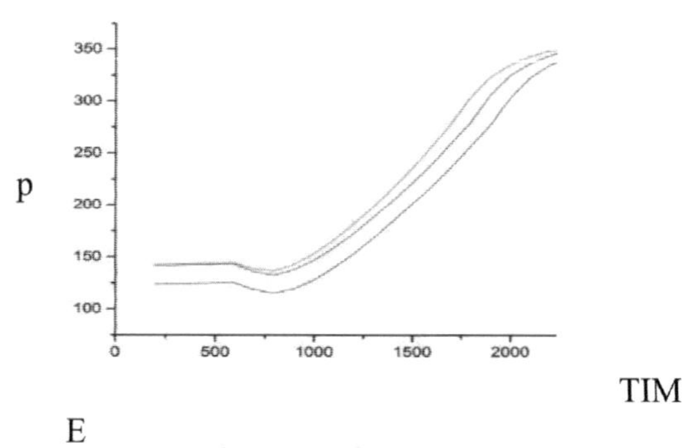

Figure 13. Comparison of total stress at different peaking depths

Table 3. Stress of steam water separator in different operating conditions

Operating condition	Cold start	warm start	hot start	Low load operation peak shaving (peaking depth)		
				65%	60%	55%
Maximum stress (MPa)	366	366	366	362	361	362
Minimum stress (MPa)	3	55	94	121	139	140
Stress amplitude (MPa)	363	311	272	241	222	222

It can be seen from Table 3 that during the process of peak load regulation at low load, the difference in stress amplitude is small, and the stress amplitude during peak regulation is smaller than the stress amplitude during startup.

The larger the peaking amplitude, the larger the generated stress amplitude. At the end of the peak load peaking operation, the maximum stress value is the same. The peaking depth is larger, the smaller the minimum stress, the larger the stress amplitude. In various starting modes, the order of stress amplitude from large to small is: cold start, warm start, hot start.

4. Analysis of the effect of peak regulation on life

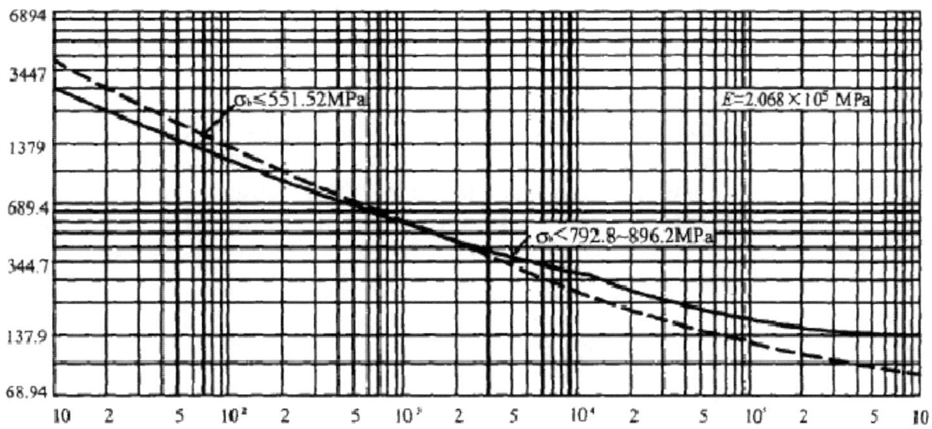

Figure 14. American ASME fatigue design curve

In this paper, the US ASME standard is used for life evaluation. For the steam-water separator, the maximum total stress value and the minimum total stress value of the steam-water separator under different working conditions can be obtained by numerical simulation. The difference between the two is the cyclic stress amplitude.

$$\sigma_{aij} = \frac{1}{2}\Delta\sigma_{aij,max} = \frac{1}{2}\left(\sigma_{aij,max} - \sigma_{aij,min}\right) \tag{4-1}$$

$\sigma_{aij,max}$ —Main stress maximum,

$\sigma_{aij,min}$ —Main stress minimum,

σ_{aij} —Maximum cyclic stress amplitude,

$\Delta\sigma_{aij,max}$ —Maximum cyclic stress amplitude fluctuation range;

After calculating the maximum cyclic stress amplitude σ_{aij}, it is necessary to modify the elastic modulus to convert it into the corresponding cyclic stress amplitude σ_a in the fatigue design curve,then we can obtain the allowable number of cycles N.

$$\sigma_a = \frac{E_d}{E}\sigma_{aij} \tag{4-2}$$

E_d —Elastic Modulus $E_d = 2.068\times10^5 MPa$,

E —Elastic modulus of the material of the life assessment component;

In this paper, the finite element analysis is used to calculate the stress values during the different starting modes and peaking depth of the once-through boiler. The ASME fatigue life curve is then used to evaluate the life of different operating conditions.

Table 4. Different working conditions life loss of Steam water separator

Different operating conditions	Cold start and stop	Warm start and stop	Hot start and stop	Peak shaving depth 65%	Peak shaving depth 60%	Peak shaving depth 55%
Stress amplitude	363	311	272	241	222	222
Corrected stress amplitude	383.6	314.2	294.9	292.3	220.76	219.68
Allowed cycle times	1.1×10^3	3.4×10^3	5×10^3	4.95×10^3	5.8×10^3	5.8×10^3
One cycle life loss	0.91	0.28	0.2	0.2	0.17	0.17

It can be seen from the above table that in various start-stop modes, the steam-water separator has the smallest stress amplitude during the hot start-stop process, and the hot-state start-stop mode has the smallest life loss for the steam-water separator. Under low load operation peaking mode, the stress amplitude at peaking depth of 65% is bigger than the thermal stress during hot start process, the stress amplitude at the peaking depth of 60% and 55% is smaller than the thermal stress during the hot start process.

5. Conclusions

For 670MW once-through boiler, when the peaking depth is lower than 65%, it is recommended to select the low-load operation mode for peak shaving; when the peaking depth of the boiler is greater than or equal to 65%, it is recommended to use the two-shift mechanism to adjust the peak.

When the peaking depth is lower than 65%, the life damage of the steam separator is less than the restart after the peak load of the low load operation. It is recommended to use the low load operation for peak shaving. In the actual peaking operation process, the crew needs frequent start and stop when using the two-shift mechanism for peak shaving, which is difficult to operate and has a large workload.

At the same time, the boiler needs to be ignited and fired when starting the boiler, and the investment cost is high .Therefore, when the peaking depth is higher than or equal to 65%, the low-load operation mode can be selected for peak shaving, or the two-shift mechanism can be selected for peak shaving.

References

[1] Clark K J, Hara K G, Lee C Q, et al. Method for extending the useful life of boiler tubes: EP, US5050108[P]. 1991.
[2] Brown T B. Assessing the effect of thermal transients on the life of boiler plant[C]// International Conference on Life Management of Power Plants. IEEE Xplore, 1994:137-143.
[3] Na Li, Wei Ping Yan. Finite Element Analysis of Transient Stress of Steam-Water Separator of 1000MW Boiler [J]. Boiler technology,2013,44(01):6-9.
[4] Yu Bing Zhao. Development of on-line monitoring system for life loss of super (super) critical boiler steam separator [D]. North China Electric Power University,2012.
[5] Li D, Fu Z, Zhang Q, et al. Mechanical Stress Analysis of a 600MW Supercritical Boiler Superheater Outlet Header[J]. Engineering, 2013, 05(1):66-68.
[6] Dou H, Li D Y, Dai Y, et al. Finite Element Stress Analysis of a 600 MW Supercritical Boiler Superheater Outlet Header[J]. Advanced Materials Research, 2012, 601:144-148.
[7] Ray A K, Tiwari Y N, Sahu J K, et al. Effect of Oxide Scale Deposition on the Residual Life of Service Exposed Reheater Tubes in Boiler of a Thermal Power Plant[J]. Corrosion Science, 2001.

IOP Conf. Series: Materials Science and Engineering **721** (2020) 012063 doi:10.1088/1757-899X/721/1/012063

Analysis of Influence of Atmospheric Parameter Fluctuation on Condenser Pressure

Pingyang Zi, Youliang Chen, Jiangang Hao and Daxing Xie

Huadian Electric Power Research Institute Co., Ltd, Hangzhou, Zhejiang, 310000, China

*Corresponding author's e-mail: dazi190026@163.com

Abstract. The influence of atmospheric temperature, humidity and pressure fluctuation on the condenser pressure is analyzed by numerical simulation. The seasons are mainly divided according to the atmospheric humidity range, and the atmospheric data are classified according to the seasons, and the fluctuation range of atmospheric temperature, humidity and pressure in specific seasons is determined, and a simplified condenser pressure coupling calculation model is established. The hypothetical iterative procedure calculates the condenser pressure value within a certain seasonal range and a certain atmospheric parameter fluctuation range. The results show that the atmospheric temperature fluctuations have a great influence on the pressure of the condenser in the spring and autumn, atmospheric humidity and pressure fluctuations have a great influence on the condenser pressure in summer. The primary and secondary analyses were carried out on several parameters of the atmosphere. The influence of temperature on the pressure of the condenser was 93%, the weight of the humidity was 6.2%, and the weight of the pressure was 0.8%, which provided a basis for optimizing the operation mode of the cold end system.

1. Introduction

In thermal power and nuclear power generator sets, the cold end system is an important part of the thermal power unit, which has an important impact on the safe and economic operation of the unit [1-5], and the condenser is an important equipment for the cold end system, and its vacuum is directly related to the power generation efficiency of the thermal power plant. The pressure of the condenser is affected by the steam exhaust volume, atmospheric parameters, circulating water flow, condenser cleanliness, vacuum pump efficiency, cooling tower air volume, etc. Its solution is a complex calculation [6-9], at the same time, the atmospheric temperature and humidity conditions vary greatly in different seasons, the pump operation mode, fan operation mode, and cycle rate are also different. These factors are coupled to each other and affect the condenser pressure, thus affecting the economics of the unit [10-15]. This requires optimization of the algorithm in order to quantitatively analyze the influence of atmospheric parameters on the condenser pressure, in order to find the economic operation mode of the pump and the fan under certain atmospheric conditions in a certain season, and provide guidance for the economic operation of the unit.

2. Unit overview

2.1Circulating water system

The circulating water system of the power plant unit adopts a secondary circulation expansion unit water supply system with mechanical ventilation cooling tower, one machine and two pumps. The pump operating conditions are: under pure coagulation conditions, one machine two pumps (high speed) are used in summer. The cooling rate is 60 times. The circulating water flow rate of each unit is about 7.61m^3/s. The flow rate of each pump is about 3.81m^3/s and the pump head is about 27.2m. In the spring and autumn, it is operated by one machine two (low speed). The cooling rate is 50.06 times, the cooling water flow rate of each unit is about 6.47m^3/s, the flow rate of each pump is 3.24m^3/s, the pump head is 24.8m, and one pump (high speed) is operated in winter, the cooling rate is 33.48 times. The circulating water flow rate of each unit is about 4.57m^3/s, the flow rate of each pump is 4.57m^3/s, and the head is 22.1m.

2.2Atmospheric statistics

According to the meteorological statistics, the atmospheric parameters are statistically calculated according to the season, so that the fluctuation range of the calculation parameters can be narrowed, so that the influence of the fluctuation of the atmospheric parameters can be studied. The specific fluctuation range is shown in tab 1.

Table 1. Atmospheric parameter fluctuation range

Season	(Spring and Autumn) April to May, October to November	(Winter) December, January to March	(Summer) June to September
Temperature fluctuation range /℃	10-25	1-15	21-31
Air pressure fluctuation range /Pa	100.8-102.5	101.7-103	100.2-101.5
Humidity fluctuation range /%	76-80	74-77	80-83

3. Condenser pressure calculation

3.1Calculation condition

The atmospheric parameters were calculated by season. The effects of atmospheric temperature, atmospheric pressure and atmospheric humidity on the condenser pressure were calculated for each season. As shown in Table 2, a total of nine operating conditions were calculated. Taking winter as an example, working conditions 1-3 are respectively calculating the influence of fluctuations in atmospheric temperature, atmospheric pressure and atmospheric humidity on the condenser pressure in the range shown in Table 1. Working conditions 4-6 correspond to spring and autumn, working conditions 7- 9 corresponds to summer.

Table 2. Calculating working condition parameters

Calculation condition	atmospheric temperature/℃	atmospheric pressure/Pa	atmospheric humidity/%	Cooling water flow /m^3·h^{-1}	Cooling ratio
1	1-15	102.3	75.5	16452	33.48
2	8	101.5-103	75.5	16452	33.48
3	8	102.3	74-77	16452	33.48
4	10-25	101.6	78	23292	50.06
5	17.5	101-102.5	78	23292	50.06
6	17.5	101.6	76-80	23292	50.06
7	21-31	100.8	81.5	27396	60

8	26	100-101.5	81.5	27396	60
9	26	100.8	80-83	27396	60

3.2Power tower calculation

The known data are atmospheric pressure P_a, inlet air dry temperature t_a and inlet air wet bulb temperature t_{au}, circulating water quantity Q, turbine exhaust volume D_c, condenser parameters, and power tower parameters. The amount to be determined is the fan air volume G, export water temperature of the power tower t_{w2}, the condenser temperature t_s.

The cooling tower capacity N_c is calculated as follows:

$$N_c = a \cdot lambda^b \tag{1}$$

In the middle: *lambda* — Gas to water ratio, calculated according to the following formula.

$$lambda = \frac{r\,\mathrm{hom} \cdot G}{A_f \cdot q} \tag{2}$$

In the middle:*rhom* — Air density, kg·m^{-3};

G— Tower air volume, m^3·s^{-1};

A_f— Mechanical tower ventilation cross section, m^2;

q— Water density, kg·m^{-2}·s^{-1}.

The number of cooling tower tasks N_t is calculated as follows:

$$N_c = \frac{C_w \cdot \Delta t}{6}\left(\frac{1}{i_2'' - i_1} + \frac{4}{i_m'' - i_m} + \frac{1}{i_1'' - i_2}\right) \tag{3}$$

In the middle:C_w —Circulating water specific heat capacity, kJ·kg^{-1}·℃$^{-1}$;

Δt—Circulating water temperature difference, ℃;

i_2'', i_1—Enthalpy of saturated gas and corresponding gas of the outlet water temperature, kJ·kg^{-1};

i_1'', i_2—Enthalpy of saturated gas and corresponding gas of the inlet water temperature, kJ·kg^{-1};

i_m'', i_m—Enthalpy of saturated gas and corresponding gas of the average inlet and outlet water temperature, kJ·kg^{-1}.

Assume that the initial flow rate of the power tower fan G is 500 m^3·s^{-1}, and the air density can be obtained by entering the air dry and wet bulb temperature of the tower. Through the calculation of the resistance and the correlation curve, the wind turbine resistance H_z and the total pressure of the fan H_c can be obtained. Along with the air volume increases, the total pressure of the fan decreases, and the fan resistance rises, therefore there is a certain air volume G_0, so that H_z is equal to H_c, and the air volume is the required air volume.

Further, the cooling capacity N_c of the cooling tower and the number of cooling tasks N_t can be obtained by the formulas (1) and (3), assuming that the water temperature of the tower is t_{w1}', and there is also a tower water temperature t_{w2}', so that N_c is equal to N_t, and the outlet temperature t_{w2}' is under the condition of the inlet water temperature t_{w1}'.

3.3Condenser calculation

Assuming that the steam exhaust steam temperature is t_{s1}, the temperature difference dt of the inlet and outlet water can be calculated. It is considered that the temperature difference between the inlet and outlet of the circulating water is equal to the temperature difference between the inlet and outlet of the cooling tower. There is a certain water temperature hypothesis value t_{w1}, so that the temperature difference between the inlet and outlet towers is equal to dt,t_{w1} is the inlet water temperature, then the turbine exhaust steam temperature t_s can be obtained, and there is a certain assumed value t_{s1}, so that t_s is equal to t_{s1}, this temperature t_{s1} is the condenser temperature, and its corresponding saturated vapor pressure P_a is the condenser pressure.

4. Analysis and conclusion

4.1Influence of atmospheric temperature on condenser pressure

As shown in Figure 1, in winter (December, January to March), the relative humidity of the atmosphere increases from 1 °C to 15 °C (the relative humidity and pressure of the atmosphere are unchanged), and the condenser pressure is increases from 6.0 kPa to 8.2 kPa, increased by 2.2 kPa; spring and autumn (April to May, October to November), atmospheric temperature increases from 10 ° C to 25 ° C (atmospheric relative humidity and pressure unchanged), condenser pressure by 6.08kPa increased to 8.8kPa, increased by 2.72kPa; in summer (June to September), atmospheric relative humidity increases from 21 °C to 31 °C (atmospheric relative humidity and pressure unchanged), condenser pressure from 6.85 The kPa increases to 9.4 kPa, which is increased by 2.55 kPa.

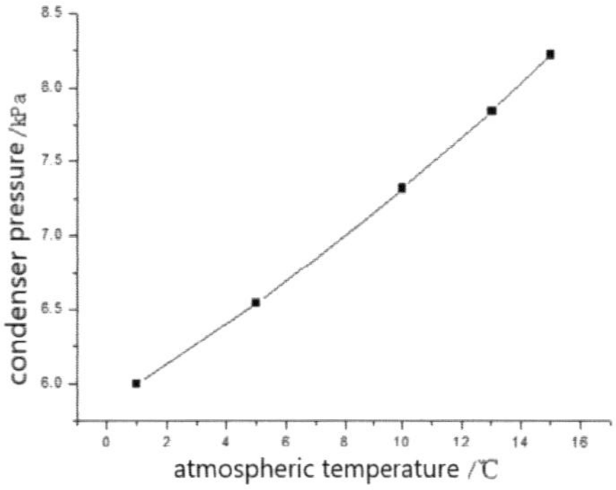

a)Winter

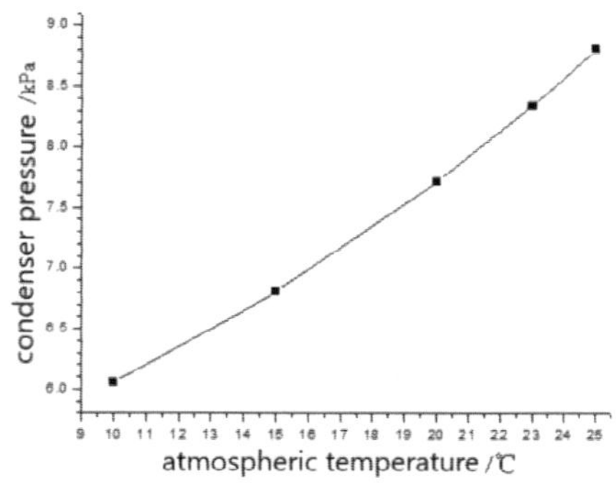

b)Spring and Autumn

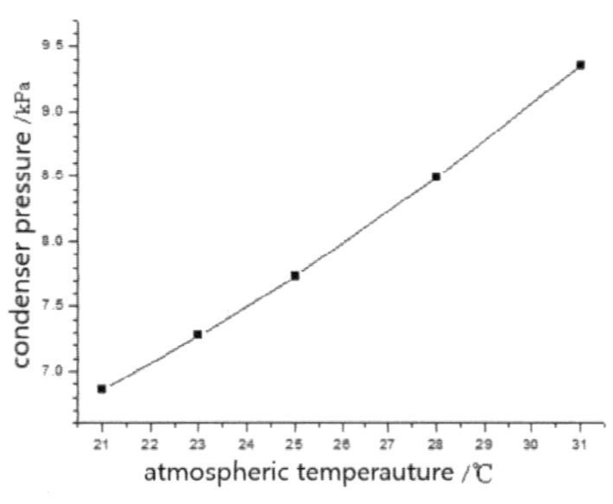

c)Summer

Fig.1 Effect of atmospheric temperature on condenser pressure

The air temperature fluctuation affects the temperature of the inlet and outlet water, which in turn affects the condenser pressure. The calculation results show that the condenser pressure increases with the inlet air temperature rises. This is because the average temperature difference between the air and the cooling water in the cooling tower decreased as the temperature of the inlet air rises, resulting in an increase in the temperature of the inlet and outlet water and an increase in the condenser pressure. At the same time, it can be seen from the graph that the condenser pressure with the variation of the atmospheric temperature (curve slope) is the largest in summer, followed by spring and autumn, and the smallest in winter. From the perspective of fluctuation range, the fluctuation of condenser pressure caused by atmospheric temperature in spring and autumn is greater than that in summer. Therefore, it is necessary to focus on the influence of temperature difference between day and night in spring and autumn on condenser pressure.

4.2Influence of atmospheric humidity on condenser pressure

As shown in Figure 2, in winter (December, January to March), the atmospheric relative humidity increases from 74% to 77% (the atmospheric dry bulb temperature and pressure did not change), and the condenser pressure increases from 6.976 kPa to 7.014 kPa, increased by 0.038 kPa; in spring and autumn (April to May, October to November), atmospheric relative humidity increases from 76% to 80% (atmospheric dry bulb temperature and pressure unchanged), condenser pressure increases from 7.195 kPa to 7.277 kPa, increased by 0.082 kPa; in summer (June to September), the atmospheric relative humidity increases from 80% to 83% (the atmospheric dry bulb temperature and pressure are unchanged), condenser pressure increases from 7.918 kPa to 8.03 kPa, an increase of 0.112 kPa.

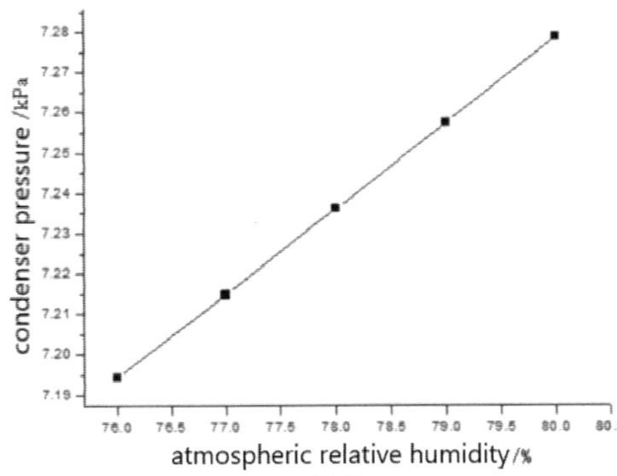

a)Winter

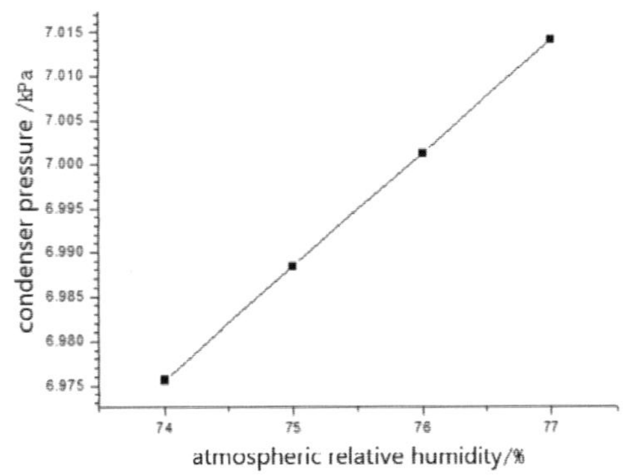

b)Spring and Autumn

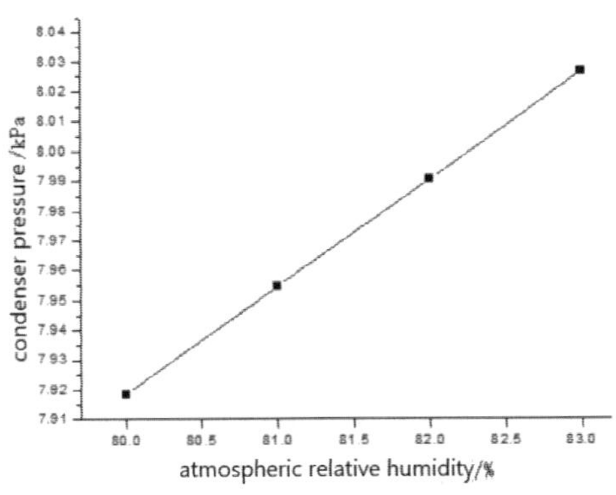

c)Summer

Fig.2 Influence of atmospheric humidity on condenser pressure

Atmospheric humidity affects the temperature of the inlet and outlet water by affecting the density and enthalpy of the incoming tower air, resulting in changes in condenser temperature and pressure. The calculation shows that the condenser pressure increases with the increase of atmospheric humidity. This is because the air enthalpy difference of import and export air decreased as humidity increases, the tower cooling task increases, results in the tower water temperature and condenser pressure rising. From the curve, the influence of atmospheric humidity on the condenser (curvature slope) sorted from big to small is: summer, spring and autumn, winter. From the perspective of fluctuation range, the condenser pressure fluctuation caused by atmospheric humidity in summer is higher than that in spring and autumn, the fluctuation range in winter is smaller. Therefore, it is necessary to focus on the atmospheric humidity fluctuation in summer.

4.3Influence of atmospheric pressure on condenser pressure

It can be seen from Figure 3 that in winter (December, January to March), the atmospheric pressure increases from 101.5 kPa to 103 kPa (the atmospheric dry bulb temperature and humidity do not change), and the condenser pressure increases from 6.989 kPa to 7.001. kPa, increased by 0.012kPa; in spring and autumn (April to May, October to November), the atmospheric pressure increases from 101kPa to 102.5kPa (the atmospheric dry bulb temperature and humidity are unchanged), the condenser pressure increases from 7.23kPa to 7.245 kPa, increased by 0.015 kPa; in summer (June to September), the atmospheric pressure increases from 100 kPa to 101.5 kPa (the atmospheric dry bulb temperature and humidity are constant), and the condenser pressure increases from 7.964 kPa to 7.780 kPa, an increase of 0.016 kPa.

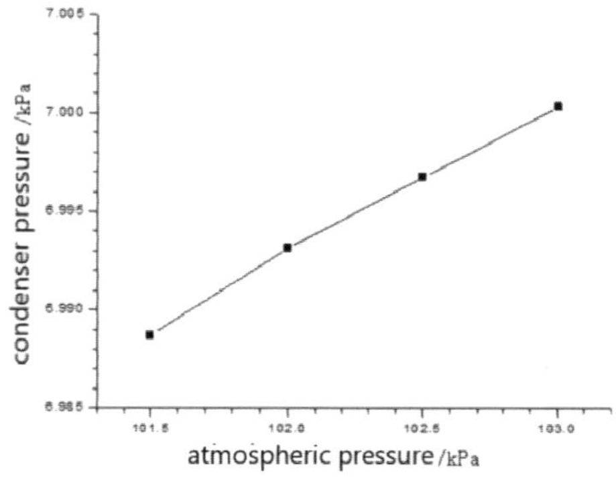

a)Winter

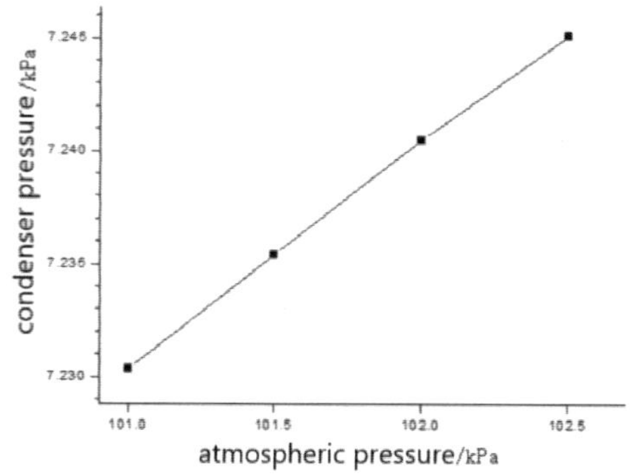

b)Spring and Autumn

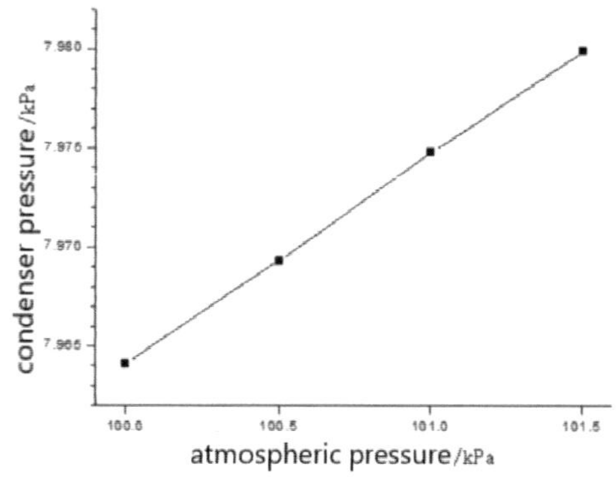

c)Summer

Figure 3. Influence of atmospheric pressure on condenser pressure

Atmospheric pressure fluctuations affect the air density, which affects the gas-water ratio of the cooling tower, and then affects the cooling task of the cooling tower, resulting in fluctuations in the temperature of the inlet and outlet towers, finally the condenser pressure changes. The calculation results show that the condenser pressure increases with the increase of atmospheric pressure. This is because as the atmospheric pressure increases, the difference between the outlet air of the cooling tower and the inlet air decreases, resulting in an increase in the cooling task and the condenser pressure. It can be obtained from the curve that the fluctuation range of the condenser pressure with atmospheric pressure (the slope of the curve) from high to low is: summer, spring and autumn, winter. From the perspective of fluctuation range, the condenser pressure fluctuation caused by atmospheric pressure in summer is higher than that in spring and autumn, and the fluctuation range in winter is smallest. Therefore, it is necessary to focus on the atmospheric pressure fluctuation in summer.

5. Conclusion

1) Through the above analysis, within the calculation range, the atmospheric temperature fluctuation has the greatest influence on the condenser pressure, accounting for 93% of the total atmospheric impact, atmospheric humidity accounting for 6.2%, and atmospheric pressure accounting for only 0.8%. From a seasonal perspective, the order of influence of atmospheric parameter changes on the

condenser pressure is ranked from high to low: summer, spring and autumn, winter, from the perspective of fluctuations, atmospheric humidity and pressure have the greatest influence on condenser pressure in summer, and atmospheric temperature has the greatest influence on condenser pressure in spring and autumn.

2) In the optimization of cold-end system, it is necessary to closely combine the fluctuation characteristics of atmospheric temperature and humidity, such as temperature difference between day and night, temperature and humidity during rainfall. In summer, we should focus on the impact of precipitation on temperature and humidity, and we should focus on the influence of temperature difference between day and night in other seasons.

Reference

[1] YU Jian,YANGJianming.Effect of water ring vacuum pump on the economics of condenser[J].Thermal Power Generation, 2009,38(10):29-32.

[2] CAO Lihua, SI Heyong, DONG Enfu, HU Pengfei, NING Zhe, LIU Wei.Aerodynamic Performance Optimization of Steam Turbine Exhaust Passage and Condenser Heat Transfer Characteristics Analysis[J].Journal of China Electrical Engineering, 2018,38(12):3597-3604.

[3] YI Jiuqing.Analysis on Energy Saving Optimization of Condenser for 330 MW Unit[J].Energy Research and Management,2018(04):102-105.

[4] ZHANG Yu,SHAORui,LILing.Optimization Calculation of Condenser Back Pressure of Generator Sets[J].Journal of University of Shanghai For Science and Technology,2018,40(05):429-434.

[5] QIU Ruifei.Discussion on the Influence of Condenser Vacuum on Steam Turbine Economy[J].Applied Energy Technology,2018(03):9-12.

[6] TAN Jiangfeng.Analysis and improvement measures of abnormally high vacuum value of condenser in a power plant[J].Electromechanical Information,2017(30):71-72.

[7] ZHANG Guoqiang, ZHANG Heng, LI Yongyi, LIU Hengping, SONG Xiaotong, YANG Zhiping, YANG Yongping.Study on Performance Optimization of Gas-Steam Combined Cycle Based on Variable Back Pressure Modification[J].Thermal Power Generation,2018,47(04):92-98.

[8] MA Shanjun, LI Shaohua, ZHANG Li, ZENG Zhendong, CHENG Haoran. Optimization and Analysis of Cold End System of Thermal Power Units[J].Thermal Power Generation,2017,46(03):82-86.

[9] Yingying Yao, Jinfeng Zhao, Shouheng Sun, Hang Wang, Shan Yang, Wenbo Lv, Yan Jiang, Zhipeng Jiang. Analysis of Heat Transfer Characteristics of Shell-and-Tube Condenser[J]. IOP Conference Series: Materials Science and Engineering,2019,490(6).

[10] Hydro-Thermal Corporation; Patent Issued for Molecular Sieve Depressurization System Utilizing A Low Pressure Inductor Type Vapor Condenser (USPTO 10,220,371)[J]. Biotech Business Week,2019.

[11] John Tomczyk. Condensers Can Offer Valuable Insights Into System Operation[J]. Air Conditioning, Heating & Refrigeration News,2019,266(13).

[12] QU Binbin, ZHANG Li, WANG Fuhua. Numerical calculation of condenser performance considering the influence of vacuum pump variable working conditions[J]. Journal of Power Engineering, 2018, 38(08): 640-644.

[13] CHENG Dongtao,ZOUYang,MATingshan,JUWenping.Study on Performance Evaluation Method of Power Plant Condenser Reconstruction[J].Journal of Power Engineering,2016,36(11):889-893.

[14] ZHAO Wensheng, SONG Baichuan, GUO Hao.Optimization of Circulating Water System for Dual Pressure Condenser[J].Technology of Steam Turbines,2017,59(02):145-147.

[15] JIANG Wen,HU Qing,NING Yuqin,ZHU Wenzhong.Optimization and Application of Cold Junction System of a 600MW Unit[J].Technology of Turbine,2017,59(01):63-64.

Experiment and Simulation of Coriolis Mass Flowmeter with Multi-Medium and Multi-Working Conditions

Hui Qiu[1], Tao Zhou[1], Zhipeng Xu[2*]

[1]West Gas East Pipeline Branch of PetroChina, Shanghai, 200122, China

[2]College of metrology & measurement engineering, China Jiliang University, Hangzhou, Zhejiang, 310027, China

[*]Corresponding author's e-mail: xuzhipeng@cjlu.edu.cn

Abstract. Coriolis mass flowmeter can directly measure mass flow, high measurement accuracy, which can be used for gas, liquid, high viscosity media and other fluids. Thus it is one of the most important directions of flow measurement. Coriolis mass flowmeters currently used in natural gas pipeline networks are usually calibrated using water. The effects of differences in the detection medium and pressure on the results are lacking in relevant studies. This article analyzes the influence of fluid pressure on the test results by statics simulation, and different pressure was carried out on the DN80 flow meter to simulate different calibration media. Then real flow calibration on several brands and models of flowmeters to compare the influence of water and natural gas respectively. The experimental results show that the error of water is less than ±0.2%, and the error of natural gas is less than ±01%. The influence brought by pressure is close to the simulation analysis, and the results of different media have good consistency.

1. Overview

The most prominent feature of the Coriolis mass flow meter is its ability to directly measure the fluid quality. It is not only accurate, repeatable and stable, but also has no choke or movable elements in the fluid passageway, making it reliable, lasting and capable of measuring the flow rate of high-viscosity fluid and high-pressure gas [1-3].

The Coriolis mass flow meter consists of two parts, namely, the mechanical sensor through which the fluid flows and the flow meter transmitter. The sensor consists of a shell, a micro-vibration measuring tube, a vibrator, a signal detector, and a temperature compensating element. When the fluid passes through the measuring tube, the measuring tube is slightly distorted due to the Coriolis force; as a result, the phase difference at the time of vibration is converted into a linear electrical signal output. The transmitter receives and processes the signals to show the mass of the metered fluid.

Since the Coriolis mass flow meter can directly measure the flow mass of natural gas, it is becoming more widely used in metering the company's internal compressor fuel gas and self-use gas. But problems still exist. For example, no conclusion has been made in the domestic research in the aspect of natural gas flow meter application and performance evaluation. In addition, there are few academic reference for the performance test of natural gas mass flow meter in the aspects of whether water or air can be used as medium instead of natural gas, whether the measurement results of mass flow meter is reliable under different pressure conditions, whether the flow meter is stable during use, and whether drift and other influencing factors will occur [4, 5].

It is necessary to analyze and evaluate the results of tests in which water and natural gas are used as the medium under different pressures, and verify the feasibility of using water instead of natural gas to calibrate the mass flow meter based on the performance indicators and measurement results [6].

This paper uses three Coriolis flow meters of different models and different calibers to obtained the test results according to the verification procedure. Since the flow meters are different in age, the test results cannot fully characterize the corresponding brand performance. To avoid misunderstanding, the serial numbers of the flow meters are replaced by letters.

2. Water flow test

This paper uses the static mass method water flow standard device according to the verification procedure, and conducts the test in the Institute of Metrology, Zhejiang Province. The test flow range of the device is (0.2~220) t/h, and the relative expansion uncertainty is 0.05% (k=2). After connecting the device, the supporting instrument and the circuit of the flow meter, select the pulse output mode, manually start the water flow device so that the pipe is full of static water medium; then close the valves at both ends, and set the instrument to zero. The test is performed according to the flow point order of Q_{max}, 0.5 Q_{max}, 0.2 Q_{max}, Q_{min}, and Q_{max} and the data is recorded respectively. In addition, a total of three tests are carried out in different seasons so as to verify the repeatability of the flow meter and compare the metering performance of flow meters under different temperatures.

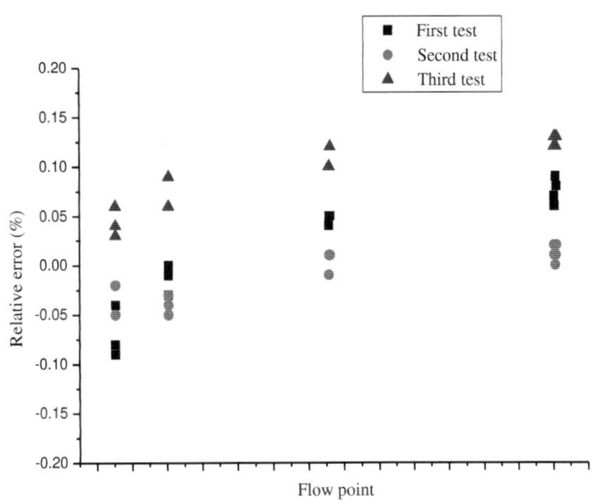

Figure 1. Water flow test results of A flow meter

The range ratio of A flow meter is approximately 10:1. It can be seen that the results of the three tests have good repeatability, with the single point repeatability higher than 0.05% and the majority higher than 0.02%. The single point errors of the first two tests are all less than 0.1%, indicating high accuracy; the average single point error of the third test is about 0.1%.

For B flow meter, its diameter is DN=80mm, its test flow range is (6~60) t/h, and the maximum-minimum flow ratio is about 10:1. From the test data, it can be seen that the results of B water flow have good repeatability, with single point repeatability higher than 0.1% and the majority higher than 0.05%. Relatively speaking, the accuracy of the first test is poorer, with its single point error of about (0.10~0.15) %, while the accuracy of the second and third tests is higher, with the single point error of about 0.05%.

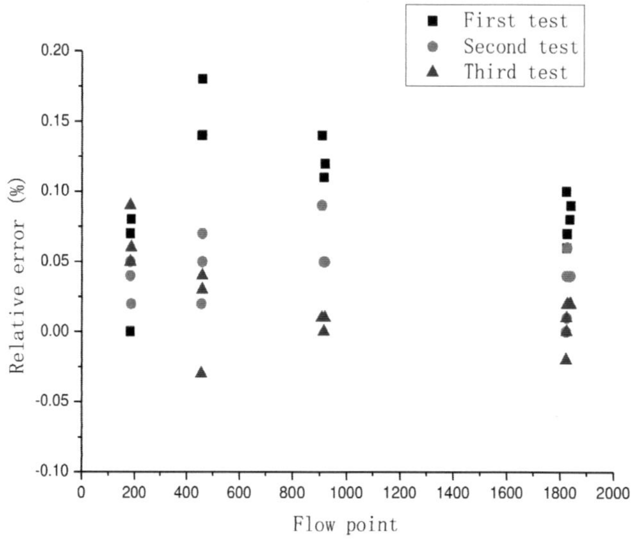

Figure 2. Water flow test results of B flow meter

The maximum-minimum flow ratio of C flow meter is approximately 5:1. The repeatability of each flow point is higher than 0.05% and the single point error of the all flow points is less than 0.1% except for the minimum flow point. In addition, in the third test, the error also shows a significant trend of turning from negative to positive. The error of the second test is generally 0.1% smaller than that of the third one.

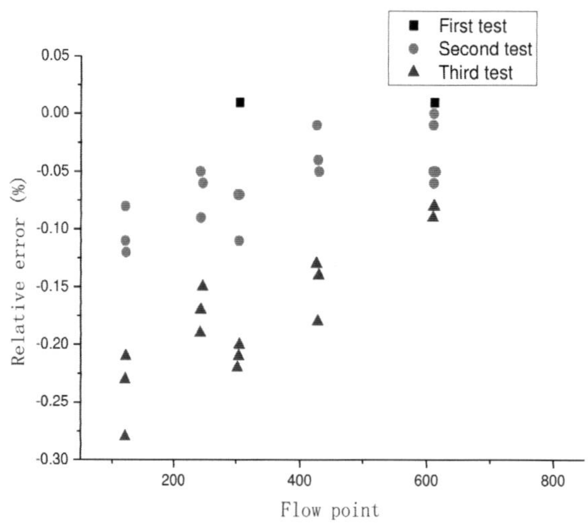

Figure 3. Water flow test results of C flow meter

3. Natural gas test

The natural gas tests are carried out at the Guangzhou Sub-station of the National Oil and Gas Large Flow Metering Station. The tests adopt standard meter method and the gas flow device, with the test flow range set as (8~15000) m3/h, the relative expansion uncertainty, 0.29% (k=2), and the gas pressure, 7MPa. The test procedure is the same as above after turning off the pressure compensation function of the flow meter. The test results of the three flow meters are shown in Figures 4~6.

The measured natural gas flow range of A flow meter is about (2~20) t/h. despite the obvious differences in temperature and pressure of the three tests, the overall repeatability is higher than 0.1%, and most test errors are close to 0.2%.

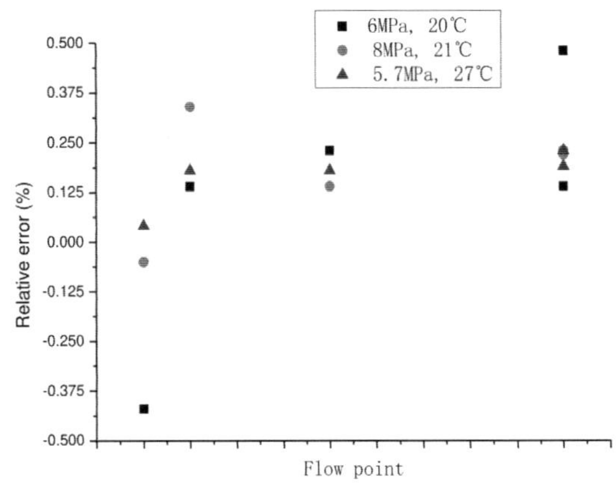

Figure 4. Natural gas flow test results of A flow meter

The measured natural gas flow range of B flow meter is about (2~20) t/h. C flow meter shows a certain fluctuation in three tests, with the error mean being -0.1%, 0% and -0.5%, respectively. The error of the third test is relatively larger, with the minimum flow point of -0.94%, and the single point repeatability fluctuating significantly from 0.01% to 0.32%

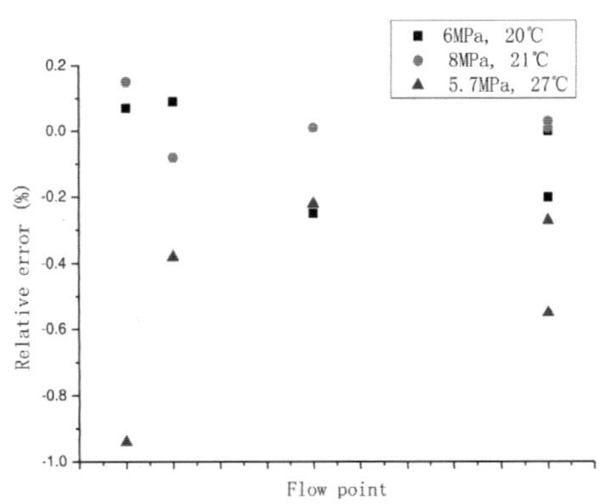

Figure 5. Natural gas flow test results of B flow meter

For C flow meter, its natural gas flow range is about (2~20) t/h, the overall repeatability is good, with the majority higher than 0.05%, and its errors are mainly negative.

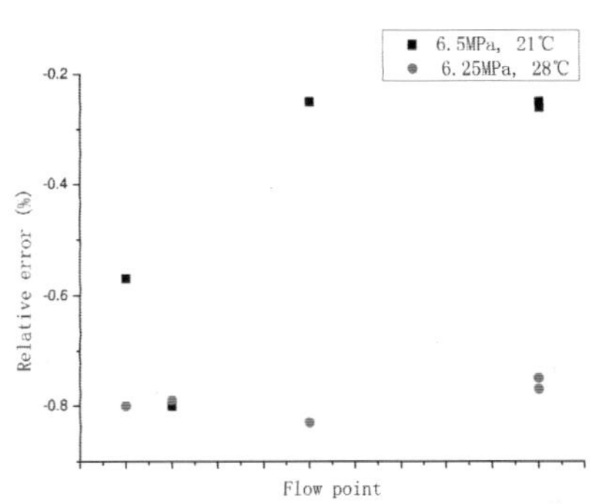

Figure 6. Natural gas flow test results of C flow meter

IOP Conf. Series: Materials Science and Engineering **721** (2020) 012064 doi:10.1088/1757-899X/721/1/012064

It can be seen from the comparison results that when the medium is water, the error curve is smoother with less fluctuation; when the medium is natural gas, the error curve shows stronger fluctuations. This phenomenon is related to the lower density of the gas (which is 1/20 or so that of the water). In addition, the uncertainty of the natural gas flow standard device itself is relatively low, which will also affect the results. The error curves of actual gas tests are generally below that of the water flow tests, and the average offset amplitude of natural gas tests is (-0.5~-0.3) %. Since pressure compensation is turned off before the test, the phenomenon is supposed to be related to the influence of the medium pressure. Therefore, this paper intends to carry out the corresponding analysis.

4. Simulation analysis

In order to analyze the influence of medium pressure on the metering characteristics, this paper establishes a simulation model of the Coriolis mass flow meter in SolidWorks. The structure and size are set according to the actual flow meter, and the non-flow pipeline section is ignored. The author selects the stress analysis in the Simulation plug-in, sets the material to stainless steel, fixes the pipe section, and generates the mesh for calculation and analysis [7, 8].

In the stimulation test of the DN80 flow meter, 0.1 MPa, 0.5 MPa, and 6 MPa are applied to respectively simulate the case where the medium is air, water, and natural gas. From the simulation results in Fig. 10, it can be seen that under the medium pressure, the maximum deformation is concentrated at the center of the U-tube since the interfaces at both ends are fixed. When the pressure is 0.1 MPa, 0.5 MPa, and 6 MPa, the maximum deformation in the diameter direction is 7.634×10-3 mm, 7.108×10-3 mm, and 6.356×10-2 mm, respectively, which indicates that greater pressure has a greater influence on the deformation of the U-shaped tube.

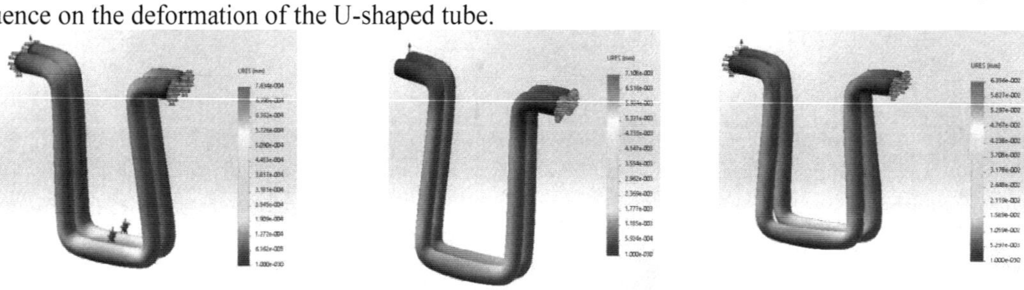

(a) p=0.1 MPa (b) p=0.5 MPa (c) p=6 MPa

Figure 7. Simulation results under different pressures

The expansion coefficient is a physical quantity that characterizes the expansion property of an object, that is, a physical quantity that characterizes the expansion degree of length, area, and volume. Its calculation formula is as follows:

$$K = \frac{R_1^2}{R_0^2} \tag{1}$$

Where: R_0 is the size before expansion and R_1 is the size after expansion.

According to the above formula, K1, the variation of the cross-sectional area of the pipeline during the natural gas test of the three Coriolis mass flow meters, and K2, the average value of the ratio of the actual flow rate to the detected flow rate can be calculated. It is known that K2 approximates the pressure compensation coefficient of the flow meter. The simulation coefficient K1 and experimental coefficient K2 are shown in Table 1. It can be seen that there is good consistency between K1 and K2 in terms of the numerical value and the variation direction. However, since K2 is calculated from the error detection, which is greatly affected by the performance of the flow meter and the detection process, it provides limited reference in quantitative evaluation [9, 10].

Table 1. Comparison of simulation and experimental results

No.	Simulation coefficient K_1	Experimental coefficient K_2
A	1.0089	1.00610
B	1.0089	1.00416
C	1.0038	1.00058

5. Conclusion

Based on the experimental comparison of three Coriolis mass flow meters, it can be seen that the detection results of tests in which water and natural gas are respectively used as medium are generally consistent. When the medium is water, the error is between -0.25% and 0.2%. When the medium is natural gas, the error is slightly larger because of low device uncertainty and the lower density of natural gas. Errors of two flow meters exceed -0.5%. Since the pressure compensation is turned off, the natural gas detection result has a significant negative deviation. Static simulation is carried out targeting the influence of test pressure on the flow meter, and the result shows that the pressure increase will cause the expansion of the metering tube, resulting in negative deviation, which is more consistent than the experimental data.

It is preliminarily determined that water can be used to replace natural gas in the Coriolis mass flow meter, but pressure compensation should be applied to the medium in its application. The author will conduct further tests to compare Coriolis mass flow meters.

Acknowledgements
This work was supported by the Zhejiang Provincial Natural Science Foundation of China (No.LY18E050009) and the research project of West-east Gas Transmission Branch of China Petroleum Pipeline Co., LTD. (No. XQDSPO123).

References

[1] Mu, J.Y. (2011) Development and Application Status of Coriolis Mass Flow Meter. China Instrumentation. 7: 60-62.
[2] Xue, H.M., Zhang, D.F. (2013) Application of Coriolis Mass Flowmeter in Natural Gas Metering. Petroleum Instruments, 27: 36-38.
[3] Liang, G.W. (1998) The Limitations Coriolis Mass Flowmeters of the in its Applications. Journal of China Institute of Metrology, 1: 61-68.
[4] Zhang, N.X. (2002) Principle of coriolis mass flowmeter and its application to gas measurement. Petroleum Instruments, 3: 18-20, 61-62.
[5] Chen, B. (2017) The Performance Test and Study of Coriolis Mass Flowmeter Used for Crude Oil Trade Measurement. Oil-Gas Field Surface Engineering, 36: 61-63.
[6] Wang, H.W., Li, G.F., Shi, Z.D. (2016) The Influence of Medium on The Verification Results Of Coriolis Mass Flowmeter. China Metrology, 12: 68-69.
[7] Mole, N., Bobovonk, G., Kutin, J., et al. (2008) An improved three-dimensional coupled fluid structure model for Coriolis flowmeters. Journal of Fluids and Structures, 24: 559-575.
[8] Ji. A.M. (2001) Modeling and Simulation Analysis of Coriolis Mass Flowmeter. University of Science and Technology of China.
[9] Chen. X.C., Zhou. T., Zhang. K. (2015) Numerical simulation of flow field of turbine flowmeters and accuracy analysis. Journal of China University of Metrology, 26: 406-410.
[10] Cai. T.Q., Liu. Q., Tang. G.J., Zhu. Q.Y. (2002) Finite Element Analysis of Ω Tube Coriolis Mass Flow Meter. Chinese Journal of Scientific Instrument, 23: 520-521.

IOP Conf. Series: Materials Science and Engineering **721** (2020) 012065 doi:10.1088/1757-899X/721/1/012065

Numerical Simulation Study on Radiation Section of Dehydrogenation Furnace

Linlin Duan[1], Cenfan Liu[1]*, Sheng Chen[1], Fang Ji[1], Guide Deng[1]

[1]China Special Equipment Inspection and Research Institute, Heping Street, Chaoyang District, Beijing, 100029, China

*Corresponding author's e-mail: liucenfan288@163.com

Abstract. The three-dimensional numerical simulation of the radiation section of a dehydrogenation furnace was carried out by CFD, the thermal-structure coupling analysis was performed. The full-scale meshing of the geometric model was used in this paper, the temperature field and flow field of the furnace and the furnace tube were simulated. The thermal radiation values and temperature values of the simulation results were compared with the design values, it shows that the simulated values are in good agreement with the designed values. The corresponding numerical calculation can be performed for half or 4/1 of the geometry under the condition with the geometric symmetry of the furnace and the furnace tube being good. The simulation results can provide a reference for the design and optimization of the dehydrogenation furnace.

1. Introduction

The furnace is the core device for propane dehydrogenation, and the radiant section is the main place for heat exchange and chemical reaction in the furnace. It is the most important component of the reformer. However, the furnace tube in the radiant section may not be able to long-term safe operation due to crack and creep with local overheating, so focus on the radiant section of the furnace [1]. The traditional research method for radiant section has limitations for the current large-scale dehydrogenation furnace, with the rapid development of computers, the CFD numerical simulation method can obtain a large amount of simulation data under less design conditions, it can provide the necessary information of the heat transfer situation and fluid flow in the furnace to optimize the furnace structure [2,3].

For the combustion and heat transfer problems in industrial furnaces[4], the computational fluid dynamics can be used to simulate the combustion and heat transfer conditions of the furnace under various working conditions in the case of determining structure of the dehydrogenation furnace and the amount of fuel intake. According to the position of the burner and the difference of reformer, many scholars have conducted corresponding research [5-7]. After establishing a three-dimensional model of geometry, generally, the simulation is divided into two parts: the numerical simulation of flame combustion and the fluid flow heat transfer simulation in the furnace tube, which will obtain accurate and reliable simulation results with less time and cost [8]. It needs the heat flux density or temperature, the heat flux distribution of the outer wall of the furnace tube in the combustion field, to calculate the coupling of steady-state combustion and thermal response of the fluid in the furnace tube. The heat flux density on the outer wall surface of the furnace tube is used as a coupling condition to further calculate the flow and heat transfer inside the furnace tube. In this paper, the calculation results are compared with the design values to adjust the parameters of the furnace and fuel to achieve long-term efficient and safe operation.

2. The setting of numerical simulation for dehydrogenation furnace

2.1 Control equation[9]

1) Mass conservation equation

$$\frac{\partial \rho_g}{\partial t} + \nabla \cdot (\rho_g \mathbf{v}_g) = 0 \tag{1}$$

2) Momentum conservation equation

$$\frac{\partial}{\partial t}(\rho_g \mathbf{v}_g) + \nabla \cdot (\rho_g \mathbf{v}_g \mathbf{v}_g) = -\nabla p + \nabla \cdot \boldsymbol{\tau}_g + \rho_g \mathbf{g} + \mathbf{F} \tag{2}$$

3) Energy conservation equation

$$\frac{\partial}{\partial t}(\rho_g h_g) + \nabla \cdot (\rho_g h_g \mathbf{v}_g) = \nabla \cdot \left[\left(\kappa_g + \kappa_{g,t} \right) \nabla T_g \right] + S_g \tag{3}$$

4) The calculation formula of the mixed fraction

$$f = \frac{Z_i - Z_{i,ox}}{Z_{i,fuel} - Z_{i,ox}} \tag{4}$$

In this study, the heat flux of the outer wall of the furnace tube is first obtained by combustion simulation. But the heat flux density at the interface of the solid domain fluid should be regarded as part of the calculation result, rather than the known condition. This is a typical conjugate heat transfer problem.

2.2 Geometric model

This paper mainly studies the radiant section of the dehydrogenation furnace, which includes 24 circular burners and 16 square burners. There are 24 tubes in the furnace radiant section. The radiant section is 11m long, 8m wide and 9m high, as shown in Figure 1. The burner is in the bottom, venting outlet is at the top, and the entry and exit method of mediums in the conversion tube are middle tubes to entry, other tubes to exit.

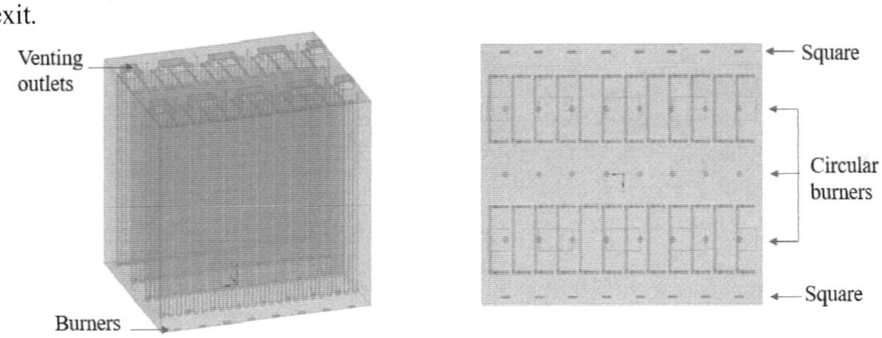

Figure 1. Three-dimensional geometric model and bottom of the radiant section

2.3. Meshing

The meshing method is poly-hexcore meshing, as shown in figure 2. The grid-independent verification is performed. The number of dehydrogenation furnace grids is 22.6 million for, and the conversion tubes is 18.2 million.

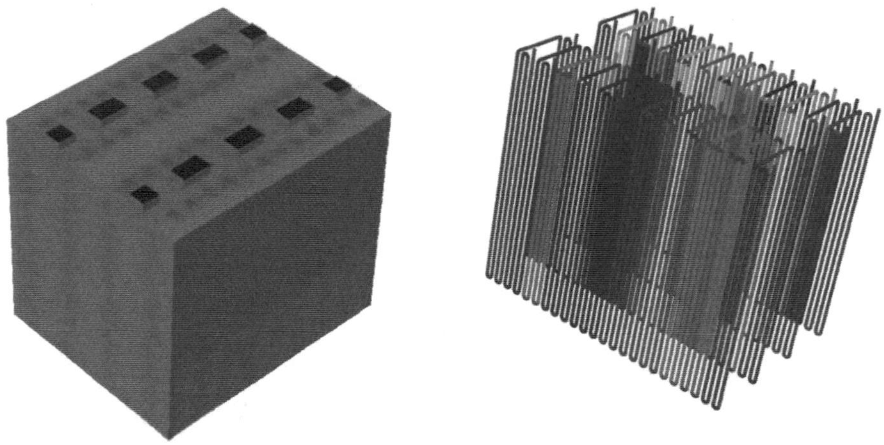

Figure 2. Meshing of dehydrogenation furnace and conversion tube

2.4 Boundary conditions and simulation settings

Under the design condition, the main component of the fuel of the dehydrogenation furnace is hydrogen. The air inlet flow rate of a single circular burner is 0.78kg/s, the gas inlet flow rate is 0.034kg/s; the air inlet flow rate of a single square burner is 0.39kg/s, gas inlet flow rate is 0.017k/s. The main component of the medium in the furnace tube is propane, the inlet flow rate is 89kg/s, the inlet and outlet temperatures are respectively 462°C and 620°C, the inlet and outlet pressures are respectively 0.34MPaA and 0.12MPaA.

The specific heat capacity, viscosity, and thermal conductivity of a fuel such as hydrogen are defined by a temperature equation, and a piecewise-polynomial is selected. The wall material is stainless steel.

Steady state calculation is adopted for the furnace combustion, the combustion model is non-premixed combustion, the radiation is discrete coordinate model, the turbulence model is Realisable k-ε, the velocity pressure coupling is coupled algorithm, and the discrete format pressure is PRESTO[10]. The simulation in the furnace tube is the steady-state calculation of the single-phase flow and the component transport equation.

3. Simulation results

3.1 Comparison of simulation results and design values

Through the simulation calculation, the relevant numerical results are obtained. The design values of the dehydrogenation furnace are compared with the simulation values, as shown in table 1. Overall, the simulation results agree well with the design values.

Table 1. The table of Comparison design values with simulation values

Category	Design Value	Simulation Value
The average temperature of the flue gas outlet	812°C	841°C
Average heat flux of the tube	32883 W/m^2	32342W/m^2
Average pressure of furnace tube inlet	0.34MPaA	0.32MPaA
The average inlet speed of the tube	About 80m/s	93m/s
Furnace tube outlet average temperature	620°C	624°C
The average outlet speed of the tube	About 250m/s	264m/s

3.2 Temperature and velocity distribution

The temperature distribution of the X-axis section of the center circular, side circular, and square burners are shown in Fig. 3. The flame shape of the circular burner is circular, and the center flame height is 1/3 of the total furnace height, that is highly uniform with design condition. The flame shape of square burners is square, and the flame height is about half of the circular flame height. The temperature distribution in the middle and upper sections of circular burner is relatively uniform and the temperature is high. The temperature of the side circular burner near the furnace wall is lower, and the temperature of intermediate furnace is relatively uniform. The square burners have uneven temperature distribution in the middle and upper regions, and the middle part is high temperature zone, and the temperature on both sides is low.

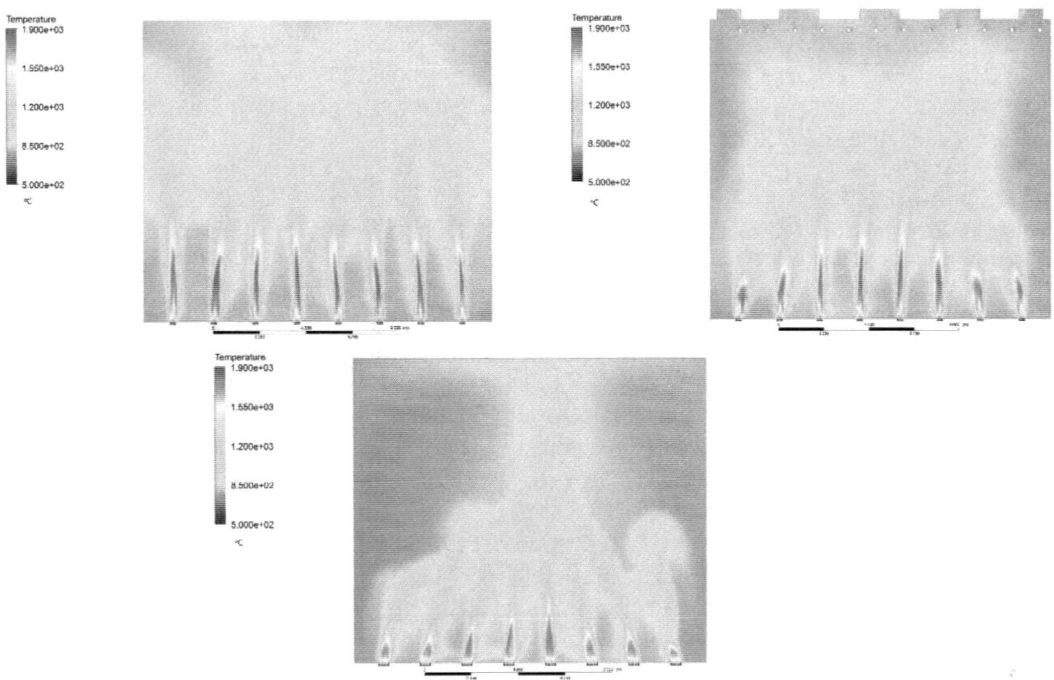

Figure 3. Temperature distribution of the X-axis section of the center circular, side circular and square burners

Temperature distribution of the inner wall of the furnace tube, and velocity distribution inside the conversion tubes are shown in Fig.4. The temperature distribution on the inner wall of conversion tubes is low on both sides, but in the middle tube wall is high, so the temperature gradient is large, and the high temperature area is concentrated in the middle of the furnace. The temperature and velocity distribution of the internal flow field of the conversion tubes are consistent. The flow velocity and temperature of the medium on the outlet side are significantly higher than that on the inlet side. The medium passes through the upper and lower tubes to reach the outlet, the temperature and velocity of the upper part of the outlet tube are also the largest. The temperature of reforming tubes in the middle of the furnace is higher than that of the outside.

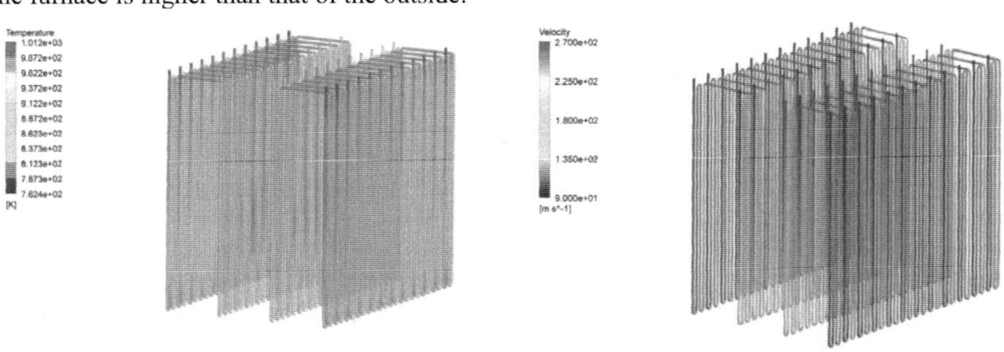

Figure 4. Temperature distribution of the inner wall of the furnace tube, velocity distribution inside the conversion tubes

4. Conclusion

The thermo-mechanical coupling characteristics of the radiant section of the dehydrogenation furnace are numerically simulated based on the CFD method. The temperature and velocity distribution of the fluid in the furnace radiant section and the conversion tubes are obtained, and the simulated values and design values of the parameters are compared well.

Overall, considering the calculation amount, the model can be numerically simulated with half or 1/4 of the geometric model in case of good symmetry of the furnace and the conversion tubes.

The simulation results are consistent with the experimental results, and the simulation method can be applied to similar furnace combustion and conversion tube design.

Acknowledgements

The authors are supported financially by Internal Project of China Special Equipment Inspection and Research Institute (2019Youth12).

References

[1] Pei Chen. Numerical Simulation of Radiation Section of Natural Gas Steam Reforming Hydrogen Reforming Furnace [D]. China University of Petroleum (Beijing), 2016.
[2] CFD simulation of the sidewall fired tubular reforming furnace. Nielsen P,Christiansen L J. ASME 2002 Pressure Vessels and Piping Conference . 2002
[3] Development of a numerical model for natural gas steam reforming and coupling with a furnace model[J] . C. Ventura,J.L.T. Azevedo. International Journal of Hydrogen Energy . 2010 (18)
[4] Tubular heating furnace [M]. China Petrochemical Press, edited by Qian Jialin, 2003
[5] Xiaojun Zhang.Analysis of Temperature Field and Smoke Flow Field of Large Hydrogen Reforming Furnace[J].Petrochemical Equipment Technology,2010,31(03):37-40+71.
[6] The progress in water gas shift and steam reforming hydrogen production technologies – A review[J] . Trevor L. LeValley,Anthony R. Richard,Maohong Fan. International Journal of Hydrogen Energy . 2014 (30)
[7] Yu Long, Xiangping Yang, Enjie Yin, Yanping Liu.Study on Comprehensive Numerical Simulation in Hydrogen Reforming Furnace[J].Refining Technology & Engineering,2010,40(08):22-25.
[8] Xiaojiao Li. Simulation of temperature distribution in furnace and hydrogenation stress analysis of furnace tube [D]. Beijing University of Chemical Technology, 2012.
[9] Fujun Wang. Computational fluid dynamics analysis [M]. Beijing: Tsinghua University Press, 2004.52
[10] Effects of geometry and operating conditions on hydrogen productivity of a fuel cell reformer[J] . Jae Seong Lee,Juhyeong Seo,Ho Young Kim,Jaewon Chung,Sam S. Yoon. International Journal of Heat and Mass Transfer . 2014

Numerical study on condensation of steam/air mixture inside corrugated low finned tubes

Bin Ren[1,2*], Hongliang Lu [1,2], Xiaoying Tang[1,2], Yannan Du [1], Jun Si[1], Facai Ren[1]

[1] Shanghai Institute of Special Equipment Inspection and Technical Research, Putuo District, Shanghai, 200062, China

[2] National Heat exchanger Product Quality Supervision and Inspection Center, Jinshan District, Shanghai, 201518, China

[*]Corresponding author's e-mail: renbin_580912@163.com

Abstract. Corrugated low finned tubes are the specical spirally corrugated tubes, which can enhance the condensation heat transfer with and without noncondensable gas. In this paper, a numerical model on the basis of diffusion layer theory was established to investigate the local heat transfer characteristic inside corrugated low finned tubes. The local parameters contain bulk temperatures, gas mass fraction, heat transfer coefficient and heat flux. The results showed that the bulk temperatures decreased along the condenser tube but the noncondensable gas mass fraction displayed the opposite trend. The local heat transfer coefficient and heat flux decrease along the condenser tube especially near the outlet. Of the five enhanced tubes, #5 tube had the best overall heat transfer performance, but it also had the fastest rate of decline as condensation progressed, indicating that the rib was the main influence factor compared with pitch.

1. Introduction

According to Bergles's perspective, the heat transfer enhancement techniques can be classified either as passive, which require no direct application of external power, or as active, which require external power. And the enhancement techniques have been developed to the third generation [1]. The common used enhanced tubes for in-tube condensation are micro-fin tubes and corrugated tubes [2]. Corrugated low finned tube (CLFT) is a kind of spirally corrugated tube with trapezoidal profile. Due to the wider spiral protrusion, its heat transfer performance is better than that of ordinary spirally corrugated tube.

Ren et al experimentally studied the condensation heat transfer and flow resistance characteristics inside corrugated low finned tubes with/without noncondensable gas [3, 4]. It was found that the heat transfer performance inside corrugated low finned tubes was greater than that inside smooth tubes even when the noncondensable gas existed. However, because of the difficulty in measuring wall temperature, the mechanism of condensation inside corrugated low finned tubes with noncondensable gas has not been understood thoroughly.

In this study, a theoretical model on the basis of diffusion layer theory was established. The theoretical model didn't need constant wall temperature or heat flux as the solving condition. The wall temperature distribution and outlet temperature of cooling water could be evaluated by coupling the in-tube condensation and out-tube forced convection flow. The profiles of these parameters along the condenser tube was showed, including bulk temperature, noncondensable gas mass fraction, heat transfer coefficient and heat transfer flux. The structural parameters of these tubes and the inlet conditions can be obtained from Ren's work [3].

2. Numerical model

2.1Basic principle

The numerical model utilizes the diffusion layer method to study the condensation characteristic of steam/air mixture. The diffusion layer model is on the basis of heat balance at the liquid–air interface, in which the heat transferred from the steam–air boundary layer is equal to the heat transferred through the condensate film [5]. The heat flux transferred from the bulk mixture to the interface can be described as the sum of sensible and latent heat fluxes:

$$q = q_s + q_{cd} = h_s(T_b^s - T_i) + m_c'' h_{fg} \tag{1}$$

In the gas region, the latent heat transfer coefficient can be expressed as:

$$h_{cd} = \frac{m_c'' h_{fg}}{(T_b^s - T_i)} \tag{2}$$

So the heat flux from the gas mixture to the interface can be described as:

$$q = h_s(T_b^s - T_i) + h_{cd}(T_b^s - T_i) = (h_s + h_{cd})(T_b^s - T_i) \tag{3}$$

The heat flux is equal to that transferred through the liquid film:

$$q = h_f(T_i - T_w) \tag{4}$$

The overall condensation heat transfer coefficient can be expressed as:

$$h_c = \frac{q}{(T_b^s - T_w)} \tag{5}$$

Substitute Eq. (3) and Eq. (4) into Eq. (5), the overall condensation heat transfer coefficient is descripted as:

$$h_c = [\frac{1}{h_s + h_{cd}} + \frac{1}{h_f}]^{-1} \tag{6}$$

From Eq. (6), it is found that, to get the overall condensation heat transfer coefficient (h_c), the sensible heat transfer coefficient (h_s), the gas region condensation heat transfer coefficient (h_{cd}) and the film heat transfer coefficient (h_f) must be firstly calculated.

The sensible heat transfer coefficient (h_s) can be obtained from:

$$h_s = \frac{Nu \cdot k_m}{d_i} \tag{7}$$

Here, the Nusselt number can be calculated from:

$$Nu = 1.121 \cdot Re^{0.638} \cdot Pr^{0.33} \cdot (\frac{p}{d_i})^{-0.239} \cdot (\frac{e}{d_i})^{0.461} \tag{8}$$

And the gas region condensation heat coefficient can be obtained from [6]:

$$h_{cd} = \frac{Sh \cdot k_c}{d_i} \tag{9}$$

Here, the Sherwood number can be get from the same formulation showed in Eq. (8) by substituting the Prandtl number with the Schmidt number. And k_c is the condensation thermal conductivity defined as:

$$k_c = \frac{1}{\phi}\left(\frac{h_{fg}^{\ 2} p_t M_s^{\ 2} D_{ab}}{R^2 T^3}\right) \tag{10}$$

For the annular flow, the simplest method to calculate heat transfer coefficient is the two-phase multiplier approach, where the formula can be expressed as the two-phase multiplier form:

$$h_f = 0.096 \cdot Re_{eq}^{\ 0.989} \cdot Pr^{0.33} \cdot \left(\frac{p}{d_i}\right)^{-0.239} \cdot \left(\frac{e}{d_i}\right)^{0.668} \cdot \left(\frac{\lambda}{d_i}\right) \tag{11}$$

2.2Calculation procedure

The numerical model requests inlet parameters as the solving condition. The inlet parameters contain steam flow rate ($\dot{m}_{s,in}$), air flow rate ($\dot{m}_{g,in}$), gas mixture temperature ($T_{s,in}$), gas mixture pressure (p_t), coolant flow rate ($\dot{m}_{co,in}$) and coolant temperature ($T_{co,in}$). The calculation procedure consists of the following steps.

Step 1: Obtain the structure parameters of condenser tube and inlet parameters of gas mixture and coolant.

Step 2: Assume the outlet temperature of coolant $T_{co,out}$, which corresponds to the boundary condition at the inlet of condenser tube.

Step 3: Divide the condenser tube into n control cells along the axial direction, using the converging criterion of 0.05 m.

Step 4: Assume the heat transfer rate Q of each control part represented by "i". Calculate physical property parameters and flow conditions such as quality, Reynolds number etc.

Step 5: Determine the interface temperature T_i. This step is made up of three steps:

Step 5-1: Estimate the interface temperature T_i and the condensation mass flux m_c".

Step 5-2: Obtain the sensible and condensation heat transfer coefficients in the gas resion from Eq. (7) and Eq. (9).

Step 5-3: Obtain the heat transfer rate Q_g from Eq. (12). Calculate the relative error between Q_g and Q in step 4. If this value exceeds a threshold (0.05), adjust the assumed T_i until the relative error is less than the threshold.

$$Q_g = (h_s + h_{cd}) \cdot (T_s - T_i) \cdot (\pi \cdot d_{in} \cdot L/n) \tag{12}$$

Step 6: Obtain the heat transfer rate Q_f through the liquid film from Eq. (13). The heat transfer coefficient h_f and inner wall temperature $T_{w,in}$ can be respectively obtained from Eq. (11) and Eq. (14). Calculate the relative error between Q_f and Q in step 4. If this value exceeds a threshold (0.05), adjust the assumed heat transfer rate Q in step 4, until the relative error is less than the threshold.

$$Q_f = h_f \cdot (T_i - T_{w,in}) \cdot (\pi d_{in}\ L/n) \tag{13}$$

$$T_{w,in} = T_{co} + \frac{Q/(\pi \cdot d_o \cdot L/n)}{h_{co} + 1/R_w} \tag{14}$$

Step 7: Calculate the steam mass flow rate at the outlet of this control part from Eq. (15). If the mass flow rate is smaller than zero, the heat transfer mode transforms into forced convection. Otherwise go to step 8. Calculate the coolant temperature at outlet of current control part from Eq. (16)

$$\dot{m}_s(i+1) = \dot{m}_s(i) - Q/h_{fg} \tag{15}$$

IOP Conf. Series: Materials Science and Engineering **721** (2020) 012066 doi:10.1088/1757-899X/721/1/012066

$$T_{co}(i+1) = T_{co}(i) - Q / (c_{p,co} \cdot \dot{m}_{co}) \tag{16}$$

Step 8: If $i<n$, repeat step 2 to step 7 for the rest control parts. Otherwise go to step 9.

Step 9: Compare the calculated inlet temperature of coolant with the measured value. If the gap exceeds 0.1°C, adjust the assumed outlet temperature of coolant in step 2 until the temperature gap is less than 0.1℃.

3. Results and discussion

Figure 1 shows the distribution of steam/air bulk temperatures along the corrugated low finned tubes at the same inlet condition. Obviously, the bulk temperatures decrease along the condenser tube. Because as the condensation proceeds, the steam flow rate decreases, leading to the reduction of steam pressure and saturated temperature. In this figure, Tube #1, #2 and #3 have the same rib height but different pitch and Tube #2, #4 and #5 have the same pitch height but different rib. It can be found that the rate of descent for tube #5 is the fastest, followed by tube #1, #2, #4 and #3.

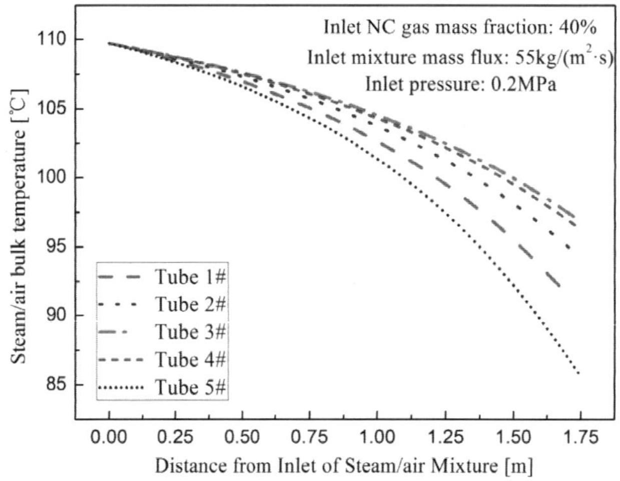

Figure 1. Variation of bulk temperature along the CLFTs

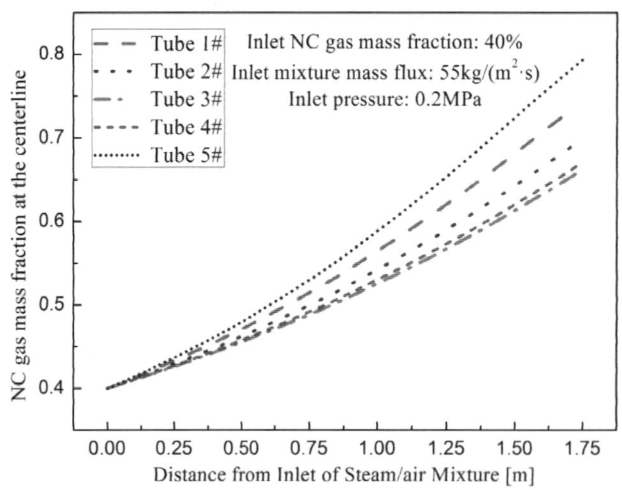

Figure 2. Variation of noncondensable gas mass fraction along the CLFTs

Figure 2 displays the variation of noncondensable gas mass fraction along the corrugated low finned tubes. The variation tendency of noncondensable gas mass fraction is opposite to that of bulk temperature. The noncondensable gas mass fraction increases along the condenser tube because of the reduction of steam flow rate. The growth rate for tube #5 is the highest, followed by tube #1, #2, #4 and #3. This indicates that # 5 tube has the best heat transfer performance during pure steam condensation, which is consistent with the results in Ren's work [4].

Figure 3 displays the distribution of condensation heat transfer coefficient along the corrugated low finned tubes. Obviously, the heat transfer coefficient decreases along the condenser tube, due to the decrease of gas mixture Reynolds number and the increase of condensate film thickness and local noncondensable fraction. The heat transfer coefficient of #5 tube is the largest at the beginning of condenser tube. But as the condensation progresses, the heat transfer coefficient decreases rapidly, and at the outlet of condenser tube it is even smaller than that of # 1, # 2 and # 3 tubes.

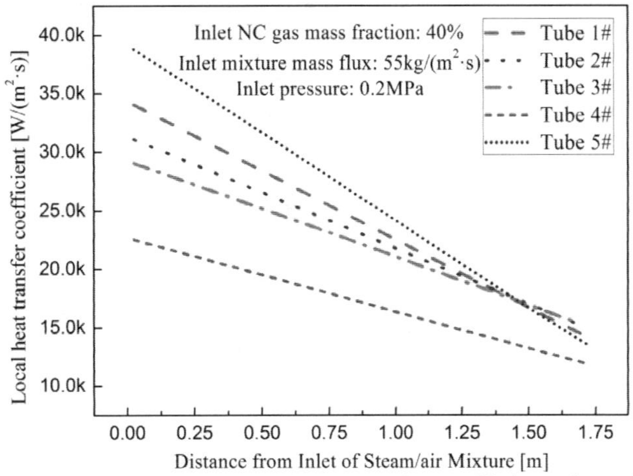

Figure 3. Variation of condensation heat transfer coefficient along the CLFTs

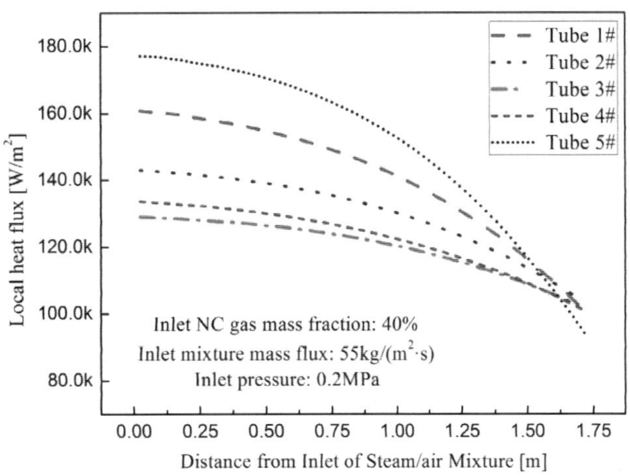

Figure 4. Variation of local heat flux along the CLFTs

Figure 4 illustrates the variation of local heat flux along the corrugated low finned tubes. As expected, the local heat flux of #5 tube is the highest at the entrance. This indicates that the rib is the main factor to enhance condensation heat transfer even in the presence of noncondensable gas. It can also be found that the heat flux deteriorates remarkably near the outlet of condenser tube. This is caused by two factors, one is the decrease of heat transfer coefficient and the other is the reduction of temperature difference of heat transfer.

4. Conclusions

In this study, a numerical model on the basis of diffusion layer theory was established. The distribution of these parameters along the condenser tube was analyzed, including bulk temperatures, gas mass fraction, heat transfer coefficient and heat flux. The results showed that the bulk temperatures decreased along the condenser tube while the noncondensable gas mass fraction displayed the opposite trend. The local heat transfer coefficient and heat flux decrease along the condenser tube especially near the outlet. Among the five enhanced tubes, #5 tube had the best overall heat transfer performance, but it also had the fastest rate of decline as condensation progressed, indicating that the rib was the main influence factor compared with pitch.

Acknowledgments

This work was funded by the 2016 Shanghai Leading Talents Training Program for Professor Xiaoying Tang.

References

[1] Bergles, A.E. (2002) ExHFT for fourth generation heat transfer technology. Experimental Thermal and Fluid Science, 26(2-4): 335-344.

[2] Cavallinia, A., Censi G., Del Col, D., et al. (2003) Condensation inside and outside smooth and enhanced tubes-a review of recent research. International Journal of Refrigeration, 26(4): 373-392.

[3] Ren, B., Tang, X. Y., Lu, H. L., et al. (2017) Experimental investigation on condensation in corrugated low finned tubes in presence of noncondensable gas. In: ASME 2017 Pressure Vessels and Piping Conference. Hawaii. pp. V03AT03A068.

[4] Ren, B., Tang, X. Y., Lu, H. L., et al. (2018) Experimental study on characteristics of condensation and flow resistance inside horizontal corrugated low finned tubes. In: ASME 2018 Pressure Vessels and Piping Conference. Prague. pp. V03AT03A041.

[5] Ren, B., Zhang, L., Cao, J., et al. (2015) Experimental and theoretical investigation on condensation inside a horizontal tube with noncondensable gas. International Journal of Heat and Mass Transfer, 82: 588-603.

[6] Liao, Y., Vierow, K. (2007) A generalized diffusion layer model for condensation of vapor with noncondensable gases. Journal of Heat Transfer-Transaction of ASME, 129: 988-994.

Effect of Flow Arrangement On Energy Efficiency Index

Hongliang Lu[1], Bin Ren[1], Xiaoying Tang[1], Xiaolong Xue[1], Aini He[1], Zhenhua Li[2]

[1] Shanghai Institute of Special Equipment Inspection and Technical Research, Shanghai, 200062, China

[2] National Heat exchanger Product Quality Supervision and Inspection Center, Jinshan District, Shanghai, 201518, China

*Corresponding author's e-mail: liholu@163.com

Abstract. An energy efficiency index (EEI=k/▽p^n) was proposed by Zhang et al to rate the energy saving performance of plate heat exchanger, it was also considered to be independent on the heat transfer area and weakly correlated or not correlated with the flow velocity. However, whether EEI as an ideal index is correlated with the structure such as the flow arrangement or not was important and remained to be confirmed by the theoretical or experimental result. The present work was to analyze EEI for 5 heat exchangers with different flow arrangement: parallel flow, counter flow and combined flow by theoretical method. They all were composed of the same 37 pieces of BP type plate with Nu=0.091Re$^{0.73}$Pt^n, Eu=42400Re$^{-0.545}$. Their inlet velocities between plates and the inlet temperatures for hot and cold fluid were the same and 5 mathematical models were developed. The EEI took into consideration and weighted quantitatively the heat transferred as well as the pressure drop. The comparison verified that the proposed EEI for 5 heat exchangers were different with each other, but 5 heat exchangers with similar flow arrangement were close to each other, less than 5%.

1. Introduction

The heat transfer and flow pressure drop are two different aspects of heat exchanger, and how to review their influences is of great importance on designing and application of heat exchangers. So far there has been several energy efficiency rating indexes such as effectiveness-NTU[1], exergy[2], entropy production[3] and entransy dissipation[4,5,6] on the basis of different thermodynamic conceptions, however they are not used presently due to their inherent shortcomings, for example, their magnitude difference by pressure drop is 2 orders of magnitude less than that by heat transfer. An energy efficiency index (EEI=k/▽p^n) was proposed last year by Zhang et al[7] to rate the energy saving performance of plate heat exchanger, it was also considered to be independent on the heat transfer area and weakly correlated or not correlated with the flow velocity. However, whether EEI as an ideal efficiency index is correlated with the structure such as the flow arrangement or not was important and remained to be confirmed by the theoretical or experimental result. The present work was to analyze EEI for 3 plat heat exchangers with different flow arrangement: parallel flow, counter flow and combined flow at the same inlet temperature and velocity by theoretical method.

2. Configuration and boundary conditions

To review the effect of flow arrangement on EEI, 5 plat heat exchangers are composed with the same 37 sheets and operated at the same inlet temperature and fluid velocity. 3 different flow arrangements include 2 parallel flow heat exchangers shown in Figure 1 A, 2 counter flow heat exchangers Figure 1

B and 1 combined flow heat exchanger Figure 1 C. Both parallel flow and counter flow heat exchangers have two configurations: 37 plates in 1 shell pass and 1 tube pass and in 18 shell passes and 18 tube passes. The combined flow plate heat exchanger is composed of 3 shell passes in the hot water side and 6 tube passes in the cold water side.

The heat exchanger is water-water heat transfer process without phase-change. The inlet temperature for hot water is 110°C and 35°C for the cold water. The outlet temperature and pressure drop changes differently when the water velocity between plates increases from 0.05 m/s to 1.0 m/s.

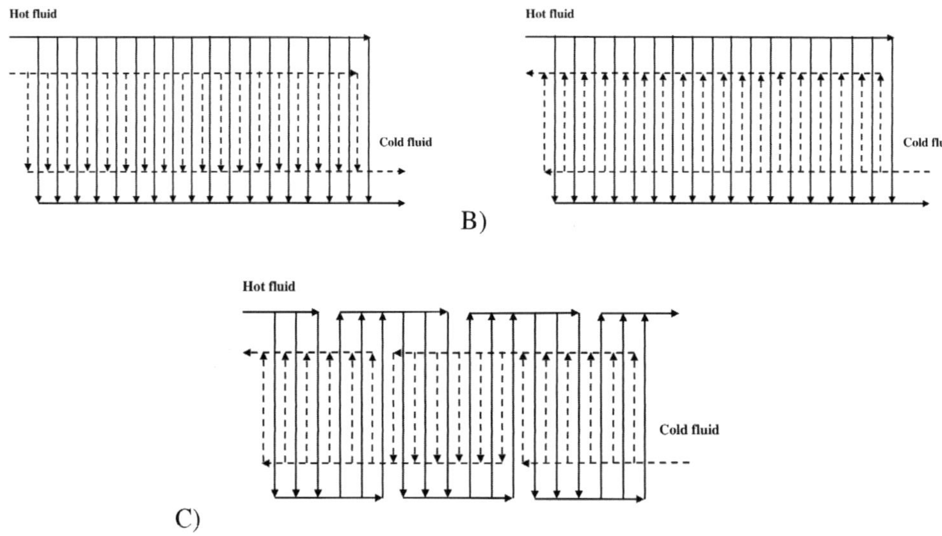

Figure 1. Heat exchanger's flow arrangement: a) parallel flow, b) counter flow, c) cross flow of multiple shell and multiple tube passes

3. Heat transfer and pressure drop model

The heat transfer and pressure drop fitting correlations by experimental data for BP type steel plate are written as:

$$Nu=0.091Re^{0.73}Pt^{n} \tag{1}$$

$$Eu=42400Re^{-0.545} \tag{2}$$

The effectiveness for heat exchanger of parallel flow is written as follows:

$$\varepsilon = \frac{1-\exp\left[-\frac{UA}{C_{min}}\left(1+\frac{C_{min}}{C_{max}}\right)\right]}{1+\frac{C_{min}}{C_{max}}} \tag{3}$$

The effectiveness for heat exchanger of counter flow is written as follows:

$$\varepsilon = \frac{1-\exp\left[-\frac{UA}{C_{min}}\left(1-\frac{C_{min}}{C_{max}}\right)\right]}{1-\frac{C_{min}}{C_{max}}\exp\left[-\frac{UA}{C_{min}}\left(1-\frac{C_{min}}{C_{max}}\right)\right]} \tag{4}$$

The effectiveness for multiple shell passes is rather complicated. The plate heat exchanger with multiple passes is mathematically processed like a shell and tube heat exchanger[8]: n shell passes and 2n tube passes The configuration of 3 shell passes and 6 tube passes is divided into two steps: first,εp for 1 shell pass and 2 tube passes; second, $\varepsilon$for 3 shell passes and 6 tube passes. The effectiveness for heat exchanger of combined flow is written as follows:

IOP Conf. Series: Materials Science and Engineering **721** (2020) 012067 doi:10.1088/1757-899X/721/1/012067

$$\varepsilon_p = \frac{2}{1+\frac{C_{min}}{C_{max}}+\sqrt{1+\left(\frac{C_{min}}{C_{max}}\right)^2}\,\frac{1+\exp\left[-\frac{UA}{C_{min}}\sqrt{1+\left(\frac{C_{min}}{C_{max}}\right)^2}\right]}{1-\exp\left[-\frac{UA}{C_{min}}\sqrt{1+\left(\frac{C_{min}}{C_{max}}\right)^2}\right]}} \tag{5}$$

$$\varepsilon = \frac{\left(\frac{1-\frac{C_{min}}{C_{max}}\varepsilon_p}{1-\varepsilon_p}\right)^n - 1}{\left(\frac{1-\frac{C_{min}}{C_{max}}\varepsilon_p}{1-\varepsilon_p}\right)^n - \frac{C_{min}}{C_{max}}} \tag{6}$$

The definition of energy efficiency index for plate heat exchanger is as follows:

$$EEI = k/\nabla p^{1/3} \tag{7}$$

4. NUMERICAL RESULT

The cold fluid velocity between plates increases from 0.05 m/s to 1.0 m/s simultaneously with the hot water velocity, 5 plate heat exchangers have different variations.

Figure 2 gives EEI distribution under the same operating conditions, obviously EEI for 5 plate heat exchangers are near each other, EEI for the counter flow of 18 shell passes and 18 tube passes is the maximum and that for the combined flow heat exchanger is the minimum, and the former is 4.26% in average greater than the latter. Furthermore, EEI for the counter flow is slightly greater than that for the parallel flow with the same 18 shell passes and 18 tube passes, and so are the configurations with the same 1 shell pass and 1 tube pass. So it can be concluded from the given 5 heat exchangers that the flow arrangement makes little effect on EEI.

In other way, EEI increases with fluid velocity between plates, and approaches the maximum when the fluid velocity is in the range of 0.65 m/s to 0.95 m/s.

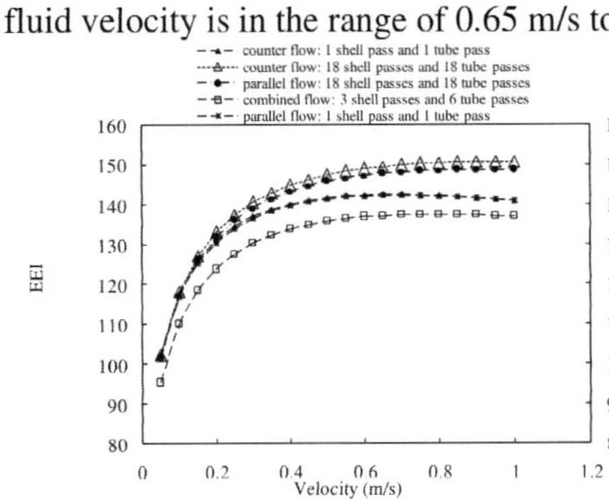

Figure 2. EEI of 5 heat exchangers

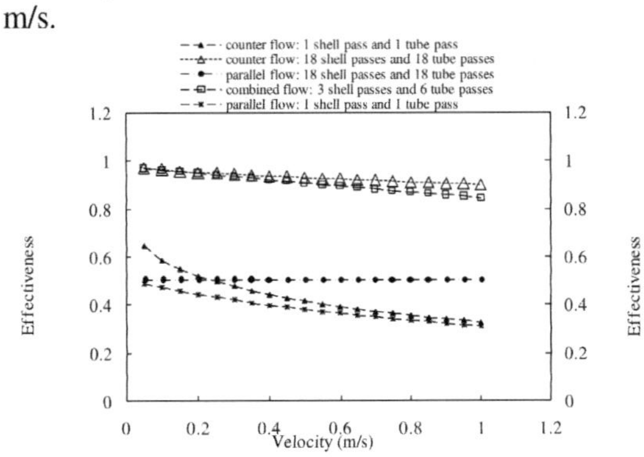

Figure 3. Effectiveness of 5 heat exchangers

Figure 3 gives the effectiveness distribution for 5 plate heat exchangers. The effectiveness for the combined flow is close to the counter flow of 18 shell passes and 18 tube passes.

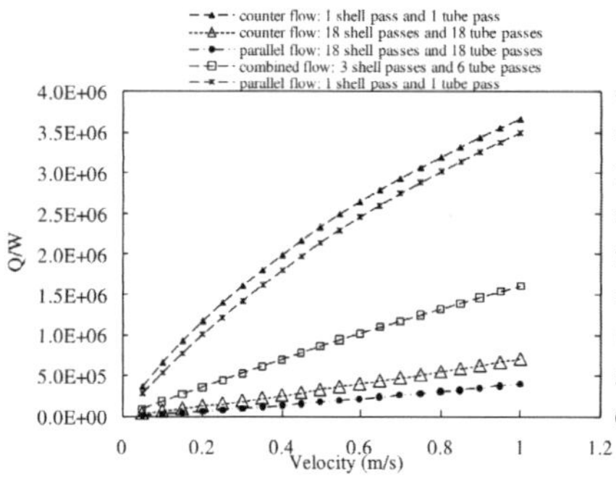

Figure 4. Heat capacity of 5 heat exchangers

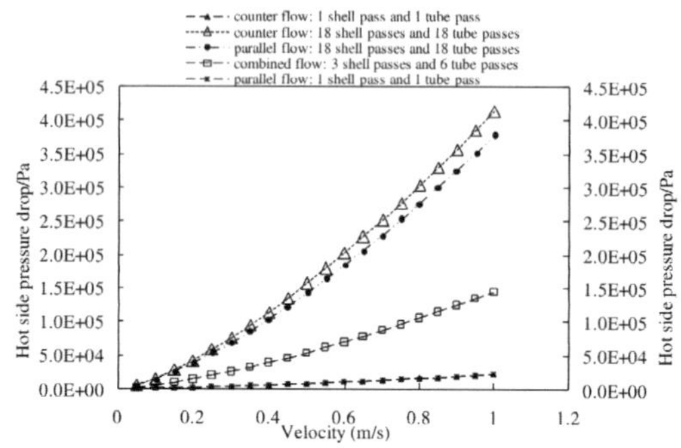

Figure 5. Pressure drop of 5 heat exchangers

From Figure 4, the heat exchangers with 1 shell pass and 1 tube pass are greater in the heat capacity than those with 18 shell passes and 18 tube passes, and the heat exchanger with 3 shell passes and 6 tube passes is between the former two. The heat capacity for the counter flow is slightly greater than that for the parallel flow with the same 1 shell pass and 1 tube pass, and so are the configurations with the same 18 shell passes and 18 tube passes.

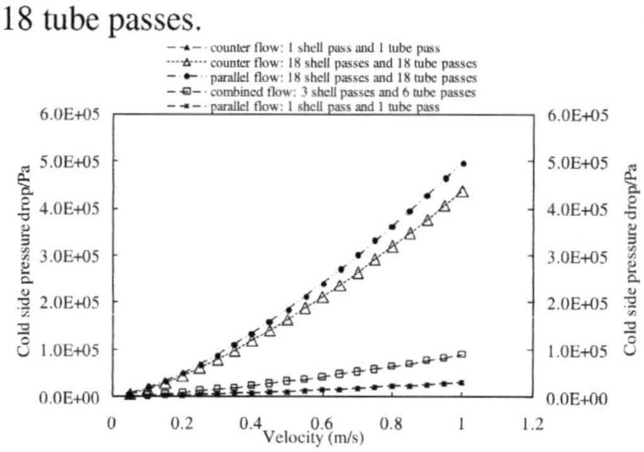

Figure 6. Cold side pressure drop of 5 heat exchangers

Figure 5, Figure 6 gives out the hot side and cold side pressure drop distribution for 5 plate heat exchangers. The heat exchangers with 18 shell passes and 18 tube passes pass are greater in the pressure drop than those with 1 shell pass and 1 tube, and the heat exchanger with 3 shell passes and 6 tube passes is between the former two.

5. CONCLUSION

The EEI takes into consideration and weighted quantitatively the heat transferred as well as the pressure drop. EEI for 5 plate heat exchangers are near each other, EEI for the counter flow of 18 shell passes and 18 tube passes is the maximum and that for the combined flow heat exchanger is the minimum, and the former is 4.26% in average greater than the latter. Furthermore, EEI for the counter flow is slightly greater than that for the parallel flow with the same 18 shell passes and 18 tube passes, and so are the configurations with the same 1 shell pass and 1 tube pass. So it can be concluded from the given 5 heat exchangers that the flow arrangement makes little effect on EEI. In other way, EEI increases with fluid velocity between plates, and approaches the maximum when the fluid velocity is near 0.7 m/s to 0.8 m/s.

Acknowledgment

The work was funded by the 2016 Shanghai Leading Talents Training Program for Professor Xiaoying Tang.

References

[1] Kays, W.M.; London, A.L. Compact Heat Exchangers. 3rd ed. McGraw-Hill Book Company. 1984.

[2] Kotas T J. The Exergy Method of Thermal Plant Analysis. London: Butterworths, 1985.
[3] Bejan, A., 1988, Advanced Engineering Thermodynamics, Wiley, New York.
[4] Z.Y. Guo, H.Y. Zhu, X.G. Liang, Entransy – a physical quantity describing heat transfer ability, Int. J. Heat Mass Transfer 50 (13–14) (2007) 2545–2556.
[5] XU Ming-Tian, CHENG Lin, GUO Jiang-Feng An application of entransy dissipation theory to heat exchanger design. Journal of Engineering Thermophysics, 2009, 30(12): 2090-2092.
[6] X.D. Qian, Z.X. Li, Analysis of entransy dissipation in heat exchangers, Int. J.Therm. Sci. 50 (4) (2011) 608–614.
[7] Zhang Y F, Jiang C, Shou B N, et al. Energy efficiency evaluation method for plate heat exchanger (in Chinese). Chin Sci Bull, 2016, 61: 802–808, doi: 10.1360/N972015-00753.
[8] J.P.Holman： Heat Transfer, 9 Edition,McGraw-Hill Companies.Inc, pp； 536. 2001.11

IOP Conf. Series: Materials Science and Engineering **721** (2020) 012068 doi:10.1088/1757-899X/721/1/012068

Transient Evaporation Simulation of the Forced Circulation Hot Water Boiler

Junjie Bao[1], Di Wang[2], Feng Liu[3], Zhiguo Qu[4], Hongtao Xu[1*]

[1]Shanghai Key Laboratory of Multiphase Flow and Heat Transfer in Power Engineering, School of Energy and Power Engineering, University of Shanghai for Science and Technology, Shanghai, 200093, China.

[2]Shanghai Institute of Special Equipment Inspection and Technical Research, Shanghai, 200062, China.

[3]China Special Equipment Inspection and Research Institute, Beijing, 10013, China.

[4]Moe Key Laboratory of Thermo-Fluid Science and Engineering, Xi'an Jiaotong University, Xi'an, 710049, China.

[*]Corresponding author's e-mail: htxu@usst.edu.cn

Abstract. This paper presents a comprehensive analysis of the water vaporization in a hot water boiler system. The Computational Fluid Dynamics (CFD) technology was adopted to simulate the transient water evaporation in the boiler termed as DZL14-1.25/115/70-AII under the assumption of extreme conditions such as the water pump failure. The results indicated that the average pressure in the pipeline increased from 0.57Mpa to 0.66Mpa (the setting pressure of the safety valve) within 1050s. The pressure increased rapidly with the time increase. The average temperature of hot water changed linearly with time, and the internal thermal deviation of the hot water pipeline in the boiler decreased with time. The internal vaporization phenomenon of the hot water boiler became significant under the extreme condition. The steam flowed upward and accumulated in the upper part of the boiler shell. At 1050s, the vapor volume fraction in the hot boiler was 1.05%, which indicated that the safety valve above the stream shell need to be opened in time after 1050s to ensure the safety of the boiler.

1. Introduction

The hot water boiler is a widely used power equipment in the industrial field. The thermal efficiency and safety of the hot water boiler have achieved more concerns with the fast growing economy in China. However, in the practical operation, water vaporization in the hot water boiler results in the water hammer and pipe burst hazard under some extreme conditions. For examples, some reasons causing water vaporization are as follows: (1) The heating system continually operated after the water circulation was interrupted. Accordingly, the hot water temperature increased rapidly and exceeded the saturation temperature at the corresponding working pressure which caused the water vaporization [1]. (2) The water pressure dropped due to the failure of the constant pressure device, which caused the boiler water vaporization [2].

Many researchers investigated the water vaporization mechanism. Kuang et al. [3] simulated the boiling flow behaviour in the evaporator of separate type heat pipe, and studied the flow patterns and flow behaviours. Under conditions of low mass and heat flux, the nucleate boiling was the dominant mechanism. Hasanpoor et al. [4] investigated the evaporation considering the energy and mass transfer

IOP Conf. Series: Materials Science and Engineering **721** (2020) 012068 doi:10.1088/1757-899X/721/1/012068

during phase change based on the VOF (volume of fluid) model. The results indicated that the surface heat flux and heat transfer coefficient decreased as the wall temperature increased. Noureddine and Ben [5] investigated the mechanism of heat and mass transfer during water evaporation in a horizontal channel. They found that the local coefficient of heat transfer was independent of the free stream vapour concentration. An analogy between heat and mass transfer occurred only at the low free stream temperature and vapour concentration.

At present stage, the previous researches mainly focused on the study of partial water vaporization. The comprehensive analysis of the overall water vaporization in the hot water boiler was imperative to prevent the water hammer hazard and local tube explosion. In this paper, the comprehensive numerical simulation work were carried out to study the transient water evaporation in the hot-water boiler and the overall steam distribution characteristics were quantitatively presented under the extreme condition of the failure of water pump.

2. Numerical simulation

2.1Physical model

A hot water boiler termed as DZL14-1.25/115/70-A II was studied. Fig. 1 shows the onsite picture and the overall physical model of the hot water boiler. The boiler shell was arranged longitudinally along the top centre of the boiler. The bottom of the boiler shell and water pipe bundle formed the furnace radiant heating surface, and the tube bundle between the rear part of the shell and the header box constituted the convection heating surface. The hot water boiler was divided into two systems: the combustion system and the hot water system, and presented in Fig. 2. Coal was burned in the chain type boiler and the hot flue gas first flow upward along the furnace arch. When it met the smoke damper, the flue gas turned to the left side and then passed below the boiler shell to the right smoke box. Finally, the flue gas went through the smoke pipe in the shell and was discharged to the chimney. The feed water after the economizer entered the boiler through backwater inlet in the boiler shell. The water then followed downward through the four coarse down-tubes, and accumulated in the headers. When the water absorbed the heat, it flowed upward and passed through the middle header. Finally, the hot water reached the boiler shell and the separation of water and stream was proceeded in it.

(a) Picture onsite (b) Physical model

Figure 1. Picture and physical model of the hot water boiler

IOP Conf. Series: Materials Science and Engineering **721** (2020) 012068 doi:10.1088/1757-899X/721/1/012068

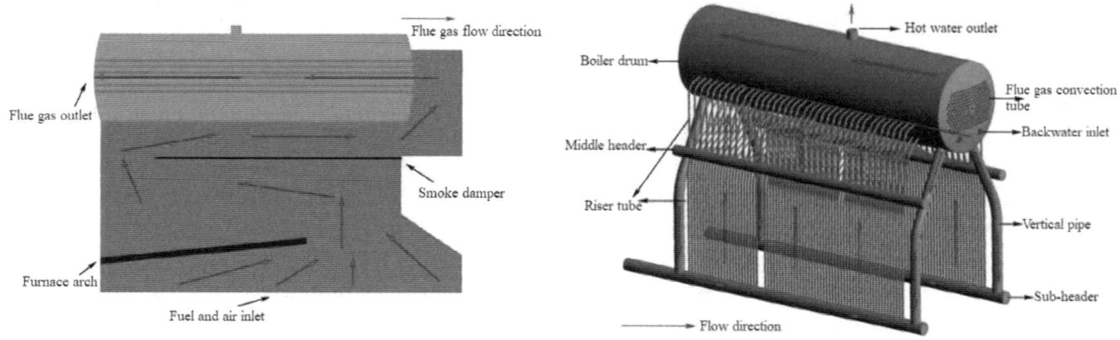

(a)Flue gas flow (b)Hot water flow

Figure 2. The schematic of flue gas and water flow in the boiler

2.2Mathematical model

The numerical simulations were carried out by Ansys Fluent v18.0, and the standard k - ε turbulence model was adopted. The eddy-dissipative combustion model and the discrete coordinate radiation model were used to calculate the combustion reaction rate and radiation respectively. The mathematical descriptive equations are expressed as:

Continuity equation:

$$\frac{\partial \rho}{\partial t}+\frac{\partial(\rho u_j)}{\partial x_j}=0 \tag{1}$$

Equation for conservation of momentum:

$$\rho\frac{\partial u_i}{\partial t}+\rho u_j=-\frac{\partial p}{\partial x_j}+\mu\frac{\partial}{\partial x_j}(\frac{\partial u_i}{\partial x_j}+\frac{\partial u_j}{\partial x_i})+S \tag{2}$$

Energy conservation equation:

$$\rho\frac{\partial T}{\partial t}+\rho\frac{\partial u_i T}{\partial x_i}=-p\frac{\partial u_i}{\partial x_i}+\lambda\frac{\partial}{\partial x_j}(\frac{\partial u_i}{\partial x_j}+\frac{\partial u_j}{\partial x_i}) \tag{3}$$

Turbulent kinetic equation:

$$\frac{\partial}{\partial t}(\rho k)+\frac{\partial}{\partial x_i}(\rho k u_i)=\frac{\partial}{\partial x_j}[(\mu+\frac{\mu_t}{\sigma_k})\frac{\partial k}{\partial x_j}]-\rho\varepsilon \tag{4}$$

Turbulent dissipation rate equation:

$$\frac{\partial}{\partial t}(\rho\varepsilon)+\frac{\partial}{\partial x_i}(\rho\varepsilon u_i)=\frac{\partial}{\partial x_i}[(\mu+\frac{\mu_t}{\sigma_\varepsilon})\frac{\partial_\varepsilon}{\partial x_j}]-C_{2\varepsilon}\rho\frac{\varepsilon^2}{k} \tag{5}$$

where i is 1 to 3, j=1~3, $C_{2\varepsilon}$ is 1.92, C_μ is 0.09, σ_k is 1.0, σ_ε is 1.3, S is source term.

The VOF model was used to simulate the two-phase flow of water and stream. The mathematical equations are expressed as:

Physical parameter equation:

$$\rho=f\rho_1+(1-f)\rho_v \tag{6}$$

Equation for conservation of momentum:

$$\frac{\partial}{\partial t}(\rho v)+\nabla(\rho vv)=-\nabla p+\nabla[\mu(\nabla v+\nabla v^{\mathrm{T}})]+\rho g+F \tag{7}$$

Energy conservation equation:

$$\frac{\partial}{\partial t}(\rho E)+\nabla[v(\rho E+p)]=\nabla p+\nabla(\mathrm{k}_{eff}\nabla \mathrm{T}) \tag{8}$$

where ρ_l is liquid density kg/m^3, ρ_v is gas density kg/m^3, μ is coefficient of kinetic viscosity, T is time s, P is pressure pa, g is acceleration of gravity m/s^2, F is surface tension represented by unit fluid N/m^3.

The combustion of coal generally goes through the following stages: drying and heating of coal [6], volatile decomposition [7], the ignition and combustion of volatiles [8], the formation and burning of coke until the formation of ash [9]. Therefore, in the numerical calculation, the combustion process of coal is divided into three stages: volatile pyrolysis, volatile combustion and coke combustion. In this paper, the appropriate models were selected for these three stages. Two competing rates model proposed by Kobayshi et al. [10] was used to simulate pyrolysis process of coal in the furnace, and the mathematical equations are expressed as:

$$C(Coal)\xrightarrow{k_1}(1-\alpha_1)s_1(cokingcoal)+\alpha_1 v_1(volatile) \quad (9)$$

$$C(Coal)\xrightarrow{k_1}(1-\alpha_2)s_2(cokingcoal)+\alpha_2 v_2(volatile) \quad (10)$$

where α_1, and α_2, and E_1, and E_2 are volatile pyrolysis kinetic parameters, K_1 and K_2 are Arrhenius rate.

Mixture fraction/PDF model proposed by Cheng [11] was used to simulate volatile combustion, and the mathematical equations is expressed as:

$$\frac{\partial}{\partial t}(\rho\bar{f})+\nabla\cdot(\rho v\bar{f})=\nabla\cdot(\frac{\mu_t}{\sigma_t}\nabla\bar{f})+S_m+S_{user} \quad (11)$$

where S_m is the mass of the reaction particles which input into the gas phase, S_{user} is source term.

Dynamic/dynamic control reaction rate model proposed by Baum and Street [12] was used to simulate the burning of coke, and the mathematical equations is expressed as:

$$\frac{dm_p}{dt}=-\pi d_p{}^2\frac{\rho RTY_{ox}}{M_{w,ox}}\frac{D_0 R}{D_0+R} \quad (12)$$

where D_0 is diffusion rate constant, R is chemical reaction rate constant.

2.3Boundary conditions under the rated working situation

The rated pressure of the water boiler DZL14-1.25/115/70-A IIhot is 1.25Mpa, and the boiler outlet temperature is 115°C based on the information from the boiler manufacturer. The detailed parameters under the rated working condition are shown in Table 1.

Table 1. Rated parameters of the hot water boiler

Item	Value
Rated thermal power	14MW
Work pressure	1.25Mpa
Flow temperature	115°C
Design thermal Efficiency	81.2%
Coal consumption	3508kg/h
Backwater temperature	70°C
Exhaust gas temperature	157°C
Rated circulating water	267T/h
Hydrostatic test pressure	1.65Mpa
Design coal	IIbituminous coal

In this work, only the extreme condition with the water pump out of service was investigated. When the water supply pump stopped completely, the boiler continuously absorbed heat and the temperature of water in pipeline increased and vaporization occurred finally. The detailed boundary parameters in the extreme condition are shown in Table 2.

Table 2. Extreme boundary condition

Boundary	Type	Value
Backwater inlet	Wall	-

Hot water outlet	Wall	-
Air inlet	Velocity inlet	0.61122m/s, 180°C
Coal inlet	Mass flow inlet	0.974kg/s, 50°C
Heat exchange tube	Couple Wall	Fluid-solid coupling boundary
Vertical pipe and header	Wall	Adiabatic boundary
Flue gas outlet	Pressure-outlet	-

3. Results and discussions

Fig. 3. presents the phase transition temperature of water under different pressures. The middle section of heating tube was selected to monitor the water pressure. The pressure of 0.66MPa was adopted as the simulation termination because it was the actual highest operation pressure of the safety relief valve.

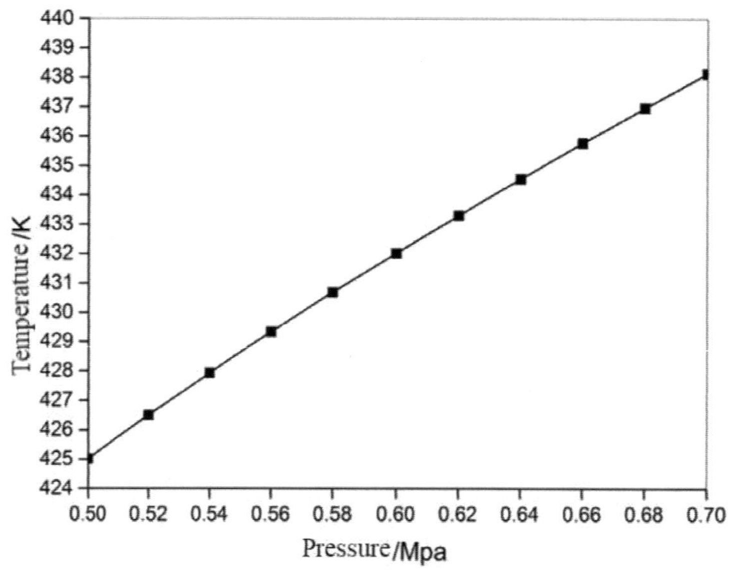

Figure 3. Vaporization temperature at different pressure

Fig. 4 presents the simulated pressure variations with time. It was founded that the pressure overall increases with time and 0.66Mpa was reached at 1050s. During the period of 0s~120s, the internal pressure of the boiler was basically unchanged at about 0.57Mpa, and the pressure increased sharply after 700s.

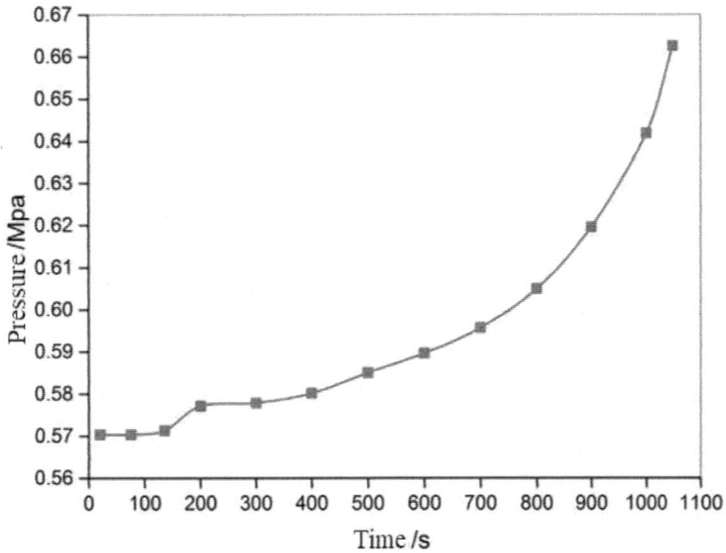

Figure 4. Internal pressure verse time curve

Fig. 5. presents the overall pressure variation trends of the water system in the boiler at different time. The results indicated that the boiler pressure increases continuously with time. The high pressure occurred at the bottom and the low pressure was at the top of the boiler shell.

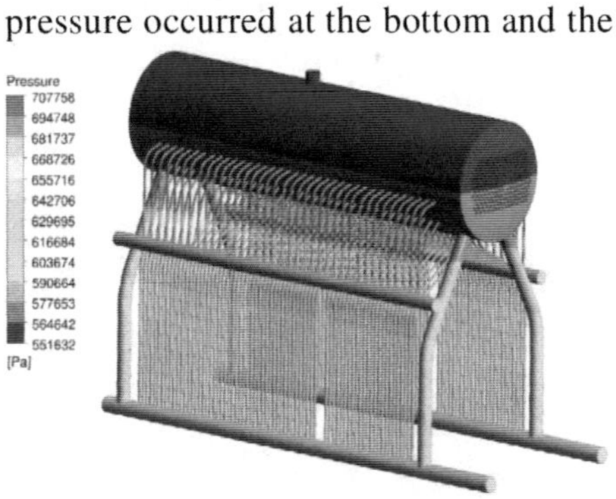

(a) t=20s

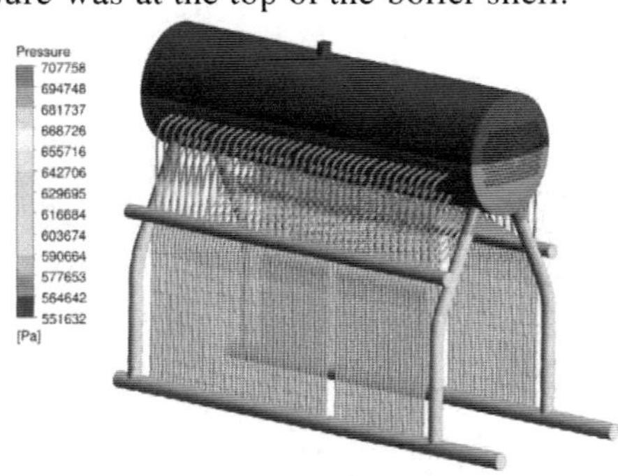

(b) t=75s

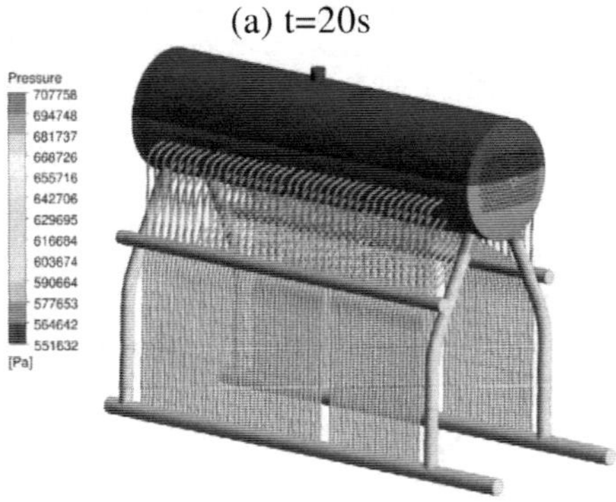

(c) t=135s

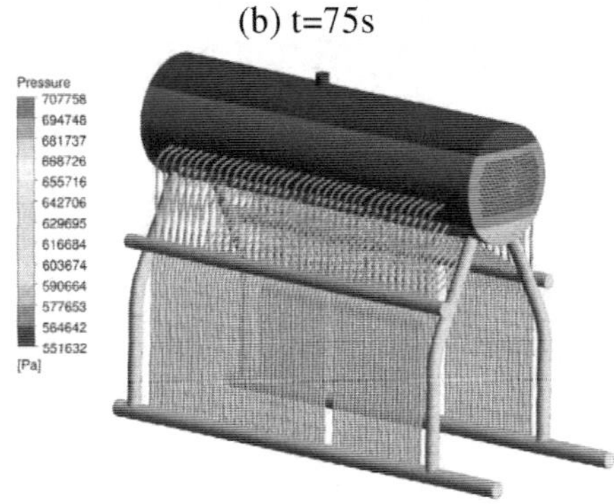

(d) t=200s

IOP Conf. Series: Materials Science and Engineering **721** (2020) 012068 doi:10.1088/1757-899X/721/1/012068

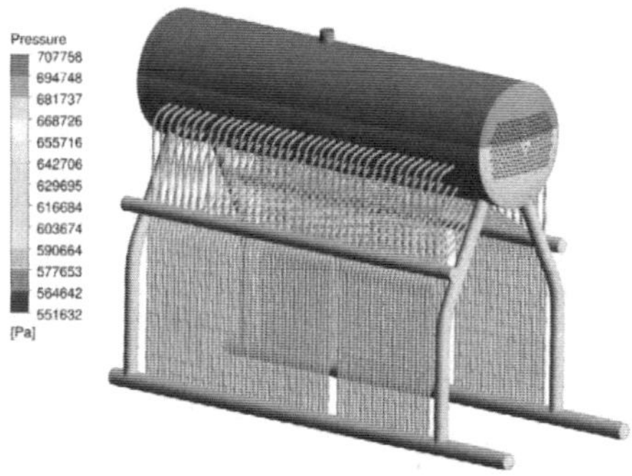

(e) t=500s

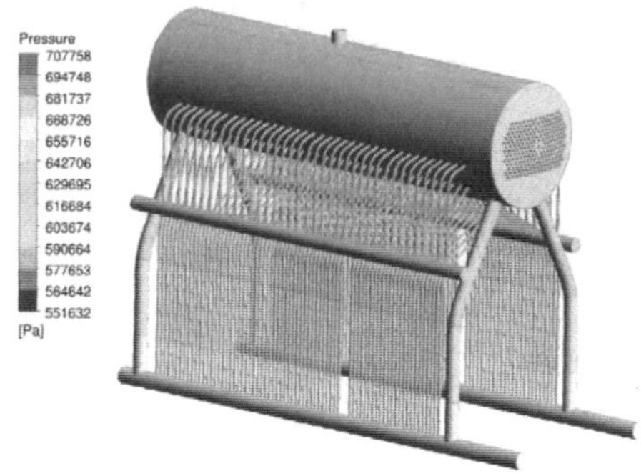

(f) t=900s

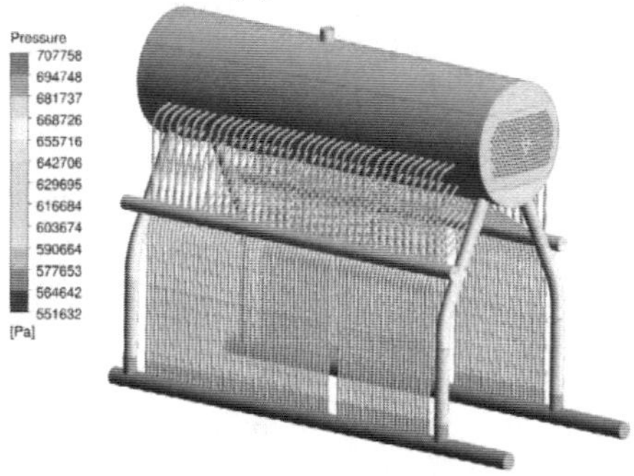

(g) t=1000s

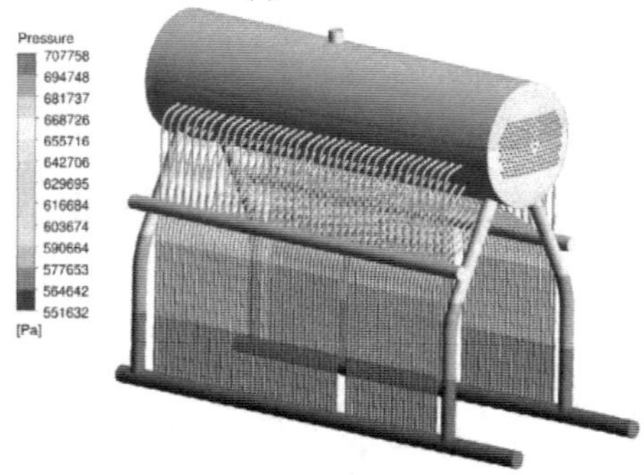

(h) t=1050s

Figure 5. Overall pressure distribution of boiler at different time

Fig. 6 shows the average temperature variation at the middle section of heating tube. The temperature of the boiler almost linearly increased from 366K to 417K.

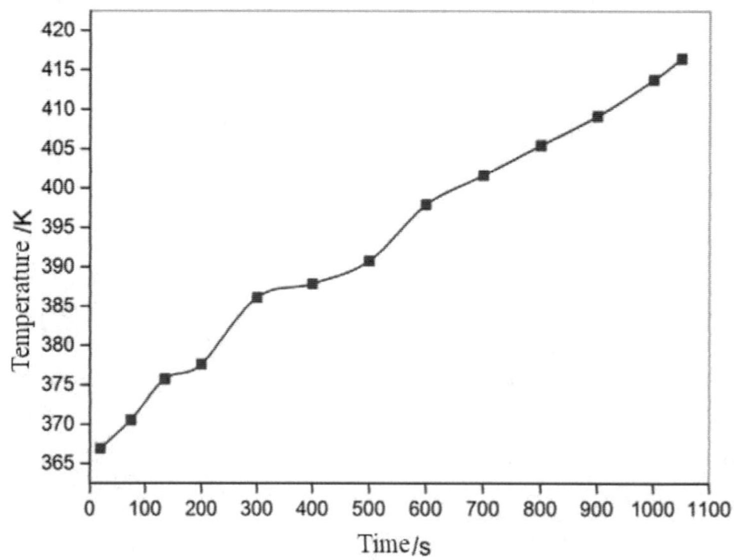

Figure 6. Internal temperature verse time curve

Fig. 7 presents the detailed internal temperature distributions of the hot water boiler pipe with respect to time. The water temperature in the ascension tubes is higher than the water in the boiler shell, indicating that the serious over-temperature phenomenon occurred in the ascension pipes. As the flue gas outlet is located on the left side of the shell, the right part of the boiler shell has a relatively higher water temperature. Though the water temperature inside the boiler is overall low when the water pump is shut down due to some reasons, the phenomenon of undercooling boiling occurs on the wall surface of the pipeline. It should be known that the over-temperature is not harmful to the boiler immediately. The thermal deviation becomes more significant with respect to time.

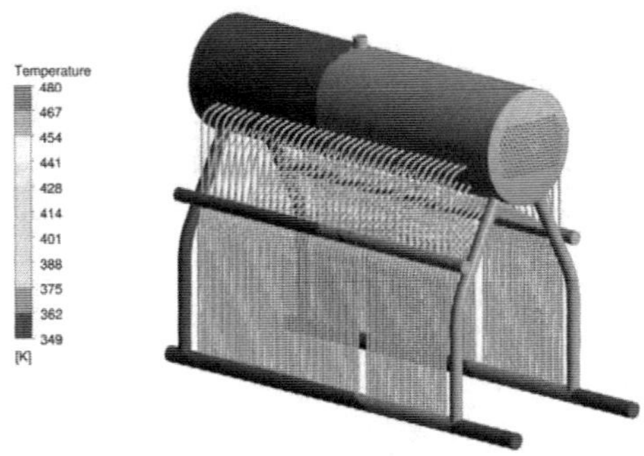

(a) t=20s

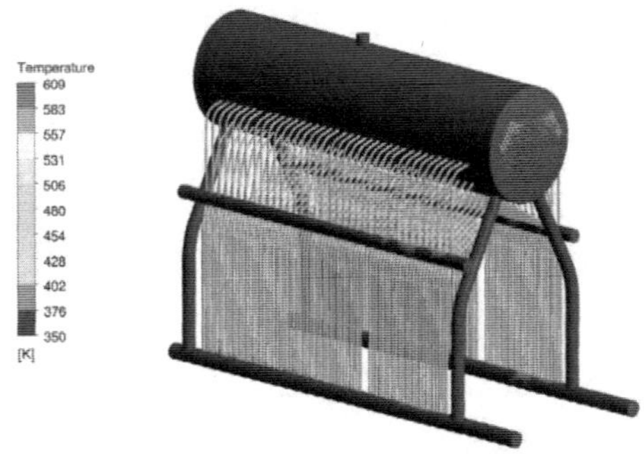

(b) t=75s

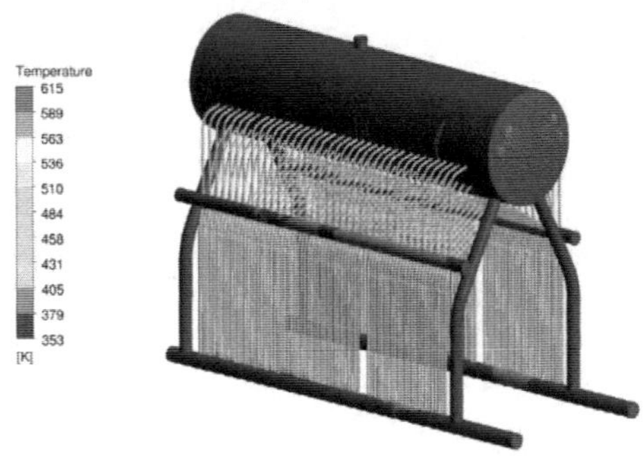

(c) t=135s

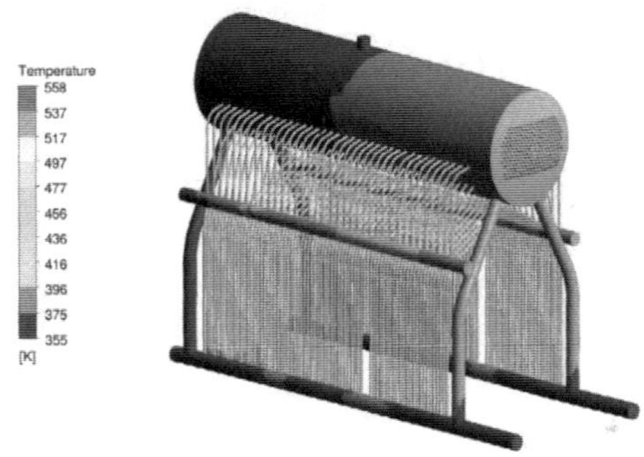

(d) t=200s

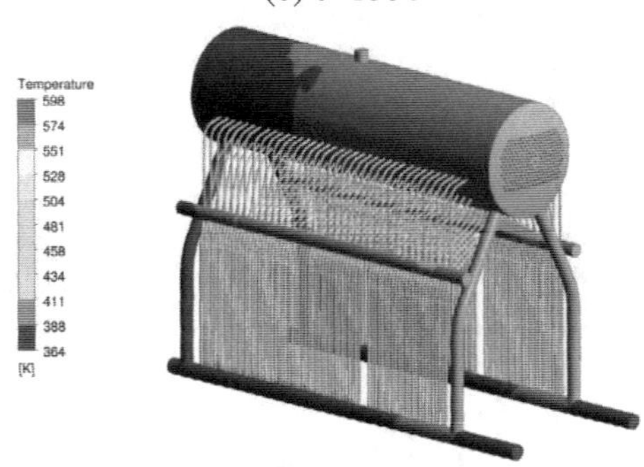

(e) t=500s

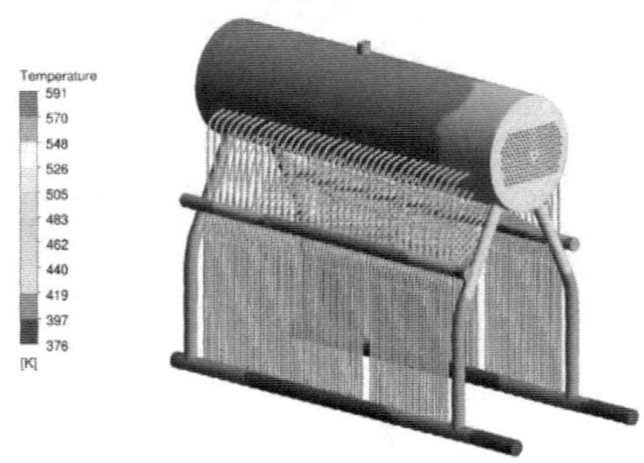

(f) t=900s

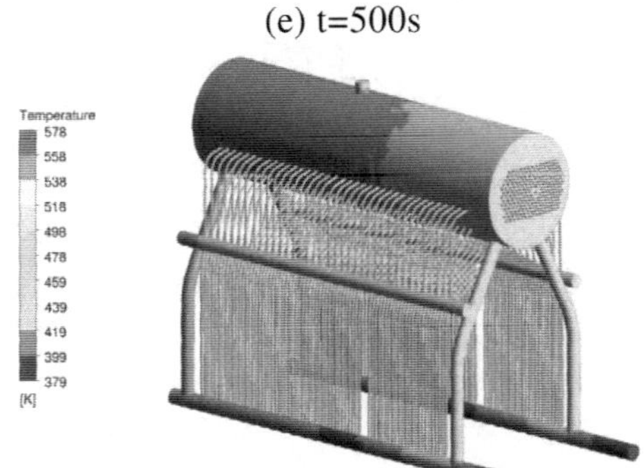

(g) t=1000s

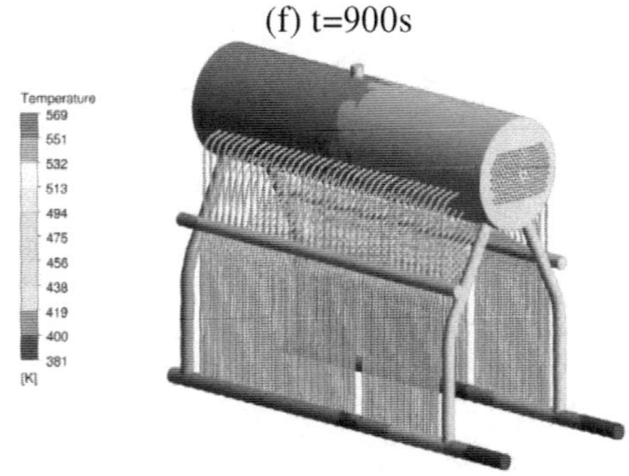

(h) t=1050s

Figure 7. Overall temperature distribution of boiler at different time

Fig. 8. shows the volume of steam generated inside the hot water boiler at different time in the entire water system. It can be seen that the volume fraction of steam increases with time and becomes fast-growing after 500s. The steam volume fraction is 0.045% at t=200s, 0.1% at t=500s, 1.05% at t=1050s, respectively.

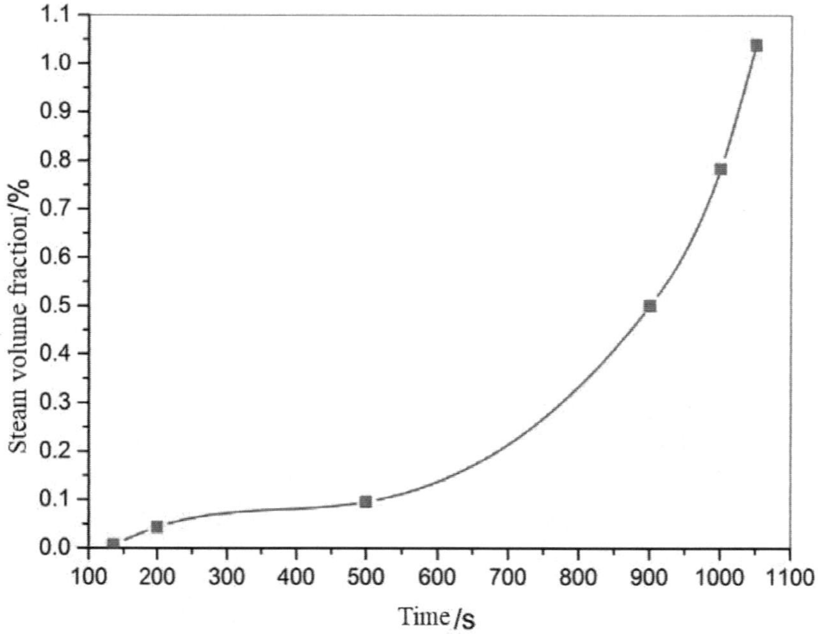

Figure 8. Internal steam volume fraction verse time curve

Fig. 9 shows the internal detailed vaporization distributions in the water pipeline and boiler shell. The vaporization distribution is not significant before135s. After 135s, the vaporization part inside the pipeline becomes larger and all the water vapor accumulated in the upper part of the boiler shell. The water vapor causes the increase of the internal pressure of the boiler continuously. Since the steam accumulates in the upper part of the boiler shell and the hot water outlet, and the over-temperature vaporization steam easily enters the outside pipe from top outlet. Once the pipeline encounters the cold water and condensation occurred, which causes accidents such as water hammer in outside water pipeline. Therefore, in the event of a water pump stop accident, and the hot water outlet must be closed in time, and the safety relief valve should be opened to release the pressure, or explosion may happen.

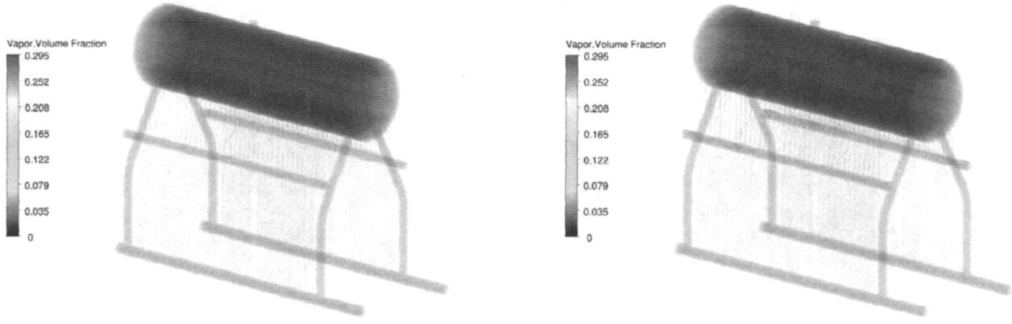

(a) t=135s (b) t=200s

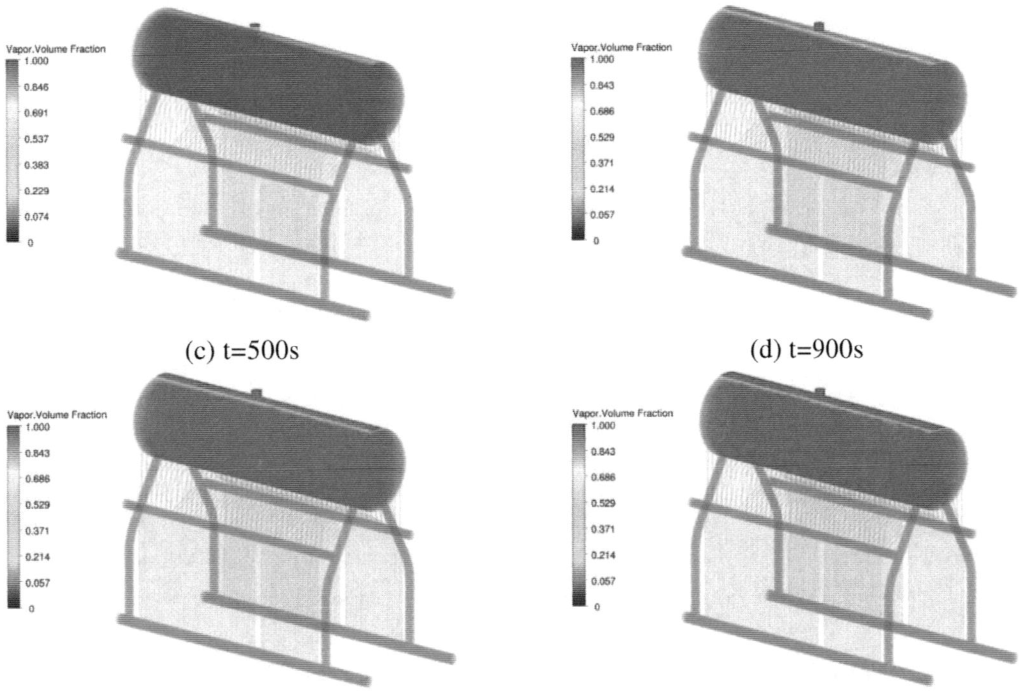

(c) t=500s (d) t=900s

(e) t=1000s (f) t=1050s

Figure 9. Overall vaporization distribution of boiler at different time

4. Conclusions

This paper presents a comprehensive simulation work for the transient evaporation phenomena of the forced circulation hot water boiler. The extreme condition is assumed to occur when the water pump does not work and the coal is still burning in the boiler. With the continuous temperature increase, water evaporation starts to happen. The detailed distributions of temperature, pressure, steam volume fraction and vaporization distribution are presented. When the pressure in the boiler shell reached 0.66Mpa within 1050s, the safety relief valve shall be open timely to release the pressure, or explosion may happen.

Acknowledgments

This work was financially supported by National Key Research and Development Project(2016YFC0801904), Project of Shanghai Municipal Bureau of Quality and Technical Supervision (2017-14), and Dongguan Social Science and Technology Development (General) Project (20185071641113).

References

[1] MENG J.W. (2009) Harm of vaporization of water in hot water boiler and preventive measure. Shanxi Architecture, 35: 187–188.
[2] XIU L.H. (2008) Shallow discuss the boiler of hot water the steam of stove melt. Gansu Metallurgy, 30: 45-46.
[3] KUANG Y.W, WANG W, ZHUAN R (2015) Simulation of boiling flow in evaporator of separate type heat pipe with low heat flux. Annals of Nuclear Energy, 75: 158-167.
[4] Hasanpoor B, Javaherdeh K, Kouhikamali R (2011) Numerical simulation of two phase flow in high pressure and low pressure evaporator pipes in recovery boiler. In: Thermal Power Plants (CTPP). Tehran. pp. 1-2.

[5] Noureddine B, Ben N.S (2001) Mass and heat transfer during water evaporation in laminar flow inside a rectangular channel-validity of heat and mass transfer analogy. International Journal of Thermal Sciences, 40: 67-81.
[6] ZHOU J.H, SONG G.L, CHEN Y.B. (2003) Test and research of flue gas temperature deviation at the furnace exit of a 2008t/h boiler with tangential firing. Thermal Power Generation, 32: 31-35.
[7] DONG C, DOU W.Y, ZHOU Q.L. (2008) Numerical simulation of the thermal load and deformation tendency of a 1100 t/h tower furnace water wall. Journal of Engineering for Thermal Energy and Power, 23: 649-654.
[8] Morsi S.A. (1972) An investigation of particle trajectories in two-phase flow systems. J. of Fluids Mech, 01: 55.
[9] ZHAO J.G. (2014) Numerical simulation of coal combustion optimization for a 600MW supercritical pressure boiler. North China Electric Power University.
[10] Kobayashi H, Howard J.B, Sarofim A.F. (1977) Coal devolatilization at high temperatures. Symposium (International) on Combustion, 16: 411-425.
[11] P. Cheng. (1964) Two-Dimensional radiating gas flow by a moment method. AIAA Journal, 2:1662-1664.
[12] Baum M.M, Street P.J. (1971) Predicting the combustion behaviour of coal particles. Combustion Science and Technology, 3: 231-243.

IOP Conf. Series: Materials Science and Engineering **721** (2020) 012069 doi:10.1088/1757-899X/721/1/012069

Numerical Study of High Performance Ablative Thermal Protect System by Adding Phase Change Material

Lei Qiao, Ri-tian Ji, Jun-qiang Bai*, Hui Wang

School of Aeronautics, Northwestern Polytechnical University, Xi' an, Shaanxi 710072, China

*Corresponding author: junqiang@nwpu.edu.cn

Abstract. Under extreme thermal environment, ablation process of thermal protection system is a key issue in atmospheric re-entry and hypersonic flight vehicles; it displays a strong coupling between ablation mass, velocity and heat transfer. This article presents a method for combining the ablating and melting process, building a right triangle carbon-carbon model for ablation mixed with octadecane phase change material to absorb the heat of combustion delaying the movement of burning interface. For the moving of boundary, it is treated by the modified Stefan model coupled with chemical reaction and thermal radiation at the front leading boundary. The governing equation of heat transfer considering the melting process of phase change material is given by referring the enthalpy method. The effect of volume ratio and aerodynamic heating on the ablation interface and net heat flux on the interface are investigated. Result shows that high volume ratio of the phase change material delays the movement of the interface, and decreases the net heat flux of the interface; on the other hand, high aerodynamic heat accelerates the movement of the interface and increases the interface net heat flux at a given time point. This study can give a rule on how the volume ratio of phase change material influence the ablative process quantitatively.

1. Introduction

With the development of the flight vehicles and spacecraft, the velocity of them has reached hypersonic field. The harsh environment for re-entry into the atmosphere and the high speed bring the severe aerodynamic heat problem. Thus, the effective thermal protection system should be considered. According to the different thermal protection mechanisms, the basic forms of heat protection schemes can be divided into endothermic thermal protection system, mass ejector thermal protection system, radiant heat protection system, mass transfer heat treatment thermal protection system, and integrated thermal protection system (ITPS)[1], among which ablative thermal protection system has been widely adopted in the thermal protect system, because it has the advantage on tuneable density, lower cost and higher heat shock resistance.

For ablative thermal protection system. There are two sides can be considered on the system, new materials and new methods for ablation. For ablative materials, rhenium, molybdenum, and tungsten was considered in early time. Graphite carbon-carbon (C-C) composite material was considered for its high heat capacity and high energy of vaporization [2]. Thus, nowadays C-C composite material is commonly used in ablative thermal protection system (TPS). What's more, with the development of material science, new composite material has been developed. For the new methods for ablation, during the study of ablation process, the nonlinear phase change problem is the key point: the moving interface at which the phase change occurs. For boundary moving process, Chu and Zhang [3] combined the moving boundary and finite volume method to calculate the heat exchanger problem, showing that the coupling algorithm is reliable in the heat exchanger modelling with the run time much shorter than only using finite volume method. Ahmed et al.[4] got a new hybrid grid and meshless method for the numerical simulation result of phase change material with moving boundary, and well compared with the analytical solution. Li et al.[5] used the Stefan moving boundary models for simulation the phase change of heat-dissociation hydrate, getting the laws of temperature and the frontier location. Turkyilmazoglu et al.[6] analysed the Stefan problems of phase changing bar using single and double phase models, and show a unique observation is that the solid-liquid interface location grows when the slab moves in the direction of melt or freeze. Abhishek et al.[7] describe a Stefan problem with temperature dependent Thermal coefficients and moving phase change material, also consider the convective boundary condition using spectral tau method to approach; and they show the accuracy of the result by comparing the result with analytical solution and get that the solidification process is affected by Peclet number (Pe) and constant γ, and the process becomes fast as the parameters Pe and γ increase. The study of phase change material melting process varies from different scales, for example, Hasadi et al.[8] showed the melting process of nanostructured phase change materials colloidal suspensions, and observed that for the case of diameter of particle equal to 2 nm, as the mass fraction of the particles increases the solid-liquid interface changes from a planar to unstable morphology during melting. Some materials are widely used for the study of phase changing process such as paraffin wax[9], TiO_2 [10], open-cell metal foams[11], octadecane[12]. We can choose the octadecane for our simulation. Besides, for hypersonic vehicles subjected to severe aerodynamic heating, carbon composites ablative materials are commonly used in the thermal protection systems (TPS) for the integrity of the structure[13].

From the study above, we can see the advantage of phase change materials, but nowadays we can see there are a lot of ablative TPS studies, but few of them combine the ablative TPS with phase change materials or take the notice on the model building on the leading edge of the hypersonic vihicles. This paper presents a method for combining the ablating and phase change material and the model building on the leading edge, by building a right triangle carbon-carbon model of Stefan problem coupled with chemical reaction and thermal radiation at the front leading boundary mixed with octadecane composite phase change material under the condition of constant aerodynamic heat flux, and the effect of the aerodynamic heat and volume ratio of the phase change material on the ablative process is studied.

2. Description of physical problem and mathematical model

2.1Physical problem descriptions

For the supersonic and hypersonic vehicles, the wing is very important, always with large sweep angle and thin wing as shown in **Figure 1**. And for the serious heating condition on leading edge, we should consider the thermal protection on that. The heat energy will get stored with the cruising speed larger than sonic speed, which cause the shape of the leading edge deformed without effective protection, resulting in the changing in aerodynamic force, even causing the aircraft to be unstable and destroyed. So, it is very important for thermal conduction at the leading edge. Some studies have got some other mechanisms: Zhu et al. [14] presented a typical actively-cooled thermal protection system cooling the leading edge by impingement jet of gaseous helium and get effectively decrease the temperature on the leading edge shown in **Figure 2**. It is an actively-cooled TPS on which the thermal stress cannot decrease significantly, so mostly we use the passive thermal protection by adding the ablative material on the leading edge.

Figure 1 structure for Concorde

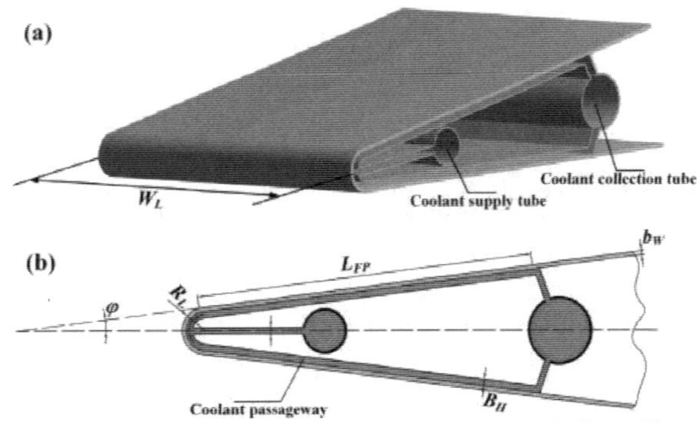

Figure 2 Schematics of impingement-cooled leading edge: (a) 3D overview; (b) Detail of the geometric dimensions[14].

In this paper, the complex leading-edge ablative material was simplified to a right triangle, and the right triangle model is treated as a one dimensional physical model for Stefan problem with chemical reaction at the leading edge ablation leading interface. Heat is only transmitted along the long right side. The degree of small acute angle is set to $\psi = 0.3rad$. Due to the aerodynamic heat cannot be transmitted from the leading point, the triangle model must remove the point, so L is length of the long right side after removed the point. We suppose that the limited size of ablation model is fixed with the length of long right side L=0.5 m. When the point is removed, the physical model is turned into a right trapezoid model. The transmitted model is described in **Figure 3**. Assuming that the left edge is the boundary of model and length of the left edge is initial thermal conductive area of ablation body. Which the initial area is expressed as A_0, setting the initial area $A_0 = 0.01m^2$. Originally, the ablation body stays its initial temperature. Supposing that the ambient temperature. As the time continued, the solid region where the temperature exceeded the temperature of gasification point. Then, the ablation body of front layer is turned into the gas phase and the front layer will gradually move away from its initial position. Meanwhile, the heat dissipates from the left ablation interface to the internal material. Along with the increase of temperature, the whole ablation body will be preheated after a period of time, for there is also phase change material which will begin to change the phase from solid to fluid with the temperature get to its solid-phase temperature, and all phase change material get to fluid entirely when temperature go to fluid-phase temperature.

On the above states of the physical model, the following assumptions could be proposed: (1) Oxygen density at the leading edge during the chemical reaction period is treated as changeless; (2) Convective heat transfer between air and the ablation body on the surface layer is neglected; (3) Effect of the oxygen partial pressure is neglected; (4) Local atmospheric environmental thermal radiation is neglected.

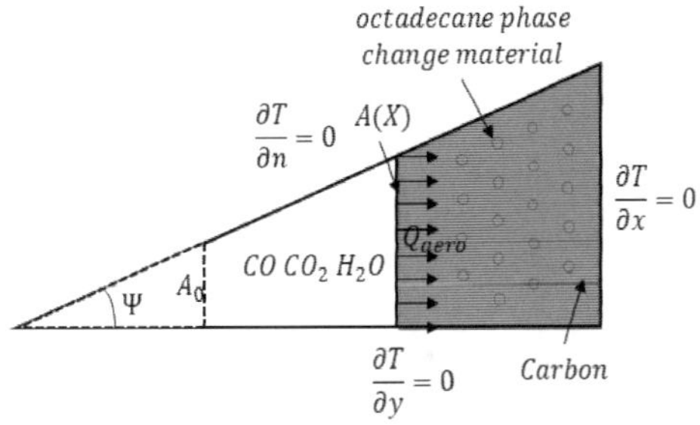

Figure 3. Description of physical model

2.2Mathematical descriptions

Obviously, the calculation area is limited to the problem based on the above physical descriptions and assumptions, and the governing equations of thermal conductive area are presented.

Instantaneous heat conductive area:

$$A(X)=A_0+\tan(\psi)S(t) \qquad \tau\geq 0 \tag{1}$$

For thermal governing equation, we use the enthalpy method as followed

$$\rho\frac{\partial H}{\partial t}=\lambda\frac{\partial^2 T}{\partial x^2} \tag{2}$$

H is the total enthalpy as shown in equation (3). This value consists of the sensible enthalpy (h') and latent enthalpy (h):

$$H=h+h'=\int cdT+f(\mathrm{T})\cdot h_e \tag{3}$$

Where h_e is the effective latent heat of phase change material. f is the fluid phase fraction [16], can be expressed as equation(4):

$$f(\mathrm{T})=\begin{cases}0 & ,\ T<T_s\\ \dfrac{T-T_s}{T_f-T_s} & ,\ T_s\leq T\leq T_f\\ 1 & ,\ T>T_f\end{cases} \tag{4}$$

Then substitute equation (4) into equation (2), and we can get the governing equation of temperature, as shown in equation (5).

$$\rho c\frac{\partial T}{\partial t}=\lambda\frac{\partial^2 T}{\partial x^2}-\rho h_e\frac{df(T)}{dt} \tag{5}$$

Initial condition:

$$T(X,0)=T_0 \qquad \tau=0,\ S(t)=0 \tag{6}$$

Boundary conditions:

$$Q_{\text{net}}=-\lambda A(X)\frac{\partial T}{\partial x}=Q_{\text{aero}}-\varepsilon A(X)\sigma\left(T^4-T_{amb}^4\right)+A(X)K_0\mathrm{e}^{-\frac{E}{RT}}P_{O_2}^{\frac{1}{2}}\mathrm{h_c} \qquad X=S(\tau) \tag{7}$$

$$\frac{\partial T(X,t)}{\partial X}=0 \qquad X=L \tag{8}$$

Net heat flux:

$$q_{net}=\frac{Q_{net}}{A(X)} \qquad 0\leq X\leq L \tag{9}$$

Some thermal physical properties of graphite are listed in **Table 1**. Additionally, to calculate the ablation front layer recession, the modified moving boundary controlling equations similar to that in classic Stefan problem are initially given below. Moving boundary controlling equations:

$$\rho\frac{dS(t)}{dt}=k_0e^{-\frac{E}{RT}}P_{O_2}^{\frac{1}{2}} \qquad X=S(t);\ S(t)=0 \qquad t=0 \tag{10}$$

equation (10) accounts for the heat balance at the ablation interface. When time $\delta t\to 0$, the recession distant is exclusively determined by mass loss of TPS material suffered from chemical ablation. Two dimensionless variables can be introduced in the above equations. Expressing as following:

$$\eta=\frac{X-S(t)}{L-S(t)} \quad (0\leq\eta\leq 1); \quad \xi=\frac{S(t)}{L} \quad (0\leq\xi\leq 1) \tag{11}$$

The η represents the dimensionless moving coordinates, and the ξ represents the dimensionless coordinate of ablation interface location. Then, plugged the above two dimensionless parameters into the equations in Eqs.(1), (2) and (5)-(11) respectively. Instantaneous Heat Conduction Area:

$$A = A_0 + \tan(\psi)\xi L \tag{12}$$

Thermal governing equation:

$$\frac{\partial T}{\partial t} = \frac{\lambda}{L^2 \rho c(1-\xi)^2}\frac{\partial^2 T}{\partial \eta^2} - \left[\frac{\lambda}{(1-\eta)L^2 \rho c_p (1-\xi)^2} + \frac{\lambda tg(\psi)}{(1-\eta)L\rho c(1-\xi)(A_0 + tg(\psi)\xi L)}\right]\frac{\partial T}{\partial \eta} - \frac{h_e}{c}\frac{df(T)}{dt} \tag{13}$$

Initial condition:

$$T(\eta, 0) = T_0 \tag{14}$$

Boundary conditions:

$$Q_{net} = -\lambda(A_0 + \mathrm{tg}(\psi)\xi L)\frac{1}{L(1-\xi)}\frac{\partial T}{\partial \eta}$$
$$= Q_{aero} - (A_0 + \mathrm{tg}(\psi)\xi L)\varepsilon\sigma(T^4 - T_{amb}^4) + (A_0 + \mathrm{tg}(\psi)\xi L)k_0 e^{-\frac{E}{RT}} p_{O_2}^{\frac{1}{2}} h_c \quad \eta = 0 \tag{15}$$

$$-\lambda_c \frac{\partial T(\eta, t)}{\partial \eta} = 0 \qquad \eta = 1 \tag{16}$$

Net heat flux:

$$q_{net} = \frac{Q_{net}}{\left(A_0 + \tan(\psi)\xi L\right)} \qquad 0 \le \xi \le 1 \tag{17}$$

Moving boundary controlling equations:

$$\rho L \frac{d\xi(t)}{dt} = k_0 e^{-\frac{E}{RT}} p_{O_2}^{\frac{1}{2}} \qquad \eta = 0\ ; \quad \xi = 0 \quad t = 0 \tag{18}$$

2.3Discrete process of equation

we use finite difference method (FDM) to discrete the thermal governing equation, for divergence can probably occur when the liquid fraction is directly calculated because of the strong nonlinearity of this problem. Therefore, the modified discretized equation suggested by Voller et al.[17] is adopted as

$$f_i^{\overset{n+1}{m+1}} = f_i^{\overset{n+1}{m}} + \frac{df(T)}{dT}\left[T_i^{\overset{n+1}{m}} - f^{-1}\left(f_i^{\overset{n+1}{m}}\right)\right] \tag{19}$$

In equation (19) $T_i^{\overset{n}{m}}$ means the temperature of the mth iteration at the nth time level for the i node, and we can use T_i^n to replace the convergence result of the nth time level for the i node. This equation is used to weaken the nonlinearity. When the second term on the right part of equation (19) is equal to 0, the calculation is converged. Then governing equation can change to:

$$\frac{T_i^{\overset{n+1}{m+1}} - T_i^{\overset{n}{m+1}}}{\Delta t} = G\frac{T_{i+1}^{\overset{n+1}{m+1}} - 2T_i^{\overset{n+1}{m+1}} + T_{i-1}^{\overset{n+1}{m+1}}}{\Delta \eta^2} - F\frac{T_{i+1}^{\overset{n+1}{m+1}} - T_{i-1}^{\overset{n+1}{m+1}}}{2\Delta \eta} - \frac{he}{c\Delta t}\left(f_i^{\overset{n+1}{m}} + \frac{df(T)}{dT}\left[T_i^{\overset{n+1}{m}} - f^{-1}\left(f_i^{\overset{n+1}{m}}\right)\right] - f_i^{\overset{n}{m+1}}\right) \tag{20}$$

$$G = \frac{\lambda}{\rho c}\frac{1}{L^2(1-\xi)^2} \qquad F = \frac{\lambda}{\rho c}\left(\frac{1}{(1-\eta)L^2(1-\xi)^2} + \frac{tg(\psi)}{(1-\eta)L(1-\xi)(A_0 + tg(\psi)\xi L)}\right) \tag{21}$$

Boundary conditions:

IOP Conf. Series: Materials Science and Engineering **721** (2020) 012069 doi:10.1088/1757-899X/721/1/012069

$$BT_1^{n+1} = CT_2^{n+1} + D \quad \eta = 0 \quad B = C = (A_0 + \mathrm{tg}(\psi)\xi L)\frac{\lambda}{L\Delta\eta(1-\xi)} \tag{22}$$

$$D = Q_{aero} - (A_0 + \mathrm{tg}(\psi)\xi L)\varepsilon\sigma\left[\left(T_1^{n+1}\right)^4 - \left(T_{amb}\right)^4\right] + (A_0 + \mathrm{tg}(\psi)\xi L)k_0 e^{-\frac{E}{R\left(T_1^{n+1}\right)}} p_{O_2}^{\frac{1}{2}} h_c \tag{23}$$

$$-\lambda_c \frac{T_{M1}^{n+1} - T_{M2}^{n+1}}{\Delta\eta} = 0 \qquad \eta = 1 \tag{24}$$

Net heat flux:

$$q_{net} = \frac{Q_{net}}{\left(A_0 + \tan(\psi)\xi L\right)} \tag{25}$$

Moving boundary controlling equations:

$$\frac{\xi^{n+1} - \xi^n}{\Delta t} = \frac{1}{\rho L} k_0 e^{-\frac{E}{RT_1^{(n+1)}}} p_{O_2}^{\frac{1}{2}} \quad \eta = 0\ ;\ \xi = 0 \qquad t = 0 \tag{26}$$

2.4Numerical method

The specific calculation steps are specifically listed as follows:

1. Using the initial temperature and dimensionless ablation front interface location distribution T^0, ξ^0 respectively, update the temperature distribution T_1^1 by TDMA algorithm. The superscript and subscript letter represent time layer and number of the inner iteration at specific time.

2. Using the updated the value of temperature, calculate the interface ξ_1^1 via equation (26).

3. Finding the maximum value relative error, $\Delta T_1^1 = \left(T_1^1 - T^0\right)/T^0$, $\Delta\xi_1^1 = \left(\xi_1^1 - \xi^0\right)/\xi^0$. If $\Delta T_1^1 \le \varepsilon_1$ and $\Delta\xi_1^1 \le \varepsilon_2$, ($\varepsilon_1 = \varepsilon_2 = 5\times10^{-8}$), then the present values are considered convergent and go to Step1. If either of the judgmental condition fails, repeat step 1 until satisfied.

4. Saving the convergent values and apply them as the initial conditions of the next time layer, repeat Step 1.

Table 1. Some thermal physical properties of C-C material

Parameters	C-C	Octadecane[15]
Oxidation rate of graphite, k_0 /kg·m^{-2}·s^{-}	3.28×10^9	3.28×10^9
Thermal conductivity, λ_c /W·m^{-1}·K^{-1}	200	0.23
Activation energy, E /cal·mol^{-1}	9.66×10^4	9.66×10^4
Heat value of combustion, h_c /J·kg^{-1}	3.28×10^7	4.75×10^7
Density, ρ /kg·m^{-3}	2000	600
Capacity, c/ J·kg^{-1}·K^{-1}	806	1850

3. Results and discussion

3.1The grid independence check

In the course of the simulation, the discrete equations are solved numerically by TDMA algorithm. The grid independence is checked as shown in **Figure 4**. It can be seen that with increase of grid number, the temperature trends a constant value. The final adopted grid number is chosen as 101.

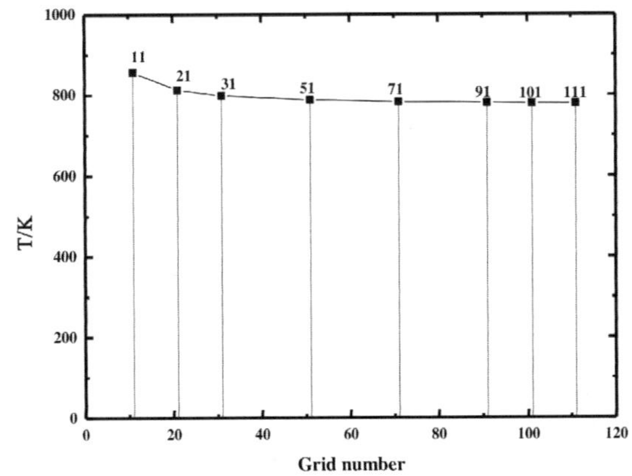

Figure 4. Front layer temperature variations with grid number

3.2Effect of volume ratio of phase change material on ablation interface location and net heat flux of ablation front layer

From **Figure 5** we can see that with the volume ratio of phase change material increasing, the moving speed of ablation interface recession location will go down. From the volume ratio of phase change material going up from 0 to 0.01, the moving speed of ablation interface location has dropped by 50 percent. And when the volume ratio of phase change material goes up to 0.05, the ablation interface location almost keeps constant. This appearance is caused from the volume ratio changing, for the volume ratio of the phase change material will affect the thermal properties: with the phase change material mixed some heat can be absorbed by the phase changing, so the moving of the interface will be slowed down.

As shown in **Figure 6**, it presents the effects of different volume ratio of phase change material on the performance of net heat flux on ablation front layer nodes. We can see that the net heat flux decreasing trend is similar with volume ratio varying. But with the volume ratio going up, the decreasing speed of the heat flux will significantly go down. This trend can be discussed for the interface moving, for energy will be released by radiation and materials absorbing, and the releasing speed has relationship with the boundary moving. What's more, the boundary moving speed means the ablative TPS material consumption rate. Slowing down the rate cases the net heat flux decreasing slowly.

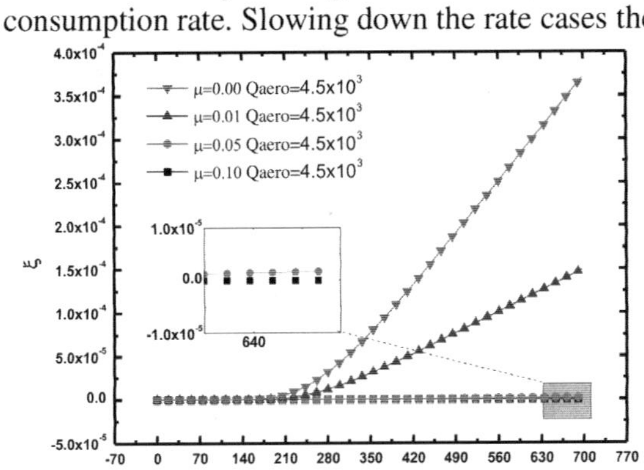

Figure 5. Effect of volume ratio of phase change material on ablation interface location

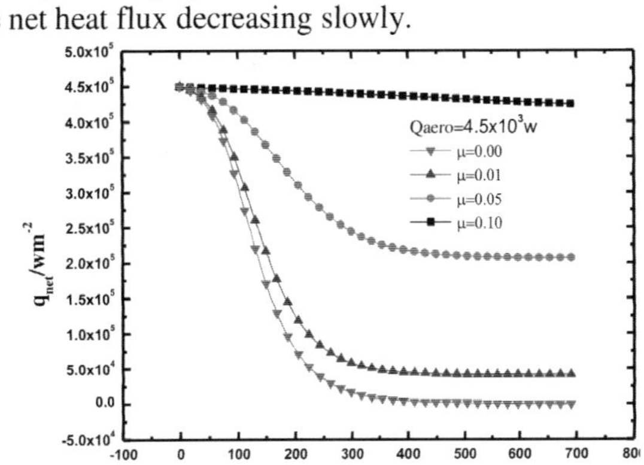

Figure 6. Effect of volume ratio on net heat flux of ablation front layer

3.3Effect of aerodynamic heat on ablation interface location and net heat flux of ablation front layer

Figure 7 presents the effects of different aerodynamic heat on the location distribution of ablation front layer on different times. There is no matter for a given condition the trend of the interface moving will not change, and under a high aerodynamic heat condition, the moving of the interface of the ablation will be quicker. That is because the much heat means more temperature and the high speed of the ablation reaction. We can consider under the high speed, the higher aerodynamic heat, the leading edge of the aircraft will get easily ablative deformation.

From **Figure 8** we can get that with the aerodynamic heat changing, the heat flux distribution on the interface of the ablative process. The trend of the heat flux with time is the same as shown in **Figure 6**: with time flowing, the interface will radiate energy to the surrounding environment, and with temperature rising more, the radiate energy will increase sharply. Thus, the heat flux of the interface will go down. Furthermore, with the aerodynamic heat increasing, we can see the heat flux increase firstly, which is because the input energy is larger, and in a short time the energy output is with no big difference. Then with time goes to 200s, the condition begins to show a opposite trend for larger aero heat get a less heat flux. The reason for this condition is that under the high aerodynamic heat, the moving of the interface will be accelerated and temperature of that will be a higher, so it will get more energy radiated. In other word it also can be expressed as: more aero heat will accelerate the entire process, which means the speed of going down will also be accelerated. That is why after around 200s the higher aero heat will result in a lower heat flux.

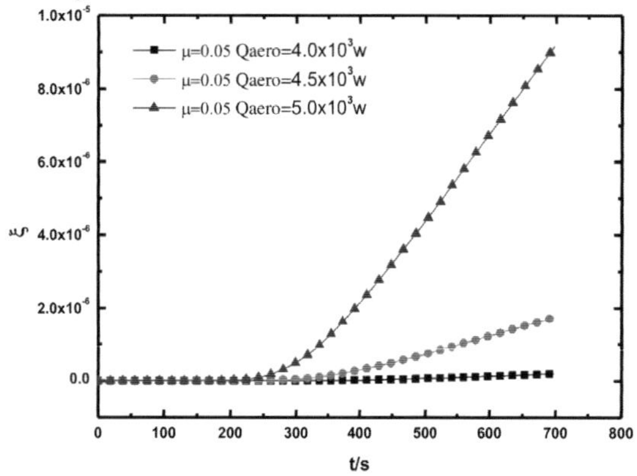

Figure 7. Effects of aerodynamic heat on ablation interface recession location

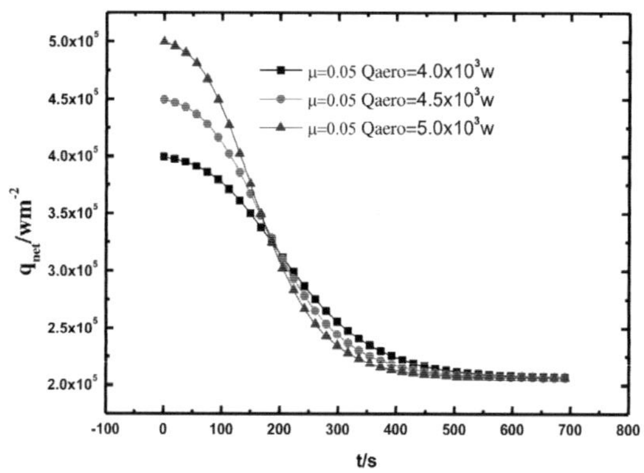

Figure 8. Effects of aerodynamic heat on net heat flux of front layer

4. Conclusions

In this paper, a modified Stefan model is proposed, considering the chemical reaction of ablation showing the movement of the ablating interface, phase change absorbing the burning heat with fluid phase fraction changing, and heated boundary radiation for temperature rising too high. Considering the effect of volume ratio of phase change material and aerodynamic heat to the physical and chemical properties, we use TDMA method to present the temperature distributions and simulate how the burning interface move. From what we get we can derive the heat flux, movement of the interface with considering on the influence of volume fraction and aero heat. We can see the influence of the aerodynamic heat and volume ratio of the phase change material is very significant: high volume ratio of the phase change material delays the movement of the interface, and decreases the net heat flux of the interface; on the other hand, high aerodynamic heat accelerates the movement of the interface and increases the interface net heat flux at a given time point. So the effective phase change material can offset the effects of high aerodynamic heat and is necessary for future application on ablative and phase change TPS.

References

[1] Fan X.J. (2004) Aerodynamic Heating and Thermal Protection System. Science Press. Beijing.
[2] Natali M, Torre L. (2011) Ablative materials. In: Wiley encyclopedia of composites. New York: Wiley.
[3] Chu Z.H., Zhang W.P. (2019) Moving-boundary and finite volume coupling algorithm for heat exchanger with fluid phase change. International Journal of Heat and Mass Transfer. 131: 313-

328.
[4] Ahmed S.G.. (2019) Numerical solution of moving boundary problems using a new hybrid grid and Meshless method. Engineering Analysis with Boundary Elements. 103: 22-31.
[5] Li M.C., Fan S.S., Su Y.L., Xu F.H., Yan Li, Lu M.J., Sheng G.L., Yan K. (2018) The Stefan moving boundary models for the heat-dissociation hydrate with a density difference. Energy.160: 1124-1132.
[6] Turkyilmazoglu M. (2018) Stefan problems for moving phase change materials and multiple solutions. International Journal of Thermal Sciences; 126: 67-73.
[7] Singh A.K., Kumar A., Rajeev. (2018) A Stefan problem with variable thermal coefficents and moving phase change material. Journal of King Saud University Science.
[8] Yousef M.F. El Hasadi. (2019) Numerical simulation of the melting process off nanostructured based colloidal suspensions phase change materials including the effect of the transport of the particles. Journal of Molecular Liquids. 287.
[9] Yadav A.K., Donepudi T., Sriram S.B. (2019) Numerical and experimental investigation of melting characteristics of phase change material RT58. Thermal Science and Engineering Progress.
[10] Motahar S., Alemrajabi A.A., Khodabandeh R. (2017) Experimental investigation on heat transfer characteristics during melting of a phase change material with dispersed TiO2 nanoparticles in a rectangular enclosure. International Journal of Heat and Mass Transfer.109: 134-146.
[11] Guo Z.X., Bai Q.S., Hou J.L., Yang X.H., Liu Y.H., Sun Y.J. (2018) Experimental investigation on the melting behavior of phase change materials in open-cell metal foams in an inclied rectangular enclosure. Energy Procedia. 152: 215-220.
[12] Zhu Y.L., Liang S.E., Wang H., Zhang K., Jia X.R., Tian C.R., Zhou Y.L., Wang J.H. (2016) Morphological control and thermal properties of nanoencapsulated n-octadecane phase change material with organosilica shell materials. Energy Conversion and Management. 119: 151-162.
[13] Riccio A., Raimondo F., Sellitto A., Carandente V., Scigliano R., Tescione D. (2017) Optimum design of ablative thermal protection systems for atmospheric entry vehicles. Applied Thermal Engineering. 119: 541-552.
[14] Zhu X.W., Zhao J.Q. (2016) Study on helium impingement cooling for a sharp leading edge subject to aerodynamic heating. Applied Thermal Engineering. 107: 253-163.
[15] Cai D., Li J., Jiao N.X. (2019) Preparation and thermolphysical properties of grapheme nanoplateets-octadecane phase change composite materials. Acta Phys. Sin. 68: No.10.
[16] Zhang J.J., Qu Z.G., Liu Y. (2016) Numerical study on the melting themal characteristics of a microencapsulated phase change plate. Numerical Heat Transfer, Part A: Application. 70: 399-419.
[17] Voller V.R. and Swaninathan C.R. (1991) ERAL source-based method for solidification phase change, Numer. Heat Transfer, Part B Fundam. 28: 175-189.

IOP Conf. Series: Materials Science and Engineering **721** (2020) 012070 doi:10.1088/1757-899X/721/1/012070

Numerical modelling of a counter-flow regenerative evaporative cooler based on modified effectiveness-NTU method

Zhiyin Duan[1*], Sikai Wen[1], Minghui Wang[1], Junjun Li[1]

[1]Beijing Key Lab of Heating, Gas Supply, Ventilating and Air Conditioning Engineering, Beijing University of Civil Engineering and Architecture, Beijing, 100044, China

[*]Corresponding author's e-mail: duanzhiyin@bucea.edu.cn

Abstract. This study develops a numerical model for a counter-flow regenerative evaporative cooler in Engineering Equation Solver environment by using modified effectiveness-NTU method. The overall performance of the counter-flow heat exchanger was predicted by varying inlet air temperature and humidity.

1. Introduction

Regenerative evaporative cooler, as the IECs with closed-loop configuration, has the ability to provide supply air at a temperature below the wet-bulb towards the dew-point temperature of inlet air theoretically [1]. According to the estimates [2], a counter-flow REC could reduce 13-58% of the total electrical power annually consumed by conventional compression air conditioners for various climates in China. Predominantly, there are two types of regenerative evaporative coolers including single-stage counter-flow and multi-stage M-cycle cross-flow configurations. More recently, there have been growing interests in developing the single-stage counter-flow regenerative evaporative coolers by numerical methods [3-5]. However, the effectiveness-NTU method hasn't been developed in the modelling heat and mass exchanger. To fill the gap, this paper developed a dedicated numerical computer model for the proposed counter-flow heat and mass exchanger based on modified effectiveness-NTU method. The model was validated with published experimental data. We used the model to simulate the effect of inlet air temperature and humidity on the performance of counter-flow heat and mass exchanger.

2. Numerical modelling of the heat and mass exchanger

2.1Physical model of the heat and mass exchanger

Fig. 1 shows the schematic of geometry and the operating conditions for the counter-flow plate heat and mass exchanger. In the dry channels of the heat exchanger, product air at P_{pi}= 0.1 MPa, t_{pi}= 38.9 °C, $t_{wb,pi}$=21.11 enters one set of channels. The total volumetric flow rate of air passing through all of the dry channels is V_p= 2737 m^3/h. Portions of product air, as working air enters the other set of channels (wet channels) at P_{wi}=P_{po} and t_{wi}=t_{po} through the perforations at the end of dry channels. R_{ratio} is the ratio of working to intake airflow rate. Each wall plate is δ_{wall}= 0.375 thick and is composed of plastic film and cellulose fibre sheet. The water film thickness is assumed to be δ_{water}= 0.3 mm. The plates are L=

0.86 long in the flow direct and W= 0.458 wide. The plate separation distances, i.e., the channel height and width are 3.0 mm and 45.8 mm respectively. There are N=320 pairs of channels.

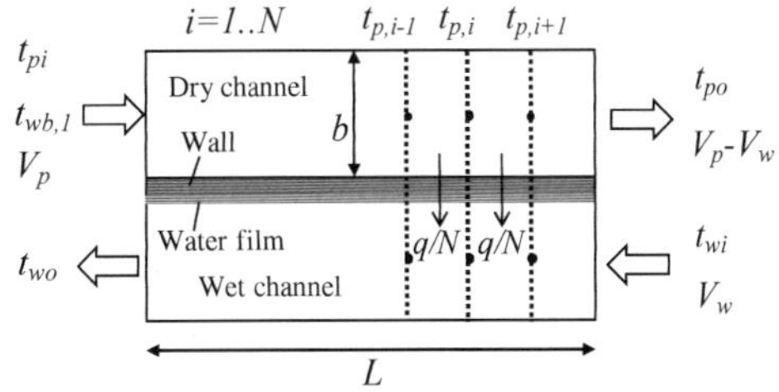

Fig. 1 The plate type heat and mass exchanger in a counter-flow configuration

2.2Governing equations and numerical solution method

An efficient numerical model of the heat exchanger can be obtained by applying the modified ε-NTU solutions to each sub-heat exchanger and requiring that the boundary conditions associated with each sub-heat exchanger are consistent with the adjacent sub-heat exchangers. The following equations are coded in Engineering Equation Solver (EES) software and solved by Newton's alternative method. It is found that the heat conductance approach constant rapidly when the number of computational elements is larger than 5, therefore, 30 computational elements number was used to provide sufficient accuracy. A few assumptions are made to simplify the solutions: 1) the heat exchanger is well insulated; 2) axial heat conduction and fouling resistance of heat exchanger is ignored; 3) In each computational sub-heat exchanger, heat and mass transfer coefficients as well as fluid properties are fixed; 4) the Lewis number is unity. 5) the temperature of air and water interface is saturated at the temperature of water film.

The governing mathematical equations for each sub-heat exchanger, or computational element, are given as follows. The heat transfer rate for product air in dry channels is expressed by Eq. (1)

$$dq = {}_{p}V_{p}(i_{p1} \quad i_{p2}) \tag{1}$$

The heat transfer rate for working air in wet channels is given by Eq. (2)

$$dq = {}_{w}V_{w}(i_{w1} \quad i_{w2}) \tag{2}$$

For the regenerative configuration, the outlet air conditions in dry channel is equivalent to the inlet air conditions in wet channels, i.e., $i_{p,o}=i_{w,i}$ and $t_{p,o}=t_{w,i}$

The heat transfer between product air and working air is given by Eq. (3)

$$dq = UdA(i_{s}(t_{p}) \quad i(t_{w})) \tag{3}$$

Where, the overall conductance of the heat exchanger is determined by Eq. (4)

$$UdA = [a(\frac{1}{h_{p}} + \frac{wall}{k_{wall}} + \frac{water}{k_{water}}) + \frac{1}{h_{m}}]^{1} NWdx \tag{4}$$

It is assumed that the air enthalpy has a linear relation with saturation temperature, which is defined as Eq. (5).

$$i_{s}(t_{p}) = a(t_{p}) + c \tag{5}$$

The heat capacity rate of product air and working air can be redefined by establishing the heat balance on product air and working air, which are expressed as the following two equations.

$${}_{p}V_{p}c_{p,p}(t_{p1} \quad t_{p2}) = C_{p}(i_{s}(t_{p1}) \quad i_{s}(t_{p2})) \tag{6}$$

$${}_{w}V_{w}c_{p,w}(t_{w1} \quad t_{w2}) = C_{w}(i(t_{w1}) \quad i(t_{w2})) \tag{7}$$

Rearranging Eq. (6) gives:

$$C_{p} = \frac{{}_{p}V_{p}c_{p,p}(t_{p1} \quad t_{p2})}{i_{s}(t_{p1}) \quad i_{s}(t_{p2})} = \frac{{}_{p}V_{p}c_{p,p}}{a} \tag{8}$$

IOP Conf. Series: Materials Science and Engineering **721** (2020) 012070 doi:10.1088/1757-899X/721/1/012070

Rearranging Eq. (7) gives:

$$C_w = \frac{\ _w V_w c_{p,w}(t_{w1} \quad t_{w2})}{i(t_{w1}) \quad i(t_{w2})} = \quad _w V_w \tag{9}$$

As the overall conductance approaches infinity, the heat exchanger could obtain the maximum possible heat transfer rate:

$$dq_{max} = C_{min}(i(t_p) \quad i(t_w)) \tag{10}$$

Where, $C_{min} = \min(C_p, C_w)$

The effectiveness of sub-exchanger is the ratio of the actual heat transfer rate to the maximum possible heat transfer rate:

$$= \frac{dq}{dq_{max}} \tag{11}$$

The number of transfer units is given by Eq. (12):

$$NTU = \frac{UdA}{C_{min}} \tag{12}$$

For the heat exchanger in counter-flow configuration, the effectiveness is expressed by Eq. (13):

$$= \frac{1 \quad e^{[\quad NTU(1 \quad C_r)]}}{1 \quad C_r e^{[\quad NTU(1 \quad C_r)]}} \tag{13}$$

Where, $C_r = \frac{C_{min}}{C_{max}}$ □and $C_{max} = \max(C_p, C_w)$

For each computational element, the local convective heat transfer coefficients from product air and working air to exchanger walls are determined by Eqs. (14) and (15) respectively:

$$h_p = \frac{Nu_p k_p}{D_p} \tag{14}$$

$$h_w = \frac{Nu_w k_w}{D_w} \tag{15}$$

□(18)

For a uniform wall temperature, the local Nusselt number for a laminar, hydrodynamically and thermally fully developed flow in a rectangular duct (b/e=0.066) is given by Eq. (16).

$$Nu = 7.54 \tag{16}$$

The mass transfer coefficient between working air and water film can be determined by the following formula:

$$h_m = \frac{h_w}{\ _w c_{p,w}} \tag{17}$$

For laminar flow in rectangular duct, the average friction factor of each element is given by:

$$f = \frac{96}{\mathrm{Re}} \tag{18}$$

Wet-bulb effectiveness of the heat exchanger is defined as the following equation:

$$_{wb} = \frac{t_{pi} \quad t_{wo}}{t_{pi} \quad t_{wb,pi}} \tag{19}$$

The sensible cooling capacity is determined by Eq. (20):

$$Q_{\text{cooling}} = \frac{\rho c_p (V_p - V_w)(t_{pi} - t_{po})}{3.6} \tag{20}$$

Coefficient of performance (COP) is defined by the ratio of sensible cooling capacity to total power consumption by fan and water pump of the cooler as below:

$$COP = \frac{Q_{\text{cooling}}}{P_{\text{fan}} + P_{\text{pump}}} \tag{21}$$

3. Results and analyses of numerical modelling

Fig. 2 plots the effects of inlet air temperature on wet-bulb effectiveness, product air outlet temperature, cooling capacity and COP of the exchanger under three different humidity ratio values. The prediction results show that, for the fixed air humidity ratio, the product air outlet temperature, cooling capacity and COP of the cooler increased with increasing the product air inlet temperature from 25 to 45 °C. However, the wet-bulb effectiveness didn't alter significantly as the inlet air temperature varies. As the humidity ratio of inlet air reduces from 16 to 6 g/kg, the cooler's cooling capacity and COP were increased significantly. Meanwhile, the wet-bulb effectiveness didn't change too much. The findings reveal that the cooler would be more suitable for applications in hot and arid regions than hot and humid regions. For applications in hot and humid regions, air dehumidification unit should be added prior to the evaporative cooler to reduce air humidity.

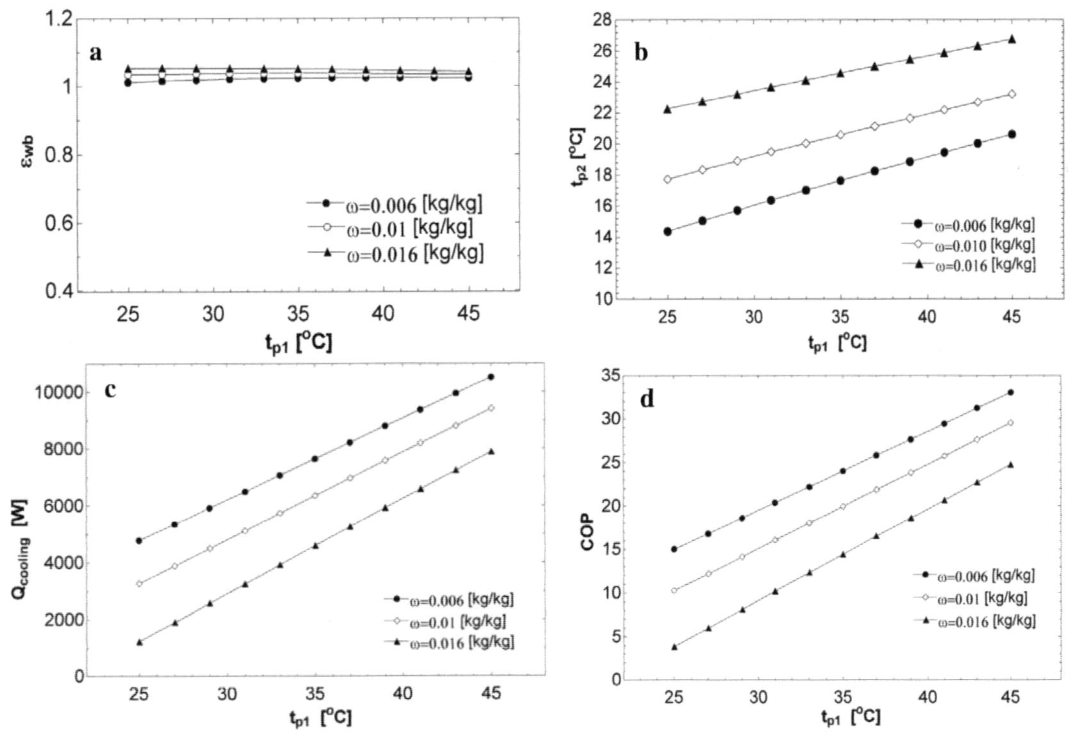

Fig. 2 The effect of inlet air temperature on (a) wet-bulb effectiveness, (b) product air outlet temperature, (c) cooling capacity and (d) coefficient of performance of the counter-flow heat and mass exchanger

4. Conclusions

This paper develops a numerical mode of regenerative evaporative cooler in counter-flow configuration based on modified ε-NTU method. The effects of inlet airflow temperature and humidity were predicted by the model. A few conclusions are summarised as follows: The inlet air temperature has positive impacts on cooling capacity and COP. As the inlet humidity increases, both the cooling capacity and

IOP Conf. Series: Materials Science and Engineering **721** (2020) 012070 doi:10.1088/1757-899X/721/1/012070

COP reduce while the outlet temperature of product air increases; The wet-bulb effectiveness changes minorly as either inlet air temperature or inlet air humidity vary significantly.

Acknowledgments
This work has been funded by National Nature Science Foundation of China No. 51908020 and Beijing Natural Science Foundation No. 3184044.

Nomenclature:

a—the slope of the temperature-enthalpy saturation line
c—constant in equation
c_p—specific heat capacity at constant pressure, J/(kg·K)
C—heat capacity rate, (J/K)
dq—heat transfer rate between product and working air for each computational element, W
D—hydrometric diameter of dry or wet channel, m
h—convective heat coefficent, W(m^2·K)
h_m—mass transfer coefficicent, kg/(m^2·s)
i—enthalpy of airflow, J/kg
k—thermal conductivity, W(m·K)
NTU—number of transfer units
Nu —Nusselt number
P—power consumption, W
Q—cooling capacity of the cooler, W
UdA—overall conductance of the heat exchanger, W/K
V—volumetric airflow rate, m3/s

Greek symbols
ε—wet-bulb effectiveness
ρ—density of airflow, kg/m^3
σ—surface wettability factor, 0.8

subscripts
i—inlet
o—outlet
p—product air
w—working air
wb—wet-bulb
s—saturation
1—inlet for each computation element
2—outlet for each computation element
max—maximum
min—minimum
fan—air fan
pump—water pump

References

[1] Zhao X, Li J, Riffat SB. (2008) Numerical study of a novel counter-flow heat and mass exchanger for dew point evaporative cooling. Appl Therm Eng, 28: 1942-51.
[2] Duan Z, Zhao X, Zhan C, Dong X, Chen H. (2017) Energy saving potential of a counter- flow regenerative evaporative cooler for various climates of China: experiment-based evaluation. Energy Build, 148: 199-210.
[3] adi Jr M, Riffat S. (2014) Experimental and numerical investigation of a dew-point cooling system for thermal comfort in buildings. Appl Energy, 132: 524-35.
[4] Heidarinejad G, Moshari S. (2015) Novel modeling of an indirect evaporative cooling system with cross-flow configuration. Energy Build;92: 351-62.
[5] Cui X, Chua KJ, Yang WM. (2014) Numerical simulation of a novel energy-efficient dew-point evaporative air cooler. Appl Energy,136: 979-88.

IOP Conf. Series: Materials Science and Engineering **721** (2020) 012071 doi:10.1088/1757-899X/721/1/012071

Apparatus for investigation of heat transfer to H_2O/CO_2 mixtures in both near-critical and supercritical regions

Haomin Wu[1], Hanlin Zhang[1], Sha Li, Dong Liu[*], Qiang Li

MIIT Key Laboratory of Thermal Control of Electronic Equipment, School of Energy and Power Engineering, Nanjing University of Science and Technology, Nanjing 210094, China.

[1]These authors contributed equally to this work

[*]Email address: liudong15@njust.edu.cn

Abstract. Supercritical H_2O/CO_2 mixtures are the working fluid in a novel power generation system with coal gasified in supercritical water, so their heat transfer behaviors are important to design heat transfer devices. However, heat transfer to supercritical mixtures has received few attention. Here, we designed and established an apparatus to measure the forced convection coefficients and constant pressure heat capacities of H_2O/CO_2 mixtures in both near-critical and supercritical regions. The highest mixture temperature is designed to be 873 K with the highest pressure of 25 MPa and with the CO_2 mass fraction up to 25%. Results show that our apparatus is stable in both near-critical and supercritical regions, which is challenging for high-temperature and high-pressure experiments. Furthermore, our apparatus has lower uncertainties than those in previous studies.

1. Introduction

Our collaborators and we have been investigating a novel power generation system with coal gasified in supercritical water [1]. The coal is gasified in supercritical water in the gasification reactor with the organic matter converted to H_2 and CO_2, while with N, S, P, Hg and other elements deposited as inorganic salt ash or slag attributed to the unique properties of supercritical water. Therefore, pollution production including NO_X, SO_X and PM2.5 are intrinsically eliminated in this system compared to conventional coal-fired power plants. The working fluid in this system is supercritical H_2O/CO_2 mixtures (~ 600°C, 25 MPa), so the forced convection coefficients and constant pressure heat capacities are important thermophysical properties to design heat exchangers or other heat transfer processes. However, heat transfer to supercritical mixtures has received few attention [2-4]. Therefore, in this paper, an apparatus has been set up to measure these properties of H_2O/CO_2 mixtures in both near-critical and supercritical regions. To reach steady state for measurements are challenging in apparatus design, especially in the near-critical region.

2. Apparatus Design

2.1Process configuration

As shown in Figure 1, de-ionized water is pressurized to 25 MPa by an OBL plunger metering pump and then heated to target temperature by Preheaters 1 and 2. High-purity carbon dioxide from gas cylinders is liquefied in the storage tank, pressurized to 25 MPa by an OBL plunger metering pump, and then heated to the same temperature as water by Preheater 3. H_2O and CO_2 are mixed at a buffer vessel and then heated or cooled to a prescribed temperature at the test section inlet by Preheater 4. After passing the test section, the mixture flows through a condenser and a pressure regulating valve. Then, the H_2O/CO_2 mixture at low temperature and pressure is separated and flow out of the apparatus. The apparatus was designed to be an open system instead of a closed loop to achieve constant component mass fraction. The H_2O/CO_2 mixture has two phases with different component mass fractions in liquid and vapor phases if the mixture is cooled to subcritical temperature before circulated. Considering security and stability, the regenerative heat exchanger that pre-heats the water using the high temperature mixture coming from the test section is not used in the apparatus.

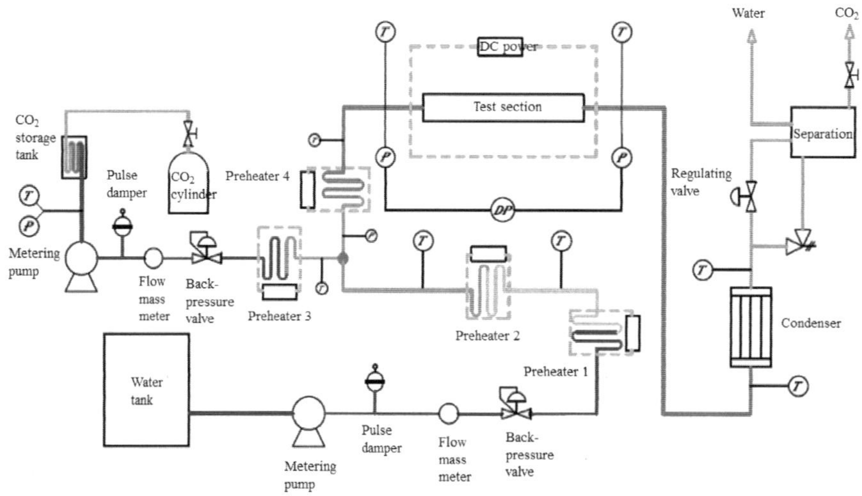

Figure 1. Schematic of apparatus.

To reach steady state for measurements are challenging in apparatus design, especially in the near-critical regions, because the thermophysical properties of H_2O vary significantly in this region. Therefore, H_2O and CO_2 feeding systems, and automatic control system are specifically designed. H_2O and liquefied CO_2 flow through the metering plunger pump, pulse damper, mass flow meter and back-pressure valve in the feeding system before preheaters. The pulse damper is used to eliminate flow oscillation generated from the pump. The metering pump can be automatically adjusted with a feedback controller according to the output data of the mass flow meter. Therefore, the flow rate throughout the apparatus is expected to be stable. The pressure in the test section is controlled by the regulating valve at the outlet of the condenser. This valve can be automatically adjusted with a feedback controller according to the output data of the pressure transducer at the test section inlet. Therefore, the pressure throughout the test section is expected to be stable. Preheater and test tubes are electrically heated by direct current power with low voltages and high currents. Temperatures at preheater outlets and test section inlet are controlled by the power supply which can be automatically adjusted with feedback controllers according to the output data of thermocouples at each location. Therefore, the temperatures throughout the apparatus are expected to be stable.

2.2Data reduction

Figure 2 shows the schematic diagram of the vertical test section, which is a Φ6.35×1.65 mm smooth tube. The heated length of the test section is 2 m. An unheated, 250 mm long tube is installed at the inlet of the test section to assure fully developed flow in the test section. The thermocouple arrangement in test section is also shown in Figure 2. Totally 30 K-type thermocouples are used to measure the tube outer wall temperatures, T_{ow} in 10 locations with three thermocouples spot welded on the top, bottom and side surfaces for each location. The bulk fluid temperatures were measured at the test section inlet and outlet with thermocouples inserted inside the fluid through special customized union tee fittings. The local fluid enthalpies can be linearly interpolated from the known inlet and outlet bulk enthalpies

IOP Conf. Series: Materials Science and Engineering **721** (2020) 012071 doi:10.1088/1757-899X/721/1/012071

[5] with the local fluid temperatures, $T_{b,x}$, referenced from NIST REFPROP and the results reported by our collaborators [6]. The electrical power was defined as a product of the effective value of the voltage and current. The fluid pressure at the outlet of the test section was measured by the EJA 530E pressure transducer. As summarized in Table 1, the temperatures of the supercritical H_2O/CO_2 mixtures range from 300 to 500 °C, the test section inlet pressure is 25 MPa and the CO_2 mass fractions range from 0 to 0.25 in our designed experiments.

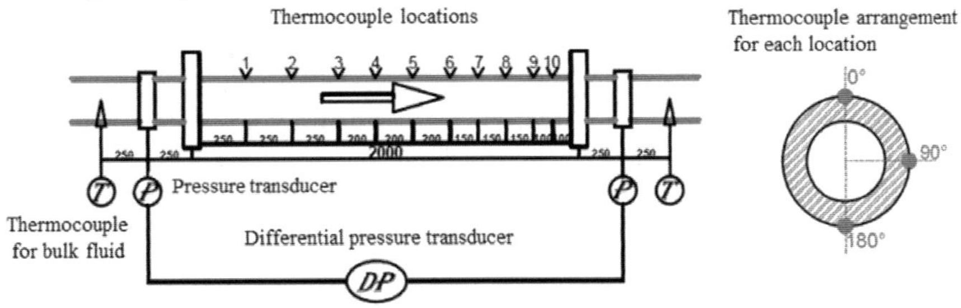

Figure 2. Schematic of test section.

Table 1. Experimental parameters

Fluid temperature (°C)	Pressure (MPa)	CO_2 mass fraction (%)	Mass velocity (kg m^{-2} s^{-1})	Inner diameter (mm)	Heat flux (kW m^{-2})	Flow direction
350-450	24	0-0.25	800-1600	3.05	20-200	Vertical upward and downward Horizontal

The heat absorbed by fluid in the test section is calculated by

$$Q = U \times I \times \eta \tag{1}$$

where U is the voltage and I is the current applied on the test tube measured by transducers, and η is electrical heating efficiency which is calibrated by the bulk fluid enthalpy gains between test section inlet and outlet in supercritical water experiments. The heat flux on the inner wall is

$$q = \frac{Q}{\pi d L} \tag{2}$$

where d and L are the inner diameter and length of the test tube. Then, the local heat transfer coefficient is expressed as

$$HTC = \frac{q}{T_{iw} - T_{b,x}} \tag{3}$$

where the tube inner wall temperatures T_{iw} is determined from the one-dimensional heat conduction equation with internal heat source for fully-circumferential uniform heating condition [7]. All the data were monitored and collected by a computer equipped with a data acquisition board connected with all the sensors used in the experiments. The relative estimated uncertainties in the experiment were summarized in Table 2.

Table 2. Relative estimated uncertainties.

Parameter	**Uncertainties**
Pressure (MPa)	0.1%
Pressure difference (kPa)	0.1%
Mass flux (kg m^{-2} s)	2.3%
Mass fraction (%)	2.9%
Electrical power (kW)	1.4%
Fluid temperature (°C)	0.6°C

Heat flux (kW m^{-2})	2.2%
Heat transfer coefficient (kW m^{-2} K)	18.1%

3. Results and Discussion

3.1Stability test

Figure 3 shows the appearance of our established apparatus. Figure 4(a) shows the variations of the test section pressure and inlet temperature with time in the supercritical region for the condition that the CO_2 mass fraction is 9.9%, the mass velocity is 1270 kg m^{-2} s^{-1} and the heat flux is 60 kW m^{-2}. Results show that the pressure oscillation is below ±0.03 MPa and the temperature oscillation is below ±0.2°C for 24 MPa and supercritical temperatures. Figure 4(b) shows the variations of the test section pressure and inlet temperature with time in the near-critical region for the condition that the CO_2 mass fraction is 10.2%, the mass flux is 1596 kg m^{-2} s^{-1} and the heat flux is 60 kW m^{-2}. Results show that the pressure oscillation is below ±0.03 MPa and the temperature oscillation is below ±0.3°C for 24 MPa and near-critical temperatures.

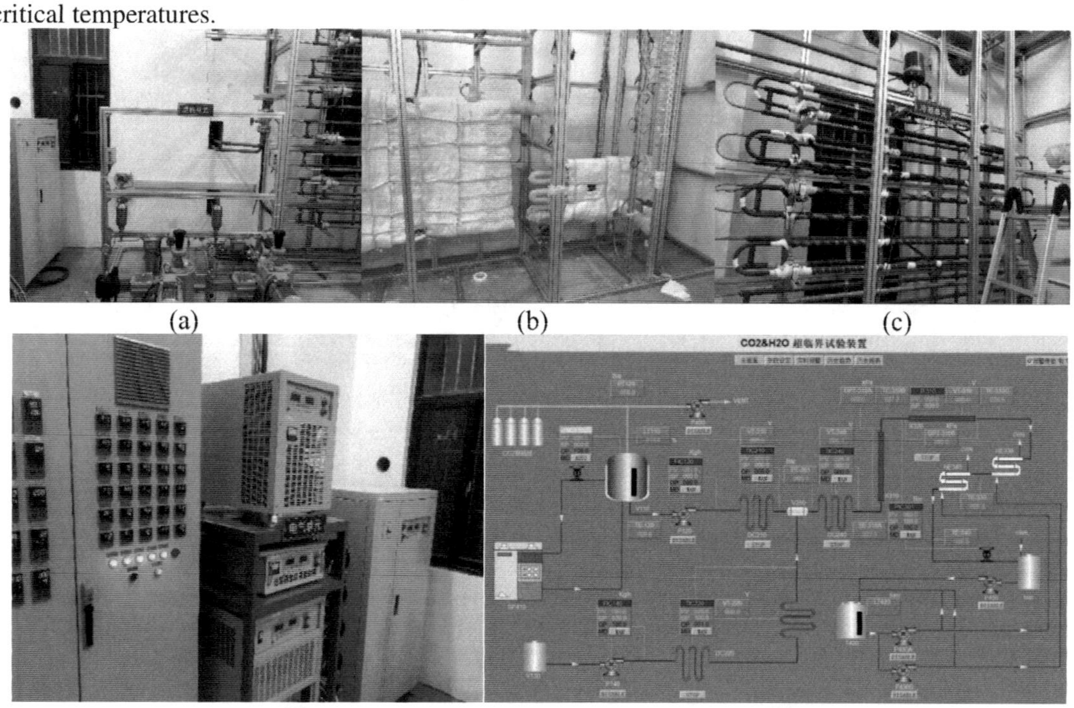

(a) (b) (c)

(d) (e)

Figure 3. Photographs of (a) pumps and mass flow meters, (b) preheaters and test section, (c) condenser, (d) electric controlling and power supplies and (d) automatic control interface.

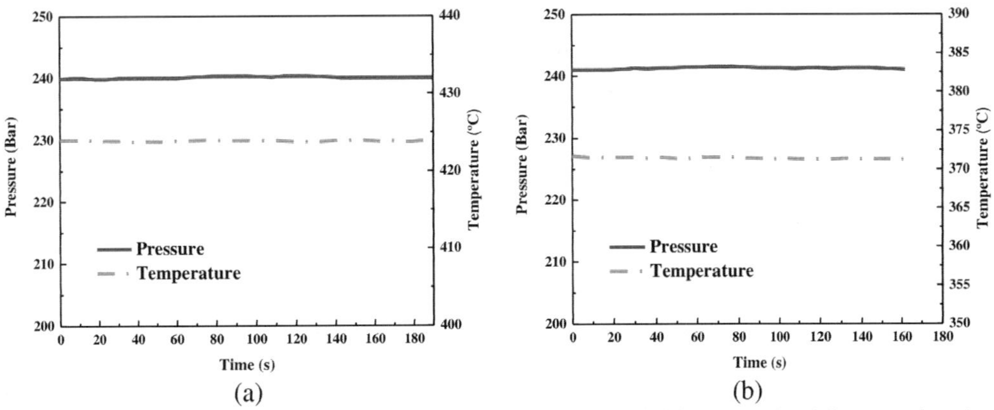

Figure 4. Stability test (a) in the supercritical region for the condition that the CO_2 mass fraction is 9.9%, the mass flux is 1122 kg m^{-2} s^{-1} and the heat flux is 60 kW m^{-2}; (b) in the near-critical region for the condition that the CO_2 mass fraction is 10.2%, the mass flux is 1596 kg m^{-2} s^{-1} and the heat flux is 60 kW m^{-2}.

3.2Accuracy validation

The measured forced convection heat transfer coefficients were compared with those calculated using the Mokry correlation [8]. Figure 5 shows that the measured and correlated results agree well demonstrating that the experimental data in our experiments was reliable.

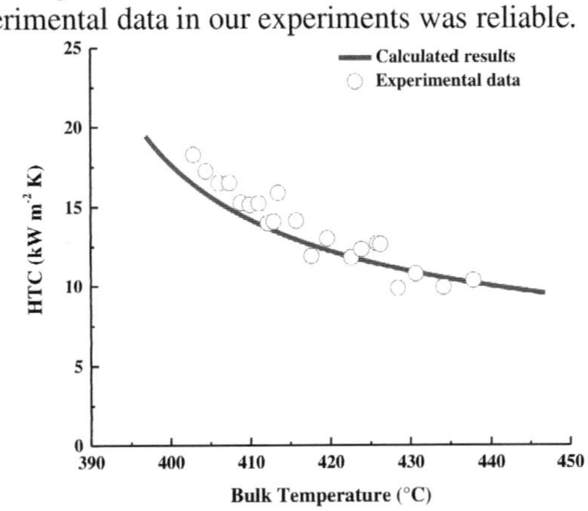

Figure 5. Accuracy validation for the condition that pure water with 1122 kg m^{-2} s^{-1} mass velocity and 60 kW m^{-2} heat flux.

4. Conclusion

In summary, we established an apparatus to investigate the heat transfer behavior of H_2O/CO_2 mixtures in both near-critical and supercritical regions. The apparatus is stable and the measurement uncertainties of forced convection heat transfer coefficients were analyzed and are expected to be below 20%.

References

[1] Guo L., Jin H., Ge Z., Lu Y., Cao C., (2015) Industrialization Prospects for Hydrogen Production by Coal Gasification in Supercritical Water and Novel Thermodynamic Cycle Power Generation System with No Pollution Emission. Science China Technological Sciences, 58: 1989-2002.

[2] Rogak S. N., Faraji D., (2004) Heat Transfer to Water-Oxygen Mixtures at Supercritical Pressure. ASME Journal of Heat Transfer, 126: 419-424.

[3] Huang D., Wu Z., Sunden B., Li W., (2016) A Brief Review on Convection Heat Transfer of Fluids at Supercritical Pressures in Tubes and the Recent Progress. Applied Energy, 162: 494-

505.

[4] Jackson J.D., (2013) Fluid Flow and Convective Heat Transfer to Fluids at Supercritical Pressure. Nuclear Engineering and Design, 264: 24-40.

[5] Lei X., Li H., Zhang W., Dinh N.T., Guo Y., Yu S., (2017) Experimental Study on the Difference of Heat Transfer Characteristics between Vertical and Horizontal Flows of Supercritical Pressure Water. Applied Thermal Engineering, 113: 609-620.

[6] Yang X., Xu J., Wu S., Yu M., Hu B., Cao B., Cao J., (2018) A Molecular Dynamics Simulation Study of PVT Properties for $H_2O/H_2/CO_2$ Mixtures in near-Critical and Supercritical Regions of Water. International Journal of Hydrogen Energy, 43: 10980-10990.

[7] Li Z., Wu Y., Lu J., Zhang D., Zhang H., (2014) Heat Transfer to Supercritical Water in Circular Tubes with Circumferentially Non-Uniform Heating. Applied Thermal Engineering, 70: 190-200.

[8] Mokry S., Pioro I., Farah A., King K., Gupta S., Peiman W., Kirillov P., (2011) Development of supercritical water heat-transfer correlation for vertical bare tubes. Nuclear Engineering and Design, 241: 1126-1136.

IOP Conf. Series: Materials Science and Engineering **721** (2020) 012072 doi:10.1088/1757-899X/721/1/012072

Experimental Analysis of Thermal Efficiency of Biomass-burning Hot Blast Furnace

BingJin, LinWang, ChenLiu

Liaoning Inspection, Examination&Certification Centre (Liaoning Academy of Safety Science), Shenyang 110004, Liaoning Province, China

Tel: 024-23891311, Email: 463267841@qq.com

Abstract: This paper analyzes the main factors affecting the thermal efficiency of biomass hot blast furnaces, and compares the theoretical factors with the experimental ones that influence the thermal efficiency so as to provide suggestions for the improvement of the biomass hot blast furnace.

1. Introduction

In some areas of northern China, a hot blast furnace is needed in the process of grain drying. The hot blast furnace heats the air, and the hot blast can directly passes through the grain drying equipment to dry the grain. The materials used to produce energy for hot blast furnace installation include coal, gas, biomass and fuel [1]. The working principle of the hot blast furnace is: first, the fuel is fully burned in the combustion chamber of the hot blast furnace, and the hot gas after combustion enters the heat exchanger to exchange heat with the cold air outside; the gas after heat exchange is discharged into the atmosphere through the ash removal system, and the hot blast outside is sent to the grain drying tower through the pipeline to dry the grain. The hot blast furnace is mainly composed of three systems, namely, the fuel combustion system, the heat exchange system, and the gas exhaust system. The efficiency of the combustion system and the exhaust system can be analyzed according to the efficiency of the boiler. At present, there are many domestic studies on boiler efficiency [2]. Tubular hot blast furnace is mainly used domestically [3, 4], and its efficiency is related to the efficiency of combustion and heat exchanging.

2. Thermal efficiency of hot blast furnace

According to the literature and standards [5,6], the thermal efficiency of the hot blast furnace is related to the hot blast flow rate, the diameter of the air flue, the hot blast temperature and the inlet air temperature. It can be calculated with the heat of boiler and the heat of heat exchanger taken into consideration.

2.1 Thermal efficiency algorithm of hot blast furnace

2.1.1 Output heat

$$Qc = q\rho \times (\triangle CT) \quad (1)$$

In the formula, q is the hot blast flow, ρ is the hot blast density, and the ΔCT is the difference between the products of the average constant-pressure mass specific heat capacity and the temperature.

2.1.2 Input heat

$$Qr= Qnet,v,arB \tag{2}$$

In the formula, Q is the input heat; Qnet, v, and ar is the low calorific value of the fuel; B is the average fuel consumption per hour.

2.1.3 Thermal efficiency

$$\eta= Qc/ Qr\times 100\% \tag{3}$$

In the formula, η is the heat transfer efficiency of the hot blast furnace

2.2 Heat calculation formula of heat exchanger

$$A=Q/K(Tr-\Delta t) \tag{4}$$

In the formula, A is the heat exchange area, Q is the total heat transfer, and K is the thermal conductivity. The thermal conductivity varies from material to material and from medium to medium. If the same material is used in the heat exchangers with different structures, the K value will be different. Tr is the average temperature of the hotter medium, and Δt is the average temperature of the second hottest medium.

Theoretically, the main factors influencing the thermal efficiency of the boiler include the heat loss from exhaust gas, from incomplete solid combustion, from incomplete gas combustion, from heat dissipation and from physical heat loss of ash.

The area and the average temperature of the heat exchange surface are related to the calculation of heat exchanger. The heat transfer coefficient directly affects the overall efficiency of the hot blast furnace. In order to increase the efficiency of the heat exchanger, it is necessary to enlarge the heat transfer area, expand the average temperature difference between the hot and cold fluids, and increase the heat transfer coefficient.

Therefore, the selection calculation of the boiler and the heat exchanger is particularly important in the aspects of reducing heat loss of the boiler and improving the heat exchange efficiency so as to improve the thermal efficiency of the hot blast furnace.

3. Test and analysis of the hot furnace

3.1 Basic situation of the hot furnace

The tested hot blast furnace is a biomass hot blast furnace with a rated power of 4.2 MW. It is reconstructed from the original coal hot blast furnace, with the chain grate stoker unchanged. The combustion space inside the furnace is rebuilt and the original furnace body is re-insulated. The original heat exchanger belongs to the built-up type, and some convective heat exchange tubes are replaced after the ash is removed. Part of the gas is drained from the gas exhaust system and mixed with the fresh air before entering the combustion chamber.

3.2 Test situation

The use of the hot blast furnace is tested during its operation. The measure point map is shown in Figure 1 and the test results are shown in Table 1.

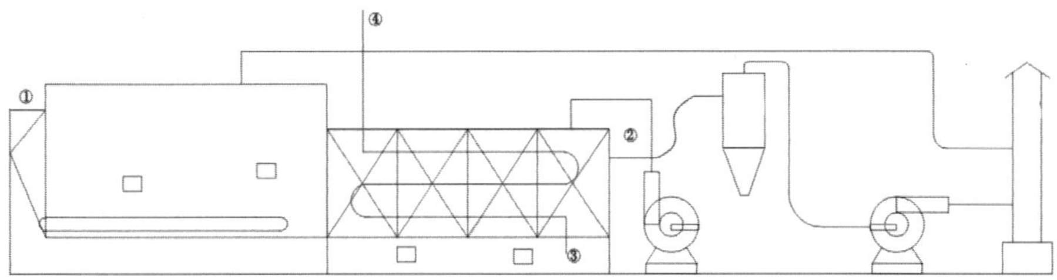

Figure 1 Measure point map of biomass hot blast furnace

1- measure point of biomass consumption; 2- measure point of exhaust gas; 3- measure point of heat exchanging;4- measure point of hot blast temperature and flow; Note: slag and ash sampling and gas pressure test are not shown in the map

For the hot blast furnace and the heat exchanger, the heat that can be obtained mainly include the heat loss of the gas, the recovered waste heat of gas, and the heat used by the hot blast furnace.

Table 1 Test data

Item	Before rebuilding	After rebuilding
Temperature /°C	-5	-6
Humidity /%	32%	35%
Hot blast temperature /°C	120	135
Hot blast pressure /Pa	2000	2000
Unit fuel quantity /Kg/h	950	950
Low calorific value of biomass /KJ/Kg	14280	14280
Exhaust gas temperature /°C	140.90	109.45
Exhaust gas temperature /%	1.66	1.30
Carbon content of ash /%	10%	7%

The average exhaust gas temperature is reduced from 140.90°C to 109.45°C after building. The average temperature of the exhaust gas is reduced by 31.45°C or 22%. When the amount of fuel in the hot blast furnace and the air volume of the heat exchanger are kept substantially constant, reducing the exhaust gas temperature can greatly reduce the heat loss of the exhaust gas.

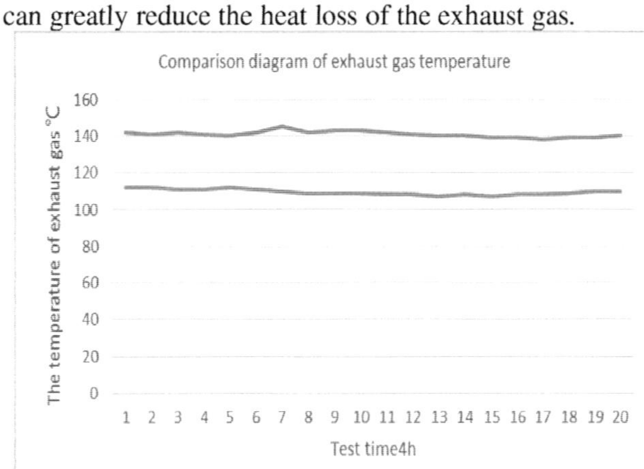

Figure 2 Comparison of exhaust gas temperature before and after rebuilding

The heat loss curves obtained before and after the hot blast furnace rebuilding are shown in the figure below. It can be seen from the figure that after rebuilding, the heat loss of exhaust gas is reduced by 34.71%, the heat loss of incomplete solid combustion is reduced by 25.30%, and the heat dissipation is reduced by 24.47%, while the physical heat loss of ash remains the same. The heat loss of incomplete gas combustion sees an increase of 50.00% after rebuilding. But the heat brought by gas is not enough to offset the heat loss of incomplete gas combustion.

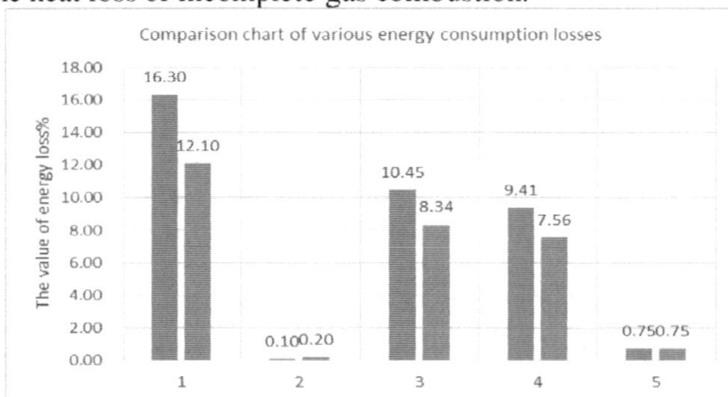

Figure 3 Comparison of various heat losses

After rebuilding, the average hot blast temperature rises from 120.60°C to 135.70°C, an increase of 15.10°C or 12.53%. When the amount of fuel entering the hot blast furnace and the air volume of the heat exchanger are kept substantially constant, higher hot blast temperature leads to higher efficiency of the hot blast furnace.

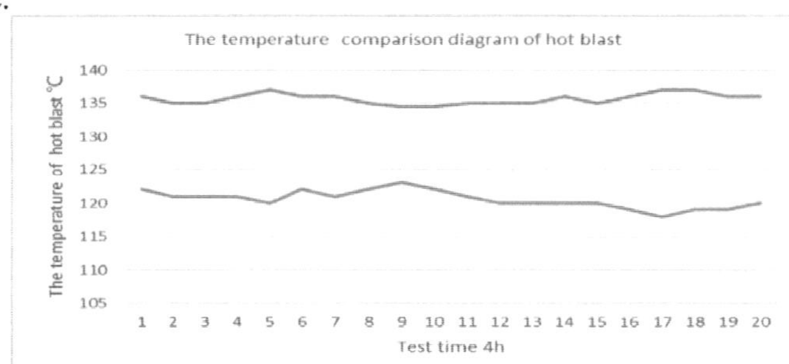

Figure 4 Comparison of hot blast temperature before and after rebuilding

4. Conclusion

Based on the analysis and measurement of a nominal 4.2MW hot blast furnace in a grain depot, this paper draws the following conclusions.

(1) When the amount of fuel in the whole hot blast furnace and the air volume of the heat exchanger are kept basically constant, for the hot blast furnace after rebuilding and ash removing, its average exhaust gas temperature is reduced from 140.90°C to 109.45°C, a reduction of 31.45°C or 22%. Reducing the exhaust gas temperature can greatly reduce the heat loss of the gas, thereby improving the thermal efficiency of the hot blast furnace.

(2) When the amount of fuel in the whole hot blast furnace and the air volume of the heat exchanger are kept substantially constant, for the hot blast furnace after rebuilding and ash removing, its average hot blast temperature jumps from 120.60°C to 135.70°C, an increase of 15.10°C or 12.53%. It can be proved that the ash removing can significantly improve the heat transfer coefficient of heat exchangers, especially for those using the original coal-fired furnace as a heating source.

(3) When the amount of fuel in the whole hot blast furnace and the air volume of the heat exchanger are kept substantially constant, the heat lost of the incomplete gas combustion increases by 50.00% after rebuilding when the exhaust gas is partially introduced into the combustor. But the heat brought by gas is not enough to offset the heat loss of incomplete gas combustion. It can be seen that the introduction of exhaust gas for combustion reaction has no positive effect on the improvement of the thermal efficiency of the entire system.

References

[1] Luo Ping.Application of Biomass Hot Blast Furnace in Greenhouse in Northeast China.Enterprise and Technology.2011（16）：200

[2] Qi Guoli,Guan Jian,Leng Hao.Discussion on the Test Methods of Industrial Boiler Thermal Effiiciency for Approved Products.Energy Conservation Technology,march.2013,No.2

[3] Zhang Baoliang,Wang Lihong.Design and Structure of Coal-fired Hot Blast Furnace.Journal of A gricultural Mechanization Research,2019（8）

[4] Hu Jingchuan.The Study of Hot Blast Furnace with Thermal Radiant Panel.Journal of Tea Machinery,1995（10）

[5] Ministry of Industry and Information Technology.JB/T 6672-2011.4. Coal-burnning furnace[S]

[6] Wang Dehua,Zhou Yun,Ren Jinxiang,Zhou Xianguang,Shen Jianhong.Suggestions about enhangcing coal-burning air heater's effiviency in grain drying system.Science and Technology of Cereals,Oils and Foods.2011(16):200

IOP Conf. Series: Materials Science and Engineering **721** (2020) 012073 doi:10.1088/1757-899X/721/1/012073

Heat Transfer Characteristics and Energy Efficiency Analysis of Finned Tube Heat Exchangers

Zhenchuan Wang, Chao Liu, Songsong Zhang

China Special Equipment Inspection and Research Institute

Wangzhenchuan@csei.org.cn

Abstract. Flow and heat transfer characteristics of the integral finned tube heat exchanger are studied by the numerical simulation method in the present paper. The energy efficiency levels of two types of finned tube heat exchangers (aligned finned tube and staggered finned tube) are evaluated by PEC method. The heat transfer intensity of different finned tube heat exchangers was evaluated by the field synergy principle. When the tube spacing ratio ($S_1/S_2=1$) is small, the field synergy of staggered finned tube is similar to the aligned finned tube, and the PEC index is less than 1; when the spacing ratio ($S_1/S_2=3$) is large, the field synergy of the staggered finned tube get much stronger than the aligned finned tube, and the PEC index is larger than 1 as the $Re \geq 8\times10^4$. Considering both the change of heat transfer intensity and the increase of resistance, it is recommended to evaluate the energy efficiency level of finned tube heat exchanger by PEC method.

1. Introduction

Heat exchangers are one of the common equipment for various industrial processes. Heat transfer performance research and energy efficiency evaluation are of great significance for the efficient use of heat exchangers. Heat exchanger research includes heat transfer enhancement, control and simulation and so on. In order to study the heat transfer enhancement of heat exchangers, scholars have proposed methods to expand heat transfer surfaces, fluid induction and surface treatment. The fin structure can effectively improve the heat exchange capacity of the pipe, and the fluid disturbance can be enhanced by the pipeline arrangement and the fin treatment, thereby further enhancing the heat transfer capability of the heat exchanger. For the heat transfer performance calculation of various fin-and-tube heat exchangers, scholars have proposed many heat transfer coefficient correlation formulas [1-4]. Zhukauskas[1] summarized a highly applicable heat transfer correlation for the heat transfer of the fluid across the tube bundle, with the Pr number ranged from 0.6 to 500.

For the study of flow heat transfer characteristics in heat exchangers, many scholars use numerical methods for analysis and discussion. Mon [5] numerical studied the effects of fin spacing on four rows of annular staggered finned tube bundles by three-dimensional numerical studies. Kahalerras [6] used numerical simulation methods to improve the heat transfer performance of the double-tube heat exchanger by using porous fins. He [7] conducted a three-dimensional numerical simulation study on the laminar heat transfer and flow characteristics of plate-fin heat exchangers. From the field synergy principle, it is analyzed that reducing the angle between velocity and temperature gradient is the basic principle of enhancing convective heat transfer. Suárez [8] proposed a novel one-dimensional model to predict the heat transfer rate of a continuous finned tube heat exchanger. Ephraïm [9] carried out three-dimensional numerical simulation of four rows of elliptical finned tube heat exchangers by LES numerical simulation method, and analyzed the flow evolution in the whole heat exchanger and the instability of complex vortex structures. The results show that as the Reynolds number increases, flow instability occurs faster downstream of the heat exchanger. Zhang [10] conducted a CFD numerical simulation study on the performance of the finned tube heat exchanger in the air cooler, and predicted the air flow, temperature curve and heat transfer characteristics under different operating conditions.

The heat transfer research of finned tube heat exchangers is mainly focused on the study of heat transfer enhancement. Guo [11] first proposed the field synergy principle and carried out numerical verification. The field synergy number, Fc, is defined as the degree of synergy between the velocity field and the temperature gradient field over the entire region. The results show that the field synergy will be much less than 1 under most heat transfer conditions, and there is a greater potential for heat transfer enhancement. Webb [12] divides heat transfer enhancement into three purposes, reducing surface area, increasing heat load and reducing power consumption, and proposed PEC method to evaluate the effect of enhanced heat transfer structure on heat transfer.

Many scholars have obtained empirical correlations for convective heat transfer of fin-and-tube heat exchangers through experimental methods, numerical simulations, theoretical analysis, etc. The research on the enhanced heat transfer of fin-and-tube heat exchangers is also sufficient. While for the evaluation of the energy efficiency level of the heat exchanger is insufficient. In this paper, the field synergy theory and PEC method are used to evaluate the enhanced heat transfer effect and energy efficiency level of the finned tube heat exchanger in the staggered finned tube and the aligned finned tube. It is recommended to evaluate the energy efficiency level of the finned tube heat exchanger by PEC method. The field synergy principle could be used to evaluate the heat transfer intensity of the finned tube heat exchanger.

2. Physical model and numerical method

2.1Physical model

In this paper, flow and heat transfer characteristics of the staggered finned tube and the aligned finned tube with different tube pitch was calculated by the numerical method. Finned tube bundle contains 16 rows of tubes. A two-dimensional assumption is made to the finned tube heat exchanger, ignoring the flow and heat transfer along the length of the tube. The working fluid is air, assuming the tube bundle is at a constant temperature of 80°C and the inlet air temperature is 25°C. The inlet boundary adopts the velocity boundary condition, the outlet adopts the pressure outlet boundary, and the other boundary adopt the periodic boundary condition. The Reynolds number is determined by the average flow velocity at the smallest section of the bundle, and the feature length is the outer diameter of the tube. The Reynolds number ranges from 10^4-8×10^4. The tube spacing is divided into two types, $S_1=1\times S_2=1.5\times D$, and $S_1=3\times S_2=4.5\times D$.

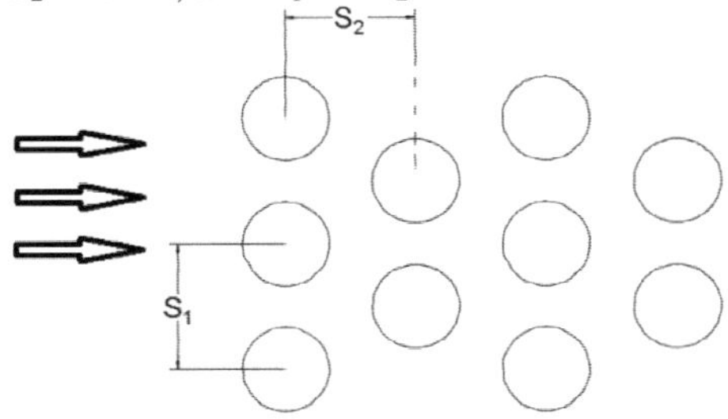

(a) staggered finned tube

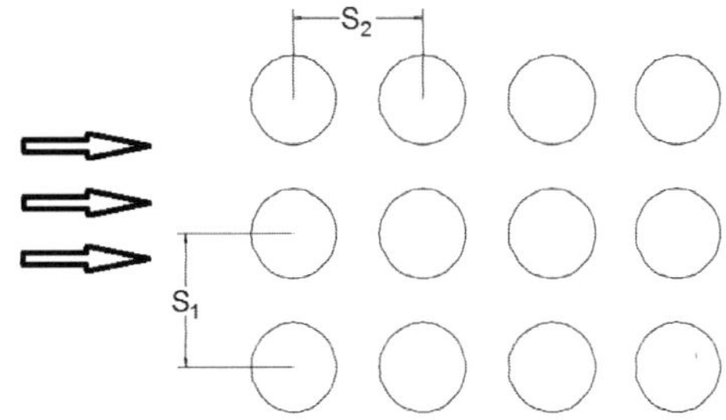

(b) aligned finned tube

Figure 1 Tube bundle model

2.2Numerical method

The computational study was conducted using the Ansys Fluent 13.0 [13]. The flow and heat transfer was calculated by the standard k-ε turbulence model. The continuity, momentum and energy equation are as below:

Continuity:

$$\frac{\partial U}{\partial x}+\frac{\partial V}{\partial y}=0 \tag{1}$$

U-momentum:

$$\frac{\partial U^2}{\partial x}+\frac{\partial (VU)}{\partial y}=-\frac{1}{\rho}\frac{\partial p}{\partial x}+2\frac{\partial}{\partial x}[(\vartheta+\vartheta_t)(\frac{\partial U}{\partial x})]+\frac{\partial}{\partial y}[(\vartheta+\vartheta_t)\left(\frac{\partial U}{\partial y}+\frac{\partial V}{\partial x}\right)] \tag{2}$$

V-momentum:

$$\frac{\partial V^2}{\partial x}+\frac{\partial (UV)}{\partial y}=-\frac{1}{\rho}\frac{\partial p}{\partial x}+2\frac{\partial}{\partial y}[(\vartheta+\vartheta_t)(\frac{\partial V}{\partial y})]+\frac{\partial}{\partial x}[(\vartheta+\vartheta_t)\left(\frac{\partial V}{\partial x}+\frac{\partial U}{\partial y}\right)] \tag{3}$$

Energy:

$$\frac{\partial (Uh)}{\partial x}+\frac{\partial (Vh)}{\partial y}=\frac{\partial}{\partial x}\left[\left(\frac{\vartheta}{Pr}+\frac{\vartheta_t}{Pr_t}\right)\frac{\partial h}{\partial x}\right]+\frac{\partial}{\partial y}\left[\left(\frac{\vartheta}{Pr}+\frac{\vartheta_t}{Pr_t}\right)\frac{\partial h}{\partial y}\right] \tag{4}$$

Turbulent kinetic energy (k)

$$\left[\frac{\partial (Uk)}{\partial x}+\frac{\partial (Vk)}{\partial y}\right]=\frac{\partial}{\partial x}\left[\left(\vartheta+\frac{\vartheta_t}{\sigma_k}\right)\frac{\partial k}{\partial x}\right]+\frac{\partial}{\partial y}\left[r\left(\vartheta+\frac{\vartheta_t}{\sigma_k}\right)\frac{\partial k}{\partial y}\right]+P_k+G_k-\rho\varepsilon \tag{5}$$

Dissipation rate (ε)

$$\left[\frac{\partial (U\varepsilon)}{\partial x}+\frac{\partial (V\varepsilon)}{\partial y}\right] =\frac{\partial}{\partial x}\left[\left(\vartheta+\frac{\vartheta_t}{\sigma_\varepsilon}\right)\frac{\partial \varepsilon}{\partial x}\right]+\frac{\partial}{\partial y}\left[\left(\vartheta+\frac{\vartheta_t}{\sigma_\varepsilon}\right)\frac{\partial \varepsilon}{\partial y}\right]+C_{\varepsilon 1}f_1\frac{\varepsilon}{k}(P_k+G_k) - C_{\varepsilon 2}f_2\frac{\varepsilon^2}{k} \tag{6}$$

The complete computational domain was discretized using Gambit 2.4 into unconstructed mesh. The mesh was refined around the tube where the heating commenced which significantly influences the prediction of the wall temperature. The SIMPLEC scheme was used to couple the pressure and velocity fields. The QUICK scheme was used to approximate the convection terms in the momentum equations with the UPWIND scheme used for the other transport equations. The convergence criterion for the normalized residual of each equation was set to be less than 10^{-5}.

2.3Validation of the numerical simulation

The predicted results are compared with the Zhukauskas experimental correlation in the following figures. Figure 2 compares the Nusselt number with Re number for different tube arrangement, and the tube pitch S_1 is equal to S_2. The correlation is as below:

$$Nu_f=0.27Re_f^{0.63}Pr_f^{0.36}({}^{Pr_f}/_{Pr_w})^{0.25} \text{ for aligned finned tube} \tag{7}$$

$$Nu_f=0.35({}^{S_1}/_{S_2})Re_f^{0.6}Pr_f^{0.36}({}^{Pr_f}/_{Pr_w})^{0.25}, {}^{S_1}/_{S_2}\le 2$$

$$Nu_f=0.40Re_f^{0.6}Pr_f^{0.36}({}^{Pr_f}/_{Pr_w})^{0.25}, {}^{S_1}/_{S_2}>2 \text{ for staggered finned tube} \tag{8}$$

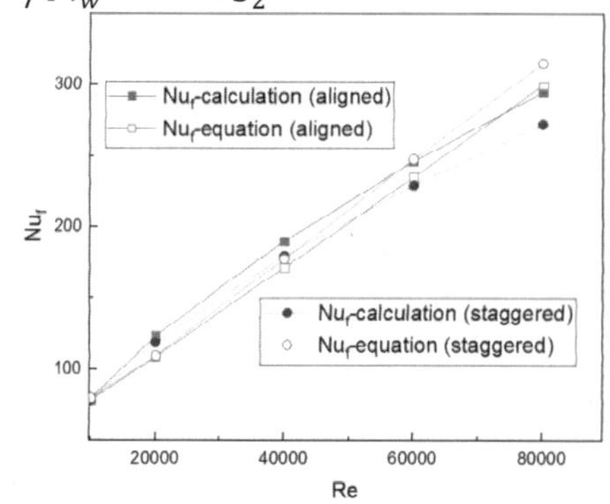

Figure 2 Nusselt number

All the difference of the results between the numerical calculation and the Zhukauskas equation is within 15%, and most of them are less than 10%. The results show that the numerical model is accurate, and the model could be used to get the flow and heat transfer characteristics for the finned tube.

3. Results and discussion

Figure 3 shows the temperature field distribution of the two kinds of finned tube heat exchangers whit the tube spacing $S_1 = S_2$ and Re = 6×10^4. Combined with the results of the Fig. 2, it can be seen that when the pipe spacing S_1 is relatively small, the convective heat transfer intensity of the staggered finned tube heat exchanger is higher than that of the heat exchanger, but the change of the heat exchange intensity is not large. When the pipe spacing is small, the turbulence intensity is similar, so that the heat exchange of the two types of finned tube heat exchangers is similar.

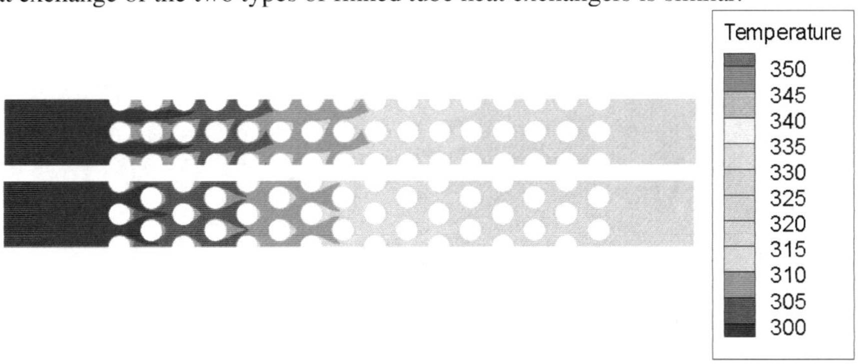

Figure 3 temperature field for finned tube

Figure 4 shows the PEC evaluation index of the staggered finned tube heat exchanger obtained based on the evaluation of the aligned finned tube heat exchanger. The formulas of PEC and f are as below:

$$\text{PEC} = \frac{Nu/_{Nu_0}}{(f/_{f_0})^{1/3}} \tag{9}$$

$$f = \frac{\Delta P}{2\rho u_{max}} \times \frac{D}{L} \tag{10}$$

As shown in the fig. 4, the PEC index of the two kinds of heat exchanger increases with the increase of Reynolds number under two different pipe spacing forms. Except the condition of the Reynolds number equals to 8×10^4, and S_1 equals to three times of S_2, the other PEC indexes are less than 1. The results show that the energy efficiency level of the staggered finned tube heat exchanger is higher than that of the heat exchanger when the Reynolds number is large. When the Reynolds number is small or the fin tube spacing is similar, the energy efficiency level of the aligned finned tube heat exchanger is higher than that of the staggered finned tube heat exchanger.

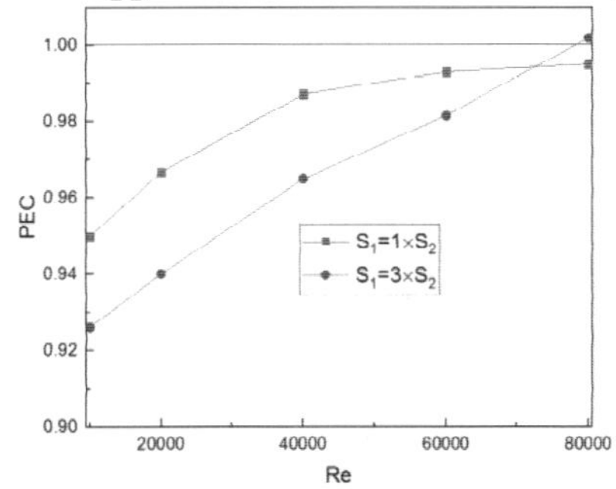

Figure 4 PEC evaluation

IOP Conf. Series: Materials Science and Engineering **721** (2020) 012073 doi:10.1088/1757-899X/721/1/012073

Figure 5 shows the angle distribution of the velocity direction and the temperature gradient direction in the flow field region for the two kinds of heat exchanger with the Reynolds number equals to 8×10^4. The angle between the velocity field and the temperature gradient field is defined as follows:

$$cos\emptyset = \frac{\bar{U}\cdot\overline{\nabla T}}{|\bar{U}|\cdot|\overline{\nabla T}|} \tag{11}$$

It can be seen from the angle distribution cloud diagram that the flow path disturbance is enhanced due to the piping arrangement of the staggered finned tube heat exchanger, and the region of the angle less than 90°is larger than that of the aligned finned tube heat exchanger. According to the field synergy principle, when the other quantities are constant, the closer the angle is to 90°, the smaller the heat transfer intensity is, and the closer to 0°, the greater the heat transfer intensity. It can be seen from the degree of synergy between the flow field and the temperature field that the heat exchange intensity of the staggered finned tube heat exchanger is enhanced relative to the aligned finned tube heat exchanger.

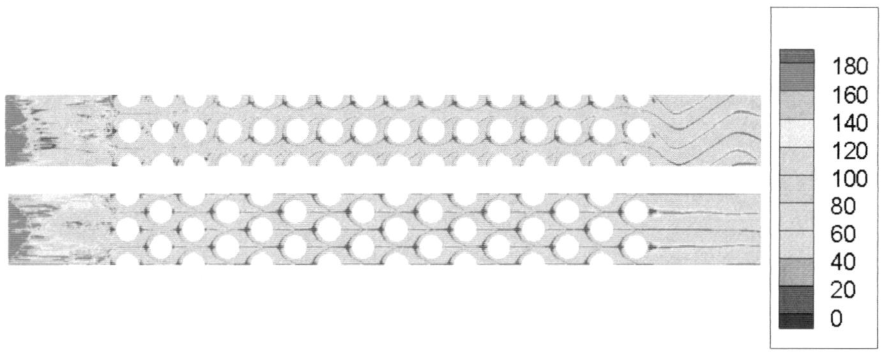

Figure 5 Field synergy angle

Figure 6 shows the variation of field synergy number with Reynolds number in two kinds of heat exchangers under different pipe spacing conditions. As shown in Fig.6, the field synergy number decreases with the increase of Reynolds number, which indicates that with the increase of Reynolds number, the effect of fluid impingement pipeline is reduced, and the effect of fluid flow and dissociation is enhanced.

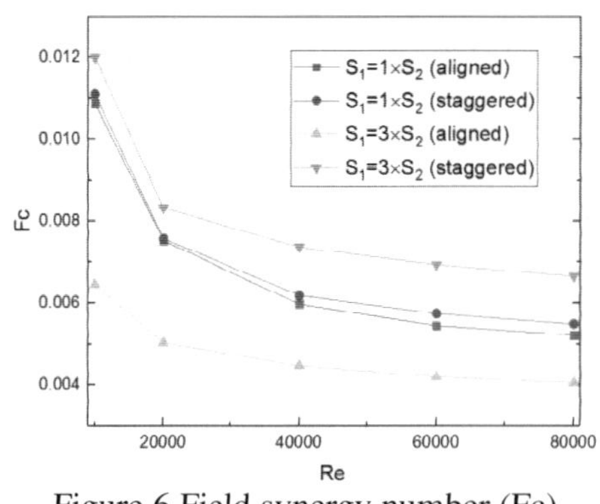

Figure 6 Field synergy number (Fc)

From the perspective of field synergy, increasing the fluid velocity reduces the degree of synergy between the flow field and the temperature field. Under the two tube spacing, the field synergy number of the staggered finned tube heat exchanger is larger than that of the aligned finned tube heat exchanger, but when the tube spacing S_1 equals to S_2, the field synergy number of the two kinds of heat exchanger field is very close. When the pipe spacing is small, the resistance of the staggered finned tube is larger than that of the aligned finned tube heat exchanger. In practice application, the staggered finned tube heat exchanger is more prone to fouling and dust accumulation, so it is recommended to prioritize use the aligned finned tube heat exchanger when the pipe spacing is small. When the pipe spacing is increased, the field synergy number of the staggered finned tube heat exchanger field is larger than that of the aligned finned tube heat exchanger, indicating that the larger the pipe spacing, the better of the field synergy and the greater heat exchange intensity. When comparing the heat transfer degree of the fin-and-tube heat exchanger, the field synergy can be used for quantitative evaluation. Under the same Reynolds number and Prandtl number, the larger the field synergy, the higher the heat transfer intensity.

4. Conclusion

In this paper, the flow and heat transfer characteristics of the aligned finned tube heat exchanger and staggered finned tube heat exchanger are numerical simulated with different tube pitch. The enhanced heat transfer level and energy efficiency of the two heat exchangers were evaluated by PEC method and field synergy principle. When considering heat transfer strength and resistance changes, it is recommended to use the PEC method for the energy efficiency on the finned tube heat exchangers. When evaluating the enhanced heat transfer level of fin-and-tube heat exchangers, it is recommended to use the field synergy number for quantitative evaluation.

Acknowledgements: This project was supported by the CSEI Research Program (No. 201806).

References

[1] Zhukauskas A A. Convection heat transfer in heat exchanger[M], beijing, CSPM, 1986.
[2] Kang haijun, Li huizhen et al. Experimental study on heat transfer and pressure drop for plane fin-and-tube heat exchanger. Journal of Xi'an Jiaotong University, 1994, 28:91-98.
[3] Li Wu, Li huizhen, Kang haijun. Study on the characteristics of heat transfer and pressure drop of heat exchanger with low density fins. Journal of Xi'an Jiaotong University, 1994, 28:81-86.
[4] Qu Z G , Tao W Q , He Y L . Three-Dimensional Numerical Simulation on Laminar Heat Transfer and Fluid Flow Characteristics of Strip Fin Surface With X-Arrangement of Strips[J]. Journal of Heat Transfer, 2004, 126(5):697.
[5] Mon M S , Gross U . Numerical study of fin-spacing effects in annular-finned tube heat exchangers[J]. International Journal of Heat and Mass Transfer, 2004, 47(8-9):1953-1964.
[6] Kahalerras H , Targui N . Numerical analysis of heat transfer enhancement in a double pipe heat exchanger with porous fins[J]. International Journal of Numerical Methods for Heat & Fluid Flow, 2008, 18(5):593-617.
[7] He Y L , Tao W Q , Song F Q , et al. Three-dimensional numerical study of heat transfer characteristics of plain plate fin-and-tube heat exchangers from view point of field synergy principle[J]. International Journal of Heat and Fluid Flow, 2005, 26(3):459-473.
[8] Suárez, Felipe, Keegan S D , Mariani, Néstor J, et al. A novel one-dimensional model to predict fin efficiency of continuous fin-tube heat exchangers[J]. Applied Thermal Engineering, 2019, 149:1192-1202.
[9] Ephraïm Toubiana , Rémi Gautier , Daniel Bougeard , et al. Large Eddy Simulation of transitional flows in an elliptical finned-tube heat exchanger[J]. International Journal of Thermal Sciences, 2019, 144: 158-172.
[10] Zhang X.Y, Ge Y. T., Sun J. N., et al. CFD Modelling of Finned-tube CO2 Gas Cooler for Refrigeration Systems[J]. Energy Procedia, 2019, 161: 275-282.
[11] Guo Z Y , Tao W Q , Shah R K . The field synergy (coordination) principle and its applications in enhancing single phase convective heat transfer[J]. International Journal of Heat and Mass Transfer, 2005, 48(9):1797-1807.
[12] Ralph L. Webb, Nae-Hyun Kim, Principles of enhanced heat transfer [M]. New York: Taylor & Francis Group，2005：55-74.
[13] ANSYS Inc. ANSYS Fluent Theory Guide Release 13.0.; ANSYS Inc., USA, 2011.

IOP Conf. Series: Materials Science and Engineering **721** (2020) 012074 doi:10.1088/1757-899X/721/1/012074

The Study of Metallographic Testing of Welding Joint of Welding in High Alloy Steel Installation of Power Station Boiler

Bing Zhang[1*], Peng He[2#], Jinqing Wang[3], Zhangmin Jin[1] and Xinmin Pan[1]

[1]Wenzhou Special Equipment Inspection and Research Institute, Wenzhou, Zhejiang, 325007, China

[2]Shanxi Boiler and Pressure Vessel Supervision Inspection Institute, Taiyuan Shanxi, 030012, China

[3]School of Metrology and Measurement Engineering, China Metrology University, Hangzhou, Zhejiang, 310018, China

[*]Corresponding author's e-mail: 326109580@qq.com

[#]These two authors contribute equally to this work

Abstract. Two disputes about the metallographic testing of welding joints of the high alloy steel on-site installation welding of the current power plant boiler are discussed. This thesis analyzes the reasons for the two dispute. On this basis, the accordance for the detection, detection method, qualification assessment standard and the ratio of detected parts to quantity are defined. This paper can provide reference for strengthening the quality of welding joints for high alloy steel installation welding of power station boilers.

1.Introduction

The author has found in the process of monitoring and inspection of the boiler installation of the power station that the installation organization does not perform metallographic testing on TP347H (equivalent to 07Cr18Ni11Nb in GB5310, Aussyta body heat-resistant steel) material's screen-mounted superheater site-mounted butt welding joints and TP347H and 12Cr1MoV welding joints of installation welding. The reason is that in the past similar boiler installation process no metallographic testing of the TP347H material installation welding joint is required. Metallographic testing is required by no supervision and inspection agency. The installation organization says that they are not familiar with accordance for testing, testing methods, qualification assessment criteria, etc. and the relevant system documents also makes no requirements.

2.The current situation of metallographic testing of welding joints of high alloy steel installation welding

The investigation made by the author shows at present, some of the domestic power plant boiler installation organization performs no metallographic testing in the installation process. Some supervision and inspection agencies also do not require metallographic testing of welded joints for the installation welding of high alloy steel. There are two disputes about the metallographic testing.

2.1 No need for metallographic testing
Main reasons for no need to perform metallographic testing are as follows.

2.1.1 Boiler Safety Technical Supervision Procedure（TSG G0001-2012）article 4.5 stipulates quality inspection of boiler pressurized components and their welding joints includes appearance inspection, ball test, chemical composition analysis, non-destructive test, mechanical performance test, hydraulic test, etc[1].No metallographic testing is included.

2.1.2 Grade A boiler installation standard Technical Specifications for Electrical Power Construction Part 2: Boiler Unit （DL5190.2）. No metallographic testing is included[2].

2.1.3 Welding Technical Regulations of Thermal Power Plants (DL/T869-2012) article 6.5.3 stipulates that the on-site micro metallographic testing of welded joints should be carried out in accordance with the dL/T884 regulations only when required by contract or design documentation or verification[3].Metallographic testing does not have to be performed when it is not required by contract or design documents or verification.

2.1.4 Boiler Supervision and Inspection Rules (TSG G7001-2015) item 3.2.5 article 10 stipulates that it is necessary to review the metallographic testing report of welding joint of high alloy steel material installation welding whether it meets the requirements of relevant standards[4]. The rules are formulated for inspection institutions in order to standardize the supervision and inspection work of inspection institutions, and the installation organizations are allowed to make no implement of the rules.

2.2 Need for metallographic examination
Main reasons for need to perform metallographic testing are as follows.

2.2.1 Boiler Supervision and Inspection Rules (TSG G7001-2015) item 3.2.5 article 10 stipulates it is necessary to review the metallographic testing report of welding joint of high alloy steel material installation welding whether it meets the requirements of relevant standards[4].

2.2.2 Welding Technical Regulations of Thermal Power Plants（DL/T869-2012）article 6.5.3 stipulates that on-site micro metallographic testing of welded joints should be carried out when required by contract or design requirements or verification[3].From the point of view of the systematic structure of special equipment laws and regulations in China, Boiler Supervision and Inspection Rules belongs to the safety technical specification. Welding Technical Regulations of Thermal Power Plants belongs to industry recommended standards. When there is a discrepancy between the two, the provisions of the safety technical specifications should be enforced.

2.2.3 Relevant standards set testing requirements on welding joints (tubes) metallographic testing. Regulations on Safety Supervision of Boiler Pressure Vessels in the Power Industry （DL/T612-2017） article 9.5.1（c）speculates welding quality inspection of pressurized components includes on-site metallographic examination[5].Metal Technical Supervision Procedure of Thermal Power Plants （DL/T438-2009）article 9.1.1(b) speculates that the inspection items in the pipe quality inspection supervision report include low-fold inspection and metallographic testing organizations. Article 9.3.8 speculates the superheater tube and reheater tube with wall temperature greater than 450 degrees Celsius during the operation of the unit should be sampled for aging and deterioration of the mechanical properties of the metallographic testing organization, and the quality of the pipe wall thickness, pipe diameter, metallographic testing organization, decarbonization layer and mechanical properties should be checked. Article 9.3.9 stipulates that the superheater tube, reheater tube and the heterogeneous steel welding joints connected to the aucotic stainless steel shall be sampled for aging of the metallographic testing organization and the deterioration of the mechanical properties[6].Seamless Steel Pipes for High-Pressure Boilers (GB/T5310-2017) table 11 has set the requirements of the inspection items, sampling quantity, sampling methods, test methods, etc of metallographic testing[7].Certificates of Acceptance, Packaging, Marking and Quality of Steel Pipes (GB/T2102-2006) article 4.5.2 stimulates that re-examination is not permitted when the low-fold tissue defects have white dot, micro-tissue in the metallographic testing, the grain size, and decarbonization layer are not qualified in the first test[8].Rules for the Acceptance of Boiler Materials into the Plant (JB/T3375-2002) article 6.1.2.4 (c) states that high-quality carbon structure steel pipes and alloy structural steel pipes supplied according to GB5310 shall be required for the actual grain size determination and microtissue inspection. Article 6.1.2.4 (d) stipulates cold-drawn (rolled) tubes supplied by GB5310 with outer diameter of not more than 76mm should be required for decarbonization layer inspection[9].Guidelines for the Supervision of Electrical Equipment (DL/T586-2008) table A1 has set requirements for the inspection of the heterogeneous steel joints of superheater and reheater [10].Guidelines for the Supervision and Inspection of Thermal Surface Tubes in Boilers in Thermal Power Plants (DL/T939-2005）article 6.6.6(n) stipulates after running time reaches 80,000 hours, the heterogeneous steel joints connected to Auschwitz stainless steel should be examined for the metallographic testing if necessary[11].

3.Analysis & understanding

3.1 Analysis and understanding of legal and regulatory standards

Welding Technical Regulations of Thermal Power Plants（DL/T869-2012）standard applicable steel includes carbon steel, low alloy steel and high alloy steel[3].Welding of carbon steel and low alloy steel is relatively easy. Control welding quality is generally not easy to form cracks, superheated tissue and hardened mastosome tissue. Therefore, if there is no contract or design requirements or verification needs, welding joints of on-site installation welding are not required for testing. High alloy steel (Auschwitz body heat-resistant steel) alloy element content is high, and the selection of welding process is not proper, and it is easy to have defects such as intercrystalline corrosion and thermal crack. Requirements for welding specifications (e.g. low heat input, multi-layer multi-channel welding, strict control of interlayer temperature, the welding torch shall not swing in the transverse direction, etc.) and inspection process are high. Therefore, in order to strictly control the welding quality and prevent the early failure of high alloy steel welding joints, it is necessary to perform metallographic testing.

Boiler Supervision and Inspection Rules（TSG G7001-2015）item 3.2.5 article 10 stipulates it is necessary to review the metallographic testing report of welding joint of high alloy steel material installation welding whether it meets the requirements of relevant standards[4].This article does not have any premises, such as no contract or the design document, or the need for verification. On the basis of this article, the following three points can be drawn.

(a) Metallographic testing shall be carried out for the installation of welded joints of high alloy steel material at the boiler installation site, the methods of metallographic inspection of installation organization, the proportion and qualification standards shall conform to the regulations of Boiler Safety Technical Supervision Procedure and related standards[12];

(b) Boiler installation supervision and inspection personnel examines metallographic testing report;

(c) The content of the examination is whether the metallographic test report (metallographic testing items and test results) meets the requirements of the relevant standards.

The metallographic testing is a follow-up test. In order to evaluate the aging and aging speed of the components and materials accurately, the microstructure of the on-site welding shall be the source basis, and the later organizational morphology change, metallographic composition change, carbide coarsening and metallographic structure change shall have the reference standard.

As a formal boiler installation organization licensed by the state administration, the quality assurance system documents shall meet the requirements of the special equipment safety technical specification, and shall also implement the requirements of item 3.2.5 article 10 of Boiler Supervision and Inspection Rules (TSG G7001-2015).The standards mentioned in 2.2.3 of this article have relevant rules for metallographic testing requirements; therefore, it has good reason that the high alloy steel installation welding joint requires that the metallographic examination. There are many metallographic examination techniques and methods applicable for power plant components. The Technical Guide for Metallographic Testing and Evaluation of Thermal Power Plants (DL/T884)[13] specifies the basic requirements and main operating steps for the on-site and laboratory inspection of high temperature components in thermal power plants by metallographic methods. The basic process and evaluation standard of metallographic analysis are also stipulated and metallographic testing can be carried out according to the standards. The specific method of metallographic detection and analysis is selected by the installation organization according to the purpose of the analysis, and it is clearly stipulated in the quality assurance system document.

3.2 Theoretical analysis and understanding

In the welding of the high alloy steel and the low alloy steel, the austenitic stainless steel and the ferritic heat-resistant steel, the failure problem is very easy to occur at the welding joint[14]; Because of the two materials have great differences in the mechanical property, the chemical composition, the metallographic structure and the thermal expansion coefficient, the welding of the austenitic steel and the 12Cr1MoV results in a crack source[15], as shown in figure 1,2; In the hardened fusion zone, there is an engineering example. Before the water pressure test made after the installation of an ultra supercritical unit is completed, it is found that there is a crack parallel to the weld bead in welding joint between the high temperature superheater tube screen SA-213T91 and HR3C high alloy dissimilar steel[16], as shown in figure 3.

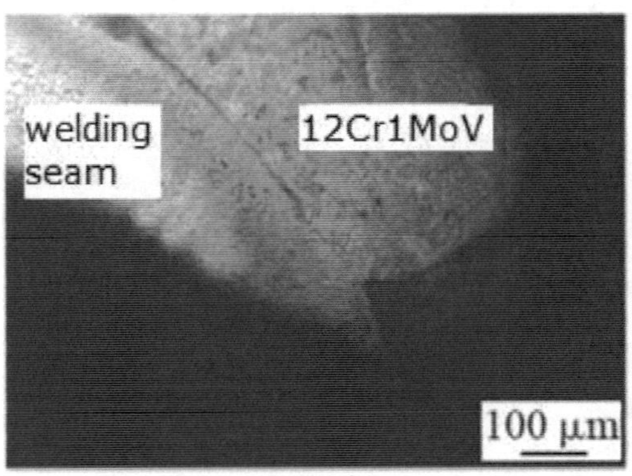

Fig.1 Fracture initiation position

Fig.2 Macroscopic morphology of the broken fracture

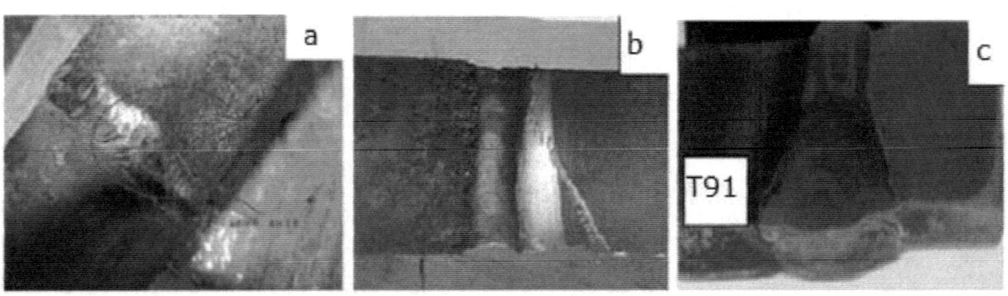

Fig.3 Macro morphology of crack and the pit on the wglding zone
(a)crack on outer surface (b)crack on inner surface (c)crack on deep direction

[17] It is pointed out in relevant literature that in the process of post-weld heat treatment, carbon will migrate and spread between the welding seam and the base metal in the welded joint of TP347H and 12Cr1MoV dissimilar steel. Because the chromium with high nickel-based weld content is easy to form carbides with carbon, the free carbon concentration in the weld metal near the fusion line decreases, and the carbon in the low alloy base metal diffuses to the weld and further forms carbides in the weld. A certain carbon-poor zone is formed at the base metal near the fusion line, resulting in a great difference in carbon content between the two sides of the fusion line. The low alloy steel side hardness of fusion line decreases (softens), and the weld side hardness of fusion wire increases (hardened), which makes the fusion line become a weak part with large hardness difference, and it is easy to form microcracks at a higher residual stress level, which leads to the early failure of welded joint.

The metallographic examination of welded joints with high alloy steel can obtain the information of carbide precipitation growth and spheroidizing degree, creep porosity and cracking tendency, and eliminate the existence of microcracks, overheated microstructure and hardened martensite structure, which plays a positive role in preventing the early failure of welded joints.

3.3 Other analysis and understanding

*3.3.1*qualification criteria. The conformity standard of metallographic structure of welded joint shall comply with the following regulations: 1. No cracks, no looseness. 2. There is no overheated tissue. 3. There is no hardened martensite structure. 4. δ-Ferrite must not exist in the metallographic structure of single Austenite steel weld metal.

*3.3.2*according to the relevant literature [15-19], table 1 shows the common failure locations of high alloy steel welded joints.

Table 1. Common failure locations of high alloy steel welded joints.

Serial number	Base metal	Welding consumables	Failure position
1	12Cr1MoV+ZGCr20Ni14Si2Mn1.5	ER316L	Near fusion line
2	SA-213 T91+HR3C	ERNiCr-3	Near fusion line
3	TP347H+12Cr1MoV	Inconel82	Near fusion line
4	TP347H+06Cr19Ni10	ER308	Near fusion line

As seen from table 1, the failure of welded joints of high alloy steel mainly occurs near fusion line, so metallographic testing should focus on the metallographic structure near fusion line.

*3.3.3.*the current relevant standards do not stipulate the proportion of metallographic testing of welded joints in installation welding. Before the introduction of the relevant standards, the installation organization shall define the proportion of metallographic testing in the quality assurance system documents.

4.Conclusion

The material of the high-temperature component of the power station boiler is easily aged and damaged, which is of great importance to the safety operation of the boiler, and the aging damage condition of the welding joint can be effectively checked and tracked by the metallographic examination method. At present, some installation organizations and supervision and inspection organizations do not make requirements for metallographic testing of welded joints for high alloy steel installation of power station boilers, which need to be paid great attention to in all aspects. It is suggested that the newly revised Boiler Safety Technical Supervision Procedure and other safety technical specifications and related standards should clarify the specific requirements for metallographic testing of welded joints installed with high alloy steel for power station boilers, so that all parties can comply with the implementation and ensure the installation and welding quality and safe operation of high alloy steel for power station boilers.

References

[1] TSG G0001-2012 Supervision Regulation of Boiler Safety Technology[S].
[2] DL5190.2 Technical Specification for Construction of Electric Power Construction Part 2: Boiler Unit [S].
[3] DL/T869-2012 Technical Specification for Welding of Thermal Power Plant[S].
[4] TSG G7001-2015 Boiler Supervision and Inspection Rule[S].
[5] DL/T612-2017 Safety Supervision Regulation for Boiler Pressure Vessel in Electric Power Industry[S].
[6] DL/T438-2009 Technical Supervision Regulation for Thermal Power Plant[S].
[7] GB/ T5310-2017 Seamless Steel Tube for High-Pressure Boiler[S].
[8] GB/ T2102-2006 Acceptance, Packaging, Marking and Quality Certificate of Steel Pipe[S].
[9] JB/ T3375-2002 Acceptance Rule of Material for Boiler[S].
[10] DL/ T586-2008 Technical Guidelines for Supervision of Power Equipment[S].
[11] DL/ T939-2005 Technical Guidelines for Supervision and Inspection of Boiler Heating Surface of Thermal Power Plant[S].
[12] Qian,G. (2015) Interpretation of Boiler Supervision and Inspection Rules. Tianjin Peoples Publishing House, Tianjin.
[13] DL/T 884 Technical Guidelines for Metallographic Inspection and Evaluation of Thermal Power Plants [S].
[14] Khalid R H,Kishore B N,Phanikumar G,et al.(2013) Microstructural Evolution during Friction Surfacing of Austenitic Stainless Steel AISI 304 on Low Carbon Steel. J. Metallurgical and Materials Transactions A, 44(1): 345-350.
[15] Liu D, Ning Y, Zhang S. (2018) Failure Analysis of Welded Joint of Screen-Type Superheater Dissimilar Steel Welding. J. Hot Working Process, 47 (19): 258-262.
[16] Yang S, Liu S. (2011) Crack Analysis of Welded Joints of Dissimilar Steel Welding. J. Metal Heat Treatment, 36: 273-276.
[17] Cai W, Zhao W, Yang F. (2009) Study on Early Failure of Welded Joints of TP347H and 12Cr1MoV Dissimilar Steel Welding. J. China Power, 42 (4): 22-24.
[18] Chen Z, Gao M, Xie X. (2019) Fracture Behavior of Welded Joints of TP347H Steel under High Temperature and High Stress, J. Welding Technology, 48 (9): 118-121.
[19] Zhang J. (2007) High Temperature Deformation and Fracture of Materials. Science Publishing House, Beijing.

IOP Conf. Series: Materials Science and Engineering **721** (2020) 012075 doi:10.1088/1757-899X/721/1/012075

Research Status and Analysis of Biomass Pyrolysis Gasification Technology

Zhongwei Wang[1], Bing Zhang[2] and Guoli Qi[1*]

[1]China Special Equipment Inspection and Research Institute, Beijing, 100029, China

[2]Wenzhou Special Equipment Inspection and Research Institute, Wenzhou, Zhejiang, 325000, China

[*]Corresponding author's e-mail: wenyuhe0616@126.com

Abstract: This paper analyzes the main methods and principles of biomass thermochemical conversion worldwide, introduces common equipment for biomass pyrolysis and gasification, and discusses the current research and numerical simulation of biomass pyrolysis gasification experiments. This paper also analyzes the application status of biomass pyrolysis gasification technology and summarizes the bottleneck problems encountered so as to preliminarily discuss the ways and methods of using biomass pyrolysis gasification to solve tar problems in the future.

1.Introduction

A useful measure to curb global warming is to use renewable energy to replace fossil fuels. Biomass is a unique renewable energy source as it can be stored and converted into fuel or chemical raw materials. In the use of biomass energy, biomass gasification technology can convert solid biomass into gaseous fuel used for power generation, gas supply, and central heating through thermochemical reaction.

From a biological point of view, the main components of biomass are cellulose, hemicellulose, lignin, and a small amount of extracts and ash, as shown in Table 1. Woody plants with a dense fibrous structure and high lignin content grow relatively slowly, while herbaceous plants with loose fibers, low lignin content and high cellulose content are generally annual plants. Different biomass materials differ in composition, and the relative ratio of cellulose to lignin is a factor in the determination of biomass type and the way to handle biomass. The pyrolysis products of cellulose include non-condensable gas, coke and a large amount of aldehydes, ketones and organic acids. In comparison, hemicellulose produces less tar but mor non-condensable gas during pyrolysis. Wood produces much more coke than cellulose and hemicellulose during pyrolysis. The pyrolysis process of biomass can be seen as a linear superposition of the pyrolysis process of cellulose, hemicellulose, and lignin [1].

Table 1. Chemical composition of different biomass

Type	Cellulose	Hemicellulose	Lignin	Extract	Ash
Cork	41	24	28	2	0.4
Hardwood	39	35	20	3	0.3
Pine bark	34	16	34	14	2

Wheat straw	40	28	17	11	7
Rice husk	30	25	12	18	16
Peat	10	32	44	11	6

The biomass is converted into biomass energy through two main pathways, namely, thermochemical conversion and biochemical conversion. Generally speaking, thermochemical conversion is more efficient in terms of reaction time and is stronger in converting organic matter. For example, lignin is a typical non-fermenting material; therefore it cannot be thermochemically but biochemically degraded. The thermochemical conversion mainly includes the steps of direct combustion, pyrolysis, gasification and liquidation [1], as shown in Figure 1.

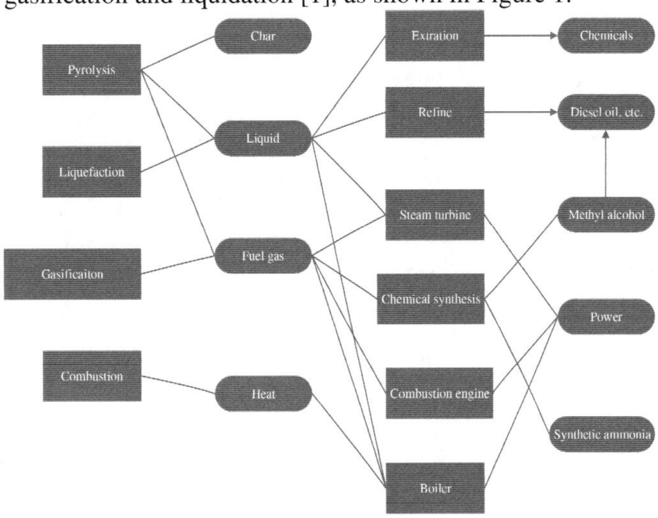

Fig.1 Thermo-chemical processes for bioenergy production and the corresponding products

Combustion is the most widely used in biomass conversion. Biomass combustion equipment mainly includes fixed bed, moving bed, fluidized bed and rotary kiln [2]. Pyrolysis refers to the thermal decomposition process of converting biomass into solid carbon, liquid (biomass oil and tar) and non-condensable gases under high temperature anoxic or anaerobic conditions. Nowadays, pyrolysis research is becoming increasingly important, which is because pyrolysis technique is considered to be an industrially achievable biomass conversion process and pyrolysis is the most important first step in gasification and combustion [3]. Biomass gasification is a process in which low oxygen or a suitable oxidant (such as water vapor and CO_2) converts carbon into a kind of combustible gas. When air or oxygen is used as the gasification medium, the process of gasification is similar to that of combustion. However, gasification is the process of local biomass combustion, which mainly uses fixed bed/moving bed, fluidized bed, circulating fluidized bed, etc. [4].

Direct liquidation is a thermochemical process carried out at high pressure and low temperature conditions, in which the biomass is cracked into small fragments and dissolved in water or other suitable solvent. Lignocellulosic materials are rich in hydroxyl groups, thus able to be liquefied into intermediate products for biopolymer production, such as epoxy resin, polyurethane foam, plywood adhesives, etc. [5].

2. Research status of biomass pyrolysis gasification

The downdraft gasifier is the most widely used in current fixed bed gasification systems. His et al. [6] studied the biomass gasifier using the downdraft fixed bed gasifier with air as the gasification agent, aming to explore the effect of operating parameters on biomass gas production. But the experiment did not mention the production rate of tar. Garcı'a-Bacaicoa et al. [7] studied the gasification characteristics of biomass and high-density polyethylene in a downdraft gasifier, aiming to investigate the feasibility of gasification using the mix of the two substances. In 2009, Sharma [8] studied 75 kW downdraft biomass gasifier system, and investigated the effect of pressure decrease on the combustion of the latter turbine. The whole system adopted water washing and sand bed to purify biomass gas, but failed to give the content of tar in the biomass gas. Ratnadhariya et al. [9] studied the molar distribution of CO/CO_2 and CO/H2 along the length of the downdraft gasifier, suggesting that the molar distribution equations of CO/CO2 and CO/H2 can be used to develop the actual gasification status in the downdraft gasifier. Compared with the fixed bed gasifier, the biggest advantage of the fluidized bed gasifier is that the uniform temperature distribution in the window and larger particle size. Dual fluidized bed gasifier is the current research focus. This kind of gasifier was pioneered by Professor Kunii [10], followed by Bartelle-Columbus and FERCO in the United States, TNEE in France, and AVSA in Belgium. The current research is concentrated in Europe (Austria and the Netherlands), Japan, and China [11]. Studies suggested that high tar content in biomass gas remains the main problem for this gasifier. Capucine Dupont et al. [12] completed the experimental study of biomass pyrolysis gas on a fluidized bed reactor with a temperature range of 800-1000 °C and particle diameters of 0.4 mm and 1.1 mm. The experimental results is closely related to the size of the biomass particles. Biomass pretreatment and feedstock are the biggest difficulties facing the application of fluidized bed. Recently, the University of Western Kentucky in the US developed a new biomass gasification gas production technology, which can reduce tar content and improve carbon conversion rate and gasification efficiency. The US National Renewable Energy Laboratory conducted studies on high-pressure combined gasification using coal biomass fluidized bed, and partial gas recirculation combustion was used to supplement energy and reduce tar content [13, 14].

Domestic research on biomass gas production has a late start but a rapid development. Wei [15] selected 13 kinds of biomass of agricultural and forestry wastes and plants such as straw, rice husk, corn cob, sawdust, the branches and leaves of salix mongolica and salix matsudana, alfalfa, reed, and alkali grass to make the biomass mixture and uses thermal gravimetric analyzer to study the co-pyrolysis characteristics of the biomass mixture and a typical lignite and the temperature of the lignite precipitation. Mi et al. [16] used a thermogravimetric analyzer to study the reactivity of biomass semi-coke in a CO2 atmosphere. The reactivity experiment of peanut shell coke under different partial pressures of CO2 shows that the reactivity of the coke sample is positively related to the concentration of the reaction gas. Cao et al. [17] conducted a thermal experimental study of a high-temperature air generator and obtained the its performance parameters, thereby preparing for the further study of high-temperature air gasification system. Wei et al. [18] studied biomass gasification by using a fixed bed updraft gasifier to explore the effects of operating conditions such as furnace temperature, raw material characteristics and fuel layer thickness. Zhou et al. [19] studied the effects of rice husk, Korean pine and fraxinus mandshurica ongas production under different parameters of oxygen, temperature, biomass ratio. The result shows that the gasification temperature is the main influencing factor for biomass entrained-flow bed. Zhao et al. [20] conducted a rapid pyrolysis gasification study of biomass under high-temperature and high-energy conditions generated by thermal plasma. Hu et al. [21] conducted a comparative analysis on the difference in tar yield rate, carbon conversion rate and gasification efficiency between coupled gasification and conventional coupled gasification by adopting upper and lower reactors of the fixed bed and taking vinasse as the gasification fuel. Duan et al. [22] carried out an experimental study on the NO reduction using simulated biomass gas, thereby obtaining the reburning characteristics of biomass gas.

3.Biomass pyrolysis gasification application status

Foreign biomass gasification research mainly focuses on the fields including district heating using biomass gas, cement plant gas supply and power generation, power generation using biomass gasification, ammonia production using biomass gasification, and dimethyl ether or methanol production using biomass gasification. Compared with equipment used in China, the biomass gasification equipment used at abroad is generally larger with higher automation degree. Its design raw material is often wood waste and is mostly used for power generation and heating. According to statistics, the ratio of commercial gasification equipment types manufactured in Europe and America is: updraft fixed bed gasifier: 2.5%; downdraft fixed bed gasifier: 75%; fluidized bed: 20%; others: 2.5% [23]. Updraft fixed bed gasification unit designed and developed by Moore Corporation of Canada can accept biomass feedstock with a higher water content (up to 60%); however, since the tar formed during pyrolysis does not pass through the combustion zone, the equipment will produce gas with a higher tar content. The production of most of the updraft gasifiers has been suspended due to its pollution (water pollution caused by tar residues). The gasification efficiency of the fluidized bed gasification unit designed and manufactured by Canada General Fuel Gasification Equipment Co., Ltd., the carbonization-gasification gas generation system manufactured by American Standard Solid Fuel Company, and the downdraft gasification gas generation system manufactured by German Army Belt Energy Co., Ltd is 60%-70%. The USA has invested much in biomass gasification research and development, such as the US Bttelle (63MW) and Hawaii (6MW) projects, which are mainly based on the biomass gasification-steam combined cycle power generation (IGCC) with high power generation efficiency. Tokyo Institute of Technology in Japan has developed a biomass high-temperature air gasification system for municipal waste treatment [24, 25]. Sweden's pressurized biomass gasification power generation project (4MW) adopts high-temperature air and water vapor as a medium for gasification study of high-density biofuels, suggesting that both gas production and low calorific value increase with increasing temperature [26]. The power generation efficiency of Italy's 12MW biomass gasification power generation BIGCC project is about 31.7%. But its construction cost reached 25,000 yuan/kwh, so the power generation cost is about 1.2 yuan/(kwh), making it difficult to be applied on a large scale [27]. In Europe, the UK (8 MW), Finland (6 MW) and the EU established three 7-12MW biomass gasification power generation BIGCC demonstration project, and France, Germany, as well as the Netherlands also carried out research on biomass gasification [28-30].

Domestic biomass gasification technology has developed rapidly since the 1980s. Guangzhou Institute of Energy, Chinese Academy of Sciences has made progress in the field of circulating fluidized bed gasification power generation, and has built and operated several gasification power generation systems [31, 32]. China Agricultural Machinery Research Institute has developed and promoted a wood drying technology using fixed-bed gasifiers [33]. Shandong Energy Research Institute successfully developed key equipment in straw gasification unit and centralized gas supply system during the Seventh Five-Year Plan and Eighth Five-Year Plan, thus helping to complete the set of technologies concerning gas generation, transmission and distribution. After the completion of the first pilot project for centralized gas supply in 1994, China has established a number of biomass gasification centralized gas supply projects nationwide [34, 35], for example, Liaoning Energy Research Institute has developed a downdraft biomass gasifier [36].

The most prominent problems restricting the commercialization of biomass pyrolysis gasification technology in China are high tar content in biomass gas and low automation degree [37]. Professor Beenackers of the Netherlands pointed out in his article commenting on European moving bed biomass gasification technology that the two problems above are common to see in the amplification design of current updraft and downdraft gasifiers, which undoubtedly hinders the promotion and application of biomass gasification process [38]. The advantages and disadvantages of various gasifiers are shown in Table 2.

Table 2. Advantages and disadvantages of selected types of gasifiers

	Advantage	**Disadvantage**
Updraft fixed bed/ moving bed	Simple, low price; outlet temperature of about 250 °C; good operation under pressure; high carbon conversion efficiency; low dust content in gas; high thermal efficiency	High tar content; easy to burn empty, be overhead, and produce slag; small size of feedstock; much time and effort for packing

Downdraft fixed bed / moving bed	Simple process; tar only exists in the produced gas	Small principle size; the ash content of the feedstock should not be too high; the scale is limited; easy to be overhead and produce slag; tar still exists
Fluidized bed	Flexible feeding speed and feedstock composition; can accept high ash fuel; can undertake pressure; high CH_4 content in gas; scale can be large or small; temperature is easy to control	Operating temperature is limited by ash sintering; high gas temperature; high tar and particulate content in gas; high carbon content in fly ash
Circulating fluidized bed	Flexible process; operating temperature up to 850 °C	Corrosion and wear; poor control of operation with biomass as fuel; tar problem still exists

4.Research status of numerical simulation of biomass pyrolysis

In 1989 and 1991, Koufopanos et al. established the pyrolysis kinetic model of biomass and its components and the pyrolysis model based on the biological particles affected by kinetics, heat and heat transfer [39, 40]. The chemical reaction kinetics model mainly studies the pyrolysis kinetics of lignocellulosic materials, aiming to provide a simple kinetic model for engineering application. The particle size is below 1 mm, thus the mass and heat transfer can be neglected and the reaction is considered to be controlled by pyrolysis kinetics. The pyrolysis rate of a kind of biomass can be described by the accumulation of the corresponding pyrolysis rates of cellulose, lignin, and hemicellulose. The kinetics of each component is modeled by a kinetic flow diagram that can predict the pyrolysis rate and ultimate weight loss, with a wide range of pyrolysis parameters (including heating conditions). The main disadvantage of this model is the failure to consider the transfer phenomenon (heat transfer and mass transfer) that plays a controlling role in the pyrolysis process of large particles.

In 1995, Srivastava et al. [41] predicted the content of biomass materials during the pyrolysis process by selecting different stages of the first-stage reaction and the second-order reaction under isothermal and non-isothermal conditions, respectively. The results show that this model can be used to predict the concentration curve of various biomasses, but the main disadvantage is the failure to combine with the transfer phenomenon. In 1999, Jalan et al. [42] studied the kinetics and heat transfer effects of pyrolysis of a single biological particle in a cylindrical coordinate system. The model of the single biological particle is based on physical and chemical changes that are both controlled by heat transfer phenomena. In addition, chemical changes include primary and secondary pyrolysis reactions. The proposed energy balance model equation considers the non-isothermal reaction of biological particles.

In 2000, Janse et al. [43] established a model for the flash pyrolysis of individual wood particles. This model makes the following assumptions: the model is one-dimensional; the ideal gas criterion is applied; local heat balance exists inside the particles between the gas phase and the solid phase; the macroscopic size of particles do not change during the reaction; wood and charcoal have changeable densities; cellulose and hemicellulose include macroscopic pores, and the diameter of impervious microscopic pores is less than 100 angstroms; porosity remains constant during conversion; Darcy gas model is used to describe the process during which the steam passes through the porous matrix of wood and charcoal to the particle surface; there is no restriction of external mass transfer; water evaporation of is not considered. The simulation results show that there is no need for the detailed description of the internal mass transfer phenomenon in the flash pyrolysis model, while correct information on chemical kinetics and heat transfer parameters is essential. This model only targets at one wood species, and makes no special explanation of tar and biomass oil conversion. In 2000, Larfeldt et al. [44, 45] conducted model and measurement studies on heat transfer of charcoal pyrolyzed from large-particle wood. The result shows that: the pyrolysis rate of large-particle biomass restricts heat transfer through the surface charcoal layer; thermal diffusion of charcoal is constant (approximately 0.7 mm^2/s). The main problem of the model is that it fails to consider the effects of kinetics.

From 2003 to 2004, Babu et al. [46, 47] studied the pyrolysis model of solid particles based on chemical reaction kinetics and heat transfer effects. Later, Babu simulated the biomass pyrolysis process on a fixed bed [48]. The main disadvantage of the above models is that they are only applicable to a single wood species without taking into account the effects of porosity [49]. The simulation result shows that the current model can deliver better-matched results with the experiment than the earlier models. However, Babu's model shows no improvement in the aspect of physical and chemical mechanism compared to Koufopanos' model, but only applies different numerical solutions.

In 2007, Zabaniotou et al. [50] studied the model of internal particle transfer and chemical reactions based on rapid pyrolysis of olive residues. The study selects spherical particles with a radius of 175 μm, the temperature range of 573 to 873 K, and the heating rate of 200 °C/s. The model considers both chemical reaction kinetics and heat transfer, and the simulation results are consistent with the test results (temperature, conversion process, and product distribution) obtained from the wire mesh reactor. However, the model does not consider shrinkage, so it can only be applied to a single material with woody physical parameters. In 2008, Gerun et al. [50] studied the local oxidation process of CFD simulation of two-stage downdraft gasifier, aiming to investigate the tar cracking situation of each reaction zone. This model is more inclined to simulate the combustion process of tar than its secondary thermal cracking. In 2009, Papadikis et al. [51] studied the rapid pyrolysis of biomass using CFD targeting the entrained-flow bed, using the two-stage hemispherical model.

In addition, Ratnadhariya et al. [52] studied the thermodynamic equilibrium and kinetic free model of the new biomass gasifier. Sharma et al. [7] also studied the biomass pyrolysis model based on thermodynamic equilibrium. Di Blasi et al. [53-55] have worked on the simulation of biomass pyrolysis gasification since 1992, but he focused more on the biomass pyrolysis and gasification process from the aspects of heat transfer, mass transfer and flow, but less on chemical reaction kinetic models. Based on the definition of tar and biomass oil, the liquid produced by biomass pyrolysis can be divided into two categories: tar and biomass oil. This classification can simulate not only the biomass pyrolysis reaction for oil production, but the relationship between tar production and temperature during biomass pyrolysis gasification under high temperature conditions [56]. The details are shown in Figure 2.

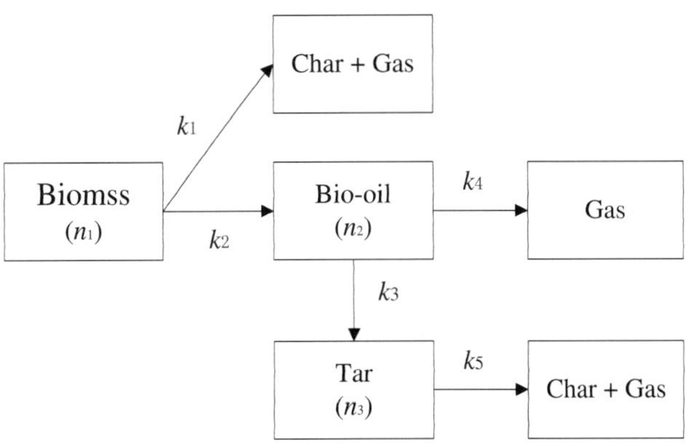

Fig.2 Chemical kinetic scheme of biomass pyrolysis

Biomass pyrolysis simulation is mostly carried out under low temperature conditions. The kinetic models of chemical reaction are mostly competitive models. Biomass after pyrolysis produces non-condensable gas products, liquid products (condensable gas products) and solid products. However, the liquid products produce biomass oil (below 500 °C) and tar (above 500 °C). The composition of the two condensable gas products is greatly different, so their excitation of pyrolysis under different temperature conditions is quite different, meaning that the application of a uniform excitation may affect simulation accuracy.

5.Conclusion

(1) At present, the main problems in biomass gasification include high tar content in biomass gas, large amount of waste water pollution due to biomass gas purification, low biomass calorific value and low energy utilization efficiency. High tar content is the most prominent problem, which causes pipe blocking and equipment shutdown. The removal of tar by water washing leads to a waste of energy (about 3% of total biomass energy) and environmental pollution due to the difficulty of waste water treatment. The catalytic cracking method can pyrolyze most of the tar using a catalyst under certain temperature conditions, but it brings the trouble of replacing the deactivated catalysts such as nickel-based catalysts, novel catalysts, dolomite and CaO. More importantly, the tar is often pyrolyzed to hard tar, which is more difficult to handle, thereby causing greater difficulty to subsequent tar processing. In addition, catalytic cracking requires complicated techniques and large-scale equipment. High-pressure super-critical gasification is mainly used for hydrogen production, while the technique for syngas production is more complicated. Therefore, it is difficult to realize industrial production for dispersed biomass resources. The high-temperature thermal cracking method of tar is of engineering application prospects and is simple and easy to operate. However, the conventional biomass gasifier has low gasification temperature and short residence time, which is not conducive to the full cracking of tar.

(2) The current research on high-temperature thermal cracking of biomass tar focuses mainly on the pyrolysis using aromatic compounds such as naphthalene to simulate tar. The composition of biomass tar is extremely complicated. So far, the researchers have not been able to provide sufficient pyrolysis data under high-temperature gas conditions.

(3) Due to the complexity of the physical and chemical properties of biomass, some current models ignore the internal temperature gradients of the particles or only simulate the pyrolysis of one chemical component. In addition, most models stimulate the high-temperature pyrolysis under low heating intensity and slow temperature rising conditions, and the model of biomass pyrolysis is not good enough under high temperature conditions. From the engineering point of view, it is great significance to understand the secondary cracking of tar, temperature distribution, degree of conversion, amount of gaseous products, solid product accumulation during high-temperature particle pyrolysis for the design and operation of the reactor. However, direct measurement of the above data is quite difficult. Therefore, it is necessary to establish a model that can be used under high temperature conditions to reflect the high-temperature pyrolysis of biomass.

Acknowledgments

This study was Supported by the National Key Research and Development Project (No. 2018YFF0216000) and the Science and Technology Planning Project of AQSIQ (No. 2017QK178).

References

[1] Bridgwater A. V., G. Peacocke V. C. (2000) Fast Pyrolysis Processes for Biomass. Renewable & Sustainable Energy Reviews, 4(1): 1~73.

[2] Energy Society of Japan. Handbook of biomass and Bioenergy. Translated by Shi Z.P, Hua Z.Z. Chemical Industry Press, 2007:97~103.

[3] Babu B. V. (2008) Biomass Pyrolysis: A State-of-the-Art Review. Biofuels Bioproducts & Biorefining-Biofpr, 2(5): 393~414.

[4] Zhang L. H., Xu C. B. (2010) Champagne. Overview of Recent Advances in Thermo-Chemical Conversion of Biomass. Energy Conversion and Management, 51(5): 969~982.

[5] Demirbas A. (2001) Biomass Resource Facilities and Biomass Conversion Processing for Fuels and Chemicals. Energy Conversion and Management, 42(11): 1357~1378.

[6] Hsi C. L., Wang T. Y., Tsai C. H., Chang C. Y., Liu C. H., Chang Y. C., Kuo J. T. (2008) Characteristics of an Air-Blown Fixed-Bed Downdraft Biomass Gasifier. Energy & Fuels, 22(6): 4196~4205.

[7] Garcia-Bacaicoa P., Mastral J. F., Ceamanos J., Berrueco C., Serrano S. (2008) Gasification of

Biomass/High Density Polyethylene Mixtures in a Downdraft Gasifier. Bioresource Technology, 99(13): 5485~5491.
[8] Sharma A. K. (2008) Equilibrium Modeling of Global Reduction Reactions for a Downdraft (Biomass) Gasifier. Energy Conversion and Management, 49(4): 832~842.
[9] Ratnadhariya J. K., Channiwala S. A. (2010) Experimental Studies On Molar Distribution of Co/Co2 and Co/H-2 Along the Length of Downdraft Wood Gasifier. Energy Conversion and Management, 51(3): 452~458.
[10] Igarashi M., Hayafune Y., Sugamiya R., Nakagawa Y., Makishima K., Kunii D. (1984) Pyrolysis of Municipal Solid Waste in Japan. J. Energy Resour. Technol, 106: 377~382.
[11] Corella J., Toledo J. M., Molina G. (2007) A Review on Dual Fluidized-Bed Biomass Gasifiers. Industrial & Engineering Chemistry Research, 46(21): 6831~6839.
[12] Dupont C., Commandre J. M., Gauthier P., Boissonnet G., Salvador S., Schweich D. (2008) Biomass Pyrolysis Experiments in an Analytical Entrained Flow Reactor Between 1073 K and 1273 K. Fuel, 87(7): 1155~1164
[13] Mclendon T. R., Lui A. P., Pineault R. L., Beer,S. K. Richardson S. W. (2004) High-Pressure Co-Gasification of Coal and Biomass in a Fluidized Bed. Biomass & Bioenergy, 26(4): 377~388
[14] Cao Y., Wang Y., Riley J. T., Pan W. P. (2006) A Novel Biomass Air Gasification Process for Producing Tar-Free Higher Heating Value Fuel Gas. Fuel Processing Technology, 87(4): 343~353
[15] Yan Wei-ping, LU Xu-ao (2009) Investigation of effect of the blending ratio of biomass and lignite on characteristic of co-gasification. Proceedings of the CSEE, 29(Supplement):150-155.
[16] Mi Tie, Chen Han-ping, Tang Ru-jiang, Wu Chuang-zhi, Ma Long-long, Shao Jing-ai, Wang Xian-hua, He Qiao, Liu De-chang (2005) Zheng Chu-guang. Reactivity study on biomass chars in CO_2 atmosphere. Acta Energiae Solaris Sinica, 26(6):766~771.
[17] Cao Xiao-ling, Jiang Shao-jian, Wu Chuang-zhi, Ai Yuan-fang (2005) Hot-state experiment research of high temperature air generator. Proceedings of the CSEE, 25(2):109~113.
[18] Wei Dun-song, Li Fang-qin, Li Lian-min (2006) Test study on biomass gasification in fixed beds. Journal of Tongji University (Natural Science), 34(2):254~259.
[19] Zhou Jin-song, Zhao Hui, Cao Xiao-wei, Luo Zhong-yang, Cen Ke-fa (2008) Syngas production with biomass entrained flow gasification. Acta Energiae Solaris Sinica, 29(11):1406~1413.
[20] Zhao Zeng-li, Li Hai-bin, Wu Chuang-zhi, Chen Yong (2005) The study on the plasma gasification of biomass. Acta Energiae Solaris Sinica, 26(4):468~472
[21] Hu Jing-hui, Wang Yin, Liu Xin-hua, Jiang Deng-gao, Xu Guang-wen (2009) Characteristics of Biomass Gasification by Decoupling Drying/Pyrolysis and Char Gasification. The Chinese Journal of Process Engineering, 9(4):731~737.
[22] Duan Jia, Luo Yong-hao, Yan Nai-qiang, Chen Yi, Lu Fang, Wang Qing-chen (2007) Experimental study on characteristics of biomass gasification-reburning. Journal of Fuel Chemistry and Technology, 35(2):245~248.
[23] Chen Guan-yi, Gao Wen-xue, Yan Bei-bei, Jia Jia-ni (2006) Present Research Status and Development of Biomass Gasification Technologies. Gas & Heat, 26(07):20~26.
[24] T. Hanaoka, S. Inoue, S. Uno, T. Ogi, T. Minowa. Effect of Woody Biomass Components On Air-Steam Gasification. Biomass & Bioenergy. 2005, 28(1): 69~76
[25] C. C. P. Pian, K. Yoshikawa. Development of a High-Temperature Air-Blown Gasification System. Bioresource Technology. 2001, 79(3): 231~241
[26] C. Lucas, D. Szewczyk, W. Blasiak, S. Mochida. High-Temperature Air and Steam Gasification of Densified Biofuels. Biomass & Bioenergy. 2004, 27(6): 563~575
[27] R. S, F. P. U. Catalytic Gasification of Biomass to Produce Hrydrogen Rich Gas. Hydrogen Energy. Hydrogen Energy, 1998, 23(7): 551~557
[28] P. D'Jesus, N. Boukis, B. Kraushaar-Czarnetzki, E. Dinjus. Gasification of Corn and Clover Grass

in Supercritical Water. Fuel. 2006, 85(7-8): 1032~1038

[29] G. Chen, J. Andries, H. Spliethoff. Catalytic Pyrolysis of Biomass for Hydrogen Rich Fuel Gas Production. Energy Conversion and Management. 2003, 44(14): 2289~2296

[30] Y. Calzavara, C. Joussot-Dubien, G. Boissonnet, S. Sarrade. Evaluation of Biomass Gasification in Supercritical Water Process for Hydrogen Production. Energy Conversion and Management. 2005, 46(4): 615~631

[31] Chen Ping, Yin Xiu-li, Zhou Zhao-qiu, Wu Chuang-zhi, Chen Yong. Operational performance of MW-scale biomass gasification and power generation plant. Acta Energiae Solaris Sinica, 28(4):389~393.

[32] WU Zheng-shun, WU Chuang-zhi, ZHENG Shun-peng, MA Long-long (2003) A design study of demonstration project of 4MW power generation system by biomass integrated gasification combined cycle. ENERGY ENGINEERING, (3):14~17.

[33] LIU Bao-liang, JIANG Jian-chun (2006) Research and Development of Biomass Gasification Technique for Power Generation in China. Biomass Chemical Engineering, 40(4):47~52.

[34] DONG Yu-ping, WANG Li-peng, DENG Bo, LU Ping, SHEN Shu-yun (2007) A review of the research and development of biomass energy technology. JOURNAL OF SHANDONG UNIVERSITY (ENGINEERING SCIENCE), 37(3):64~69.

[35] DONG Yu-ping, SUN Yu-quan. (2000) Development of Stalks Gasification Unit with Wet Purge. Transactions of The Chinese Society of Agricultural Machinery, 31(05):48~49.

[36] MI Tie, TANG Ru-jiang, CHEN Han-ping, LIU De-chang (2004) Utilization Technology of Biomass Energy and Its Research Progress. Gas & Heat, 24(12):701~705.

[37] LAI Yan-hua (2002) Investigation on Heat and Mass Transfer during Biomass Pyrolysis and Low-Tar Gasification Technology. Doctoral thesis of Southeast University, 4-90.

[38] Beenackers A. A. C. M. (1999) Biomass Gasification in Moving Beds, a Review of European Technologies. Renewable Energy, 16(1-4): 1180~1186.

[39] Koufopanos C. A., Papayannakos N., Maschio G., Lucchesi A. (1991) Modelling of the Pyrolysis of Biomass Particles. Studies On Kinetics, Thermal and Heat Transfer Effects, 69(4): 907~915.

[40] Koufopanos C. A., Maschio G., Lucchesi A. (1989) Kinetic Modelling of the Pyrolysis of Biomass and Biomass Components. Can. J. Chem. Eng., 67: 75~84.

[41] Srivastava V. K., Sushil, Jalan R. K. (1996) Prediction of Concentration in the Pyrolysis of Biomass Material. Energy Conversion and Management, 37(4): 473~483.

[42] Jalan R. K., Srivastava V. K. (1999) Studies on Pyrolysis of a Single Biomass Cylindrical Pellet - Kinetic and Heat Transfer Effects. Energy Conversion and Management, 40(5): 467~494.

[43] Janse A. M. C., Westerhout R. W. J., Prins W. (2000) Modelling of Flash Pyrolysis of a Single Wood Particle. Chemical Engineering and Processing, 39(3): 239~252.

[44] Larfeldt J., Leckner B., Melaaen M. C. (2000) Modelling and Measurements of Heat Transfer in Charcoal From Pyrolysis of Large Wood Particles. Biomass & Bioenergy, 18(6): 507~514.

[45] Larfeldt J., Leckner B., Melaaen M. C. (2000) Modelling and Measurements of the Pyrolysis of Large Wood Particles. Fuel, 79(13): 1637~1643.

[46] Babu B. V., Chaurasia A. S. (2003) Modeling for Pyrolysis of Solid Particle: Kinetics and Heat Transfer Effects. Energy Conversion and Management, 44(14): 2251~2275.

[47] Babu B. V., Chaurasia A. S. (2004) Heat Transfer and Kinetics in the Pyrolysis of Shrinking Biomass Particle. Chemical Engineering Science, 59(10): 1999~2012.

[48] Babu B. V., Sheth P. N. (2006) Modeling and Simulation of Reduction Zone of Downdraft Biomass Gasifier: Effect of Char Reactivity Factor. Energy Conversion and Management, 47(15-16): 2602~2611.

[49] Zabaniotou A., Damartzis T. (2007) Modelling the Intra-Particle Transport Phenomena and [50] Chemical Reactions of Olive Kernel Fast Pyrolysis. Journal of Analytical and Applied Pyrolysis, 80(1): 187~194.

[50] Gerun L., Paraschiv M., Vijeu R., Bellettre J., Tazerout M., Gobel B., Henriksen U. (2008)

Numerical Investigation of the Partial Oxidation in a Two-Stage Downdraft Gasifier. Fuel, 87(7): 1383~1393.

[51] Papadikis K., Gu S., Bridgwater A. V., Gerhauser H. (2009) Application of CFD to Model Fast Pyrolysis of Biomass. Fuel Processing Technology, 90(4): 504~512.

[52] Ratnadhariya J. K., Channiwala S. A. (2009) Three Zone Equilibrium and Kinetic Free Modeling of Biomass Gasifier - A Novel Approach. Renewable Energy, 34(4): 1050~1058.

[53] Di Blasi C. (1993) Modeling and Simulation of Combustion Processes of Charring and Non-Charring Solid Fuels. Progress in Energy and Combustion Science, 19(1): 71~104.

[54] Blasi C. Di. (2000) Modelling the Fast Pyrolysis of Cellulosic Particles in Fluid-Bed Reactors. Chemical Engineering Science, 55(24): 5999~6013

[55] Blasi C. Di. (2008) Modeling Chemical and Physical Processes of Wood and Biomass Pyrolysis. Progress in Energy and Combustion Science. 2008, 34(1): 47~90

[56] Qi GL, Dong P, Guan J. (2018) Pyrolysis simulation of granular biomass considering darcy flow, Acta Energiae Solaris Sinica, 39(2):292-298

Fuel characteristics and explosiveness analysis of pulverized coal industrial boilers in China

Zhongwei Wang[1], Bing Zhang[2] and Guoli Qi[1*]

[1]China Special Equipment Inspection and Research Institute, Beijing, 100029, China

[2]Wenzhou Special Equipment Inspection and Research Institute, Wenzhou, Zhejiang, 325000, China

[*]Corresponding author's e-mail: wenyuhe0616@126.com

Abstract. Pulverized coal industrial boilers are widely used in China in recent years for high thermal efficiency and suitable for contract energy management, especially the in service amount is growing fast in regions like Shanxi, Gansu, Shandong, Tianjin, Liaoning, Hebei, etc. Considering the existing operational safety issues of pulverized coal fired industrial boiler, this article researched the relationship between fuel characteristics including calorific value, volatile, ash, moisture and powder explosion in order to explore and analysis in service pulverized coal industrial boilers explosiveness.

1. Introduction

Currently, developed countries take oil-fired or gas-fired industrial boilers as primary, coal-fired industrial boilers are still applied with high standard technology and management in developed countries and regions such as America and Europe. Germany shares a similar type of energy composition with China, in which the chain-grate boiler has a significant proportion[1]. America has 3195 coal-fired industrial boilers. The chain-grate boilers mainly consume block coals selected by washing. The new industrial diagnosis and automatic control technology leads chain-grate boilers' automatically adjusts the running and reaches the boiler efficiency up to 78%[2]. The adoption of low-sulfur and low-ash coal, the high-efficiency dust removal, desulphurization equipment, the natural gas re-burning and SCR equipment can reduce the emission of NOx and reach the standard of coal pollutants emission[3].

Small and Medium pulverized coal industrial boiler technology be mature in the late 1990s and widely applied in Germany and France. After twenty years of developing, Germany has wholly commercialized pulverized coal industrial boilers and developed many technique standards[4].

China only has a ten-year history of pulverized coal industrial boiler from public documents. Relevant code and standards are incomplete. Current safety standard TSG G0001-2012 *Boiler Safety Technology Supervisory Regulations* cites GB/T5203 *Technical Code for Explosion Proof Design of Coal and Pulverizing System in Thermal Power Plant*, covers little about pulverized coal industrial boiler and mainly about the pulverized coal power plant boiler, especially about the powder-making system. Explosion-proof standard of pulverized coal industrial boiler follows T/CEEIA 221-2013 *Pulverized Coal Industrial Boiler and Its System*. The standard provides the burner ignition sequence, and furnace safeguard supervisory system (blowing before the fire, fire alarm, and blowing after the fire) standards above are formed before 2013 and mainly about power plant boilers. Standards and regulations about pulverized coal industrial boiler are little[5, 6].

2. The explosiveness of coal power

The explosiveness of coal power means the difficulty of detonation and the intention of explosion pressure. To the boilers that consume inflammable coal, their explosiveness depends on the properties of coal powder. Explosive accident of pulverized coal industrial boiler usually happens at the start, rundown, and discontinue of powder. The explosiveness of coal power depends on the flammability, ash content, humidity, fineness, temperature, concentration, and the oxygen level of the primary mixture. The characteristic ignition index generally values the explosiveness of the coal powder[1].

2.1Explosiveness of coal powder and Explosive index

Relevance between explosiveness of coal powder and Explosive index is formulated in DL/T5145-2012 Technical Regulation for Design and Calculation of Pulverizing System in Thermal Power Plant.

Explosive index is calculated according to equation as below.

$$K_d = \frac{V_d}{V_{\text{vol.que}}} \tag{1}$$

$$V_{\text{vol.que}} = \frac{V_{\text{vol}}\left(1+\dfrac{100-V_d}{V_d}\right)}{100+V_{\text{vol}}\dfrac{100-V_d}{V_d}} \tag{2}$$

$$V_{\text{vol}} = 1260\times 4.187 / Q_{\text{vol}} \tag{3}$$

$$Q_{\text{vol}} = \left(Q_{\text{net,v,daf}} - 7850\times 4.187 FC_{\text{daf}}\right) / V_{\text{daf}} \tag{4}$$

$$FC_{\text{daf}} = 1 - V_{\text{daf}} \tag{5}$$

K_d ——Explosive index;

V_d ——Dry basis volatile matter of coal, %;

$V_{\text{vol.que}}$ ——Lower limit of combustible volatiles required for combustion (Ash and fixed carbon are considered), %;

V_{vol} ——Lower limit of combustible volatiles required for combustion (Ash and fixed carbon are not considered), %;

Q_{vol} ——Calorific value of volatile matter, kJ/kg;

$Q_{\text{net,v,daf}}$ ——Low calorific value of coal on dry ash-free basis, kJ/kg;

FC_{daf} ——Fixed carbon content of coal on dry ash-free basis, %;

V_{daf} ——Dry ash-free volatile content of coal, %.

Table 1. Relevance between explosiveness of coal powder and Explosive index

Explosive index	extreme low explosivity
$K_d \le 1.0$	extreme low explosivity
$1.0 < K_d \le 3.0$	very low explosivity
$3.0 < K_d \le 7.0$	low explosivity
$7.0 < K_d \le 12.0$	medium explosivity
$12.0 < K_d < 17.0$	high explosivity
$K_d \ge 17.0$	very high explosivity

2.2Explosive index and Ignition Temperature of Pulverized Coal Gas Flow

Ignition Temperature of Pulverized Coal Gas Flow (IT) combines the influence of flammability and ash content, can broadly represent Explosive index under certain humidity, fineness, temperature, and concentration. Volatile matter and explosibility shares relativity, so volatile content of coal is an approximate judgment of Explosive index. The table2 lists DL/T5145-2012 Technical Regulation for Design and Calculation of Pulverizing System in Thermal Power Plant, gives reference value between coal explosiveness and Explosive index.

Table 2. The relationship between the explosiveness of pulverized coal and the ignition temperature of pulverized coal air flow and dry ash-free volatile matter

Explosive index of pulverized coal	Ignition temperature of pulverized coal gas flow IT (°C)	Dry ash-free volatile V_{daf} (%)
$K_d \le 1.0$	$IT \ge 800$	$V_{daf} \le 6.0$
$1.0 < K_d \le 3.0$	$800 > IT \ge 780$	$6.0 < V_{daf} \le 10.0$
$3.0 < K_d \le 7.0$	$780 > IT \ge 750$	$10.0 < V_{daf} \le 15.0$
$7.0 < K_d \le 12.0$	$750 > IT > 720$	$15.0 < V_{daf} < 20.0$
$12.0 < K_d < 17.0$	$720 > IT > 680$	$20.0 < V_{daf} < 25.0$
$K_d \ge 17.0$	$IT \le 680$	$V_{daf} \ge 25.0$

2.3Explosive index

Russian all-Russian Thermal Engineering Research Institute has a Long term experimental study about explosion characteristics of pulverized coal, proposed an engineering evaluation method for explosiveness of pulverized coal for power fuel different that China's explosive index, and formed a standard.

This explosiveness calculated according to equation as below.

$$K_T = \frac{V_d}{\mu_{V.A}^{L}} \tag{6}$$

$$\mu_{V.A}^{L} = \frac{\mu_V^L\left(1+\dfrac{100-V_d}{V_d}\right)100}{100+\mu_V^L\dfrac{100-V_d}{V_d}} \tag{7}$$

$$\mu_{V.A}^{L} = \frac{1260}{Q_V} \tag{8}$$

$$Q_V = \frac{Q_{net,daf} - 329\left(C_{daf} + H_{daf} + O_{daf} - V_{daf}\right)}{V_{daf}} \tag{9}$$

V_d ——Volatile content of coal on dry basis, %;

$\mu_{V.A}^{L}$ ——the lower limit concentration of the flame spread in the combustion state when the volatiles containing ash and coke are mixed with air, % (In addition to volatile components, there are ash and coke in the pulverized coal. They hardly participate in the reaction in the process of rapid explosion, but the ash consumes part of the heat, so it will increase the lower limit of flame expansion.);

μ_V^L ——the lower limit concentration of the flame spread in the combustion state when the volatiles without ash and coke are mixed with air, %;

Q_V ——Calorific value of volatile matter, kJ/kg;

$Q_{net,daf}$ ——Calorific value of dry ash-free base coal, kJ/kg;

C_{daf} ——Carbon content of dry ash free base coal, %;

H_{daf} ——Hydrogen content of dry ash free base coal, %;

O_{daf} ——Oxygen content of dry ash free base coal, %;

V_{daf} ——Volatile content of coal on dry ash-free basis, %.

3. Pulverized coal analysis data

36 boilers in Shanxi, Hebei, Shandong, Sichuan, Shanghai, Gansu, Liaoning, Jiangsu and other provinces were tested for coal quality, the test data are as below.

3.1Analysis of calorific value Analysis of coal used in pulverized coal industrial boilers in China

As can be seen from Fig.1, above the red line is the third-class bituminous coal, and below the red line is the second-class bituminous coal. In China, the coal used in pulverized coal industrial boilers is mainly AIII bituminous coal, which accounts for more than 90%, AII bituminous coal accounts for 5% and anthracite accounts for 5%. The main range of calorific value is 22000kJ/kg-28000kJ/kg.

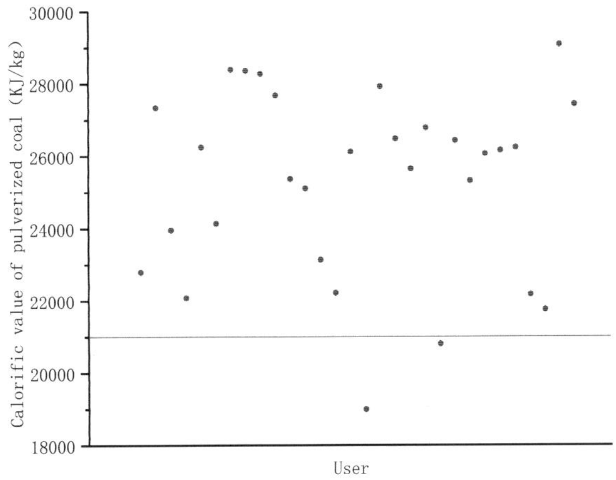

Fig.1 Distribution of calorific value of pulverized coal used in boilers in different areas

3.2Ash analysis of coal

Fig.2 is a distribution chart of AR (as received basis) ash of pulverized coal used in boilers in different areas of China. The highest AR ash content of powdered coal used in pulverized coal industrial boilers in China is 34.2% and the lowest AR ash is 5.44%. The AR ash content which is below 15% accounts for 87% of the total coal used. Because of the low ash content of the coal, the thermal efficiency of boiler is relatively high.

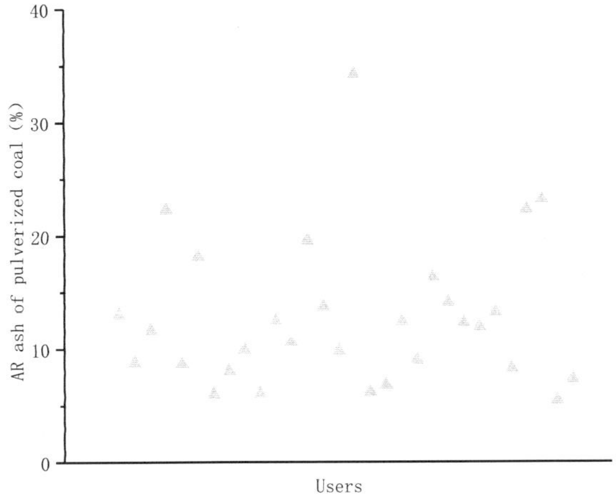

Fig.2 Distribution of AR ash of pulverized coal used in boilers in different areas

3.3Analysis of VDAF of coal

Fig.3 is a distribution chart of VDAF (dry ash-free basis volatiles) of pulverized coal used in boilers in different areas of China. The highest VDAF of powdered coal used in China's pulverized coal industrial boilers is 33.31% and the lowest VDAF is 20.50%. The VDAF content of all the pulverized coals used is above 20%, and 77% of the used coals have the VDAF content of more than 28%, which are the typical three types of bituminous coal.

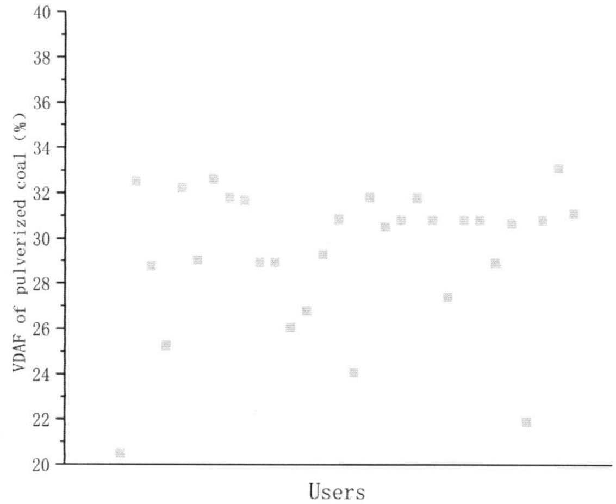

Fig.3 Distribution of VDAF of pulverized coal used in boilers in different areas

3.4Analysis of total moisture distribution of coal

Fig.4 shows the total moisture distribution of pulverized coal used in boilers in different areas of China. The total water content of coal-fired industrial boilers in China is 15.5% at most and 4% at least. The proportion of total water below 5% is 57%, and the proportion of total water below 10% is over 90%.

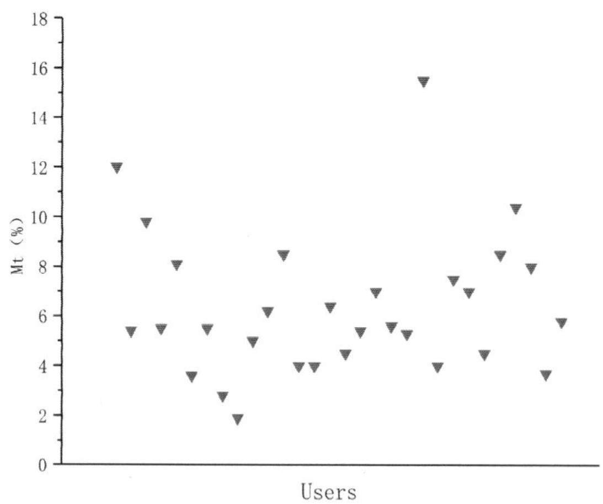

Fig.4 Total water distribution of pulverized coal used in boilers in different Areas

4. Results and discussion

The quality of pulverized coal used in industrial boilers of China has been analyzed and more than 300 pulverized coal industrial boilers have been involved. Only one of the same type industrial boilers of pulverized coal has been selected for the analysis.

K_d is used to analyze the explosivity of pulverized coal and please see Table 3 for the details. From the table, it can be seen that only 6% of the pulverized coal is of medium explosivity, 15% of the pulverized coal is of high explosivity, and over 78% of the pulverized coal is of extreme explosivity.

Table 3. Relationship between explosivity of pulverized coal and explosion Index K_d

explosivity index of pulverized coal	explosivity	Proportion（%）
$K_d \le 1.0$	extreme low explosivity	0
$1.0 < K_d \le 3.0$	low explosivity	0
$3.0 < K_d \le 7.0$	very low explosivity	0
$7.0 < K_d \le 12.0$	medium explosivity	6.25
$12.0 < K_d < 17.0$	high explosivity	15.63
$K_d \ge 17.0$	$IT \le 680$	$V_{daf} \ge 25.0$

The explosivity of pulverized coal is analyzed by KT, as shown in Table 4. From the table, it can be seen that only 6% of the powered coal is of medium explosivity, 90% of the powdered coal is of high explosivity, and 3% of the pulverized coal is of extreme explosivity. Comparing Table 3 and Table 4, it is found that the two are different in evaluation of explosion index of powdered coal. K_T and K_d are consistent in evaluation of the medium explosivity of pulverized coal, both accounting for 6.25%. However, in terms of high explosivity, there is a big difference between them, which K_d is 15.63%, and K_T is 90.63%. The main reason for the difference is that K_T takes the influence of moisture and ash into the consideration.

Table 4. Relationship between explosivity of pulverized coal and explosion index K_T

explosivity index of pulverized coal	explosivity	Proportion（%）
$K_T \le 1.0$	low	0
$1.0 < K_T \le 1.5$	medium	6.25
$1.5 < K_T \le 3.5$	high	90.63
$K_T > 3.5$	extreme	3.12
explosivity index of pulverized coal	explosivity	Proportion（%）
$K_T \le 1.0$	low	0

5. Conclusion

This article has comprehensively sorted out the safety regulations and standards for the powder supply and boiler explosion of pulverized coal industrial boilers in China, and selected the powdered coal used in 36 representative boiler models. Through analysis and test, it compares the differences between the explosion index of pulverized coal used in industrial boilers of China and that of Russia. According to the Chinese standard, more than 90% of pulverized coal used in industrial boilers is extremely explosive, but according to the Russian standard, the explosivity of pulverized coal is only high explosive as the influences of moisture and ash are mainly considered in the Russian standard.

Acknowledgments

This study was Supported by the Science and Technology Planning Project of AQSIQ (No. 2017QK178) and the National Key Technology R&D Program (No.2014BAA07B05).

IOP Conf. Series: Materials Science and Engineering **721** (2020) 012076 doi:10.1088/1757-899X/721/1/012076

References

[1] Zhang JW, Li WD, Xu CK (2017) Selection of combustion equipment for pulverized coal boiler in power plant. China Power Press, Beijing.
[2] Rosin P.O., Fehling H.R. (1937) The Ignition of Coal on a Grate. Journal of the Institute of Fuel, 11:102-117.
[3] Luo C, Luo K, Wang YZ, Ma ZT, Gong YL (2017) The effect analysis of thermal efficiency and optimal design for boiler system. Energy Procedia, 105:3045-3050.
[4] Han YF, Shen B, Zhang T (2017) A techno-economic assessment of fuel switching options of addressing environmental challenges of coal-fired industrial boiler: An analytical work for China. Energy Procedia, 142:3083-3087
[5] Regulagadda P., Dinecer I. (2010) Naterer G. Exergy analysis of a thermal power plant with measured boiler and turbine losses. Appl Therm Eng, 30(8):970-976
[6] Kilicaslan I., Ozdemir E. (2005) Energy economy with a variable speed drive in an oxygen trim controlled boiler house. J Energy Resour Technol, 127(1):59-65

IOP Conf. Series: Materials Science and Engineering **721** (2020) 012077 doi:10.1088/1757-899X/721/1/012077

Design of LNG gasification and thermal energy comprehensive utilization system for vehicles

Haomin Zhang[1*], Shihao Wang[1], Shiqiang Jing[1]

[1]School energy and power engineering, Wuhan university of technology, Wuhan, Hubei, 430063, China

[*]Corresponding author's e-mail: 1350612523@qq.com

Abstract: The development of new energy vehicles is one of the important measures for national energy conservation and emission reduction, and natural gas is an important part of new energy. At present, most natural gas vehicles on the market are CNG vehicles, and LNG has many advantages, mainly solving the problem of vehicle life. This work has designed a vehicle LNG gasification system, including the design of stable self-pressurization and new gasifier, so that LNG can be used instead of CNG in the car, and it can be combined with the designed thermal energy comprehensive utilization system, which can utilize LNG. Cold energy can further reduce energy consumption and achieve energy saving and emission reduction. We first carried out the theoretical check calculation, then completed the piping design of the system, and carried out the ANSYS simulation analysis of the key component gasifier, and verified the simulation results through experiments, and further simulated the system. Analyze, design and optimize the control mode of the system. On this basis, the system control is programmed based on the single-chip microcomputer. We also made the system model. It uses the single-chip microcomputer to control and can switch and display various working modes.

1. Introduction

As the country's economic development is growing faster, the demand for resources is also increasing. According to relevant statistics, 60% of urban atmospheric pollution comes from exhaust emissions from motor vehicles. The use of natural gas as a fuel for vehicles instead of traditional gasoline and diesel is one of the important measures to reduce vehicle exhaust pollution. In the long run, the country has decided to set a timetable for the ban on fuel vehicles, which will promote the development of the natural gas automobile industry.

At present, most of the natural gas vehicles at home and abroad use compressed natural gas CNG, and LNG has many advantages. Its energy density is high (about 3 times that of CNG), the vehicle has long cruising range, low storage pressure and safer. All countries in the world are vigorously promoting LNG vehicles, of which the United States is in a leading position in LNG vehicle technology.

Compared with CNG, a gasification system is required to use LNG in a car. Since the combustion concentration range of LNG is small, the LNG engine has high requirements for gas supply, so the supply pressure control of the LNG gasification system for the vehicle is Very important. In addition, the gasification of LNG at the gas station is mainly based on seawater gasification or air heating, and the cold energy of LNG is not well utilized [3]. Our project solves the problem of gasification control through the corresponding piping structure design and control design, and cooperates with the designed thermal energy comprehensive utilization system, which can utilize the cold energy of LNG to further reduce energy consumption and achieve energy saving and emission reduction. effect.

IOP Conf. Series: Materials Science and Engineering **721** (2020) 012077 doi:10.1088/1757-899X/721/1/012077

2. Design plan

2.1. Piping design

The system consists of LNG bottles, gasifiers, natural gas engines, radiators, water air conditioners, surge tanks, cooling water tanks, air conditioning water tanks, cooling water pumps, air conditioning water pumps, and valve components and related control devices. The LNG bottle includes a filling port, a filter, a safety valve, a pressure gauge, a liquid level gauge, a discharge valve, an economic valve, a self-pressurizing port, etc., when the pressure in the bottle is high, the economic valve will automatically open, and the bottle will be automatically opened. The natural gas is used to avoid the excessive pressure from leaking from the safety valve and causing waste. The specific design is shown in Figure 1.

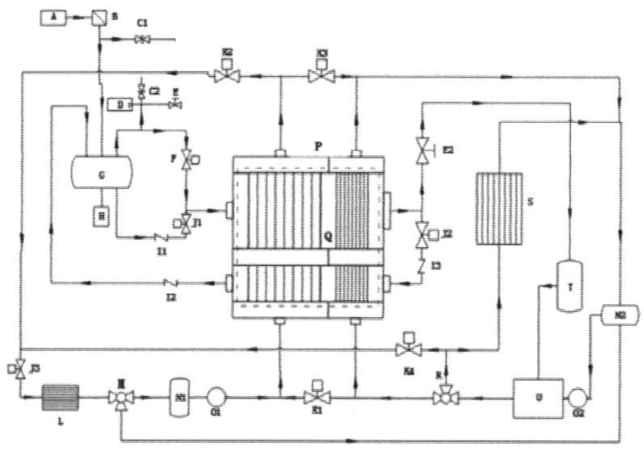

Figure 1. System piping design

2.2. Gasifier design

The gasifier uses a smaller, lighter plate-fin heat exchanger. In the gasifier, LNG flows in from the inlet, flows out from the outlet 2 after gasification, and is divided into two parts. One part is supplied to the engine after being stabilized by the surge tank, and the other part flows into the gasifier from the 3 inlet again, and is heated. After boosting, it flows back to the LNG bottle from the 4 outlet to complete self-pressurization. The engine cooling water or air conditioning water enters the gasifier from the inlets 5, 6 and flows out from the 7 and 8 outlets. In order to prevent the liquid from solidifying and blocking the flow path when the temperature is too low, the surface of the gasifier is also equipped with an electric heating layer, which can also assist the warm-up and improve the engine life. The gasifier is shown in Figure 2.

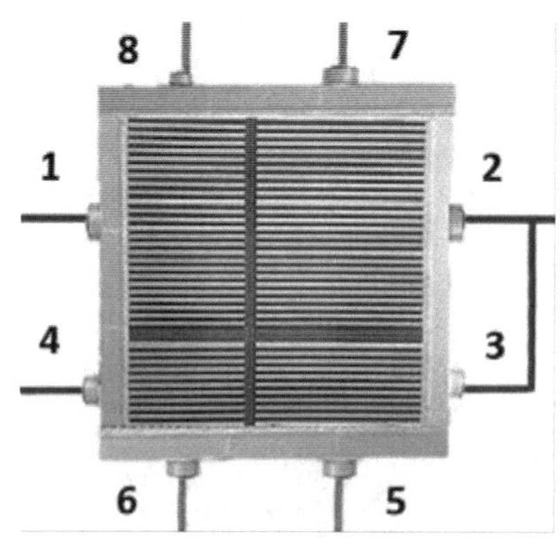

Figure 2. Gasifier

3. Gasifier numerical simulation

In Since the LNG gasification process in the plate-fin heat exchanger belongs to the boiling heat transfer in the narrow flow channel, it has the characteristics of high heat transfer coefficient and complicated process, and its heat transfer efficiency is closely related to the size and shape of the flow channel. In order to design the gasifier structural parameters suitable for LNG gasification, we use the FLUENT module in ANSYS software to analyze the LNG gasification process of a single narrow channel.

4. System and Control Simulink Simulation

In order to design a reasonable control method and verify the feasibility of the system, we established a feedback control function for the opening of each valve by establishing the Simulink model of the system, and stabilized and accelerated under the two modes of air conditioning not working and cooling. The four working conditions of step, step and sine were simulated. The established Simulink model is shown in Figure 3.

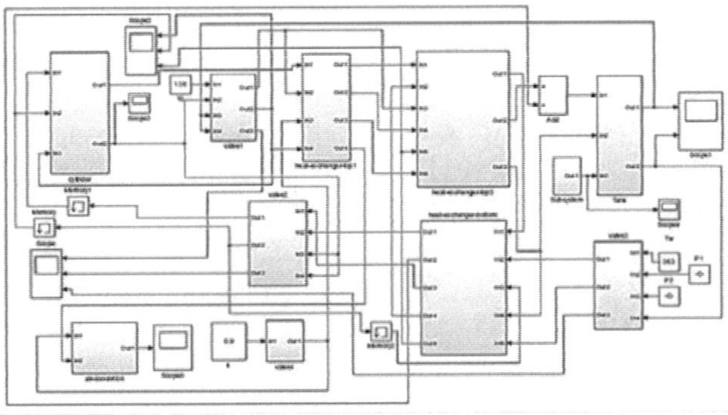

Figure 3. System Simulink model

In the various working conditions of our simulation, although the pressure and temperature fluctuate, they are all within the condition range, which can meet the requirements of the engine air supply. It can be seen that the opening control method and control function of the system and each valve are reasonable. The system can solve the problem of LNG gasification and cold energy utilization for vehicles.

5. Model making

We made the model of the system and placed the equipment on a self-made car chassis, and the layout of each device was also designed. The model uses transparent tubing and uses different colors of water to distinguish between different fluids. The model is controlled by a single-chip microcomputer and can switch and display various working modes. The model is shown in Figure 4.

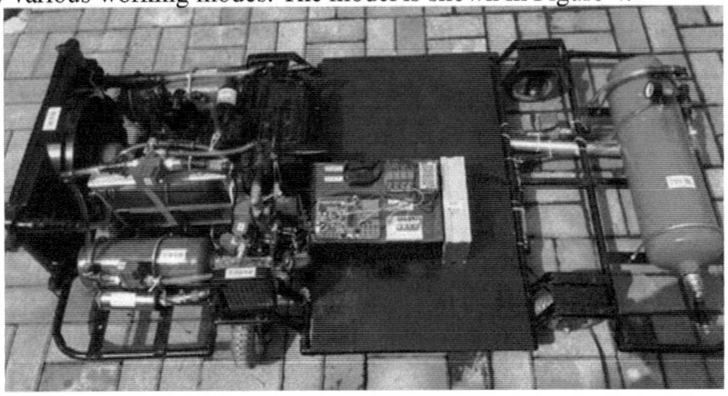

Figure 4. Model photo

6. Application prospect

Inquiring about the Shanghai Stock Exchange stock code 601857 annual performance report, we can see that the gasoline sales in 2017 is 5173.39 kilolitres. The gasoline price is based on the 2018 easy car network data, taking the average of different cities and different types of gasoline in the country, which is 6.64 yuan / liter, the car The number is based on the statistical data of the 2017 National Statistical Report on National Economic and Social Development of the People's Republic of China, which is

21.992 million. It can be calculated that the average fuel consumption per vehicle per year is 2.35 kiloliters, and the annual fuel consumption is about 15620 yuan. The average selling price of natural gas is 3.60 yuan / m^3, while driving the same mileage, the consumption of 1L of gasoline is equivalent to the consumption of 1.1$m^3$of natural gas. The calculation shows that the average annual savings per vehicle is about 6,304 yuan, which is about 40%.

According to the information provided by BP China Carbon Emissions Calculator, each natural gas vehicle can reduce carbon dioxide emissions by 90% compared with gasoline vehicles, and reduce emissions by 80%. Each vehicle can reduce CO2 emissions by about 60 tons per year. Inquire about the 2017 annual reports of companies such as Poussin, Jetta and BYD. Currently, there are about 500,000 CNG vehicles on the market. If 10% of the current cars are replaced or converted into natural gas vehicles, the annual CO2 emissions will be reduced by 131.97 million tons. . Due to our water-air conditioning technology, there is no need to use a compressor for air-conditioning in the summer. If the average air-conditioning is used for 3 hours every year from June to September, and the cooling power is calculated at 2kW, the engine efficiency is considered. It can further save 130.11kg of natural gas per year, about 652.5 yuan, and further reduce emissions by 7.3%.

The natural gas vehicle itself has the advantages of economy and environmental protection, which makes it highly competitive in the market. With the support of national policies, as long as the technical problems of the CNG vehicle's battery life difference are solved, the penetration rate of natural gas vehicles will be greatly increased. Improvement, and our project has solved this problem and has broad market prospects.

7. Conclusion

LNG is one of the clean energy sources. It has broad application prospects in transportation facilities such as vehicles and ships. This project studies key issues such as regulation, stabilization and full utilization of energy in the gas phase of LNG from liquid phase to gas phase. The middle and frontier issues have important theoretical and applied research significance for related research.

References

[1] Zhong Wei, Liao Changwei, Huang Xiaomei, Peng Shini. Development Status and Market Prospect of Vehicle LNG[J]. Gas and Heat,2013,33(01):10-13.

[2] Zhou Shuhui, Yang Yi, Li Bo. Development Status, Problems and Market Space of China's LNG Industry[J]. International Petroleum Economics, 2013, 21(06): 5-15+109-110.

[3] Li,L.(2018) Let water cutting technology play a more efficient role in the Chinese market Hyundai Manufacturing 2018 Issue 16th P16 1671-9395.

[4] Hao Wensheng, Shi Shaobo, Zhou Dan, Wang Lichun. Selection of LNG liquefaction and gasification process equipment[J]. China Chemical Industry, 2017, 19(02): 37-44.

[5] Wu Runhan, Xia Junzhong, Meng Linkun, Gu Xuesong, Yu Mingqi. Analysis of Vaporization Quantity of LNG Automobile Water Bath Vaporizer[J]. Automobile Practical Technology, 2016, (05): 75-77.9

AUTHOR INDEX